Nineteenth Century Science

Nineteenth Century Science

A Selection of Original Texts

edited and introduced by
A.S. Weber

Broadview Press

Canadian Cataloguing in Publication Data

Main entry under title:

Nineteenth century science

Includes index.
ISBN 1-55111-165-9

1. Science – History – 19th century – Sources. I. Weber, A.S. (Alan S.), 1963– .

Q158.N56 1999 S09'.034 C99-930706-1

Broadview Press Ltd., is an independent, international publishing house, incorporated in 1985

North America
Post Office Box 1243, Peterborough, Ontario, Canada K9J 7H5
3576 California Road, Orchard Park, NY 14127
Tel: (705) 743-8990; Fax: (705) 743-8353;
e-mail: customerservice@broadviewpress.com

United Kingdom:
Turpin Distribution Services, Ltd., Blackhorse Rd., Letchworth, Hertfordshire SG6 1HN
Tel: (1462) 672555; Fax: (1462) 480947; e-mail: turpin@rsc.org

Australia:
St. Clair Press, P.O. Box 287, Rozelle, NSW 2039
Tel: (02) 818-1942; Fax: (02) 418-1923

www.broadviewpress.com

Broadview Press gratefully acknowledges the financial support of the Book Publishing Industry Development Program, Ministry of Canadian Heritage, Government of Canada.

Typesetting and assembly: True to Type Inc., Mississauga, Canada.

PRINTED IN CANADA

Contents

List of Illustrations

List of Illustrations

ACKNOWLEDGEMENTS

I would like to acknowledge the assistance of the libraries of The Pennsylvania State University, Cornell University, Yale University, The University of Pittsburgh and the State University of New York, Binghamton, and the British Library, Library of Congress, and National Library of Medicine for obtaining rare materials. This book could not have appeared without the support and advice of Frederick Amrine, Isobel Armstrong, Joseph Bizup, Peter Bowler, K.K. Collins, Brian P. Dolan, Andrew Draudt, Thomas Drucker, Priscilla Finley, Barbara Gates, Jan Golinski, Thomas W. Goodhue, Pamela Gossin, Ann-Barbara Graff, Sandra Herbert, M.J.S. Hodge, Mark Horan, George Levine, A. Edward Manier, Lawrence McCauley, Ken McKay, Peter Morgan, David Musson, M.D., Marcia Nelson, Richard Olson, Stuart Peterfreund, Mary Pickering, Ed Potter, Ellen Rosenman, Britt Salvesen, David Shaw, Irene Stubb, and Michael Wolff. I regret that I was unable to incorporate all of their suggestions. Bernard Lightman kindly read the entire manuscript and offered valuable suggestions. I would additionally like to thank the President and editors at Broadview Press for their recommendations.

INTRODUCTION

This volume of excerpts from original nineteenth-century scientific texts provides a brief overview of scientific activity during what Alfred Russel Wallace called "The Wonderful Century." When looking backward at the close of the nineteenth century—when the discovery of X-rays and radium dominated the headlines of the popular press—it is difficult not to adopt the prevailing view of the positivists that European and American science had advanced considerably throughout the century and had accumulated increasingly complex and powerful bodies of knowledge. It was during this period that the term "scientist" was first adopted and that the sciences developed the professional and institutional structures familiar to us today.

This extraordinary scientific activity—if indeed the nineteenth century differed significantly from previous periods—may have been brought about by the following factors: cheaper and more widespread access to printed materials, the recognition of science as a profession, greater government support for scientific endeavour, a stronger relationship between industry and the sciences, and greater literacy rates which brought about increased public knowledge and support of science in Europe and America. Although it was generally accepted in the nineteenth century that advances in scientific knowledge had ushered in tremendous technological and economic development, an alternate hypothesis should be considered: the view that the sciences arose instead as a *consequence* of the great increase in wealth from Western colonization and industrialization which supplied Europe with the leisure and resources to devote to scientific speculation and praxis. In some respects this represents a chicken-and-egg dilemma, a search for ultimate origins which most nineteenth-century researchers themselves refused to delve into, but the predominant flow of cultural energy may have been from socio-economic sources towards scientific spheres of activity.

Any anthology represents an interpretation at some level, and although the choice of texts has been guided by the desire to present a broad range of concepts and discoveries which in their own day were recognized as scientifically significant, the selection of works will necessarily remain idiosyncratic. Science in the nineteenth century was certainly not a monolithic structure; and as the selections below illustrate, there was no agreed upon methodology, or even consensus on what should be counted as science. The editor, for example, is conscious of the many omissions (which another editor may have included) that a one-volume anthology must entail, and it is regrettable that figures such as Hertz, Buckland, the Herschels, Davy, Mill, Cannizzaro, Riemann, etc. could not be included for reasons of space. It is hoped that students of the nineteenth-century sciences will investigate these authors on their own and thereby discover how limited a single-volume treatment of the topic must be. It is also hoped that readers will take more than an antiquary interest in the nineteenth-century sciences—many of the questions originally posed by these sciences have still not been adequately answered, and some of the most revolutionary scientific developments of the twentieth century originated from researchers who knew their immediate precursors well. Furthermore, the selections presented here have not lost their value as models of rhetorical persuasion, precise observation, and clear exposition of the perennial problems of natural philosophy.

In most cases, complete chapters of first editions of original texts have been used for the excerpts, unless a later revised edition holds some historical interest. Annotation of the entries has been kept to a minimum, and if readers encounter questions about individual writers, scientific concepts, or specific terminology, they should consult standard reference works such as *The Dictionary of Scientific Biography*, *The Dictionary of National Biography*, *The Encyclopedia of Philosophy*, and *The Dictionary of the History of Ideas*, as well as the many dictionaries of scientific terms published in both the nineteenth and twentieth centuries. Scientific articles from nineteenth-century general encyclopedias may also be profitably consulted. *The Companion to the History of Modern Science*, *The Dictionary of the History of Science*, *Information Sources in the History of Science*, *The History of Science in Western Civilization* (volume 3), and *Science in the Nineteenth Century* (volume 3 of *The History of Science*, edited by René Taton) are also recommended. The history of science journal *Isis* publishes an annual bibliography with a special section devoted to books and articles published on the nineteenth century. Most of the books and articles used for this anthology have been reprinted in facsimile or modern editions which are listed in the "further reading" sections following the entries. In many instances, original editions or nineteenth-century reprints of the selected texts may still be consulted in the open stacks of many public and research libraries.

1. BENJAMIN BANNEKER

(1731–1806)

Benjamin Banneker was descended from an English indentured servant named Molly Welsh and her manumitted slave Bannka (or possibly Banneky), who claimed to be the son of an African chieftain. Banneker's parents were free residents of Baltimore County, Maryland, and Benjamin was born a free man into a culture where African-American servitude was the norm. He received some formal schooling and advanced rapidly in mathematics. Banneker became well known in his community for designing and building a wooden clock, calculating and carving the gears himself. His interest in time and astronomy was further stimulated by his friendship with George Ellicott of Ellicott's Lower Mills, who lent him instruments and introductory textbooks on astronomical calculation, such as James Ferguson's *An Easy Introduction to Astronomy* and Charles Leadbetter's *A Compleat System of Astronomy*. Banneker also assisted the Ellicotts in surveying and planning the new District of Columbia.

Banneker grew so proficient in his mathematical and astronomical studies that he was able to calculate an ephemeris for 1791. Banneker's ephemerides subsequently appeared in a series of popular almanacs printed from 1792 to 1797 bearing his name and portrait. An ephemeris consists of tables listing eclipses, the movable feasts, settings and risings of the sun and moon, etc. and was primarily used by American colonists as a calendar and guide to the weather (although completely inaccurate in this respect). Most almanacs also contained short stories, lists of fairs and major roads, meetings of the judicial courts, and medical advice excerpted from other works. Some of the medical receipts added by the printer to one of Banneker's almanacs are reproduced below. The cures are representative of European, not African, medicine. Note that bleeding (phlebotomy) is recommended—this refers to the originally Greek medical practice of purging excess humours (such as blood) from the body in order to restore physiological balance. Due to their use as a home medical guide and yearly newspaper, almanacs greatly influenced popular opinion on science, technology, and medicine.

Also reprinted below is Banneker's letter to then Secretary of State Thomas Jefferson, taking him to task for his views on slavery. The letter and Jefferson's reply appeared in Banneker's 1793 almanac and were also reprinted separately by the firm of Daniel Lawrence in 1792. Banneker had included a manuscript copy of his first almanac in his letter to Jefferson, who immediately forwarded it to Condorcet in France. Banneker's letter was frequently reprinted in the nineteenth century and he became a symbol of the scientific potential of African-Americans who were able to fully exercise their gifts. Banneker's work inspired other nineteenth-century African-American inventors and scientists, such as Norbert Rillieux (1806–94), who contributed to vacuum evaporation technology; and the agricultural chemist George Washington Carver (1864–1943) of the Tuskegee Institute, best known for his championing of the peanut as a foodstuff and raw material for industrial products.

Banneker's last surviving almanac appeared in 1797, although he continued to calculate complete ephemerides until 1802. In 1854, the Alexandrian Academy of Philadelphia, a young man's improvement society, was renamed the Banneker Institute. Silvio Bedini's biography of Banneker is the most extensive and reliable to date. Much of our biographical knowledge of Banneker derives from the memoirs of Martha Tyson, the daughter of George Ellicott. Banneker's surviving almanacs are available in the Readex Microprint series of Early American Imprints.

Benjamin Banneker's Pennsylvania, Delaware, Maryland and Virginia Almanack and Ephemeris, For the Year of Our Lord 1793.

Baltimore: Goddard and Angell, 1792, pp. 2–6.

Baltimore County, (Maryland) near Ellicott's Mills, August 19, 1791.

To THOMAS JEFFERSON, Esq; Secretary of State.

SIR,

I am fully sensible of the greatness of that freedom which I take with you on the present occasion: a liberty which seemed to me scarcely allowable, when I reflected on that distinguished and dignified station in which you stand, and the almost general prejudice and prepossession which is so prevalent in the world against those of my complexion.

I suppose it is a truth too well attested to you, to need a proof here, that we are a race of beings who have laboured under the abuse and censure of the world; that we have long been considered rather as brutish than human, and scarcely capable of mental endowments.

Sir, I hope I may safely admit, in consequence of that report which hath reached me, that you are a man far less inflexible in sentiments of this nature, than many others; that you are measurably friendly and well-disposed towards us, and that you are willing and ready to lend your aid and assistance to our relief, from those many distresses and numerous calamities to which we are reduced.

Now, Sir, if this is founded in truth, I apprehend you will readily embrace every opportunity, to eradicate that train of absurd and false ideas and opinions, which so generally prevails with respect to us; and that your sentiments are concurrent with mine, which are, that one universal Father hath given being to us all, and that he hath not only made us all of one flesh, but that he hath also, without partiality, afforded us all the same sensations, and endued us all with the same faculties, and that however variable we may be in society or religion, however diversified in situation or colour, we are all of the same family, and stand in the same relation to him.

Sir, if these are sentiments of which you are fully persuaded, I hope you cannot but acknowledge, that it is the indispensable duty of those who maintain for themselves the rights of human nature, and profess the obligations of Christianity, to extend their power and influence to the relief of every part of the human race, from whatever burthen or oppression they may unjustly labour under, and to this I apprehend a full conviction of the truth and obligation of these principles should lead all.

Sir, I have long been convinced that if your love for yourselves, and for these inestimable laws which preserve to you the rights of human nature, was founded on sincerity, you could not but be solicitous, that every individual, of whatever rank or distinction, might with you enjoy the blessings thereof; neither could you rest satisfied, short of the most active diffusion of your exertions, in order to their promotion from any state of degradation, to which the unjustifiable cruelty and barbarism of men may have reduced them.

Sir, I freely and cheerfully acknowledge, that I am of the African race, and in that colour which is natural to them of the deepest dye, and it is under a sense of the most profound gratitude to the Supreme Ruler of the Universe that I confess to you, that I am not under that state of tyrannical thraldom, and inhuman captivity, to which too many of my brethren are doomed; but that I have abundantly tasted of the fruition of those blessings, which proceed from that free and unequalled liberty with which you are favoured, and which I hope you will willingly allow you have received from the immediate hand of that Being, from whom proceedeth every good and perfect gift.

Sir, suffer me to recall to your mind that time, in which the arms and tyranny of the British crown were exerted with every powerful effort, in order to reduce you to a state of servitude; look back, I entreat you, on the variety of dangers to which you were exposed; reflect on that time in which every human aid appeared unavailable, and in which even hope and fortitude wore the aspect of inability to the conflict, and you cannot but be led to a serious and grateful sense of your miraculous and providential preservation; you cannot but acknowledge, that the present freedom and tranquillity which you enjoy, you have mercifully received, and that it is the peculiar blessing of Heaven.

This, Sir, was a time in which you clearly saw into the injustice of a state of Slavery, and in which you had just apprehensions of the horrors of its condition; it was now, Sir, that your abhorrence thereof was so excited, that you publicly held forth this true and invaluable doctrine, which is worthy to be recorded and remembered in all succeeding ages, "We hold these truths to be self-evident, that all men are created equal, and they are endowed by their Creator with certain unalienable rights; that among these are life, liberty, and the pursuit of happiness."

Here, Sir, was a time in which your tender feelings for yourselves had engaged you thus to declare; you were then impressed with proper ideas of the great valuation of liberty, and the free possession of those blessings to which you were intitled by Nature; but pitiable is it to reflect, that although you were so fully convinced of the benevolence of the Father of Mankind, and of his equal and impartial distribution of those rights and privileges which he had conferred upon them, that you should, at the same time, counteract his mercies, in detaining, by fraud and violence, so numerous a part of my brethren under groaning captivity and cruel oppression; that you should, at the same time, be found guilty of that most criminal act which you professedly detested in others, with respect to yourselves.

Sir, I suppose that your knowledge of the situation of my brethren is too extensive to need a recital here; neither shall I presume to prescribe methods by which they may be relieved, otherwise than by recommending to you and all others, to wean yourselves from those narrow prejudices which you have imbibed with respect to them; and, as Job proposed to his friends, "put your souls in their souls' stead," thus shall your hearts be enlarged with kindness toward them, and thus shall you need neither the direction of myself or others in what manner to proceed herein.

And now, although my sympathy and affection for my brethren hath caused my enlargement thus far, I ardently hope that your candour and generosity will plead with you in my behalf, when I make known to you, that it was not originally my design; but having taken up my pen in order to direct to you, as a present, a copy of an Almanack, which I have calculated for the succeeding year, I was unexpectedly and unavoidably led thereto.

This calculation is the production of my arduous study in this my advanced stage of life, for having long had unbounded desires to become acquainted with the secrets of Nature, I had to gratify my curiosity herein, through my own assiduous application to astronomical study, in which I need not to recount to you the many difficulties and disadvantages I have had to encounter.

And although I had almost declined to make my calculation for the ensuing year, in consequence of that time which I had alloted therefor being taken up at the federal territory, by the request of Mr. Andrew Ellicott, yet finding myself under several engagements to Printers of this state, to whom I had communicated my design, on my return to my place of residence, I industriously applied myself thereto, which I hope I have accomplished with correctness and accuracy, a copy of which I have taken the liberty to direct to you, and which I humbly request you will favourably receive, and although you may have the opportunity of perusing it after its publication, yet I chose to send it to you in manuscript previous thereto, that thereby you might not only have an earlier inspection, but that you might view it in my own hand-writing.

Now, Sir, I shall conclude, and subscribe myself,
With the most profound respect,
Your most obedient humble Servant,
BENJAMIN BANNEKER

[Jefferson's Reply]

Philadelphia, August 23, 1791.

SIR,
I thank you sincerely for your letter of the 19th instant, and for the Almanack it contained—Nobody wishes more than I do to see such proofs as you exhibit, that Nature has given our black brethren talents equal to those of the other colour of men, and that the appearance of a want of them is owing merely to the degraded condition of their existence, both in Africa and America. I can add, with truth, that nobody wishes more ardently to see a good system commenced for raising the condition both of their body and mind to what it ought to be, as fast as the imbecility of their present existence, and other circumstances, which cannot be neglected, will admit. I have taken the liberty of sending your Almanack to Monsieur De Condorcet, Secretary of the Academy of Sciences at Paris, and Member of the Philanthropic Society, because I considered it as a document to which your whole colour had a right, for their justification against the doubts which have been entertained of them.

I am, with great esteem, Sir,
Your most obedient humble Servant,
THOMAS JEFFERSON.

Benjamin Banneker,
near Ellicott's lower-mills, Baltimore-County

Banneker's New-Jersey, Pennsylvania, Delaware, Maryland, and Virginia Almanac, or Ephemeris, For the Year of Our Lord 1795.

Wilmington: S. and J. Adams, 1795, p. 25.

USEFUL RECEIPTS

Cure for the Bloody Flux.
Take a bottle cork, wrap it up in a bit of paper, and put it in the embers, and let it roast; then crumble it in half a pint of good spirits, and set it on fire, and reduce it to a gill; drink it at a draught, and if necessary repeat it the next day.

For the Cholic. (from Dr. Quincey)
Take two ounces of Daffy's elixir, and repeat it as occasion may require; or half a drachm of powder of rhubarb, toasted a little before the fire.

For the Quinsey, or Sore Throat.
(from *Philosophical Transactions*)
Bleeding is sometimes necessary, and cooling physic; but often jelly of black currants, swallowed down leisurely, in small quantities, effects a cure, without the assistance of any other medicine.

A Cure for the Ague and Fever.
Take one table spoonful of flower of brimstone, mixed in wine, just before the fit comes on. This has effectually cured many.

To kill Worms in Children.
Take six, eight, or ten red earth-worms, and let them purge in bay salt, then split them open and

wash them in fair water; then dry them in an earthen dish, and beat them to powder, and give them to the child in the morning fasting, for three or four mornings, but let them eat nothing for an hour after. Or, let the child eat some raisons every morning fasting, without any other food for an hour.

FURTHER READING

Allen, William G. *Wheatley, Banneker, and Horton.* Boston: Daniel Laing, Jr., 1849.

Allen, Will W. and Daniel Murray. *Banneker, The Afro-American Astronomer.* 1921. Reprint, Freeport, NY: Books for Libraries Press, 1971.

Bedini, Silvio A. *The Life of Benjamin Banneker.* New York: Charles Scribner's Sons, 1972.

A Chronology of the Life of Benjamin Banneker 1731–1806. Baltimore: African-American Bicentennial Center, 1976.

Graham, Shirley. *Your Most Humble Servant.* New York: Julian Messner, 1949.

Haber, Louis. *Black Pioneers of Science and Invention.* New York: Harcourt, Brace and World, Inc., 1970.

Latrobe, John H.B. *Memoir of Benjamin Banneker.* Publications of the Maryland Historical Society, no. 3. Baltimore: John D. Toy, 1845.

Patterson, Lillie. *Benjamin Banneker: Genius of Early America.* Nashville: Abingdon, 1978.

Tyson, Martha E. *Banneker, the Afric-American Astronomer. From the Posthumous Papers of Martha E. Tyson, Edited by Her Daughter.* Philadelphia: Friends' Book Association, 1884.

2. XAVIER BICHAT

(1771–1802)

Mais toi, Bichat, par tes paroles
Tu nous guéris et nous consoles....[1]
E.B. (Coquerelle 177)

During his short but productive life, Xavier Bichat made important contributions to surgery, the classification of tissues, and to the vitalism debate in physiology. After studying with Marc-Antoine Petit in Lyons, he worked with Pierre Joseph Desault in Paris and later edited Desault's *Journal de chirurgie* after Desault's death in 1795. He also edited the *Oeuvres chirurgicales de Desault* in honour of his friend and teacher. Bichat served briefly as a medical surgeon in the Alps with the Army of the Republic in 1793.

Bichat, drawing on Pinel's *Nosographie philosophique* and his own work on the synovial membranes, suggested in his *Traité des membranes* (1800) and later in the *Anatomie générale* (1801) that organs and their diseases should be classified according to tissue structure, an important step in the development of modern histology. Bichat's elaboration of Pinel's observations won widespread and immediate support, as Coquerelle observes, "the *Traité des membranes* enjoyed the greatest success. As soon as it appeared (January and February of 1800), it was regarded as a basic and classic text. It was cited in a host of other works, and almost all thinking men placed it with honor in their libraries" (67). Bichat developed a view of the body as a textured web composed of over 21 simple, irreducible tissues which formed the 'elements' of organs and larger structures. For Bichat, each tissue possessed its own vital properties, leading some to refer to him as a 'plurivitaliste' (Dobo and Role 269) to distinguish his views from the more generalized vitalism of his predecessors. In Bichat's system, therapeutics consisted simply of restoring the natural mode of operation of the vital properties of these tissues. Bichat did not want to imply, however, that the body could be simplistically reduced to an anatomic summation of its tissues; as Michel Foucault has observed, Bichat's tissues are "the elements of organs, but they traverse the organs, connect them into relations, and above them constitute vast systems where the human body finds the concrete forms of its unity" (129).

Unlike some physiologists of the time, Bichat was not hostile to the moral or metaphysical sciences as a means of understanding the body. In his lecture "Discours sur l'étude de la phisiologie," he spoke of the "physiologist-metaphysician" who "knows not only the senses, but also understands that their ability to receive impressions varies; he also understands the laws of sensory impression and he unites this knowledge with metaphysics, which is enriched by this knowledge" (Arène, "Appendix" vii).

In 1799 or 1800 (an VIII), Bichat published *Les Recherches physiologiques sur la vie et la mort*, excerpted below. The *Recherches*, in its basic division of life into an animal life (governed by

---

1 You, Bichat, heal and console us with your words.

animate or vital principles) and organic life (governed by physical laws), instantly sparked opposition, such as Jean Philippe Bardenat's *Les Recherches physiologiques de Xavier Bichat refutées dans leurs doctrines* (1824). Bichat's use of the concept "vie animale" recalls the original Latin root anima or soul, the governor of movement, growth, nutrition and reason in the body in classical thought. Bichat's division is not new, and closely parallels the Platonic and later Christian division of body and soul, and the animism of Paracelsus, van Helmont, Georg Stahl and the Montpellier school of medicine. Bichat's vital principles were the agents of life, in constant conflict with external physical forces, the agents of death. Bichat expressed this view in his famous opening lines "la vie est l'ensemble des fonctions qui résistent à la mort" (life is the sum of the forces which resist death). The richness of eighteenth- and nineteenth-century vitalist philosophies, recently surveyed in the collection of essays *Vitalisms from Haller to the Cell Theory*, has been obscured by nineteenth-century opponents and historians who simplistically reduced vitalist theories to the realm of the occult, the anti-mechanical, and the metaphysical—in other words, outside of the domain of physiology proper.

The selection below contains the notes of François Magendie, who edited Bichat's work, later translated into English by F. Gold. An earlier American translation by Tobias Watkins had appeared in 1809 and Bichat's works were frequently reprinted in the United States. Magendie's lengthy discussions are often critical of Bichat, and one wonders why Magendie chose to edit the volume, unless to refute its vitalism, or perhaps because Bichat had uncovered a nagging and unsolved question of physiology: what ultimately distinguishes the living from the dead body? Claude Bernard, who also seemed haunted by Bichat's work, which he knew well, rejected out of hand the validity of vitalism as a physiological question, arguing "it is enough to agree on the word life to employ it; but above all it is necessary for us to know that it is illusory and chimerical and contrary to the very spirit of science to seek an absolute definition of it" (Bernard, *Lectures* 19). Bernard's "First Lecture" in his *Leçons sur les phénomènes de la vie communs aux animaux et aux végétaux* (1878) provides an excellent outline and history of the doctrines of vitalism up to his time. The question 'what is life' is still with us, especially when we consider organized and self-organizing structures such as crystals, viruses, and the concept of artificial intelligence.

Bichat succumbed to an unknown illness in 1802 after preparing infectious laboratory specimens. Napoléon personally ordered the Minister of the Interior to erect a monument to Desault and Bichat at the Hôtel-Dieu.

Physiological Researches on Life and Death.
Translated by F. Gold. Boston: Richardson and Lord, 1827, pp. 9–24.

Chapter 1.

General Division of Life.[2]

The definition of life is usually sought for in abstract considerations; it will be found, if I mistake not, in the following general expression:—Life consists in the sum of the functions, by which death is resisted.[3]

In living bodies, such in fact is the mode of existence, that whatever surrounds them, tends to their destruction. They are influenced incessantly by inorganic bodies; they exercise themselves, the one upon the other, as constant an action; under such circumstances they could not long subsist, were they not possessed in themselves of a permanent principle of reaction. This principle is that of life; unknown in its nature, it can be only appreciated by its phenomena: an habitual alternation of action and reaction between exterior bodies, and the living body, an alternation, of which the proportions vary according to the age of the latter, is the most general of these phenomena.

There is a superabundance of life in the child: In the child, the reaction of the system is superior to the action, which is made upon it from without. In the adult, action and reaction are on a balance; the turgescence of life is gone. In the old man, the reaction of the inward principle is lessened, the action from without remaining unaltered; it is then that life languishes, and insensibly advances towards its natural term, which ensues when all proportion ceases.

The measure, then, of life in general, is the difference which exists between the effort of exterior power, and that of interior resistance. The

2 The form adopted by Bichat, in this work, has been much blamed by some, and extravagantly praised by others. The blame and the praise appear to me to be equally misplaced. His object was to exhibit the various phenomena of life; the order in which this was to be done was a matter of indifference. If Bichat gave a preference to this form it was because it was conformable to the nature of his mind; and he accomplished his task in a very happy manner. The division that he has adopted is not new, it may be found, with slight modifications in writers of different periods, and even in Aristotle. Besides, it is not necessary in the sciences to attach a very great importance to classification. All these contrivances have been invented only to aid the memory; and the functions of living bodies are not so numerous, as to render it necessary in studying them to lean upon systematic divisions.

3 The word *life* has been employed by physiologists in two different senses. With some, it means an imaginary being, the sole principle of all the functions which living bodies exhibit; with others, it means only the assemblage of these functions. It is in this last sense that Bichat employs it. This is what he means to say in the following sentence. *Life is the assemblage of the functions which resist death.* He is wrong only in allowing the idea of death to enter into it; for this idea necessarily supposes that of life. There is then really a bad circle in this definition; but in putting aside what is defective in the expression, it may be seen that Bichat considers life as a result, not as a cause.

Before and since the time of Bichat, a great number of definitions of life has been given, which are either false or incomplete. It should not be required of a definition, that it should give all the properties of the thing which it is designed to make known, this would be a description; but we have a right to expect that it should assign to this thing certain characters which belong to it alone, and thus distinguish it from every thing else.

Let us examine by this principle the definition adopted in a modern work. *Life,* it is said, *is the assemblage of the phenomena which succeed each other, for a limited time, in an organized being.* This is no doubt true of life; but, if it can also be applied to another state, it ceases to be a definition. An animal has just died; its organs from that moment are subject to the action of chemical affinities only; decomposition takes place, gases are disengaged, fluids flow out and new solid aggregates are formed. After a time every molecular motion ceases; there remains only a certain number of binary, ternary combinations, etc. Here then is an *assemblage of phenomena taking place for a limited time in an organized body,* and yet it is not life.

excess of the former is an indication of its weakness; the predominance of the latter an index of its force.

I. *Division of Life into Animal and Organic Life.*[4]

Such is life considered in the aggregate; examined more in detail it offers us two remarkable modifications, the one common to the vegetable and the animal; the other belonging exclusively to the latter. In comparing two individuals from each of the living kingdoms, the one will be seen existing only within itself, having with what surrounds it the relations only of nutrition, attached to the soil, in which its seed has been implanted, born there, growing there, and perishing there. The other will be observed combining with this interior life, which in the highest degree it enjoys, an exterior life by which it acquires a very numerous series of relations with all surrounding bodies, a life, which couples it to the existence of every other being, by which it is approximated, or removed from the objects of its desires or its fears, and seems in appropriating every thing in nature to itself, to consider every thing with regard to its individual existence only.[5]

Thus it might be said, that the vegetable is only

4 This distinction of the two lives is bad, inasmuch as it tends to separate phenomena which have a very intimate connexion, which relate to a common object, and which are often produced by means in every respect similar. Why should I rank among the organs of animal life the muscular apparatus which carries the alimentary mass from the mouth into the oesophagus, and among those of the other life, that which takes it from the cardiac orifice to the anus? Is not the action of the first apparatus in relation with nutrition as well as the action of the last, and does not the muscular apparatus of the oesophagus act upon a body which is foreign to us, as well as that of the tongue and the pharynx? Do the motions of mastication differ in their object from those of which we have just spoken, and as to the means of execution, does not the muscular action still perform the principal part?

We might in the same way bring near each other the motions by means of which we seize our food. The action itself of the senses, which directs these motions, is, with nutrition, in a relation more remote, but not less necessary, and we see in the various classes of animals that their apparatus is modified according to the different kinds of nourishment. If the distinction of the two lives be wanting in justice, as to the object of the functions it separates, we shall soon see that the characters attached to the organs of one and the other do not establish this division in a more striking manner.

5 This division between vegetables and animals is far from being so striking as is here supposed; these two classes of beings, so different when we examine them in the individuals endowed with a very complicated organization, approximate each other in a remarkable degree, when we descend to those species whose structure is most simple; it is even remarkable that the most constant character which distinguishes one from the other, is not found in the organs of animal life, but in those of vegetable or organic life. The senses are one after the other found wanting; for in an individual in whom we can discover no nervous system, there is no more reason to suppose the existence of the sense of touch as a sensation, than to suppose it in the sensitive plant, the dionaea muscipula, and other similar plants; we see only action and reaction. The motions of the arms of certain polypi no more suppose volition than the motion of the root which follows a wet sponge, or that of the branches which turn towards the light; the only very constant character is the absence or presence of a digestive cavity. To speak of an animal as a vegetable clothed with an external apparatus of organs of relation, is a more brilliant than profound view of the subject. Buisson, who, in his division of the physiological phenomena, avoids this inaccuracy, has himself fallen into error; he pretends that respiration belongs exclusively to animals; and that thus the division of Bichat was not only unfounded but also incomplete, since this function, which is neither of vegetation nor of relation, could be ranked under neither life. Buisson was not well informed; no doubt the respiration of vegetables does not exhibit the most apparent phenomena of the respiration of the mammalia, but every thing, which essentially constitutes the function, is found in the one as well as in the other; absorption of the atmospheric air, and the formation and exhalation of a new gas; the rest is only accidental and is not an appendage but in certain classes of animals. In some reptiles, though we find a particular organ for respiration, this organ is not indispensable; it may be removed, and the skin becomes the only respiratory organ; and when finally we come to consider animals with *tracheae*, we see that the conformity becomes more and more evident.

the sketch, or rather the ground-work of the animal; that for the formation of the latter, it has only been requisite to clothe the former with an apparatus of external organs, by which it might be connected with external objects.

From hence it follows, that the functions of the animal are of two very different classes. By the one (which is composed of an habitual succession of assimilation and excretion) it lives within itself, transforms into its proper substance the particles of other bodies, and afterwards rejects them when they are become heterogeneous to its nature. By the other, it lives externally, is the inhabitant of the world, and not as the vegetable of a spot only; it feels, it perceives, it reflects on its sensations, it moves according to their influence, and frequently is enabled to communicate by its voice its desires, and its fears, its pleasures, and its pains.

The aggregate of the functions of the first order, I shall name the organic life, because all organized beings, whether animal or vegetable, enjoy it more or less, because organic texture is the sole condition necessary to its existence. The sum of the functions of the second class, because it is exclusively the property of the animal, I shall denominate the animal life.

The series of the phenomena of these two lives, relate to the individual. Generation, as a function, regards the species, and thus has no place among them. Its connections with the greater number of the other functions are but very indirect; it commences a long time after them, it is extinct a long time before them. In the greater number of animals the periods of its activity are separated by long intervals of time, and during these, it is absolutely null. Even in man, with whom the remissions of its impulses, are much less durable, it has not a much more extensive connexion with the rest of the system. Castration is almost always marked by a general increase of the nutritive process; the eunuch, enjoying indeed a less degree of vital energy, but the phenomena of his life being displayed with a greater exuberance. We shall here, then, lay aside the consideration of the laws which give us existence, and occupy ourselves alone on those which maintain us in existence. Of the former we shall speak hereafter.

II. *Subdivision of each of the two lives into two orders of functions.*

The animal and the organic life, are each of them composed of two orders of functions, which succeed each other, and are concatenated in an inverse direction.

In the animal life, the first order is established from the exterior of the body, towards the brain; the second from the brain towards the organs of locomotion and the voice. The impression of objects successively affects the senses, the nerves and the brain. The first receive, the second transmit, the third perceives the impression. The impression, in such way, received, transmitted, and perceived, constitutes sensation.

The animal, in the first order of these functions, is almost passive; in the second, he becomes active.—This second order is the result of the successive actions of the brain (where volition has been produced in consequence of the previous sensation) of the nerves, which transmit such volition, and of the locomotive organs and voice, which are the agents of volition. External bodies act upon the animal by means of the first order of these functions, the animal reacts upon them by means of the second.

In general there exists between the two orders a rigorous proportion; where the one is very marked, the other is put forth with energy. In the series of living beings, the animal, which feels the most, moves also the most. The age of lively perception, is that also of vivacity of motion; in sleep, where the first order is suspended, the second ceases, or is exercised only with irregularity. The blind man, who is but half alive to what surrounds him, moves also with a tardiness which would very soon be lost, were his exterior communications to be enlarged.

A double movement is also exercised in the organic life; the one composes, the other decomposes the animal. Such is the mode of existence in the living body, that what it was at one time it ceases to be at another. Its organization remains unaltered, but its elements vary every moment. The molecules of its nutrition by turns absorbed and rejected, from the animal pass to the plant, from the plant to inorganic matter, return to the animal, and so proceed in an endless revolution.

To such revolution the organic life is well adapted. One order of its functions assimilates to the animal the substances which are destined to nourish him; another order deprives him of these substances, when, after having for some time made a part of it, they are become heterogeneous to his organization.

The first, which is that of assimilation, results from the functions of digestion, circulation, respiration, and nutrition. Every particle, which is foreign to the body before it becomes an element of it, is subject to the influence of these four functions.

When it has afterwards concurred for some time to the formation of the organs, the absorbents seize on it, and throw it out into the circulatory torrent, where it is carried on anew, and from whence it issues by the pulmonary or cutaneous exhalations, or by the different secretions by which the fluids are ejected from the body.

The second order, then, of the functions of the organic life, or that of decomposition, is formed of those of absorption, circulation, exhalation, and secretion.

The sanguiferous system, in consequence, is a middle system, the centre of the organic life, as the brain is the centre of the animal life. In this system the particles, which are about to be assimilated, are circulated and intermixed with those, which having been already assimilated, are destined to be rejected; so that the blood itself is a fluid composed of two parts; the one, the pabulum of all the parts of the body, and derived from the aliment; the other, excrementitious, composed of the wrecks and residue of the organs, and the source of the exterior secretions and exhalations.—Nevertheless these latter functions serve also, at times, the purpose of transmitting without the body, the products of digestion, although such products may not have concurred to the nourishments of the parts. This circumstance may be observed when urine and sweat are secreted after copious drinking. The skin and the kidneys being at such times the excreting organs, not of the matter of the nutritive, but of that of the digestive process; the same also may be said of the milk of animals, for this is a fluid which certainly has never been assimilated.[6]

There does not exist between the two orders of the functions of the organic life the same relation, which takes place between those of the animal life. The weakness of the first by no means renders absolutely necessary a decrease of action in the second. Hence proceed marasmus and leanness, states, in which the assimilating process ceases in part, the process of excretion remaining unaltered.

Let us leave, then, to other sciences, all artificial method, but follow the concatenation of the phenomena of life, for connecting the ideas which we form of them, and we shall perceive, that the greater part of the present physiological divisions, afford us but uncertain bases for the support of any thing like a solid edifice of science.

These divisions I shall not recapitulate; the best method of demonstrating their inutility will be, if I mistake not, to prove the solidity of the division, which I have adopted. We shall now examine the great differences, which separate the animal existing without, from the animal existing within,

6 Bichat seems here to adopt the generally received opinion that it is the chyle which furnishes to the mammary gland the materials of which the milk is composed. We know not whence this opinion arises, if it be not from the gross resemblance which the chyle and milk often exhibit. This resemblance, if it were very great, would be a poor reason for admitting, without anatomical proof, so singular a fact; but it is very far from being perfect. The chyle in fact does not exhibit the milky appearance and the white opake colour, only when the animal from whom it is taken, has fed upon substances containing fat; in all other cases, it is almost transparent; its odour and taste, under all circumstances, differ entirely from those of milk; if these two fluids are left to themselves, the milk remains a long time without coagulating, but the chyle almost immediately coagulates, and then separates into three parts. The solid portion soon exhibits cells, and an appearance of organization; nothing similar is seen in the cogulum of milk; the serum of the milk remains colourless when exposed to the simple contact of the air, that of the chyle assumes a rosy tint, often very vivid. Finally, if we examine the chemical composition of these two fluids, we shall find in them differences still more striking. (See for farther details, my Elements of Physiology, Vol. 2d.)

and wearing itself away in a continual vicissitude of assimilation and excretion.

Chapter 2.

General Differences of the Two Lives with Regard to the Outward Form of Their Respective Organs.

The organs of the animal life are symmetrical, those of the organic life irregular in their conformation; in this circumstance consists the most essential of their differences. Such character, however, to some animals, and among the fish, to the sole and turbot especially, is not applicable; but in man it is exactly traced, as well as in all the genera which are nearest to him in perfection. In them alone am I about to examine it.

I. *Symmetry of the external forms of the animal life.*[7]

Two globes in every respect the same, receive the impressions of light. Sounds and odours, have also their double analogous organ. A single membrane is affected to savours, but the median line is manifest upon it, and the two segments, which are indicated by it, are exactly similar. This line indeed is not every where to be seen in the skin, but it is every where implied. Nature, as it were, has forgotten to describe it, but from space to space she has laid down a number of points, which mark its passage. The cleft at the extremity of the nose, of the chin, and the middle of the lips, the umbilicus, the seam of the perineum, the projection of the spinous apophyses of the back, and the hollow at the posterior part of the neck are the principal points at which it is shewn.

The Nerves, which transmit the impressions received by the senses, are evidently assembled in symmetrical pairs.

The brain, the organ (on which the impressions of objects are received) is remarkable also for the regularity of its form. Its double parts are exactly alike, and even those which are single, are all of them symmetrically divided by the median line.

The Nerves again, which transmit to the agents of loco-motion and of the voice, the volitions of the brain, the locomotive organs also, which are formed in a great degree of the muscular system, of the bony system, and its dependencies, these together with the larynx and its accessaries, composing the double agents of volition, have all of them a regularity, a symmetry, which are invariable.

Such even is the truth of the character which I am now describing, that the muscles and the nerves immediately cease to be regular, as soon as they cease to appertain to the animal life. The heart, and the muscular fibres of the intestines are proofs of this assertion in the muscles; in the nerves, the great sympathetic, is an evidence of its truth.

We may conclude then from simple inspection, that Symmetry is the essential character of the organs of the animal life of man.

II. *Irregularity of the exterior forms of the organic life.*

If at present we pass to the viscera of the organic life, we shall perceive a character directly the contrary of the former. The stomach, the intestines, the spleen, the liver, etc. are all of them irregularly disposed.

---

7 It is rather to the external forms that symmetry appears to have been primitively attached, and it is in some measure accidentally and because the nature of their functions requires in general that they should be placed on the exterior, that the organs of relation are found modified in virtue of this law. In the example cited, of fishes without a bladder, the eyes, to lose nothing of their utility, must be differently placed, and on the face, which alone is in relation with the light; yet even in this case, the symmetry of external forms has been displaced rather than destroyed, and at the first examination it seems complete. When the organs of relation are found placed on the interior, they frequently exhibit some irregularity, and to take an example of a known animal, the organ of voice, in the male duck, is a very remarkable one; in man even, the wind-pipe is not symmetrical, after it arrives at the first division of the bronchia. On the contrary, among the organs of the other life, those which are prominent on the exterior, constantly present the symmetrical character, as the thyroid gland, the mammary glands, etc.

In the system of the circulation, the heart and the large vessels, such as the upper divisions of the aorta, the vena azygos, the vena portae, and the arteria innominata have no one trace of symmetry. In the vessels of the extremities continual varieties are also observed, and when they occur, it is particularly remarkable that their existence on one side in no way affects the other side of the body.

The apparatus of respiration appears indeed at first to be exactly regular; nevertheless, the bronchi are dissimilar in length, diameter, and direction; three lobes compose one of the lungs, two the other: between these organs also, there is a manifest difference of volume; the two divisions of the pulmonary artery resemble each other neither in their course, nor in their diameter; and the mediastinum is sensibly directed to the left. We shall thus perceive that symmetry is here apparent only, and that the common law has no exception.

The organs of exhalation and absorption, the serous membranes, the thoracic duct, the great right lymphatic vessel, and the secondary absorbents of all the parts have a distribution universally un-equal and irregular.

In the glandular system also we see the crypts, or mucous follicles disseminated in a disorderly manner in every part; the pancreas, the liver, the salivary glands themselves, though at first sight more symmetrical, are not exactly submitted to the median line; added to this, the kidneys differ from each other in their situation, in the length and size of their artery and vein, and in their frequent varieties more especially.[8]

From considerations so numerous we are led to a result exactly the reverse of the preceding one; namely, that the especial attribute of the organs of the interior life is irregularity of exterior form.

III. *Consequences resulting from the difference of exterior form in the organs of the two lives.*

It follows from the preceding description, that the animal life is as it were double; that its phenomena performed as they are at the same time on the two sides of the body, compose a system in each of them independent of the opposite system; that there is a life to the right, a life to the left; that the one may exist, the other ceasing to do so, and that they are doubtless intended reciprocally to supply the place of each other.

The latter circumstance we may frequently observe in those morbid affections so common, where the animal sensibility and mobility are enfeebled, or annihilated on one side of the body, and capable of no affection whatever; where the man on one side is little more than the vegetable, while on the other he preserves his claim to the animal character. Undoubtedly those partial palsies, in which the median line, is the limit where the faculties of sensation and motion finish, and the origin from whence they begin can never be remarked so invariably in animals, which, like the oyster, have an irregular exterior.

On the contrary the organic life is a single system, in which every thing is connected and concatenated; where the functions on one side cannot be interrupted, and those on the other subsist. A diseased liver influences the state of the stomach; if the colon on one side cease to act, that upon the other side cannot continue in action: the same attack, which arrests the circulation in the right side of the heart, will annihilate it also in the left side of the heart. Hence it follows, the internal organs on one side being supposed to suspend their functions, that those on the other must remain inactive, and death ensue.

This assertion, however, is a general one; it is

8 If we deny symmetry to the kidneys, because they are not uniformly composed of the same number of lobes in children, we must deny it also to the brain, the two lobes of which never exhibit the same arrangement in their circumvolutions; if we deny it to the salivary glands, because one is larger than the other, we must deny it to the extremities, because the right is usually more developed than the left. If these examples are not enough, a host of others might be cited; such as, the atrabiliary capsules, the bladder, the different organs of generation and lactation, and the very regular arrangement of the mucous follicles in certain parts situated upon the median line, etc. As to the anomalies that are observed in the distribution of the blood-vessels, they are also observed very frequently, though in a less evident manner, in the distribution of the nervous branches.

only applicable to the sum of the organic life, and not to its isolated phenomena. Some of them in fact are double, and their place may be supplied—the kidneys and lungs are of this description.

I shall not enquire into the cause of this remarkable difference, which in man, and those animals which approach him the nearest, distinguishes the organs of the two lives. I shall only observe, that it enters essentially into the nature of their phenomena, and that the perfection of the animal functions is so connected with the general symmetry observed in their respective organs, that every thing which troubles such symmetry, will more or less impair the functions.

It is from thence, no doubt, that proceeds this other difference of the two lives, namely, that nature very rarely varies the usual conformation of the organs of the animal life. Grimaud has made this observation, but has not shewn the principle on which it depends.

It is a fact, which cannot have escaped any one the least accustomed to dissection, that the spleen, the liver, the stomach, the kidneys, the salivary glands, and others of the internal life, are frequently various in form, size, position, and direction. Such in the vascular system are these varieties, that scarcely will any two subjects be found exactly alike under the scalpel of the anatomist: the organs of absorption, the lymphatic glands in particular, are rarely the same either in number or volume, neither do the mucous glands in any way affect a fixed and analogous situation.

And not only is each particular system subject to frequent aberrations, but the whole of the organs of the internal life are sometimes found in the inverse of the natural order. Of this I have lately seen an instance.

Let us now consider the organs of the animal life, the senses, the brain, the voluntary muscles, and the larynx: here every thing is exact, precise, and rigourously determined. In these there is scarcely ever seen a variety of conformation; if there do exist any, the functions are troubled, disturbed, or destroyed: they remain unaltered in the organic life, whatever may be the disposition of the parts.

The difference with respect to action, in the organs of the two lives, depends, undoubtedly, upon the symmetry of the one, whose functions the least change of conformation would have disturbed, and on the irregularity of the other, with which these different changes very well agree.

The functions of every organ of the animal life are immediately connected with the resemblance of the organ to its fellow on the opposite side if double, or if single to its similarity of conformation in its two halves: from hence the influence of organic changes upon the derangement of the functions may be well conceived.

But this assertion will become more sensible, when I shall have pointed out the relations which exist between the symmetry and the irregularity of the organs, and the harmony and the discordance of their functions.

FURTHER READING

Arène, A. *Essai sur la philosophie de Xavier Bichat.* Lyon: A. Rey, 1911.

Benton, E. "Vitalism in Nineteenth-Century Thought: A Typology and Reassessment." *Studies in History and Philosophy of Science* 5 (1974): 17–48.

Bernard, Claude. *Lectures on the Phenomena of Life Common to Animals and Plants.* Translated by H. Hoff, R. Guillemin and L. Guillemin. Springfield, IL: Charles C. Thomas, 1974.

Bichat, Xavier. *Physiological Researches on Life and Death.* Translated by F. Gold. 1827. Reprint, Significant Contributions to the History of Psychology 1750–1920. Edited by Daniel N. Robinson. Series E. Washington, D.C.: University Publications of America, Inc., 1978.

—. *Recherches physiologiques sur la vie et la mort, précédée d'une notice sur la vie et les travaux de Bichat et suivie de notes par le Docteur Cerise.* 1800. Reprint, Paris: Charpentier, 1866.

—. *Recherches physiologiques sur la vie et la mort.* Edited by André Pichot. 1800. Reprint, Paris: Flammarion, 1994.

Blanchard, R. *Centenaire de la mort de Xavier Bichat.* Paris: Librairie Scientifique et Littéraire, 1903.

Cimino, Guido and François Duchesneau, eds. *Vitalisms from Haller to the Cell Theory.* Firenze: Leo S. Olschki, 1997.

Coquerelle, Jules. *Xavier Bichat: Sa Vie, ses travaux, son apothéose.* Paris: A. Maloine, 1902.

Dezeimeris, J.E. "Bichat." *Dictionnaire historique de la médecine, ancienne et moderne.* Paris: Béchet Jeune, 1828–39, pp. 385–96.

Dobo, Nicolas and André Role. *Bichat: La vie fulgurante d'un génie.* Paris: Perrin, 1989.

Driesch, Hans. *Geschichte des Vitalismus.* Leipzig: Johann Ambrosius Barth, 1922.

Entralgo, P.L. "Sensualism and Vitalism in Bichat's 'Anatomie Générale.'" *Journal of the History of Medicine and Allied Sciences* 3 (1948): 47–64.

Foucault, Michel. *Naissance de la clinique: Une archéologie du regard médical.* Paris: Presses Universitaires de France, 1963.

Haigh, Elizabeth. *Xavier Bichat and the Medical Theory of the Eighteenth Century.* Medical History Supplement no. 4. London: Wellcome Institute for the History of Medicine, 1984.

Laignel-Lavastine, M. "Sources, principes, sillage et critique de l'oeuvre de Bichat." *Bulletin de la Société Française de Philosophie* 46 (1953): 1–38.

Lesch, John E. *Science and Medicine in France: The Emergence of Experimental Physiology, 1790–1855.* Cambridge, MA: Harvard UP, 1984.

A member of the Siphonophorae, *Forskalia tholoides*. There were two theories about these sea animals: either they represented a single animal with differentiated organs, or colonies of different animals. Ernst Haeckel, who classified the Siphonophorae and drew this plate, advanced his Medusome theory which combined aspects of both of these theories. From *Report on the Scientific Results of the Voyage of H.M.S. Challenger During the Years 1873–76* (London, 1888), vol. 28.

3. WILLIAM PALEY

(1743–1805)

William Paley exerted a profound influence over late eighteenth-century and early nineteenth-century scientific and theological thinking: his final work *Natural Theology* (1802) enjoyed numerous reprintings and his *Evidences of Christianity* (1794) and *Principles of Moral and Political Philosophy* (1785) were required reading when the young Charles Darwin entered Christ's College, Cambridge University in 1828. Paley himself had graduated as Senior Wrangler from Cambridge in 1763, and rose steadily in the Church of England as a clergyman, eventually being appointed Archdeacon of Carlisle in 1782.

His ethical and theological works mainly forgotten, Paley is now primarily associated with the idea of natural theology, the notion that the proof for an intelligent creator (God) can be deduced from nature itself, which manifests an order and design which can only have been the product of a conscious craftsman or designer. For Paley, the complexity of the human eye, for example, and the way in which its form is perfectly adapted to its function of seeing, cannot have been an accident. Paley was not the originator of this idea—he was directly influenced by two earlier works, John Ray's *The Wisdom of God Manifested in the Works of Creation* (1691) and William Derham's *Physico-Theology* (1713).

John Ray believed that the simple fact of the obvious complexity of the natural world provided directly verifiable proof of God's forethought to both philosopher and layperson alike: "for you may hear illiterate Persons, of the lowest Rank of the Commonalty, affirming, that they need no Proof of the Being of a God, for that every Pile of Grass or Ear of Corn sufficiently proves that; for, say they, all the Men of the World cannot make such a Thing as one of these; and if they cannot do it, who can or did make it but God? To tell them that it made itself, or sprung up by Chance, would be as ridiculous as to tell the greatest Philosopher so" (B1v–B2r). Despite his obvious antecedents, Paley, however, did popularize and appropriate for natural theology the central watch metaphor of eighteenth-century Deism. Deism was an eighteenth-century idea in Christianity that proposed that God had created the world like a watch, wound it up, and then left it to operate by its own mechanical principles. But instead of devaluing nature as a simple mechanistic aggregation of physical laws, natural theology led Paley to a reverend awe of the created world and its creator, as Paley himself explains: "a way and habit of remarking and contemplating the works and mysteries of nature is a delightful, and reasonable, and pious exercise of our thoughts, and is often the very first thing that leads to a religious disposition" (Edmund Paley, *Life* 319).

The basic argument of *Natural Theology*, reprinted below, can be summarized as follows: 1) nature contains abundant evidence of structure and purposeful design; 2) design cannot occur without an agent, a designer; 3) therefore, nature must have been purposefully designed by a designer (God). Most of the book is devoted to demonstrating examples of design in nature, primarily in the anatomy of animals, insects and plants, in an attempt to clinch Paley's sub-argument of the non-randomness of nature and to undermine the possibility that a disembodied "natural law" or a demiurge (Nature), acting as a deputy, instead of a creator, has produced the world. Paley argues that it is illogical for a "principle of order" to actively design something. The rhetorical skills displayed in *Natural Theology* are immense, as Paley systematically dismantles negative criticisms of his argument. Paley, however, extended the arguments for natural theology to their logical limits and some of his examples of the miracles of purposeful design are almost self-parodic, as in this passage: "there is wisdom in making the heart an involuntary muscle. Had the action of the heart been subject to the command of the will, and we could have stopped its motion at pleasure, there would have been no end of suicides, and as there would have been no external marks, they would have been undiscoverable" (Edmund Paley, *Life* 328).

Natural theology, however, despite its excesses of application, must be taken seriously if we are to understand early nineteenth-century science. The philosophy was given renewed impetus in 1825 by the commissioning of the *Bridgewater Treatises*, as prescribed in the will of Francis Henry Egerton, 8th Earl of Bridgewater. Bridgewater treatises by such noted authors as Peter Mark Roget, John Kidd, William Buckland, Thomas Chalmers, and William Whewell appeared throughout the 1830s with the express purpose of demonstrating design in nature and God's goodness as discoverable by the various natural and physical sciences. The philosophy of natural theology also framed many of the questions which were central to Charles Darwin's evolutionary theories. In the *Origin of Species*, Darwin, like Paley, wondered how existing animals seemed so well adapted to their environments, but he proposed that a physical law called natural selection was the cause of these adaptations, not the intervention or planning of a divine being.

The passage below, outlining the central argument of *Natural Theology*, is taken not from the original edition of 1802, but from an annotated version published by Henry Lord Brougham and Sir Charles Bell, author of the 1833 *Bridgewater Treatise* entitled *The Hand, Its Mechanism and Vital Endowments, as Evincing Design*. Except for the editorial notes, the passage selected follows substantially Paley's original text of 1802. Bell and Brougham's second note is especially interesting as a response to the growing acceptance of uniformitarianism in geology, the theory that geological formations are formed by natural processes acting over eons rather than by divine catastrophes such as the Biblical flood. Although Richard Dawkins in *The Blind Watchmaker* (1986) asserted that Paley in his belief in a cosmic Designer was "gloriously and utterly wrong," the astronomer Owen Gingerich, pointing to Kepler as an example, has suggested that a belief in natural theology is not incompatible with creative scientific endeavour.

Paley's Natural Theology, With Illustrative Notes by Henry Lord Brougham and Sir Charles Bell.
2 vols. London: Charles Knight, 1836, pp. 1.1–17.

Chapter I. State of the Argument.[1]

In crossing a heath, suppose I pitched my foot against a *stone*, and were asked how the stone came to be there, I might possibly answer, that, for anything I knew to the contrary, it had lain there for ever; nor would it, perhaps, be very easy to show the absurdity of this answer.[2] But suppose I had found a *watch* upon the ground, and it should be inquired how the watch happened to be in that place, I should hardly think of the answer which I had before given—that, for anything I knew, the watch might have always been there. Yet why should not this answer serve for the watch as well as for the stone? why is it not as admissible in the second case as in the first? For this reason, and for no other, viz., that, when we come to inspect the watch, we perceive (what we could not discover in the stone) that its several parts are framed and put together for a purpose, *e.g.* that they are so formed and adjusted as to produce motion, and that motion so regulated as to point out the hour of the day; that, if the different parts had been differently shaped from what they are, of a different size from what they are, or placed after any other manner, or in any other order than that in which they are placed, either no motion at all would have been carried on in the machine, or none which would have answered the use that is now served by it. To reckon up a few of the plainest of these parts, and of their offices, all tending to one result:—We see a cylindrical box containing a coiled elastic spring, which, by its endeavour to relax itself, turns round the box. We next observe a flexible chain (artificially wrought for the sake of flexure) communicating the action of the spring from the box to the fusee. We then find a series of wheels, the teeth of which catch in, and apply to, each other, conducting the motion from the fusee to the balance, and from the balance to the pointer, and, at the same time, by the size and shape of those wheels, so regulating that motion as to terminate in causing an index, by an equable and measured progression, to pass over a given space in a given time. We take notice that the wheels are made of

---

1 The last note of the Appendix describes the mechanism of a watch, and illustrates the elementary principles of mechanics. Contrasted with the mere mechanism, there is another essay on the mechanism of the animal body. These may be perused either before or after reading the present chapter.

2 The argument is here put very naturally. But a considerable change has taken place of late years in the knowledge attained even by common readers, and there are few who would be without reflection "how the stone came to be there." The changes which the earth's surface has undergone, and the preparation for its present condition, have become a subject of high interest; and there is hardly any one who now would, for an instant, believe that the stone was formed where it lay. On lifting it, he would find it rounded like gravel in a river: he would see that its asperities had been worn off, by being rolled from a distance in water: he would perhaps break it, look to its fracture, and survey the surrounding heights, to discover whence it had been broken off, or from what remote region it had been swept hither: he would consider the place where he stood, in reference to the level of the sea or the waters; and, revolving all these things in his mind, he would be impressed with the conviction, that the surface of the earth had undergone some vast revolution.

Such natural reflections lead an intelligent person to seek for information in the many beautiful and interesting works on geology that have been published in our country of late years. And by these he will be led to infer, that the fair scene before him, so happily adapted for the abode of man, was a condition of the earth resulting from many successive revolutions taking place at periods incalculably remote; and that the variety of mountain and valley, forest and fertile plain, promontory and shallow estuary, formed a world suited to his capacities and enterprise.

So true is the observation of Sir J. Herschel, "that the situation of a pebble may afford him evidence of the state of the globe he inhabits myriads of ages ago, before his species became its denizens."

brass, in order to keep them from rust; the springs of steel, no other metal being so elastic; that over the face of the watch there is placed a glass, a material employed in no other part of the work, but in the room of which, if there had been any other than a transparent substance, the hour could not be seen without opening the case. This mechanism being observed, (it requires indeed an examination of the instrument, and perhaps some previous knowledge of the subject, to perceive and understand it; but being once, as we have said, observed and understood,) the inference, we think, is inevitable, that the watch must have had a maker: that there must have existed, at some time, and at some place or other, an artificer or artificers who formed it for the purpose which we find it actually to answer; who comprehended its construction, and designed its use.

I. Nor would it, I apprehend, weaken the conclusion, that we had never seen a watch made; that we had never known an artist capable of making one; that we were altogether incapable of executing such a piece of workmanship ourselves, or of understanding in what manner it was performed; all this being no more than what is true of some exquisite remains of ancient art, of some lost arts, and, to the generality of mankind, of the more curious productions of modern manufacture. Does one man in a million know how oval frames are turned?[3] Ignorance of this kind exalts our opinion of the unseen and unknown artist's skill, if he be unseen and unknown, but raises no doubt in our minds of the existence and agency of such an artist, at some former time, and in some place or other. Nor can I perceive that it varies at all the inference, whether the question arise concerning a human agent, or concerning an agent of a different species, or an agent possessing, in some respect, a different nature.

II. Neither, secondly, would it invalidate our conclusion, that the watch sometimes went wrong, or that it seldom went exactly right. The purpose of the machinery, the design, and the designer, might be evident, and, in the case supposed, would be evident, in whatever way we accounted for the irregularity of the movement, or whether we could account for it or not. It is not necessary that a machine be perfect, in order to show with what design it was made: still less necessary, where the only question is, whether it were made with any design at all.

III. Nor, thirdly, would it bring any uncertainty into the argument, if there were a few parts of the watch, concerning which we could not discover, or had not yet discovered, in what manner they conduced to the general effect; or even some parts, concerning which we could not ascertain whether they conduced to that effect in any manner whatever. For, as to the first branch of the case, if by the loss, or disorder, or decay of the parts in question, the movement of the watch were found in fact to be stopped, or disturbed, or retarded, no doubt would remain in our minds as to the utility or intention of these parts, although we should be unable to investigate the manner according to which, or the connexion by which, the ultimate effect depended upon their action or assistance; and the more complex is the machine, the more likely is this obscurity to arise. Then, as to the second thing supposed, namely, that there were parts which might be spared without prejudice to the movement of the watch, and that he had proved this by experiment, these superfluous parts, even if we were completely assured that they were such, would not vacate the reasoning which we had instituted concerning other parts. The indication of contrivance remained, with respect to them, nearly as it was before.

---

3 It is certainly a thing not easily expressed in words. The nave of a circular wheel moves on a single pivot; but there are here two pivots, and grooves in the wheel to correspond with them. These two grooves cross each other, and play upon the pivots in such a manner that the centre of motion varies, and the rim of the wheel moves in an ellipsis. It is exactly on the same principle that we draw an oval figure, by driving two nails into a board, and throwing a band round them, and then running the pencil round within the band. These two nails are in the points called by mathematicians the *foci* of the oval or ellipse; and accordingly, a fundamental property of the curve is, that the sum of any two lines whatever, drawn from the two *foci* to any point in the curve, is always the same. These points are called *foci, fires*, because light reflected from the surface of an oval mirror is concentrated there and produces heat.

IV. Nor, fourthly, would any man in his senses think the existence of the watch, with its various machinery, accounted for, by being told that it was one out of possible combinations of material forms; that whatever he had found in the place where he found the watch, must have contained some internal configuration or other; and that this configuration might be the structure now exhibited, viz., of the works of a watch, as well as a different structure.

V. Nor, fifthly, would it yield his inquiry more satisfaction, to be answered, that there existed in things a principle of order, which had disposed the parts of the watch into their present form and situation. He never knew a watch made by the principle of order; nor can he even form to himself an idea of what is meant by a principle of order, distinct from the intelligence of the watchmaker.

VI. Sixthly, he would be surprised to hear that the mechanism of the watch was no proof of contrivance, only a motive to induce the mind to think so:

VII. And not less surprised to be informed, that the watch in his hand was nothing more than the result of the laws of *metallic* nature. It is a perversion of language to assign any law as the efficient, operative cause of anything. A law presupposes an agent; for it is only the mode according to which an agent proceeds: it implies a power; for it is the order according to which that power acts. Without this agent, without this power, which are both distinct from itself, the *law* does nothing, is nothing. The expression, "the law of metallic nature," may sound strange and harsh to a philosophic ear; but it seems quite as justifiable as some others which are more familiar to him, such as "the law of vegetable nature," "the law of animal nature," or, indeed, as "the law of nature" in general, when assigned as the cause of phenomena, in exclusion of agency and power, or when it is substituted into the place of these.[4]

VIII. Neither, lastly, would our observer be driven out of his conclusion, or from his confidence in its truth, by being told that he knew nothing at all about the matter. He knows enough for his argument: he knows the utility of the end: he knows the subserviency and adaptation of the means to the end. These points being known, his ignorance of other points, his doubts concerning other points, affect not the certainty of his reasoning. The consciousness of knowing little need not beget a distrust of that which he does know.

Chapter II. State of the Argument Continued.

Suppose, in the next place, that the person who found the watch should, after some time, discover that, in addition to all the properties which he had hitherto observed in it, it possessed the unexpected property of producing, in the course of its movement, another watch like itself (the thing is conceivable); that it contained within it a mechanism, a system of parts, a mould, for instance, or a complex adjustment of lathes, files, and other tools, evidently and separately calculated for this purpose; let us inquire what effect ought such a discovery to have upon his former conclusion.

I. The first effect would be to increase his admiration of the contrivance, and his conviction of the consummate skill of the contriver. Whether he regarded the object of the contrivance, the distinct apparatus, the intricate, yet in many parts intelligible mechanism by which it was carried on, he would perceive, in this new observation, nothing but an additional reason for doing what

4 When philosophers and naturalists observe a certain succession in the phenomena of the universe, they consider the uniformity to exist through a *law of nature*. If they discover the order of events, or phenomena, they say they have discovered the law: for example, the law of affinities, of gravitation, etc. It is a loose expression; for to obey a law supposes an understanding and a will to comply. The phrase also implies that we know the nature of the governing power which is in operation, and in the present case both conditions are wanting.

The "law" is the mode in which the power acts, and the term should infer, not only an acquiescence in the existence of the power, but of Him who has bestowed the power and enforced the law.

The term "force" is generally used instead of power, when the intensities are measurable in their mechanical results.

he had already done—for referring the construction of the watch to design, and to supreme art. If that construction *without* this property, or which is the same thing, before this property had been noticed, proved intention and art to have been employed about it, still more strong would the proof appear, when he came to the knowledge of this further property, the crown and perfection of all the rest.

II. He would reflect, that though the watch before him were, *in some sense*, the maker of the watch which was fabricated in the course of its movements, yet it was in a very different sense from that in which a carpenter, for instance, is the maker of a chair—the author of its contrivance, the cause of the relation of its parts to their use. With respect to these, the first watch was no cause at all to the second; in no such sense as this was it the author of the constitution and order, either of the parts which the new watch contained, or of the parts by the aid and instrumentality of which it was produced. We might possibly say, but with great latitude of expression, that a stream of water ground corn; but no latitude of expression would allow us to say, no stretch of conjecture could lead us to think, that the stream of water built the mill, though it were too ancient for us to know who the builder was. What the stream of water does in the affair is neither more nor less than this; by the application of an unintelligent impulse to a mechanism previously arranged, arranged independently of it, and arranged by intelligence, an effect is produced, viz., the corn is ground. But the effect results from the arrangement. The force of the stream cannot be said to be the cause or the author of the effect, still less of the arrangement. Understanding and plan in the formation of the mill were not the less necessary for any share which the water has in grinding the corn; yet is this share the same as that which the watch would have contributed to the production of the new watch, upon the supposition assumed in the last section. Therefore,

III. Though it be now no longer probable that the individual watch which our observer had found was made immediately by the hand of an artificer, yet doth not this alteration in anywise affect the inference, that an artificer had been originally employed and concerned in the production. The argument from design remains as it was. Marks of design and contrivance are no more accounted for now than they were before. In the same thing, we may ask for the cause of different properties. We may ask for the cause of the colour of a body, of its hardness, of its heat; and these causes may be all different. We are now asking for the cause of that subserviency to a use, that relation to an end, which we have remarked in the watch before us. No answer is given to this question, by telling us that a preceding watch produced it. There cannot be design without a designer; contrivance, without a contriver; order, without choice; arrangement, without anything capable of arranging; subserviency and relation to a purpose, without that which could intend a purpose; means suitable to an end, and executing their office in accomplishing that end, without the end ever having been contemplated, or the means accommodated to it. Arrangement, disposition of parts, subserviency of means to an end, relation of instruments to a use, imply the presence of intelligence and mind. No one, therefore, can rationally believe, that the insensible, inanimate watch, from which the watch before us issued, was the proper cause of the mechanism we so much admire in it;—could be truly said to have constructed the instrument, disposed its parts, assigned their office, determined their order, action, and mutual dependency, combined their several motions into one result, and that also a result connected with the utilities of other beings. All these properties, therefore, are as much unaccounted for as they were before.

IV. Nor is anything gained by running the difficulty farther back, *i.e.*, by supposing the watch before us to have been produced from another watch, that from a former, and so on indefinitely. Our going back ever so far, brings us no nearer to the least degree of satisfaction upon the subject. Contrivance is still unaccounted for. We still want a contriver. A designing mind is neither supplied by this supposition, nor dispensed with. If the difficulty were diminished the farther we went back, by going back indefinitely we might exhaust it. And this is the only case to which this sort of reasoning applies. Where there is a tendency, or, as we increase the number of terms, a continual approach towards a limit, *there*, by supposing the number of terms to be what is called infinite, we may conceive the limit to be attained; but where there is no such tendency or approach, nothing is effected by lengthening the series. There is no difference as to the point in question, (whatever there may be as to many points,) between one series and another; between a series which is finite, and a series which is infinite. A chain, composed of an infinite number of links, can no more support itself than a chain composed of a finite number of links. And of this we are assured; (though we never *can* have tried the experiment,) because, by increasing the number of links, from ten for instance to a hundred, from a hundred to a thousand, etc., we make not the smallest approach, we observe not the smallest tendency towards self-support. There is no difference in this respect (yet there may be a great difference in several respects,) between a chain of a greater or less length, between one chain and another, between one that is finite and one that is infinite. This very much resembles the case before us. The machine which we are inspecting demonstrates, by its construction, contrivance and design. Contrivance must have had a contriver; design, a designer; whether the machine immediately proceeded from another machine or not. That circumstance alters not the case. That other machine may, in like manner, have proceeded from a former machine: nor does that alter the case; the contrivance must have had a contriver. That former one from one preceding it: no alteration still; a contriver is still necessary. No tendency is perceived, no approach towards a diminution of this necessity. It is the same with any and every succession of these machines; a succession of ten, of a hundred, of a thousand; with one series, as with another; a series which is finite, as with a series which is infinite. In whatever other respects they may differ, in this they do not. In all, equally, contrivance and design are unaccounted for.

The question is not simply, How came the first watch into existence? which question, it may be pretended, is done away by supposing the series of watches thus produced from one another to have been infinite, and consequently to have had no such *first*, for which it was necessary to provide a cause. This, perhaps, would have been nearly the state of the question, if nothing had been before us but an unorganized, unmechanized substance, without mark or indication of contrivance. It might be difficult to show that such substance could not have existed from eternity, either in succession (if it were possible, which I think it is not, for unorganized bodies to spring from one another,) or by individual perpetuity. But that is not the question now. To suppose it to be so, is to suppose that it made no difference whether he had found a watch or a stone. As it is, the metaphysics of that question have no place: for, in the watch which we are examining, are seen contrivance, design; an end, a purpose; means for the end, adaptation to the purpose. And the question which irresistibly presses upon our thoughts, is, Whence this contrivance and design? The thing required is the intending mind, the adapted hand, the intelligence by which that hand was directed. This question, this demand, is not shaken off, by increasing a number or succession of substances, destitute of these properties; nor the more, by increasing that number to infinity. If it be said, that, upon the supposition of one watch being produced from another in the course of that other's movements, and by means of the mechanism within it, we have a cause for the watch in my hand, viz., the watch from which it proceeded,—I deny, that for the design, the contrivance, the suitableness of means to an end, the adaptation of instruments to a use, (all of which we discover in the watch,) we have any cause whatever. It is in vain, therefore, to assign a series of such causes, or to allege that a series may be carried back to infinity; for I do not admit that we have yet any cause at all for the phenomena, still less any series of causes either finite or infinite. Here is contrivance, but no contriver; proofs of design, but no designer.

V. Our observer would further also reflect, that the maker of the watch before him was, in truth and reality, the maker of every watch produced from it: there being no difference (except that the latter manifests a more exquisite skill,) between the making of another watch with his own hands, by the mediation of files, lathes, chisels, etc., and the disposing, fixing, and inserting of these instruments, or of others equivalent to them, in the body of the watch already made in such a manner, as to form a new watch in the course of the movements which he had given to the old one. It is only working by one set of tools instead of another.

The conclusion which the *first* examination of the watch, of its works, construction, and movement, suggested, was, that it must have had, for cause and author of that construction, an artificer who understood its mechanism, and designed its use. This conclusion is invincible. A *second* examination presents us with a new discovery. The watch is found, in the course of its movement, to produce another watch, similar to itself; and not only so, but we perceive in it a system, or organization, separately calculated for that purpose. What effect would this discovery have, or ought it to have, upon our former inference? What, as hath already been said, but to increase, beyond measure, our admiration of the skill which had been employed in the formation of such a machine? Or shall it, instead of this, all at once turn us round to an opposite conclusion, viz., that no art or skill whatever has been concerned in the business, although all other evidences of art and skill remain as they were, and this last and supreme piece of art be now added to the rest? Can this be maintained without absurdity? Yet this is atheism.[5]

FURTHER READING

Barker, Sir Ernest. "Paley and his Political Philosophy." *Traditions of Civility.* Cambridge: Cambridge UP, 1948, pp. 193–262.

Clark, F. Le Gros. *Paley's Natural Theology: Revised to Harmonize with Modern Science.* London: Society for Promoting Christian Knowledge, 1890.

Clarke, M.L. *Paley: Evidences for the Man.* Toronto: University of Toronto Press, 1974.

Crabbe, George. *An Outline of a System of Natural Theology.* London: William Pickering, 1840.

Dawkins, Richard. *The Blind Watchmaker: Why the Evidence of Evolution Reveals a Universe Without Design.* New York: W.W. Norton, 1996.

Ferré, Frederick, ed. *William Paley: Natural Theology, Selections.* Indianapolis: Bobbs-Merrill Co., Inc., 1963.

Fyfe, Aileen. "The Reception of William Paley's Natural Theology in the University of Cambridge." *British Journal for the History of Science* 30 (1997): 321–35.

Gillespie, Neal C. "Divine Design and the Industrial Revolution: William Paley's Abortive Reform of Natural Theology." *Isis* 81 (1990): 214–229.

Gingerich, Owen. "Is There a Role for Natural Theology Today?" In Murray Rae, Hilary Regan, and John Stenhouse, eds. *Science and Theology: Questions at the Interface.* Grand Rapids, MI: William B. Eerdmans Publishing Company, 1994, pp. 20–71.

Gould, Stephen Jay. "Darwin and Paley Meet the Invisible Hand." *Natural History* 11 (1990): 8–16.

LeMahieu, D.L. *The Mind of William Paley: A Philosopher and His Age.* Lincoln: University of Nebraska Press, 1976.

Meadley, George Wilson. *Memoirs of William Paley, D.D.* Sunderland, 1809.

Ospovat, Dov. *The Development of Darwin's Theory: Natural History, Natural Theology, and Natural Selection, 1838–1859.* Cambridge: Cambridge UP, 1980.

Paley, Edmund, ed. *An Account of the Life and Writings of William Paley.* 1825. Reprint, Westmead, England: Gregg International Publishers Limited, 1970.

5 We must leave this logical and satisfactory argument untouched. In this chapter our author is laying the foundation for a course of reasoning on the mechanism displayed in the animal body. The argument in favour of a creating and presiding Intelligence may be drawn from the study of the laws of physical agency:—such as the properties of heat, light, and sound; of gravitation, and chemical combination; the structure of the globe, the divisions of land and sea, the distribution of temperature; nay, the mind may rise to the contemplation of the sun and planets, their mutual dependence, and their revolutions; but, as affording proofs obvious not only to cultivated reason but to plain sense, almost to ignorance, there is nothing to be compared with that for which our author is preparing the reader in this chapter, the mechanism of the animal body, and the adaptations which affect the well-being of living creatures.

—. *The Works of William Paley.* 7 vols. London, 1825.

Ray, John. *The Wisdom of God Manifested in the Works of Creation.* 1691. Reprint, London: J. Rivington, 1759.

Topham, Jonathan R. "Beyond the Common Context: The Production and Reading of the Bridgewater Treatises." *Isis* 89, no. 2 (1998): 233–62.

Webb, Clement C.J. *Studies in the History of Natural Theology.* Oxford: Clarendon Press, 1915.

Whately, Richard. *Dr. Paley's Works: A Lecture.* London: John W. Parker and Son, 1859.

4. ERASMUS DARWIN

(1731–1802)

Erasmus Darwin, grandfather of the author of *The Origin of Species*, was born in Nottinghamshire in 1731. He studied medicine at Cambridge and Edinburgh and began practising in Nottingham in 1756. He removed immediately thereafter to Lichfield, the former home of another eighteenth-century giant, Dr. Samuel Johnson. There was scarcely room enough in the town for the two men, and Darwin and Johnson were never on intimate terms. Darwin could apparently be imperious, dismissive, and sarcastic, as his not entirely reliable biographer, the writer Anna Seward of Lichfield, has remarked: "he became, early in life, sore upon opposition, whether in argument or conduct, and always revenged it by sarcasm of very keen edge" (*Memoirs* 2).

While practising medicine, Darwin filled his commonplace book with sketches of mechanical inventions, some of which he brought to fruition: a carriage steering system, a vertical windmill, an improved pantograph, canal locks, and a bellows-powered speaking head with leather lips and "vocal cords" made of ribbon which could pronounce such words as "mama, papa, and map."

Elected to the Royal Society in 1761, Darwin published several contributions in the *Philosophical Transactions*, most notably on the adiabatic expansion of air (gas expansion without loss or gain of heat). He founded the Lunar Society in Birmingham in 1766 along with William Small and Matthew Boulton. Through this society, as well as his friendships with the leading engineers, scientists, and industrialists of the age such as James Watt and Josiah Wedgwood, Darwin became a major force behind Britain's industrial revolution. Darwin translated two of Linnaeus's treatises on plants as *A System of Vegetables* (1783) and *The Families of Plants* (1787). This laborious work inspired two poems, *The Loves of Plants* (1789) and *The Economy of Vegetation* (1791), known collectively as *The Botanic Garden*. The poems were extremely successful when issued together in 1791 by J. Johnson. Drawing on Linnaeus's classification of plants according to their sexual organs, Darwin humorously recounts in heroic couplets the love lives of plants and their sexual habits. In his notes throughout the poems, Darwin provides detailed scientific explanations for his poetic metaphors. It is clear that Darwin wished his poetry ultimately to be read as scientific fact grounded in observations of nature. Some of Darwin's critics, however, attributed his popularity more to his flashy prose and manipulation of rhetorical techniques, than to the scientific validity of his theories. John T. Rees, for example, who wrote a lengthy review of Darwin's *Zoonomia*, commented that "Darwin argues ingeniously, extends his principles far, and by turning and twisting them, often makes them explain even opposite facts; and when detected and routed in one quarter, rallies in another, and again defends himself" (*Remarks* 66).

Darwin's stated purpose in all of his poetic works was "to inlist Imagination under the banner of Science, and to lead her votaries from the looser analogies, which dress out the imagery of poetry, to the stricter ones, which form the ratiocination of philosophy" (*The*

Botanic Garden, II, "Advertisement"). *The Botanic Garden*, along with Darwin's two-volume medical treatise *Zoonomia, or the Laws of Organic Life* (1794–96), influenced the romantic poets Coleridge, Shelley, Blake, and Wordsworth. They were undoubtedly initially attracted to his radicalism and support of the French revolution, although some of them like Coleridge later outgrew or repudiated his influence.

Zoonomia briefly outlines a theory of biological evolution which may have influenced Charles Darwin. Charles wrote in the third edition of *The Origin of Species* (1861): "It is curious how largely my grandfather, Dr. Erasmus Darwin, anticipated the erroneous grounds of opinion, and the views of Lamarck, in his 'Zoonomia' (vol. i, p. 500–510), published in 1794" (13). Charles wrote a biography of his grandfather as a preface to a German work by Ernst Krause. While distancing himself from Erasmus's scientific theories, Charles nevertheless praised his grandfather for the "vividness of his imagination" and "his great originality of thought, his prophetic spirit both in science and in the mechanical arts, and ... his overpowering tendency to theorise and generalise" (Krause 48). Although he claims to have been unaffected by reading his grandfather's works, Charles Darwin was accused by such antagonists as Bishop Wilberforce of resurrecting Erasmus's ideas in *The Origin of Species* (1859).

Erasmus Darwin believed the earth to be of great antiquity and all species to be constantly improving in a state of progressive development. He also speculated that all animals were related to an original material "filament" which arose by chemical attraction and repulsion and gradually self-organized itself into "lymphatic tubes," glands, and nerves: "would it be too bold to imagine, that all warm-blooded animals have arisen from one living filament, which THE GREAT FIRST CAUSE endued with animality, with the power of acquiring new parts, attended with new propensities, directed by irritations, sensations, volitions, and associations; and thus possessing the faculty of continuing to improve by its own inherent activity, and of delivering down those improvements by generation to its posterity, world without end!" (*Zoonomia* 1.505). One of Darwin's final treatises, *Phytologia; or, The Philosophy of Agriculture and Gardening* (1800), contains reflections on agriculture and plant physiology, drawing on the work of Hales, Grew, Buffon, Spallanzani, Bonnet, Duhamel, and Malpighi.

Darwin's final work, *The Temple of Nature or, The Origin of Society* (1803), describes the origins of life, reproduction, and the progress of the mind, continuing the ideas on evolution first advanced in the chapter on "Generation" in *Zoonomia*. Erasmus Darwin fathered two illegitimate children known as the Misses Parker. He raised them along with his other children, and wrote *A Plan for the Conduct of Female Education in Boarding Schools* (1797) as a guide for the day school he established for them at Ashbourne. He died in 1802 near Derby in the home of his son Erasmus junior, who had committed suicide.

The Temple of Nature.
London: J. Johnson, 1803, pp. 3–39.

CANTO I. PRODUCTION OF LIFE.

I. By firm immutable immortal laws
Impress'd on Nature by the GREAT FIRST CAUSE,
Say, Muse! how rose from elemental strife
Organic forms, and kindled into life;
How Love and Sympathy with potent charm
Warm the cold heart, and lifted hand disarm;
Allure with pleasures, and alarm with pains,
And bind Society in golden chains.

Four past eventful Ages then recite,
And give the fifth, new-born of Time, to light;
The silken tissue of their joys disclose,
Swell with deep chords the murmur of their woes;
Their laws, their labours, and their loves proclaim,
And chant their virtues to the trump of Fame.

IMMORTAL LOVE! who ere the morn of Time,
On wings outstretch'd, o'er Chaos hung sublime;
Warm'd into life the bursting egg of Night,
And gave young Nature to admiring Light!—
You! whose wide arms, in soft embraces hurl'd
Round the vast frame, connect the whirling world!
Whether immers'd in day, the Sun your throne,
You gird the planets in your silver zone;
Or warm, descending on ethereal wing,
The Earth's cold bosom with the beams of spring;
Press drop to drop, to atom atom bind,
Link sex to sex, or rivet mind to mind;
Attend my song!—With rosy lips rehearse,
And with your polish'd arrows write my verse!—
So shall my lines soft-rolling eyes engage,
And snow-white fingers turn the volant page;
The smiles of Beauty all my toils repay,
And youths and virgins chant the living lay.

II. WHERE EDEN's sacred bowers triumphant sprung,
By angels guarded, and by prophets sung,
Wav'd o'er the east in purple pride unfurl'd,
And rock'd the golden cradle of the World;[1]
Four sparkling currents lav'd with wandering tides
Their velvet avenues, and flowery sides;
On sun-bright lawns unclad the Graces stray'd,
And guiltless Cupids haunted every glade;
Till the fair Bride, forbidden shades among,
Heard unalarm'd the Tempter's serpent-tongue;
Eyed the sweet fruit, the mandate disobey'd,
And her fond Lord with sweeter smiles betray'd.
Conscious awhile with throbbing heart he strove,
Spread his wide arms, and barter'd life for love!—
Now rocks on rocks, in savage grandeur roll'd,
Steep above steep, the blasted plains infold;
The incumbent crags eternal tempest shrouds,
And livid light'nings cleave the lambent clouds;
Round the firm base loud-howling whirlwinds blow,
And sands in burning eddies dance below.

Hence ye profane!—the warring winds exclude
Unhallow'd throngs, that press with footstep rude;
But court the Muse's train with milder skies,
And call with softer voice the good and wise.
—Charm'd at her touch the opening wall divides,
And rocks of crystal form the polish'd sides;
Through the bright arch the Loves and Graces tread,
Innocuous thunders murmuring o'er their head;
Pair after pair, and tittering, as they pass,
View their fair features in the walls of glass;
Leave with impatient step the circling bourn,
And hear behind the closing rocks return.

HERE, high in air, unconscious of the storm,
Thy temple, NATURE, rears it's mystic form;
From earth to heav'n, unwrought by mortal toil,
Towers the vast fabric on the desert soil;
O'er many a league the ponderous domes extend,
And deep in earth the ribbed vaults descend;
A thousand jasper steps with circling sweep
Lead the slow votary up the winding steep;
Ten thousand piers, now join'd and now aloof,
Bear on their branching arms the fretted roof.

Unnumber'd ailes connect unnumber'd halls,
And sacred symbols crowd the pictur'd walls;[2]
With pencil rude forgotten days design,
And arts, or empires, live in every line.
While chain'd reluctant on the marble ground,
Indignant TIME reclines, by Sculpture bound;
And sternly bending o'er a scroll unroll'd,
Inscribes the future with his style of gold.
—So erst, when PROTEUS on the briny shore,[3]
New forms assum'd of eagle, pard, or boar;
The wise ATRIDES bound in sea-weed thongs
The changeful god amid his scaly throngs;
Till in deep tones his opening lips at last
Reluctant told the future and the past.

HERE o'er piazza'd courts, and long arcades,
The bowers of PLEASURE root their waving shades;
Shed o'er the pansied moss a checker'd gloom,
Bend with new fruits, with flow'rs successive bloom.
Pleas'd, their light limbs on beds of roses press'd,
In slight undress recumbent Beauties rest;
On tiptoe steps surrounding Graces move,
And gay Desires expand their wings above.

HERE young DIONE arms her quiver'd Loves,
Schools her bright Nymphs, and practices her doves;
Calls round her laughing eyes in playful turns,
The glance that lightens, and the smile that burns;
Her dimpling cheeks with transient blushes dies,
Heaves her white bosom with seductive sighs;
Or moulds with rosy lips the magic words,
That bind the heart in adamantine cords.

Behind in twilight gloom with scowling mien
The demon PAIN, convokes his court unseen;
Whips, fetters, flames, pourtray'd on sculptur'd stone,
In dread festoons, adorn his ebon throne;
Each side a cohort of diseases stands,
And shudd'ring Fever leads the ghastly bands;
O'er all Despair expands his raven wings,
And guilt-stain'd Conscience darts a thousand stings.

Deep-whelm'd beneath, in vast sepulchral caves,
OBLIVION dwells amid unlabell'd graves;
The storied tomb, the laurell'd bust o'erturns,
And shakes their ashes from the mould'ring urns.—
No vernal zephyr breathes, no sunbeams cheer,
Nor song, nor simper, ever enters here;
O'er the green floor, and round the dew-damp wall,
The slimy snail, and bloated lizard crawl;
While on white heaps of intermingled bones
The muse of MELANCHOLY sits and moans;
Showers her cold tears o'er Beauty's early wreck,
Spreads her pale arms, and bends her marble neck.

So in rude rocks, beside the Aegean wave,
TROPHONIUS scoop'd his sorrow-sacred cave;[4]
Unbarr'd to pilgrim feet the brazen door,
And the sad sage returning smil'd no more.

SHRIN'D in the midst majestic NATURE stands,
Extends o'er earth and sea her hundred hands;
Tower upon tower her beamy forehead crests,
And births unnumber'd milk her hundred breasts;
Drawn round her brows a lucid veil depends,
O'er her fine waist the purfled woof descends;
Her stately limbs the gather'd folds surround,
And spread their golden selvage on the ground.

From this first altar fam'd ELEUSIS stole[5]
Her secret symbols and her mystic scroll;
With pious fraud in after ages rear'd
Her gorgeous temple, and the gods rever'd.
—First in dim pomp before the astonish'd throng,
Silence, and Night, and Chaos, stalk'd along;
Dread scenes of Death, in nodding sables dress'd,
Froze the broad eye, and thrill'd the unbreathing breast.
Then the young Spring, with winged Zephyr, leads
The queen of Beauty to the blossom'd meads;
Charm'd in her train admiring Hymen moves,
And tiptoe Graces hand in hand with Loves.
Next, while on pausing step the masked mimes
Enact the triumphs of forgotten times,
Conceal from vulgar throngs the mystic truth,
Or charm with Wisdom's lore the initiate youth;
Each shifting scene, some patriot hero trod,
Some sainted beauty, or some saviour god.

III. Now rose in purple pomp the breezy dawn,
And crimson dew-drops trembled on the lawn;
Blaz'd high in air the temple's golden vanes,
And dancing shadows veer'd upon the plains.—
Long trains of virgins from the sacred grove,
Pair after pair, in bright procession move,
With flower-fill'd baskets round the altar throng,
Or swing their censers, as they wind along.
The fair URANIA leads the blushing bands,
Presents their offerings with unsullied hands;
Pleas'd to their dazzled eyes in part unshrouds
The goddess-form;—the rest is hid in clouds.

"PRIESTESS OF NATURE! while with pious awe
Thy votary bends, the mystic veil withdraw;
Charm after charm, succession bright, display,
And give the GODDESS to adoring day!
So kneeling realms shall own the Power divine,
And heaven and earth pour incense on her shrine.

"Oh grant the MUSE with pausing step to press
Each sun-bright avenue, and green recess;
Led by thy hand survey the trophied walls,
The statued galleries, and the pictur'd halls;[6]
Scan the proud pyramid, and arch sublime,
Earth-canker'd urn, medallion green with time,
Stern busts of Gods, with helmed heroes mix'd,
And Beauty's radiant forms, that smile betwixt.

"Waked by thy voice, transmuted by thy wand,
Their lips shall open, and their arms expand;
The love-lost lady, and the warrior slain,
Leap from their tombs, and sigh or fight again.
—So when ill-fated ORPHEUS tuned to woe
His potent lyre, and sought the realms below;
Charm'd into life unreal forms respir'd,
And list'ning shades the dulcet notes admir'd.—

"LOVE led the Sage through Death's tremendous porch,[7]
Cheer'd with his smile, and lighted with his torch;—
Hell's triple Dog his playful jaws expands,
Fawns round the GOD, and licks his baby hands;[8]
In wondering groups the shadowy nations throng,
And sigh or simper, as he steps along;
Sad swains, and nymphs forlorn, on Lethe's brink,
Hug their past sorrows, and refuse to drink;
Night's dazzled Empress feels the golden flame
Play round her breast, and melt her frozen frame;
Charms with soft words, and sooths with amorous wiles,
Her iron-hearted Lord,—and PLUTO smiles.—
His trembling Bride the Bard triumphant led
From the pale mansions of the astonish'd dead;
Gave the fair phantom to admiring light,—
Ah, soon again to tread irremeable night!"

IV. HER snow-white arm, indulgent to my song,
Waves the fair Hierophant, and moves along.—

High plumes, that bending shade her amber hair,
Nod, as she steps, their silver leaves in air;
Bright chains of pearl, with golden buckles brac'd,
Clasp her white neck, and zone her slender waist;
Thin folds of silk in soft meanders wind
Down her fine form, and undulate behind;
The purple border, on the pavement roll'd,
Swells in the gale, and spreads its fringe of gold.

"FIRST, if you can, celestial Guide! disclose
From what fair fountain mortal life arose,
Whence the fine nerve to move and feel assign'd,
Contractile fibre, and ethereal mind:

"How Love and Sympathy the bosom warm,
Allure with pleasure, and with pain alarm,
With soft affections weave the social plan,
And charm the listening Savage into Man."

"GOD THE FIRST CAUSE!—in this terrene abode[9]
Young Nature lisps, she is the child of GOD.[10]
From embryon births her changeful forms improve,
Grow, as they live, and strengthen as they move.

"Ere Time began, from flaming Chaos hurl'd
Rose the bright spheres, which form the circling world;
Earths from each sun with quick explosions burst,[11]
And second planets issued from the first.
Then, whilst the sea at their coeval birth,
Surge over surge, involv'd the shoreless earth;
Nurs'd by warm sun-beams in primeval caves
Organic Life began beneath the waves.

"First HEAT from chemic dissolution springs,[12]
And gives to matter its eccentric wings;
With strong REPULSION parts the exploding mass,
Melts into lymph, or kindles into gas.
ATTRACTION next, as earth or air subsides,[13]
The ponderous atoms from the light divides,
Approaching parts with quick embrace combines,
Swells into spheres, and lengthens into lines.
Last, as fine goads the gluten-threads excite,
Cords grapple cords, and webs with webs unite;
And quick CONTRACTION with ethereal flame[14]
Lights into life the fibre-woven frame.—
Hence without parent by spontaneous birth[15]
Rise the first specks of animated earth;
From Nature's womb the plant or insect swims,
And buds or breathes, with microscopic limbs.

"IN earth, sea, air, around, below, above,
Life's subtle woof in Nature's loom is wove;
Points glued to points a living line extends,
Touch'd by some goad approach the bending ends;
Rings join to rings, and irritated tubes
Clasp with young lips the nutrient globes or cubes;
And urged by appetencies new select,
Imbibe, retain, digest, secrete, eject.
In branching cones the living web expands,[16]
Lymphatic ducts, and convoluted glands;
Aortal tubes propel the nascent blood,
And lengthening veins absorb the refluent flood,[17]
Leaves, lungs, and gills, the vital ether breathe
On earth's green surface, or the waves beneath.
So Life's first powers arrest the winds and floods,
To bones convert them, or to shells, or woods;
Stretch the vast beds of argil, lime, and sand,
And from diminish'd oceans form the land![18]

"Next the long nerves unite their silver train,
And young SENSATION permeates the brain;[19]
Through each new sense the keen emotions dart,
Flush the young cheek, and swell the throbbing heart.
From pain and pleasure quick VOLITIONS rise,
Lift the strong arm, or point the inquiring eyes;
With Reason's light bewilder'd Man direct,
And right and wrong with balance nice detect.
Last in thick swarms ASSOCIATIONS spring,

Thoughts join to thoughts, to motions motions cling;
Whence in long trains of catenation flow
Imagined joy, and voluntary woe.

"So, view'd through crystal spheres in drops saline,
Quick-shooting salts in chemic forms combine;
Or Mucor-stems, a vegetative tribe,[20]
Spread their fine roots, the tremulous wave imbibe.
Next to our wondering eyes the focus brings
Self-moving lines, and animated rings;
First Monas moves, an unconnected point,
Plays round the drop without a limb or joint;
Then Vibrio waves, with capillary eels,
And Vorticella whirls her living wheels;
While insect Proteus sports with changeful form
Through the bright tide, a globe, a cube, a worm.
Last o'er the field the Mite enormous swims,
Swells his red heart, and writhes his giant limbs.

V. "ORGANIC LIFE beneath the shoreless waves[21]
Was born and nurs'd in Ocean's pearly caves;
First forms minute, unseen by spheric glass,[22]
Move on the mud, or pierce the watery mass;
These, as successive generations bloom,
New powers acquire, and larger limbs assume;
Whence countless groups of vegetation spring,
And breathing realms of fin, and feet, and wing.

"Thus the tall Oak, the giant of the wood,
Which bears Britannia's thunders on the flood;
The Whale, unmeasured monster of the main,
The lordly Lion, monarch of the plain,
The Eagle soaring in the realms of air,
Whose eye undazzled drinks the solar glare,
Imperious man, who rules the bestial crowd,
Of language, reason, and reflection proud,
With brow erect who scorns this earthy sod,
And styles himself the image of his God;
Arose from rudiments of form and sense,
An embryon point, or microscopic ens![23]

"Now in vast shoals beneath the brineless tide,[24]
On earth's firm crust testaceous tribes reside;
Age after age expands the peopled plain,
The tenants perish, but their cells remain;
Whence coral walls and sparry hills ascend[25]
From pole to pole, and round the line extend.

"Next when imprison'd fires in central caves
Burst the firm earth, and drank the headlong waves;[26]
And, as new airs with dread explosion swell,
Form'd lava-isles, and continents of shell;
Pil'd rocks on rocks, on mountains mountains raised,
And high in heaven the first volcanoes blazed;
In countless swarms an insect-myriad moves[27]
From sea-fan gardens, and from coral groves;
Leaves the cold caverns of the deep, and creeps
On shelving shores, or climbs on rocky steeps.
As in dry air the sea-born stranger roves,
Each muscle quickens, and each sense improves;
Cold gills aquatic form respiring lungs,
And sounds aerial flow from slimy tongues.

"So Trapa rooted in pellucid tides,[28]
In countless threads her breathing leaves divides,
Waves her bright tresses in the watery mass,
And drinks with gelid gills the vital gas;
Then broader leaves in shadowy files advance,
Spread o'er the crystal flood their green expanse;
And, as in air the adherent dew exhales,
Court the warm sun, and breathe ethereal gales.

"So still the Tadpole cleaves the watery vale[29]
With balanc'd fins, and undulating tail;
New lungs and limbs proclaim his second birth,
Breathe the dry air, and bound upon the earth.
So from deep lakes the dread Musquito springs,[30]
Drinks the soft breeze, and dries his tender wings,
In twinkling squadrons cuts his airy way,
Dips his red trunk in blood, and man his prey.

"So still the Diodons, amphibious tribe,[31]
With two-fold lungs the sea or air imbibe;
Allied to fish, the lizard cleaves the flood
With one-cell'd heart, and dark frigescent blood;
Half-reasoning Beavers long-unbreathing dart
Through Erie's waves with perforated heart;
With gills and lungs respiring Lampreys steer,
Kiss the rude rocks, and suck till they adhere;
The lazy Remora's inhaling lips,
Hung on the keel, retard the struggling ships;
With gills pulmonic breathes the enormous Whale,
And spouts aquatic columns to the gale;
Sports on the shining wave at noontide hours,[32]
And shifting rainbows crest the rising showers.

"So erst, ere rose the science to record
In letter'd syllables the volant word;
Whence chemic arts, disclosed in pictured lines,
Liv'd to mankind by hieroglyphic signs;
And clustering stars, pourtray'd on mimic spheres,
Assumed the forms of lions, bulls, and bears;
—So erst, as Egypt's rude designs explain,[33]
Rose young DIONE from the shoreless main;[34]
Type of organic Nature! source of bliss!
Emerging Beauty from the vast abyss!
Sublime on Chaos borne, the Goddess stood,
And smiled enchantment on the troubled flood;
The warring elements to peace restored,
And young Reflection wondered and adored."

Now paused the Nymph,—The Muse responsive cries,
Sweet admiration sparkling in her eyes,
"Drawn by your pencil, by your hand unfurl'd,
Bright shines the tablet of the dawning world;
Amazed the Sea's prolific depths I view,
And VENUS rising from the waves in YOU!

"Still Nature's births enclosed in egg or seed
From the tall forest to the lowly weed,
Her beaux and beauties, butterflies and worms,
Rise from aquatic to aerial forms.
Thus in the womb the nascent infant laves
Its natant form in the circumfluent waves;
With perforated heart unbreathing swims,
Awakes and stretches all its recent limbs;[35]
With gills placental seeks the arterial flood,[36]
And drinks pure ether from its Mother's blood.
Erewhile the landed Stranger bursts his way,
From the warm wave emerging into day;
Feels the chill blast, and piercing light, and tries
His tender lungs, and rolls his dazzled eyes;[37]
Gives to the passing gale his curling hair,
And steps a dry inhabitant of air.

"Creative Nile, as taught in ancient song,
So charm'd to life his animated throng;
O'er his wide realms the slow-subsiding flood
Left the rich treasures of organic mud;
While with quick growth young Vegetation yields
Her blushing orchards, and her waving fields;
Pomona's hand replenish'd Plenty's horn,
And Ceres laugh'd amid her seas of corn.—
Bird, beast and reptile, spring from sudden birth,
Raise their new forms, half-animal, half-earth;
The roaring lion shakes his tawny mane,
His struggling limbs still rooted in the plain;
With flapping wings assurgent eagles toil
To rend their talons from the adhesive soil;
The impatient serpent lifts his crested head,
And drags his train unfinish'd from the bed.—
As Warmth and Moisture blend their magic spells,[38]
And brood with mingling wings the slimy dells;
Contractile earths in sentient forms arrange,
And Life triumphant stays their chemic change."

Then hand in hand along the waving glades
The virgin Sisters pass beneath the shades;
Ascend the winding steps with pausing march,
And seek the Portico's susurrant arch;
Whose sculptur'd architrave on columns borne
Drinks the first blushes of the rising morn,
Whose fretted roof an ample shield displays,
And guards the Beauties from meridian rays.
While on light step enamour'd Zephyr springs,
And fans their glowing features with his wings,
Imbibes the fragrance of the vernal flowers,
And speeds with kisses sweet the dancing Hours.

Urania, leaning with unstudied grace,
Rests her white elbow on a column's base;
Awhile reflecting takes her silent stand,
Her fair cheek press'd upon her lily hand;
Then, as awaking from ideal trance,
On the smooth floor her pausing steps advance,
Waves high her arm, upturns her lucid eyes,
Marks the wide scenes of ocean, earth, and
skies;
And leads, meandering as it rolls along
Through Nature's walks, the shining stream of
Song.

First her sweet voice in plaintive accents
chains
The Muse's ear with fascinating strains;
Reverts awhile to elemental strife,
The change of form, and brevity of life;
Then tells how potent Love with torch sublime
Relights the glimmering lamp, and conquers
Time.
—The polish'd walls reflect her rosy smiles,
And sweet-ton'd echoes talk along the
ailes.

Notes

1 *Cradle of the World*, l. 36. The nations, which possess Europe and a part of Asia and of Africa, appear to have descended from one family; and to have had their origin near the banks of the Mediterranean, as probably in Syria, the site of Paradise, according to the Mosaic history. This seems highly probable from the similarity of the structure of the languages of these nations, and from their early possession of similar religions, customs, and arts, as well as from the most ancient histories extant. The two former of these may be collected from Lord Monboddo's learned work on the Origin of Language, and from Mr. Bryant's curious account of Ancient Mythology.

The use of iron tools, of the bow and arrow, of earthen vessels to boil water in, of wheels for carriages, and the arts of cultivating wheat, of coagulating milk for cheese, and of spinning vegetable fibres for clothing, have been known in all European countries, as long as their histories have existed; besides the similarity of the texture of their languages, and of many words in them; thus the word sack is said to mean a bag in all of them, as *σαχχον* in Greek, saccus in Latin, sacco in Italian, sac in French, and sack in English and German.

Other families of mankind, nevertheless, appear to have arisen in other parts of the habitable earth, as the language of the Chinese is said not to resemble those of this part of the world in any respect. And the inhabitants of the islands of the South-Sea had neither the use of iron tools, nor of the bow, nor of wheels, nor of spinning, nor had learned to coagulate milk, or to boil water, though the domestication of fire seems to have been the first great discovery that distinguished mankind from the bestial inhabitants of the forest.

2 *Pictur'd walls*, l. 76. The application of mankind, in the early ages of society, to the imitative arts of painting, carving, statuary, and the casting of figures in metals, seems to have preceded the discovery of letters; and to have been used as a written language to convey intelligence to their distant friends, or to transmit to posterity the history of themselves, or of their discoveries. Hence the origin of the hieroglyphic figures which crowded the walls of the temples of antiquity; many of which may be seen in the tablet of Isis in the works of Montfaucon; and some of them are still used in the sciences of chemistry and astronomy, as the characters for the metals and planets, and the figures of animals on the celestial globe.

3 *So erst, when Proteus*, l. 83. It seems probable that Proteus was the name of a hieroglyphic figure representing Time; whose form was perpetually changing, and who could discover the past events of the world, and predict the future. Herodotus does not doubt but that Proteus was an Egyptian king or deity; and Orpheus calls him the principle of all things, and the most ancient of the gods; and adds, that he keeps the keys of Nature, *Danet's Dict.* all which might well accord with a figure representing Time.

4 *Trophonius scoop'd*, l. 126. Plutarch mentions, that prophecies of evil events were uttered from the cave of Trophonius; but the allegorical story, that whoever entered this cavern were never again seen to smile, seems to have been designed to warn the contemplative from considering too much the dark side of nature. Thus an ancient poet is said to have written a poem on the miseries of the world, and to have thence become so unhappy as to destroy himself. When we reflect on the perpetual destruction of organic life, we should also recollect, that it is perpetually renewed in other forms by the same materials, and thus the sum total of the happiness of the

world continues undiminished; and that a philosopher may thus smile again on turning his eyes from the coffins of nature to her cradles.

5 *Fam'd Eleusis stole,* l. 137. The Eleusinian mysteries were invented in Egypt, and afterwards transferred into Greece along with most of the other early arts and religions of Europe. They seem to have consisted of scenical representations of the philosophy and religion of those times, which had previously been painted in hieroglyphic figures to perpetuate them before the discovery of letters; and are well explained in Dr. Warburton's divine legation of Moses; who believes with great probability, that Virgil in the sixth book of the Aeneid has described a part of these mysteries in his account of the Elysian fields.

In the first part of this scenery was represented Death, and the destruction of all things; as mentioned in the note on the Portland Vase in the Botanic Garden. Next the marriage of Cupid and Psyche seems to have shown the reproduction of living nature; and afterwards the procession of torches, which is said to have constituted a part of the mysteries, probably signified the return of light, and the resuscitation of all things.

Lastly, the histories of illustrious persons of the early ages seem to have been enacted; who were first represented by hieroglyphic figures, and afterwards became the gods and goddesses of Egypt, Greece, and Rome. Might not such a dignified pantomime be contrived, even in this age, as might strike the spectators with awe, and at the same time explain many philosophical truths by adapted imagery, and thus both amuse and instruct?

6 *The statued galleries,* l. 176. The art of painting has appeared in the early state of all societies before the invention of the alphabet. Thus when the Spanish adventurers, under Cortez, invaded America, intelligence of their debarkation and movements was daily transmitted to Montezuma, by drawings, which corresponded with the Egyptian hieroglyphics. The antiquity of statuary appears from the Memnon and sphinxes of Egypt; that of casting figures in metals from the golden calf of Aaron; and that of carving in wood from the idols or household gods, which Rachel stole from her father Laban, and hid beneath her garments as she sat upon the straw. Gen. c. xxxi. v. 34.

7 *Love led the Sage,* l. 189. This description is taken from the figures on the Barbarini, or Portland Vase, where Eros, or Divine Love, with his torch precedes the manes through the gates of Death, and reverting his smiling countenance invites him into the Elysian fields.

8 *Fawns round the God,* l. 192. This idea is copied from a painting of the descent of Orpheus, by a celebrated Parisian artist.

9 *God the first cause,* l. 223.

> A Jove principium, musae! Jovis omnia plena.
> VIRGIL.
>
> In him we live, and move, and have our being.
> ST. PAUL.

10 *Young Nature lisps,* l. 224. The perpetual production and increase of the strata of limestone from the shells of aquatic animals; and of all those incumbent on them from the recrements of vegetables and of terrestrial animals, are now well understood from our improved knowledge of geology; and show, that the solid parts of the globe are gradually enlarging, and consequently that it is young; as the fluid parts are not yet all converted into solid ones. Add to this, that some parts of the earth and its inhabitants appear younger than others; thus the greater height of the mountains of America seems to show that continent to be less ancient than Europe, Asia, and Africa; as their summits have been less washed away, and the wild animals of America, as the tigers and crocodiles, are said to be less perfect in respect to their size and strength; which would show them to be still in a state of infancy, or of progressive improvement. Lastly, the progress of mankind in arts and sciences, which continues slowly to extend, and to increase, seems to evince the youth of human society; whilst the unchanging state of the societies of some insects, as of the bee, wasp, and ant, which is usually ascribed to instinct, seems to evince the longer existence, and greater maturity of those societies. The juvenility of the earth shows, that it has had a beginning or birth, and is a strong natural argument evincing the existence of a cause of its production, that is of the Deity.

11 *Earths from each sun,* l. 229. See Botan. Garden, Vol. I Cant. I. l. 107.

12 *First Heat from chemic,* l. 235. The matter of heat is an ethereal fluid, in which all things are immersed, and which constitutes the general power of repulsion; as appears in explosions which are produced by the sudden evolution of combined heat, and by the expansion of all bodies by the slower diffusion of it in its uncombined state. Without heat all the matter of the world would be condensed into a point by the power of attraction; and neither fluidity nor life could exist. There are also particular powers of repul-

sion, as those of magnetism and electricity, and of chemistry, such as oil and water; which last may be as numerous as the particular attractions which constitute chemical affinities; and may both of them exist as atmospheres round the individual particles of matter; see Botanic Garden, Vol. I. additional note VII. on elementary heat.

13 *Attraction next*, l. 239. The power of attraction may be divided into general attraction, which is called gravity; and into particular attraction, which is termed chemical affinity. As nothing can act where it does not exist, the power of gravity must be conceived as extending from the sun to the planets, occupying that immense space; and may therefore be considered as an ethereal fluid, though not cognizable by our senses like heat, light, and electricity.

Particular attraction, or chemical affinity, must likewise occupy the spaces between the particles of matter which they cause to approach each other. The power of gravity may therefore be called the general attractive ether, and the matter of heat may be called the general repulsive ether; which constitute the two great agents in the changes of inanimate matter.

14 *And quick Contraction*, l. 245. The power of contraction, which exists in organized bodies, and distinguishes life from inanimation, appears to consist of an ethereal fluid which resides in the brain and nerves of living bodies, and is expended in the act of shortening their fibres. The attractive and repulsive ethers require only the vicinity of bodies for the exertion of their activity, but the contractive ether requires at first the contact of a goad or stimulus, which appears to draw it off from the contracting fibre, and to excite the sensorial power of irritation. These contractions of animal fibres are afterwards excited or repeated by the sensorial powers of sensation, volition, or association, as explained at large in Zoonomia, Vol. I.

There seems nothing more wonderful in the ether of contraction producing the shortening of a fibre, than in the ether of attraction causing two bodies to approach each other. The former indeed seems in some measure to resemble the latter, as it probably occasions the minute particles of the fibre to approach into absolute or adhesive contact, by withdrawing from them their repulsive atmospheres; whereas the latter seems only to cause particles of matter to approach into what is popularly called contact, like the particles of fluids; but which are only in the vicinity of each other, and still retain their repulsive atmospheres, as may be seen in riding through shallow water by the number of minute globules of it thrown up by the horses feet, which roll far on its surface; and by the difficulty with which small globules of mercury poured on the surface of a quantity of it can be made to unite with it.

15 *Spontaneous birth*, l. 247. See additional Note, No. I.

16 *In branching cones*, l. 259. The whole branch of an artery or vein may be considered as a cone, though each distinct division of it is a cylinder. It is probable that the amount of the areas of all the small branches from one trunk may equal that of the trunk, otherwise the velocity of the blood would be greater in some parts than in others, which probably only exists when a part is compressed or inflamed.

17 *Absorb the refluent flood*, l. 262. The force of the arterial impulse appears to cease, after having propelled the blood through the capillary vessels; whence the venous circulation is owing to the extremities of the veins absorbing the blood, as those of the lymphatics absorb the fluids. The great force of absorption is well elucidated by Dr. Hales's experiment on the rise of the sap-juice in a vine-stump; see Zoonomia, Vol. I. Sect. XXIII.

18 *And from diminish'd oceans*, l. 268. The increase of the solid parts of the globe by the recrements of organic bodies, as limestone rocks from shells and bones, and the beds of clay, marl, coals, from decomposed woods, is now well known to those who have attended to modern geology; and Dr. Halley, and others, have endeavoured to show, with great probability, that the ocean has decreased in quantity during the short time which human history has existed. Whence it appears, that the exertions of vegetable and animal life convert the fluid parts of the globe into solid ones; which is probably effected by combining the matter of heat with the other elements, instead of suffering it to remain simply diffused amongst them, which is a curious conjecture, and deserves further investigation.

19 *And young Sensation*, l. 270. Both sensation and volition consist in an affection of the central part of the sensorium, or of the whole of it; and hence cannot exist till the nerves are united in the brain. The motions of a limb of any animal cut from the body, are therefore owing to irritation, not to sensation or to volition. For the definitions of irritation, sensation, volition, and association, see additional Note II.

20 *Or Mucor-stems*, l. 283. Mucor or mould in its early state is properly a microscopic vegetable, and is spontaneously produced on the scum of all

decomposing organic matter. The Monas is a moving speck, the Vibrio an undulating wire, the Proteus perpetually changes its shape, and the Vorticella has wheels about its mouth, with which it makes an eddy, and is supposed thus to draw into its throat invisible animalcules. These names are from Linneus and Muller; see Appendix to Additional Note I.

21 *Beneath the shoreless waves*, l. 295. The earth was originally covered with water, as appears from some of its highest mountains, consisting of shells cemented together by a solution of part of them, as the limestone rocks of the Alps; Ferber's Travels. It must be therefore concluded, that animal life began beneath the sea.

Nor is this unanalogous to what still occurs, as all quadrupeds and mankind in their embryon state are aquatic animals; and thus may be said to resemble gnats and frogs. The fetus in the uterus has an organ called the placenta, the fine extremities of the vessels of which permeate the arteries of the uterus, and the blood of the fetus becomes thus oxygenated from the passing stream of the maternal arterial blood; exactly as is done by the gills of fish from the stream of water, which they occasion to pass through them.

But the chicken in the egg possesses a kind of aerial respiration, since the extremities of its placental vessels terminate on a membranous bag, which contains air, at the broad end of the egg; and in this the chick in the egg differs from the fetus in the womb, as there is in the egg no circulating maternal blood for the insertion of the extremities of its respiratory vessels, and in this also I suspect that the eggs of birds differ from the spawn of fish; which latter is immersed in water, and which has probably the extremities of its respiratory organ inserted into the soft membrane which covers it, and is in contact with the water.

22 *First forms minute*, l. 297. See Additional Note I. on Spontaneous Vitality.

23 *An embryon point*, l. 314. The arguments showing that all vegetables and animals arose from such a small beginning, as a living point or living fibre, are detailed in Zoonomia, Sect. XXXIX. 4. 8. on Generation.

24 *Brineless tide*, l. 315. As the salt of the sea has been gradually accumulating, being washed down into it from the recrements of animal and vegetable bodies, the sea must originally have been as fresh as river water; and as it is not saturated with salt, must become annually saline. The sea-water about our island contains at this time from about one twenty-eighth to one thirtieth part of sea salt, and about one eightieth of magnesian salt; Brownrigg on Salt.

25 *Whence coral walls*, l. 319. An account of the structure of the earth is given in Botanic Garden, Vol. I. Additional Notes, XVI. XVIII. XIX. XX. XXIII. XXIV.

26 *Drank the headlong waves*, l. 322. See Additional Note III.

27 *An insect-myriad moves*, l. 327. After islands or continents were raised above the primeval ocean, great numbers of the most simple animals would attempt to seek food at the edges or shores of the new land, and might thence gradually become amphibious; as is now seen in the frog, who changes from an aquatic animal to an amphibious one; and in the gnat, which changes from a natant to a volant state.

At the same time new microscopic animalcules would immediately commence wherever there was warmth and moisture, and some organic matter, that might induce putridity. Those situated on dry land, and immersed in dry air, may gradually acquire new powers to preserve their existence; and by innumerable successive reproductions for some thousands, or perhaps millions of ages, may at length have produced many of the vegetable and animal inhabitants which now people the earth.

As innumerable shell-fish must have existed a long time beneath the ocean, before the calcareous mountains were produced and elevated; it is also probable, that many of the insect tribes, or less complicate animals, existed long before the quadrupeds or more complicate ones, which in some measure accords with the theory of Linneus in respect to the vegetable world; who thinks, that all the plants now extant arose from the conjunction and reproduction of about sixty different vegetables, from which he constitutes his natural orders.

As the blood of animals in the air becomes more oxygenated in their lungs, than that of animals in water by their gills; it becomes of a more scarlet colour, and from its greater stimulus the sensorium seems to produce quicker motions and finer sensations; and as water is a much better vehicle for vibrations or sounds than air, the fish, even when dying in pain, are mute in the atmosphere, though it is probable that in the water they may utter sounds to be heard at a considerable distance. See on this subject, Botanic Garden, Vol. I. Canto IV. l. 176, Note.

28 *So Trapa rooted*, l. 335. The lower leaves of this plant grow under water, and are divided into

minute capillary ramifications; while the upper leaves are broad and round, and have air bladders in their footstalks to support them above the surface of the water. As the aerial leaves of vegetables do the office of lungs, by exposing a large surface of vessels with their contained fluids to the influence of the air; so these aquatic leaves answer a similar purpose like the gills of fish, and perhaps gain from water a similar material. As the material thus necessary to life seems to be more easily acquired from air than from water, the subaquatic leaves of this plant and of sisymbrium, ocnanthe, ranunculus aquatilis, water crow-foot, and some others, are cut into fine divisions to increase the surface, whilst those above water are undivided; see Botanic Garden, Vol. II. Canto IV. l. 204, Note.

Few of the water plants of this country are used for economical purposes, but the ranunculus fluviatilis may be worth cultivation; as on the borders of the river Avon, near Ringwood, the cottagers cut this plant every morning in boats, almost all the year round, to feed their cows, which appear in good condition, and give a due quantity of milk; see a paper from Dr. Pultney in the Transactions of the Linnean Society, Vol. V.

29 *So still the Tadpole,* l. 343. The transformation of the tadpole from an aquatic animal into an aerial one is abundantly curious, when first it is hatched from the spawn by the warmth of the season, it resembles a fish; it afterwards puts forth legs, and resembles a lizard; and finally losing its tail, and acquiring lungs instead of gills, becomes an aerial quadruped.

The rana temporaria of Linneus lives in the water in spring, and on the land in summer, and catches flies. Of the rana paradoxa the larva or tadpole is as large as the frog, and dwells in Surinam, whence the mistake of Merian and of Seba, who call it a frog fish. The esculent frog is green, with three yellow lines from the mouth to the anus; the back transversely gibbous, the hinder feet palmated; its more frequent croaking in the evenings is said to foretell rain. Linnei Syst. Nat. Art. rana.

Linneus asserts in his introduction to the class Amphibia, that frogs are so nearly allied to lizards, lizards to serpents, and serpents to fish, that the boundaries of these orders can scarcely be ascertained.

30 *The dread Musquito springs,* l. 347. See Additional Note IV.

31 *So still the Diodons,* l. 351. See Additional Note V.

32 *At noontide hours,* l. 363. The rainbows in our latitude are only seen in the mornings or evenings, when the sun is not much more than forty-two degrees high. In the more northern latitudes, where the merdian sun is not more than forty-two degrees high, they are also visible at noon.

33 *As Egypt's rude designs,* l. 371. See Additional Note VI.

34 *Rose young Dione,* l. 372. The hieroglyphic figure of Venus rising from the sea supported on a shell by two tritons, as well as that of Hercules armed with a club, appear to be remains of the most remote antiquity. As the former is devoid of grace, and of the pictorial art of design, as one half of the group exactly resembles the other; and as that of Hercules is armed with a club, which was the first weapon.

The Venus seems to have represented the beauty of organic Nature rising from the sea, and afterwards became simply an emblem of ideal beauty; while the figure of Adonis was probably designed to represent the more abstracted idea of life or animation. Some of these hieroglyphic designs seem to evince the profound investigations in science of the Egyptian philosophers, and to have outlived all written language; and still constitute the symbols, by which painters and poets give form and animation to abstracted ideas, as to those of strength and beauty in the above instances.

35 *Awakes and stretches,* l. 392. During the first six months of gestation, the embryon probably sleeps, as it seems to have no use for voluntary power; it then seems to awake, and to stretch its limbs, and change its posture in some degree, which is termed quickening.

36 *With gills placental,* l. 393. The placenta adheres to any side of the uterus in natural gestation, or of any other cavity in extra-uterine gestation; the extremities of its arteries and veins probably permeate the arteries of the mother, and absorb from thence through their fine coats the oxygen of the mother's blood; hence when the placenta is withdrawn, the side of the uterus, where it adhered, bleeds; but not the extremities of its own vessels.

37 *His dazzled eyes,* l. 398. Though the membrana pupillaris described by modern anatomists guards the tender retina from too much light; the young infant nevertheless seems to feel the presence of it by its frequently moving its eyes, before it can distinguish common objects.

38 *As warmth and moisture,* l. 417.
In eodem corpore saepe
Altera pars vivit; rudis est pars altera tellus.

Quippe ubi temperiem sumpsêre humorque calorque,
Concipiunt; et ab his oriuntur, cuncta duobus.
OVID. MET. 1. 1. 430.

This story from Ovid of the production of animals from the mud of the Nile seems to be of Egyptian origin, and is probably a poetical account of the opinions of the magi or priests of that country; showing that the simplest animations were spontaneously produced like chemical combinations, but were distinguished from the latter by their perpetual improvement by the power of reproduction, first by solitary, and then by sexual generation; whereas the products of natural chemistry are only enlarged by accretion, or purified by filtration.

FURTHER READING

Brandl, Leopold. *Erasmus Darwins Botanic Garden.* Wiener Beiträge zur Englischen Philologie. Bd. 30. Wien: Wilhelm Braumüller, 1909.

—. *Erasmus Darwin's Temple of Nature.* Wiener Beiträge zur Englischen Philologie. Bd. 16. Wien: Wilhelm Braumüller, 1902.

Colp, Ralph Jr. "The Relationship of Charles Darwin to the Ideas of his Grandfather, Dr. Erasmus Darwin." *Biography* 9 (1986): 1–24.

Darwin, Erasmus. *The Essential Writings of Erasmus Darwin.* Edited by Desmond King-Hele. London: MacGibbon and Kee, 1968.

—. *The Golden Age. The Temple of Nature, or the Origin of Society.* Introduced by Donald H. Reiman. New York: Garland Publishing, Inc., 1978.

Garfinkle, Norton. "Science and Religion in England, 1790–1800: The Critical Response to the Work of Erasmus Darwin." *Journal of the History of Ideas* 16 (1955): 376–88.

Harrison, James. "Erasmus Darwin's View of Evolution." *Journal of the History of Ideas* 32 (1971): 247–64.

Hassler, Donald M. *The Comedian as Letter D: Erasmus Darwin's Comic Materialism.* The Hague: Martinus Nijhoff, 1973.

King-Hele, Desmond. *Doctor of Revolution: The Life and Genius of Erasmus Darwin.* London: Faber and Faber, 1977.

—. *Erasmus Darwin.* London: Macmillan, 1963.

—. *Erasmus Darwin and the Romantic Poets.* London: Macmillan, 1986.

—. *The Letters of Erasmus Darwin.* Cambridge: Cambridge UP, 1981.

Krause, Ernst. *Erasmus Darwin. With a Preliminary Notice by Charles Darwin.* Translated by W.S. Dallas. London: John Murray, 1879.

Logan, James Venable. *The Poetry and Aesthetics of Erasmus Darwin.* Princeton: Princeton UP, 1936.

McNeil, Maureen. *Under the Banner of Science: Erasmus Darwin and His Age.* Manchester: Manchester UP, 1987.

Pearson, Hesketh. *Doctor Darwin.* London: J.M. Dent and Sons, Ltd., 1930.

Primer, Irwin. "Erasmus Darwin's *Temple of Nature:* Progress, Evolution, and the Eleusinian Mysteries." *Journal of the History of Ideas* 25 (1964): 58–76.

Rees, John T. *Remarks on the Medical Theories of Brown, Cullen, Darwin, and Rush.* Philadelphia: Robert Carr, 1805.

Rey, Roselyne. "Érasme Darwin et la théorie de la génération." In Claude Blanckaert, Jean-Louis Fischer, and Roselyne Rey, eds. *Nature, histoire, société: Essais en hommage à Jacques Roger.* Paris: Klincksieck, 1995.

Ross, Robert N. "'To charm thy curious eye': Erasmus Darwin's Poetry at the Vestibule of Knowledge." *Journal of the History of Ideas* 32 (1971): 379–94.

Seward, Anna. *Memoirs of the Life of Dr. Darwin, Chiefly During His Residence in Lichfield.* 1804. Reprint, Philadelphia: William Poyntell, and Co., 1804.

5. JOHN DALTON

(1766–1844)

John Dalton, the unassuming middle-class Quaker who provided chemistry with its first coherent and quantitative atomic theory, was born in Eaglesfield, Cumberland where he taught in the village school at age 12. Early in life he came under the influence of the Quaker naturalists John Fletcher and Elihu Robinson. In 1781 he joined his brother Jonathan as a teacher at another Quaker school in Kendal, where he met the blind philosopher John Gough, who appears in William Wordsworth's poem *The Excursion*. Dalton began a meteorological journal at this time, which he continued until his death; from this journal he published his *Meteorological Observations and Essays* (1793). His interest in atmospheric gases, water vapour, and gas solubility led to his formulation of the atomic theory in the opening decade of the nineteenth century. He independently derived Charles's Law (given an equal increase of temperature at constant pressure, all gases expand in the same proportions). He was also the first to state the Law of Partial Pressures (in a mixture of gases at equilibrium in a closed vessel, each gas exerts pressure independently as if it were the only gas present).

Dalton successfully diagnosed his own colourblindness (originally known as "Daltonism"), and theorized that the vitreous or aqueous humours of his eye were bluish. He rejected Thomas Young's three-receptor theory of colour perception as well as Young's now vindicated hypothesis that Dalton's red-receiving receptors were absent or malfunctioning. Dalton's requested post-mortem of his eye revealed no bluish humours nor any other abnormalities. DNA analysis of Dalton's eye tissue, still preserved, by John D. Mollon *et al.* in 1995 determined that Dalton was probably suffering from a form of colourblindness called deuteranopia as a result of a defective gene coding for one variety of cone cell in his retina.

His personal life was uneventful: he never married and lived frugally on his earnings as a professor of mathematics at the New College in Manchester, an academy for dissenters, and then as a private tutor in his own Mathematical Academy. His main work was carried out in Manchester, which boasted of an active scientific community fostered by the Manchester Literary and Philosophical Society (The Lit and Phil), presided over by Dalton from 1817–44. He reluctantly became a member of the Royal Society, which awarded him their first Royal Medal in 1826.

Atomic theories had been discussed for centuries, first by the Greeks Leucippus and Democritus of Abdera (fifth century B.C.E.) and later by Epicurus of Samos, whose Roman disciple Lucretius produced the longest and most lucid account of classical atomism in his poem *De rerum natura* of about 50 B.C.E. Lucretius attributed all causation in the universe to indivisible (*atomos* [Greek] = uncuttable) and invisible particles colliding mechanically in a void space. Epicurus's alleged atheism and Aristotle's rejection of atomism impeded the acceptance of atomic theories in the Christian Middle Ages, although natural philosophers did discuss *minima*, or the smallest theoretical units of bodies. The dominant view of matter in Western thought from the time of Aristotle to the seventeenth century was Aristotle's theory of the four primary elements (Earth, Air, Fire, and Water—originally borrowed from Empedocles) which were believed to be transmutable one into the other. Aristotle, the Stoics, and later Christians believed that these elements completely filled the world (the *plenum*), as opposed to the empty space (*kenon, inane*) of the Epicurean and Lucretian atomists.

In the Renaissance, atomism was revived by Nicholas Hill, Daniel Sennert, Pierre Gassendi, and Robert Boyle. Isaac Newton likewise developed a corpuscular theory of matter and suggested in his *Opticks* that chemical forces between corpuscles or atoms were the same as those operating between gross bodies. Two contemporaries of Dalton, the Irish chemists Bryan and William Higgins, extended the observations of Newton, although their work was probably not known directly to Dalton while developing his theory. While eighteenth-century chemistry was characterized more by an interest in the forces between particles of matter (chemical "affinities"), Dalton's work helped to shift the focus of chemistry towards an interest in the matter itself, specifically its weight.

In developing his atomic theory, Dalton was therefore not operating without important philosophical and scientific precedents. The precise origin of Dalton's atomic theory has been obscured by an inaccurate early biography by W.C. Henry, the destruction of Dalton's manuscripts and notebooks in a 1940 German bombing raid, and Dalton's own faulty statements later in life about his theory. While studying the mechanisms of the solubility of gases in water, Dalton calculated the relative weights of several gases and appended them to a paper read to the Manchester Society in 1803. Again Dalton was not without precedent, as quantitative chemistry had been developing on the Continent. J.L. Proust (1754–1826) and C.L. Berthollet (1748–1822) were engaged in a debate over whether chemical compounds combined in definite or indefinite proportions. In formulating his theory, Dalton may also have drawn on the work of J.B. Richter (1762–1807), who developed a table of equivalent weights of acids and bases.

Dalton's atoms can be characterized as unchangeable ultimate particles of various weights. Each atom is surrounded by an atmosphere of caloric or heat, which causes atoms of the same element to repel one another. One confusing aspect of Dalton's work is that he called both elementary particles and combinations of particles (compounds) "atoms." In the selection below, Dalton summarizes the simple rules he developed for the chemical combination of these atoms, now formulated as the Law of Reciprocal Proportions, the Law of Constant Composition, and the Law of Multiple Proportions. Thomas Thomson, a Scottish chemist, published an account of Dalton's system in 1807 in his *System of Chemistry*, and Dalton soon after published his first volume of *A New System of Chemical Philosophy* in 1808. Dalton proceeded to publish part 2 of Volume I of *A New System* in 1810 and part 1 of Volume II in 1827. In addition to his *Elements of English Grammar* (1801) and several short pamphlets on chemical topics, Dalton also collected some of his Lit and Phil lectures into *Philosophical Essays from Vol. V, Part ii of the Memoirs of the Literary and Philosophical Society, Manchester* (1802).

A New System of Chemical Philosophy.
2 vols. London: R. Bickerstaff, 1808–27, pp. 1.208–16, 2.347–51.

On the Constitution of Solids.

A solid body is one, the particles of which are in a state of equilibrium betwixt two great powers, attraction and repulsion, but in such a manner, that no change can be made in their distances without considerable force. If an approximation of the particles is attempted by force, then the heat resists it; if a separation, then the attraction resists it. The notion of Boscovich of alternating planes of attraction and repulsion seems unnecessary; except that upon forcibly breaking the cohesion of any body, the newly exposed surface must receive such a modification in its atmosphere of heat, as may prevent the future junction of the parts, without great force.

The essential distinction between liquids and solids, perhaps consists in this, that heat changes the figure of arrangement of the ultimate particles of the former continually and gradually, whilst they retain their liquid form; whereas in the latter, it is probable, that change of temperature does no more than change the size, and not the arrangement of the ultimate particles.

Notwithstanding the *hardness* of solid bodies, or the difficulty of moving the particles one amongst another, there are several that admit of such motion without fracture, by the application of proper force, especially if assisted by heat. The ductility and malleability of the metals, need only to be mentioned. It should seem the particles glide along each others surface, somewhat like a piece of polished iron at the end of a magnet, without being at all weakened in their cohesion. The absolute force of cohesion, which constitutes the *strength* of bodies, is an enquiry of great practical importance. It has been found by experiment, that wires of the several metals beneath, being each 1/10 of an inch in diameter, were just broken by the annexed weights.

Lead	29 1/4	Pounds.
Tin	49 1/4	
Copper	299 1/4	
Brass	360	
Silver	370	
Iron	450	
Gold	500	

A piece of good oak, an inch square and a yard long, will just bear in the middle 330 lbs. But such a piece of wood should not in practice be trusted, for any length of time, with above 1/3 or 1/4 of that weight. Iron is about 10 times as strong as oak, of the same dimensions.

One would be apt to suppose that *strength* and *hardness* ought to be found proportionate to each other; but this is not the case. Glass is harder than iron, yet the latter is much the stronger of the two.

Crystallization exhibits to us the effects of the natural arrangement of the ultimate particles of various compound bodies; but we are scarcely yet sufficiently acquainted with chemical synthesis and analysis to understand the rationale of this process. The rhomboidal form may arise from the proper position of 4, 6, 8 or 9 globular particles, the cubic form from 8 particles, the triangular form from 3, 6 or 10 particles, the hexahedral prism from 7 particles, etc. Perhaps, in due time, we may be enabled to ascertain the number and order of elementary particles, constituting any given compound element, and from that determine the figure which it will prefer on crystallization, and *vice versâ*; but it seems premature to form any theory on this subject, till we have discovered from other principles the number and order of the primary elements which combine to form some of the compound elements of most frequent occurrence; the method for which we shall endeavour to point out in the ensuing chapter.

On Chemical Synthesis.

When any body exists in the elastic state, its ultimate particles are separated from each other to a much greater distance than in any other state; each particle occupies the centre of a comparatively large sphere, and supports its dignity by keeping all the rest, which by their gravity, or otherwise are disposed to encroach up it, at a respectful distance. When we attempt to conceive the *number* of particles in an atmosphere, it is somewhat like attempting to conceive the number of stars in the universe; we are confounded with the thought. But if we limit the subject, by taking a given volume of any gas, we seem persuaded that, let the divisions be ever so minute, the number of particles must be finite; just as in a given space of the universe, the number of stars and planets cannot be infinite.

Chemical analysis and synthesis go no farther than to the separation of particles one from another, and to their reunion. No new creation or destruction of matter is within the reach of chemical agency.

We might as well attempt to introduce a new planet into the solar system, or to annihilate one already in existence, as to create or destroy a particle of hydrogen. All the changes we can produce, consist in separating particles that are in a state of cohesion or combination, and joining those that were previously at a distance.

In all chemical investigations, it has justly been considered an important object to ascertain the relative *weights* of the simples which constitute a compound. But unfortunately the enquiry has terminated here; whereas from the relative weights in the mass, the relative weights of the ultimate particles or atoms of the bodies might have been inferred, from which their number and weight in various other compounds would appear, in order to assist and to guide future investigations, and to correct their results. Now it is one great object of this work, to shew the importance and advantage of ascertaining *the relative weights of the ultimate particles, both of simple and compound bodies, the number of simple elementary particles which constitute one compound particle, and the number of less compound particles which enter into the formation of one more compound particle.*

If there are two bodies, A and B, which are disposed to combine, the following is the order in which the combinations may take place, beginning with the most simple: namely,

1 atom of A + 1 atom of B = 1 atom of C, binary.
1 atom of A + 2 atoms of B = 1 atom of D, ternary.
2 atoms of A + 1 atom of B = 1 atom of E, ternary.
1 atom of A + 3 atoms of B = 1 atom of F, quaternary.
3 atoms of A + 1 atom of B = 1 atom of G, quaternary. etc. etc.

The following general rules may be adopted as guides in all our investigations respecting chemical synthesis.

1st. When only one combination of two bodies can be obtained, it must be presumed to be a *binary* one, unless some cause appear to the contrary.

2d. When two combinations are observed, they must be presumed to be a *binary* and a *ternary*.

3d. When three combinations are obtained, we may expect one to be a *binary*, and the other two *ternary*.

4th. When four combinations are observed, we should expect one *binary*, two *ternary*, and one *quaternary*, etc.

5th. A *binary* compound should always be specifically heavier than the mere mixture of its two ingredients.

6th. A *ternary* compound should be specifically heavier than the mixture of a binary and a simple, which would, if combined, constitute it; etc.

7th. The above rules and observations equally apply, when two bodies, such as C and D, D and E, etc. are combined.

From the application of these rules, to the chemical facts already well ascertained, we deduce the following conclusions; 1st. That water is a binary compound of hydrogen and oxygen, and the relative weights of the two elementary atoms are as 1 : 7, nearly; 2d. That ammonia is a binary compound of hydrogen and azote, and the relative weights of the two atoms are as 1 : 5, nearly; 3d. That nitrous gas is a binary compound of azote and oxygen, the atoms of which weigh 5 and 7 respectively; that nitric acid is a binary or ternary compound according as it is derived, and consists of one atom of azote and two of oxygen, together weighing 19; that nitrous oxide is a compound similar to nitric acid, and consists of one atom of oxygen and two of azote, weighing 17; that nitrous acid is a binary compound of nitric acid and nitrous gas, weighing 31; that oxynitric acid is a binary compound of nitric acid and oxygen, weighing 26; 4th. That carbonic oxide is a binary compound, consisting of one atom of charcoal, and one of oxygen, together weighing nearly 12; that carbonic acid is a ternary compound, (but sometimes binary) consisting of one atom of charcoal, and two of oxygen, weighing 19; etc. etc. In all these cases the weights are expressed in atoms of hydrogen, each of which is denoted by unity.

In the sequel, the facts and experiments from which these conclusions are derived, will be detailed; as well as a great variety of others from which are inferred the constitution and weight of the ultimate particles of the principal acids, the alkalis, the earths, the metals, the metallic oxides and sulphurets, the long train of neutral salts, and in short, all the chemical compounds which have hitherto obtained a tolerably good analysis. Several of the conclusions will be supported by original experiments.

From the novelty as well as importance of the ideas suggested in this chapter, it is deemed expedient to give plates, exhibiting the mode of combination in some of the more simple cases. A specimen of these accompanies this first part. The elements or atoms of such bodies are conceived at present to be simple, are denoted by a small circle, with some distinctive mark; and the combinations consist in the juxta-position of two or more of these; when three or more particles of elastic fluids are combined together in one, it is to be supposed that the particles of the same kind repel each other, and therefore take their stations accordingly.

On the Principles of the Atomic System of Chemistry.

It is generally allowed that the great objects of the atomic system are, 1st to determine the relative weights of the simple elements; and 2d to determine the *number*, and consequently the weight, of simple elements that enter into combination to form compound elements. The greatest *desideratum* at the present time is the exact relative weight of the element hydrogen. The small weight of 100 cubic inches of hydrogen gas, the important modifications of that weight by even very minute quantities of common air and aqueous vapour, and the difficulties in ascertaining the proportions of air and vapour in regard to hydrogen, are circumstances sufficient to make one distrust results obtained by the most expert and scientific operator. The specific gravity of hydrogen gas was formerly estimated at 1/10 that of common air; it descended to 1/12.5, which is the ratio we adopted in the Table at the end of Vol. 1. It is now commonly taken to be 1/14.5, and whether it may not in the sequel be found to be 1/16.5 is more than any one at present, I believe, has sufficient data to determine. The other factitious gases have mostly undergone some material alterations in their specific gravities in the last twenty years, several of which I have no doubt are improvements; but when we see these specific gravities extended to the 3rd, 4th, and 5th places of decimals, it appears to me to require a credit far greater than any one of us is entitled to. In the mean time, it may be thought a fortunate circumstance, that the weight of common air has undergone no change for the last thirty or forty years; 100 cubic inches being estimated to weigh 30.5 grains at the temperature of 60 degrees, and pressure of 30 inches of mercury: (whether this is exclusive of the moisture I do not recollect.) It is also a fortunate circumstance, (provided it be correct) that this weight is nearly free from decimal figures. I may be allowed to add, that according to my experience, the weight of 100 cubic inches of air is more nearly 31 grains than 30.5. I apprehend these observations are sufficient to shew that something more remains to be done before we obtain a tolerably correct table of the specific gravities of gases; the importance of this object can not be too highly estimated.

The combinations of gases in equal volumes, and in multiple volumes, is naturally connected with this subject. The cases of this kind, or at least approximations to them, frequently occur; but no principle has yet been suggested to account for the phenomena; till that is done I think we ought to investigate the facts with great care, and not

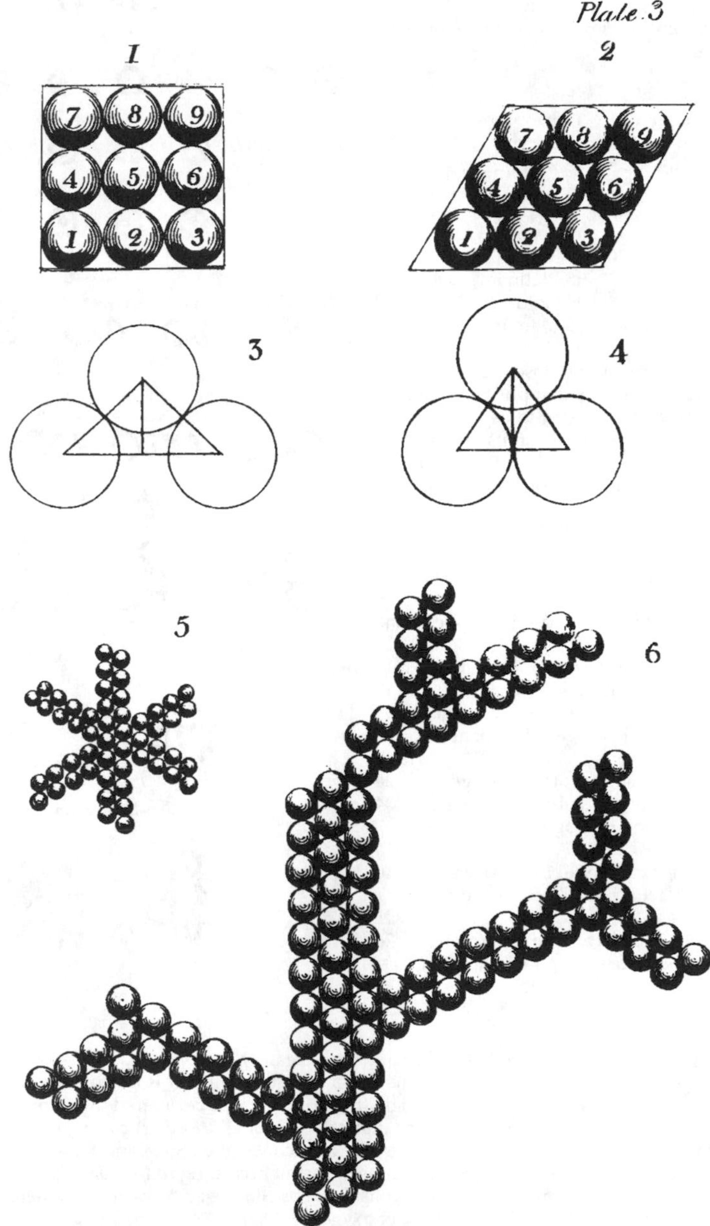

PLATE III. See page 135. The balls in Fig. 1 and 2 represent particles of water: in the former, the square form denotes the arrangement in water, the rhomboidal form in the latter, denotes the arrangement in ice. The angle is always 60° or 120°.

Fig. 3. represents the perpendicular section of a ball resting upon two others, as 4 and 8, Fig. 1.

Fig. 4 represents the perpendicular section of a ball resting upon two balls, as 7 and 5, Fig. 2. The perpendiculars of the triangles shew the heights of the strata in the two arrangements.

Fig. 5 represents one of the small spiculae of ice formed upon the sudden congelation of water cooled below the freezing point...

Fig. 6 represents the shoots or ramifications of ice at the commencement of congelation. The angles are 60° and 120°.

PLATE IV. This plate contains the arbitrary marks or signs chosen to represent the several chemical elements or ultimate particles.

Fig.		Fig.	
1 Hydrog. its rel. weight	1	11 Strontites	46
2 Azote,	5	12 Barytes	68
3 Carbone or charcoal,	5	13 Iron	38
4 Oxygen,	7	14 Zinc	56
5 Phosphorus,	9	15 Copper	56
6 Sulphur,	13	16 Lead	95
7 Magnesia,	20	17 Silver	100
8 Lime,	23	18 Platina	100
9 Soda,	28	19 Gold	140
10 Potash,	42	20 Mercury	167

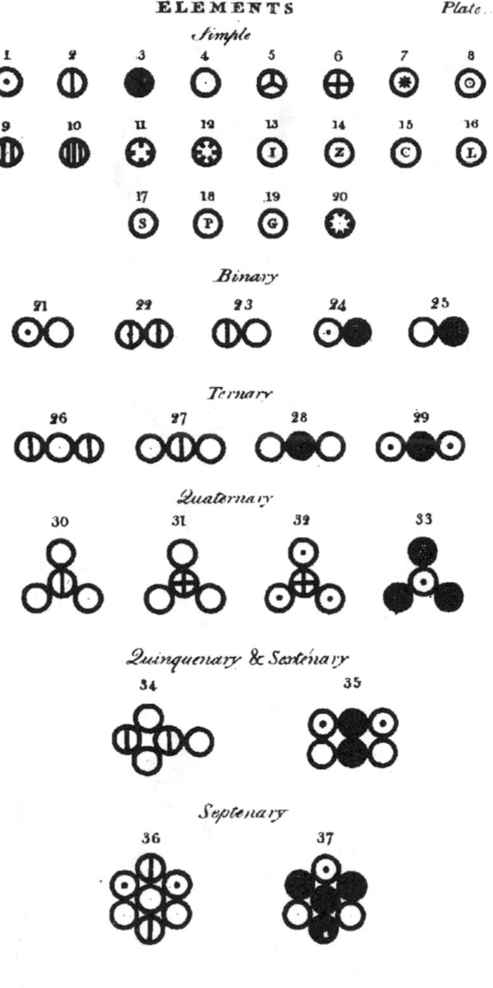

21. An atom of water or steam, composed of 1 of oxygen and 1 of hydrogen, retained in physical contact by a strong affinity, and supposed to be surrounded by a common atmosphere of heat; its relative weight = . . . 8
22. An atom of ammonia, composed of 1 of azote and 1 of hydrogen . . . 6
23. An atom of nitrous gas, composed of 1 of azote and 1 of oxygen . . . 12
24. An atom of olefiant gas, composed of 1 of carbone and 1 of hydrogen . . . 6
25. An atom of carbonic oxide composed of 1 of carbone and 1 of oxygen . . . 12
26. An atom of nitrous oxide, 2 azote + 1 oxygen . . . 17
27. An atom of nitric acid, 1 azote + 2 oxygen . . . 19
28. An atom of carbonic acid, 1 carbone + 2 oxygen . . 19
29. An atom of carburetted hydrogen, 1 carbone + 2 hydrogen . . . 7
30. An atom of oxynitric acid, 1 azote + 3 oxygen . . . 26
31. An atom of sulphuric acid, 1 sulphur + 3 oxygen . 34
32. An atom of sulphuretted hydrogen, 1 sulphur + 3 hydrogen . . . 16
33. An atom of alcohol, 3 carbone + 1 hydrogen . . . 16
34. An atom of nitrous acid, 1 nitric acid + 1 nitrous gas . . . 31
35. An atom of acetous acid, 2 carbone + 2 water . . . 26
36. An atom of nitrate of ammonia, 1 nitric acid + 1 ammonia + 1 water . . . 33
37. An atom of sugar, 1 alcohol + 1 carbonic acid . . . 35

Enough has been given to shew the method; it will be quite unnecessary to devise characters and combinations of them to exhibit to view in this way all the subjects that come under investigation; nor is it necessary to insist upon the accuracy of all these compounds, both in number and weight; the principle will be entered into more particularly hereafter, as far as respects the individual results. It is not to be understood that all those articles marked as simple substances, are necessarily such by the theory; they are only necessarily of such weights. Soda and Potash, such as they are found in combination with acids, are 28 and 42 respectively in weight; but according to Mr. Davy's very important discoveries, they are metallic oxides; the former then must be considered as composed of an atom of metal, 21, and one of oxygen, 7; and the latter, of an atom of metal, 35, and one of oxygen, 7. Or, soda contains 75 per cent. metal and 25 oxygen; potash, 83.3 metal and 16.7 oxygen. It is particularly remarkable, that according to the above-mentioned gentleman's essay on the Decomposition and Composition of the fixed alkalies, in the Philosophical Transactions (a copy of which essay he has just favoured me with) it appears that "the largest quantity of oxygen indicated by these experiments was, for potash 17, and for soda, 26 parts in 100, and the smallest 13 and 19."

suffer ourselves to be led to adopt these analogies till some reason can be discovered for them.

The 2d object of the atomic theory, namely that of investigating the *number* of atoms in the respective compounds, appears to me to have been little understood, even by some who have undertaken to expound the principles of the theory.

When two bodies, A and B, combine in multiple proportions; for instance, 10 parts of A combine with 7 of B, to form one compound, and with 14 to form another, we are directed by some authors to take the smallest combining proportion of one body as representative of the elementary particle or atom of that body. Now it must be obvious to any one of common reflection, that such a rule will be more frequently wrong than right. For, by the above rule, we must consider the first of the combinations as containing 1 atom of B, and the second as containing 2 atoms of B, with 1 atom or more of A; whereas it is equally probable by the same rule, that the compounds may be 2 atoms of A to 1 of B, and 1 atom of A to 1 of B respectively; for, the proportions being 10 A to 7 B, (or, which is the same ratio, 20 A to 14 B,) and 10 A to 14 B; it is clear by the rule, that when the numbers are thus stated, we must consider the former combination as composed of 2 atoms of A, and the latter of 1 atom of A, united to 1 or more of B. Thus there would be an *equal* chance for right or wrong. But it is possible that 10 of A, and 7 of B, may correspond to 1 atom A, and 2 atoms B; and then 10 of A, and 14 of B, must represent 1 atom A, and 4 atoms B. Thus it appears the rule will be more frequently wrong than right.

It is necessary not only to consider the combinations of A with B, but also those of A with C, D, E, etc.; as well as those of B with C, D, etc., before we can have good reason to be satisfied with our determinations as to the *number* of atoms which enter into the various compounds. Elements formed of azote and oxygen appear to contain portions of oxygen, as the numbers 1, 2, 3, 4, 5, successively, so as to make it highly improbable that the combinations can be effected in any other than one of two ways. But in deciding which of those two we ought to adopt, we have to examine not only the compositions and decompositions of the several compounds, of these two elements, but also compounds which each of them forms with other bodies. I have spent much time and labour upon these compounds, and upon others of the primary elements carbone, hydrogen, oxygen, and azote, which appear to me to be of the greatest importance in the atomic system; but it will be seen that I am not satisfied on this head, either by my own labour or that of others, chiefly through the want of an accurate knowledge of combining proportions.

FURTHER READING

Brockbank, E.M. *John Dalton: Experimental Physiologist and Would-Be Physician.* Manchester: Falkner, 1929.

—. *John Dalton: Some Unpublished Letters of Personal and Scientific Interest with Additional Information About his Colour-Vision and Atomic Theories.* Manchester: Manchester UP, 1944.

Cardwell, D.S.L. *John Dalton and the Progress of Science.* Manchester: Manchester UP, 1968.

Dalton, John, *et al. Foundations of the Atomic Theory Comprising Papers and Extracts by John Dalton, William Hyde Wollaston, M.D. and Thomas Thomson, M.D. (1802–1808).* Edinburgh: The Alembic Club, 1911.

—. *A New System of Chemical Philosophy.* 2 vols. 1808–27. Reprint, London: W. Dawson, 1960.

—. *A New System of Chemical Philosophy.* 1808–27. Reprint, New York: Citadel Press, 1964.

—. *A New System of Chemical Philosophy.* Introduced by Alexander Joseph. 1808–27. Reprint, New York: Philosophical Library, 1960.

Dickinson, Christine, Ian Murray, and David Carden, eds. *John Dalton's Colour Vision Legacy.* Manchester: Taylor and Francis, 1997.

Donovan, Arthur, ed. "The Chemical Revolution." Special issue of *Osiris*, 2d ser., 4 (1988).

Foundations of the Atomic Theory Comprising Papers and Extracts by John Dalton, William Hyde Wollaston, M.D. and Thomas Thomson, M.D. Alembic Club Reprints 2. Edinburgh: E. and S. Livingstone Ltd., 1948.

Greenaway, Frank. *John Dalton and the Atom.* Ithaca: Cornell UP, 1966.

Heilbron, J.L. "Weighing Imponderables and Other Quantitative Science Around 1800."

Special issue of *Historical Studies in the Physical and Biological Sciences*, supplement to vol. 24, part 1 (1993): 1–227.

Henry, W.C. *Memoirs of the Life and Scientific Researches of John Dalton*. London: Cavendish Society, 1854.

Knight, D.M. *Atoms and Elements: A Study of Theories of Matter in England in the Nineteenth Century*. London: Hutchinson, 1967.

Lonsdale, Henry. *John Dalton*. The Worthies of Cumberland. London: George Routledge and Sons, 1874.

Meldrum, Andrew N. *Avogadro and Dalton: The Standing in Chemistry of their Hypotheses*. Aberdeen: Greyfriars Press, 1904.

Partington, J.R. *A History of Chemistry*. Vol. 3. London: Macmillan, 1962.

Patterson, Elizabeth C. *John Dalton and the Atomic Theory*. Garden City, NY: Doubleday and Co., 1970.

Randall, Wyatt W., ed. *The Expansion of Gases by Heat: Memoirs by Dalton, Gay-Lussac, Regnault, and Chappuis*. New York: American Book Co., 1902.

Rocke, Alan J. *Chemical Atomism in the Nineteenth Century From Dalton to Cannizzaro*. Columbus: Ohio State UP, 1984.

Roscoe, Henry E. *John Dalton and the Rise of Modern Chemistry*. New York: Macmillan and Co., 1895.

Roscoe, Henry E. and Arthur Harden. *A New View of the Origin of Dalton's Atomic Theory*. Introduced by Arnold Thackray. 1896. Reprint, New York: Johnson Reprint Corporation, 1970.

Smyth, A.L. *John Dalton 1766–1844: A Bibliography of Works By and About Him*. Manchester: Manchester UP, 1966.

Thackray, Arnold. *Atoms and Powers: An Essay on Newtonian Matter-Theory and the Development of Chemistry*. Cambridge, MA: Harvard UP, 1970.

—. *John Dalton: Critical Assessments of His Life and Science*. Harvard Monographs in the History of Science. Cambridge, MA: Harvard UP, 1972.

van Melsen, A.G. *From Atomos to Atom: The History of the Concept Atom*. Pittsburgh: Duquesne UP, 1949.

Whitt, L.A. "Atoms or Affinities? The Ambivalent Recepton of Daltonian Theory." *Studies in History and Philosophy of Science* 21, no. 1 (1990): 57–89.

6. JEAN-BAPTISTE LAMARCK

(1744–1829)

Jean-Baptiste Pierre Antoine de Monet de Lamarck was neglected by his French contemporaries, rebuffed by Napoléon, and disparaged by Georges Cuvier in a eulogy, in spite of the fact that his views on species change were widely read in the early nineteenth century. First educated for the priesthood, Lamarck abandoned a religious career for military adventure, and fought in the Seven Years War. After illness forced him to abandon the army, he studied medicine, and during excursions with Jean-Jacques Rousseau, developed an interest in botany, which resulted in *Flore Françoise* of 1779, revised and updated by A.P de Candolle in 1805. The majority of Lamarck's scientific career was spent in systematizing and categorizing first plants and then the lower animals. He also contributed to the *Dictionnaire de botanique* (1783–95) and *Histoire naturelle des végétaux* (1803), completed by B. de Mirbel. The scientific articles which Lamarck wrote for the *Nouveau dictionnaire d'histoire naturelle* (1816–19) have recently been collected by J. Roger and G. Laurent (Éditions Belin, 1991). From 1788–93, Lamarck worked at the Jardin du Roi, an appointment secured for him by Buffon. Lamarck was probably the first scientist to use the term "biology" in its modern sense.

Lamarck wrote on meteorology, chemistry, and hydrogeology, although his lasting contributions came in the area of invertebrate zoology. In 1793, Lamarck was placed in charge of insects and worms at the newly organized Muséum National d'Histoire Naturelle. Two of his works on invertebrates resulting from this appointment include *Système des animaux sans vertèbres* (1801) and *Histoire naturelle des animaux sans vertèbres* (1815–22). After 1818, he was completely blind.

It was Lamarck's interest in fossil and living shells, his observation of a scale of increasing complexity among animals in nature, and his belief in a slowly changing earth over a long geological timespan that probably convinced him that species also changed over time. Most naturalists of his day maintained that species had remained unchanged since their creation by God, allowing for some obvious variation within individuals, with new species possibly arising by special acts of creation. However, highly speculative and brief accounts of species transformism or evolution (Lamarck did not use these terms) had previously appeared in Buffon, Erasmus Darwin, Maupertuis, Diderot, and in Benoît de Maillet's *Telliamed* (1748). The species for Lamarck was a stable entity only until the environment changed; thus species were only relative, not absolute, categories for him. As L. Szyfman explains "[for Lamarck], species therefore exist and one can speak of their constancy, in the sense of their invariability, as long as the conditions of the environment in which they live do not change. On the other hand, it is impossible to admit the existence of species forever fixed in the form once supposedly created by God, and never having changed either in number or state" (219). Lamarck believed moreover that simple species were created by spontaneous generation.

Lamarck's first exposition of evolution appeared in 1801 in the preface to *Système des animaux sans vertèbres*, and was repeated in *Recherches sur l'organisation des corps vivants* (1802), more fully in 1809 in the *Philosophie zoologique*, and finally in *Histoire naturelle des animaux sans vertèbres* (1815–22).

Lamarck hypothesized that new organs arose in species from use and habit (*habitude*), and that traits that had been acquired by an animal during its lifetime were passed to its offspring. This portion of Lamarck's evolutionary theory, not original with him, became known as the inheritance of acquired characters. Lamarck also asserted in the *Philosophie zoologique* (1809), excerpted below, that a primal natural life force propelled animals toward increasing complexity. In Lamarck's scheme, the fluids of an animal's external environment—electricity, magnetism, and caloric (heat)—caused internal vital movements in animals and the internal fluids were driven to seek canals, openings, and exits which eventually resulted in new organs. The external environment thus shapes the internal organs, and along with the habit or use which increases and strengthens the organs, precipitates the formation of new structures. The two major examples of structural change which Lamarck used were the lengthening of the legs of shore birds as a result of stretching their feet to avoid getting wet, and the lengthening of the necks of giraffes stretching to reach higher foliage in trees. Later English thinkers on evolution, particularly Charles Darwin and Charles Lyell (whose *Principles of Geology* contains a lengthy discussion of Lamarck), ridiculed and dismissed Lamarck's theories, perhaps because he was seen as an atheistic materialist embodying the ideals of revolutionary France. They particularly attacked his idea of the *sentiment intérieur*, mistakenly believing that Lamarck was proposing that animals developed their own organs through a conscious act of desire or will. Darwin maintained, however, that acquired characteristics could be one factor in species change.

August Weismann discredited the idea of acquired characters at the end of the nineteenth century by cutting the tails off successive generations of mice and demonstrating that this characteristic was never inherited. He further demonstrated the independence of the somatic tissues of the body and the germ plasm (reproductive cells) and concluded that somatic changes did not affect the germ substance of the nucleus, which remained constant and controlled inheritance. Weismann's work is generally confirmed by later knowledge of DNA and inheritance, although drugs and environmental radiation cause genetic mutations which can be expressed in an organism, and recently J.J. Cairns ("The Origin of Mutants," *Nature*, 1988) has theorized that mutations in bacteria may occur in response to the needs of the organism.

Neo-Lamarckian movements arose in several countries in the late nineteenth and early twentieth centuries. The Austrian scientist Paul Kammerer produced evidence for acquired characters in toads, but committed suicide in 1926 after being accused of doctoring his specimens. Trofim Lysenko, a Russian follower of Lamarck, claimed to be able to produce certain desirable plant characteristics by exposing seeds to artificial environmental influences, a process he called "vernalization." With the help of Stalin, he persecuted "capitalist" geneticists who opposed his theories and tragically dominated Soviet biology from the 1930s to 1960s.

Zoological Philosophy: An Exposition with Regard to the Natural History of Animals. Translated by Hugh Elliot. London: Macmillan and Co., 1914, pp. 106–27.

OF THE INFLUENCE OF THE ENVIRONMENT ON THE ACTIVITIES AND HABITS OF ANIMALS, AND THE INFLUENCE OF THE ACTIVITIES AND HABITS OF THESE LIVING BODIES IN MODIFYING THEIR ORGANISATION AND STRUCTURE.

We are not here concerned with an argument, but with the examination of a positive fact—a fact which is of more general application than is supposed, and which has not received the attention that it deserves, no doubt because it is usually very difficult to recognise. This fact consists in the influence that is exerted by the environment on the various living bodies exposed to it.

It is indeed long since the influence of the various states of our organisation on our character, inclinations, activities and even ideas has been recognised; but I do not think that anyone has yet drawn attention to the influence of our activities and habits even on our organisation. Now since these activities and habits depend entirely on the environment in which we are habitually placed, I shall endeavour to show how great is the influence exerted by that environment on the general shape, state of the parts and even organisation of living bodies. It is, then, with this very positive fact that we have to do in the present chapter.

If we had not had many opportunities of clearly recognising the result of this influence on certain living bodies that we have transported into an environment altogether new and very different from that in which they were previously placed, and if we had not seen the resulting effects and alterations take place almost under our very eyes, the important fact in question would have remained for ever unknown to us.

The influence of the environment as a matter of fact is in all times and places operative on living bodies; but what makes this influence difficult to perceive is that its effects only become perceptible or recognisable (especially in animals) after a long period of time.

Before setting forth to examine the proofs of this fact, which deserves our attention and is so important for zoological philosophy, let us sum up the thread of the discussions that we have already begun.

In the preceding chapter we saw that it is now an unquestionable fact that on passing along the animal scale in the opposite direction from that of nature, we discover the existence, in the groups composing this scale, of a continuous but irregular degradation in the organisation of animals, an increasing simplification in their organisation, and, lastly, a corresponding diminution in the number of their faculties.

This well-ascertained fact may throw the strongest light over the actual order followed by nature in the production of all the animals that she has brought into existence, but it does not show us why the increasing complexity of the organisation of animals from the most imperfect to the most perfect exhibits only an *irregular gradation,* in the course of which there occur numerous anomalies or deviations with a variety in which no order is apparent.

Now on seeking the reason of this strange irregularity in the increasing complexity of animal organisation, if we consider the influence that is exerted by the infinitely varied environments of all parts of the world on the general shape, structure and even organisation of these animals, all will then be clearly explained.

It will in fact become clear that the state in which we find any animal, is, on the one hand, the result of the increasing complexity of organisation tending to form a regular gradation; and, on the other hand, of the influence of a multitude of very various conditions ever tending to destroy the regularity in the gradation of the increasing complexity of organisation.

I must now explain what I mean by this statement: *the environment affects the shape and organisation of animals,* that is to say that when the environment becomes very different, it produces in course of time corresponding modifications in the shape and organisation of animals.

It is true if this statement were to be taken literally, I should be convicted of an error; for, whatever the environment may do, it does not work any direct modification whatever in the shape and organisation of animals.

But great alterations in the environment of animals lead to great alterations in their needs, and these alterations in their needs necessarily lead to others in their activities. Now if the new needs become permanent, the animals then adopt new habits which last as long as the needs that evoked them. This is easy to demonstrate, and indeed requires no amplification.

It is then obvious that a great and permanent alteration in the environment of any race of animals induces new habits in these animals.

Now, if a new environment, which has become permanent for some race of animals, induces new habits in these animals, that is to say, leads them to new activities which become habitual, the result will be the use of some one part in preference to some other part, and in some cases the total disuse of some part no longer necessary.

Nothing of all this can be considered as hypothesis or private opinion; on the contrary, they are truths which, in order to be made clear, only require attention and the observation of facts.

We shall shortly see by the citation of known facts in evidence, in the first place, that new needs which establish a necessity for some part really bring about the existence of that part, as a result of efforts; and that subsequently its continued use gradually strengthens, develops and finally greatly enlarges it; in the second place, we shall see that in some cases, when the new environment and the new needs have altogether destroyed the utility of some part, the total disuse of that part has resulted in its gradually ceasing to share in the development of the other parts of the animal; it shrinks and wastes little by little, and ultimately, when there has been total disuse for a long period, the part in question ends by disappearing. All this is positive; I propose to furnish the most convincing proofs of it.

In plants, where there are no activities and consequently no habits, properly so-called, great changes of environment none the less lead to great differences in the development of their parts; so that these differences cause the origin and development of some, and the shrinkage and disappearance of others. But all this is here brought about by the changes sustained in the nutrition of the plant, in its absorption and transpiration, in the quantity of caloric, light, air and moisture that it habitually receives; lastly, in the dominance that some of the various vital movements acquire over others.

Among individuals of the same species, some of which are continually well fed and in an environment favourable to their development, while others are in an opposite environment, there arises a difference in the state of the individuals which gradually becomes very remarkable. How many examples I might cite both in animals and plants which bear out the truth of this principle! Now if the environment remains constant, so that the condition of the ill-fed, suffering or sickly individuals becomes permanent, their internal organisation is ultimately modified, and these acquired modifications are preserved by reproduction among the individuals in question, and finally give rise to a race quite distinct from that in which the individuals have been continuously in an environment favourable to their development.

A very dry spring causes the grasses of a meadow to grow very little, and remain lean and puny; so that they flower and fruit after accomplishing very little growth.

A spring intermingled with warm and rainy days causes a strong growth in this same grass, and the crop is then excellent.

But if anything causes a continuance of the unfavourable environment, a corresponding variation takes place in the plants: first in their general appearance and condition, and then in some of their special characters.

Suppose, for instance, that a seed of one of the meadow grasses in question is transported to an elevated place on a dry, barren and stony plot much exposed to the winds, and is there left to germinate; if the plant can live in such a place, it will always be badly nourished, and if the individuals reproduced from it continue to exist in this bad environment, there will result a race fundamentally different from that which lives in the meadows and from which it originated. The individuals of this new race will have small and meagre parts; some of their organs will have developed more than others, and will then be of unusual proportions.

Those who have observed much and studied large collections, have acquired the conviction that according as changes occur in environment, situation, climate, food, habits of life, etc., corresponding changes in the animals likewise occur in size, shape, proportions of the parts, colour, consistency, swiftness and skill.

What nature does in the course of long periods we do every day when we suddenly change the environment in which some species of living plant is situated.

Every botanist knows that plants which are transported from their native places to gardens for purposes of cultivation, gradually undergo changes which ultimately make them unrecognisable. Many plants, by nature hairy, become glabrous or nearly so; a number of those which used to lie and creep on the ground, become erect; others lose their thorns or excrescences; others again whose stem was perennial and woody in their native hot climates, become herbaceous in our own climates and some of them become annuals; lastly, the size of their parts itself undergoes very considerable changes. These effects of alterations of environment are so widely recognised, that botanists do not like to describe garden plants unless they have been recently brought into cultivation.

Is it not the case that cultivated wheat (*Triticum sativum*) is a plant which man has brought to the state in which we now see it? I should like to know in what country such a plant lives in nature, otherwise than as the result of cultivation.

Where in nature do we find our cabbages, lettuces, etc., in the same state as in our kitchen gardens? and is not the case the same with regard to many animals which have been altered or greatly modified by domestication?

How many different races of our domestic fowls and pigeons have we obtained by rearing them in various environments and different countries; birds which we should now vainly seek in nature?

Those which have changed the least, doubtless because their domestication is of shorter standing and because they do not live in a foreign climate, none the less display great differences in some of their parts, as a result of the habits which we have made them contract. Thus our domestic ducks and geese are of the same type as wild ducks and geese; but ours have lost the power of rising into high regions of the air and flying across large tracts of country; moreover, a real change has come about in the state of their parts, as compared with those of the animals of the race from which they come.

Who does not know that if we rear some bird of our own climate in a cage and it lives there for five or six years, and if we then return it to nature by setting it at liberty, it is no longer able to fly like its fellows, which have always been free? The slight change of environment for this individual has indeed only diminished its power of flight, and doubtless has worked no change in its structure; but if a long succession of generations of individuals of the same race had been kept in captivity for a considerable period, there is no doubt that even the structure of these individuals would gradually have undergone notable changes. Still more, if instead of a mere continuous captivity, this environmental factor had been further accompanied by a change to a very different climate; and if these individuals had by degrees been habituated to other kinds of food and other activities for seizing it, these factors when combined together and become permanent would have unquestionably given rise imperceptibly to a new race with quite special characters.

Where in natural conditions do we find that multitude of races of dogs which now actually exist, owing to the domestication to which we have reduced them? Where do we find those bull-dogs, grey-hounds, water-spaniels, spaniels, lap-dogs, etc., etc.; races which show wider differences than those which we call specific when they occur among animals of one genus living in natural freedom?

No doubt a single, original race, closely resembling the wolf, if indeed it was not actually the wolf, was at some period reduced by man to domestication. That race, of which all the individuals were then alike, was gradually scattered with man into different countries and climates; and after they had been subjected for some time to the influences of their environment and of the various habits which had been forced upon them in each country, they underwent remarkable alterations and formed various special races. Now man travels about to very great distances, either for trade or any other purpose; and thus brings into thickly populated places, such as a great capital, various races of dogs formed in very distant countries. The crossing of these races by reproduction then gave rise in turn to all those that we now know.

The following fact proves in the case of plants how the change of some important factor leads to alteration in the parts of these living bodies.

So long as *Ranunculus aquatilis* is submerged in the water, all its leaves are finely divided into minute segments; but when the stem of this plant reaches the surface of the water, the leaves which develop in the air are large, round and simply lobed. If several feet of the same plant succeed in growing in a soil that is merely damp without any immersion, their stems are then short, and none of their leaves are broken up into minute divisions, so that we get *Ranunculus hederaceus*, which botanists regard as a separate species.

There is no doubt that in the case of animals, extensive alterations in their customary environment produce corresponding alterations in their parts; but here the transformations take place much more slowly than in the case of plants; and for us therefore they are less perceptible and their cause less readily identified.

As to the conditions which have so much power in modifying the organs of living bodies, the most potent doubtless consist in the diversity of the places where they live, but there are many others as well which exercise considerable influence in producing the effects in question.

It is known that localities differ as to their character and quality, by reason of their position, construction and climate: as is readily perceived on passing through various localities distinguished by special qualities; this is one cause of variation for animals and plants living in these various places. But what is not known so well and indeed what is not generally believed, is that every locality itself changes in time as to exposure, climate, character and quality, although with such extreme slowness, according to our notions, that we ascribe to it complete stability.

Now in both cases these altered localities involve a corresponding alteration in the environment of the living bodies that dwell there, and this again brings a new influence to bear on these same bodies.

Hence it follows that if there are extremes in these alterations, there are also finer differences: that is to say, intermediate stages which fill up the interval. Consequently there are also fine distinctions between what we call species.

It is obvious then that as regards the character and situation of the substances which occupy the various parts of the earth's surface, there exists a variety of environmental factors which induces a corresponding variety in the shapes and structure of animals, independent of that special variety which necessarily results from the progress of the complexity of organisation in each animal.

In every locality where animals can live, the conditions constituting any one order of things remain the same for long periods: indeed they alter so slowly that man cannot directly observe it. It is only by an inspection of ancient monuments that he becomes convinced that in each of these localities the order of things which he now finds has not always been existent; he may thence infer that it will go on changing.

Races of animals living in any of these localities must then retain their habits equally long: hence the apparent constancy of the races that we call species,—a constancy which has raised in us the belief that these races are as old as nature.

But in the various habitable parts of the earth's surface, the character and situation of places and climates constitute both for animals and plants environmental influences of extreme variability. The animals living in these various localities must therefore differ among themselves, not only by reason of the state of complexity of organisation attained in each race, but also by reason of the habits which each race is forced to acquire; thus when the observing naturalist travels over large portions of the earth's surface and sees conspicuous changes occurring in the environment, he invariably finds that the characters of species undergo a corresponding change.

Now the true principle to be noted in all this is as follows:

1. Every fairly considerable and permanent alteration in the environment of any race of

animals works a real alteration in the needs of that race.

2. Every change in the needs of animals necessitates new activities on their part for the satisfaction of those needs, and hence new habits.

3. Every new need, necessitating new activities for its satisfaction, requires the animal, either to make more frequent use of some of its parts which it previously used less, and thus greatly to develop and enlarge them; or else to make use of entirely new parts, to which the needs have imperceptibly given birth by efforts of its inner feeling; this I shall shortly prove by means of known facts.

Thus to obtain a knowledge of the true causes of that great diversity of shapes and habits found in the various known animals, we must reflect that the infinitely diversified but slowly changing environment in which the animals of each race have successively been placed, has involved each of them in new needs and corresponding alterations in their habits. This is a truth which, once recognised, cannot be disputed. Now we shall easily discern how the new needs may have been satisfied, and the new habits acquired, if we pay attention to the two following laws of nature, which are always verified by observation.

FIRST LAW.
In every animal which has not passed the limit of its development, a more frequent and continuous use of any organ gradually strengthens, develops and enlarges that organ, and gives it a power proportional to the length of time it has been so used; while the permanent disuse of any organ imperceptibly weakens and deteriorates it, and progressively diminishes its functional capacity, until it finally disappears.

SECOND LAW.
All the acquisitions or losses wrought by nature on individuals, through the influence of the environment in which their race has long been placed, and hence through the influence of the predominant use or permanent disuse of any organ; all these are preserved by reproduction to the new individuals which arise, provided that the acquired modifications are common to both sexes, or at least to the individuals which produce the young.

Here we have two permanent truths, which can only be doubted by those who have never observed or followed the operations of nature, or by those who have allowed themselves to be drawn into the error which I shall now proceed to combat.

Naturalists have remarked that the structure of animals is always in perfect adaptation to their functions, and have inferred that the shape and condition of their parts have determined the use of them. Now this is a mistake: for it may be easily proved by observation that it is on the contrary the needs and uses of the parts which have caused the development of these same parts, which have even given birth to them when they did not exist, and which consequently have given rise to the condition that we find in each animal.

If this were not so, nature would have had to create as many different kinds of structure in animals, as there are different kinds of environment in which they have to live; and neither structure nor environment would ever have varied.

This is indeed far from the true order of things. If things were really so, we should not have race-horses shaped like those in England; we should not have big draught-horses so heavy and so different from the former, for none such are produced in nature; in the same way we should not have basset-hounds with crooked legs, nor grey-hounds so fleet of foot, nor water-spaniels, etc.; we should not have fowls without tails, fantail pigeons, etc.; finally, we should be able to cultivate wild plants as long as we liked in the rich and fertile soil of our gardens, without the fear of seeing them change under long cultivation.

A feeling of the truth in this respect has long existed; since the following maxim has passed into a proverb and is known by all, *Habits form a second nature.*

Assuredly if the habits and nature of each animal could never vary, the proverb would have been false and would not have come into existence, nor been preserved in the event of any one suggesting it.

If we seriously reflect upon all that I have just set forth, it will be seen that I was entirely justified when in my work entitled *Recherches sur les corps vivants* (p. 50), I established the following proposition:

"It is not the organs, that is to say, the nature and shape of the parts of an animal's body, that have given rise to its special habits and faculties; but it is, on the contrary, its habits, mode of life and environment that have in course of time controlled the shape of its body, the number and state of its organs and, lastly, the faculties which it possesses."

If this proposition is carefully weighed and compared with all the observations that nature and circumstances are incessantly throwing in our way, we shall see that its importance and accuracy are substantiated in the highest degree.

Time and a favourable environment are as I have already said nature's two chief methods of bringing all her productions into existence: for her, time has no limits and can be drawn upon to any extent.

As to the various factors which she has required and still constantly uses for introducing variations in everything that she produces, they may be described as practically inexhaustible.

The principal factors consist in the influence of climate, of the varying temperatures of the atmosphere and the whole environment, of the variety of localities and their situation, of habits, the commonest movements, the most frequent activities, and, lastly, of the means of self-preservation, the mode of life and the methods of defence and multiplication.

Now as a result of these various influences, the faculties become extended and strengthened by use, and diversified by new habits that are long kept up. The conformation, consistency and, in short, the character and state of the parts, as well as of the organs, are imperceptibly affected by these influences and are preserved and propagated by reproduction.

These truths, which are merely effects of the two natural laws stated above, receive in every instance striking confirmation from facts; for the facts afford a clear indication of nature's procedure in the diversity of her productions.

But instead of being contented with generalities which might be considered hypothetical, let us investigate the facts directly, and consider the effects in animals of the use or disuse of their organs on these same organs, in accordance with the habits that each race has been forced to contract.

Now I am going to prove that the permanent disuse of any organ first decreases its functional capacity, and then gradually reduces the organ and causes it to disappear or even become extinct, if this disuse lasts for a very long period throughout successive generations of animals of the same race.

I shall then show that the habit of using any organ, on the contrary, in any animal which has not reached the limit of the decline of its functions, not only perfects and increases the functions of that organ, but causes it in addition to take on a size and development which imperceptibly alter it; so that in course of time it becomes very different from the same organ in some other animal which uses it far less.

The permanent disuse of an organ, arising from a change of habits, causes a gradual shrinkage and ultimately the disappearance and even extinction of that organ.

Since such a proposition could only be accepted on proof, and not on mere authority, let us endeavour to make it clear by citing the chief known facts which substantiate it.

The vertebrates, whose plan of organisation is almost the same throughout, though with much variety in their parts, have their jaws armed with teeth; some of them, however, whose environment has induced the habit of swallowing the objects they feed on without any preliminary mastication, are so affected that their teeth do not develop. The teeth then remain hidden in the bony framework of the jaws, without being able to appear outside; or indeed they actually become extinct down to their last rudiments.

In the right-whale, which was supposed to be completely destitute of teeth, M. Geoffroy has nevertheless discovered teeth concealed in the jaws of the foetus of this animal. The professor has moreover discovered in birds the groove in which the teeth should be placed, though they are no longer to be found there.

Even in the class of mammals, comprising the most perfect animals, where the vertebrate plan of organisation is carried to its highest completion, not only is the right-whale devoid of teeth, but the ant-eater (*Myrmecophaga*) is also found to be in the same condition, since it has acquired a habit of carrying out no mastication, and has long preserved this habit in its race.

Eyes in the head are characteristic of a great number of different animals, and essentially constitute a part of the plan of organisation of the vertebrates.

Yet the mole, whose habits require a very small use of sight, has only minute and hardly visible eyes, because it uses that organ so little.

Olivier's *Spalax* (*Voyage en Égypte et en Perse*), which lives underground like the mole, and is apparently exposed to daylight even less than the mole, has altogether lost the use of sight: so that it shows nothing more than vestiges of this organ. Even these vestiges are entirely hidden under the skin and other parts, which cover them up and do not leave the slightest access to light.

The *Proteus*, an aquatic reptile allied to the salamanders, and living in deep dark caves under the water, has, like the *Spalax*, only vestiges of the organ of sight, vestiges which are covered up and hidden in the same way.

The following consideration is decisive on the question which I am now discussing,

Light does not penetrate everywhere; consequently animals which habitually live in places where it does not penetrate, have no opportunity of exercising their organ of sight, if nature has endowed them with one. Now animals belonging to a plan of organisation of which eyes were a necessary part, must have originally had them. Since, however, there are found among them some which have lost the use of this organ and which show nothing more than hidden and covered up vestiges of them, it becomes clear that the shrinkage and even disappearance of the organ in question are the results of a permanent disuse of that organ.

This is proved by the fact that the organ of hearing is never in this condition, but is always found in animals whose organisation is of the kind that includes it: and for the following reason.

The substance of sound,[1] that namely which, when set in motion by the shock or the vibration of bodies, transmits to the organ of hearing the impression received, penetrates everywhere and passes through any medium, including even the densest bodies: it follows that every animal, belonging to a plan of organisation of which hearing is an essential part, always has some opportunity for the exercise of this organ wherever it may live. Hence among the vertebrates we

1 Physicists believe and even affirm that the atmospheric air is the actual substance of sound, that is to say, that it is the substance which, when set in motion by the shocks or vibrations of bodies, transmits to the organ of hearing the impression of the concussions received.

That this is an error is attested by many known facts, showing that it is impossible that the air should penetrate to all places to which the substance producing sound actually does penetrate.

See my memoir *On the Substance of Sound*, printed at the end of my *Hydrogéologie*, p. 225, in which I furnished the proofs of this mistake.

Since the publication of my memoir, which by the way is seldom cited, great efforts have been made to make the known velocity of the propagation of sound in air tally with the elasticity of the air, which would cause the propagation of its oscillations to be too slow for the theory. Now, since the air during oscillation necessarily undergoes alternate compressions and dilatations in its parts, recourse has been had to the effects of the caloric squeezed out during the sudden compressions of the air and of the caloric absorbed during the rarefactions of that fluid. By means of these effects, quantitatively determined by convenient hypotheses, geometricians now account for the velocity with which sound is propagated through air. But this is no answer to the fact that sound is also propagated through bodies which air can neither traverse nor set in motion.

These physicists assume forsooth a vibration in the smallest particles of solid bodies; a vibration of very dubious existence, since it can only be propagated through homogeneous bodies of equal density, and cannot spread from a dense body to a rarefied one or *vice versâ*. Such a hypothesis offers no explanation of the well-known fact that sound is propagated through heterogeneous bodies of very different densities and kinds.

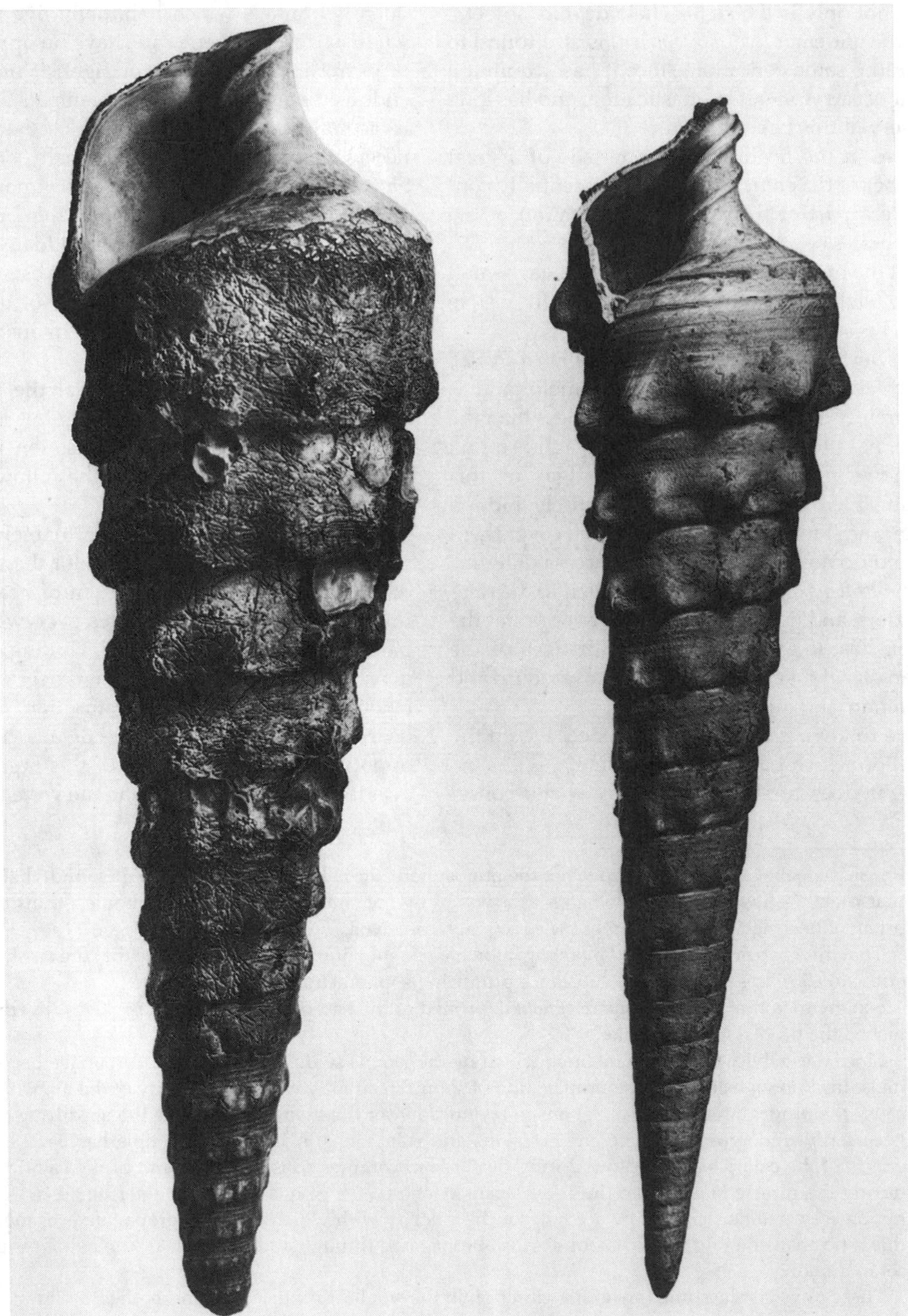

Fossil mollusks (*Cerithium giganteum*) from the extensive collection of J.B. Lamarck housed at the Muséum d'Histoire Naturelle in Geneva. In addition to his theories of species evolution, Lamarck made numerous taxonomic contributions to conchology.

do not find any that are destitute of the organ of hearing; and after them, when this same organ has come to an end, it does not subsequently recur in any animal of the posterior classes.

It is not so with the organ of sight; for this organ is found to disappear, re-appear and disappear again according to the use that the animal makes of it.

In the acephalic molluscs, the great development of the mantle would make their eyes and even their head altogether useless. The permanent disuse of these organs has thus brought about their disappearance and extinction, although molluscs belong to a plan of organisation which should comprise them.

Lastly, it was part of the plan of organisation of the reptiles, as of other vertebrates, to have four legs in dependence on their skeleton. Snakes ought consequently to have four legs, especially since they are by no means the last order of the reptiles and are farther from the fishes than are the batrachians (frogs, salamanders, etc.).

Snakes, however, have adopted the habit of crawling on the ground and hiding in the grass; so that their body, as a result of continually repeated efforts at elongation for the purpose of passing through narrow spaces, has acquired a considerable length, quite out of proportion to its size. Now, legs would have been quite useless to these animals and consequently unused. Long legs would have interfered with their need of crawling, and very short legs would have been incapable of moving their body, since they could only have had four. The disuse of these parts thus became permanent in the various races of these animals, and resulted in the complete disappearance of these same parts, although legs really belong to the plan of organisation of the animals of this class.

Many insects, which should have wings according to the natural characteristics of their order and even of their genus, are more or less completely devoid of them through disuse. Instances are furnished by many Coleoptera, Orthoptera, Hymenoptera and Hemiptera, etc., where the habits of these animals never involve them in the necessity of using their wings.

But it is not enough to give an explanation of the cause which has brought about the present condition of the organs of the various animals,—a condition that is always found to be the same in animals of the same species; we have in addition to cite instances of changes wrought in the organs of a single individual during its life, as the exclusive result of a great mutation in the habits of the individuals of its species. The following very remarkable fact will complete the proof of the influence of habits on the condition of the organs, and of the way in which permanent changes in the habits of an individual lead to others in the condition of the organs, which come into action during the exercise of these habits.

M. Tenon, a member of the Institute, has notified to the class of sciences, that he had examined the intestinal canal of several men who had been great drinkers for a large part of their lives, and in every case he had found it shortened to an extraordinary degree, as compared with the same organ in all those who had not adopted the like habit.

It is known that great drinkers, or those who are addicted to drunkenness, take very little solid food, and eat hardly anything; since the drink which they consume so copiously and frequently is sufficient to feed them.

Now since fluid foods, especially spirits, do not long remain either in the stomach or intestine, the stomach and the rest of the intestinal canal lose among drinkers the habit of being distended, just as among sedentary persons, who are continually engaged on mental work and are accustomed to take very little food; for in their case also the stomach slowly shrinks and the intestine shortens.

This has nothing to do with any shrinkage or shortening due to a binding of the parts which would permit of the ordinary extension, if instead of remaining empty these viscera were again filled; we have to do with a real shrinkage and shortening of considerable extent, and such that these organs would burst rather than yield at once to any demand for the ordinary extension.

Compare two men of equal ages, one of whom has contracted the habit of eating very little, since his habitual studies and mental work have made digestion difficult, while the other habitually takes much exercise, is often out-of-doors, and eats well; the stomach of the first will have very little capacity left and will be filled up by a very small quantity of food, while that of the second will have preserved and even increased its capacity.

Here then is an organ which undergoes profound modification in size and capacity, purely on account of a change of habits during the life of the individual.

The frequent use of any organ, when confirmed by habit, increases the functions of that organ, leads to its development and endows it with a size and power that it does not possess in animals which exercise it less.

We have seen that the disuse of any organ modifies, reduces and finally extinguishes it. I shall now prove that the constant use of any organ, accompanied by efforts to get the most out of it, strengthens and enlarges that organ, or creates new ones to carry on functions that have become necessary.

The bird which is drawn to the water by its need of finding there the prey on which it lives, separates the digits of its feet in trying to strike the water and move about on the surface. The skin which unites these digits at their base acquires the habit of being stretched by these continually repeated separations of the digits; thus in course of time there are formed large webs which unite the digits of ducks, geese, etc., as we actually find them. In the same way efforts to swim, that is to push against the water so as to move about in it, have stretched the membranes between the digits of frogs, sea-tortoises, the otter, beaver, etc.

On the other hand, a bird which is accustomed to perch on trees and which springs from individuals all of whom had acquired this habit, necessarily has longer digits on its feet and differently shaped from those of the aquatic animals that I have just named. Its claws in time become lengthened, sharpened and curved into hooks, to clasp the branches on which the animal so often rests.

We find in the same way that the bird of the water-side which does not like swimming and yet is in need of going to the water's edge to secure its prey, is continually liable to sink in the mud. Now this bird tries to act in such a way that its body should not be immersed in the liquid, and hence makes its best efforts to stretch and lengthen its legs. The long-established habit acquired by this bird and all its race of continually stretching and lengthening its legs, results in the individuals of this race becoming raised as though on stilts, and gradually obtaining long, bare legs, denuded of feathers up to the thighs and often higher still. (*Système des Animaux sans vertèbres*, p.14.)

We note again that this same bird wants to fish without wetting its body, and is thus obliged to make continual efforts to lengthen its neck. Now these habitual efforts in this individual and its race must have resulted in course of time in a remarkable lengthening, as indeed we actually find in the long necks of all water-side birds.

If some swimming birds like the swan and goose have short legs and yet a very long neck, the reason is that these birds while moving about on the water acquire the habit of plunging their head as deeply as they can into it in order to get the aquatic larvae and various animals on which they feed; whereas they make no effort to lengthen their legs.

If an animal, for the satisfaction of its needs, makes repeated efforts to lengthen its tongue, it will acquire a considerable length (ant-eater, green-woodpecker); if it requires to seize anything with this same organ, its tongue will then divide and become forked. Proofs of my statement are found in the humming-birds which use their tongues for grasping things, and in lizards and snakes which use theirs to palpate and identify objects in front of them.

Needs which are always brought about by the environment, and the subsequent continued efforts to satisfy them, are not limited in their results to a mere modification, that is to say, an increase or decrease of the size and capacity of organs; but they may even go so far as to extinguish organs, when any of these needs make such a course necessary.

Fishes, which habitually swim in large masses of water, have need of lateral vision; and, as a matter of fact, their eyes are placed on the sides of their head. Their body, which is more or less flattened according to the species, has its edges perpendicular to the plane of the water; and their eyes are placed so that there is one on each flattened side. But such fishes as are forced by their habits to be constantly approaching the shore, and especially slightly inclined or gently sloping beaches, have been compelled to swim on their flattened surfaces in order to make a close approach to the water's edge. In this position, they receive more light from above than below and stand in special need of paying constant attention to what is passing above them; this requirement has forced one of their eyes to undergo a sort of displacement, and to assume the very remarkable position found in the soles, turbots, dabs, etc. (*Pleuronectes* and *Achirus*). The position of these eyes is not symmetrical, because it results from an incomplete mutation. Now this mutation is entirely completed in the skates, in which the transverse flattening of the body is altogether horizontal, like the head. Accordingly the eyes of skates are both situated on the upper surface and have become symmetrical.

Snakes, which crawl on the surface of the earth, chiefly need to see objects that are raised or above them. This need must have had its effect on the position of the organ of sight in these animals, and accordingly their eyes are situated in the lateral and upper parts of their head, so as easily to perceive what is above them or at their sides; but they scarcely see at all at a very short distance in front of them. They are, however, compelled to make good the deficiency of sight as regards objects in front of them which might injure them as they move forward. For this purpose they can only use their tongue, which they are obliged to thrust out with all their might. This habit has not only contributed to making their tongue slender and very long and contractile, but it has even forced it to undergo division in the greater number of species, so as to feel several objects at the same time; it has even permitted of the formation of an aperture at the extremity of their snout, to allow the tongue to pass without having to separate the jaws.

Nothing is more remarkable than the effects of habit in herbivorous mammals.

A quadruped, whose environment and consequent needs have for long past inculcated the habit of browsing on grass, does nothing but walk about on the ground; and for the greater part of its life is obliged to stand on its four feet, generally making only few or moderate movements. The large portion of each day that this kind of animal has to pass in filling itself with the only kind of food that it cares for, has the result that it moves but little and only uses its feet for support in walking or running on the ground, and never for holding on, or climbing trees.

From this habit of continually consuming large quantities of food material, which distend the organs receiving it, and from the habit of making only moderate movements, it has come about that the body of these animals has greatly thickened, become heavy and massive and acquired a very great size: as is seen in elephants, rhinoceroses, oxen, buffaloes, horses, etc.

The habit of standing on their four feet during the greater part of the day, for the purpose of browsing, has brought into existence a thick horn which invests the extremity of their digits; and since these digits have no exercise and are never moved and serve no other purpose than that of support like the rest of the foot, most of them have become shortened, dwindled and, finally, even disappeared.

Thus in the pachyderms, some have five digits on their feet invested in horn, and their hoof is consequently divided into five parts; others have only four, and others again not more than three; but in the ruminants, which are apparently the oldest of the mammals that are permanently confined to the ground, there are not more than two digits on the feet and indeed, in the solipeds, there is only one (horse, donkey).

Nevertheless some of these herbivorous animals, especially the ruminants, are incessantly exposed to the attacks of carnivorous animals in the desert countries that they inhabit, and they can only find safety in headlong flight. Necessity has in these cases forced them to exert themselves in swift running, and from this habit their body has become more slender and their legs much finer; instances are furnished by the antelopes, gazelles, etc.

In our own climates, there are other dangers, such as those constituted by man, with his continual pursuit of red deer, roe deer and fallow deer; this has reduced them to the same necessity, has impelled them into similar habits, and had corresponding effects.

Since ruminants can only use their feet for support, and have little strength in their jaws, which only obtain exercise by cutting and browsing on the grass, they can only fight by blows with their heads, attacking one another with their crowns.

In the frequent fits of anger to which the males especially are subject, the efforts of their inner feeling cause the fluids to flow more strongly towards that part of their head; in some there is hence deposited a secretion of horny matter, and in others of bony matter mixed with horny matter, which gives rise to solid protuberances: thus we have the origin of horns and antlers, with which the head of most of these animals is armed.

It is interesting to observe the result of habit in the peculiar shape and size of the giraffe (*Camelopardalis*): this animal, the largest of the mammals, is known to live in the interior of Africa in places where the soil is nearly always arid and barren, so that it is obliged to browse on the leaves of trees and to make constant efforts to reach them. From this habit long maintained in all its race, it has resulted that the animal's fore-legs have become longer than its hind legs, and that its neck is lengthened to such a degree that the giraffe, without standing up on its hind legs, attains a height of six metres (nearly 20 feet).

Among birds, ostriches, which have no power of flight and are raised on very long legs, probably owe their singular shape to analogous circumstances.

The effect of habit is quite as remarkable in the carnivorous mammals as in the herbivores; but it exhibits results of a different kind.

Those carnivores, for instance, which have become accustomed to climbing, or to scratching the ground for digging holes, or to tearing their prey, have been under the necessity of using the digits of their feet: now this habit has promoted the separation of their digits, and given rise to the formation of the claws with which they are armed.

But some of the carnivores are obliged to have recourse to pursuit in order to catch their prey: now some of these animals were compelled by their needs to contract the habit of tearing with their claws, which they are constantly burying deep in the body of another animal in order to lay hold of it, and then make efforts to tear out the part seized. These repeated efforts must have resulted in its claws reaching a size and curvature which would have greatly impeded them in walking or running on stony ground: in such cases the animal has been compelled to make further efforts to draw back its claws, which are so projecting and hooked as to get in its way. From this there has gradually resulted the formation of those peculiar sheaths, into which cats, tigers, lions, etc. withdraw their claws when they are not using them.

Hence we see that efforts in a given direction, when they are long sustained or habitually made by certain parts of a living body, for the satisfaction of needs established by nature or environment, cause an enlargement of these parts and the acquisition of a size and shape that they would never have obtained, if these efforts had not become the normal activities of the animals exerting them. Instances are everywhere furnished by observations on all known animals.

Can there be any more striking instance than that which we find in the kangaroo? This animal, which carries its young in a pouch under the abdomen, has acquired the habit of standing upright, so as to rest only on its hind legs and tail; and of moving only by means of a succession of leaps, during which it maintains its erect attitude in order not to disturb its young. And the following is the result:

1. Its fore legs, which it uses very little and on which it only supports itself for a moment on abandoning its erect attitude, have never acquired a development proportional to that of the other parts, and have remained meagre, very short and with very little strength.

2. The hind legs, on the contrary, which are almost continually in action either for supporting the whole body or for making leaps, have acquired a great development and become very large and strong.

3. Lastly, the tail, which is in this case much used for supporting the animal and carrying out its chief movements, has acquired an extremely remarkable thickness and strength at its base.

These well-known facts are surely quite sufficient to establish the results of habitual use on an organ or any other part of animals. If on observing in an animal any organ particularly well-developed, strong, and powerful, it is alleged that its habitual use has nothing to do with it, that its continued disuse involves it in no loss, and finally, that this organ has always been the same since the creation of the species to which the animal belongs, then I ask, Why can our domestic ducks no longer fly like wild ducks? I can, in short, cite a multitude of instances among ourselves, which bear witness to the differences that accrue to us from the use or disuse of any of our organs, although these differences are not preserved in the new individuals which arise by reproduction: for if they were their effects would be far greater.

I shall show in Part II., that when the will guides an animal to any action, the organs which have to carry out that action are immediately stimulated to it by the influx of subtle fluids (the nervous fluid), which become the determining factor of the movements required. This fact is verified by many observations, and cannot now be called in question.

Hence it follows that numerous repetitions of these organised activities strengthen, stretch, develop and even create the organs necessary to them. We have only to watch attentively what is happening all around us, to be convinced that this is the true cause of organic development and changes.

Now every change that is wrought in an organ through a habit of frequently using it, is subsequently preserved by reproduction, if it is common to the individuals who unite together in fertilisation for the propagation of their species. Such a change is thus handed on to all succeeding individuals in the same environment, without their having to acquire it in the same way that it was actually created.

Furthermore, in reproductive unions, the crossing of individuals who have different qualities or structures is necessarily opposed to the permanent propagation of these qualities and structures. Hence it is that in man, who is exposed to so great a diversity of environment, the accidental qualities or defects which he acquires are not preserved and propagated by reproduction. If, when certain peculiarities of shape or certain defects have been acquired, two individuals who are both affected were always to unite together, they would hand on the same peculiarities; and if successive generations were limited to such unions, a special and distinct race would then be formed. But perpetual crossings between individuals, who have not the same peculiarities of shape, cause the disappearance of all peculiarities acquired by special action of the environment. Hence, we may be sure that if men were not kept apart by the distances of their habitations, the crossing in reproduction would soon bring about the disappearance of the general characteristics distinguishing different nations.

If I intended here to pass in review all the classes, orders, genera and species of existing animals, I should be able to show that the conformation and structure of individuals, their organs, faculties, etc., etc., are everywhere a pure result of the environment to which each species is exposed by its nature, and by the habits that the individuals composing it have been compelled to acquire; I should be able to show that they are not the result of a shape which existed from the beginning, and has driven animals into the habits they are known to possess.

It is known that the animal called the *ai* or sloth (*Bradypus tridactylus*) is permanently in a state of such extreme weakness that it only executes very slow and limited movements, and walks on the ground with difficulty. So slow are its movements that it is alleged that it can only take fifty steps in a day. It is known, moreover, that the organisation of this animal is entirely in harmony with its state of feebleness and incapacity for walking; and that if it wished to make other movements than those which it actually does make it could not do so.

Hence on the supposition that this animal had received its organisation from nature, it has been asserted that this organisation forced it into the habits and miserable state in which it exists.

This is very far from being my opinion; for I am convinced that the habits which the ai was originally forced to contract must necessarily have brought its organisation to its present condition.

If continual dangers in former times have led

the individuals of this species to take refuge in trees, to live there habitually and feed on their leaves, it is clear that they must have given up a great number of movements which animals living on the ground are in a position to perform. All the needs of the ai will then be reduced to clinging to branches and crawling and dragging themselves among them, in order to reach the leaves, and then to remaining on the tree in a state of inactivity in order to avoid falling off. This kind of inactivity, moreover, must have been continually induced by the heat of the climate; for among warm-blooded animals, heat is more conducive to rest than to movement.

Now the individuals of the race of the ai have long maintained this habit of remaining in the trees, and of performing only those slow and little varied movements which suffice for their needs. Hence their organisation will gradually have come into accordance with their new habits; and from this it must follow:

1. That the arms of these animals, which are making continual efforts to clasp the branches of trees, will be lengthened;
2. That the claws of their digits will have acquired a great length and a hooked shape, through the continued efforts of the animal to hold on;
3. That their digits, which are never used in making independent movements, will have entirely lost their mobility, become united and have preserved only the faculty of flexion or extension all together;
4. That their thighs, which are continually clasping either the trunk or large branches of trees, will have contracted a habit of always being separated, so as to lead to an enlargement of the pelvis and a backward direction of the cotyloid cavities;
5. Lastly, that a great many of their bones will be welded together, and that parts of their skeleton will consequently have assumed an arrangement and form adapted to the habits of these animals, and different from those which they would require for other habits.

This is a fact that can never be disputed; since nature shows us in innumerable other instances the power of environment over habit and that of habit over the shape, arrangement and proportions of the parts of animals.

Since there is no necessity to cite any further examples, we may now turn to the main point elaborated in this discussion.

It is a fact that all animals have special habits corresponding to their genus and species, and always possess an organisation that is completely in harmony with those habits.

It seems from the study of this fact that we may adopt one or other of the two following conclusions, and that neither of them can be verified.

Conclusion adopted hitherto: Nature (or her Author) in creating animals, foresaw all the possible kinds of environment in which they would have to live, and endowed each species with a fixed organisation and with a definite and invariable shape, which compel each species to live in the places and climates where we actually find them, and there to maintain the habits which we know in them.

My individual conclusion: Nature has produced all the species of animals in succession, beginning with the most imperfect or simplest, and ending her work with the most perfect, so as to create a gradually increasing complexity in their organisation; these animals have spread at large throughout all the habitable regions of the globe, and every species has derived from its environment the habits that we find in it and the structural modifications which observation shows us.

The former of these two conclusions is that which has been drawn hitherto, at least by nearly everyone: it attributes to every animal a fixed organisation and structure which never have varied and never do vary; it assumes, moreover, that none of the localities inhabited by animals ever vary; for if they were to vary, the same animals could no longer survive, and the possibility of finding other localities and transporting themselves thither would not be open to them.

The second conclusion is my own: it assumes that by the influence of environment on habit, and thereafter by that of habit on the state of the parts and even on organisation, the structure and organisation of any animal may undergo modifications, possibly very great, and capable of accounting for the actual condition in which all animals are found.

In order to show that this second conclusion is baseless, it must first be proved that no point on the surface of the earth ever undergoes variation as to its nature, exposure, high or low situation, climate, etc., etc.; it must then be proved that no part of animals undergoes even after long periods of time any modification due to a change of environment or to the necessity which forces them into a different kind of life and activity from what has been customary to them.

Now if a single case is sufficient to prove that an animal which has long been in domestication differs from the wild species whence it sprang, and if in any such domesticated species, great differences of conformation are found between the individuals exposed to such a habit and those which are forced into different habits, it will then be certain that the first conclusion is not consistent with the laws of nature, while the second, on the contrary, is entirely in accordance with them.

Everything then combines to prove my statement, namely: that it is not the shape either of the body or its parts which gives rise to the habits of animals and their mode of life; but that it is, on the contrary, the habits, mode of life and all the other influences of the environment which have in course of time built up the shape of the body and of the parts of animals. With new shapes, new faculties have been acquired, and little by little nature has succeeded in fashioning animals such as we actually see them.

Can there be any more important conclusion in the range of natural history, or any to which more attention should be paid than that which I have just set forth?

Let us conclude this Part I. with the principles and exposition of the natural order of animals.

FURTHER READING

Barsanti, G. *Dalla storia naturale alla storia della natura: Saggio su Lamarck.* Milan: Feltrinelli, 1979.

Barthélemy–Madaule, Madeleine. *Lamarck the Mythical Precursor: A Study of the Relations Between Science and Ideology.* Translated by M.H. Shank. Cambridge, MA: M.I.T. Press, 1982.

Bowler, Peter. *The Eclipse of Darwinism: Anti-Darwinian Evolution Theories in the Decades around 1900.* Baltimore: Johns Hopkins UP, 1983.

Burckhardt, Richard W. *The Spirit of System: Lamarck and Evolutionary Biology. Now With "Lamarck in 1995."* Cambridge, MA: Harvard UP, 1995.

Butler, Samuel. *Evolution Old and New.* London, 1879.

Cannon, H. Graham. *Lamarck and Modern Genetics.* Manchester: Manchester UP, 1959.

C.E.R.I.C. *Lamarck et son temps, Lamarck et notre temps.* Paris: J. Vrin, 1981.

Corsi, Pietro. *The Age of Lamarck: Evolutionary Theories in France 1790–1830.* Translated by Jonathan Mandelbaum. Berkeley: University of California Press, 1988.

Daudin, Henri. *Cuvier et Lamarck: Les Classes zoologiques et l'idée de série animale (1790–1830).* 2 vols. Paris: Félix Alcan, 1926.

Faure, Jean-Pierre. *Le Cas Lamarck.* Paris: Albert Blanchard, 1978.

Gillispie, C.C. "The Formation of Lamarck's Evolutionary Theory." *Archives internationales d'histoire des sciences* 9 (1956): 323–338.

Glass, B., O. Temkin, and W.L. Straus, eds. *Forerunners of Darwin: 1745–1859.* Baltimore: Johns Hopkins Press, 1959.

Hodge, M.J.S. "Lamarck's Science of Living Bodies." *British Journal for the History of Science* 5, no. 20 (1971): 323–352.

Jordanova, L.J. *Lamarck.* Oxford: Oxford UP, 1984.

Kühner, F. *Lamarck: Die Lehre vom Leben.* Jena: Eugen Diederichs, 1913.

Lamarck, J.-B. *Hydrogeology.* Translated by Albert V. Carozzi. Urbana: University of Illinois Press, 1964.

—. *Inédites de Lamarck.* Edited by Max Vachon. Paris: Masson et Compagnie, 1972.

—. *Recherches sur l'organisation des corps vivants.* Edited by Jean-Marc Drouin. 1802. Reprint, Paris: Fayard, 1986.

—. *Zoological Philosophy: An Exposition with Regard to the Natural History of Animals.* Translated by Hugh Elliot. 1914. Reprint, New York: Hafner Publishing Company, 1963.

—. *Zoological Philosophy: An Exposition with*

Regard to the Natural History of Animals. Translated by Hugh Elliot. Edited by Richard W. Burckhardt, Jr. 1914. Reprint, Chicago: University of Chicago Press, 1984.

Lamarck e il Lamarckismo: Atti del convegno Napoli, 1-3 dicembre 1988. Napoli: La Città del Sole, 1995.

Landrieu, Marcel. *Lamarck, le fondateur du transformisme: Sa vie, son oeuvre.* Paris: Société Zoologique de France, 1909.

Laurent, Goulven. *Paléontologie et évolution en France de 1800 à 1860: Une histoire des ideés de Cuvier et Lamarck à Darwin.* Paris: Éditions du C.T.H.S., 1987.

Livingstone, David N. "Natural Theology and Neo-Lamarckism: The Changing Context of Nineteenth-century Geography in the United States and Great Britain." *Annals of the Association of American Geographers* 74, no. 1 (1984): 9–28.

Lovejoy, A.O. *The Great Chain of Being: The Study of the History of an Idea.* Cambridge, MA: Harvard UP, 1936.

Métraux, Alexandre. "Jean-Baptiste Lamarck's Quest for Natural Species." *Science in Context* 9 (1996): 541–53.

Packard, A.S. *Lamarck, the Founder of Evolution: His Life and Work.* London: Longmans, Green, and Co., 1901.

Roger, J. *Les sciences de la vie dans la pensée française du XVIIIe siècle.* Paris: Armand Colin, 1963.

Roule, Louis. *Lamarck et l'interprétation de la nature.* Paris: Ernest Flammarion, 1927.

Schiller, J., ed. *Colloque internationale "Lamarck."* Paris: A. Blanchard, 1971.

Szyfman, L. *Jean-Baptiste Lamarck et son époque.* Paris: Masson, 1982.

Wagner, Adolf. *Geschichte des Lamarckismus als Einführung in die psycho-biologische Bewegung der Gegenwart.* Stuttgart: Franckh'sche Verlagshandlung, 1908.

7. JOHANN WOLFGANG GOETHE

(1749–1832)

Gott, heißt es, schied die Finsternis vom Licht,
Doch mocht' es ihm nicht ganz gelingen;
Denn wenn das Licht in Farben sich erbricht,
Mußt' es vorher die Finsternis verschlingen
(Goethe, *Zahme Xenien* VIII).[1]

Universally recognized as Germany's greatest poet, the author of *Faust* and *Die Leiden des jungen Werthers*, Goethe also spent a substantial part of his career in scientific pursuits and even believed that his scientific discoveries would outlast his poetic achievements. Goethe's scientific and literary work cannot be neatly separated, as evidenced by his novel *Die Wahlverwandtschaften* (*The Elective Affinities*) with its controlling metaphor of chemical affinity applied to human relationships.

Goethe studied law in Strasbourg and was invited to Karl August's court at Weimar in 1775, where he superintended roadworks, mining, and the state theatre. In 1784, Goethe claimed to have established the existence of the intermaxillary bone in humans. His discovery was based partly on comparative osteology, and partly on his belief in archetypes, the Platonic-inspired idea that primal or archetypal phenomena (*Urphänomene*) underlie all of nature. In the botanical world, Goethe viewed individual plant species as realized variations of the archetypal plant, the *Urpflanze*, described in the *Metamorphose der Pflanzen* (1790). Similarly in geology, Goethe believed that most rock strata were chemically precipitated on top of a primal granite (*Urgebirge*).

Goethe's work on colours, not surprisingly therefore, was founded on the belief in the polarity of lightness and darkness as the primal phenomenon underlying the creation of all colour. His studies of coloured afterimages in the eye convinced him that colour represented a struggle of opposing forces—light and dark—and that the eye "demanded" missing colours in order to compose a harmony. This belief in a fundamental cosmic duality (*Zweiheit*) expressed paradoxically as a unity demonstrates the influence of Spinoza on his thought. Goethe's interest in colours began with the *Beiträge zur Optik* (1791–92), and culminated in *Zur Farbenlehre* (1810), usually translated as "Theory of Colours" but perhaps better rendered as "teachings" on colour.

Zur Farbenlehre details Goethe's experiments and observations on light, explains his polarity theory, and provides an historical sketch of ideas on colour from antiquity to the late eighteenth century. Goethe divided the *Farbenlehre* into a Didactic, Polemical, and Historical

1 God, it is written, divided the Darkness from the Light
But he could not completely divide it;
For when he emanated light into colours
He first had to interlace the darkness.

part, and in the Polemical section launched numerous intemperate attacks on Newton's *Opticks* (1704). Newton had separated white light into component colours with a glass prism and theorized that white light constituted a mixture of coloured rays, all exhibiting the property of differential refrangibility (each coloured ray is refracted at different angles by the prism causing the separation of white light into monochromatic rays). For Goethe, on the other hand, white light represented an unanalyzable, primary element. It is difficult to find a winner and loser in the Newton-Goethe debate since they were not investigating exactly the same phenomena, with Newton concerned with the physical nature of light and Goethe more interested in the interaction of light and the human eye and mind. Colour theories based on the Newtonian model have had much greater explanatory value, but Goethe's work raised some still unsolved questions about colour perception which bridge the modern disciplines of psychology, physiology, physics, and chemistry.

Hermann von Helmholtz, who studied the physiological effects of colour in the mid-nineteenth century, wrote in his essay "Goethe's Scientific Researches" (1853) that Goethe's poetic nature was incompatible with the mathematical sciences. In another essay, however, "Goethe's Presentiment of Coming Scientific Ideas" (1892), Helmholtz praised Goethe's intuitions about the flaws in the Newtonian view of light; Helmholtz, however, could not fathom how a mind of Goethe's breadth could have misunderstood optical effects clearly accounted for by Newton's theory.

The twentieth-century physicist Werner Heisenberg in his *Philosophic Problems* felt that Goethe's battle with Newton over colour was based on science's traditional division of reality into the subjective (our sense of colour) and the objective (numerical measurement of wavelength), but concluded that this dichotomy was too simplistic. Heisenberg, like Goethe, understood that the experimenter (the subject) in the act of artificially observing nature (the object), necessarily changes that nature, an observation which Heisenberg quantified in his Uncertainty Principle. The experimenter, according to Goethe, acts as a mediator between the subject and object of the experiment, and the subject/object dichotomy in scientific practice becomes a blurry one.

There has been some interest in this century in Goethe's work as a precursor of the modern science of colour perception (see for example the experiments of Edwin Land, who created the full colour spectrum from two rotating monochromatic lights). Goethe's work also stands as a reminder to modern researchers that colour perception is not simply a function of the wavelength of light (i.e., the physical or objective nature of light) acting on light receptors in the retina, which is implied in the Newtonian and later Young-Helmholtzian three receptor model. As Wolfgang Buchheim points out, "colour theory, according to modern understanding, is a collective term for the contributions of many disciplines on colour phenomena, and the perceptual conditions of these phenomena inside and outside of the eye" (1). Recent historical studies by Dennis L. Sepper and Frederick Burwick demonstrate that the Newton-Goethe debate, far from simply illustrating the triumph of Newton's colour theory, is still a fruitful area of enquiry which can teach us about scientific method and the construction of scientific theories. The following selection is from Charles Lock Eastlake's standard translation of the first Didactic section of the *Farbenlehre*, which has been reprinted by M.I.T. Press (1970) and by Rupprecht Matthaei in facsimile.

Goethe's Theory of Colours.

Translated by Charles Lock Eastlake. London: John Murray, 1840, pp. 273–82, 316–20.

PART IV.

GENERAL CHARACTERISTICS.

688.
We have hitherto, in a manner forcibly, kept phenomena asunder, which, partly from their nature, partly in accordance with our mental habits, have, as it were, constantly sought to be reunited. We have exhibited them in three divisions. We have considered colours, first, as transient, the result of an action and re-action in the eye itself; next, as passing effects of colourless, light-transmitting, transparent, or opaque mediums on light; especially on the luminous image; lastly, we arrived at the point where we could securely pronounce them as permanent, and actually inherent in bodies.

689.
In following this order we have as far as possible endeavoured to define, to separate, and to class the appearances. But now that we need no longer be apprehensive of mixing or confounding them, we may proceed, first, to state the general nature of these appearances considered abstractedly, as an independent circle of facts, and, in the next place, to show how this particular circle is connected with other classes of analogous phenomena in nature.

THE FACILITY WITH WHICH COLOUR APPEARS.

690.
We have observed that colour under many conditions appears very easily. The susceptibility of the eye with regard to light, the constant re-action of the retina against it, produce instantaneously a slight iridescence. Every subdued light may be considered as coloured, nay, we ought to call any light coloured, inasmuch as it is seen. Colourless light, colourless surfaces, are, in some sort, abstract ideas; in actual experience we can hardly be said to be aware of them.—Note Z.

691.
If light impinges on a colourless body, is reflected from it or passes through it, colour immediately appears; but it is necessary here to remember what has been so often urged by us, namely, that the leading conditions of refraction, reflection, etc., are not of themselves sufficient to produce the appearance. Sometimes, it is true, light acts with these merely as light, but oftener as a defined, circumscribed appearance, as a luminous image. The semi-opacity of the medium is often a necessary condition; while half, and double shadows, are required for many coloured appearances. In all cases, however, colour appears instantaneously. We find, again, that by means of pressure, breathing heat (432, 471), by various kinds of motion and alteration on smooth clean surfaces (461), as well as on colourless fluids (470), colour is immediately produced.

692.
The slightest change has only to take place in the component parts of bodies, whether by immixture with other particles or other such effects, and colour either makes its appearance or becomes changed.

THE FORCE OF COLOUR.

693.
The physical colours, and especially those of the prism, were formerly called *"colores emphatici,"* on account of their extraordinary beauty and force. Strictly speaking, however, a high degree of effect may be ascribed to all appearances of colour, assuming that they are exhibited under the purest and most perfect conditions.

694.
The dark nature of colour, its full rich quality, is what produces the grave, and at the same time fascinating impression we sometimes experience, and as colour is to be considered a condition of light, so it cannot dispense with light as the co-operating cause of its appearance, as its basis or

ground; as a power thus displaying and manifesting colour.

THE DEFINITE NATURE OF COLOUR.

695.
The existence and the relatively definite character of colour are one and the same thing. Light displays itself and the face of nature, as it were, with a general indifference, informing us as to surrounding objects perhaps devoid of interest or importance; but colour is at all times specific, characteristic, significant.

696.
Considered in a general point of view, colour is determined towards one of two sides. It thus presents a contrast which we call a polarity, and which we may fitly designate by the expressions *plus* and *minus*.

Plus.	*Minus.*
Yellow.	Blue.
Action.	Negation.[2]
Light.	Shadow.
Brightness.	Darkness.
Force.	Weakness.
Warmth.	Coldness.
Proximity.	Distance.
Repulsion.	Attraction.
Affinity with acids.	Affinity with alkalis.

COMBINATION OF THE TWO PRINCIPLES

697.
If these specific, contrasted principles are combined, the respective qualities do not therefore destroy each other: for if in this intermixture the ingredients are so perfectly balanced that neither is to be distinctly recognised, the union again acquires a specific character; it appears as a quality by itself in which we no longer think of combination. This union we call green.

698.
Thus, if two opposite phenomena springing from the same source do not destroy each other when combined, but in their union present a third appreciable and pleasing appearance, this result at once indicates their harmonious relation. The more perfect result yet remains to be adverted to.

AUGMENTATION TO RED.

699.
Blue and yellow do not admit of increased intensity without presently exhibiting a new appearance in addition to their own. Each colour, in its lightest state, is a dark; if condensed it must become darker, but this effect no sooner takes place than the hue assumes an appearance which we designate by the word reddish.

700.
This appearance still increases, so that when the highest degree of intensity is attained it predominates over the original hue. A powerful impression of light leaves the sensation of red on the retina. In the prismatic yellow-red which springs directly from the yellow, we hardly recognise the yellow.

701.
This deepening takes place again by means of colourless semi-transparent mediums, and here we see the effect in its utmost purity and extent. Transparent fluids, coloured with any given hues, in a series of glass-vessels, exhibit it very strikingly. The augmentation is unremittingly rapid and constant; it is universal, and obtains in physiological as well as in physical and chemical colours.

JUNCTION OF THE TWO AUGMENTED EXTREMES.

702.
As the extremes of the simple contrast produce a beautiful and agreeable appearance by their union, so the deepened extremes on being united, will present a still more fascinating colour; indeed, it might naturally be expected that we

2 Wirkung, Beraubung; the last would be more literally rendered *privation*. The author has already frequently made use of the terms *active* and *passive* as equivalent to *plus* and *minus*.–T.

should here find the acme of the whole phenomenon.

703.
And such is the fact, for pure red appears; a colour to which, from its excellence, we have appropriated the term "purpur."[3]

704.
There are various modes in which pure red may appear. By bringing together the violet edge and yellow-red border in prismatic experiments, by continued augmentation in chemical operations, and by the organic contrast in physiological effects.

705.
As a pigment it cannot be produced by intermixture or union, but only by arresting the hue in substances chemically acted on, at the high culminating point. Hence the painter is justified in assuming that there are *three* primitive colours from which he combines all the others. The natural philosopher, on the other hand, assumes only *two* elementary colours, from which he, in like manner, developes and combines the rest.

COMPLETENESS THE RESULT OF VARIETY IN COLOUR.

706.
The various appearances of colour arrested in their different degrees, and seen in juxtaposition, produce a whole. This totality is harmony to the eye.

707.
The chromatic circle has been gradually presented to us; the various relations of its progression are apparent to us. Two pure original principles in contrast, are the foundation of the whole; an augmentation manifests itself by means of which both approach a third state; hence there exists on both sides a lowest and highest, a simplest and most qualified state. Again, two combinations present themselves; first that of the simple primitive contrasts, then that of the deepened contrasts.

HARMONY OF THE COMPLETE STATE.

708.
The whole ingredients of the chromatic scale, seen in juxtaposition, produce an harmonious impression on the eye. The difference between the physical contrast and harmonious opposition in all its extent should not be overlooked. The first resides in the pure restricted original dualism, considered in its antagonizing elements; the other results from the fully developed effects of the complete state.

709.
Every single opposition in order to be harmonious must comprehend the whole. The physiological experiments are sufficiently convincing on this point. A development of all the possible contrasts of the chromatic scale will be shortly given.[4]

FACILITY WITH WHICH COLOUR MAY BE MADE TO TEND EITHER TO THE PLUS OR MINUS SIDE.

710.
We have already had occasion to take notice of the mutability of colour in considering its so-called augmentation and progressive variations round the whole circle; but the hues even pass and repass from one side to the other, rapidly and of necessity.

711.
Physiological colours are different in appearance as they happen to fall on a dark or on a light ground. In physical colours the combination of the objective and subjective experiments is very remarkable. The epoptical colours, it appears, are contrasted according as the light shines through or upon them. To what extent the chemical

3 Wherever this word occurs incidentally it is translated *pure red*, the English word *purple* being generally employed to denote a colour similar to violet.–T.

4 No diagram or table of this kind was ever given by the author.–T.

colours may be changed by fire and alkalis, has been sufficiently shown in its proper place.

EVANESCENCE OF COLOUR.

712.
All that has been adverted to as subsequent to the rapid excitation and definition of colour, immixture, augmentation, combination, separation, not forgetting the law of compensatory harmony, all takes place with the greatest rapidity and facility; but with equal quickness colour again altogether disappears.

713.
The physiological appearances are in no wise to be arrested; the physical last only as long as the external condition lasts; even the chemical colours have great mutability, they may be made to pass and repass from one side to the other by means of opposite re-agents, and may even be annihilated altogether.

PERMANENCE OF COLOUR.

714.
The chemical colours afford evidence of very great duration. Colours fixed in glass by fusion, and by nature in gems, defy all time and re-action.

715.
The art of dyeing again fixes colour very powerfully. The hues of pigments which might otherwise be easily rendered mutable by reagents, may be communicated to substances in the greatest permanency by means of mordants.

COMPLETENESS AND HARMONY.

803.
We have hitherto assumed, for the sake of clearer explanation, that the eye can be compelled to assimilate or identity itself with a single colour; but this can only be possible for an instant.

804.
For when we find ourselves surrounded by a given colour which excites its corresponding sensation on the eye, and compels us by its presence to remain in a state identical with it, this state is soon found to be forced, and the organ unwillingly remains in it.

805.
When the eye sees a colour it is immediately excited, and it is its nature, spontaneously and of necessity, at once to produce another, which with the original colour comprehends the whole chromatic scale. A single colour excites, by a specific sensation, the tendency to universality.

806.
To experience this completeness, to satisfy itself, the eye seeks for a colourless space next every hue in order to produce the complemental hue upon it.

807.
In this resides the fundamental law of all harmony of colours, of which every one may convince himself by making himself accurately acquainted with the experiments which we have described in the chapter on the physiological colours.

808.
If, again, the entire scale is presented to the eye externally, the impression is gladdening, since the result of its own operation is presented to it in reality. We turn our attention therefore, in the first place, to this harmonious juxtaposition.

809.
As a very simple means of comprehending the principle of this, the reader has only to imagine a moveable diametrical index in the colorific circle.[5] The index, as it revolves round the whole circle, indicates at its two extremes the complemental colours, which, after all, may be reduced to three contrasts.

5 Plate 1, fig. 3.

810.

Yellow demands Red-blue,
Blue " " Red-yellow,
Red " " Green,
and contrariwise.

811.

In proportion as one end of the supposed index deviates from the central intensity of the colours, arranged as they are in the natural order, so the opposite end changes its place in the contrasted gradation, and by such a simple contrivance the complemental colours may be indicated at any given point. A chromatic circle might be made for this purpose, not confined, like our own, to the leading colours, but exhibiting them with their transitions in an unbroken series. This would not be without its use, for we are here considering a very important point which deserves all our attention.[6]

812.

We before stated that the eye could be in some degree pathologically affected by being long confined to a single colour; that, again, definite moral impressions were thus produced, at one time lively and aspiring, at another susceptible and anxious—now exalted to grand associations, now reduced to ordinary ones. We now observe that the demand for completeness, which is inherent in the organ, frees us from this restraint; the eye relieves itself by producing the opposite of the single colour forced upon it, and thus attains the entire impression which is so satisfactory to it.

813.

Simple, therefore, as these strictly harmonious contrasts are, as presented to us in the narrow circle, the hint is important, that nature tends to emancipate the sense from confined impressions by suggesting and producing the whole, and that in this instance we have a natural phenomenon immediately applicable to aesthetic purposes.

814.

While, therefore, we may assert that the chromatic scale, as given by us, produces an agreeable impression by its ingredient hues, we may here remark that those have been mistaken who have hitherto adduced the rainbow as an example of the entire scale; for the chief colour, pure red, is deficient in it, and cannot be produced, since in this phenomenon, as well as in the ordinary prismatic series, the yellow-red and blue-red cannot attain to a union.

815.

Nature perhaps exhibits no general phenomenon where the scale is in complete combination. By artificial experiments such an appearance may be produced in its perfect splendour. The mode, however, in which the entire series is connected in a circle, is rendered most intelligible by tints on paper, till after much experience and practice, aided by due susceptibility of the organ, we become penetrated with the idea of this harmony, and feel it present in our minds.

NOTE Z.—Par. 690.

The author appears to mean that a degree of brightness which the organ can bear at all, must of necessity be removed from dazzling, white light. The slightest tinge of colour to this brightness, implies that it is seen through a medium, and thus, in painting, the lightest, whitest surface should partake of the quality of depth. Goethe's view here again accords, it must be admitted, with the practice of the best colourists, and with the precepts of the highest authorities.—See Note C. [Eastlake's Note.]

NOTE C—Par. 50.

Every treatise on the harmonious combination of colours contains the diagram of the chromatic circle more or less elaborately constructed. These diagrams, if intended to exhibit the contrasts pro-

6 See Note C.

duced by the action and re-action of the retina, have one common defect. The opposite colours are made equal in intensity; whereas the complemental colour pictured on the retina is always less vivid, and always darker or lighter than the original colour. This variety undoubtedly accords more with harmonious effects in painting.

The opposition of two pure hues of equal intensity, differing only in the abstract quality of colour, would immediately be pronounced crude and inharmonious. It would not, however, be strictly correct to say that such a contrast is too violent; on the contrary, it appears the contrast is not carried far enough, for though differing in colour, the two hues may be exactly similar in purity and intensity. Complete contrast, on the other hand, supposes dissimilarity in all respects.

In addition to the mere difference of hue, the eye, it seems, requires difference in the lightness or darkness of the hue. The spectrum of a colour relieved as a dark on a light ground, is a light colour on a dark ground, and *vice versâ*. Thus, if we look at a bright red wafer on the whitest surface, the complemental image will be still lighter than the white surface; if the same wafer is placed on a black surface, the complemental image will be still darker. The colour of both these spectra may be called greenish, but it is evident that a colour must be scarcely appreciable as such, if it is lighter than white and darker than black. It is, however, to be remarked, that the white surface round the light greenish image seems tinged with a reddish hue, and the black surface round the dark image becomes slightly illuminated with the same colour, thus in both cases assisting to render the image apparent (58).

The difficulty or impossibility of describing degrees of colour in words, has also had a tendency to mislead, by conveying the idea of more positive hues than the physiological contrast warrants. Thus, supposing scarlet to be relieved as a dark, the complemental colour is so light in degree and so faint in colour, that it should be called a pearly grey; whereas the theorists, looking at the quality of colour abstractedly, would call it a green-blue, and the diagram would falsely present such a hue equal in intensity to scarlet, or as nearly equal as possible.

Even the difference of mass which good taste requires may be suggested by the physiological phenomena, for unless the complemental image is suffered to fall on a surface precisely as near to the eye as that on which the original colour was displayed, it appears larger or smaller than the original object (22), and this in a rapidly increasing proportion. Lastly, the shape itself soon becomes changed (26).

That vivid colour demands the comparative absence of colour, either on a lighter or darker scale, as its contrast, may be inferred again from the fact that bright colourless objects produce strongly coloured spectra. In darkness, the spectrum which is first white, or nearly white, is followed by red: in light, the spectrum which is first black, is followed by green (39–44). All colour, as the author observes (259), is to be considered as half-light, inasmuch as it is in every case lighter than black and darker than white. Hence no contrast of colour with colour, or even of colour with black or white, can be so great (as regards lightness or darkness) as the contrast of black and white, or light and dark abstractedly. This distinction between the differences of degree and the differences of kind is important, since a just application of contrast in colour may be counteracted by an undue difference in lightness or darkness. The mere contrast of colour is happily employed in some of Guido's lighter pictures, but if intense darks had been opposed to his delicate carnations, their comparative whiteness would have been unpleasantly apparent. On the other hand, the flesh-colour in Giorgione, Sebastian del Piombo (his best imitator), and Titian, was sometimes so extremely glowing[7] that the deepest colours, and black, were indispensable accom-

7 "Ardito veramente alquanto, sanguigno, e quasi fiammeggiante."–*Zanetti della Pittura Veneziana*, Ven. 1771, p. 90. Warm as the flesh colour of the colourists is, it still never approaches a positive hue, if we except some examples in frescoes and other works intended to be seen at a great distance. Zanetti, speaking of a fresco by Giorgione, now almost obliterated, compares the colour to "un vivo raggio di cocente sole."–*Varie Pitture a fresco dei Principali Maestri Veneziani*. Ven. 1760.

paniments. The manner of Titian as distinguished from his imitation of Giorgione, is golden rather than fiery, and his biographers are quite correct in saying that he was fond of opposing red (lake) and blue to his flesh.[8] The correspondence of these contrasts with the physiological phenomena will be immediately apparent, while the occasional practice of Rubens in opposing bright red to a still cooler flesh-colour, will be seen to be equally consistent.

The effect of white drapery (the comparative absence of colour) in enhancing the glow of Titian's flesh-colour, has been frequently pointed out:[9] the shadows of white thus opposed to flesh, often present, again, the physiological contrast, however delicately, according to the hue of the carnation. The lights, on the other hand, are not, and probably never were, quite white, but from the first, partook of the quality of depth, a quality assumed by the colourists to pervade every part of a picture more or less.[10]

It was before observed that the description of colours in words may often convey ideas of too positive a nature, and it may be remarked generally that the colours employed by the great masters are, in their ultimate effect, more or less subdued or broken. The physiological contrasts are, however, still applicable in the most comparatively neutral scale.

Again, the works of the colourists show that these oppositions are not confined to large masses (except perhaps in works to be seen only at a great distance); on the contrary, they are more or less apparent in every part, and when at last the direct and intentional operations of the artist may have been insufficient to produce them in their minuter degrees, the accidental results of glazing and other methods may be said to extend the contrasts to infinity. In such productions, where every smallest portion is an epitome of the whole, the eye still appreciates the fascinating effect of contrast, and the work is pronounced to be true and complete, in the best sense of the words.

The Venetian method of scumbling and glazing exhibits these minuter contrasts within each other, and is thus generally considered more refined than the system of breaking the colours, since it ensures a fuller gradation of hues, and produces another class of contrasts, those, namely, which result from degrees of transparence and opacity. In some of the Flemish and Dutch masters, and sometimes in Reynolds, the two methods are combined in great perfection.

The chromatic diagram does not appear to be older than the last century. It is one of those happy adaptations of exacter principles to the objects of taste which might have been expected from Leonardo da Vinci. That its true principle was duly felt is abundantly evident from the works of the colourists, as well as from the general observations of early writers.[11] The more practical directions occasionally to be met with in the treatises of Leon Battista Alberti, Leonardo da Vinci and others, are conformable to the same system. Some Italian works, not written by painters, which pretend to describe this harmony, are, however, very imperfect.[12] A passage in Lodovico Dolce's Dialogue on Colours is perhaps the only one worth quoting. "He," says that writer, "who wishes to combine colours that are agreeable to the eye, will put grey next dusky orange; yellow-green next rose-colour; blue next orange; dark purple, black, next dark-green;

8 Ridolfi.

9 Zanetti, l. ii.

10 Two great authorities, divided by more than three centuries, Leon Battista Alberti and Reynolds, have recommended this subdued treatment of white. "It is to be remembered," says the first, "that no surface should be made so white that it cannot be made more so. In white dresses again, it is necessary to stop far short of the last degree of whiteness."–*Della Pittura*, l. ii., compare with Reynolds, vol. i. dis. 8.

11 Vasari observes, "L'unione nella pittura è una discordanza di colori diversi accordati insième."–Vol. i. c. 18. This observation is repeated by various writers on art in nearly the same words, and at last appears in Sandrart: "Concordia, potissimum picturae decus, in discordiâ consistit, et quasi litigio colorum."–P. i. c. 5. The source, perhaps, is Aristotle: he observes, "We are delighted with harmony, because it is the union of contrary principles having a ratio to each other."–*Problem*.

12 See "Occolti Trattato de' Colori." Parma, 1568.

white next black, and white next flesh-colour."[13] The Dialogue on Painting, by the same author, has the reputation of containing some of Titian's precepts: if the above passage may be traced to the same source, it must be confessed that it is almost the only one of the kind in the treatise from which it is taken. [Eastlake's Note.]

FURTHER READING

Amrine, Frederick, Francis J. Zucker and Harvey Wheeler, eds. *Goethe and the Sciences: A Reappraisal.* Vol. 97. Boston Studies in the Philosophy of Science. Dordrecht: D. Reidel Publishing Company, 1987.

—. *Goethe in the History of Science.* 2 vols. New York: Peter Lang, 1996.

Bjerke, André. *Neue Beiträge zu Goethes Farbenlehre.* Translated by Louise Funk. Stuttgart: Verlag Freies Geistesleben, 1961.

Buchheim, Wolfgang. *Der Farbenlehrstreit Goethes mit Newton in wissenschaftsgeschichtlicher Sicht.* Berlin: Akademie Verlag, 1991.

Burwick, Frederick. *The Damnation of Newton: Goethe's Color Theory and Romantic Perception.* Berlin: Walter de Gruyter, 1986.

Élie, Maurice. *Lumière, couleurs et nature: L'Optique et la physique de Goethe et de la Naturphilosophie.* Paris: Librairie Philosophique J. Vrin, 1993.

Fink, Karl J. *Goethe's History of Science.* Cambridge: Cambridge UP, 1991.

Glockner, Hermann. *Das Philosophische Problem in Goethes Farbenlehre.* Heidelberg: Carl Winter Universitätsbuchhandlung, 1924.

Heisenberg, Werner. *Philosophic Problems of Nuclear Science.* Translated by F.C. Hayes. London: Faber and Faber, 1952.

Höpfner, Felix. *Wissenschaft wider die Zeit: Goethes Farbenlehre aus rezeptionsgeschichtlicher Sicht, mit einer Bibliographie zur Farbenlehre.* Heidelberg: Carl Winter Universitätsverlag, 1990.

Land, Edwin. "The Retinex Theory of Color Vision." *Scientific American* 237, no. 6 (1977): 108–28.

—. "Experiments in Color Vision." *Scientific American* 200, no. 5 (1959): 84–99.

Lobeck, Fritz. *Erfahrungen mit Goethes Farbenlehre an Iris, Halo, Hof.* Leipzig: Heitz und Co., 1937.

Lütkehaus, Ludger, ed. *Arthur Schopenhauer: Der Briefwechsel mit Goethe und Andere Dokumente Zur Farbenlehre.* Zürich: Haffmans Verlag, 1992.

MacAdam, David L., ed. *Sources of Color Science.* Cambridge, MA: M.I.T. Press, 1970.

Martin, M. *Die Kontroverse um die Farbenlehre: Anschauliche Darstellung der Forschungwege von Newton und Goethe.* Schaffhausen: Novalis-Verlag, 1979.

Matthei, Rupprecht. *Goethe's Color Theory.* Translated by Herb Aach. New York: Van Nostrand Reinhold Company, 1971.

Rehbock, Theda. *Goethe und die "Rettung der Phänomene": Philosophische Kritik des naturwissenschaftlichen Weltbilds am Beispiel der Farbenlehre.* Konstanz: Verlag am Hockgraben, 1995.

Schmidt, Peter. *Goethes Farbensymbolik: Untersuchungen zu Verwendung und Bedeutung der Farben in den Dichtungen und Schriften Goethes.* Berlin: Erich Schmidt Verlag, 1965.

Schöne, Albrecht. *Goethes Farbentheologie.* München: C.H. Beck, 1987.

Schönherr, Hartmut R. *Einheit und Werden: Goethes Newton-Polemik als systematische Konsequenz seiner Naturkonzeption.* Würzburg: Königshausen und Neumann, 1993.

Seamon, David, and Arthur Zajonc, eds. *Goethe's Way of Science: A Phenomenology of Nature.* Albany: State University of New York Press, 1998.

Sepper, Dennis L. *Goethe contra Newton: Polemics and the Project for a New Science of Color.* Cambridge: Cambridge UP, 1988.

Westphal, Otto. *Die Weltgeschichte im Spiegel von Goethes Farbenlehre.* Stuttgart: W. Kohlhammer, 1957.

13 "Volendo l'uomo accoppiare insième colori che all'occhio dilettino–porrà insième il berrettino col leonato; il verde-giallo con l'incarnato e rosso; il turchino con l'arangi; il morello col verde oscuro; il nero col bianco; il bianco con l'incarnato."–*Dialogo di M. Lodovico Dolce nel quale si ragiona della qualità, diversità, e proprietà de' colori.* Venezia, 1565.

8. ALEXANDER VON HUMBOLDT

(1769–1859)

Humboldt was born in Prussia during the reign of Frederick the Great. His father was a Prussian soldier and his mother a wealthy heiress from a family of French Huguenot plate-glass manufacturers. His brother Wilhelm became a noted philologist and diplomat. In 1787, Humboldt began university studies in political economy at Frankfurt and then moved on to Göttingen where he studied chemistry and physics for one year. His early interest in world travel was stimulated by his friendship with Georg Forster, who had accompanied his father on Cook's second world voyage.

Humboldt studied mining in Freiberg with A.G. Werner, the founder of the geological school of Geognosy, which believed that rocks had originated as precipitates from a primeval ocean. Humboldt entered civil service in the Prussian Mining Department in 1792. He designed mining lamps and reorganized the mines to the extent of setting up mining schools for the labourers in 1793. Humboldt's mother died in 1796, providing him with the financial independence which allowed him to resign from the mining service and to devote himself entirely to science and travel.

Humboldt was invited to participate in an Egyptian journey with Lord Bristol, and Bougainville also asked him to accompany him on a world voyage, but both projected journeys were abandoned due to military hostilities in Europe and the arrest of Lord Bristol for suspected spying. Humboldt instead crossed the Pyrenees to Madrid with the botanist Aimé Bonpland, making astronomical observations as they went. In 1799, they set sail from Spain with unprecedented permission from King Carlos IV to investigate the Spanish new world possessions. Humboldt and Bonpland visited Cuba, Venezuela, Ecuador, Colombia, Peru, and Mexico while Humboldt filled page after page of his *Reisetagebücher*, and collected animal, plant, and geological specimens. These travel books have been published with commentary by the Akademie der Wissenschaften der DDR. In South America, Humboldt confirmed that the Amazon and Orinoco Rivers were connected and he ascended Mount Chimborazo in the Andes to a height of 19,286 feet, a world record. Humboldt also made extensive notes on the economic structure and social practices of the colonies he visited, including the institution of slavery. One of Humboldt's first observations when he landed in Cuba in 1800 was the increasing uneasiness of the plantation owners and colonists as the number of imported African slaves rose in the colony (Beck, *Alexander von Humboldt* 1.174). His insights and records provide us with invaluable witness to the Spanish colonial system.

Humboldt left South America in 1804 and travelled north to the United States, where he was elected to the American Philosophical Society in Philadelphia. He also met with President Jefferson, himself an avid amateur natural philosopher. Returning to Europe in 1804, Humboldt began the publication of his journals as *Voyage aux régions équinoxiales du nouveau continent* (1805–34), eventually reaching 35 volumes. Humboldt was aided in the publication and classification of botanical specimens by Bonpland and C.S. Kunth, who brought out the notable *Nova genera et species plantarum* (1815–25). In 1829, Humboldt travelled through Siberia on another scientific expedition. In his later life, he was best known for his popular and encyclopedic *Kosmos* (1845–1862), translated into English by Elsie C. Otté. Humboldt's works enjoyed great popularity in Europe and several English translations of his books appeared in his lifetime: *Personal Narrative of Travels to the Equinoctial Regions of the New Continent During the Years 1799–1804, By Alexander de Humboldt and Aimé Bonpland* (1814–29) and *Researches, Concerning the Institutions and Monuments of the Ancient Inhabitants of America, with Descriptions and Views of Some of the Most Striking Scenes in the Cordilleras!* (1814) both translated by Helen Maria Williams. Humboldt's *Views of Nature: or, Contemplations on the Sublime Phenomena of Creation; with Scientific Illustrations*, translated by Elsie C. Otté and Henry G. Bohn, appeared in 1850.

With one foot in the rationalism and Newtonianism of the eighteenth century and the other in early nineteenth-century romanticism, Humboldt was influenced by German Romantic science and the work of Goethe, with its emphasis on totalities and harmonies in nature. W. Petri has called Humboldt's scientific philosophy a "kosmischer Humanismus," or Cosmic Humanism. For Humboldt, cosmos is "the total nature in which humankind is not only just a part, but also in which humans can and will find the meaning of their existence and the way to realize it" (Petri 480).

Humboldt's influence on travel literature and scientific travel cannot be underestimated, especially in Britain and France. As W.H. Brock points out, "Humboldt's writings, much as Darwin's *Journal of Researches* later, conveyed urban Britons on a magic carpet to places of extraordinary colour, beauty and strangeness" (Brock 365). The scientific journey narrative became an important literary genre in the nineteenth century, with contributions from Lewis and Clark, Mungo Park, Alfred Russel Wallace, Francis Galton, Charles Lyell, Charles Darwin, and John McDouall Stuart, who explored the interior of Australia. In addition, Henry Stanley's controversial relief in 1871 near Lake Tanganyika of Dr. David Livingstone, who was searching for the source of the Nile and carrying out missionary work, captured the imagination of the British public. The selection below illustrates Humboldt's method of observation and his frustrations in collecting and preserving specimens in *terra incognita*. Because of his thorough and precise measurements and observations of the agriculture, geography, climate, economics, and sociology of the island of Cuba, Humboldt was called "el segundo descubridor de Cuba" (the second discoverer of Cuba) by the Cuban José de la Luz y Caballero. This dubiously honorary title is inscribed on the base of Humboldt's statue in front of the Humboldt-Universität in Berlin. Humboldt was also designated "el segundo Colón" (the second Columbus) by Manuel Nicolás Corpancho.

The Island of Cuba.

Translated by J.S. Thrasher. New York: Derby and Jackson, 1856, pp. 351–61.

Toward the close of April,[1] 1801, Monsieur Bonpland and myself, having completed the series of observations we had proposed making on the extreme northern limit of the torrid zone, were about to depart for Vera Cruz with the squadron of Admiral Aristizabal; but the false intelligence contained in the public gazettes, relative to the expedition of Captain Baudin, induced us to abandon the project we had entertained, of crossing Mexico on our way to the Philippine Islands. Many papers, and particularly those of the United States, announced that two French corvettes, the *Géographe* and the *Naturaliste*, had sailed for Cape Horn, and would run along the coasts of Chili and Peru, from whence they were to proceed to New Holland.

This news excited me greatly, for it again filled my imagination with the projects I had formed during my stay in Paris, when I had not ceased for a moment to urge the ministry of the Directory to hasten the departure of Captain Baudin. While on the point of leaving Spain, I had promised to join the expedition wherever I might be able to reach it. When one desires a thing that may produce untoward results, he easily persuades himself that a sense of obligation is the only motive that influences his determination. Monsieur Bonpland, always enterprising and confident in our good fortune, determined at once to divide our collection of plants into three parts.

In order not to expose all that we had collected, with so much labor, on the banks of the Orinoco, Atabapo, and Rio Negro, to the chances of a long sea voyage, we sent one part to Germany by way of England, another to France by way of Cadiz; and left the third at Havana. We afterwards had reason to congratulate ourselves on the adoption of this course, which prudence counselled. Each part contained, with slight difference, the same species and classes, and no precaution was omitted to secure the remission of the cases to Sir Joseph Banks, or to the directors of the Museum of Natural History at Paris, in case they should fall into the hands of English or French cruisers.

Fortunately, the manuscripts which I had at first intended to send with the portion sent to Cadiz, were not placed in charge of our friend and fellow traveller, friar Juan Gonzalez. This estimable young man, of whom I have often had occasion to speak, had accompanied us to Havana, on his way to Spain, and sailed from Cuba shortly after our departure; but the vessel in which he embarked was lost with all her passengers and freight, in a tempest on the coast of Africa. By this shipwreck we lost one of the duplicates of our collection of plants; and also, which was a greater misfortune for the cause of science, all the insects that Bonpland had gathered, under a thousand difficulties, during our voyage to the Orinoco and Rio Negro.

By an extraordinary fatality we remained two years in the Spanish colonies without receiving a single letter from Europe, and those which reached us in the three subsequent years, gave no information in regard to the collections we had sent. One will readily conceive how anxious I was to learn the fate of a diary which contained all our astronomical observations, and barometrical readings of altitudes, and which I had so patiently copied out in full. It was only after having traversed New Granada, Mexico, and Peru, and when I was on the point of leaving the New World, that in the public library at Philadelphia, I accidentally ran my eye over the table of contents of a scientific review, and there saw these words, "Arrival of the Manuscripts of M. Humboldt, at the residence of his brother, in Paris, by way of Spain." With difficulty I suppressed the expression of my joy, and it seemed to me that no table of contents had ever before been so well arranged.

1 Thus, in the original, but it is undoubtedly a slip of the pen, and should read February instead of "April." Baron Humboldt arrived at Havana, on his first visit to Cuba, on the 19th December, 1800, and sailed from Trinidad on the 16th March, 1801.

While M. Bonpland labored night and day, dividing and arranging our collections, I had the ungracious task of meeting a thousand obstacles that presented themselves to our sudden and unforeseen departure. There was no vessel in the harbor of Havana that would convey us to Porto Bello or Carthagena, and the persons whom I consulted took a pleasure in exaggerating the inconveniences that attended the crossing of the isthmus, and the delays incident to a voyage southward, from Panama to Guayaquil, and thence to Lima or Valparaiso.

They censured me, and perhaps with reason, for not continuing to explore the vast and rich countries of Spanish America, which had been closed for half a century to foreign travellers. The vicissitudes of a voyage round the world, touching only at a few islands, or the arid coasts of a continent, did not seem to them preferable to studying the geological constitution of New Spain, which alone contributed five-eighths of the mass of silver taken yearly from all the mines of the known world. To these arguments, I opposed the wish to determine on a large scale, the inflexion of the curves of equal inclination of the decrease of the magnetic force from the pole toward the equator, and the temperature of the ocean as it varies with the latitude, the direction of the currents, and the proximity of banks and shoals.

In proportion as obstacles rose to my plans, I hastened the more to put them in execution, and not being able to find a passage in a neutral vessel, I chartered a Catalan schooner lying in the roadstead of Batabanó, to take me to Porto Bello or Carthagena, as the winds might permit. The extended relations of the prosperous commerce of Havana afforded me the means for making my pecuniary arrangements for several years. General Gonzalo de O'Farril, distinguished alike for his talents and his high character, then resided in my own country, as minister from the court of Spain. I was enabled to exchange my income in Prussia for a part of his in the island of Cuba, and the family of Don Ygnacio O'Farril y Herrera, his brother, kindly did all they could to forward my projects at the time of my unexpected departure from Havana.

On the sixth of March, we learned that the schooner I had chartered was ready for sea. The road to Batabanó lead us again through Güines, to the sugar plantation of Rio Blanco, the residence of Count de Jaruco y Mopox, which was adorned with all the luxuries that good taste and a large fortune can command. That hospitality which generally wanes as civilization advances, is still practised in Cuba with the same profusion as in the most distant countries of Spanish America. We naturalists and simple travellers accord with pleasure to the inhabitants of Havana, the same grateful acknowledgments that have been given to them by those illustrious strangers,[2] who, everywhere that I have followed their route, have left in the New World the remembrance of their noble simplicity, their ardor for learning and their love for the public weal.

From Rio Blanco to Batabanó, the road passes through an uncultivated country, a portion of which contains many springs. In the open spaces the indigo and cotton plants grow wild for want of cultivation. As the capsule of the *Gossipium* opens at that season of the year when the northern storms are most frequent, the fibre which surrounds the seed is torn from side to side, and the cotton, which in other respects is of the best quality, suffers greatly when the period of the storms coincides with its ripening. Further south we found a new species of the palm, with fan-like leaves (*corifa maritima*), having a free filament in the interstices between the leaves. This corifa abounds through a portion of the southern coast, and takes the place of the majestic royal palm, and the *coco crispa* of the northern shore. Porous limestone (of the Jurassic formation) appeared from time to time in the plain.

Batabanó was at this time a poor hamlet, where a church had been built a few years before. Half a league beyond it the swamp begins, which extends to the entrance of the Bay of Jagua, a distance of seventy leagues from west to east. It is

2 The young princes of the House of Orleans (the Duke d'Orleans, the Duke de Montpensier, and the Count de Beaujolois), who visited the United States and Havana, descending the Ohio and Mississippi rivers, and remained a year in the island of Cuba.—H.

supposed at Batabanó that the sea continues its encroachments upon the land, and that the oceanic irruption has been observed particularly at the time of the great upheaving at the close of the eighteenth century, when the tobacco mills near Havana were destroyed, and the course of the river Chorrera was changed. Nothing can be more gloomy than the view of the marshes around Batabanó, for not a tree breaks the monotony of the scene, and the decaying trunks of a few palms only rise, like broken masts, in the midst of great thickets of running vines and purple flag flowers.

As we remained only one night at Batabanó, I regretted that I could not obtain exact information relative to the two species of *cocodrilos* that infest the swamp. The inhabitants call one the *cayman*, and the other the *cocodrilo*, which name is generally applied to both. We were assured that the latter is the most agile, and the tallest when on its feet; that its snout runs to a much sharper point than that of the cayman, with which it never associates. It is very fearless, and is even said to leap on board of vessels when it can find a support for its tail. The great daring of this animal was noticed during the early expeditions of Diego Velasquez. At the river Cauto, and along the marshy coast of Jagua, it will wander a league from the sea-shore to devour the hogs in the fields. Some attain a length of fifteen feet, and the most savage of them will, it is said, chase a man on horseback like the wolves of Europe—while those that are known as *caymanes* at Batabanó, are so timid that the people do not fear to bathe in waters where they dwell in droves.

These habits, and the name of *cocodrilo*, which is given in Cuba to the most dangerous of the carnivorous saurians, seem to me to indicate a different species from the great animals of the Orinoco and Magdalena rivers, and St. Domingo. The colonists in all other parts of Spanish America, deceived by the exaggerated tales of the ferocity of the Egyptian crocodile, affirm that there are no true crocodiles except in the Nile; while zoologists have found in America the *cayman*, with obtuse snout and no scales on his legs, and the *cocodrilo*, with pointed snout and with scales on his legs. At the same time we find on the old continent, the common crocodile, and those of the Ganges, with rounded snout.

The *crocodilus acutus* of St. Domingo, which I cannot now undertake to class specifically, and the *cocodrilo* of the great Orinoco and Magdalena rivers, have, in the words of Cuvier, so admirable a resemblance to the crocodile of the Nile, that it has been necessary to examine minutely every part, in order to show that the law of Buffon, relative to the distribution of species in the tropical regions of the two continents, was not defective.

As on my second visit to Havana, in 1804, I could not revisit the marshes of Batabanó, I procured at a great expense specimens of the two species, which the inhabitants call cayman and cocodrilo. Two of the latter reached Havana alive, the oldest being about four feet three inches long. Their capture had been very difficult, and they were brought to the city muzzled, tied upon the back of a jack-mule. They were strong and ferocious, and in order to observe their habits and movements, we put them in a large room, where, from the top of a high table, we could see them attacked by dogs.

Having been for six months on the Orinoco, Apure, and Magdalena rivers, in the midst of cocodrilos, we observed with renewed pleasure, before our return to Europe, these singular animals, that pass with an astonishing rapidity from a state of complete immobility to the most impetuous motion. Those which were sent to us from Batabanó as cocodrilos, had the snout as pointed as those of the Orinoco and Magdalena (*Crocodilus acutus*, Cuv.); their color was somewhat darker, being a blackish green on the back, and white on the belly, with yellow spots on the sides. I counted thirty-eight teeth in the upper, and thirty in the lower jaw, as in the true crocodile. Of the upper teeth, the ninth and tenth, and of the lower, the first and fourth, were the largest. The description which Bonpland and myself made on the spot at Costa Firma, expressly states that the fourth lower tooth projects freely over the upper jaw; the posterior extremities were flattened. These cocodrilos of Batabanó, seemed to us, specifically the same with the crocodilus acutus, although it is true that what we were told of its habits, does not accord with what we ourselves had observed on the Orinoco; but the carnivorous saurians of like species, and in the same river, are mild and timid, or ferocious and fearless, according to the nature of the locality.

The animal called cayman at Batabanó, died on the way to Havana, and those in charge had not the foresight to bring the body to us, so that we were not able to compare the two species. Are there, perhaps, on the south side of Cuba true caymans, with the rounded snout, and the fourth under tooth entering the upper jaw; and another species (alligators), like those of Florida? In view of the assertions of the colonists relative to the more pointed head of the cocodrilo of Batabanó, this is almost certain. If this is the case, the people of the island have made, by a happy instinct, a distinction between the cocodrilo and the cayman, with all the exactitude now used by zoologists in separating families that belong to the same genera, and bear the same name.

FURTHER READING

Alexander von Humboldt-Kommission. *Alexander von Humboldt 14.9.1769–6.5.1859: Gedenkschrift zur 100. Wiederkehr Seines Todestages.* Berlin: Akademie-Verlag, 1959.

Beck, Hanno. *Alexander von Humboldt.* 2 vols. Wiesbaden: Franz Steiner, 1959–61.

—. *Alexander von Humboldts Amerikanische Reise.* Stuttgart: K. Thienemann, 1985.

—. *Alexander von Humboldts Reise durchs Baltikum nach Russland und Sibirien 1829.* Stuttgart: K. Thienemann, 1983.

de Bopp, Marianne O. *et al. Ensayos sobre Humboldt.* México: Universidad Nacional Autónoma de México, 1962.

Botting, Douglas. *Humboldt and the Cosmos.* New York: Harper and Row, 1973.

Brann, Edward R. *Alexander von Humboldt: Patron of Science.* Madison, WI: Littel Printing Co., 1954.

Brock, W.H. "Humboldt and the British: A Note on the Character of British Science." *Annals of Science* 50, no. 4 (1993): 365–72.

Bruhns, Karl *et al. Life of Alexander von Humboldt.* Translated by Jane and Caroline Lassell. 2 vols. London: Longmans, Green, and Co., 1873.

Cannon, Susan Faye. "Humboldtian Science." In *Science in Culture: The Early Victorian Period.* New York: Science History Publications, 1978.

De Terra, Helmut. *Humboldt: the Life and Times of Alexander von Humboldt, 1769–1859.* New York: Alfred A. Knopf, 1955.

Gascar, Pierre. *Humboldt l'explorateur.* Paris: Gallimard, 1985.

Heikenroth, H., and Inga Deters, eds. *Alexander-von-Humboldt-Ehrung in der DDR.* Berlin: Akademie-Verlag Berlin, 1986.

Humboldt, Alexander. *Aus Meinem Leben.* Edited by Kurt-R. Biermann. München: C.H. Beck, 1987.

—. *The Island of Cuba.* Translated by J.S. Thrasher. 1856. Reprint, New York: Negro Universities Press, 1969.

—. *Political Essay on the Kingdom of New Spain.* Translated by John Black. Edited and abridged by Mary Maples Dunn. New York: Alfred A. Knopf, 1972.

—. *Relation historique du voyage aux régions équinoxiales du nouveau continent.* Edited by Hanno Beck. 3 vols. Stuttgart: F.A. Brockhaus, 1970.

Kellner, Charlotte. *Alexander von Humboldt.* London: Oxford UP, 1963.

Leitner, Ulrike *et al.*, eds. *Studia Fribergensia: Vorträge des Alexander-von-Humboldt-Kolloquiums in Freiberg vom 8. bis 10. November 1991.* Berlin: Akademie-Verlag, 1994.

Meyer-Abich, Adolf, ed. *Alexander von Humboldt: 1769/1969.* Bonn: Inter Nationes, 1969.

—, ed. *Alexander von Humboldt: Vom Orinoko zum Amazonas.* Wiesbaden: F.A. Brockhaus, 1958.

Minguet, Charles. *Alexandre de Humboldt, historien et géographe de l'Amérique espagnole, 1799–1804.* Paris: François Maspero, 1969.

—, ed. *Cartas Americanas.* Caracas, Venezuela: Biblioteca Ayacucho, 1980.

Miranda, José. *Humboldt y México.* México: Universidad Nacional Autónoma de México, 1962.

Ortega y Medina, Juan A. *Humboldt desde México.* México: Universidad Nacional Autónoma de México, 1960.

Pereyra, Carlos. *Humboldt en América.* Madrid: Editorial-América, 1917.

Petri, W. "Alexander von Humboldt als Patriarch eines kosmischen Humanismus." *Philosophia naturalis* 17 (1979): 479–93.

Pfeiffer, Heinrich. *Alexander von Humboldt: Werk und Weltgeltung.* München: R. Piper und Co., 1969.

Santner, Reinhold, ed. *Alexander von Humboldt: Abenteuer eines Weltreisenden.* Wien: Prisma Verlag, 1980.

Schleucher, Kurt. *Alexander von Humboldt: Der Mensch, der Forscher, der Schriftsteller.* Darmstadt: Eduard Roether Verlag, 1984.

—. *Alexander von Humboldt.* Berlin: Stapp Verlag, 1988.

Schultze, Joachim H., ed. *Alexander von Humboldt: Studien zu seiner universalen Geisteshaltung.* Berlin: Walter de Gruyter und Co., 1959.

Scurla, Herbert. *Im Lande der Kariben: Reisen deutscher Forscher des 19. Jahrhunderts in Guayana.* Berlin: Verlag der Nation, 1965.

Selsam, Millicent E. *Stars, Mosquitoes, and Crocodiles: the American Travels of Alexander von Humboldt.* New York: Harper and Row, 1962.

Stafford, R.A. *Scientist of Empire: Sir Roderick Murchison, Scientific Exploration and Victorian Imperialism.* Cambridge: Cambridge UP, 1989.

Stearn, W.T., ed. *Humboldt, Bonpland, Kunth and Tropical American Botany.* Stuttgart: J. Cramer, 1968.

von Hagen, Victor Wolfgang. *South America Called Them: Explorations of the Great Naturalists La Condamine, Humboldt, Darwin, Spruce.* New York: Duell, Sloan and Pearce, 1945.

Zeuske, M., and Bernd Schröter, eds. *Alexander von Humboldt und das neue Geschichtsbild von Lateinamerika.* Leipzig: Leipziger Universitätsverlag, 1992.

Zuñiga, Neptali. *Diario Inedito del Viaje de Humboldt por la Provincia de Guayaquil.* Guayaquil, Ecuador: Universidad de Guayaquil, 1983.

9. CHARLES BABBAGE

(1791–1871)

The mathematician Charles Babbage is best known today as the pioneer of the modern digital computer, although he made contributions to many areas of science. At Cambridge University, where he was eventually elected Lucasian Professor of Mathematics, he successfully campaigned for the replacement of Newton's calculus notation with the more flexible system of Liebniz. He also advanced actuarial science in his *A Comparative View of the Various Institutions for the Assurance of Lives* (1826). Some of his other pursuits included cryptography, economics (*On the Economy of Machinery and Manufactures*, 1832), and even natural theology (*The Ninth Bridgewater Treatise; A Fragment*, 1837).

From 1822-34, Babbage developed his Difference Engine, a machine that would not only calculate mathematical tables useful for astronomy and navigation (logarithms and sines), but would also print out the results directly, greatly reducing transcription errors. Direct governmental support of scientific endeavour was uncommon at this time, but Babbage secured a £1500 grant in 1823 to build a working model. Babbage suspended work on the engine after cost overruns, the loss of government support, and a quarrel with his chief engineer Joseph Clement. Meanwhile, Babbage learned that the Swedish engineers Georg and Edvard Scheutz had built a working Difference Engine in 1853 on the basis of Babbage's design reported in a popular article written by Dionysius Lardner. The Science Museum in London eventually produced a full-scale working model of Babbage's Difference Engine No. 2 in 1991.

Abandoning the Difference Engine, Babbage instead devoted his energies to a similar but more powerful machine, the Analytical Engine, in many respects a prototype of the modern computer. Lardner prophesied that Babbage's engines would provide "numerical tables, unlimited in quantity and variety, restricted to no particular species, and limited by no particular law; extending not merely to the boundaries of existing knowledge, but spreading their powers over the undefined regions of future discovery" (Morrison and Morrison 164). This dream of a knowledge-producing pansophic machine has antecedents in such systems as medieval philosopher Ramon Llull's *Ars magna*, a device of rotating wheels for determining logical permutations, or the Universal Character of Bishop John Wilkins, an artificial scientific language of the seventeenth century. Although based on decimal rather than binary principles, the Analytical Engine anticipated the major components of the modern computer—it consisted of an "operating system" or mill regulated by a music box-type barrel with pins; "software programs," which employed the punched card system of the Jacquard loom (similar to the paper punch cards of early IBM computers); and "memory," stored numbers also regulated by punch cards for calculating functions. Babbage was unable to build the Analytical Engine for financial and engineering reasons, but his son Henry built a small model after Babbage's death. Byron's daughter Ada Lovelace translated a description of Babbage's Analytical Engine written by L. Menabrea, which brought greater exposure to

Babbage's ideas. Babbage's work may have inspired the research of the modern computer scientists Howard Aiken and Alan Turing.

Throughout his life, Babbage was concerned with the reform and development of British science and he helped to found the Analytical Society (1812), Astronomical Society (1820), and the British Association for the Advancement of Science (1831). In 1830, he published *Reflections on the Decline of Science in England and On Some of Its Causes*, a vitriolic, sardonic, and alarmist attack on both the Royal Society and Babbage's personal enemies. Babbage complained of a number of minor abuses (such as the falsification of Royal Society minutes) and major outrages: the incompetent governance of the Royal Society, the lack of government support for science, the neglect of scientific studies in the universities, and Britain's loss of scientific pre-eminence to continental researchers. Reform was in the air in the 1820s and 1830s; in the face of popular pressure, Parliament passed the Great Reform Bill of 1832, expanding the electorate, and Charles Lyell had anticipated Babbage's concerns about reforming university curricula in his 1827 *Quarterly Review* article entitled "State of the Universities." Babbage hoped that the Royal Society could be reformed into an institution along the lines of the professional, state-supported societies which he had visited in Paris and Berlin after the death of his wife Georgina. Babbage suggested many useful changes in British science, including establishing a scientific "order of merit" and limiting Royal Society membership to working scientists, but few of his suggestions were taken up.

A work so polemical, closely reasoned, and well documented was bound to spawn a variety of private and public debates, which have become known collectively as "The Declinist Movement." David Brewster supported Babbage's conclusions in an 1830 *Quarterly Review* article, while Gerard Moll, with the backing of Faraday, attacked Babbage in *On the Alleged Decline of Science in England* (1831). In the same year that Babbage's critique of British science appeared, Augustus Bozzi-Granville published *Science Without a Head: or the Royal Society Dissected*, which also complained of the deplorable state of science in Britain, in part caused by the great number of non-scientific members of the Royal Society and the publication of papers in the *Philosophical Transactions* on grounds other than scientific merit. If in fact British science was in decline in 1830 as the declinists warned, and this is a highly debatable proposition, it was nevertheless thriving by mid-century. Babbage failed to emphasize in his book the unprecedented growth during this time of English scientific societies and mechanics' institutes for popular scientific education, as well as the development of University College London with its scientific curricula.

Babbage published his autobiography *Passages from the Life of a Philosopher* in 1864, which has recently been augmented by the appearance of a memoir by H.W. Buxton (written c. 1872–80). In his later years, Babbage led an ill-tempered campaign to outlaw street musicians.

Reflections on the Decline of Science in England and On Some of Its Causes.
London: B. Fellowes, 1830, pp. 1–39.

Introductory Remarks.

It cannot have escaped the attention of those, whose acquirements enable them to judge, and who have had opportunities of examining the state of science in other countries, that in England, particularly with respect to the more difficult and abstract sciences, we are much below other nations, not merely of equal rank, but below several even of inferior power. That a country, eminently distinguished for its mechanical and manufacturing ingenuity, should be indifferent to the progress of inquiries which form the highest departments of that knowledge on whose more elementary truths its wealth and rank depend, is a fact which is well deserving the attention of those who shall inquire into the causes that influence the progress of nations.

To trace the gradual decline of mathematical, and with it of the highest departments of physical science, from the days of Newton to the present, must be left to the historian. It is not within the province of one who, having mixed sufficiently with scientific society in England to see and regret the weakness of some of its greatest ornaments, and to see through and deplore the conduct of its pretended friends, offers these remarks, with the hope that they may excite discussion,—with the conviction that discussion is the firmest ally of truth,—and with the confidence that nothing but the full expression of public opinion can remove the evils that chill the enthusiasm, and cramp the energies of the science of England.

The causes which have produced, and some of the effects which have resulted from, the present state of science in England, are so mixed, that it is difficult to distinguish accurately between them. I shall, therefore, in this volume, not attempt any minute discrimination, but rather present the result of my reflections on the concomitant circumstances which have attended the decay, and at the conclusion of it, shall examine some of the suggestions which have been offered for the advancement of British science.

Chapter I. On the Reciprocal Influence of Science and Education.

That the state of knowledge in any country will exert a directive influence on the general system of instruction adopted in it, is a principle too obvious to require investigation. And it is equally certain that the tastes and pursuits of our manhood will bear on them the traces of the earlier impressions of our education. It is therefore not unreasonable to suppose that some portion of the neglect of science in England, may be attributed to the system of education we pursue. A young man passes from our public schools to the universities, ignorant almost of the elements of every branch of useful knowledge; and at these latter establishments, formed originally for instructing those who are intended for the clerical profession, classical and mathematical pursuits are nearly the sole objects proposed to the student's ambition.

Much has been done at one of our universities during the last fifteen years, to improve the system of study; and I am confident that there is no one connected with that body, who will not do me the justice to believe that, whatever suggestions I may venture to offer, are prompted by the warmest feelings for the honour and the increasing prosperity of its institutions. The ties which connect me with Cambridge are indeed of no ordinary kind.

Taking it then for granted that our system of academical education ought to be adapted to nearly the whole of the aristocracy of the country, I am inclined to believe that whilst the modifications I should propose would not be great innovations on the spirit of our institutions, they would contribute materially to that important object.

It will be readily admitted, that a degree conferred by a university, ought to be a pledge to the public that he who holds it possesses a certain quantity of knowledge. The progress of society has rendered knowledge far more various in its kinds than it used to be; and to meet this variety in the tastes and inclinations of those who come to us for instruction, we have, besides the regular lectures to which all must attend, other sources of information from whence the students may acquire sound and varied knowledge in the numerous lectures on chemistry, geology, botany, history, etc. It is at present a matter of option with the student, which, and how many of these courses he shall attend, and such it should still remain. All that it would be necessary to add would be, that previously to taking his degree, each person should be examined by those Professors, whose lectures he had attended. The pupils should then be arranged in two classes, according to their merits, and the names included in these classes should be printed. I would then propose that no young man, except his name was found amongst the "List of Honours," should be allowed to take his degree, unless he had been placed in the first class of some one at least of the courses given by the professors. But it should still be imperative upon the student to possess such mathematical knowledge as we usually require. If he had attained the first rank in several of these examinations, it is obvious that we should run no hazard in a little relaxing the strictness of his mathematical trial.

If it should be thought preferable, the sciences might be grouped, and the following subjects be taken together:—

Modern History.
Laws of England.
Civil Law.

Political Economy.
Applications of Science to Arts and Manufactures.

Chemistry.
Mineralogy.
Geology.

Zoology, including Physiology and Comparative Anatomy.
Botany, including Vegetable Physiology and Anatomy.

One of the great advantages of such a system would be, that no young person would have an excuse for not studying, by stating, as is most frequently done, that the only pursuits followed at Cambridge, classics and mathematics, are not adapted either to his taste, or to the wants of his after life. His friends and relatives would then reasonably expect every student to have acquired distinction in *some* pursuit. If it should be feared that this plan would lead to too great a diversity of pursuits in the same individual, a limitation might be placed upon the number of examinations into which the same person might be permitted to enter. It might also be desirable not to restrict the whole of these examinations to the third year, but to allow the student to enter on some portion of them in the first or second year, if he should prefer it.

By such an arrangement, which would scarcely interfere seriously with our other examinations, we should, I think, be enabled effectually to keep pace with the wants of society, and retaining fully our power and our right to direct the studies of those who are intended for the church, as well as of those who aspire to the various offices connected with our academical institutions; we should, at the same time, open a field of honourable ambition to multitudes, who, from the exclusive nature of our present studies, leave us with but a very limited addition to their stock of knowledge.

Much more might be said on a subject so important to the interests of the country, as well as of our university, by my wish is merely to open it for our own consideration and discussion. We have already done so much for the improvement of our system of instruction, that public opinion will not reproach us for any unwillingness to alter. It is our first duty to be well satisfied that we can improve: such alterations ought only to be the result of a most mature consideration, and of a free interchange of sentiments on the subject, in order that we may condense upon the question the accumulated judgment of many minds.

It is in some measure to be attributed to the defects of our system of education, that scientific knowledge scarcely exists amongst the higher classes of society. The discussions in the Houses of Lords or of Commons, which arise on the occurrence of any subjects connected with science, sufficiently prove this fact, which, if I had consulted the extremely limited nature of my personal experience, I should, perhaps, have doubted.

Chapter II. Of the Inducements to Individuals to Cultivate Science.

Interest or inclination form the primary and ruling motives in this matter: and both these exert greater or less proportionate influence in each of the respective cases to be examined.

Section 1. Professional Impulses.

A large portion of those who are impelled by ambition or necessity to advance themselves in the world, make choice of some profession in which they imagine their talents likely to be rewarded with success; and there are peculiar advantages resulting to each from this classification of society into professions. The *esprit de corps* frequently overpowers the jealousy which exists between individuals, and pushes on to advantageous situations some of the more fortunate of the profession; whilst, on the other hand, any injury or insult offered to the weakest, is redressed or resented by the whole body. There are other advantages which are perhaps of more importance to the public. The numbers which compose the learned professions in England are so considerable, that a kind of public opinion is generated amongst them, which powerfully tends to repress conduct that is injurious either to the profession or to the public. Again, the mutual jealousy and rivalry excited amongst the whole body is so considerable, that although the rank and estimation which an individual holds in the profession may be most unfairly appreciated, by taking the opinion of his rival; yet few estimations will be found generally more correct than the opinion of a whole profession on the merits of any one of its body. This test is of great value to the public, and becomes the more so, in proportion to the difficulty of the study to which the profession is devoted. It is by availing themselves of it that men of sense and judgment, who have occasion for the services of professional persons, are, in a great measure, guided in their choice.

The pursuit of science does not, in England, constitute a distinct profession, as it does in many other countries. It is therefore, on that ground alone, deprived of many of the advantages which attach to professions. One of its greatest misfortunes arises from this circumstance; for the subjects on which it is conversant are so difficult, and require such unremitted devotion of time, that few who have not spent years in their study can judge of the relative knowledge of those who pursue them. It follows, therefore, that the public, and even that men of sound sense and discernment, can scarcely find means to distinguish between the possessors of knowledge, in the present day, merely elementary, and those whose acquirements are of the highest order. This remark applies with peculiar force to all the more difficult applications of mathematics; and the fact is calculated to check the energies of those who only look to reputation in England.

As there exists with us no peculiar class professedly devoted to science, it frequently happens that when a situation, requiring for the proper fulfilment of its duties considerable scientific attainments, is vacant, it becomes necessary to select from among amateurs, or rather from among persons whose chief attention has been bestowed on other subjects, and to whom science has been only an occasional pursuit. A certain quantity of scientific knowledge is of course possessed by individuals in many professions; and when added to the professional acquirements of the army, the navy, or to the knowledge of the merchant, is highly meritorious: but it is obvious that this may become, when separated from the profession, quite insignificant as the basis of a scientific reputation.

To those who have chosen the profession of medicine, a knowledge of chemistry, and of some branches of natural history, and, indeed, of several other departments of science, affords useful assistance. Some of the most valuable names which adorn the history of English science have been connected with this profession.

The causes which induce the selection of the clerical profession are not often connected with science; and it is, perhaps, a question of considerable doubt whether it is desirable to hold out to its members hopes of advancement from such acquirements. As a source of recreation, nothing can be more fit to occupy the attention of a divine; and our church may boast, in the present as in past times, that the domain of science has been extended by some of its brightest ornaments.

In England, the profession of the law is that which seems to hold out the strongest attraction to talent, from the circumstance, that in it ability, coupled with exertion, even though unaided by patronage, cannot fail of obtaining reward. It is frequently chosen as an introduction to public life. It also presents great advantages, from its being a qualification for many situations more or less remotely connected with it, as well as from the circumstance that several of the highest officers of the state must necessarily have sprung from its ranks.

A powerful attraction exists, therefore, to the promotion of a study and of duties of all others engrossing the time most completely, and which is less benefited than most others by any acquaintance with science. This is one amongst the causes why it so very rarely happens that men in public situations are at all conversant even with the commonest branches of scientific knowledge, and why scarcely an instance can be cited of such persons acquiring a reputation by any discoveries of their own.

But, however consistent other sciences may be with professional avocations, there is one which, from its extreme difficulty, and the overwhelming attention which it demands, can only be pursued with success by those whose leisure is undisturbed by other claims. To be well acquainted with the present state of mathematics, is no easy task; but to add to the powers which that science possesses, is likely to be the lot of but few English philosophers.

Section 2. Of National Encouragement.

The little encouragement which at all previous periods has been afforded by the English Government to the authors of useful discoveries, or of new and valuable inventions, is justified on the following grounds:

1. The public, who consume the new commodity or profit by the new invention, are much better judges of its merit than the government can be.
2. The reward which arises from the sale of the commodity is usually much larger than that which government would be justified in bestowing; and it is exactly proportioned to the consumption, that is, to the want which the public feel for the new article.

It must be admitted that, as general principles, these are correct: there are, however, exceptions which flow necessarily from the very reasoning from which they were deduced. Without entering minutely into these exceptions, it will be sufficient to show that all abstract truth is entirely excluded from reward under this system. It is only the application of principles to common life which can be thus rewarded. A few instances may perhaps render this position more evident. The principle of the hydrostatic paradox was known as a speculative truth in the time of Stevinus;[1] and its application to raising heavy weights has long been stated in elementary treatises on natural philosophy, as well as constantly exhibited in lectures. Yet, it may fairly be regarded as a mere abstract principle, until the late Mr. Bramah, by substituting a pump instead of the smaller column, converted it into a most valuable and powerful engine.—The principle of the convertibility of the centres of oscillation and suspension in the pendulum, discovered by Huygens more than a century and a half ago, remained, until within these few years, a sterile, though most elegant proposition; when, after being hinted at by Prony, and distinctly pointed out by Bonenberger, it was employed by Captain Kater as the foundation of a most convenient practical method of determining the length of the pendulum.—The interval which separated the discovery, by Dr. Black, of latent heat, from the beautiful and successful application of it to the steam engine, was comparatively short; but it required the efforts of two minds; and both were of the highest order.—The influence of electricity in producing decompositions, although of inestimable value as an instrument of discovery in chemical inquiries, can

1 About the year 160.

hardly be said to have been applied to the practical purposes of life, until the same powerful genius[2] which detected the principle, applied it, by a singular felicity of reasoning, to arrest the corrosion of the copper-sheathing of vessels. That admirably connected chain of reasoning, the truth of which is confirmed by its very failure as a remedy,[3] will probably at some future day supply, by its successful application, a new proof of the position we are endeavouring to establish.

Other instances might, if necessary, be adduced, to show that long intervals frequently elapse between the discovery of new principles in science and their practical application: nor ought this at all to surprise us. Those intellectual qualifications, which give birth to new principles or to new methods, are of quite a different order from those which are necessary for their practical application.

At the time of the discovery of the beautiful theorem of Huygens, it required in its author not merely a complete knowledge of the mathematical science of his age, but a genius to enlarge its boundaries by new creations of his own. Such talents are not always united with a quick perception of the details, and of the practical applications of the principles they have developed, nor is it for the interest of mankind that minds of this high order should lavish their powers on subjects unsuited to their grasp.

In mathematical science, more than in all others, it happens that truths which are at one period the most abstract, and apparently the most remote from all useful application, become in the next age the bases of profound physical inquiries, and in the succeeding one, perhaps, by proper simplification and reduction to tables, furnish their ready and daily aid to the artist and the sailor.

It may also happen that at the time of the discovery of such principles, the mechanical arts may be too imperfect to render their application likely to be attended with success. Such was the case with the principle of the hydrostatic paradox; and it was not, I believe, until the expiration of Mr. Bramah's patent, that the press which bears his name received that mechanical perfection in its execution, which has deservedly brought it into such general use.

On the other hand, for one person who is blessed with the power of invention, many will always be found who have the capacity of applying principles; and much of the merit ascribed to these applications will always depend on the care and labour bestowed in the practical detail.

If, therefore, it is important to the country that abstract principles should be applied to practical use, it is clear that it is also important that encouragement should be held out to the few who are capable of adding to the number of those truths on which such applications are founded. Unless there exist peculiar institutions for the support of such inquirers, or unless the Government directly interfere, the contriver of a thaumatrope may derive profit from his ingenuity, whilst he who unravels the laws of light and vision, on which multitudes of phenomena depend, shall descend unrewarded to the tomb.

Perhaps it may be urged, that sufficient encouragement is already afforded to abstract science in our different universities, by the professorships established at them. It is not however in the power of such institutions to create; they may foster and aid the development of genius; and, when rightly applied, such stations ought to be its fair and honourable rewards. In many instances their emolument is small; and when otherwise, the lectures which are required from the professor are not perhaps in all cases the best mode of employing the energies of those who are capable of inventing.

I cannot resist the opportunity of supporting these opinions by the authority of one of the greatest philosophers of a past age, and of

2 I am authorised in stating, that this was regarded by Laplace as the greatest of Sir Humphry Davy's discoveries.

3 It did *not* fail in producing the effect foreseen by Sir H. Davy,—*the preventing the corrosion of the copper*; but it failed as a cure of the evil, by producing one of an *opposite* character; either by preserving too perfectly from decay the surface of the copper, or by rendering it negative, it allowed marine animals and vegetables to accumulate on its surface, and thus impede the progress of the vessel.

expressing my acknowledgments to the author of a most interesting piece of scientific biography. In the correspondence which terminated in the return of Galileo to a professorship in his native country, he remarks, "But, because my private lectures and domestic pupils are a great hinderance and interruption of my studies, I wish to live entirely exempt from the former, and in great measure from the latter."—*Life of Galileo*, p. 18. And, in another letter to Kepler, he speaks with gratitude of Cosmo, the Grand Duke of Tuscany, who "has now invited me to attach myself to him with the annual salary of 1000 florins, and with the title of Philosopher and principal Mathematician to his Highness, without the duties of any office to perform, but with most complete leisure; so that I can complete my treatise on Mechanics, etc."—p. 31.[4]

Surely, if knowledge is valuable, it can never be good policy in a country far wealthier than Tuscany, to allow a genius like Mr. Dalton's, to be employed in the drudgery of elementary instruction.[5] Where would have been the military renown of England, if, with an equally improvident waste of mental power, its institutions had forced the Duke of Wellington to employ his life in drilling recruits, instead of planning campaigns?

If we look at the fact, we shall find that the great inventions of the age are not, with us at least, always produced in universities. The doctrines of "definite proportions," and of the "chemical agency of electricity,"—principles of a high order, which have immortalized the names of their discoverers,—were not produced by the meditations of the cloister: nor is it in the least a reproach to those valuable institutions to mention truths like these. Fortunate circumstances must concur, even to the greatest, to render them eminently successful. It is not permitted to all to be born, like Archimedes, when a science was to be created; nor, like Newton, to find the system of the world "without form and void;" and, by disclosing *gravitation*, to shed throughout that system the same irresistible radiance as that with which the Almighty Creator had illumined its material substance. It can happen to but few philosophers, and but at distant intervals, to snatch a science, like Dalton, from the chaos of indefinite combination, and binding it in the chains of number, to exalt it to rank amongst the exact. Triumphs like these are necessarily "few and far between;" nor can it be expected that that portion of encouragement, which a country may think fit to bestow on science, should be adapted to meet such instances. Too extraordinary to be frequent, they must be left, if they are to be encouraged at all, to some direct interference of the government.

The dangers to be apprehended from such a specific interference, would arise from one, or several, of the following circumstances:—That class of society, from whom the government is selected, might not possess sufficient knowledge either to judge themselves, or know upon whose judgment to rely. Or the number of persons devoting themselves to science, might not be sufficiently large to have due weight in the expression of public opinion. Or, supposing this class to be large, it might not enjoy, in the estimation of the world, a sufficiently high character for independence. Should these causes concur in any

4 *Life of Galileo*, published by the Society for the Diffusion of Useful Knowledge.

5 I utter these sentiments from no feelings of private friendship to that estimable philosopher, to whom it is my regret to be almost unknown, and whose modest and retiring merit, I may, perhaps, have the misfortune to offend by these remarks. But Mr. Dalton was of no party; had he ever moved in that vortex which has brought discredit, and almost ruin, on the Royal Society of England;—had he taken part with those who vote to each other medals, and, affecting to be tired of the fatigues of office, make to each other requisitions to retain places they would be most reluctant to quit; his great and splendid discovery would long since have been represented to government. Expectant mediocrity would have urged on his claims to remuneration, and those who covered their selfish purposes with the cloak of science, would have hastened to shelter themselves in the mantle of his glory.—But the philosopher may find consolation for the tardy approbation of that Society, in the applause of Europe. If he was insulted by their medal, he escaped the pain of seeing his name connected with their proceedings.

country, it might become highly injurious to commit the encouragement of science to any department of the government. This reasoning does not appear to have escaped the penetration of those who advised the abolition of the late Board of Longitude.

The question whether it is good policy in the government of a country to encourage science, is one of which those who cultivate it are not perhaps the most unbiassed judges. In England, those who have hitherto pursued science, have in general no very reasonable grounds of complaint; they knew, or should have known, that there was no demand for it, that it led to little honour, and to less profit.

That blame has been attributed to the government for not fostering the science of the country is certain; and, as far as regards past administrations, is, to a great extent, just; with respect to the present ministers, whose strength essentially depends on public opinion, it is not necessary that they should precede, and they cannot remain long insensible to any expression of the general feeling. But supposing science were thought of some importance by any administration, it would be difficult in the present state of things to do much in its favour; because, on the one hand, the higher classes in general have not a profound knowledge of science, and, on the other, those persons whom they have usually consulted, seem not to have given such advice as to deserve the confidence of government. It seems to be forgotten, that the money allotted by government to purposes of science ought to be expended with the same regard to prudence and economy as in the disposal of money in the affairs of private life.[6]

To those who measure the question of the national encouragement of science by its value in pounds, shillings, and pence, I will here state a fact, which, although pretty generally known, still, I think, deserves attention. A short time since it was discovered by government that the terms on which annuities had been granted by them were erroneous, and new tables were introduced by act of Parliament. It was stated at the time that the erroneous tables had caused a loss to the country of between two and three millions sterling. The fact of the sale of those annuities being a losing concern was long known to many; and the government appear to have been the last to be informed on the subject. Half the interest of half that loss, judiciously applied to the encouragement of mathematical science, would, in a few years, have rendered utterly impossible such expensive errors.

To those who bow to the authority of great names, one remark may have its weight. *The Mecanique Coeleste*,[7] and the *Théorie Analytique des Probabilités*, were both dedicated, by Laplace, to Napoleon. During the reign of that extraordinary man, the triumphs of France were as eminent in Science as they were splendid in arms. May the institutions which trained and rewarded her philosophers be permanent as the benefits they have conferred upon mankind!

In other countries it has been found, and is admitted, that a knowledge of science is a recommendation to public appointments, and that a man does not make a worse ambassador because he has directed an observatory, or has added by his discoveries to the extent of our knowledge of animated nature. Instances even are not wanting of ministers who have begun their career in the inquiries of pure analysis. As such examples are perhaps more frequent than is generally imagined, it may be useful to mention a few of those

---

6 Who, for instance, could have advised the government to incur the expense of printing *seven hundred and fifty* copies of the Astronomical Observations made at Paramatta, to form a third part of the Philosophical Transactions for 1829, whilst of the Observations made at the Royal Observatory at Greenwich, two hundred and fifty copies only are printed?

Of these seven hundred and fifty copies, seven hundred and ten will be distributed to members of the Royal Society, to six hundred of whom they will probably be wholly uninteresting or useless; and thus the country incurs a constantly recurring annual expense. Nor is it easy to see on what principle a similar destination could be refused for the observations made at the Cape of Good Hope.

7 The first volume of the first translation of this celebrated work into our own language, has just arrived in England from——America.

Country.	Name.	Department of Science.	Public Office.
France	Marquis Laplace*	Mathematics	President of the Conservative Senate.
France	M. Carnot†	Mathematics	Ministry of War.
France	Count Chaptal	Chemistry	Minister of the Interior.
France	Baron Cuvier‡	Comparative Anatomy, Natural History	Minister of Public Instruction.
Prussia	Baron Humboldt	Oriental languages	Ambassador to England.
Prussia	Baron Alexander Humboldt	The celebrated Traveller	Chamberlain to the King of Prussia.
Modena	Marquis Rangoni§	Mathematics	Minister of Finance and of Public Instruction, President of the Italian Academy of Forty.
Tuscany	Count Fossombroni‖	Mathematics	Prime Minister of the Grand Duke of Tuscany.
Saxony	M. Lindenau¶	Astronomy	Ambassador.

* Author of the *Mecanique Coeleste.*
† Author of *Traité de Chimie Appliqué aux Arts.*
‡ Author of *Leçons d'Anatomie Comparée—Récherches sur les Ossemens Fossiles, &c.&c.*
§ Author of *Memoria sulle Funzioni Generatrici,* Modena, 1824, and of various other memoirs on mathematical subjects.
‖ Author of several memoirs on mechanics and hydraulics, in the Transactions of the Academy of Forty.
¶ Author of *Tables Barometriques,* Gotha, 1809—*Tabulæ Veneris, novæ et correctæ,* Gothæ, 1810—*Investigatio Nova Orbitæ a Mercurio circa Solem descriptæ,* Gothæ, 1813, and of other works.

men of science who have formerly held, or who now hold, high official stations in the governments of their respective countries.

M. Lindenau, the Minister from the King of Saxony to the King of the Netherlands, commenced his career as astronomer at the observatory of the Grand Duke of Gotha, by whom he was sent as his representative at the German Diet. On the death of the late reigning Duke, M. Lindenau was invited to Dresden, and filled the same situation under the King of Saxony; after which he was appointed his minister at the court of the King of the Netherlands. Such occurrences are not to be paralleled in our own country, at least not in modern times. Newton was, it is true, more than a century since, appointed Master of the Mint; but let any person suggest an appointment of a similar kind in the present day, and he will gather from the smiles of those to whom he proposes it that the highest knowledge conduces nothing to success, and that political power is almost the only recommendation.

Sec. 3.

Of Encouragement from Learned Societies.

There are several circumstances which concur in inducing persons pursuing science, to unite together, to form societies or academies. In former times, when philosophical instruments were more rare, and the art of making experiments was less perfectly known, it was almost necessary. More recently, whilst numerous additions are constantly making to science, it has been found that those who are most capable of extending human knowledge, are frequently least able to encounter the expense of printing their investigations. It is therefore convenient, that some means should be devised for relieving them from this difficulty, and the volumes of the transactions of academies have accomplished the desired end.

There is, however, another purpose to which academies contribute. When they consist of a limited number of persons, eminent for their knowledge, it becomes an object of ambition to be admitted on their list. Thus a stimulus is applied to all those who cultivate science, which urges on their exertions, in order to acquire the wished-for distinction. It is clear that this envied position will be valued in proportion to the difficulty of its attainment, and also to the celebrity of those who enjoy it; and whenever the standard of scientific knowledge which qualifies for its ranks is lowered, the value of the distinction itself will be diminished. If, at any time, a multitude of persons having no sort of knowledge of science are admitted, it must cease to be sought after as an object of ambition by men of science, and the class of persons to whom it will become an object of desire will be less intellectual.

Let us now compare the numbers composing some of the various academies of Europe.—The Royal Society of London, the Institute of France, the Italian Academy of Forty, and the Royal Academy of Berlin, are amongst the most distinguished.

Name . Country	Population	Number of Members of its Academy	Number of Foreign Members
1. England	22,299,000	685	50
2. France.	32,058,000	75	{ 8 Mem. 100 Corr.
3. Prussia.	12,415,000	38	16
4. Italy.	12,000,000	40	8

It appears then, that in France, one person out of 427,000 is a member of the Institute. That in Italy and Prussia, about one out of 300,000 persons is a member of their Academies. That in England, every 32,000 inhabitants produces a Fellow of the Royal Society. Looking merely at these proportions, the estimation of a seat in the Academy of Berlin, must be more than nine times as valuable as a similar situation in England; and a member of the Institute of France will be more than thirteen times more rare in his country than a Fellow of the Royal Society is in England.

Favourable as this view is to the dignity of such situations in other countries, their comparative rarity is by no means the most striking difference in the circumstances of men of science. If we look at the station in society occupied by the *savans* of other countries, in several of them we shall find it high, and their situations profitable. Perhaps, at the present moment, Prussia is, of all the countries in Europe, that which bestows the greatest attention, and most unwearied encouragement on science. Great as are the merits of many of its philosophers, much of this support arises from the character of the reigning family, by whose enlightened policy even the most abstract sciences are fostered.

The maxim that "knowledge is power," can be perfectly comprehended by those only who are themselves well versed in science; and to the circumstance of the younger branches of the royal family of Prussia having acquired considerable knowledge in such subjects, we may attribute the great force with which that maxim is appreciated.

In France, the situation of its *savans* is highly respectable, as well as profitable. If we analyze the list of the Institute, we shall find few who do not possess titles or decorations; but as the value of such marks of royal favour must depend, in a great measure, on their frequency, I shall mention several particulars which are probably not familiar to the English reader.[8]

Number of the Members of the Institute of France who belong to the Legion of Honour.		Total Number of each Class of the Legion of Honour.
Grand Croix	3	80
Grand Officer	3	160
Commandeur	4	400
Officier	17	2,000
Chevalier	40	Not limited.

Number of Members of the Institute decorated with the Order of St. Michel.		Total Number of that Order.
Grand Croix	2	} 100
Chevalier	27	

Amongst the members of the Institute there are,—

Dukes	2
Marquis	1
Counts	4
Viscounts	2
Barons	14
	23
Of these there are Peers of France	5

We might, on turning over the list of the 685

8 This analysis was made by comparing the list of the Institute, printed for that body in 1827, with the *Almanach Royale* for 1823.

members of the Royal Society, find a greater number of peers than there are in the Institute of France; but a fairer mode of instituting the comparison, is to inquire how many titled members there are amongst those who have contributed to its Transactions. In 1827, there were one hundred and nine members who had contributed to the Transactions of the Royal Society; amongst these were found:—

Peer..............................1
Baronets......................5
Knights 5

It should be observed, that five of these titles were the rewards of members of the medical profession, and one only, that of Sir H. Davy, could be attributed exclusively to science.

It must not be inferred that the titles of nobility in the French list, were all of them the rewards of scientific eminence; many are known to have been such; but it would be quite sufficient for the argument to mention the names of Lagrange, Laplace, Berthollet, and Chaptal.

The estimation in which the public hold literary claims in France and England, was curiously illustrated by an incidental expression in the translation of the debates in the House of Lords, on the occasion of His Majesty's speech at the commencement of the session of 1830. The Gazette de France stated, that the address was moved by the Duc de Buccleugh, "*chef de la maison de Walter Scott*." Had an English editor wished to particularize that nobleman, he would undoubtedly have employed the term *wealthy*, or some other of the epithets characteristic of that quality most esteemed amongst his countrymen.

If we turn, on the other hand, to the emoluments of science in France, we shall find them far exceed those in our own country. I regret much that I have mislaid a most interesting memorandum on this subject, which I made several years since: but I believe my memory on the point will not be found widely incorrect. A foreign gentleman, himself possessing no inconsiderable acquaintance with science, called on me a few years since, to present a letter of introduction. He had been but a short time in London; and, in the course of our conversation, it appeared to me that he had imbibed very inaccurate ideas respecting our encouragement of science.

Thinking this a good opportunity of instituting a fair comparison between the emoluments of science in the two countries, I placed a sheet of paper before him, and requested him to write down the names of six Englishmen, in his opinion, best known in France for their scientific reputation. Taking another sheet of paper, I wrote upon it the names of six Frenchmen, best known in England for their scientific discoveries. We exchanged these lists, and I then requested him to place against each name (as far as he knew) the annual income of the different appointments held by that person. In the mean time, I performed the same operation on his list, against some names of which I was obliged to place a *zero*. The result of the comparison was an average of nearly 1200*l*. per annum for the six French *savans* whom I had named. Of the average amount of the sums received by the English, I only remember that it was very much smaller. When we consider what a command over the necessaries and luxuries of life 1200*l*. will give in France, it is underrating it to say it is equal to 2000*l*. in this country.

Let us now look at the prospects of a young man at his entrance into life, who, impelled by an almost irresistible desire to devote himself to the abstruser sciences, or who, confident in the energy of youthful power, feels that the career of science is that in which his mental faculties are most fitted to achieve the reputation for which he pants. What are his prospects? Can even the glowing pencil of enthusiasm add colour to the blank before him? There are no situations in the state; there is no position in society to which hope can point, to cheer him in his laborious path. If, indeed, he belong to one of our universities, there are some few chairs in his *own* Alma Mater to which he may at some distant day pretend; but these are not numerous; and whilst the salaries attached are seldom sufficient for the sole support of the individual, they are very rarely enough for that of a family. What then can he reply to the entreaties of his friends, to betake himself to some business in which perhaps they have power to assist him, or to choose some profession in which his talents may produce for him their fair reward? If he have no fortune, the choice is taken away: he *must* give up that line of life in which his habits of thought and his ambition qualify him to succeed eminently, and he *must* choose the bar, or some other profession, in which, amongst so many competitors, in spite of his great talents, he can be but moderately successful. The loss to him is great, but to the country it is greater. We thus, by a destructive misapplication of talent which our institutions create, exchange a profound philosopher for but a tolerable lawyer.

If, on the other hand, he possess some moderate fortune of his own; and, intent on the glory of an immortal name, yet not blindly ignorant of the state of science in this country, he resolve to make for that aspiration a sacrifice the greater, because he is fully aware of its extent;—if, so circumstanced, he give up a business or a profession on which he might have entered with advantage, with the hope that, when he shall have won a station high in the ranks of European science, he may a little augment his resources by some of those few employments to which science leads;—if he hope to obtain some situation, (at the Board of Longitude,[9] for example,) where he may be permitted to exercise the talents of a philosopher for the paltry remuneration of a clerk, he will find that other qualifications than knowledge and a love of science are necessary for its attainment. He will also find that the high and independent spirit, which usually dwells in the breast of those who are deeply versed in these pursuits, is ill adapted for such appointments; and that even if successful, he must hear many things he disapproves, and raise no voice *against* them.

Thus, then, it appears that scarcely any man can be expected to pursue abstract science unless he possess a private fortune, and unless he can resolve to give up all intention of improving it. Yet, how few thus situated are likely to undergo the labour of the acquisition; and if they do from some irresistible impulse, what inducement is there for them to deviate one step from those inquiries in which they find the greatest delight, into those which might be more immediately useful to the public?

9 This body is now dissolved.

FURTHER READING

Alborn, Timothy. "The Business of Induction: Industry and Genius in the Language of British Scientific Reform 1820–40." *History of Science* 34 (1996): 91–121.

Aspray, William, ed. *Computing Before Computers.* Ames: Iowa State UP, 1990.

Babbage, Charles. *Passages from the Life of a Philosopher.* Edited by Martin Campbell-Kelly. New Brunswick, NJ: Rutgers UP, 1994.

—. *Reflections on the Decline of Science in England and On Some of Its Causes.* 1830. Reprint, Farnborough: Gregg International, 1969.

—. *Science and Reform: Selected Works of Charles Babbage.* Edited by Anthony Hyman. Cambridge: Cambridge UP, 1989.

—. *The Works of Charles Babbage.* Edited by Martin Campbell-Kelly. 11 vols. London: William Pickering, 1989.

Babbage, Henry Prevost. *Babbage's Calculating Engines: A Collection of Papers by Henry Prevost Babbage.* Edited by Allan G. Bromley. Los Angeles: Tomash Publishers, 1982.

Buxton, H.W. *Memoir of the Life and Labours of the Late Charles Babbage Esq. F.R.S.* Edited by Anthony Hyman. Cambridge, MA: M.I.T. Press, 1988.

Cannon, Susan Faye. *Science in Culture: The Early Victorian Period.* New York: Dawson and Science History Publications, 1978.

Cardwell, D.S.L. *The Organisation of Science in England: A Retrospect.* London: William Heinemann, Ltd., 1957.

Collier, Bruce. *The Little Engines that Could've: The Calculating Machines of Charles Babbage.* New York: Garland Publishing, Inc., 1990.

Dubbey, J.M. *The Mathematical Work of Charles Babbage.* Cambridge: Cambridge UP, 1978.

Foote, George A. "The Place of Science in the British Reform Movement 1830–50." *Isis* 42, no. 129 (1951): 192–208.

Franksen, Ole Immanuel. *Mr. Babbage's Secret: The Tale of a Cypher-and APL.* Englewood Cliffs, NJ: Prentice-Hall, Inc., 1985.

Hyman, Anthony. *Charles Babbage Pioneer of the Computer.* Princeton: Princeton UP, 1982.

Lindgren, Michael. *Glory and Failure: The Difference Engines of Johann Müller, Charles Babbage and Georg and Edvard Scheutz.* Translated by Dr. Craig G. McKay. Stockholm: Royal Institute of Technology Library, 1987.

Morrell, J.B. "Individualism and the Structure of British Science in 1830." *Historical Studies in the Physical Sciences* 3 (1971): 183–204.

Morrell, Jack, and Arnold Thackray. *Gentlemen of Science: Early Years of the British Association for the Advancement of Science.* Oxford: Clarendon Press, 1981.

Morrison, Philip, and Emily Morrison, eds. *Charles Babbage and His Calculating Engines: Selected Writings by Charles Babbage and Others.* New York: Dover Publications, 1961.

Moseley, Maboth. *Irascible Genius: The Life of Charles Babbage, Inventor.* Chicago: Henry Regnery Company, 1964.

Swade, Doron. *Charles Babbage and His Calculating Engines.* London: Science Museum, 1991.

Williams, L. Pearce. "The Royal Society and the Founding of the British Association for the Advancement of Science." *Notes and Records of the Royal Society of London* 16 (1961): 221–33.

10. CHARLES LYELL

(1797–1875)

The subtitle of Sir Charles Lyell's influential *Principles of Geology* (1830–33) summarizes its primary argument: "An Attempt to Explain the Former Changes of the Earth's Surface, By Reference to Causes Now in Operation." Lyell argued that geological strata and present-day rock formations are the products of uniform, unvarying natural processes which had existed since the creation of the earth and which continue to exert their influence at the present time. Lyell was a barrister, and the *Principles* is a masterpiece of rhetoric centring on the uniformitarian theme; Stephen Jay Gould has described the work as "a passionate brief for a single, well-formed argument, hammered home relentlessly" (Gould 105).

Although sometimes hailed as the founder of modern geology, Lyell's work grows out of several late eighteenth-century geological debates. Lyell adopted the ideas of James Hutton's *Theory of the Earth* (1795)—popularized by John Playfair's *Illustrations of the Huttonian Theory of the Earth* (1802)—on the uniformity of geological processes and the volcanic origin of granite and basalt (hence the term "Plutonist" which the Huttonians have been designated). Hutton believed that rocks originated in erosion and lithification by the earth's internal heat. Similarly the "Vulcanists" Jean Etienne Guettard (1715–86) and Nicholas Desmarest (1725–1815) had argued for the igneous origin of basalt, and against the idea that volcanoes were minor and inconsequential geological phenomena arising from burning underground coal deposits. Lyell also drew on the geological and paleontological work of George Poulett Scrope and William Smith, who had correlated fossils with particular strata and produced the first accurate stratigraphic map of England.

William Whewell coined the term "uniformitarianism" as an umbrella term for a number of Lyell's related views that geological processes were uniform in rate and mode of action over time, were still acting (actualism), and proceeded without directional progression (Rudwick 164–218; Gould 117–126). As A. Hallam explains, "Lyell failed to distinguish clearly between a *methodological uniformitarianism*, comprising the assumption that natural laws are constant in space and time, and a *substantive uniformitarianism*, which postulates a uniformity of material conditions or of rate of processes. The former, which also demands that no hypothetical unknown processes be invoked if observed historical results can be explained by presently observed processes, is vital for the interpretation of past events. It is in effect synonymous with actualism and was readily conceded by such early opponents as Sedgwick" (105).

The Plutonists with whom Lyell was in sympathy were reacting against the school of "Geognosy," founded by the German Abraham Gottlob Werner (1749–1817) and championed by Robert Jamieson, who believed that most rocks had been formed not by heat, but as chemical precipitates from a universal ocean (hence their name "Neptunists," as Lyell calls them), with some later strata being mechanically re-deposited or locally distributed by alluvial sedimentation. Lyell looked critically upon the Neptunists as well as the "Catastrophists," who believed that violent catastrophes of great magnitude such as extraordinary earthquakes and floods constituted the primary geological agents of change. Lyell also looked skeptically upon the related claims of the Diluvialists, who traced all geology back to the effects of the Biblical flood. Lyell, on the other hand, stressed that the present was the key to the past, in opposition to the weighty authority of the premier geologists of his day—William Buckland, William D. Conybeare, and Adam Sedgwick—who often invoked the biblical book of Genesis and Noah's flood to explain the history of the earth. In Lyell's own selective history of geology provided in the *Principles*, his empirically-based gradualism inspired by Hutton and Playfair triumphs over a number of speculative bogeymen, including Wernerians (Neptunists), Catastrophists, and Diluvialists.

Lyell's general principle of uniformitarianism has been challenged by modern Neo-catastrophists working in the fields of cosmology, geomorphology, sedimentology, and evolutionary theory, including Gould's and Eldredge's hypothesis of "punctuated equilibrium," which proposes that species change occurs during brief periods of dramatic environmental shift, followed by longer periods of species stability. Asteroids and meteors colliding with the planets of our solar system in the past have certainly caused dramatic cosmological change.

Lyell and his wife Mary, who shared his geological interests, travelled extensively in Europe and America investigating rock strata and fossils. Lyell subsequently published a series of travel narratives entitled *Travels in North America* (1845) and *A Second Visit to the United States of North America* (1849). He also wrote a textbook, *Elements of Geology* (1838), whose name was later changed to *A Manual of Elementary Geology* in 1851. Lyell contributed to stratigraphy by subdividing the Tertiary strata into four epochs: the Eocene, Miocene, and older and newer Pliocene, names suggested to him by William Whewell.

Darwin took the first volume of the *Principles* with him on the voyage of the Beagle in 1831 and Lyell, who became Darwin's close friend after 1836, influenced Darwin's work immensely: Darwin became the spokesman for gradualism and actualism in speciation just as Lyell had been in geology. The greater part of volume II of the *Principles* is devoted to an examination and eventual dismissal of Lamarck's theories of species transmutation. Lyell also critiqued Darwin's evolutionary theories in *Geological Evidences of the Antiquity of Man* (1863). He presented Darwin's theory of natural selection and discussed Lamarck again along with Robert Chambers's *Vestiges of Natural Creation*, but it is difficult to find a full, unambiguous endorsement of natural selection or species transmutation in Lyell's work. He later expressed full support of Darwin's theory and rewrote the *Principles* to reflect this change of mind.

In the latter part of the nineteenth century, both Darwin's theory of speciation and Lyell's uniformitarianism—which both required long expanses of geological time—were challenged by Lord Kelvin's calculations of the age of the earth as a relatively young 100 million years old. Kelvin's figures were based on estimating the cooling rate of an originally molten earth. In a 1904 address, Ernest Rutherford proposed that heat generated from the decay of radioactive materials known to exist in the earth's crust would have slowed the cooling of the earth, providing a greater age for the earth than that extrapolated from Lord Kelvin's thermal data.

In the passages below, Lyell first defines geology, outlines his methodology for studying geology, and then reviews the "causes which have retarded the progress of Geology." Drawing on modern discoveries, Lyell dismisses the counterarguments and objections to his central theory that existing causes have produced the former changes of the earth's surface.

Principles of Geology.

Vol. 1. London: John Murray, 1830, pp. 1–4; 75–91.

Chapter 1.

Geology defined—Compared to History—Its relation to other Physical Sciences—Its distinctness from all—Not to be confounded with Cosmogony.

Geology is the science which investigates the successive changes that have taken place in the organic and inorganic kingdoms of nature; it enquires into the causes of these changes, and the influence which they have exerted in modifying the surface and external structure of our planet.

By these researches into the state of the earth and its inhabitants at former periods, we acquire a more perfect knowledge of its *present* condition, and more comprehensive views concerning the laws *now* governing its animate and inanimate productions. When we study history, we obtain a more profound insight into human nature, by instituting a comparison between the present and former states of society. We trace the long series of events which have gradually led to the actual posture of affairs; and by connecting effects with their causes, we are enabled to classify and retain in the memory a multitude of complicated relations—the various peculiarities of national character—the different degrees of moral and intellectual refinement, and numerous other circumstances, which, without historical associations, would be uninteresting or imperfectly understood. As the present condition of nations is the result of many antecedent changes, some extremely remote and others recent, some gradual, others sudden and violent, so the state of the natural world is the result of a long succession of events, and if we would enlarge our experience of the present economy of nature, we must investigate the effects of her operations in former epochs.

We often discover with surprise, on looking back into the chronicles of nations, how the fortune of some battle has influenced the fate of millions of our contemporaries, when it has long been forgotten by the mass of the population. With this remote event we may find inseparably connected the geographical boundaries of a great state, the language now spoken by the inhabitants, their peculiar manners, laws, and religious opinions. But far more astonishing and unexpected are the connexions brought to light, when we carry back our researches into the history of nature. The form of a coast, the configuration of the interior of a country, the existence and extent of lakes, valleys, and mountains, can often be traced to the former prevalence of earthquakes and volcanoes, in regions which have long been undisturbed. To these remote convulsions the present fertility of some districts, the sterile character of others, the elevation of land above the sea, the climate, and various peculiarities, may be distinctly referred. On the other hand, many distinguishing features of the surface may often be ascribed to the operation at a remote era of slow and tranquil causes—to the gradual deposition of sediment in a lake or in the ocean, or to the prolific growth in the same of corals and testacea. To select another example, we find in certain localities subterranean deposits of coal, consisting of vegetable matter, formerly drifted into seas and lakes. These seas and lakes have since been filled up, the lands whereon the forests grew have disappeared or changed their form, the rivers and currents which floated the vegetable masses can no longer be traced, and the plants belonged to species which for ages have passed away from the surface of our planet. Yet the commercial prosperity, and numerical strength of a nation, may now be mainly dependent on the local distribution of fuel determined by that ancient state of things.

Geology is intimately related to almost all the physical sciences, as is history to the moral. An historian should, if possible, be at once profoundly acquainted with ethics, politics, jurisprudence, the military art, theology; in a word, with all branches of knowledge, whereby any insight into human affairs, or into the moral and intellectual nature of man, can be obtained. It would be no less desirable that a geologist should be well versed in chemistry, natural philosophy, mineralogy, zoology, comparative anatomy, botany; in short, in every science relating to organic and inorganic nature. With these accomplishments the historian and geologist would rarely fail to draw correct and philosophical conclusions from the various monuments transmitted to them of former occurrences. They would know to what combination of causes analogous effects were referrible, and they would often be enabled to supply by inference, information concerning many events unrecorded in the defective archives of former ages. But the brief duration of human life, and our limited powers, are so far from permitting us to aspire to such extensive acquisitions, that excellence even in one department is within the reach of few, and those individuals most effectually promote the general progress, who concentrate their thoughts on a limited portion of the field of inquiry. As it is necessary that the historian and the cultivators of moral or political science should reciprocally aid each other, so the geologist and those who study natural history or physics stand in equal need of mutual assistance. A comparative anatomist may derive some accession of knowledge from the bare inspection of the remains of an extinct quadruped, but the relic throws much greater light upon his own science, when he is informed to what relative era it belonged, what plants and animals were its contemporaries, in what degree of latitude it once existed, and other historical details. A fossil shell may interest a conchologist, though he be ignorant of the locality from which it came; but it will be of more value when he learns with what other species it was associated, whether they were marine or fresh-water, whether the strata containing them were at a certain elevation above the sea, and what relative position they held in regard to other groups of strata, with many other particulars determinable by an experienced geologist alone. On the other hand, the skill of the comparative anatomist and conchologist are often indispensable to those engaged in geological research, although it will rarely happen that the geologist will himself combine these different qualifications in his own person.

Some remains of former organic beings, like the ancient temple, statue, or picture, may have both their intrinsic and their historical value, while there are others which can never be expected to attract attention for their own sake. A painter, sculptor, or architect, would often neglect many curious relics of antiquity, as devoid of beauty and uninstructive with relation to their own art, however illustrative of the progress of refinement in some ancient nation. It has therefore been found desirable that the antiquary should unite his labours to those of the historian, and similar co-operation has become necessary in geology. The field of inquiry in living nature being inexhaustible, the zoologist and botanist can rarely be induced to sacrifice time in exploring the imperfect remains of lost species of animals and plants, while those still existing afford constant matter of novelty. They must entertain a desire of promoting *geology* by such investigations, and some knowledge of its objects must guide and direct their studies. According to the different opportunities, tastes, and talents of individuals, they may employ themselves in collecting particular kinds of minerals, rocks, or organic remains, and these, when well examined and explained, afford data to the geologist, as do coins, medals, and inscriptions to the historian.

It was long ere the distinct nature and legitimate objects of geology were fully recognized, and it was at first confounded with many other branches of inquiry, just as the limits of history, poetry, and mythology were ill-defined in the infancy of civilization. Werner appears to have regarded geology as little other than a subordinate department of mineralogy, and Desmarest included it under the head of Physical Geography. But the identification of its objects with those of Cosmogony has been the most common and serious source of confusion. The first who endeavoured to draw a clear line of demarcation between these distinct departments, was Hutton, who declared that geology was in no ways concerned "with questions as to the origin of things." But his doctrine on this head was vehemently opposed at first, and although it has gradually gained ground, and will ultimately prevail, it is yet far from being established. We shall attempt in the sequel of this work to demonstrate that geology differs as widely from cosmogony, as speculations concerning the creation of man differ from history. But before we enter more at large on this controverted question, we shall endeavour to trace the progress of opinion on this topic, from the earliest ages, to the commencement of the present century.

Chapter V.

Review of the causes which have retarded the progress of Geology—Effects of prepossessions in regard to the duration of past time—Of prejudices arising from our peculiar position as inhabitants of the land—Of those occasioned by our not seeing subterranean changes now in progress—All these causes combine to make the former course of Nature appear different from the present—Several objections to the assumption, that existing causes have produced the former changes of the earth's surface, removed by modern discoveries.

We have seen that, during the progress of geology, there have been great fluctuations of opinion respecting the nature of the causes to which all former changes of the earth's surface are referrible. The first observers conceived that the monuments which the geologist endeavours to decipher, relate to a period when the physical constitution of the earth differed entirely from the present, and that, even after the creation of living beings, there have been causes in action distinct in kind or degree from those now forming part of the economy of nature. These views have been gradually modified, and some of them entirely abandoned in proportion as observations have been multiplied, and the signs of former mutations more skilfully interpreted. Many appearances, which for a long time were regarded as indicating mysterious and extraordinary agency, are finally recognized as the necessary result of the laws now governing the material world; and the discovery of this unlooked for conformity has induced some geologists to infer that there has never been any interruption to the same uniform order of physical events. The same assemblage of general causes, they conceive, may have been sufficient to produce, by their various combinations, the endless diversity of effects, of which the shell of the earth has preserved the memorials, and, consistently with these principles, the recurrence of analogous changes is expected by them in time to come.

Whether we coincide or not in this doctrine, we must admit that the gradual progress of opinion concerning the succession of phenomena in remote eras, resembles in a singular manner that which accompanies the growing intelligence of every people, in regard to the economy of nature in modern times. In an early stage of advancement, when a great number of natural appearances are unintelligible, an eclipse, an earthquake, a flood, or the approach of a comet, with many other occurrences afterwards found to belong to the regular course of events, are regarded as prodigies. The same delusion prevails as to moral phenomena, and many of these are ascribed to the intervention of demons, ghosts, witches, and other immaterial and supernatural agents. By degrees, many of the enigmas of the moral and physical world are explained, and, instead of being due to extrinsic and irregular causes, they are found to depend on fixed and invariable laws. The philosopher at last becomes convinced of the undeviating uniformity of secondary causes, and, guided by his faith in this principle, he determines the probability of accounts transmitted to him of former occurrences, and often rejects the fabulous tales of former ages, on the ground of their being irreconcilable with the experience of more enlightened ages.

As a belief in want of conformity in the physical constitution of the earth, in ancient and modern times, was for a long time universally prevalent, and that too amongst men who were convinced that the order of nature is *now* uniform, and has continued so for several thousand years; every circumstance which could have influenced their minds and given an undue bias to their opinions deserves particular attention. Now the reader may easily satisfy himself, that, however undeviating the course of nature may have been from the earliest epochs, it was impossible for the first cultivators of geology to come to such a conclusion, so long as they were under a delusion as to the age of the world, and the date of the first creation of animate beings. However fantastical some theories of the sixteenth century may now appear to us,—however unworthy of men of great talent and sound judgment, we may rest assured that, if the same misconceptions now prevailed in regard to the memorials of human transactions, it would give rise to a similar train of absurdities. Let us imagine, for example, that Champollion, and the French and Tuscan literati now engaged in exploring the antiquities of Egypt, had visited that country with a firm belief that the banks of the Nile were never peopled by the human race before the beginning of the nineteenth century, and that their faith in this dogma was as difficult to shake as the opinion of our ancestors, that the earth was never the abode of living beings until the creation of the present continents, and of the species now existing,—it is easy to perceive what extravagant systems they would frame, while under the influence of this delusion, to account for the monuments discovered in Egypt. The sight of the pyramids, obelisks, colossal statues, and ruined temples, would fill them with such astonishment, that for a time they would be as men spell-bound—wholly incapacitated to reason with sobriety. They might incline at first to refer the construction of such stupendous works to some superhuman powers of a primeval world. A system might be invented resembling that so gravely advanced by Manetho, who relates that a dynasty of gods originally ruled in Egypt, of whom Vulcan, the first monarch, reigned nine thousand years. After them came Hercules and other demi-gods, who were at last succeeded by human kings. When some fanciful speculations of this kind had amused the imagination for a time, some vast repository of mummies would be discovered and would immediately undeceive those antiquaries who enjoyed an opportunity of personally examining them, but the prejudices of others at a distance, who were not eye-witnesses of the whole phenomena, would not be so easily overcome. The concurrent report of many travellers would indeed render it necessary for them to accommodate ancient theories to some of the new facts, and much wit and ingenuity would be required to modify and defend their old positions. Each new invention would violate a greater number of known analogies; for if a theory be required to embrace some false principle, it becomes more visionary in proportion as facts are multiplied, as would be the case if geometers were now required to form an astronomical system on the assumption of the immobility of the earth.

Amongst other fanciful conjectures concerning the history of Egypt, we may suppose some of the following to be started. 'As the banks of the Nile have been so recently colonized, the curious substances called mummies could never in reality have belonged to men. They may have been generated by some *plastic virtue* residing in the interior of the earth, or they may be abortions of nature produced by her incipient efforts in the work of creation. For if deformed beings are sometimes born even now, when the scheme of the universe is fully developed, many more may have been "sent before their time, scarce half made up," when the planet itself was in the embryo state. But if these notions appear to derogate from the perfection of the Divine attributes, and if these mummies be in all their parts true representations of the human form, may we not refer them to the future rather than the past? May we not be looking into the womb of nature, and not her grave? may not these images be like the shades of the unborn in Virgil's Elysium—the archetypes of men not yet called into existence?'

These speculations, if advocated by eloquent writers, would not fail to attract many zealous votaries, for they would relieve men from the painful necessity of renouncing preconceived opinions. Incredible as such scepticism may appear, it would be rivalled by many systems of the sixteenth and seventeenth centuries, and among others by that of the learned Falloppio, who regarded the tusks of fossil elephants as earthy concretions, and the vases of Monte Testaceo, near Rome, as works of nature, and not of art. But when one generation had passed away, and another not compromised to the support of antiquated dogmas had succeeded, they would review the evidence afforded by mummies more impartially, and would no longer controvert the preliminary question, that human beings had lived in Egypt before the nineteenth century: so that when a hundred years perhaps had been lost, the industry and talents of the philosopher would be at last directed to the elucidation of points of real historical importance.

But we have adverted to one only of many prejudices with which the earlier geologists had to contend. Even when they conceded that the earth had been peopled with animate beings at an earlier period than was at first supposed, they had no conception that the quantity of time bore so great a proportion to the historical era as is now generally conceded. How fatal every error as to the quantity of time must prove to the introduction of rational views concerning the state of things in former ages, may be conceived by supposing that the annals of the civil and military transactions of a great nation were perused under the impression that they occurred in a period of one hundred instead of two thousand years. Such a portion of history would immediately assume the air of a romance; the events would seem devoid of credibility, and inconsistent with the present course of human affairs. A crowd of incidents would follow each other in thick succession. Armies and fleets would appear to be assembled only to be destroyed, and cities built merely to fall in ruins. There would be the most violent transitions from foreign or intestine war to periods of profound peace, and the works effected during the years of disorder or tranquillity would be alike superhuman in magnitude.

He who should study the monuments of the natural world under the influence of a similar infatuation, must draw a no less exaggerated picture of the energy and violence of causes, and must experience the same insurmountable difficulty in reconciling the former and present state of nature. If we could behold in one view all the volcanic cones thrown up in Iceland, Italy, Sicily, and other parts of Europe, during the last five thousand years, and could see the lavas which have flowed during the same period; the dislocations, subsidences and elevations caused by earthquakes; the lands added to various deltas, or devoured by the sea, together with the effects of devastation by floods, and imagine that all these events had happened in one year, we must form most exalted ideas of the activity of the agents, and the suddenness of the revolutions. Were an equal amount of change to pass before our eyes in the next year, could we avoid the conclusion that some great crisis of nature was at hand? If geologists, therefore, have misinterpreted the signs of a succession of events, so as to conclude that centuries were implied where the characters imported thousands of years, and thousands of years where the language of nature signified millions, they could not, if they reasoned logically from such false premises, come to any other conclusion, than that the system of the natural world had undergone a complete revolution.

We should be warranted in ascribing the erection of the great pyramid to superhuman power, if we were convinced that it was raised in one day; and if we imagine, in the same manner, a mountain chain to have been elevated, during an equally small fraction of the time which was really occupied in upheaving it, we might then be justified in inferring, that the subterranean movements were once far more energetic than in our own times. We know that one earthquake may raise the coast of Chili for a hundred miles to the average height of about five feet. A repetition of two thousand shocks of equal violence might produce a mountain chain one hundred miles long, and ten thousand feet high. Now, should one only of these convulsions happen in a century, it would be consistent with the order of events experienced by the Chilians from the earliest times; but if the whole of them were to occur in the next hundred years, the entire district must be depopulated, scarcely any animals or plants could survive, and the surface would be one confused heap of ruin and desolation.

One consequence of undervaluing greatly the quantity of past time is the apparent coincidence which it occasions of events necessarily disconnected, or which are so unusual, that it would be inconsistent with all calculation of chances to suppose them to happen at one and the same time. When the unlooked for association of such rare phenomena is witnessed in the present course of nature, it scarcely ever fails to excite a suspicion of the preternatural in those minds which are not firmly convinced of the uniform agency of secondary causes;—as if the death of some individual in whose fate they are interested, happens to be accompanied by the appearance of a luminous meteor, or a comet, or the shock of an earthquake. It would be only necessary to multiply such coincidences indefinitely, and the mind of every philosopher would be disturbed. Now it would be difficult to exaggerate the number of physical events, many of them most rare and unconnected in their nature, which were imagined by the Woodwardian hypothesis to have happened in the course of a few months; and numerous other examples might be found of popular geological theories, which require us to imagine that a long succession of events happened in a brief and almost momentary period.

The sources of prejudice hitherto considered may be deemed as in a great degree peculiar to the infancy of the science, but others are common to the first cultivators of geology and to ourselves, and are all singularly calculated to produce the same deception, and to strengthen our belief that the course of nature in the earlier ages differed widely from that now established. Although we cannot fully explain all these circumstances, without assuming some things as proved, which it will be the object of another part of this work to demonstrate, we must briefly allude to them in this place.

The first and greatest difficulty, then, consists in our habitual unconsciousness that our position as observers is essentially unfavourable, when we endeavour to estimate the magnitude of the changes now in progress. In consequence of our inattention to this subject, we are liable to the greatest mistakes in contrasting the present with former states of the globe. We inhabit about a fourth part of the surface; and that portion is almost exclusively the theatre of decay and not of reproduction. We know, indeed, that new deposits are annually formed in seas and lakes, and that every year some new igneous rocks are produced in the bowels of the earth, but we cannot watch the progress of their formation; and, as they are only present to our minds by the aid of reflection, it requires an effort both of the reason and the imagination to appreciate duly their importance. It is, therefore, not surprising that we imperfectly estimate the result of operations invisible to us; and that, when analogous results of some former epoch are presented to our inspection, we cannot recognise the analogy. He who has observed the quarrying of stone from a rock, and has seen it shipped for some distant port, and then endeavours to conceive what kind of edifice will be raised by the materials, is in the same predicament as a geologist, who, while he is confined to the land, sees the decomposition of rocks, and the transportation of matter by rivers to the sea, and then endeavours to picture to himself the new strata which Nature is building beneath the waters. Nor is his position less unfavourable when, beholding a volcanic eruption, he tries to conceive what changes the column of lava has produced, in its passage upwards, on the intersected strata; or what form the melted matter may assume at great depths on cooling down; or what may be the extent of the subterranean rivers and reservoirs of liquid matter far beneath the surface. It should, therefore, be remembered, that the task imposed on those who study the earth's history requires no ordinary share of discretion, for we are precluded from collating the corresponding parts of a system existing at two different periods. If we were inhabitants of another element—if the great ocean were our domain, instead of the narrow limits of the land, our difficulties would be considerably lessened; while, on the other hand, there can be little doubt, although the reader may, perhaps, smile at the bare suggestion of such an idea, that an amphibious being, who should possess our faculties, would still more easily arrive at sound theoretical opinions in geology, since he might behold, on the one hand, the decomposition of rocks in the atmosphere, and the transportation of matter by running water; and, on the other, examine the deposition of sediment in the sea, and the imbedding of animal remains in new strata. He might ascertain, by direct observation, the action of a mountain torrent, as well as of a marine current; might compare the products of volcanos on the land with those poured out beneath the waters; and might mark, on the one hand, the growth of the forest, and on the other that of the coral reef. Yet, even with these advantages, he would be liable to fall into the greatest errors when endeavouring to reason on rocks of subterranean origin. He would seek in vain, within the sphere of his observation, for any direct analogy to the process of their formation, and would therefore be in danger of attributing them, wherever they are upraised to view, to some "primeval state of nature." But if we may be allowed so far to indulge the imagination, as to suppose a being, entirely confined to the nether world—some "dusky melancholy sprite," like Umbriel, who could "flit on sooty pinions to the central earth," but who was never permitted to "sully the fair face of light," and emerge into the regions of water and of air; and if this being should busy himself in investigating the structure of the globe, he might frame theories the exact converse of those usually adopted by human philosophers. He might infer that the stratified rocks, containing shells and other organic remains, were the oldest of created things, belonging to some original and nascent state of the planet. "Of these masses," he might say, "whether they consist of loose incoherent sand, soft clay, or solid rock, none have been formed in modern times. Every year some part of them are broken and shattered by earthquakes, or melted up by volcanic fire; and, when they cool down slowly from a state of fusion, they assume a crystalline form, perfectly distinct from those inexplicable rocks which are so regularly bedded, and contain stones full of curious impressions and fantastic markings. This process cannot have been carried on for an indefinite time, for in that case all the stratified rocks would long ere this have been fused and crystallized. It is therefore probable that the whole planet once consisted of these curiously-bedded formations, at a time when the volcanic fire had not yet been brought into activity. Since that period there seems to have been a gradual development of heat, and this augmentation we may expect to continue till the whole globe shall be in a state of fluidity and incandescence."

Such might be the system of the Gnome at the very same time that the followers of Liebnitz, reasoning on what they saw on the outer surface, would be teaching the doctrine of gradual refrigeration, and averring that the earth had begun its career as a fiery comet, and would hereafter become a frozen icy mass. The tenets of the schools of the nether and of the upper world would be directly opposed to each other, for both would partake of the prejudices inevitably resulting from the continual contemplation of one class of phenomena to the exclusion of another. Man observes the annual decomposition of crystalline and igneous rocks, and may sometimes see their conversion into stratified deposits; but he cannot witness the reconversion of the sedimentary into the crystalline by subterranean fire. He is in the habit of regarding all the sedimentary rocks as more recent than the unstratified, for the same reason that we may suppose him to fall into the opposite error if he saw the origin of the igneous class only.

It is only by becoming sensible of our natural disadvantages that we shall be roused to exertion, and prompted to seek out opportunities of discovering the operations now in progress, such as do not present themselves readily to view. We are called upon, in our researches into the state of the earth, as in our endeavours to comprehend the mechanism of the heavens, to invent means for overcoming the limited range of our vision. We are perpetually required to bring, as far as possible, within the sphere of observation, things to which the eye, unassisted by art, could never obtain access. It was not an impossible contingency that astronomers might have been placed, at some period, in a situation much resembling that in which the geologist seems to stand at present. If the Italians, for example, in the early part of the twelfth century, had discovered at Amalphi, instead of the pandects of Justinian, some ancient manuscripts filled with astronomical observations relating to a period of three thousand years, and made by some ancient geometers who possessed optical instruments as perfect as any in modern Europe, they would probably, on consulting these memorials, have come to a conclusion that there had been a great revolution in the solar and sidereal systems. "Many primary and secondary planets," they might say, "are enumerated in these tables, which exist no longer. Their positions are assigned with such precision, that we may assure ourselves that there is nothing in their place at present but the blue ether. Where one star is visible to us, these documents represent several thousands. Some of those which are now single, consisted then of two separate bodies, often distinguished by different colours, and revolving periodically round a common centre of gravity. There is no analogy to them in the universe at present, for they were neither fixed stars nor planets, but stood in the mutual relation of sun and planet to each other. We must conclude, therefore, that there has occurred, at no distant period, a tremendous catastrophe, whereby thousands of worlds have been annihilated at once, and some heavenly bodies absorbed into the substance of others." When such doctrines had prevailed for ages, the discovery of one of the worlds, supposed to have been lost, by aid of the first rude telescope, would not dissipate the delusion, for the whole burden of proof would now be thrown on those who insisted on the stability of the system from the beginning of time, and these philosophers would be required to demonstrate the existence of *all* the worlds said to have been annihilated. Such popular prejudices would be most unfavourable to the advancement of astronomy; for, instead of persevering in the attempt to improve their instruments, and laboriously to make and record observations, the greater number would despair of verifying the continued existence of the heavenly bodies not visible to the naked eye. Instead of confessing the extent of their ignorance, and striving to remove it by bringing to light new facts, they would be engaged in the indolent employment of framing imaginary theories concerning catastrophes and mighty revolutions in the system of the universe.

For more than two centuries the shelly strata of the Sub-apennine hills afforded matter of speculation to the early geologists of Italy, and few of them had any suspicion that similar deposits were then forming in the neighbouring sea. They were as unconscious of the continued action of causes still producing similar effects, as the astronomers, in the case supposed by us, of the existence of certain heavenly bodies still giving and reflecting light, and performing their movements as in the olden time. Some imagined that the strata, so rich in organic remains, instead of being due to secondary agents, had been so created in the beginning of things by the fiat of the Almighty; and others ascribed the imbedded fossil bodies to some plastic power which resided in the earth in the early ages of the world. At length Donati explored the bed of the Adriatic, and found the closest resemblance between the new deposits there forming, and those which constituted hills above a thousand feet high in various parts of the peninsula. He ascertained that certain genera of living testacea were grouped together at the bottom of the sea in precisely the same manner as were their fossil analogues in the strata of the hills, and that some species were common to the recent and fossil world. Beds of shells, moreover, in the Adriatic, were becoming incrusted with calcareous rock; and others were recently enclosed in deposits of sand and clay, precisely as fossil shells were found in the hills. This splendid discovery of the identity of modern and ancient submarine operations was not made without the aid of artificial instruments, which, like the telescope, brought phenomena into view not otherwise within the sphere of human observation.

In like manner, in the Vincentin, a great series of volcanic and marine sedimentary rocks were examined in the early part of the last century; but no geologist suspected, before the time of Arduino, that these were partly composed of ancient submarine lavas. If, when these enquiries were first made, geologists had been told that the mode of formation of such rocks might be fully elucidated by the study of processes then going on in certain parts of the Mediterranean, they would have been as incredulous as geometers would have been before the time of Newton, if any one had informed them that, by making experiments on the motion of bodies on the earth, they might discover the laws which regulated the movements of distant planets.

The establishment, from time to time, of numerous points of identification, drew at length from geologists a reluctant admission, that there was more correspondence between the physical constitution of the globe, and more uniformity in the laws regulating the changes of its surface, from the most remote eras to the present, than they at first imagined. If, in this state of the science, they still despaired of reconciling every class of geological phenomena to the operations of ordinary causes, even by straining analogy to the utmost limits of credibility, we might have expected, that the balance of probability at least would now have been presumed to incline towards the identity of the causes. But, after repeated experience of the failure of attempts to speculate on different classes of geological phenomena, as belonging to a distinct order of things, each new sect persevered systematically in the principles adopted by their predecessors. They invariably began, as each new problem presented itself, whether relating to the animate or inanimate world, to assume in their theories, that the economy of nature was formerly governed by rules quite independent of those now established. Whether they endeavoured to account for the origin of certain igneous rocks, or to explain the forces which elevated hills or excavated valleys, or the causes which led to the extinction of certain races of animals, they first presupposed an original and dissimilar order of nature; and when at length they approximated, or entirely came round to an opposite opinion, it was always with the feeling, that they conceded what they were justified *à priori* in deeming improbable. In a word, the same men who, as natural philosophers, would have been greatly surprised to find any deviation from the usual course of Nature *in their own time*, were equally surprised, as geologists, not to find such deviations at every period of the past.

The Huttonians were conscious that no check could be given to the utmost licence of conjecture in speculating on the causes of geological phenomena, unless we can assume invariable constancy in the order of Nature. But when they asserted this uniformity without any limitation as to time, they were considered, by the majority of their contemporaries, to have been carried too far, especially as they applied the same principle to the laws of the organic, as well as of the inanimate world.[1]

We shall first advert briefly to many difficulties which formerly appeared insurmountable, but which, in the last forty years, have been partially or entirely removed by the progress of science; and shall afterwards consider the objections that still remain to the doctrine of absolute uniformity.

In the first place, it was necessary for the supporters of this doctrine to take for granted incalculable periods of time, in order to explain the formation of sedimentary strata by causes now in diurnal action. The time which they required theoretically, is now granted, as it were, or has become absolutely requisite, to account for another class of phenomena brought to light by more recent investigations. It must always have been evident to unbiassed minds, that successive strata, containing, in regular order of superposition, distinct beds of shells and corals, arranged in families as they grow at the bottom of the sea, could only have been formed by slow and insensible degrees in a great lapse of ages; yet, until organic remains were minutely examined and specifically determined, it was rarely possible to prove that the series of deposits met with in one country was not formed simultaneously with that found in another. But we are now able to determine, in numerous instances, the relative dates of sedimentary rocks in distant regions, and to show, by their organic remains, that they were not of contemporary origin, but formed in succession. We often find, that where an interruption in the consecutive formation in one district is indicated by a sudden transition from one assemblage of fossil species to another, the chasm is filled up, in some other district, by other important groups of strata. The more attentively we study the European continent, the greater we find the extension of the whole series of geological formations. No sooner does the calender appear to be completed, and the signs of a succession of physical events arranged in chronological order, than we are called upon to intercalate, as it were, some new periods of vast duration. A geologist, whose observations have been confined to England, is accustomed to consider the superior and newer groups of marine strata in our island as modern, and such they are, comparatively speaking; but when he has travelled through the Italian peninsula and in Sicily, and has seen strata of more recent origin forming mountains several thousand feet high, and has marked a long series both of volcanic and submarine operations, all newer than any of the regular strata which enter largely into the physical

1 Playfair, after admitting the extinction of some species, says, "The inhabitants of the globe, then, like all other parts of it, are subject to change. It is not only the individual that perishes, but whole *species*, and even perhaps *genera*, are extinguished."—"A change in the animal kingdom seems to be a *part of the order of nature*, and is visible in instances to which human power cannot have extended."—Illustrations of the Huttonian Theory, § 413.

structure of Great Britain, he returns with more exalted conceptions of the antiquity of some of those modern deposits, than he before entertained of the oldest of the British series. We cannot reflect on the concessions thus extorted from us, in regard to the duration of past time, without forseeing that the period may arrive when part of the Huttonian theory will be combated on the ground of its departing too far from the assumption of uniformity in the order of nature. On a closer investigation of extinct volcanos, we find proofs that they broke out at successive eras, and that the eruptions of one group were often concluded long before others had commenced their activity. Some were burning when one class of organic beings were in existence, others came into action when different races of animals and plants existed,—it follows, therefore, that the convulsions caused by subterranean movements, which are merely another portion of the volcanic phenomena, occurred also in succession, and their effects must be divided into separate sums, and assigned to separate periods of time; and this is not all:—when we examine the volcanic products, whether they be lavas which flowed out under water or upon dry land, we find that intervals of time, often of great length, intervened between their formation, and that the effects of one eruption were not greater in amount than that which now results during ordinary volcanic convulsions. The accompanying or preceding earthquakes, therefore, may be considered to have been also successive, and to have been in like manner interrupted by intervals of time, and not to have exceeded in violence those now experienced in the ordinary course of nature. Already, therefore, may we regard the doctrine of the sudden elevation of whole continents by paroxysmal eruptions as invalidated; and there was the greatest inconsistency in the adoption of such a tenet by the Huttonians, who were anxious to reconcile former changes to the present economy of the world. It was contrary to analogy to suppose, that Nature had been at any former epoch parsimonious of time and prodigal of violence—to imagine that one district was not at rest while another was convulsed—that the disturbing forces were not kept under subjection, so as never to carry simultaneous havoc and desolation over the whole earth, or even over one great region. If it could have been shown, that a certain combination of circumstances would at some future period produce a crisis in the subterranean action, we should certainly have had no right to oppose our experience for the last three thousand years as an argument against the probability of such occurrences in past ages; but it is not pretended that such a combination can be foreseen. In speculating on catastrophes by water, we may certainly anticipate great floods in future, and we may therefore presume that they have happened again and again in past times. The existence of enormous seas of fresh-water, such as the North American lakes, the largest of which is elevated more than six hundred feet above the level of the ocean, and is in parts twelve hundred feet deep, is alone sufficient to assure us, that the time will come, however distant, when a deluge will lay waste a considerable part of the American continent. No hypothetical agency is required to cause the sudden escape of the confined waters. Such changes of level, and opening of fissures, as have accompanied earthquakes since the commencement of the present century, or such excavation of ravines as the receding cataract of Niagra is now effecting, might breach the barriers. Notwithstanding, therefore, that we have not witnessed within the last three thousand years the devastation by deluge of a large continent, yet, as we may predict the future occurrence of such catastrophes, we are authorized to regard them as part of the present order of Nature, and they may be introduced into geological speculations respecting the past, provided we do not imagine them to have been more frequent or general than we expect them to be in time to come.

The great contrast in the aspect of the older and newer rocks, in their texture, structure, and in the derangement of the strata, appeared formerly one of the strongest grounds for presuming that the causes to which they owed their origin were perfectly dissimilar from those now in operation. But this incongruity may now be regarded as the natural result of subsequent modifications, since the difference of relative age is demonstrated to have been so immense, that, however slow and insensible the change, it must have become important in the course of so many ages. In addition to volcanic heat, to which the Vulcanists formerly attributed too much influence, we must allow for the effect of mechanical pressure, of chemical affinity, of percolation by mineral waters, of permeation by elastic fluids, and the action, perhaps, of many other forces less understood, such as electricity and magnetism. In regard to the signs of upraising and sinking, of fracture and contortion in rocks, it is evident that newer strata cannot be shaken by earthquakes, unless the subjacent rocks are also affected; so that the contrast in the relative degree of disturbance in the more ancient and the newer strata, is one of many proofs that the convulsions have happened in different eras, and the fact confirms the uniformity of the action of subterranean forces, instead of their greater violence in the primeval ages.

The popular doctrine of universal formations, or the unlimited geographical extent of strata, distinguished by similar mineral characters, appeared for a long time to present insurmountable objections to the supposition, that the earth's crust had been formed by causes now acting. If it had merely been assumed, that rocks originating from fusion by subterranean fire presented in all parts of the globe a perfect correspondence in their mineral composition, the assumption would not have been extravagant; for, as the elementary substances that enter largely into the composition of rocks are few in number, they may be expected to arrange themselves invariably in the same forms, whenever the elementary particles are freely exposed to the action of chemical affinities. But when it was imagined that sedimentary mixtures, including animal and vegetable remains, and evidently formed in the beds of ancient seas, were of homogeneous nature throughout a whole hemisphere, or even farther, the dogma precluded at once all hope of recognizing the slightest analogy between the ancient and modern causes of decay and reproduction. For we know that existing rivers carry down from different mountain-chains sediment of distinct colours and composition; where the chains are near the sea, coarse sand and gravel is swept in; where they are distant, the finest mud. We know, also, that the matter introduced by springs into lakes and seas is very diversified in mineral composition; in short, contemporaneous strata now in the progress of formation are greatly varied in their composition, and could never afford formations of homogeneous mineral ingredients co-extensive with the greater part of the earth's surface. This theory, however, is as inapplicable to the effects of those operations to which the formation of the earth's crust is due, as to the effects of existing causes. The first investigators of sedimentary rocks had never reflected on the great areas occupied by modern deltas of large rivers; still less on the much greater areas over which marine currents, preying alike on river-deltas, and continuous lines of sea-coast, might be diffusing homogeneous mixtures. They were ignorant of the vast spaces over which calcareous and other mineral springs abound upon the land and in the sea, especially in and near volcanic regions, and of the quantity of matter discharged by them. When, therefore, they ascertained the extent of the geographical distribution of certain groups of ancient strata—when they traced them continuously from one extremity of Europe to the other, and found them flanking, throughout their entire range, great mountain-chains, they were astonished at so unexpected a discovery; and, considering themselves at liberty to disregard all modern analogy, they indulged in the sweeping generalization, that the law of continuity prevailed throughout strata of contemporaneous origin over the whole planet. The difficulty of dissipating this delusion was extreme, because some rocks, formed under similar circumstances at different epochs, present the same external characters, and often the same internal composition; and all these were assumed to be contemporaneous until the contrary could be shown, which, in the absence of evidence derived from direct superposition, and in the scarcity of organic remains, was often impossible.

Innumerable other false generalizations have been derived from the same source; such, for instance, as the former universality of the ocean, now disproved by the discovery of the remains of terrestrial vegetation, contemporary with every successive race of marine animals; but we shall dwell no longer on exploded errors, but proceed at once to contend against weightier objections, which will require more attentive consideration.

FURTHER READING

Albritton, Claude C. *The Abyss of Time.* San Francisco: Freeman, Cooper, 1980.

Bailey, Sir Edward. *Charles Lyell.* Garden City, NY: Doubleday and Co., 1963.

Beringer, Carl Cristophe. *Geschichte der Geologie und des geologischen Weltbildes.* Stuttgart: Ferdinand Enke, 1954.

Bowler, Peter J. *Fossils and Progress: Paleontology and the Idea of Progressive Evolution in the Nineteenth Century.* New York: Science History Publications, 1976.

Challinor, John. *The History of British Geology: A Bibliographical Study.* Newton Abbot: David and Charles, 1971.

Charles Lyell Centenary Symposium: Programme and Abstracts. London: International Union of Geological Sciences, 1975.

Eiseley, L. *Charles Lyell.* Scientific American Reprint. San Francisco: W.H. Freeman, 1959.

Geike, Sir Archibald. *The Founders of Geology.* London: Macmillan and Co., 1897.

Gillispie, Charles Coulston. *Genesis and Geology: A Study in the Relations of Scientific Thought, Natural Theology, and Social Opinion in Great Britain, 1790–1850.* Foreward by Nicolaas A. Rupke. Cambridge, MA: Harvard UP, 1996.

Gould, Stephen Jay. *Time's Arrow Time's Cycle: Myth and Metaphor in the Discovery of Geological Time.* Cambridge, MA: Harvard UP, 1987.

Greene, M.T. *Geology in the Nineteenth Century: Changing Views of a Changing World.* Ithaca, NY: Cornell UP, 1982.

Hallam, A. *Great Geological Controversies.* Oxford: Oxford UP, 1983.

Hooykaas, R. *Natural Law and Divine Miracle: A Historical-Critical Study of the Principle of Uniformity in Geology, Biology, and Theology.* Leiden: E.J. Brill, 1959.

Klaver, J.M.I. *Geology and Religious Sentiment: The Effect of Geological Discoveries on English Society and Literature between 1829 and 1859.* Leiden: E.J. Brill, 1997.

Laudan, Rachel. *From Mineralogy to Geology: The Foundations of a Science, 1650–1830.* Chicago: University of Chicago Press, 1987.

Lyell, Charles. *Principles of Geology.* Foreword by M.J.S. Rudwick. 3 vols. 1830–33. Reprint, Chicago: The University of Chicago Press, 1990.

Lyell, Katherine, ed. *Life Letters and Journals of Sir Charles Lyell, Bart.* 2 vols. London: John Murray, 1881.

Oldroyd, David. *Thinking About the Earth: A History of Ideas in Geology.* London: Athlone Press, 1996.

North, John F. *Sir Charles Lyell: Interpreter of the Principles of Geology.* London: A. Barker, 1965.

Porter, Roy. *The Making of Geology: Earth Science in Britain, 1660–1815.* Cambridge: Cambridge UP, 1977.

Rudwick, M.J.S. *Georges Cuvier, Fossil Bones, and Geological Catastrophes: New Translations and Interpretations of the Primary Texts.* Chicago: Chicago UP, 1997.

—. *The Great Devonian Controversy: The Shaping of Scientific Knowledge Among Gentlemanly Specialists.* Chicago: Chicago UP, 1985.

—. *The Meaning of Fossils: Episodes in the History of Palaeontology.* 2nd ed. New York: Science History Publications, 1976.

Rupke, Nicolaas A. "'The End of History' in the Early Picturing of Geological Time." *History of Science* 36 (1998): 61–90.

—. *The Great Chain of History: William Buckland and the English School of Geology (1814–1849).* Oxford: Clarendon Press, 1983.

Schneer, Cecil J. *Toward a History of Geology.* Cambridge, MA: M.I.T. Press, 1969.

Secord, James A. *Controversy in Victorian Geology: The Cambrian-Silurian Dispute.* Princeton: Princeton UP, 1986.

Skinner, Hubert C. *Charles Lyell on North American Geology.* New York: Arno Press, 1978.

Wilson, Leonard G. *Charles Lyell. The Years to 1841: The Revolution in Geology.* New Haven, CT: Yale UP, 1972.

—, ed. *Sir Charles Lyell's Scientific Journals on the Species Question.* New Haven, CT: Yale UP, 1970.

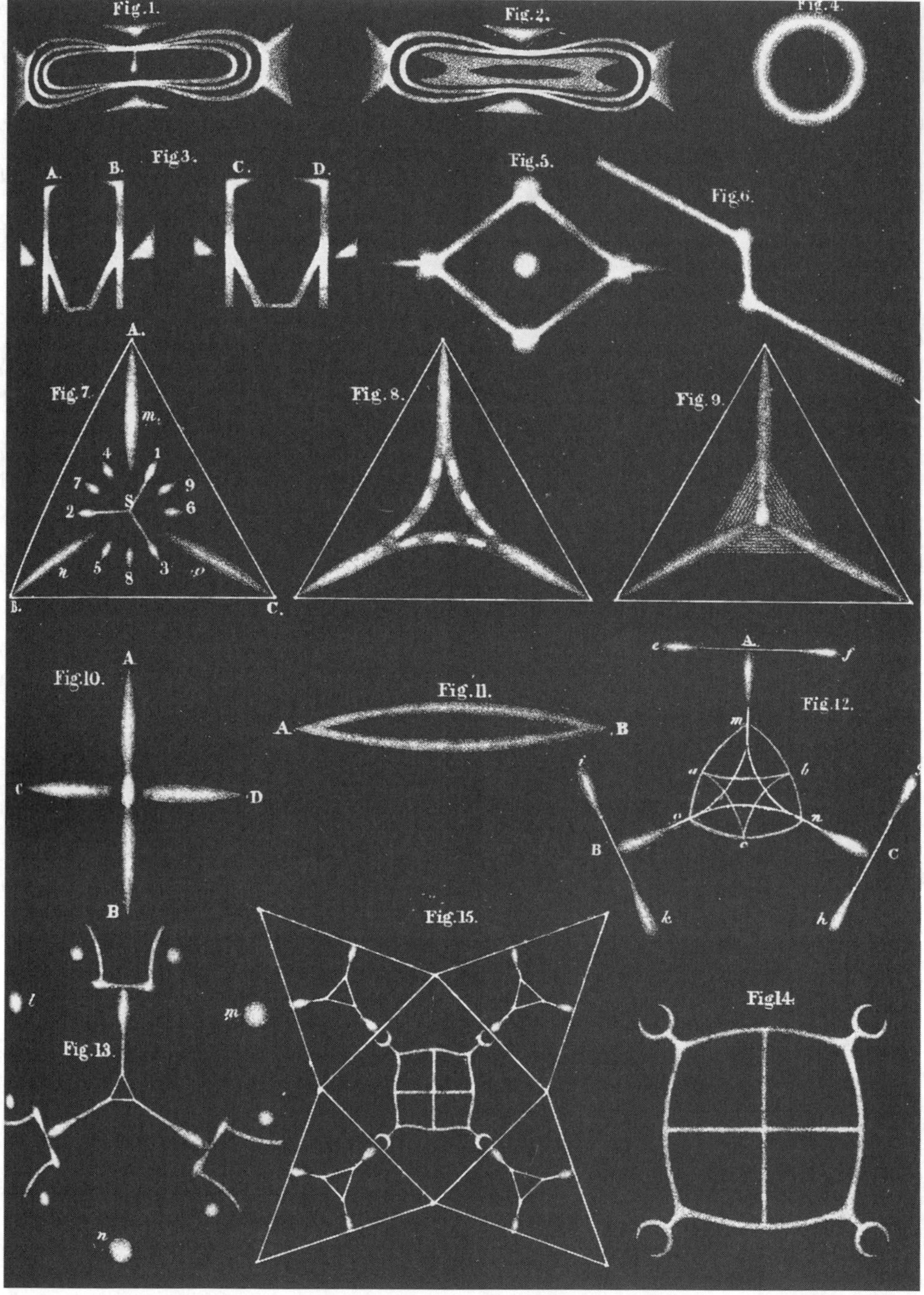

Optical figures produced by the disintegrated surfaces of crystals from the crystallographic research of Sir David Brewster. Published in *Transactions of the Royal Society of Edinburgh*, vol. 14 (1840): 166.

11. MARY FAIRFAX SOMERVILLE

(1780–1872)

Mary Somerville, born in Scotland, became a leading figure in the English and French scientific communities through her translations, popular accounts of scientific discoveries, and original scientific contributions on ultra-violet radiation. Her father, an English naval officer, discovered her studying algebra on her own as a teenager, and he forbade her from reading advanced mathematics, a subject he felt inappropriate for women. Her first husband, Samuel Greig, was equally unsympathetic to her self-education, but she nevertheless persisted and received welcome support from her second husband, Dr. William Somerville, who became a Royal Society Fellow in 1817. With these new scientific connections and encouragement, Somerville continued her studies in astronomy, mathematics, and the natural sciences which had been supported by her friends William Wallace and John Playfair.

In 1826, Somerville published a paper in the *Philosophical Transactions of the Royal Society* (116, p. 132) on magnetism and solar radiation entitled "On the Magnetizing Power of the More Refrangible Solar Rays." She wrote two other original papers on solar radiation, published by the Royal Society (1845) and in the *Comptes rendus* of the French Académie des Sciences (1836). Her work *The Mechanism of the Heavens* (1831), a translation of Laplace's *Mécanique céleste*, a treatise on Newtonian celestial mechanics, was widely acclaimed and was used as a textbook in British and American universities. In 1835, she and Caroline Herschel, the famed observational astronomer and sister of William Herschel, were elected to the Royal Astronomical Society.

Somerville's second book, *On the Connexion of the Physical Sciences* (1834), sought to demonstrate to a popular audience the recent developments in mathematical astronomy, sound, electricity, magnetism, and light. An interesting tension in the work—also present in the scientific community at the time—arises between her desire to define and categorize the physical sciences, while at the same time to demonstrate how some phenomena transcended disciplinary boundaries. Elizabeth Patterson locates this tension in Somerville's phenomenal approach: "Her chief interest centers always on phenomena and the explanations for them. Her search is for connections between the physical sciences, not for categories" (*Cultivation* 129). Sir David Brewster reviewed *On the Connexion* favourably in the *Edinburgh Review* in 1834, although he complained of the lack of diagrams, which he saw as a serious flaw in a popular scientific exposition. He obviously recognized her rare mathematical gifts, and exhorted her to turn towards original scientific research: "Although," he wrote, "the manner in which the work is executed would justify us in expressing a wish that Mrs Somerville would undertake a series of separate treatises on the Physical Sciences, yet we feel some reluctance in making such a demand upon one whose intellectual capacities are fitted for higher labours. Mrs. Somerville's great mathematical acquirements, her correct and profound knowledge of the principles of physical science, and the talent for original enquiry which she has already evinced in her paper on the magnetism of the violet rays, induce us to urge her to original investigation in some of the more elegant departments of science" (170–71).

In *On the Connexion*, Somerville stresses the interconnection of the physical sciences and the universality of physical law, specifically gravitation. Although she omits complicated mathematical formulae, the book is more than a popularization of scientific material and was used by practising scientists to keep abreast of new discoveries outside their own fields. Charles Lyell, for example, requested an advance copy of the *Connexion* from Somerville's publisher John Murray while he was revising his own *Principles of Geology*. Michael Faraday, William Whewell, and Charles Lyell were all consulted about the volume and Faraday provided a detailed critique of her sections on electricity and magnetism. The sections reprinted below clearly reveal Somerville's general thoughts on science and her philosophical influences: Newtonian mechanics, Deism, and natural theology. Her specific comments on gravity in the last section also reveal her desire that a unified theory of force will one day be discovered.

Somerville's third book, a textbook on *Physical Geography* (1848), brought widespread international attention, and many rewards. Her final book, *On Molecular and Microscopic Science* (1869), published when she was 89, was unfortunately outdated and not well regarded at the time. Somerville opened up many professional avenues for women in science during the nineteenth century, and Somerville College at Oxford was named in 1879 to honour her achievements.

On the Connexion of the Physical Sciences.
London: John Murray, 1834, pp. 1–5; 407–14.

Section I.

All the knowledge we possess of external objects is founded upon experience, which furnishes facts; and the comparison of these facts establishes relations, from which induction, the intuitive belief that like causes will produce like effects, leads to general laws. Thus, experience teaches that bodies fall at the surface of the earth with an accelerated velocity, and with a force proportional to their masses. By comparison, Newton proved that the force which occasions the fall of bodies at the earth's surface, is identical with that which retains the moon in her orbit; and induction led him to conclude that, as the moon is kept in her orbit by the attraction of the earth, so the planets might be retained in their orbits by the attraction of the sun. By such steps he was led to the discovery of one of those powers with which the Creator has ordained that matter should reciprocally act upon matter.

Physical astronomy is the science which compares and identifies the laws of motion observed on earth with the motions that take place in the heavens; and which traces, by an uninterrupted chain of deduction from the great principle that governs the universe, the revolutions and rotations of the planets, and the oscillations of the fluids at their surfaces; and which estimates the changes the system has hitherto undergone, or may hereafter experience—changes which require millions of years for their accomplishment.

The accumulated efforts of astronomers, from the earliest dawn of civilization, have been necessary to establish the mechanical theory of astronomy. The courses of the planets have been observed for ages with a degree of perseverance that is astonishing, if we consider the imperfection and even the want of instruments. The real motions of the earth have been separated from the apparent motions of the planets; the laws of the planetary revolutions have been discovered; and the discovery of these laws has led to the knowledge of the gravitation of matter. On the other hand, descending from the principle of gravitation, every motion in the solar system has been so completely explained, that the account of no astronomical phenomenon can now be transmitted to posterity of which the laws have not been determined.

Science, regarded as the pursuit of truth, which can only be attained by patient and unprejudiced investigation, wherein nothing is too great to be attempted, nothing so minute as to be justly disregarded, must ever afford occupation of consummate interest and subject of elevated meditation. The contemplation of the works of creation elevates the mind to the admiration of whatever is great and noble; accomplishing the object of all study,—which, in the elegant language of Sir James Mackintosh, 'is to inspire the love of truth, of wisdom, of beauty, especially of goodness, the highest beauty, and of that supreme and eternal Mind, which contains all truth and wisdom, all beauty and goodness. By the love or delightful contemplation and pursuit of these transcendent aims, for their own sake only, the mind of man is raised from low and perishable objects, and prepared for those high destinies which are appointed for all those who are capable of them.'

The heavens afford the most sublime subject of study which can be derived from science. The magnitude and splendour of the objects, the inconceivable rapidity with which they move, and the enormous distances between them, impress the mind with some notion of the energy that maintains them in their motions with a durability to which we can see no limit. Equally conspicuous is the goodness of the great First Cause, in having endowed man with faculties by which he can not only appreciate the magnificence of His works, but trace, with precision, the operation of his laws; use the globe he inhabits as a base wherewith to measure the magnitude and distance of the sun and planets, and make the diameter of the earth's orbit the first step of a scale by which he may ascend to the starry firmament. Such pursuits, while they ennoble the mind, at the same time inculcate humility, by showing that there is a barrier which no energy, mental or physical, can ever enable us to pass: that however profoundly we may penetrate the depths of space, there still remain innumerable systems, compared with which those apparently so vast must dwindle into insignificance, or even become invisible; and that not only man, but the globe he inhabits,—nay, the whole system of which it forms so small a part,—might be annihilated, and its extinction be unperceived in the immensity of creation.

Although it must be acknowledged that a complete acquaintance with physical astronomy can be attained by those only who are well versed in the higher branches of mathematical and mechanical science, and that they alone can appreciate the extreme beauty of the results, and of the means by which these results are obtained, it is nevertheless true that a sufficient skill in analysis to follow the general outline,—to see the mutual dependence of the different parts of the system, and to comprehend by what means some of the most extraordinary conclusions have been arrived at,—is within the reach of many who shrink from the task, appalled by difficulties, which, perhaps, are not more formidable than those incident to the study of the elements of every branch of knowledge; and who possibly overrate them from disregarding the distinction between the degree of mathematical acquirement necessary for making discoveries, and that which is requisite for understanding what others have done. That the study of mathematics, and their application to astronomy, are full of interest, will be allowed by all who have devoted their time and attention to these pursuits; and they only can estimate the delight of arriving at the truths they disclose, whether it be in the discovery of a world or of a new property of numbers.

Section XXXVII.

The known quantity of matter bears a very small proportion to the immensity of space. Large as the bodies are, the distances which separate them are immeasurably greater; but as design is manifest in every part of creation, it is probable that, if the various systems in the universe had been nearer to one another, their mutual disturbances would have been inconsistent with the harmony and stability of the whole. It is clear that space is not pervaded by atmospheric air, since its resistance would, long ere this, have destroyed the velocity of the planets; neither can we affirm it to be a void, since it is replete with ether, and traversed in all directions by light, heat, gravitation, and possibly by influences whereof we can form no idea.

Whatever the laws may be that obtain in the more distant regions of creation, we are assured that one alone regulates the motions not only of our own system, but also the binary systems of the fixed stars; and as general laws form the ultimate object of philosophical research, we cannot conclude these remarks without considering the nature of gravitation—that extraordinary power whose effects we have been endeavouring to trace through some of their mazes. It was at one time imagined that the acceleration in the moon's mean motion was occasioned by the successive transmission of the gravitating force; but it has been proved that, in order to produce this effect, its velocity must be about fifty millions of times greater than that of light, which flies at the rate of 200000 miles in a second: its action, even at the distance of the sun, may therefore be regarded as instantaneous; yet so remote are the nearest of the fixed stars, that it may be doubted whether the sun has any sensible influence on them.

The curves in which the celestial bodies move by the force of gravitation are only lines of the second order; the attraction of spheroids, according to any other law of force than that of gravitation, would be much more complicated; and as it is easy to prove that matter might have been moved according to an infinite variety of laws, it may be concluded that gravitation must have been selected by Divine Wisdom out of an infinity of others, as being the most simple, and that which gives the greatest stability to the celestial motions.

It is a singular result of the simplicity of the laws of nature, which admit only of the observation and comparison of ratios, that the gravitation and theory of the motions of the celestial bodies are independent of their absolute magnitudes and distances; consequently, if all the bodies of the solar system, their mutual distances, and their velocities, were to diminish proportionally, they would describe curves in all respects similar to those in which they now move; and the system might be successively reduced to the smallest sensible dimensions, and still exhibit the same appearances. We learn by experience that a very different law of attraction prevails when the particles of matter are placed within inappreciable distances from each other, as in chemical and capillary attraction and the attraction of cohesion: whether it be a modification of gravity, or that some new and unknown power comes into action, does not appear; but as a change in the law of the force takes place at one end of the scale, it is possible that gravitation may not remain the same throughout every part of space. Perhaps the day may come when even gravitation, no longer regarded as an ultimate principle, may be resolved into a yet more general cause, embracing every law that regulates the material world.

The action of the gravitating force is not impeded by the intervention even of the densest substances. If the attraction of the sun for the centre of the earth, and of the hemisphere diametrically opposite to him, were diminished by a difficulty in penetrating the interposed matter, the tides would be more obviously affected. Its attraction is the same also, whatever the substances of the celestial bodies may be; for if the action of the sun upon the earth differed by a millionth part from his action upon the moon, the difference would occasion a periodical variation in the moon's parallax whose maximum would be the 1/15 of a second, and also a variation in her longitude amounting to several seconds, a supposition proved to be impossible, by the agreement of theory with observation. Thus all matter is pervious to gravitation, and is equally attracted by it.

As far as human knowledge extends, the intensity of gravitation has never varied within the limits of the solar system; nor does even analogy lead us to expect that it should; on the contrary, there is every reason to be assured that the great laws of the universe are immutable, like their

Author. Not only the sun and planets, but the minutest particles, in all the varieties of their attractions and repulsions,—nay, even the imponderable matter of the electric, galvanic, or magnetic fluid,—are all obedient to permanent laws, though we may not be able in every case to resolve their phenomena into general principles. Nor can we suppose the structure of the globe alone to be exempt from the universal fiat, though ages may pass before the changes it has undergone, or that are now in progress, can be referred to existing causes with the same certainty with which the motions of the planets, and all their periodic and secular variations, are referable to the law of gravitation. The traces of extreme antiquity perpetually occurring to the geologist give that information as to the origin of things in vain looked for in the other parts of the universe. They date the beginning of time with regard to our system; since there is ground to believe that the formation of the earth was contemporaneous with that of the rest of the planets; but they show that creation is the work of Him with whom 'a thousand years are as one day, and one day as a thousand years.'

It thus appears that the theory of dynamics, founded upon terrestrial phenomena, is indispensable for acquiring a knowledge of the revolutions of the celestial bodies and their reciprocal influences. The motions of the satellites are affected by the forms of their primaries, and the figures of the planets themselves depend upon their rotations. The symmetry of their internal structure proves the stability of these rotatory motions, and the immutability of the length of the day, which furnishes an invariable standard of time; and the actual size of the terrestrial spheroid affords the means of ascertaining the dimensions of the solar system, and provides an invariable foundation for a system of weights and measures. The mutual attraction of the celestial bodies disturbs the fluids at their surfaces, whence the theory of the tides and the oscillations of the atmosphere. The density and elasticity of the air, varying with every alternation of temperature, lead to the consideration of barometrical changes, the measurement of heights, and capillary attraction; and the doctrine of sound, including the theory of music, is to be referred to the small undulations of the aërial medium. A knowledge of the action of matter upon light is requisite for tracing the curved path of its rays through the atmosphere, by which the true places of distant objects are determined, whether in the heavens or on the earth. By this we learn the nature and properties of the sunbeam, the mode of its propagation through the etherial fluid, or in the interior of material bodies, and the origin of colour. By the eclipses of Jupiter's satellites, the velocity of light is ascertained, and that velocity, in the aberration of the fixed stars, furnishes the only direct proof of the real motion of the earth. The effects of the invisible rays of light are immediately connected with chemical action; and heat, forming a part of the solar ray, so essential to animated and inanimated existence, whether considered as invisible light or as a distinct quality, is too important an agent in the economy of creation not to hold a principal place in the order of physical science. Whence follows its distribution over the surface of the globe, its power on the geological convulsions of our planet, its influence on the atmosphere and on climate, and its effects on vegetable and animal life, evinced in the localities of organized beings on the earth, in the waters, and in the air. The connexion of heat with electrical phenomena, and the electricity of the atmosphere, together with all its energetic effects, its identity with magnetism and the phenomena of terrestrial polarity, can only be understood from the theories of these invisible agents, and are probably principal causes of chemical affinities. Innumerable instances might be given in illustration of the immediate connexion of the physical sciences, most of which are united still more closely by the common bond of analysis which is daily extending its empire, and will ultimately embrace almost every subject in nature in its formulæ.

These formulæ, emblematic of Omniscience, condense into a few symbols the immutable laws of the universe. This mighty instrument of human power itself originates in the primitive constitution of the human mind, and rests upon a few fundamental axioms which have eternally existed in Him who implanted them in the breast of man when He created him after His own image.

FURTHER READING

Brewster, Sir David. "On the Connexion of the Physical Sciences, by Mrs. Somerville." *Edinburgh Review* 59, no. 119 (1834): 154–71.

Brück, Mary T. "Mary Somerville, Mathematician and Astronomer of Underused Talents." *Journal of the British Astronomical Association* 106 (1996): 201–6.

Hamer, Sarah Sharp. *Mrs. Somerville and Mary Carpenter.* London: Cassell and Co., 1887.

McKinlay, Jane. *Mary Somerville, 1780–1872.* Edinburgh: University of Edinburgh Press, 1987.

Patterson, Elizabeth Chambers. "The Case of Mary Somerville: An Aspect of Nineteenth-Century Science." *Proceedings of the American Philosophical Society* 118, no. 3 (1974): 269–75.

—. "Mary Somerville." *The British Journal for the History of Science* 4, part 4, no. 16 (1969): 311–39.

—. *Mary Somerville and the Cultivation of Science, 1815–1840.* Boston: Martinus Nijhoff Publishers, 1983.

—. *Mary Somerville, 1780–1872.* Oxford: Somerville College, 1979.

Proctor, Richard A. *Light Science for Leisure Hours: Familiar Essays on Scientific Subjects, Natural Phenomenon, etc. with a Sketch of the Life of Mary Somerville.* 2d Series. London: Longmans, Green, and Co., 1882.

Somerville, Martha. *Personal Recollections from Early Life to Old Age of Mary Somerville, with Selections from her Correspondence.* London: John Murray, 1873.

Walford, L.B. *Four Biographies from 'Blackwood': Jane Taylor, Hannah More, Elizabeth Fry, Mary Somerville.* Edinburgh: William Blackwood and Sons, 1888.

Weitzenhoffer, Kenneth. "The Education of Mary Somerville." *Sky and Telescope* 73, no. 2 (1987): 138–9.

12. THEODOR SCHWANN

(1810–1882)

Theodor Schwann is credited along with Matthias Schleiden with developing the cell theory. William Coleman emphasizes the importance of this theory for the subsequent development of biology: "After mid-century the cell had become for the great majority of biologists the essential structural reference point for the interpretation of organic form" (*Biology* 17). It is difficult to imagine a time when the idea that plants and animals are composed of cells was not common, but staining techniques in the early nineteenth century were not thoroughly developed and the poor resolution of microscopes did not allow researchers to distinguish microcellular structures clearly, even though cork cells (cavities) had been observed by Robert Hooke and blood corpuscles by Swammerdam as early as the seventeenth century. Even when the cell was recognized as an elementary structure, few physiologists in Schwann's day imagined that cellular growth, structure, and metabolism played such a dominant role in the general functioning of an organism (Duchesneau 139).

Because of his retiring and modest disposition, there is not a great deal of biographical information on Theodor Schwann: from an early age he became aware of his extreme timidity and introspective nature. In 1826, Schwann entered the Jesuit College at Cologne known as the Tricoronatum. He came under the influence of the religious doctrines of Wilhelm Smets and remained extremely pious throughout his life. In 1829 he studied at the University of Bonn with Johannes Müller (1801–58), a confirmed vitalist and author and editor of the *Handbuch der Physiologie des Menschen*. Schwann eventually followed Müller to Berlin in 1833 as his doctoral student.

Schwann investigated quantitatively the physical laws of muscle contraction and also discovered the digestive enzyme pepsin. He was also the first to identify the glial cells surrounding the axons of neurons. These cells serve to facilitate nerve transmission and are now known as Schwann cells. In the 1830s he carried out experiments on infusoria and fermentation. By 1836 Schwann was convinced that alcoholic fermentation was caused by a living being (yeast) rather than by oxidation, as the prevailing theory of putrefaction suggested. In a satirical article of 1839, Friedrich Wöhler and Justus von Liebig ridiculed Schwann's views on fermentation; Liebig's own theory of chemical ferments activated by oxygen and acting on vegetable sugars held immense power in scientific circles until challenged by Pasteur. Schwann was turned down for a requested teaching post in physiology at the University of Bonn, undoubtedly due in part to the hostility to his ideas from Germany's established scientists; he instead occupied the chair of anatomy at the University of Louvain, Belgium from 1839–48.

Matthias Jakob Schleiden (1804–81), drawing on the work of the "globulist" theorists as well as Robert Brown, who described the cell nucleus in 1832, theorized that plants were composed of cells possessing a nucleus which precedes the formation of the cell. After Schwann noticed similar structures in a variety of animal cells, it was a logical step for him to argue in *Mikroskopische Untersuchungen über die Uebereinstimmung in der Struktur und dem Wachstum der Thiere und Pflanzen* (*Microscopical Researches into the Accordance in the Structure and Growth of Animals and Plants*) that all animal tissues were composed of cells just like plants. Schwann invited Schleiden to the operating theatre and demonstrated to him the resemblance between plant nuclei and nuclei in cells of the chorda dorsalis (notochord). This meeting represented a turning point in Schwann's development of the cell theory: "from that moment all my efforts were directed to prove the pre-existence of the nucleus in the cell" (Watermann 99).

Schwann, expanding Schleiden's ideas, further hypothesized that all animal tissues essentially consisted of nucleated cells, formed not from pre-existing cells, but within a fluid "cytoblastema" through a process of accretion and deposition beginning with the appearance of a nucleolus, a nucleus, and then a cell wall. Schwann further suggested that cell production was analogous to the inorganic process of crystallization. Rudolf Virchow later disproved several aspects of Schwann's cell theory in his *Die Cellularpathologie* (1858); Virchow epitomized his opposition to Schwann in the dictum *omnis cellula e cellula* (all cells from cells), after abundant proof from Robert Remak and from his own work showed that cells arise from the division (*Theilung*) of other cells, not in the way Schwann had hypothesized.

Schwann and Schleiden's cell theory also stated that cells develop according to the same universal laws which govern molecular interactions, not from a unifying life force. Rejecting the vitalism of Müller, Schwann argued that cells were formed by natural and material laws resulting from an internal physical principle, not a generalized vital force in the organism. Rudolf Virchow and Johannes Müller in his work on tumours extended cell theory into medical practice by developing the area of cellular pathology. Schwann was awarded the Royal Society's Copley Medal for 1845 for his cell work, and Henry Smith translated the *Mikroskopische Untersuchungen* for the Sydenham Society in 1847.

From 1848–80, Schwann occupied the chair of anatomy at the University of Liège. There he became involved in designing equipment for use in the mining industry. At the 1876 Health and Safety Exposition in Brussels, Schwann demonstrated a closed circuit breathing apparatus employing compressed air with hydrated calcium oxide to absorb respired carbon dioxide. The device anticipated modern breathing apparatus such as SCUBA as well as scientific instruments, including Léon Fredericq's oxygénographe, for measuring basal metabolism. Technology for the reabsorption of respired CO_2—important in anesthesiology as well as in the life support systems of space craft—has benefitted from Schwann's pioneering research. He began work on an unpublished extension of the *Mikroskopische Untersuchungen* in an attempt to unify his scientific, religious, and philosophical reflections on life, but died before its completion.

Microscopical Researches into the Accordance in the Structure and Growth of Animals and Plants.

Translated by Henry Smith. London: The Sydenham Society, 1847, pp. 186–215.

Theory of the Cells.

The whole of the foregoing investigation has been conducted with the object of exhibiting from observation alone the mode in which the elementary parts of organized bodies are formed. Theoretical views have been either entirely excluded, or where they were required (as in the foregoing retrospect of the cell-life), for the purpose of rendering facts more clear, or preventing subsequent repetitions, they have been so presented that it can be easily seen how much is observation and how much argument. But a question inevitably arises as to the basis of all these phenomena; and an attempt to solve it will be more readily permitted us, since by making a marked separation between theory and observation the hypothetical may be clearly distinguished from that which is positive. An hypothesis is never prejudicial so long as we are conscious of the degree of reliance which may be placed upon it, and of the grounds on which it rests. Indeed it is advantageous, if not necessary for science, that when a certain series of phenomena is proved by observation, some provisional explanation should be conceived that will suit them as nearly as possible, even though it be in danger of being overthrown by subsequent observations; for it is only in this manner that we are rationally led to new discoveries, which either establish or refute the explanation. It is from this point of view I would beg that the following theory of organization may be regarded; for the inquiry into the source of development of the elementary parts of organisms is, in fact, identical with the theory of organized bodies.

The various opinions entertained with respect to the fundamental powers of an organized body may be reduced to two, which are essentially different from one another. The first is, that every organism originates with an inherent power, which models it into conformity with a predominant idea, arranging the molecules in the relation necessary for accomplishing certain purposes held forth by this idea. Here, therefore, that which arranges and combines the molecules is a power acting with a definite purpose. A power of this kind would be essentially different from all the powers of inorganic nature, because action goes on in the latter quite blindly. A certain impression is followed of necessity by a certain change of quality and quantity, without regard to any purpose. In this view, however, the fundamental power of the organism (or the soul, in the sense employed by Stahl) would, inasmuch as it works with a definite individual purpose, be much more nearly allied to the immaterial principle, endued with consciousness which we must admit operates in man.

The other view is, that the fundamental powers of organized bodies agree essentially with those of inorganic nature, that they work altogether blindly according to laws of necessity and irrespective of any purpose, that they are powers which are as much established with the existence of matter as the physical powers are. It might be assumed that the powers which form organized bodies do not appear at all in inorganic nature, because this or that particular combination of molecules, by which the powers are elicited, does not occur in inorganic nature, and yet they might not be essentially distinct from physical and chemical powers. It cannot, indeed, be denied that adaptation to a particular purpose, in some individuals even in a high degree, is characteristic of every organism; but, according to this view, the source of this adaptation does not depend upon each organism being developed by the operation of its own power in obedience to that purpose, but it originates as in inorganic nature, in the creation of the matter with its blind powers by a rational Being. We know, for instance, the powers which operate in our planetary system. They operate, like all physical powers, in accordance with blind laws of necessity, and yet is the planetary system remarkable for its adaptation to a purpose. The ground of this adaptation does not lie in the powers, but in Him, who has so constituted matter with its powers, that in blindly obeying its laws it produces a whole suited to fulfil an intended purpose. We may even assume that the planetary system has an individual adaptation to a purpose. Some external influence, such as a comet, may occasion disturbances of motion, without thereby bringing the whole into collision; derangements may occur on single planets, such as a high tide, etc., which are yet balanced entirely by physical laws. As respects their adaptation to a purpose, organized bodies differ from these in degree only; and by this second view we are just as little compelled to conclude that the fundamental powers of organization operate according to laws of adaptation to a purpose, as we are in inorganic nature.

The first view of the fundamental powers of organized bodies may be called the *teleological*, the second the *physical* view. An example will show at once, how important for physiology is the solution of the question as to which is to be followed. If, for instance, we define inflammation and suppuration to be the effort of the organism to remove a foreign body that has been introduced into it; or fever to be the effort of the organism to eliminate diseased matter, and both as the result of the "autocracy of the organism," then these explanations accord with the teleological view. For, since by these processes the obnoxious matter is actually removed, the process which effects them is one adapted to an end; and as the fundamental power of the organism operates in accordance with definite purposes, it may either set these processes in action primarily, or may also summon further powers of matter to its aid, always, however, remaining itself the "primum movens." On the other hand, according to the physical view, this is just as little an explanation as it would be to say, that the motion of the earth around the sun is an effort of the fundamental power of the planetary system to produce a change of seasons on the planets, or to say, that ebb and flood are the reaction of the organism of the earth upon the moon.

In physics, all those explanations which were suggested by a teleological view of nature, as "horror vacui," and the like, have long been discarded. But in animated nature, adaptation—individual adaptation—to a purpose is so prominently marked, that it is difficult to reject all teleological explanations. Meanwhile it must be remembered that those explanations, which explain at once all and nothing, can be but the last resources, when no other view can possibly be adopted; and there is no such necessity for admitting the teleological view in the case of organized bodies. The adaptation to a purpose which is characteristic of organized bodies differs only in degree from what is apparent also in the inorganic part of nature; and the explanation that organized bodies are developed, like all the phenomena of inorganic nature, by the operation of blind laws framed with the matter, cannot be rejected as impossible. Reason certainly requires some ground for such adaptation, but for her it is sufficient to assume that matter with the powers inherent in it owes its existence to a rational Being. Once established and preserved in their integrity, these powers may, in accordance with their immutable laws of blind necessity, very well produce combinations, which manifest, even in a high degree, individual adaptation to a purpose. If, however, rational power interpose after creation merely to sustain, and not as an immediately active agent, it may, so far as natural science is concerned, be entirely excluded from the consideration of the creation.

But the teleological view leads to further difficulties in the explanation, and especially with respect to generation. If we assume each organism to be formed by a power which acts according to a certain predominant idea, a portion of this power may certainly reside in the ovum during generation; but then we must ascribe to this subdivision of the original power, at the separation of the ovum from the body of the mother, the capability of producing an organism similar to that which the power, of which it is but a portion, produced: that is, we must assume that this power is infinitely divisible, and yet that each part may perform the same actions as the whole power. If, on the other hand, the power of organized bodies reside, like the physical powers, in matter as such, and be set free only by a certain combination of the molecules, as, for instance, electricity is set free by the combination of a zinc and copper plate, then also by the conjunction of molecules to form an ovum the power may be set free, by which the ovum is capable of appropriating to itself fresh molecules, and these newly-conjoined molecules again by this very mode of combination acquire the same power to assimilate fresh molecules. The first development of the many forms of organized bodies—the progressive formation of organic nature indicated by geology—is also much more difficult to understand according to the teleological than the physical view.

Another objection to the teleological view may be drawn from the foregoing investigation. The molecules, as we have seen, are not immediately combined in various ways, as the purpose of the organism requires, but the formation of the elementary parts of organic bodies is regulated by laws which are essentially the same for all elementary parts. One can see no reason why this should be the case, if each organism be endued with a special power to frame the parts according to the purpose which they have to fulfil: it might much rather be expected that the formative principle, although identical for organs physiologically the same, would yet in different tissues be correspondingly varied. This resemblance of the elementary parts has, in the instance of plants, already led to the conjecture that the cells are really the organisms, and that the whole plant is an aggregate of these organisms arranged according to certain laws. But since the elementary parts of animals bear exactly similar relations, the individuality of an entire animal would thus be lost; and yet precisely upon the individuality of the whole animal does the assumption rest, that it possesses a single fundamental power operating in accordance with a definite idea.

Meanwhile we cannot altogether lay aside teleological views if all phenomena are not clearly explicable by the physical view. It is, however, unnecessary to do so, because an explanation, according to the teleological view, is only admissible when the physical can be shown to be impossible. In any case it conduces much more to the object of science to strive, at least, to adopt the physical explanation. And I would repeat that, when speaking of a physical explanation of organic phenomena, it is not necessary to understand an explanation by known physical powers, such, for instance, as that universal refuge electricity, and the like; but an explanation by means of powers which operate like the physical powers, in accordance with strict laws of blind necessity, whether they be also to be found in inorganic nature or not.

We set out, therefore, with the supposition that an organized body is not produced by a fundamental power which is guided in its operation by a definite idea, but is developed, according to blind laws of necessity, by powers which, like those of inorganic nature, are established by the very existence of matter. As the elementary materials of organic nature are not different from those of the inorganic kingdom, the source of the organic phenomena can only reside in another combination of these materials, whether it be in a peculiar mode of union of the elementary atoms to form atoms of the second order, or in the arrangement of these conglomerate molecules when forming either the separate morphological elementary parts of organisms, or an entire organism. We have here to do with the latter question solely, whether the cause of organic phenomena lies in the whole organism, or in its separate elementary parts. If this question can be answered, a further inquiry still remains as to whether the organism or its elementary parts possess this power through the peculiar mode of combination of the conglomerate molecules, or through the mode in which the elementary atoms are united into conglomerate molecules.

We may, then, form the two following ideas of the cause of organic phenomena, such as growth, etc. First, that the cause resides in the totality of the organism. By the combination of the molecules into a systematic whole, such as the organism is in every stage of its development, a power is engendered, which enables such an organism to take up fresh material from without, and appropriate it either to the formation of new elementary parts, or to the growth of those already present. Here, therefore, the cause of the growth of the elementary parts resides in the totality of the organism. The other mode of explanation is, that growth does not ensue from a power resident in the entire organism, but that each separate elementary part is possessed of an independent power, an independent life, so to speak; in other words, the molecules in each separate elementary part are so combined as to set free a power by which it is capable of attracting new molecules, and so increasing, and the whole organism subsists only by means of the reciprocal[1] action of the single elementary parts. So that here the single elementary parts only exert an active influence on nutrition, and totality of the organism may indeed be a condition, but is not in this view a cause.

In order to determine which of these two views is the correct one, we must summon to our aid the results of the previous investigation. We have seen that all organized bodies are composed of essentially similar parts, namely, of cells; that these cells are formed and grow in accordance with essentially similar laws; and, therefore, that these processes must, in every instance, be produced by the same powers. Now, if we find that some of these elementary parts, not differing from the others, are capable of separating themselves from the organism, and pursuing an independent growth, we may thence conclude that each of the other elementary parts, each cell, is already possessed of power to take up fresh molecules and grow; and that, therefore, every elementary part possesses a power of its own, an independent life, by means of which it would be enabled to develop itself independently, if the relations which it bore to external parts were but similar to those in which it stands in the organism. The ova of animals afford us examples of such independent cells, growing apart from the organism. It may, indeed, be said of the ova of higher animals, that after impregnation the ovum is essentially different from the other cells of the organism; that by impregnation there is a something conveyed to the ovum, which is more to it than an external condition for vitality, more than nutrient matter; and that it might thereby have first received its peculiar vitality, and therefore that nothing can be inferred from it with respect to the other cells. But this fails in application to those classes which consist only of female individuals, as well as with the spores of the lower plants; and, besides, in the inferior plants any given cell may be separated from the plant, and then grow alone. So that here are whole plants consisting of cells, which can be positively proved to have independent vitality. Now, as all cells grow according to the same laws, and consequently the cause of growth cannot in one case lie in the cell, and in another in the whole organism; and since it may be further proved that some cells, which do not differ from the rest in their mode of growth, are developed independently, we must ascribe to all cells an independent vitality, that is, such combinations of molecules as occur in any single cell, are capable of setting free the power by which it is enabled to take up fresh molecules. The cause of nutrition and growth resides not in the organism as a whole, but in the separate elementary parts—the cells. The failure of growth in the case of any particular cell, when separated from an organized body, is as slight an objection to this theory, as it is an objection against the independent vitality of a bee, that it cannot continue long in existence after being separated from its swarm. The manifestation of the power which resides in the cell depends upon conditions to which it is subject only when in connexion with the whole (organism).

The question, then, as to the fundamental power of organized bodies resolves itself into that of the fundamental powers of the individual cells. We must now consider the general phenomena attending the formation of cells, in order to discover what powers may be presumed to exist in the cells to explain them. These phenomena may be arranged in two natural groups: first, those which relate to the combination of the molecules to form a cell, and which may be denominated the *plastic* phenomena of the cells; secondly, those which result from chemical changes either in the component particles of the cell itself, or in the surrounding cytoblastema, and which may be called *metabolic* phenomena (τὸ μεταβολικὸν, implying that which is liable to occasion or to suffer change).

The general plastic appearances in the cells are, as we have seen, the following: at first a minute corpuscle is formed, (the nucleolus); a layer of substance (the nucleus) is then precipitated around it, which becomes more thickened and expanded by the continual deposition of fresh

1 The word "reciprocal action" must here be taken in its widest sense, as implying the preparation of material by one elementary part, which another requires for its own nutrition.

molecules between those already present. Deposition goes on more vigorously at the outer part of this layer than at the inner. Frequently the entire layer, or in other instances the outer part of it only, becomes condensed to a membrane, which may continue to take up new molecules in such a manner that it increases more rapidly in superficial extent than in thickness, and thus an intervening cavity is necessarily formed between it and the nucleolus. A second layer (cell) is next precipitated around this first, in which precisely the same phenomena are repeated, with merely the difference that in this case the processes, especially the growth of the layer, and the formation of the space intervening between it and the first layer (the cell-cavity), go on more rapidly and more completely. Such were the phenomena in the formation of most cells; in some, however, there appeared to be only a single layer formed, while in others (those especially in which the nucleolus was hollow) there were three. The other varieties in the development of the elementary parts were (as we saw) reduced to these—that if two neighbouring cells commence their formation so near to one another that the boundaries of the layers forming around each of them meet at any spot, a common layer may be formed enclosing the two incipient cells. So at least the origin of nuclei, with two or more nucleoli, seemed explicable, by a coalescence of the first layers (corresponding to the nucleus), and the union of many primary cells into one secondary cell by a similar coalescence of the second layers (which correspond to the cell). But the further development of these common layers proceeds as though they were only an ordinary single layer. Lastly, there were some varieties in the progressive development of the cells, which were referable to an unequal deposition of the new molecules between those already present in the separate layers. In this way modifications of form and division of the cells were explained. And among the number of the plastic phenomena in the cells we may mention, lastly, the formation of secondary deposits; for instances occur in which one or more new layers, each on the inner surface of the previous one, are deposited on the inner surface of a simple or of a secondary cell.

These are the most important phenomena observed in the formation and development of cells. The unknown cause, presumed to be capable of explaining these processes in the cells, may be called the plastic power of the cells. We will, in the next place, proceed to determine how far a more accurate definition of this power may be deduced from these phenomena.

In the first place, there is a power of attraction exerted in the very commencement of the cell, in the nucleolus, which occasions the addition of new molecules to those already present. We may imagine the nucleolus itself to be first formed by a sort of crystallization from out of a concentrated fluid. For if a fluid be so concentrated that the molecules of the substance in solution exert a more powerful mutual attraction than is exerted between them and the molecules of the fluid in which they are dissolved, a part of the solid substance must be precipitated. One can readily understand that the fluid must be more concentrated when new cells are being formed in it than when those already present have merely to grow. For if the cell is already partly formed, it exerts an attractive force upon the substance still in solution. There is then a cause for the deposition of this substance, which does not co-operate when no part of the cell is yet formed. Therefore, the greater the attractive force of the cell is, the less concentration of the fluid is required; while, at the commencement of the formation of a cell, the fluid must be more than concentrated. But the conclusion which may be thus directly drawn, as to the attractive power of the cell, may also be verified by observation. Wherever the nutrient fluid is not equally distributed in a tissue, the new cells are formed in that part into which the fluid penetrates first, and where, consequently, it is most concentrated. Upon this fact, as we have seen, depended the difference between the growth of organized and unorganized tissues (see page 169). And this confirmation of the foregoing conclusion by experience speaks also for the correctness of the reasoning itself.

The attractive power of the cells operates so as to effect the addition of new molecules in two ways,—first, in layers, and secondly, in such a manner in each layer that the new molecules are deposited between those already present. This is only an expression of the fact; the more simple law, by which several layers are formed and the molecules are not all deposited between those already present, cannot yet be explained. The formation of layers may be repeated once, twice, or thrice. The growth of the separate layers is regulated by a law, that the deposition of new molecules should be greatest at the part where the nutrient fluid is most concentrated. Hence the outer part particularly becomes condensed into a membrane both in the layer corresponding to the nucleus and in that answering to the cell, because the nutrient fluid penetrates from without, and consequently is more concentrated at the outer than at the inner part of each layer. For the same reason the nucleus grows rapidly, so long as the layer of the cell is not formed around it, but it either stops growing altogether, or at least grows much more slowly so soon as the cell-layer has surrounded it; because then the latter receives the nutrient matter first, and, therefore, in a more concentrated form. And hence the cell becomes, in a general sense, much more completely developed, while the nucleus-layer usually remains at a stage of development, in which the cell-layer had been in its earlier period. The addition of new molecules is so arranged that the layers increase more considerably in superficial extent than in thickness; and thus an intervening space is formed between each layer and the one preceding it, by which cells and nuclei are formed into actual hollow vesicles. From this it may be inferred that the deposition of new molecules is more active between those which lie side by side along the surface of the membrane, than between those which lie one upon the other in its thickness. Were it otherwise, each layer would increase in thickness, but there would be no intervening cavity between it and the previous one, there would be no vesicles, but a solid body composed of layers.

Attractive power is exerted in all the solid parts of the cell. This follows, not only from the fact that new molecules may be deposited everywhere between those already present, but also from the formation of secondary deposits. When the cavity of a cell is once formed, material may be also attracted from its contents and deposited in layers; and as this deposition takes place upon the inner surface of the membrane of the cell, it is probably that which exerts the attractive influence. This formation of layers on the inner surface of the cell-membrane is, perhaps, merely a repetition of the same process by which, at an earlier period, nucleus and cell were precipitated as layers around the nucleolus. It must, however, be remarked that the identity of these two processes cannot be so clearly proved as that of the processes by which nucleus and cell are formed; more especially as there is a variety in the phenomena, for the secondary deposits in plants occur in spiral forms, while this has at least not yet been demonstrated in the formation of the cell-membrane and the nucleus, although by some botanical writers the cell-membrane itself is supposed to consist of spirals.

The power of attraction may be uniform throughout the whole cell, but it may also be confined to single spots; the deposition of new molecules is then more vigorous at these spots, and the consequence of this uneven growth of the cell-membrane is a change in the form of the cell.

The attractive power of the cells manifests a certain form of election in its operation. It does not take up all the substances contained in the surrounding cytoblastema, but only particular ones, either those which are analogous with the substance already present in the cell (assimilation), or such as differ from it in chemical properties. The several layers grow by assimilation, but when a new layer is being formed, different material from that of the previously-formed layer is attracted: for the nucleolus, the nucleus and cell-membrane are composed of materials which differ in their chemical properties.

Such are the peculiarities of the plastic power of the cells, so far as they can as yet be drawn from observation. But the manifestations of this power presuppose another faculty of the cells. The cytoblastema, in which the cells are formed, contains the elements of the materials of which the cell is composed, but in other combinations: it is not a mere solution of cell-material, but it contains only certain organic substances in solution. The cells, therefore, not only attract materials from out of the cytoblastema, but they must have the faculty of producing chemical changes in its constituent particles. Besides which, all the parts of the cell itself may be chemically altered during the process of its vegetation. The unknown cause of all these phenomena, which we comprise under the term metabolic phenomena of the cells, we will denominate the *metabolic power*.

The next point which can be proved is, that this power is an attribute of the cells themselves, and that the cytoblastema is passive under it. We may mention vinous fermentation[2] as an instance of this. A decoction of malt will remain for a long time unchanged; but as soon as some yeast is added to it, which consists partly of entire fungi and partly of a number of single cells, the chemical change immediately ensues. Here the decoction of malt is the cytoblastema; the cells clearly exhibit activity, the cytoblastema, in this instance even a boiled fluid, being quite passive during the change. The same occurs when any simple cells, as the spores of the lower plants, are sown in boiled substances.

In the cells themselves again, it appears to be the solid parts, the cell-membrane and the nucleus, which produce the change. The contents of the cell undergo similar and even more various changes than the external cytoblastema, and it is at least probable that these changes originate with the solid parts composing the cells, especially the cell-membrane, because the secondary deposits are formed on the inner surface of the cell-membrane, and other precipitates are generally formed in the first instance around the nucleus. It may therefore, on the whole, be said that the solid component particles of the cells possess the power of chemically altering the substances in contact with them.

The substances which result from the transformation of the contents of the cell are different from those which are produced by change in the external cytoblastema. What is the cause of this difference, if the metamorphosing power of the cell-membrane be limited to its immediate neigh-

2 I could not avoid bringing forward fermentation as an example, because it is the best known illustration of the operation of the cells, and the simplest representation of the process which is repeated in each cell of the living body. Those who do not as yet admit the theory of fermentation set forth by Cagniard-Latour, and myself, may take the development of any simple cells, especially of the spores, as an example; and we will in the text draw no conclusion from fermentation which cannot be proved from the development of other simple cells which grow independently, particularly the spores of the inferior plants. We have every conceivable proof that the fermentation-granules are fungi. Their form is that of fungi; in structure they, like them, consist of cells, many of which enclose other young cells. They grow, like fungi, by the shooting forth of new cells at their extremities; they propagate like them, partly by the separation of distinct cells, and partly by the generation of new cells within those already present, and the bursting of the parent-cells. Now, that these fungi are the cause of fermentation, follows, first, from the constancy of their occurrence during the process; secondly, from the cessation of fermentation under any influences by which they are known to be destroyed, especially boiling heat, arseniate of potass, etc. ; and, thirdly, because the principle which excites the process of fermentation must be a substance which is again generated and increased by the process itself, a phenomenon which is met with only in living organisms. Neither do I see how any further proof can possibly be obtained otherwise than by chemical analysis, unless it can be proved that the carbonic acid and alcohol are formed only at the surface of the fungi. I have made a number of attempts to prove this, but they have not as yet completely answered the purpose. A long test-tube was filled with a weak solution of sugar, coloured of a delicate blue with litmus, and a very small quantity of yeast was added to it, so that fermentation might not begin until several hours afterwards, and the fungi, having thus previously settled at the bottom, the fluid might become clear. When the carbonic acid (which remained in solution) commenced to be formed, the reddening of the blue fluid actually began at the bottom of the tube. If at the beginning a rod were put into the tube, so that the fungi might settle upon it also, the reddening began both at the bottom, and upon the rod. This proves, at least, that an undissolved substance which is heavier than water gives rise to fermentation; and the experiment was next repeated on a small scale under the microscope, to see whether the reddening really proceeded from the fungi, but the colour was too pale to be distinguished, and when the fluid was coloured more deeply no fermentation ensued; meanwhile, it is probable that a reagent upon carbonic acid may be found which will serve for microscopic observation, and not interrupt fermentation. The foregoing inquiry into the process by which organized bodies are formed, may perhaps, however, serve in some measure to recommend this theory of fermentation to the attention of chemists.

bourhood merely? Might we not much rather expect that converted substances would be found without distinction on the inner as on the outer surface of the cell-membrane? It might be said that the cell-membrane converts the substance in contact with it without distinction, and that the variety in the products of this conversion depends only upon a difference between the convertible substance contained in the cell and the external cytoblastema. If it be true that the cell-membrane, which at first closely surrounds the nucleus, expands in the course of its growth, so as to leave an interspace between it and the cell, and that the contents of the cell consist of fluid which has entered this space merely by imbibition, they cannot differ essentially from the external cytoblastema. I think therefore that, in order to explain the distinction between the cell-contents and the external cytoblastema, we must ascribe to the cell-membrane not only the power in general of chemically altering the substances which it is either in contact with, or has imbibed, but also of so separating them that certain substances appear on its inner, and others on its outer surface. The secretion of substances already present in the blood, as, for instance, of urea, by the cells with which the urinary tubes are lined, cannot be explained without such a faculty of the cells. There is, however, nothing so very hazardous in it, since it is a fact that different substances are separated in the decompositions produced by the galvanic pile. It might perhaps be conjectured from this peculiarity of the metabolic phenomena in the cells, that a particular position of the axes of the atoms composing the cell-membrane is essential for the production of these appearances.

Chemical changes occur, however, not only in the cytoblastema and the cell-contents, but also in the solid parts of which the cells are composed, particularly the cell-membrane. Without wishing to assert that there is any intimate connexion between the metabolic power of the cells and galvanism, I may yet, for the sake of making the representation of the process more clear, remark that the chemical changes produced by a galvanic pile are accompanied by corresponding changes in the pile itself.

The more obscure the cause of the metabolic phenomena in the cells is, the more accurately we must mark the circumstances and phenomena under which they occur. One condition to them is a certain temperature, which has a maximum and a minimum. The phenomena are not produced in a temperature below 0 degrees or above 80 degrees R.; boiling heat destroys this faculty of the cells permanently; but the most favorable temperature is one between 10 degrees and 32 degrees R. Heat is evolved by the process itself.

Oxygen, or carbonic acid, in a gaseous form or lightly confined, is essentially necessary to the metabolic phenomena of the cells. The oxygen disappears and carbonic acid is formed, or *vice versa,* carbonic acid disappears, and oxygen is formed. The universality of respiration is based entirely upon this fundamental condition to the metabolic phenomena of the cells. It is so important that, as we shall see further on, even the principal varieties of form in organized bodies are occasioned by this peculiarity of the metabolic process in the cells.

Each cell is not capable of producing chemical changes in every organic substance contained in solution, but only in particular ones. The fungi of fermentation, for instance, effect no changes in any other solutions than sugar; and the spores of certain plants do not become developed in all substances. In the same manner it is probable that each cell in the animal body converts only particular constituents of the blood.

The metabolic power of the cells is arrested not only by powerful chemical actions, such as destroy organic substances in general, but also by matters which chemically are less uncongenial; for instance, concentrated solutions of neutral salts. Other substances, as arsenic, do so in less quantity. The metabolic phenomena may be altered in quality by other substances, both organic and inorganic, and a change of this kind may result even from mechanical impressions on the cells.

Such are the most essential characteristics of the fundamental powers of the cells, so far as they can as yet be deduced from the phenomena. And now, in order to comprehend distinctly in what the peculiarity of the formative process of a cell, and therefore in what the peculiarity of the essential phenomenon in the formation of organized bodies consists, we will compare this process with a phenomenon of inorganic nature as nearly as possible similar to it. Disregarding all that is specially peculiar to the formation of cells, in order to find a more general definition in which it may be included with a process occurring in inorganic nature, we may view it as a process in which a solid body of definite and regular shape is formed in a fluid at the expense of a substance held in solution by that fluid. The process of crystallization in inorganic nature comes also within this definition, and is, therefore, the nearest analogue to the formation of cells.

Let us now compare the two processes, that the difference of the organic process may be clearly manifest. First, with reference to the plastic phenomena, the forms of cells and crystals are very different. The primary forms of crystals are simple, always angular, and bounded by plane surfaces; they are regular, or at least symmetrical, and even the very varied secondary forms of crystals are almost, without exception, bounded by plane surfaces. But manifold as is the form of cells, they have very little resemblance to crystals; round surfaces predominate, and where angles occur, they are never quite sharp, and the polyhedral crystal-like form of many cells results only from mechanical causes. The structure too of cells and of crystals is different. Crystals are solid bodies, composed merely of layers placed one upon another; cells are hollow vesicles, either single, or several inclosed one within another. And if we regard the membranes of these vesicles as layers, there will still remain marks of difference between them and crystals; these layers are not in contact, but contain fluid between them, which is not the case with crystals; the layers in the cells are few, from one to three only; and they differ from each other in chemical properties, while those of crystals consist of the same chemical substance. Lastly, there is also a great difference between crystals and cells in their mode of growth. Crystals grow by apposition, the new molecules are set only upon the surface of those already deposited, but cells increase also by intussusception, that is to say, the new molecules are deposited also between those already present.

But greatly as these plastic phenomena differ in cells and in crystals, the metabolic are yet more different, or rather they are quite peculiar to cells. For a crystal to grow, it must be already present as such in the solution, and some extraneous cause must interpose to diminish its solubility. Cells, on the contrary, are capable of producing a chemical change in the surrounding fluid, of generating matters which had not previously existed in it as such, but of which only the elements were present in another combination. They therefore require no extraneous influence to effect a change of solubility; for if they can produce chemical changes in the surrounding fluid, they may also produce such substances as could not be held in solution under the existing circumstances, and therefore need no external cause of growth. If a crystal be laid in a pretty strong solution, of a substance similar even to itself, nothing ensues without our interference, or the crystal dissolves completely: the fluid must be evaporated for the crystal to increase. If a cell be laid in a solution of a substance, even different from itself, it grows and converts this substance without our aid. And this it is from which the process going on in the cells (so long as we do not separate it into its several acts) obtains that magical character, to which attaches the idea of Life.

From this we perceive how very different are the phenomena in the formation of cells and of crystals. Meanwhile, however, the points of resemblance between them should not be overlooked. They agree in this important point, that solid bodies of a certain regular shape are formed in obedience to definite laws at the expense of a substance contained in solution in a fluid; and the crystal, like the cell, is so far an active and positive agent as to cause the substances which are precipitated to be deposited on itself, and nowhere else. We must, therefore, attribute to it as well as to the cell a power to attract the substance held in solution in the surrounding fluid. It does not indeed follow that these two attractive powers, the power of crystallization—to give it a brief title—and the plastic power of the cells are essentially the same. This could only be admitted, if it were proved that both powers acted according to the same laws. But this is seen at the first glance to be by no means the case: the phenomena in the formation of cells and crystals, are, as we have observed, very different, even if we regard merely the plastic phenomena of the cells, and leave their metabolic power (which may possibly arise from some other peculiarity of organic substance) for a time entirely out of the question.

It is, however, possible that these distinctions are only secondary, that the power of crystallization and the plastic power of the cells are identical, and that an original difference can be demonstrated between the substance of cells and that of crystals, by which we may perceive that the substance of cells must crystallize as cells according to the laws by which crystals are formed, rather than in the shape of the ordinary crystals? It may be worth while to institute such an inquiry.

In seeking such a distinction between the substance of cells and that of crystals, we may say at once that it cannot consist in anything which the substance of cells has in common with those organic substances which crystallize in the ordinary form. Accordingly, the more complicated arrangement of the atoms of the second order in organic bodies cannot give rise to this difference; for we see in sugar, for instance, that the mode of crystallization is not altered by this chemical composition.

Another point of difference by which inorganic bodies are distinguished from at least some of the organic bodies, is the faculty of imbibition. Most organic bodies are capable of being infiltrated by water, and in such a manner that it penetrates not so much into the interspaces between the elementary tissues of the body, as into the simple structureless tissues, such as areolar tissue, etc.; so that they form an homogeneous mixture, and we can neither distinguish particles of organic matter, nor interspaces filled with water. The water occupies the infiltrated organic substances, just as it is present in a solution, and there is as much difference between the capacity for imbibition and capillary permeation, as there is between a solution and the phenomena of capillary permeation. When water soaks through a layer of glue, we do not imagine it to pass through pores, in the common sense of the term; and this is just the condition of all substances capable of imbibition. They possess, therefore, a double nature, they have a definite form like solid bodies; but like fluids, on the other hand, they are also permeable by anything held in solution. As a specifically lighter fluid poured on one specifically heavier so carefully as not to mix with it, yet gradually penetrates it, so also, every solution, when brought into contact with a membrane already infiltrated with water, bears the same relations to the membrane, as though it were a solution. And crystallization being the transition from the fluid to the solid state, we may conceive it possible, or even probable, that if bodies, capable of existing in an intermediate state between solid and fluid could be made to crystallize, a considerable difference would be exhibited from the ordinary mode of crystallization. In fact, there is nothing, which we call a crystal, composed of substance capable of imbibition; and even among organized substances, crystallization takes place only in those which are capable of imbibition, as fat, sugar, tartaric acid, etc. The bodies capable of imbibition, therefore, either do not crystallize at all, or they do so under a form so different from the crystal, that they are not recognized as such.

Let us inquire what would most probably ensue, if material capable of imbibition crystallized according to the ordinary laws, what varieties from the common crystals would be most likely to show themselves, assuming only that the solution has permeated through the parts of the crystal already formed, and that new molecules can therefore be deposited between them. The ordinary crystals increase only by apposition; but there may be an important difference in the mode of this apposition. If the molecules were all deposited symmetrically one upon another, we might indeed have a body of a certain external form like a crystal; but it would not have the structure of one, it would not consist of layers. The existence of this laminated structure in crystals presupposes a double kind of apposition of their molecules; for in each layer the newly-deposited molecules coalesce, and become continuous with those of the same layer already present; but those molecules which form the adjacent surfaces of two layers do not coalesce. This is a remarkable peculiarity in the formation of crystals, and we are quite ignorant of its cause.

We cannot yet perceive why the new molecules, which are being deposited on the surface of a crystal (already formed up to a certain point), do not coalesce and become continuous with those already deposited, like the molecules in each separate layer, instead of forming, as they do, a new layer; and why this new layer does not constantly increase in thickness, instead of producing a second layer around the crystal, and so on. In the meantime, we can do no more than express the fact in the form of a law, that the coalescing molecules are deposited rather along the surface beside each other, than in the thickness upon one another, and thus, as the breadth of the layer depends upon the size of the crystal, so also the layer can attain only a certain thickness, and beyond this, the molecules which are being deposited cannot coalesce with it, but must form a new layer.

If we now assume that bodies capable of imbibition could also crystallize, the two modes of junction of the molecules should be shown also by them. Their structure should also be laminated, at least there is no perceptible reason for a difference in this particular, as the very fact of layers being formed in common crystals shows that the molecules need not be all joined together in the most exact manner possible. The closest possible conjunction of the molecules takes place only in the separate layers. In the common crystals this occurs by apposition of the new molecules on the surface of those present and coalescence with them. In bodies capable of imbibition, a much closer union is possible, because in them the new molecules may be deposited by intussusception between those already present. It is scarcely, therefore, too bold an hypothesis to assume, that when bodies capable of imbibition crystallize, their separate layers would increase by intussusception; and that this does not happen in ordinary crystals, simply because it is impossible.

Let us then imagine a portion of the crystal to be formed: new molecules continue to be deposited, but do not coalesce with the portion of the crystal already formed; they unite with one another only, and form a new layer, which, according to analogy with the common crystals, may invest either the whole or a part of the crystal. We will assume that it invests the entire crystal. Now, although this layer be formed by the deposition of new molecules between those already present instead of by apposition, yet this does not involve any change in the law, in obedience to which the deposition of the coalescing molecules goes on more vigorously in two directions, that is, along the surface, than it does in the third direction corresponding to the thickness of the layer; that is to say; the molecules which are deposited by intussusception between those already present, must be deposited much more vigorously between those lying together along the surface of the layer than between those which lie over one another in its thickness. This deposition of molecules side by side is limited in common crystals by the size of the crystal, or by that of the surface on which the layer is formed; the coalescence of molecules therefore ceases as regards that layer, and a new one begins. But if the layers grow by intussusception in crystals capable of imbibition, there is nothing to prevent the deposition of more molecules between those which lie side by side upon the surface, even after the lamina has invested the whole crystal; it may continue to grow without the law by which the new molecules coalesce requiring to be altered. But the consequence is, that the layer becomes, in the first instance more condensed, that is, more solid substance is taken into the same space; and afterwards it will expand and separate from the completed part of the crystal so as to leave a hollow space between itself and the crystal; this space fills with fluid by imbibition, and the first-formed portion of the crystal adheres to a spot on its inner surface. Thus, in bodies capable of imbibition, instead of a new layer attached to the part of the crystal already formed, we obtain a hollow vesicle. At first this must have the shape of the body of the crystal around which it is formed, and must, therefore, be angular, if the crystal is angular. If, however, we imagine this layer to be composed of soft substance capable of imbibition, we may readily comprehend how such a vesicle must very soon become round or oval. But the first formed part of the crystal also consists of substance capable of imbibition, so that it is very doubtful whether it must have an angular form at all. In common crystals atoms of some one particular substance are deposited together, and we can understand how a certain angular form of the crystal may result if these atoms have a certain form, or if in certain axes they attract each other differently. But in bodies capable of imbibition, an atom of one substance is not set upon another atom of the same substance, but atoms of water come between; atoms of water, which are not united with an atom of solid substance, so as to form a compound atom, as in the water of crystallization, but which exist in some other unknown manner between the atoms of solid substance. It is not possible, therefore, to determine whether that part of the crystal which is first formed must have an angular figure or not.

An ordinary crystal consists of a number of laminae; when so small as to be but just discernible, it has the form which the whole crystal afterwards exhibits, at least as far as regards the angles; we must therefore suppose that the first layer is formed around a very small corpuscle, which is of the same shape as the subsequent crystal. We will call this the primitive corpuscle. It is doubtful what may be the shape of this corpuscle in the crystals which are capable of imbibition. The first layer, then, is formed around the corpuscle in the way mentioned; it grows by intussusception, and thus forms a hollow, round or oval vesicle, to the inner surface of which the primitive corpuscle adheres. As all the new molecules that are being deposited may be placed in this layer without any alteration being required in the law which regulates the coalescence of the molecules during crystallization, we must conclude that it remains the only layer, and becomes greatly expanded, so as to represent all the layers of an ordinary crystal. It is, however, a question whether there may not exist some reasons why several layers can be formed. We can certainly conceive such to be the case. The quantity of the solid substance that must crystallize in a given time, depends upon the concentration of the fluid; the number of molecules that may, in accordance with the law already mentioned, be deposited in the layer in a given time depends upon the quantity of the solution which can penetrate the membrane by imbibition during that time. If in consequence of the concentration of the fluid there must be more precipitated in the time than can penetrate the membrane, it can only be deposited as a new layer on the outer surface of the vesicle. When this second layer is formed, the new molecules are deposited in it, and it rapidly becomes expanded into a vesicle, on the inner surface of which the first vesicle lies with its primitive corpuscle. The first vesicle now either does not grow at all, or at any rate much more slowly, and then only when the endosmosis into the cavity of the second vesicle proceeds so rapidly that all that might be precipitated while passing through it, is not deposited. The second vesicle, when it is developed at all, must needs be developed relatively with more rapidity than the first; for as the solution is in the most concentrated state at the beginning, the necessity for the formation of a second layer then occurs sooner; but when it is formed, the concentration of the fluid is diminished, and this necessity occurs either later or not at all. It is impossible, however, that even a third, or fourth, and more, may be formed; but the outermost layer must always be relatively the most vigorously developed; for then the concentration of the solution is only so strong, that all that *must* be deposited in a certain time, *can* be deposited in the outermost layer, it is all applied to the increase of this layer.

Such, then, would be the phenomena under which substances capable of imbibition would probably crystallize, if they did so at all. I say probably, for our incomplete knowledge of crystallization and the faculty of imbibition, does not as yet admit of our saying anything positively *a priori*. It is, however, obvious that these are the principal phenomena attending the formation of cells. They consist always of substance capable of imbibition; the first part formed is a small corpuscle, not angular (nucleolus), around this a lamina is deposited (nucleus), which advances rapidly in its growth, until a second lamina (cell) is formed around it. This second now grows more quickly and expands into a vesicle, as indeed often happens with the first layer. In some rarer instances only one layer is formed; in others, again, there are three. The only other difference in the formation of cells is, that the separate layers do not consist of the same chemical substance, while a common crystal is always composed of one material. In instituting a comparison, therefore, between the formation of cells and crystallization, the above-mentioned differences in form, structure, and mode of growth fall altogether to the ground. If crystals were formed from the same substance as cells, they would probably, in *these* respects, be subject to the same conditions as the cells. Meanwhile the metabolic phenomena, which are entirely absent in crystals, still indicate essential distinctions.

Should this important difference between the mode of formation of cells and crystals lead us to deny all intimate connexion of the two processes, the comparison of the two may serve at least to give a clear representation of the cell-life. The following may be conceived to be the state of the matter: the material of which the cells are composed is capable of producing chemical changes in the substance with which it is in contact, just as the well-known preparation of platinum converts alcohol into acetic acid. This power is possessed by every part of the cell. Now, if the cytoblastema be so changed by a cell already formed, that a substance is produced which cannot become attached to that cell, it immediately crystallizes as the central nucleolus of a new cell. And then this converts the cytoblastema in the same manner. A portion of that which is converted may remain in the cytoblastema in solution, or may crystallize as the commencement of new cells; another portion, the cell-substance, crystallizes around the central corpuscle. The cell-substance is either soluble in the cytoblastema, and crystallizes from it, so soon as the latter becomes saturated with it; or else it is insoluble, and crystallizes at the time of its formation, according to the laws of crystallization of bodies capable of imbibition mentioned above, forming in this manner one or more layers around the central corpuscle, and so on. If we conceive the above to represent the mode of the formation of cells, we regard the plastic power of the cells as identical with the power by which crystals grow. According to the foregoing description of the crystallization of bodies capable of imbibition, the most important plastic phenomena of the cells are certainly satisfactorily explained. But let us see if this comparison agrees with all the characteristics of the plastic power of the cells. (See above, p. 194 et seq.)

The attractive power of the cells does not always operate symmetrically; the deposition of new molecules may be more vigorous in particular spots, and thus produce a change in the form of the cell. This is quite analogous to what happens in crystals; for although in them an angle is never altered, there may be much more material deposited on some surfaces than on others; and thus, for instance, a quadrilateral prism may be formed out of a cube. In this case new layers are deposited on one, or on two opposite sides of a cube. Now, if one layer in cells represent a number of layers in a common crystal, it may be easily perceived that instead of several new layers being formed on two opposite surfaces of a cell, the one layer would grow more at those spots, and thus a round cell would be elongated into a fibre; and so with the other changes of form. Division of the cells can have no analogue in common crystals, because that which is once deposited is incapable of any further change. But this phenomenon may be made to accord with the representation of crystals capable of imbibition, just as well as the coalescence of numerous cells in the manner described at page 184 does. And if we ascribe to a layer of a crystal capable of imbibition the power of producing chemical changes in organic substances, we can very well understand also the origin of secondary deposits on its inner surface as they occur in cells. For if, in accordance with the laws of crystallization, the lamina has become expanded into a vesicle, and its cavity has become filled by imbibition with a solution of organic substance, there may be materials formed by means of the converting influence of the lamina, which cannot any longer be held in solution. These may, then, either crystallize within the vesicle, as new crystals capable of imbibition under the form of cells; or if they are allied to the substance of the vesicle, they may so crystallize as to form part of the system of the vesicle itself: the latter may occur in two ways, the new matters may be applied to the increase of the vesicle, or they may form new layers on its inner surface from the same cause which led to the first formation of the vesicle itself as a layer. In the cells of plants these secondary deposits have a spiral arrangement. This is a very important fact, though the laws of crystallization do not seem to account for the absolute necessity of it. If, however, it could be mathematically proved from the laws of the crystallization of inorganic bodies, that under the altered circumstances in which bodies capable of imbibition are placed, these deposits must be arranged in spiral forms, it might be asserted without hesitation that the plastic power of cells and the fundamental powers of crystals are identical.

We come now, however, to some peculiarities in the plastic power of cells, to which we might, at first sight, scarcely expect to find anything analogous in crystals. The attractive power of the cells manifests a certain degree of election in its operation; it does not attract every substance present in the cytoblastema, but only particular ones; and here a muscle-cell, there a fat-cell, is generated from the same fluid, the blood. Yet crystals afford us an example of a precisely similar phenomenon, and one which has already been frequently adduced as analogous to assimilation. If a crystal of nitre be placed in a solution of nitre and sulphate of soda, only the nitre crystallizes; when a crystal of sulphate of soda is put in, only the sulphate of soda crystallizes. Here, therefore, there occurs just the same selection of the substance to be attracted.

We observed another law attending the development of the plastic phenomena in the cells, viz. that a more concentrated solution is requisite for the first formation of a cell than for its growth when already formed, a law upon which the difference between organized and unorganized tissues is based. In ordinary crystallization the solution must be more than saturated for the process to begin. But when it is over, there remains a mother lye, according to Thénard, which is no longer saturated at the same temperature. This phenomenon accords precisely with the cells; it shows that a more concentrated solution is requisite for the commencement of crystallization than for the increase of a crystal already formed. The fact has indeed been disputed by Thomson; but if, in the undisputed experiment quoted above, the crystal of sulphate of soda attracts the dissolved sulphate of soda rather than the dissolved nitre, and *vice versâ*, the crystal of nitre attracts the dissolved nitre more than the dissolved sulphate of soda, it follows that a crystal does attract a salt held in solution, because the experiment proves that there are degrees of this attraction. But if there be such an attraction exerted by a crystal, then the introduction of a crystal into a solution of salt, affords an efficient cause for the deposition of this salt, which does not exist when no crystal is introduced. The solution must therefore be more concentrated in the latter case than in the former, though the difference be so slight as not to be demonstrable by experiment. It would not, however, be superfluous to repeat the experiments. In the instance of crystals capable of imbibition, this difference may be considerably augmented, since the attraction of molecules may increase perhaps considerably by the penetrating of the solution between those already deposited.

We see then how all the plastic phenomena in the cells may be compared with phenomena which, in accordance with the ordinary laws of crystallization, would probably appear if bodies capable of imbibition could be brought to crystallize. So long as the object of such a comparison were merely to render the representation of the process by which cells are formed more clear, there could not be much urged against it; it involves nothing hypothetical, since it contains no explanation; no assertion is made that the fundamental power of the cells really has something in common with the power by which crystals are formed. We have, indeed, compared the growth of organisms with crystallization, in so far as in both cases solid substances are deposited from a fluid, but we have not therefore asserted the identity of the fundamental powers. So far we have not advanced beyond the data, beyond a certain simple mode of representing the facts.

The question is, however, whether the exact accordance of the phenomena would not authorize us to go further. If the formation and growth of the elementary particles of organisms have nothing more in common with crystallization than merely the deposition of solid substances from out of a fluid, there is certainly no reason for assuming any more intimate connexion of the two processes. But we have seen, first, that the laws which regulate the deposition of the molecules forming the elementary particles of organisms are the same for all elementary parts; that there is a common principle in the development of all elementary parts, namely, that of the formation of cells; it was then shown that the power which induced the attachment of the new molecules did not reside in the entire organism, but in the separate elementary particles (this we called the plastic power of the cells); lastly, it was shown that the laws, according to which the new molecules combine to form cells, are (so far as our incomplete knowledge of the laws of crystallization admits of our anticipating their probability) the same as those by which substances capable of imbibition would crystallize. Now the cells do, in fact, consist only of material capable of imbibition; should we not then be justified in putting forth the proposition, that the formation of the elementary parts of organisms is nothing but a crystallization of substance capable of imbibition, and the organism nothing but an aggregate of such crystals capable of imbibition?

To advance so important a point as absolutely true, would certainly need the clearest proof; but it cannot be said that even the premises which have been set forth have in all points the requisite force. For too little is still known of the cause of crystallization to predict with safety (as was attempted above) what would follow if a substance capable of imbibition were to crystallize. And if these premises were allowed, there are two other points which must be proved in order to establish the proposition in question: 1. That the metabolic phenomena of the cells, which have not been referred to in the foregoing argument, are as much the necessary consequence of the faculty of imbibition, or of some other peculiarity of the substance of cells, as the plastic phenomena are. 2. That if a number of crystals capable of imbibition are formed, they must combine according to certain laws so as to form a systematic whole, similar to an organism. Both these points must be clearly proved, in order to establish the truth of the foregoing view. But it is otherwise if this view be adduced merely as an hypothesis, which may serve as a guide for new investigations. In such case the inferences are sufficiently probable to justify such an hypothesis, if only the two points just mentioned can be shown to accord with it.

With reference to the first of these points, it would certainly be impossible, in our ignorance as to the cause of chemical phenomena in general, to prove that a crystal capable of imbibition must produce chemical changes in substances surrounding it; but then we could not infer, from the manner in which spongy platinum is formed, that it would act so peculiarly upon oxygen and hydrogen. But in order to render this view tenable as a possible hypothesis, it is only necessary to see that it *may* be a consequence. It cannot be denied that it may: there are several reasons for it, though they certainly are but weak. For instance, since all cells possess this metabolic power, it is more likely to depend on a certain position of the molecules, which in all probability is essentially the same in all cells, than on the chemical combination of the molecules, which is very different in different cells. The presence, too, of different substances on the inner and the outer surface of the cell-membrane (see above, page 199) in some measure implies that a certain direction of the axes of the atoms may be essential to the metabolic phenomena of the cells. I think, therefore, that the cause of the metabolic phenomena resides in that definite mode of arrangement of the molecules which occurs in crystals, combined with the capacity which the solution has to penetrate between these regularly deposited molecules (by means of which, presuming the molecules to possess polarity, a sort of galvanic pile will be formed), and that the same phenomena would be observed in an ordinary crystal, if it could be rendered capable of imbibition. And then perhaps the differences of quality in the metabolic phenomena depend upon their chemical composition.

In order to render tenable the hypothesis contained in the second point, it is merely necessary to show that crystals capable of imbibition can unite with one another according to certain laws. If at their first formation all crystals were isolated, if they held no relation whatever to each other, the view would leave entirely unexplained how the elementary parts of organisms, that is, the crystals in question, become united to form a whole. It is therefore necessary to show that crystals do unite with each other according to certain laws, in order to receive, at least, the possibility of their uniting also to form an organism, without the need of any further combining power. But there are many crystals in which a union of this kind, according to certain laws, is indisputable;

indeed they often form a whole, so like an organism in its entire form that groups of crystals are known in common life by the names of flowers, trees, etc. I need only refer to the ice-flowers on the windows, or to the lead-tree, etc. In such instances a number of crystals arrange themselves in groups around others, which form an axis. If we consider the contact of each crystal with the surrounding fluid to be an indispensable condition to the growth of crystals which are not capable of imbibition, but that those which are capable of imbibition, in which the solution can penetrate whole layers of crystals, do not require this condition, we perceive that the similarity between organisms and these aggregations of crystals is as great as could be expected with such difference of substance. As most cells require for the production of their metabolic phenomena, not only their peculiar nutrient fluid, but also the access of oxygen and the power of exhaling carbonic acid, or *vice versâ*; so, on the other hand, organisms in which there is no circulation of respiratory fluid, or in which at least it is not sufficient, must be developed in such a way as to present as extensive a surface as possible to the atmospheric air. This is the condition of plants, which require for their growth that the individual cells should come into contact with the surrounding medium in a similar manner, if not in the same degree as occurs in a crystal tree, and in them indeed the cells unite into a whole organism in a form much resembling a crystal tree. But in animals the circulation renders the contact of the individual cells with the surrounding medium superfluous, and they may have more compact forms, even though the laws by which the cells arrange themselves are essentially the same.

The view then that organisms are nothing but the form under which substances capable of imbibition crystallize, appears to be compatible with the most important phenomena of organic life, and may be so far admitted, that it is a possible hypothesis, or attempt towards an explanation of these phenomena. It involves very much that is uncertain and paradoxical, but I have developed it in detail, because it may serve as a guide for new investigations. For even if no relation between crystallization and the growth of organisms be admitted in principle, this view has the advantage of affording a distinct representation of the organic process; an indispensable requisite for the institution of new inquiries in a systematic manner, or for testing by the discovery of new facts a mode of explanation which harmonizes with phenomena already known.

FURTHER READING

Ackerknecht, Erwin H. *Rudolf Virchow: Doctor, Statesman, Anthropologist.* Madison: University of Wisconsin Press, 1953.

Baker, John R. "The Cell-theory: a Restatement, History, and Critique." *Quarterly Journal of Microscopical Science* 89 (1948): 103–25; 90 (1949): 87–108, 331; 93 (1952): 157–90; 94 (1953): 407–40; 96 (1955): 449–81.

Cappelletti, Vincenzo. *Entelechìa: Saggi sulle dottrine biologiche del secolo decimonono.* Firenze: G.C. Sansoni, 1965.

Coleman, William. *Biology in the Nineteenth Century: Problems of Form, Function, and Transformation.* Cambridge: Cambridge UP, 1977.

Duchesneau, François. *Genèse de la théorie cellulaire.* Paris: J. Vrin, 1987.

Florkin, Marcel. *Naissance et déviation de la théorie cellulaire dans l'oeuvre de Théodore Schwann.* Paris: Hermann, 1960.

—. "Théodore Schwann: 1810–1882." In *Florilège des sciences en Belgique pendant le XIXe siècle et le début du XXe.* Bruxelles: Académie Royale de Belgique, 1968, pp. 929–55.

—. *Théodore Schwann et les débuts de la médecine scientifique.* Paris: Les Conférences du Palais de la Découverte, 1956.

Fredericq, Léon. *Théodore Schwann: Sa vie et ses travaux.* Liège: Université de Liège, 1884.

Klein, M. *Histoire des origines de la théorie cellulaire.* Paris: Hermann et Compagnie, 1936.

Rádl, E. *The History of Biological Theories.* Oxford: Oxford UP, 1930.

Rather, L.J., Patricia Rather, and John B. Frerichs. *Johannes Müller and the Nineteenth-Century Origins of Tumor Cell Theory.* Canton, MA: Science History Publications, 1986.

Watermann, Rembert. *Theodor Schwann: Leben und Werk.* Düsseldorf: L. Schwann, 1960.

13. NIKOLAI IVANOVICH LOBACHEVSKY

(1792–1856)

Born in Nizhni Novgorod, Russia, Lobachevsky possessed an early aptitude for mathematics, and after studies at Kazan University (1807–11), became a professor of mathematics at that institution. Upon becoming rector in 1827, he expanded the scientific mission of the university and began publishing a *Scientific Proceedings*. His first communication of his discovery of a Non-Euclidean or Hyperbolic Geometry came in a lecture of 1826, later published as "On the Principles of Geometry" in the *Kazan Herald* (1829–30), but his ideas were not understood or well received at the time. Mathematics was experiencing rapid change at this time and the nineteenth century has been called the "Golden Age in mathematics" (Boyer 496)—even such sciences as biology and psychology grew more quantitative and statistically-based by the end of the century, and new mathematical concepts such as non-Euclidean geometries, noncommutative algebra, and n-dimensional spaces were developed.

The story of Lobachevsky's new geometry begins with Euclid's work the *Elements* (c. 300 B.C.E.) which outlined a geometrical system of definitions (primitive terms), axioms, and theorems logically deduced from the axioms. Lobachevskian geometry in fact accepts all of the Euclidean propositions, except for the fifth parallel-postulate (here expressed in simplified terms): *only one straight line can be drawn parallel to a given straight line through a point not on that line (in the same plane).* Fruitless attempts were made by Proclus and Posidonius in classical times and by eighteenth-century mathematicians Girolamo Saccheri and Johann Lambert to prove the validity of this assumption as a theorem instead of accepting it as a self-evident axiom. Lobachevsky in his attempt to prove the parallel-postulate (by demonstrating that denial of Euclid's parallel-postulate leads to contradiction) found that a new set of theorems could be formulated describing a consistent, non-contradictory and logical system which he called "imaginary geometry."

He began, in contrast to Euclid's parallel-postulate, with the axiom "through a point not lying on a given line one can draw in the plane determined by this point and line at least two lines which do not have a point of intersection with the given line" (Shirokov 12). Lobachevsky, similarly to Gauss and Wachter, further demonstrated that on the imaginary figure the horosphere (a sphere of infinite radius, comprised of an infinite number of horocycles or circles of infinite radius), both Euclidean and normal plane geometry were valid, suggesting that these geometries were subsets of Lobachevky's more powerful and more generalized system. Lobachevsky published three book-length explanations of his new geometrical system: *New Foundations of Geometry* (1835–38); *Geometrical Researches on the Theory of Parallels* (1840), of which the first section is reprinted here below; and *Pangeometry* (1855), the final expression of his complete system.

The possibility of a non-Euclidean geometry had also been envisioned by Carl Friedrich Gauss (1777–1855), the most influential mathematician of the nineteenth century, but Gauss, perhaps fearing ridicule, did not publish or systematize his insights in his lifetime. The Hungarian mathematician Janós Bolyai (1802–1860), at the same time as Gauss and Lobachevsky, also developed a non-Euclidean "Absolute Science of Space," published as an appendix to one of his father's mathematical treatises in 1832, although Gauss surprisingly withheld public support for Bolyai's views. Obviously involving a case of simultaneous discovery among researchers interested in the same questions, credit for the full elaboration of non-Euclidean geometry, however, is most often awarded to Lobachevsky.

Lobachevsky's work wrought a profound change in mathematical thinking which had relied for over a thousand years on Euclid's system as an exact description of space and everyday geometrical reality. The realization of alternate geometries helped to dislodge the notion "that Euclidean geometry is 'innate,' 'unique,' 'natural,' or 'god-given'" (Yaglom vi). Work on non-Euclidean geometries, however, was not fully appreciated before G.F.B. Riemann's *Über die Hypothesen welche der Geometrie zu Grunde liegen* (1867), in which he proposed an even more general approach to geometry, thus subsuming both Euclidean and Lobachevskian geometries as special cases of a more universalized geometry. Riemann by theorizing curved metric spaces paved the way for Einstein's relativity theory in the twentieth century.

Geometrical Researches on the Theory of Parallels.
Translated by George Bruce Halstead. New edition. Chicago:
Open Court Publishing Company, 1914, pp. 11–19.

THEORY OF PARALLELS

In geometry I find certain imperfections which I hold to be the reason why this science, apart from transition into analytics, can as yet make no advance from that state in which it has come to us from Euclid.

As belonging to these imperfections, I consider the obscurity in the fundamental concepts of the geometrical magnitudes and in the manner and method of representing the measuring of these magnitudes, and finally the momentous gap in the theory of parallels, to fill which all efforts of mathematicians have been so far in vain.

For this theory Legendre's endeavors have done nothing, since he was forced to leave the only rigid way to turn into a side path and take refuge in auxiliary theorems which he illogically strove to exhibit as necessary axioms. My first essay on the foundations of geometry I published in the Kasan *Messenger* for the year 1829. In the hope of having satisfied all requirements, I undertook hereupon a treatment of the whole of this science, and published my work in separate parts in the *"Gelehrten Schriften der Universitæt Kasan"* for the years 1836, 1837, 1838, under the title "New Elements of Geometry, with a complete Theory of Parallels." The extent of this work perhaps hindered my countrymen from following such a subject, which since Legendre had lost its interest. Yet I am of the opinion that the Theory of Parallels should not lose its claim to the attention of geometers, and therefore I aim to give here the substance of my investigations, remarking beforehand that contrary to the opinion of Legendre, all other imperfections—for example, the definition of a straight line—show themselves foreign here and without any real influence on the theory of parallels.

In order not to fatigue my reader with the multitude of those theorems whose proofs present no difficulties, I prefix here only those of which a knowledge is necessary for what follows.

1. A straight line fits upon itself in all its positions. By this I mean that during the revolution of the surface containing it the straight line does not change its place if it goes through two unmoving points in the surface: (*i.e.*, if we turn the surface containing it about two points of the line, the line does not move.)

2. Two straight lines can not intersect in two points.

3. A straight line sufficiently produced both ways must go out beyond all bounds, and in such way cuts a bounded plain into two parts.

4. Two straight lines perpendicular to a third never intersect, how far soever they be produced.

5. A straight line always cuts another in going from one side of it over to the other side: (*i.e.*, one straight line must cut another if it has points on both sides of it.)

6. Vertical angles, where the sides of one are productions of the sides of the other, are equal. This holds of plane rectilineal angles among themselves, as also of plane surface angles: (*i.e.*, dihedral angles.)

7. Two straight lines can not intersect, if a third cut them at the same angle.

8. In a rectilineal triangle equal sides lie opposite equal angles, and inversely.

9. In a rectilineal triangle, a greater side lies opposite a greater angle. In a right-angled triangle the hypothenuse is greater than either of the other sides, and the two angles adjacent to it are acute.

10. Rectilineal triangles are congruent if they have a side and two angles equal, or two sides and the included angle equal, or two sides and the angle opposite the greater equal, or three sides equal.

11. A straight line which stands at right angles upon two other straight lines not in one plane with it is perpendicular to all straight lines drawn through the common intersection point in the plane of those two.

12. The intersection of a sphere with a plane is a circle.

13. A straight line at right angles to the intersection of two perpendicular planes, and in one, is perpendicular to the other.

14. In a spherical triangle equal sides lie opposite equal angles, and inversely.

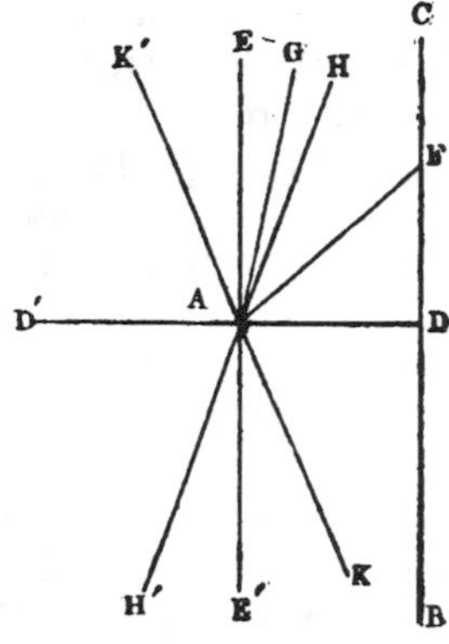

Figure 1

15. Spherical triangles are congruent (or symmetrical) if they have two sides and the included angle equal, or a side and the adjacent angles equal.

From here follow the other theorems with their explanations and proofs.

16. All straight lines which in a plane go out from a point can, with reference to a given straight line in the same plane, be divided into two classes—into *cutting* and *not-cutting*.

The *boundary lines* of the one and the other class of those lines will be called *parallel to the given line*.

From the point A (Fig. 1) let fall upon the line BC the perpendicular AD, to which again draw the perpendicular AE.

In the right angle EAD either will all straight lines which go out from the point A meet the line DC, as for example AF, or some of them, like the perpendicular AE, will not meet the line DC. In the uncertainty whether the perpendicular AE is the only line which does not meet DC, we will assume it may be possible that there are still other lines, for example AG, which do not cut DC, how far soever they may be prolonged. In passing over from the cutting lines, as AF, to the not-cutting lines, as AG, we must come upon a line AH, parallel to DC, a boundary line, upon one side of which all lines AG are such as do not meet the line DC, while upon the other side every straight line AF cuts the line DC.

The angle HAD between the parallel HA and

the perpendicular AD is called the parallel angle (angle of parallelism), which we will here designate by $\Pi(p)$ for AD = p.

If $\Pi(p)$ is a right angle, so will the prolongation AE´ of the perpendicular AE likewise be parallel to the prolongation DB of the line DC, in addition to which we remark that in regard to the four right angles, which are made at the point A by the perpendiculars AE and AD, and their prolongations AE´ and AD´, every straight line which goes out from the point A, either itself or at least its prolongation, lies in one of the two right angles which are turned toward BC, so that except the parallel EE´ all others, if they are sufficiently produced both ways, must intersect the line BC.

If $\Pi(p) < \frac{1}{2}\pi$, then upon the other side of AD, making the same angle DAK = $\Pi(p)$ will lie also a line AK, parallel to the prolongation DB of the line DC, so that under this assumption we must also make a distinction of *sides in parallelism*.

All remaining lines or their prolongations within the two right angles turned toward BC pertain to those that intersect, if they lie within the angle HAK = 2 $\Pi(p)$ between the parallels; they pertain on the other hand to the non-intersecting AG, if they lie upon the other sides of the parallels AH and AK, in the opening of the two angles EAH = $\frac{1}{2}\pi - \Pi(p)$, E´AK = $\frac{1}{2}\pi - \Pi(p)$, between the parallels and EE´ the perpendicular to AD. Upon the other side of the perpendicular EE´ will in like manner the prolongations AH´ and AK´ of the parallels AH and AK likewise be parallel to BC; the remaining lines pertain, if in the angle K´AH´, to the intersecting, but if in the angles K´AE, H´AE´ to the non-intersecting.

In accordance with this, for the assumption $\Pi(p) = \frac{1}{2}\pi$ the lines can be only intersecting or parallel; but if we assume that $\Pi(p) < \frac{1}{2}\pi$, then we must allow two parallels, one on the one and one on the other side; in addition we must distinguish the remaining lines into non-intersecting and intersecting.

For both assumptions it serves as the mark of parallelism that the line becomes intersecting for the smallest deviation toward the side where lies the parallel, so that if AH is parallel to DC, every line AF cuts DC, how small soever the angle HAF may be.

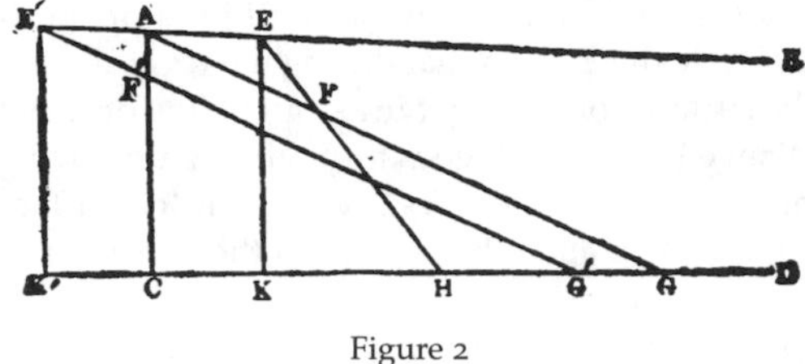

Figure 2

17. *A straight line maintains the characteristic of parallelism at all its points.*

Given AB (Fig. 2) parallel to CD, to which latter AC is perpendicular. We will consider two points taken at random on the line AB and its production beyond the perpendicular.

Let the point E lie on that side of the perpendicular on which AB is looked upon as parallel to CD.

Let fall from the point E a perpendicular EK on CD and so draw EF that it falls within the angle BEK.

Connect the points A and F by a straight line, whose production then (by Theorem 16) must cut CD somewhere in G. Thus we get a triangle ACG, into which the line EF goes; now since this latter, from the construction, can not cut AC, and can not cut AG or EK a second time (Theorem 2), therefore it must meet CD somewhere at H (Theorem 3).

Now let E´ be a point on the production of AB and E´K´ perpendicular to the production of the line CD; draw the line E´F´ making so small an angle AE´F´ that it cuts AC somewhere in F´; making the same angle with AB, draw also from A the line AF, whose production will cut CD in G (Theorem 16).

Thus we get a triangle AGC, into which goes the production of the line E´F´; since now this line can not cut AC a second time, and also can not cut AG, since the angle BAG = BE´G´, (Theorem 7), therefore must it meet CD somewhere in G´.

Therefore from whatever points E and E´ the lines EF and E´F´ go out, and however little they may diverge from the line AB, yet will they always cut CD, to which AB is parallel.

18. *Two lines are always mutually parallel.*

Let AC be a perpendicular on CD, to which AB is parallel if we draw from C the line CE making any acute angle ECD with CD, and let fall from A the perpendicular AF upon CE, we obtain a right-angled triangle ACF, in which AC, being the hypothenuse, is greater than the side AF (Theorem 9.)

Make AG = AF, and slide the figure EFAB until AF coincides with AG, when AB and FE will take the position AK and GH, such that the angle BAK = FAC, consequently AK must cut the line DC somewhere in K (Theorem 16), thus forming a triangle AKC, on one side of which the perpendicular GH intersects the line AK in L (Theorem 3), and thus determines the distance AL of the intersection point of the lines AB and CE on the line AB from the point A.

Hence it follows that CE will always intersect AB, how small soever may be the angle ECD, consequently CD is parallel to AB (Theorem 16.)

19. *In a rectilineal triangle the sum of the three angles can not be greater than two right angles.*

Suppose in the triangle ABC (Fig. 4) the sum of the three angles is equal to $\pi + a$; then choose in case of the inequality of the sides the smallest BC, halve it in D, draw from A through D the line AD and make the prolongation of it, DE, equal to AD, then join the point E to the point C by the straight line EC. In the congruent triangles ADB and CDE, the angle ABD = DCE, and BAD = DEC (Theorems 6 and 10); whence follows that also in the triangle ACE the sum of the three angles must be equal to $\pi + a$; but also the smallest angle BAC (Theorem 9) of the triangle ABC in passing over into the new triangle ACE has been cut up into the two parts EAC and AEC. Continuing this process, continually halving the side opposite the smallest angle, we must finally attain to a triangle in which the sum of the three angles is $\pi + a$, but wherein are two angles, each of which in absolute magnitude is less than $\frac{1}{2}a$; since now, however, the third angle can not be greater than π, so must a be either null or negative.

20. *If in any rectilineal triangle the sum of the three angles is equal to two right angles, so is this also the case for every other triangle.*

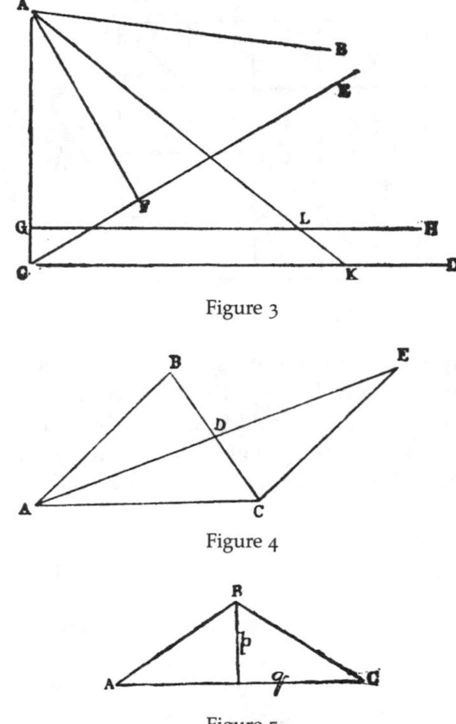

Figure 3

Figure 4

Figure 5

If in the rectilineal triangle ABC (Fig. 5) the sum of the three angles = π, then must at least two of its angles, A and C, be acute. Let fall from the vertex of the third angle B upon the opposite side AC the perpendicular p. This will cut the triangle into two right-angled triangles, in each of which the sum of the three angles must also be π, since it can not in either be greater than π, and in their combination not less than π.

So we obtain a right-angled triangle with the perpendicular sides p and q, and from this a quadrilateral whose opposite sides are equal and whose adjacent sides p and q are at right angles (Fig. 6.)

By repetition of this quadrilateral we can make another with sides np and q, and finally a quadrilateral ABCD with sides at right angles to each other, such that AB = np, AD = mq, DC = np, BC = mq, where m and n are any whole numbers. Such a quadrilateral is divided by the diagonal DB into two congruent right-angled triangles, BAD and BCD, in each of which the sum of the three angles = π.

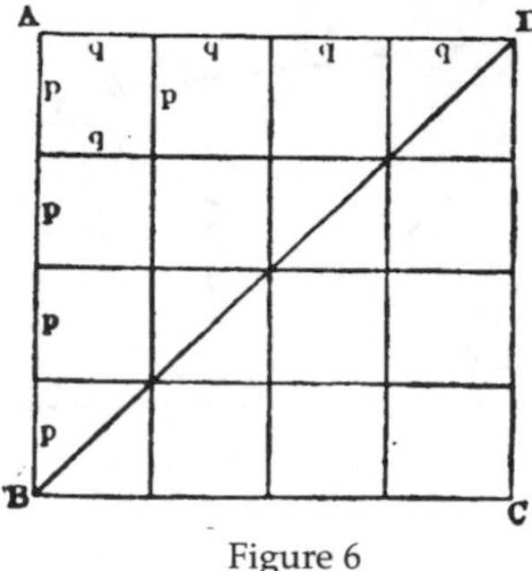

Figure 6

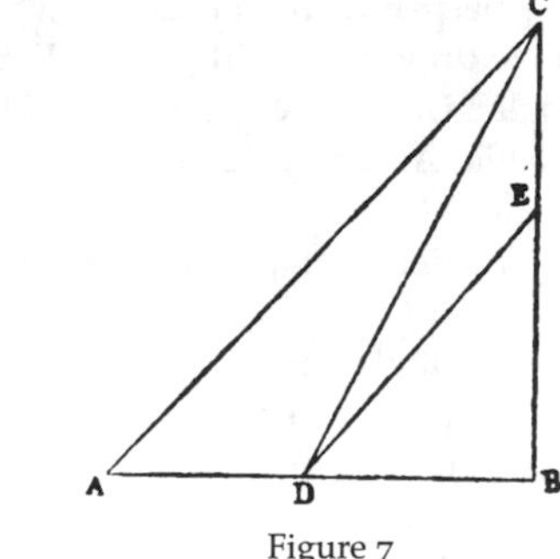

Figure 7

The numbers n and m can be taken sufficiently great for the right-angled triangle ABC (Fig. 7) whose perpendicular sides AB = np, BC = mq, to enclose within itself another given (right-angled) triangle BDE as soon as the right-angles fit each other.

Drawing the line DC, we obtain right-angled triangles of which every successive two have a side in common.

The triangle ABC is formed by the union of the two triangles ACD and DCB, in neither of which can the sum of the angles be greater than π; consequently it must be equal to π, in order that the sum in the compound triangle may be equal to π.

In the same way the triangle BDC consists of the two triangles DEC and DBE, consequently must in DBE the sum of the three angles be equal to π, and in general this must be true for every triangle, since each can be cut into two right-angled triangles.

From this it follows that only two hypotheses are allowable: Either is the sum of the three angles in all rectilineal triangles equal to π, or this sum is in all less than π.

21. *From a given point we can always draw a straight line that shall make with a given straight line an angle as small as we choose.*

Let fall from the given point A (Fig. 8) upon the given line BC the perpendicular AB; take upon BC at random the point D; draw the line AD; make DE = AD, and draw AE.

In the right-angled triangle ABD let the angle ADB = a; then must in the isosceles triangle ADE the angle AED be either ½a or less (Theorems 8 and 20). Continuing thus we finally attain to such an angle, AEB, as is less than any given angle....

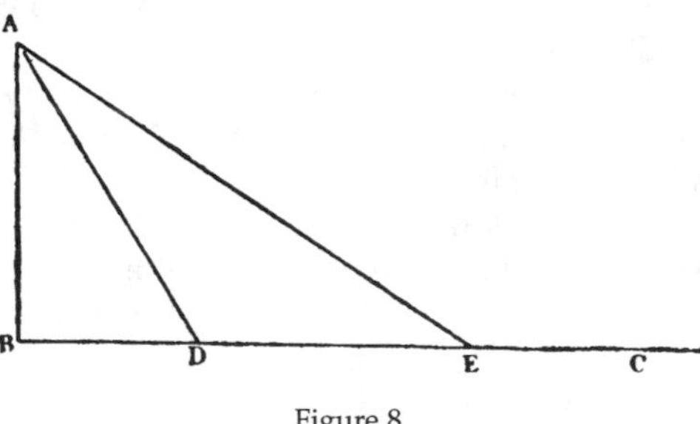

Figure 8

[Lobachevsky continues to elaborate the increasingly complex implications of his system.]

FURTHER READING

Benedetti, R. *Lectures on Hyperbolic Geometry.* Berlin: Springer-Verlag, 1992.

Bonola, Roberto. *Non-Euclidean Geometry: A Critical and Historical Study of Its Development.* Translated by H.S. Carslaw. Chicago: Open Court Publishing Company, 1912.

Boyer, Carl B. *A History of Mathematics.* Revised by Uta C. Merzbach. 2nd ed. New York: John Wiley and Sons, Inc., 1991.

Coolidge, Julian Lowell. *The Elements of Non-Euclidean Geometry.* Oxford: Oxford UP, 1909.

Coxeter, H.S.M. *Non-Euclidean Geometry.* 3rd ed. Toronto: University of Toronto Press, 1957.

Engel, Friedrich. *Nikolaj Iwanowitsch Lobatschefskij: Zwei geometrische Abhandlungen.* Leipzig: B.G. Teubner, 1898.

Engel, Friedrich, and Paul Stäckel. *Die Theorie der Parallellinien von Euklid bis auf Gauss.* Leipzig: B.G. Teubner, 1895.

Euclid. *The Thirteen Books of Euclid's Elements.* Trans. T L. Heath. Cambridge: Cambridge UP, 1926.

Gans, David. *An Introduction to Non-Euclidean Geometry.* New York: Academic Press, 1973.

Klein, Felix. *Vorlesungen über nicht-euklidische Geometrie.* Ed. W. Rosemann. Berlin: Julius Springer, 1928.

Kulczycki, Stefan. *Non-Euclidean Geometry.* Trans. Stanislaw Knapowski. New York: Pergamon Press, 1961.

Levy, Silvio, ed. *Flavors of Geometry.* Cambridge: Cambridge UP, 1997.

Lieber, Lillian R., and Hugh Gray Lieber. *Non-Euclidean Geometry or Three Moons in Mathesis.* Lancaster: Science Press Printing Company, 1931.

Meschkowski, Herbert. *Noneuclidean Geometry.* Trans. A. Shenitzer. New York: Academic Press, 1964.

Pont, Jean-Claude. *L'Aventure des parallèles. Histoire de la géométrie non euclidienne: Précurseurs et attardés.* Berne: Peter Lang, 1986.

Rédei, L. *Foundations of Euclidean and Non-Euclidean Geometries According to F. Klein.* Oxford: Pergamon Press, 1968.

Roeser, Ernst. *Die nichteuklidischen Geometrien und ihre Beziehungen untereinander.* München: R. Oldenbourg, 1957.

Rozenfeld, B.A. *A History of Non-Euclidean Geometry: Evolution of the Concept of a Geometric Space.* New York: Springer-Verlag, 1988.

Shirokov, P.A. *A Sketch of the Fundamentals of Lobachevskian Geometry.* Trans. Leo F. Boron and Ward D. Bouwsma. Groningen: P. Noordhoff, Ltd., 1964.

Smogorzhevsky, A.S. *Lobachevskian Geometry.* Trans. V. Kisin. Moscow: Mir Publishers, 1976.

Sommerville, D.M.Y. *Bibliography of Non-Euclidean Geometry.* London: Harrison and Sons, 1911.

Stäckel, Paul, ed. *Wolfgang und Johann Bolyai: Geometrische Untersuchungen.* Leipzig: B.G. Teubner, 1913.

Thurston, William P. *Three-Dimensional Geometry and Topology.* Ed. Silvio Levy. Princeton: Princeton UP, 1997.

Tresse, A. *Théorie élémentaire des géométries non euclidiennes.* Paris: Gauthier-Villars, 1957.

Trudeau, Richard J. *The Non-Euclidean Revolution.* Boston: Birkhäuser, 1987.

Turc, Albert. *Introduction élémentaire à la géométrie lobatschewskienne.* Paris: Albert Blanchard, 1967.

Yaglom, I.M. *A Simple Non-Euclidean Geometry and Its Physical Basis.* Trans. Abe Shenitzer and Basil Gordon. New York: Springer-Verlag, 1979.

AWFUL CHANGES.

MAN FOUND ONLY IN A FOSSIL STATE——REAPPEARANCE OF ICHTHYOSAURI.

A Lecture.—"You will at once perceive," continued PROFESSOR ICHTHYOSAURUS, "that the skull before us belonged to some of the lower order of animals; the teeth are very insignificant, the power of the jaws trifling, and altogether it seems wonderful how the creature could have procured food."

14. ROBERT CHAMBERS

(1802–71)

Robert Chambers was a self-educated Scot who, after a downturn in his family's fortunes, established with his brother William a successful publishing company whose hallmark became low-cost literature for a popular audience. Chambers wrote voluminously on literature (*Life of Sir Walter Scott*, 1832), history (*History of the Rebellions in Scotland*, 1828–9), and on a host of miscellaneous topics (*The Book of Days*, 1864). In 1832, the firm of W. and R. Chambers launched the successful *Chambers's Edinburgh Journal*, which contained articles on astronomy, geology, and general science. In 1835, the brothers edited a series of textbooks on natural philosophy, geometry, and chemistry entitled *Chambers's Educational Course*.

When Chambers's *Vestiges of the Natural History of Creation* appeared anonymously in 1844, the storm of controversy was immediate and widespread. Chambers was obliged to answer his critics in *Explanations, by the Author of Vestiges of the Natural History of Creation* (1845). Disraeli satirized the work in *Tancred* (1847) and Tennyson's *In Memoriam* (1833–50) was also influenced by its ideas. Based on fossil evidence, theories of embryonic development, and the nebular hypothesis of Laplace and John Nichol—which proposed that the solar system had condensed and developed from a primordial cloud of hot gas—Chambers argued that plant and animal species had also developed over time from a primeval germ according to a natural law of progression and development. Chambers believed that small changes in individual members of species over time would create, by a branching process, new species unable to breed with the original parent type. The vast majority of naturalists in Chambers's day believed that species were fixed and unchangeable creations, introduced into nature by God, with the appearance of new species or the extinction of former ones only possible through divine fiat.

Although the idea of species transmutation had been proposed earlier by Jean-Baptiste Lamarck and Benoît de Maillet in France and Erasmus Darwin in England, Chambers's *Vestiges* was the first major statement of species evolution in Britain. The *Vestiges* was immediately attacked by conservative clerics for its materialism and for its denial of the direct role of God in creating and destroying species. In a lengthy review of the *Vestiges*, Francis Bowen, for example, asked rhetorically whether the world had been created by "one all-wise and all-powerful Being, or by particles of brute matter, acting of themselves, without direction, interference, or control" (*Theory* 53). The book was equally offensive to professional scientists, such as Adam Sedgwick and T.H. Huxley, who savaged it, not only because of its unfounded superstitions and errors of scientific detail (Chambers believed, for example, in such things as the transmutation of oats into rye), but also because professional scientists may have felt upstaged by an obvious amateur who had at the very least competently assembled a wide variety of evidence necessary for consideriation of the species question. Chambers, to his credit, was well read in geology, such that the label of "amateur" may not be entirely appropriate; he was elected to the Royal Society of Edinburgh, contributed 14 scientific papers on geology to such journals as the *Edinburgh New Philosophical Journal*, and completed a geological study of *Ancient Sea Margins, As Memorials of Changes in the Relative Level of Sea and Land* (1848). William Chambers later wrote about his brother, "his patient investigation, his long and toilsome journeys, and his careful accumulation of facts employed in establishing his geological theories, indicate the true scientific spirit and enthusiasm..." (*Memoir* 351). Many of the major points on which he was attacked had also been advanced by established scientists, but Chambers lacked the detailed field knowledge and intense focus of a Huxley or a Darwin, leaving him open to attack by scientific critics.

With the publication of the *Origin of Species* in 1859, the evolution controversy shifted to Charles Darwin, and Chambers's work was gradually forgotten. In later editions of the *Origin*, Darwin grudgingly acknowledged its influence: "in my opinion it has done excellent service in calling in this country attention to the subject, in removing prejudice, and in thus preparing the ground for the reception of analogous views" (1861 edition, xvi). Chambers was revealed as the author of *Vestiges* in the final 1884 edition.

"Hypothesis of the Development of the Vegetable and Animal Kingdoms."

Vestiges of the Natural History of Creation.

London: John Churchill, 1844, pp. 191–235.

It has been already intimated, as a general fact, that there is an obvious gradation amongst the families of both the vegetable and animal kingdoms, from the simple lichen and animalcule respectively up to the highest order of dicotyledonous trees and the mammalia. Confining our attention, in the meantime, to the animal kingdom—it does not appear that this gradation passes along one line, on which every form of animal life can be, as it were, strung; there may be branching or double lines at some places; or the whole may be in a circle composed of minor circles, as has been recently suggested. But still it is incontestable that there are general appearances of a scale beginning with the simple and advancing to the complicated. The animal kingdom was divided by Cuvier into four sub-kingdoms, or divisions, and these exhibit an unequivocal gradation in the order in which they are here enumerated:—Radiata, (polypes, etc.;) mollusca, (pulpy animals;) articulata, (jointed animals;) vertebrata, (animals with internal skeleton.) The gradation can, in like manner, be clearly traced in the *classes* into which the sub-kingdoms are subdivided, as, for instance, when we take those of the vertebrata in this order—reptiles, fishes, birds, mammals.

While the external forms of all these various animals are so different, it is very remarkable that the whole are, after all, variations of a fundamental plan, which can be traced as a basis throughout the whole, the variations being merely modifications of that plan to suit the particular conditions in which each particular animal has been designed to live. Starting from the primeval germ, which, as we have seen, is the representative of a particular order of full-grown animals, we find all others to be merely advances from that type, with the extension of endowments and modification of forms which are required in each particular case; each form, also, retaining a strong affinity to that which precedes it, and tending to impress its own features on that which succeeds. This unity of structure, as it is called, becomes the more remarkable, when we observe that the organs, while preserving a resemblance, are often put to different uses. For example: the ribs become, in the serpent, organs of locomotion, and the snout is extended, in the elephant, into a prehensile instrument.

It is equally remarkable that analogous purposes are served in different animals by organs essentially different. Thus, the mammalia breathe by lungs; the fishes, by gills. These are not modifications of one organ, but distinct organs. In mammifers, the gills exist and act at an early stage of the foetal state, but afterwards go back and appear no more; while the lungs are developed. In fishes, again, the gills only are fully developed; while the lung structure either makes no advance at all, or only appears in the rudimentary form of an air-bladder. So, also, the baleen of the whale and the teeth of the land mammalia are different organs. The whale, in embryo, shews the rudiments of teeth; but these, not being wanted, are not developed, and the baleen is brought forward instead. The land animals, we may also be sure, have the rudiments of baleen in their organization. In many instances, a particular structure is found advanced to a certain point in a particular set of animals, (for instance, feet in the serpent tribe,) although it is not there required in any degree; but the peculiarity, being carried a little farther forward, is perhaps useful in the next set of animals in the scale. Such are called rudimentary organs. With this class of phenomena are to be ranked the useless mammae of the male human being, and the unrequired process of bone in the male opossum, which is needed in the female for supporting her pouch. Such curious features are most conspicuous in animals which form links between various classes.

As formerly stated, the marsupials, standing at the bottom of the mammalia, shew their affinity to the oviparous vertebrata, by the rudiments of two canals passing from near the anus to the external surfaces of the viscera, which are fully developed in fishes, being required by them for the respiration of aerated waters, but which are not needed by the atmosphere-breathing marsupials. We have also the peculiar form of the sternum and rib-bones of the lizards *represented* in the mammalia in certain white cartilaginous lines traceable among their abdominal muscles. The struphionidae (birds of the ostrich type) form a link between birds and mammalia, and in them we find the wings imperfectly or not at all developed, a diaphragm and urinary sac, (organs wanting in other birds,) and feathers approaching the nature of hair. Again, the ornithorynchus belongs to a class at the bottom of the mammalia, and approximating to birds, and in it behold the bill and web-feet of that order!

For further illustration, it is obvious that, various as may be the lengths of the upper part of the vertebral column in the mammalia, it always consists of the same parts. The giraffe has in its tall neck the same number of bones with the pig, which scarcely appears to have a neck at all.[1] Man, again, has no tail; but the notion a much-ridiculed philosopher of the last century is not altogether, as it happens, without foundation, for the bones of a caudal extremity exist in an undeveloped state in the *os coccygis* of the human subject. The limbs of all the vertebrate animals are, in like manner, on one plan, however various they may appear. In the hind-leg of a horse, for example, the angle called the hock is the same part which in us forms the heel; and the horse, and all other quadrupeds, with almost the solitary exception of the bear, walk, in reality, upon what answers to the toes of a human being. In this and many other quadrupeds the fore part of the extremities is shrunk up in a hoof, as the tail of the human being is shrunk up in the bony mass at the bottom of the back. The bat, on the other hand, has these parts largely developed. The membrane, commonly called its wing, is framed chiefly upon bones answering precisely to those of the human hand; its extinct congener, the pterodactyle, had the same membrane extended upon the fore-finger only, which in that animal was prolonged to an extraordinary extent. In the paddles of the whale and other animals of its order, we see the same bones as in the more highly developed extremities of the land mammifers; and even the serpent tribes, which present no external appearance of such extremities, possess them in reality, but in an undeveloped or rudimental state.

The same law of development presides over the vegetable kingdom. Amongst phanerogamous plants, a certain number of organs appear to be always present, either in a developed or rudimentary state; and those which are rudimentary can be developed by cultivation. The flowers

1 Daubenton established the rule, that all the viviparous quadrupeds have seven vertebrae in the neck.

which bear stamens on one stalk and pistils on another, can be caused to produce both, or to become perfect flowers, by having a sufficiency of nourishment supplied to them. So also, where a special function is required for particular circumstances, nature has provided for it, not by a new organ, but by a modification of a common one, which she has effected in development. Thus, for instance, some plants destined to live in arid situations, require to have a store of water which they may slowly absorb. The need is arranged for by a cup-like expansion round the stalk; in which water remains after a shower. Now the *pitcher*, as this is called, is not a new organ, but simply a metamorphose of a leaf.

These facts clearly shew how all the various organic forms of our world are bound up in one—how a fundamental unity pervades and embraces them all, collecting them, from the humblest lichen up to the highest mammifer, in one system, the whole creation of which must have depended upon one law or decree of the Almighty, though it did not all come forth at one time. After what we have seen, the idea of a separate exertion for each must appear totally inadmissible. The single fact of abortive or rudimentary organs condemns it; for these, on such a supposition, could be regarded in no other light than as blemishes or blunders—the thing of all others most irreconcilable with that idea of Almighty Perfection which a general view of nature so irresistibly conveys. On the other hand, when the organic creation is admitted to have been effected by a general law, we see nothing in these abortive parts but harmless peculiarities of development, and interesting evidences of the manner in which the Divine Author has been pleased to work.

We have yet to advert to the most interesting class of facts connected with the laws of organic development. It is only in recent times that physiologists have observed that each animal passes, in the course of its germinal history, through a series of changes resembling the *permanent forms* of the various orders of animals inferior to it in the scale. Thus, for instance, an insect, standing at the head of the articulated animals, is, in the larva state, a true annelid, or worm, the annelida being the lowest in the same class. The embryo of a crab resembles the perfect animal of the inferior order myriapoda, and passes through all the forms of transition which characterize the intermediate tribes of crustacea. The frog, for some time after its birth, is a fish with external gills, and other organs fitting it for an aquatic life, all of which are changed as it advances to maturity, and becomes a land animal. The mammifer only passes through still more stages, according to its higher place in the scale. Nor is man himself exempt from this law. His first form is that which is permanent in the animalcule. His organization gradually passes through conditions generally resembling a fish, a reptile, a bird, and the lower mammalia, before it attains its specific maturity. At one of the last stages of his foetal career, he exhibits an intermaxillary bone, which is characteristic of the perfect ape; this is suppressed, and he may then be said to take leave of the simial type, and become a true human creature. Even, as we shall see, the varieties of his race are represented in the progressive development of an individual of the highest, before we see the adult Caucasian, the highest point yet attained in the animal scale.

To come to particular points of the organization. The brain of man, which exceeds that of all other animals in complexity of organization and fulness of development, is, at one early period, only "a simple fold of nervous matter, with difficulty distinguishable into three parts, while a little tail-like prolongation towards the hinder parts, and which had been the first to appear, is the only representation of a spinal marrow. Now, in this state it perfectly resembles the brain of an adult fish, thus assuming *in transitu* the form that in the fish is permanent. In a short time, however, the structure is become more complex, the parts more distinct, the spinal marrow better marked; it is now the brain of a reptile. The change continues; by a singular motion, certain parts (*corpora quadragemina*) which had hitherto appeared on the upper surface, now pass towards the lower; the former is their permanent situation in fishes and reptiles, the latter in birds and mammalia. This is another advance in the scale, but more remains yet to be done. The complication of the organ increases; cavities termed *ventricles* are formed, which do not exist in fishes, reptiles, or birds; curiously organized parts, such as the corpora striata, are added; it is now the brain of the mammalia. Its last and final change alone seems wanting, that which shall render it the brain of MAN."[2] And this change in time takes place.

So also with the heart. This organ, in the mammalia, consists of four cavities, but in the reptiles of only three, and in fishes of two only, while in the articulated animals it is merely a prolonged tube. Now in the mammal foetus, at a certain early stage, the organ has the form of a prolonged tube; and a human being may be said to have then the heart of an insect. Subsequently it is shortened and widened, and becomes divided by a contraction into two parts, a ventricle and an auricle; it is now the heart of a fish. A subdivision of the auricle afterwards makes a triple-chambered form, as in the heart of the reptile tribes; lastly, the ventricle being also subdivided, it becomes a full mammal heart.

Another illustration here presents itself with the force of the most powerful and interesting analogy. Some of the earliest fishes of our globe, those of the Old Red Sandstone, present, as we have seen, certain peculiarities, as the one-sided tail and an inferior position of the mouth. No fishes of the present day, in a mature state, are so characterized; but some, at a certain stage of their existence, have such peculiarities. It occurred to a geologist to inquire if the fish which existed before the Old Red Sandstone had any peculiarities assimilating them to the foetal condition of existing fish, and particularly if they were small. The first which occur before the time of the Old Red Sandstone, are those described by Mr. Murchison, as belonging to the Upper Ludlow Rocks; *they are all rather small.* Still older are those detected by Mr. Philips, in the Aymestry Limestone, being the most ancient of the class which have as yet been discovered; *these are so extremely minute as only to be distinguishable by the microscope.* Here we apparently have very clear demonstrations of a parity, or rather identity, of laws presiding over the development of the animated tribes on the face of the earth, and that of the individual in embryo.

The tendency of all these illustrations is to make us look to *development* as the principle which has been immediately concerned in the peopling of this globe, a process extending over a vast space of time, but which is nevertheless connected in character with the briefer process by which an individual being is evoked from a simple germ. What mystery is there here—and how shall I proceed to enunciate the conception which I have ventured to form of what may prove to be its proper solution! It is an idea by no means calculated to impress by its greatness, or to puzzle by its profoundness. It is an idea more marked by simplicity than perhaps any other of those which have explained the great secrets of nature. But in this lies, perhaps, one of its strongest claims to the faith of mankind.

The whole train of animated beings, from the simplest and oldest up to the highest and most recent, are, then, to be regarded as a series of *advances of the principle of development,* which have depended upon external physical circumstances, to which the resulting animals are appropriate. I contemplate the whole phenomena as having been in the first place arranged in the counsels of Divine Wisdom, to take place, not only upon this sphere, but upon all the others in space, under necessary modifications, and as being carried on, from first to last, here and elsewhere, under immediate favour of the creative will or energy.[3] The nucleated vesicle, the fundamental form of all organization, we must regard as the meeting-point between the inorganic and the organic—the end of the mineral and beginning of the vegetable and animal kingdoms, which thence start in different directions, but in perfect parallelism and analogy. We have already seen that this nucleated vesicle is itself a type of mature and independent being in the infusory animalcules, as well as the starting point of the foetal progress of every higher individual in creation, both animal and vegetable. We have seen that it is a form of being which electric agency will produce—though not

2 Lord's Popular Physiology. It is to Tiedemann that we chiefly owe these curious observations; but ground was first broken in this branch of physiological science by Dr. John Hunter.

3 When I formed this idea, I was not aware of one which seems faintly to foreshadow it—namely, Socrates's doctrine, afterwards dilated on by Plato, that "previous to the existence of the world, and beyond its present limits, there existed certain archetypes, the embodiment (if we may use such a word) of general ideas; and that these archetypes were models, in imitation of which all particular beings were created."

perhaps usher into full life—in albumen, one of those compound elements of animal bodies, of which another (urea) has been made by artificial means. Remembering these things, we are drawn on to the supposition, that the first step in the creation of life upon this planet was *a chemico-electric operation, by which simple germinal vesicles were produced*. This is so much, but what were the next steps? Let a common vegetable infusion help us to an answer. There, as we have seen, simple forms are produced at first, but afterwards they become more complicated, until at length the life-producing powers of the infusion are exhausted. Are we to presume that, in this case, the simple engender the complicated? Undoubtedly, this would not be more wonderful as a natural process than one which we never think of wondering at, because familiar to us—namely, that in the gestation of the mammals, the animalcule-like ovum of a few days is the parent, in a sense, of the chick-like form of a few weeks, and that in all the subsequent stages—fish, reptile, etc.—the one may, with scarcely a metaphor, be said to be the progenitor of the other. I suggest, then, as an hypothesis already countenanced by much that is ascertained, and likely to be further sanctioned by much that remains to be known, that the first step was *an advance under favour of peculiar conditions, from the simplest forms of being, to the next more complicated, and this through the medium of the ordinary process of generation*.

Unquestionably, what we ordinarily see of nature is calculated to impress a conviction that each species invariably produces its like. But I would here call attention to a remarkable illustration of natural law which has been brought forward by Mr. Babbage, in his *Ninth Bridgewater Treatise*. The reader is requested to suppose himself seated before the calculating machine, and observing it. It is moved by a weight, and there is a wheel which revolves through a small angle round its axis, at short intervals, presenting to his eye successively, a series of numbers engraved on its divided circumference.

Let the figures thus seen be the series, 1,2,3,4,5, etc., of natural numbers, each of which exceeds its immediate antecedent by unity.

"Now, reader," says Mr. Babbage, "let me ask you how long you will have counted before you are firmly convinced that the engine has been so adjusted, that it will continue, while its motion is maintained, to produce the same series of natural numbers? Some minds are so constituted, that, after passing the first hundred terms, they will be satisfied that they are acquainted with the law. After seeing five hundred terms few will doubt, and after the fifty thousandth term the propensity to believe that the succeeding term will be fifty thousand and one, will be almost irresistible. That term *will* be fifty thousand and one; and the same regular succession will continue; the five millionth and fifty millionth term will still appear in their expected order, and one unbroken chain of natural numbers will pass before your eyes, from *one* up to *one hundred million*.

"True to the vast induction which has been made, the next succeeding term will be one hundred million and one; but the next number presented by the rim of the wheel, instead of being one hundred million and two, is one hundred million *ten thousand* and two. The whole series from the commencement being thus,—

	1	
	2	
	3	
	4	
	5	
	...	
	....	
	.....	
	......	
	99,999,999	
	100,000,000	
regularly as far as	100,000,001	
	100,010,002	the law changes.
	100,030,003	
	100,060,004	
	100,100,005	
	100,150,006	
	100,210,007	
	100,280,008	
	... ...	
	... ...	

"The law which seemed at first to govern this series failed at the hundred million and second term. This term is larger than we expected by

10,000. The next term is larger than was anticipated by 30,000, and the excess of each term above what we had expected forms the following table:—

10,000
30,000
60,000
100,000
150,000
... ...
... ...

being, in fact, the series of *triangular numbers*, [4] each multiplied by 10,000.

"If we now continue to observe the numbers presented by the wheel, we shall find, that for a hundred, or even for a thousand terms, they continue to follow the new law relating to the triangular numbers; but after watching them for 2761 terms, we find that this law fails in the case of the 2762d term.

"If we continue to observe, we shall discover another law then coming into action, which also is dependent, but in a different manner, on triangular numbers. This will continue through about 1430 terms, when a new law is again introduced which extends over about 950 terms, and this, too, like all its predecessors, fails, and gives place to other laws, which appear at different intervals.

"Now it must be observed that *the law that each number presented by the engine is greater by unity than the preceding number*, which law the observer had deduced from an induction of a hundred million instances, *was not the true law that regulated its action*, and that the occurrence of the number 100,010,002 at the 100,000,002nd term was *as necessary a consequence of the original adjustment, and might have been as fully foreknown at the commencement, as was the regular succession of any one of the intermediate numbers to its immediate antecedent*. The same remark applies to the next apparent deviation from the new law, which was founded on an induction of 2761 terms, and also to the succeeding law, with this limitation only—that, whilst their consecutive introduction at various definite intervals, is a necessary consequence of the mechanical structure of the engine, our knowledge of analysis does not enable us to predict the periods themselves at which the more distant laws will be introduced."

It is not difficult to apply the philosophy of this passage to the question under consideration. It must be borne in mind that the gestation of a single organism is the work of but a few days, weeks, or months; but the gestation (so to speak) of a whole creation is a matter probably involving enormous spaces of time. Suppose that an ephemeron, hovering over a pool for its one April day of life, were capable of observing the fry of the frog in the water below. In its aged afternoon, having seen no change upon them for such a long time, it would be little qualified to conceive that the external branchiae of these creatures were to decay, and be replaced by internal lungs, that feet were to be developed, the tail erased, and the animal then to become a denizen of the land. Precisely such may be our difficulty in conceiving that any of the species which people our earth is

4 The numbers 1,3,6,10,15,21,28, etc. are formed by adding the successive terms of the series of natural numbers thus:

1 = 1
1+2=3
1+2+3=6
1+2+3+4=10, etc.

They are called triangular numbers, because a number of points corresponding to any term can always be placed in the form of a triangle; for instance—

capable of advancing by generation to a higher type of being. During the whole time which we call the historical era, the limits of species have been, to ordinary observation, rigidly adhered to. But the historical era is, we know, only a small portion of the entire age of our globe. We do not know what may have happened during the ages which preceded its commencement, as we do not know what may happen in ages yet in the distant future. All, therefore, that we can properly infer from the apparently invariable production of like by like is, that such is the ordinary procedure of nature in the time immediately passing before our eyes. Mr. Babbage's illustration powerfully suggests that this ordinary procedure may be subordinate to a higher law which only *permits* it for a time, and in proper season interrupts and changes it. We shall soon see some philosophical evidence for this very conclusion.

It has been seen that, in the reproduction of the higher animals, the new being passes through stages in which it is successively fish-like and reptile-like. But the resemblance is not to the adult fish or the adult reptile, but to the fish and reptile at a certain point in their foetal progress; this holds true with regard to the vascular, nervous, and other systems alike. It may be illustrated by a simple diagram. The foetus of all the four classes may be supposed to advance in an identical condition to the point A.

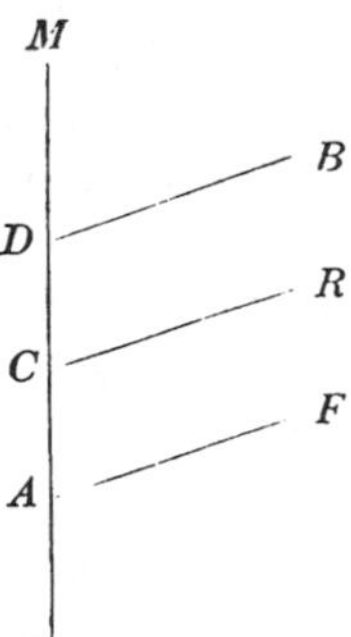

The fish there diverges and passes along a line apart, and peculiar to itself, to its mature state at F. The reptile, bird, and mammal, go on together to C, where the reptile diverges in like manner, and advances by itself to R. The bird diverges at D, and goes on to B. The mammal then goes forward in a straight line to the highest point of organization at M. This diagram shews only the main ramifications; but the reader must suppose minor ones, representing the subordinate differences of orders, tribes, families, genera, etc., if he wishes to extend his views to the whole varieties of being in the animal kingdom. Limiting ourselves at present to the outline afforded by this diagram, it is apparent that the only thing required for an advance from one type to another in the generative process is that, for example, the fish embryo should not diverge at A, but go on to C before it diverges, in which case the progeny will be, not a fish, but a reptile. To protract the *straightforward part of the gestation over a small space*—and from species to species the space would be small indeed—is all that is necessary.

This might be done by the force of certain external conditions operating upon the parturient system. The nature of these conditions we can only conjecture, for their operation, which in the geological eras was so powerful, has in its main strength been long interrupted, and is now perhaps only allowed to work in some of the lowest departments of the organic world, or under extraordinary casualties in some of the higher, and to these points the attention of science has as yet been little directed. But though this knowledge were never to be clearly attained, it need not much affect the present argument, provided it be satisfactorily shewn that there must be some such influence within the range of natural things.

To this conclusion it must be greatly conducive that the law of organic development is still daily seen at work to certain effects, only somewhat short of a transition from species to species. Sex we have seen to be a matter of development. There is an instance, in a humble department of the animal world, of arrangements being made by the animals themselves for adjusting this law to the production of a particular sex. Amongst bees, as amongst several other insect tribes, there is in each community but one true female, the queen bee, the workers being false females or neuters; that is to say, sex is carried on in them to a point where it is attended by sterility. The preparatory states of the queen bee occupy sixteen days; those of the neuters, twenty; and those of males, twenty-four. Now it is a fact, settled by innumerable observations and experiments, that the bees can so modify a worker in the larva state, that, when it emerges from the pupa, it is found to be a queen or true female. For this purpose they enlarge its cell, make a pyramidal hollow to allow of its assuming a vertical instead of a horizontal position, keep it warmer than other larvae are kept, and feed it with a peculiar kind of food. From these simple circumstances, leading to a shortening of the embryotic condition, results a creature different in form, and also in dispositions, from what would have otherwise been produced. Some of the organs possessed by the worker are here altogether wanting. We have a creature "destined to enjoy love, to burn with jealousy and anger, to be incited to vengeance, and to pass her time without labour," instead of one "zealous for the good of the community, a defender of the public rights, enjoying an immunity from the stimulus of sexual appetite and the pains of parturition; laborious, industrious, patient, ingenious, skilful; incessantly engaged in the nurture of the young, in collecting honey and pollen, in elaborating wax, in constructing cells and the like!—paying the most respectful and assiduous attention to objects which, had its ovaries been developed, it would have hated and pursued with the most vindictive fury till it had destroyed them!"[5] All these changes may be produced by a mere modification of the embryotic progress, which it is within the power of the adult animals to effect. But it is important to observe that this modification is different from working a direct change upon the embryo. It is not the different food which effects a metamorphosis. All that is done is merely to accelerate the period of the insect's perfection. By the arrangements made and the food given, the embryo becomes sooner fit for being ushered forth in its imago or perfect state. Development may be said to be thus arrested at a particular stage—that early one at which the female sex is complete. In the other circumstances, it is allowed to go on four days longer, and a stage is then reached between the two sexes, which in this species is designed to be the perfect condition of a large portion of the community. Four days more make it a perfect male. It is at the same time to be observed that there is, from the period of oviposition, a destined distinction between the sexes of the young bees. The queen lays the whole of the eggs which are designed to become workers, before she begins to lay those which become males. But probably the condition of her reproductive system governs the matter of sex, for it is remarked that when her impregnation is delayed beyond the twenty-eighth day of her entire existence, she lays only eggs which become males.

We have here, it will be admitted, a most remarkable illustration of the principle of development, although in an operation limited to the production of sex only. Let it not be said that the phenomena concerned in the generation of bees may be very different from those concerned in the reproduction of the higher animals. There is a unity throughout nature which makes the one case an instructive reflection of the other.

We shall now see an instance of development operating within the production of what approaches to the character of variety of species. It is fully established that a human family, tribe, or nation, is liable, in the course of generations, to be either advanced from a mean form to a higher one, or degraded from a higher to a lower, by the influence of the physical conditions in which it lives. The coarse features, and other structural peculiarities of the negro race only continue while these people live amidst the circumstances usually associated with barbarism. In a more temperate clime, and higher social state, the face and figure become greatly refined. The few African nations which possess any civilization also exhibit forms approaching the European; and when the same people in the United States of America have enjoyed a within-door life for several generations, they assimilate to the whites amongst whom they live. On the other hand, there are authentic instances of a people originally well-formed and good-looking, being brought, by imperfect diet and a variety of physical hardships, to a meaner form. It is remarkable that prominence of the jaws, a recession and diminution of the cranium, and an elongation and attenuation of the limbs, are peculiarities always produced by these miserable conditions, for they

5 Kirby and Spence.

indicate an unequivocal retrogression towards the type of the lower animals. Thus we see nature alike willing to go back and to go forward. Both effects are simply the result of the operation of the law of development in the generative system. Give good conditions, it advances; bad ones, it recedes. Now, perhaps, it is only because there is no longer a possibility, in the higher types of being, of giving sufficiently favourable conditions to carry on species to species, that we see the operation of the law so far limited.

Let us trace this law also in the production of certain classes of monstrosities. A human foetus is often left with one of the most important parts of its frame imperfectly developed: the heart, for instance, goes no farther than the three-chambered form, so that it is the heart of a reptile. There are even instances of this organ being left in the two-chambered or fish form. Such defects are the result of nothing more than a failure of the power of development in the system of the mother, occasioned by weak health or misery. Here we have apparently a realization of the converse of those conditions which carry on species to species, so far, at least, as one organ is concerned. Seeing a complete specific retrogression in this one point, how easy it is to imagine an access of favourable conditions sufficient to reverse the phenomenon, and make a fish mother develop a reptile heart, or a reptile mother develop a mammal one. It is no great boldness to surmise that a super-adequacy in the measure of this under-adequacy (and the one thing seems as natural an occurrence as the other) would suffice in a goose to give its progeny the body of a rat, and produce the ornithorynchus, or might give the progeny of an ornithorynchus the mouth and feet of a true rodent, and thus complete at two stages the passage from the aves to the mammalia.

Perhaps even the transition from species to species does still take place in some of the obscurer fields of creation, or under extraordinary casualties, though science professes to have no such facts on record. It is here to be remarked, that such facts might often happen, and yet no record be taken of them, for so strong is the prepossession for the doctrine of invariable like-production, that such circumstances, on occurring, would be almost sure to be explained away on some other supposition, or, if presented, would be disbelieved and neglected. Science, therefore, has no such facts, for the very same reason that some small sects are said to have no discreditable members—namely, that they do not receive such persons, and extrude all who begin to verge upon the character. There are, nevertheless, some facts which have chanced to be reported without any reference to this hypothesis, and which it seems extremely difficult to explain satisfactorily upon any other. One of these has already been mentioned—a progression in the forms of the animalcules in a vegetable infusion from the simpler to the more complicated, a sort of microcosm, representing the whole history of the progress of animal creation as displayed by geology. Another is given in the history of the Acarus Crossii, which may be only the ultimate stage of a series of similar transformations effected by electric agency in the solution subjected to it. There is, however, one direct case of a translation of species, which has been presented with a respectable amount of authority.[6] It appears that, whenever oats sown at the usual time are kept cropped down during summer and autumn, and allowed to remain over the winter, a thin crop of rye is the harvest presented at the close of the ensuing summer. This experiment has been tried repeatedly, with but one result; invariably the *secale cereale* is the crop reaped where the *avena sativa*, a recognised different species, was sown. Now it will not satisfy a strict inquirer to be told that the seeds of the rye were latent in the ground and only superseded the dead product of the oats; for if any such fact were in the case, why should the usurping grain be always rye? Perhaps those curious facts which have been stated with regard to forests of one kind of trees, when burnt down, being succeeded (without planting) by other kinds, may yet be found most explicable, as this is, upon the hypothesis of a progression of species which takes place under certain favouring conditions, now apparently of comparatively rare occurrence. The case of the oats is the more valuable, as bearing upon the suggestion as to a

6 See an article by Dr. Weissenborn, in the New Series of "Magazine of Natural History," vol. i. p. 574.

protraction of the gestation at a particular part of its course. Here, the generative process is, by the simple mode of cropping down, kept up for a whole year beyond its usual term. The type is thus allowed to advance, and what was oats becomes rye.

The idea, then, which I form of the progress of organic life upon the globe—and the hypothesis is applicable to all similar theatres of vital being—is, *that the simplest and most primitive type, under a law to which that of like-production is subordinate, gave birth to the type next above it, that this again produced the next higher, and so on to the very highest,* the stages of advance being in all cases very small—namely, from one species only to another; so that the phenomenon has always been of a simple and modest character. Whether the whole of any species was at once translated forward, or only a few parents were employed to give birth to the new type, must remain undetermined; but, supposing that the former was the case, we must presume that the moves along the line or lines were simultaneous, so that the place vacated by one species was immediately taken by the next in succession, and so on back to the first, for the supply of which the formation of a new germinal vesicle out of inorganic matter was alone necessary. Thus, the production of new forms, as shewn in the pages of the geological record, has never been anything more than a new stage of progress in gestation, an event as simply natural, and attended as little by any circumstances of a wonderful or startling kind, as the silent advance of an ordinary mother from one week to another of her pregnancy. Yet, be it remembered, the whole phenomena are, in another point of view, wonders of the highest kind, for in each of them we have to trace the effect of an Almighty Will which had arranged the whole in such harmony with external physical circumstances, that both were developed in parallel steps—and probably this development upon our planet is but a sample of what has taken place, through the same cause, in all the other countless theatres of being which are suspended in space.

This may be the proper place at which to introduce the preceding illustrations in a form calculated to bring them more forcibly before the mind of the reader. The following table was suggested to me, in consequence of seeing the scale of animated nature presented in Dr. Fletcher's *Rudiments of Physiology*. Taking that scale as its basis, it shews the wonderful parity observed in the progress of creation, as presented to our observation in the succession of fossils, and also in the foetal progress of one of the principal human organs.[7] This scale, it may be remarked, was not made up with a view to support such an hypothesis as the present, nor with any apparent regard to the history of fossils, but merely to express the appearance of advancement in the orders of the Cuvierian system, assuming, as the criterion of that advancement, "an increase in the number and extent of the manifestations of life, or of the relations which an organized being bears to the external world." Excepting in the relative situation of the annelida and a few of the mammal orders, the parity is perfect; nor may even these small discrepancies appear when the order of fossils shall have been further investigated, or a more correct scale shall have been formed. Meanwhile, it is a wonderful evidence in favour of our hypothesis, that a scale formed so arbitrarily should coincide to such a nearness with our present knowledge of the succession of animal forms upon earth, and also that both of these series should harmonize so well with the view given by modern physiologists of the embryotic progress of one of the organs of the highest order of animals.

The reader has seen physical conditions several times referred to, as to be presumed to have in some way governed the progress of the development of the zoological circle. This language may seem vague, and, it may be asked,—can any par-

7 "It is a fact of the highest interest and moment that as the brain of every tribe of animals appears to pass, during its development, in succession through the types of all those below it, so the brain of man passes through the types of those of every tribe in the creation. It represents, accordingly, before the second month of utero-gestation, that of an avertebrated animal; at the second month, that of an osseous fish; at the third, that of a turtle; at the fourth, that of a bird; at the fifth, that of one of the rodentia; at the sixth, that of one of the ruminantia; at the seventh, that of one of the digitigrada; at the eighth, that of one of the quadrumana;

ticular physical condition be adduced as likely to have affected development? To this it may be answered, that air and light are probably amongst the principal agencies of this kind which operated in educing the various forms of being. Light is found to be essential to the development of the individual embryo. When tadpoles were placed in a perforated box, and that box sunk in the Seine, light being the only condition thus abstracted, they grew to a great size in their original form, but did not pass through the usual metamorphose which brings them to their mature state as frogs. The proteus, an animal of the frog kind, inhabiting the subterraneous waters of Carniola, and which never acquires perfect lungs so as to become a land animal, is presumed to be an example of arrested development, from the same cause. When, in connexion with these facts, we learn that human mothers living in dark and close cells under ground,—that is to say, with an inadequate provision of air and light,—are found to produce an unusual proportion of defective children,[8] we can appreciate the important effects of both these physical conditions in ordinary reproduction. Now there is nothing to forbid the supposition that the earth has been at different stages of its career under different conditions, as to both air and light. On the contrary, we have seen reason for supposing that the proportion of carbonic acid gas (the element fatal to animal life) was larger at the time of the carboniferous formation than it afterwards became. We have also seen that astronomers regard the zodiacal light as a residuum of matter enveloping the sun, and which was probably at one time denser than it is now. Here we have the indications of causes for a progress in the purification of the atmosphere and in the diffusion of light during the earlier ages of the earth's history, with which the progress of organic life may have been conformable. An accession to the proportion of oxygen, and the effulgence of the central luminary, may have been the immediate prompting cause of all those advances from species to species which we have seen, upon other grounds, to be necessarily supposed as having taken place. And causes of the like nature may well be supposed to operate on other spheres of being, as well as on this. I do not indeed present these ideas as furnishing the true explanation of the progress of organic creation; they are merely thrown out as hints towards the formation of a just hypothesis, the completion of which is only to be looked for when some considerable advances shall have been made in the amount and character of our stock of knowledge.

Early in this century, M. Lamarck, a naturalist of the highest character, suggested an hypothesis of organic progress which deservedly incurred much ridicule, although it contained a glimmer of the truth. He surmised, and endeavoured, with a great deal of ingenuity, to prove, that one being

---

till at length, at the ninth, it compasses the brain of Man! It is hardly necessary to say, that all this is only an approximation to the truth; since neither is the brain of all osseous fishes, of all turtles, of all birds, nor of all the species of any one of the above order of mammals, by any means precisely the same, nor does the brain of the human foetus at any time precisely resemble, perhaps, that of any individual whatever among the lower animals. Nevertheless, it may be said to represent, at each of the above-mentioned periods, the aggregate, as it were, of the brains of each of the tribes stated; consisting as it does, about the second month, chiefly of the mesial parts of the cerebellum, the corpora quadrigemina, thalami optici, rudiments of the hemispheres of the cerebrum and corpora striata; and receiving in succession, at the third, the rudiments of the lobes of the cerebrum; at the fourth, those of the fornix, corpus callosum, and septum lucidum; at the fifth, the tubor annulare, and so forth; the posterior lobes of the cerebrum increasing from before to behind, so as to cover the thalami optici about the fourth month, the corpora quadrigemina about the sixth, and the cerebellum about the seventh. This, then, is another example of an increase in the complexity of an organ succeeding its centralization; as if Nature, having first piled up her materials in one spot, delighted afterwards to employ her abundance, not so much in enlarging old parts as in forming new ones upon the old foundations, and thus adding to the complexity of a fabric, the rudimental structure of which is in all animals equally simple."—*Fletcher's Rudiments of Physiology.*

8 Some poor people having taken up their abode in the cells under the fortifications of Lisle, the proportion of defective infants produced by them became so great, that it was deemed necessary to issue an order commanding these cells to be shut up.

226 HYPOTHESIS OF THE DEVELOPMENT OF THE VEGETABLE AND ANIMAL KINGDOMS. 227

SCALE OF ANIMAL KINGDOM. (The numbers indicate orders:)	ORDER OF ANIMALS IN	ASCENDING SERIES OF ROCKS.	FŒTAL HUMAN BRAIN RESEMBLES, IN
		1 Gneiss and Mica Slate system	1st month, that of an avertebrated animal;
RADIATA (1, 2, 3, 4, 5)	Zoophyta; Polypiaria	2 Clay Slate and Grawacke system	
MOLLUSCA (6, 7, 8, 9, 10, 11)	Conchifera; Double-shelled Mollusks	3 Silurian system	
ARTICULATA: *Annelida* (12, 13, 14); *Crustacea* (15, 16, 17, 18, 19, 20); *Arachnida* & *Insecta* (21—31)	Crustacea; Annelida; Crustaceous Fishes	4 Old Red Sandstone	
VERTEBRATA: *Pisces* (32, 33, 34, 35, 36)	True Fishes	5 Carboniferous formation	2nd month, that of a fish;
Reptilia (37, 38, 39, 40)	Piscine Saurians (ichthyosaurus, &c.); Pterodactyles; Crocodiles; Tortoises; Batrachians	6 New Red Sandstone	3rd month, that of a turtle;
Aves (41, 42, 43, 44, 45, 46)	Birds		4th month, that of a bird;
Mammalia: 47 Cetacea; 48 Ruminantia	(Bone of a marsupial animal)	7 Oolite	
		8 Cretaceous formation	
49 Pachydermata	Pachydermata (tapirs, horses, &c.)	9 Lower Eocene	
50 Edentata			
51 Rodentia	Rodentia (dormouse, squirrel, &c.)		5th month, that of a rodent;
52 Marsupialia	Marsupialia (racoon, opossum, &c.)		6th month, that of a ruminant;
53 Amphibia			
54 Digitigrada	Digitigrada (genette, fox, wolf, &c.)	10 Miocene	7th month, that of a digitigrade animal;
55 Plantigrada	Plantigrada (bear)		
	Cetacea (lamantins, seals, whales)		
56 Insectivora	Edentata (sloths, &c.)	11 Pliocene	
	Ruminantia (oxen, deer, &c.)		
57 Cheiroptera			
58 Quadrumana	Quadrumana (monkeys)		8th month, that of the quadrumana;
59 Bimana	Bimana (man)	12 Superficial deposits	9th month, attains full human character.

advanced in the course of generations to another, in consequence merely of its experience of wants calling for the exercise of its faculties in a particular direction, by which exercise new developments of organs took place, ending in variations sufficient to constitute a new species. Thus he thought that a bird would be driven by necessity to seek its food in the water, and that, in its efforts to swim, the outstretching of its claws would lead to the expansion of the intermediate membranes, and it would thus become web-footed. Now it is possible that wants and the exercise of faculties have entered in some manner into the production of the phenomena which we have been considering; but certainly not in the way suggested by Lamarck, whose whole notion is obviously so inadequate to account for the rise of the organic kingdoms, that we can only place it with pity among the follies of the wise. Had the laws of organic development been known in his time, his theory might have been of a more imposing kind. It is upon these that the present hypothesis is mainly founded. I take existing natural means, and shew them to have been capable of producing all the existing organisms, with the simple and easily conceivable aid of a higher generative law, which we perhaps still see operating upon a limited scale. I also go beyond the French philosopher to a very important point, the original Divine conception of all the forms of being which these natural laws were only instruments in working out and realizing. The actuality of such a conception I hold to be strikingly demonstrated by the discoveries of Macleay, Vigors, and Swainson, with respect to the affinities and analogies of animal (and by implication vegetable) organisms.[9] Such a regularity in the *structure*, as we may call it, of the *classification of animals*, as is shewn in their systems, is totally irreconcilable with the idea of form going on to form merely as needs and wishes in the animals themselves dictated. Had such been the case, all would have been irregular, as things arbitrary necessarily are. But, lo, the whole plan of being is as symmetrical as the plan of a house, or the laying out of an old-fashioned garden! This must needs have been devised and arranged for beforehand. And what a preconception or forethought have we here! Let us only for a moment consider how various are the external physical conditions in which animals live—climate, soil, temperature, land, water, air—the peculiarities of food, and the various ways in which it is to be sought; the peculiar circumstances in which the business of reproduction and the care-taking of the young are to be attended to—all these required to be taken into account, and thousands of animals were to be formed suitable in organization and mental character for the concerns they were to have with these various conditions and circumstances—here a tooth fitted for crushing nuts; there a claw fitted to serve as a hook for suspension; here to repress teeth and develop a bony net-work instead; there to arrange for a bronchial apparatus, to last only for a certain brief time; and all these animals were to be schemed out, each as a part of a great range, which was on the whole to be rigidly regular: let us, I say, only consider these things, and we shall see that the decreeing of laws to bring the whole about was an act involving such a degree of wisdom and device as we only can attribute, adoringly, to the one Eternal and Unchangeable. It may be asked, how does this reflection comport with that timid philosophy which would have us to draw back from the investigation of God's works, lest the knowledge of them should make us undervalue his greatness and forget his paternal character? Does it not rather appear that our ideas of the Deity can only be worthy of him in the ratio in which we advance in a knowledge of his works and ways; and that the acquisition of this knowledge is consequently an available means of our growing in a genuine reverence for him!

But the idea that any of the lower animals have been concerned in any way with the origin of man—is not this degrading? Degrading is a term, expressive of a notion of the human mind, and the human mind is liable to prejudices which prevent its notions from being invariably correct. Were we acquainted for the first time with the circumstances attending the production of an individual of our race, we might equally think them degrading, and be eager to deny them, and

9 These affinities and analogies are explained in the next chapter.

exclude them from the admitted truths of nature. Knowing this fact familiarly and beyond contradiction, a healthy and natural mind finds no difficulty in regarding it complacently. Creative Providence has been pleased to order that it should be so, and it must therefore be submitted to. Now the idea as to the progress of organic creation, if we become satisfied of its truth, ought to be received precisely in this spirit. It has pleased Providence to arrange that one species should give birth to another, until the second highest gave birth to man, who is the very highest: be it so, it is our part to admire and to submit. The very faintest notion of there being anything ridiculous or degrading in the theory—how absurd does it appear, when we remember that every individual amongst us actually passes through the characters of the insect, the fish, the reptile, (to speak nothing of others,) before he is permitted to breathe the breath of life! But such notions are mere emanations of false pride and ignorant prejudice. He who conceives them little reflects that they, in reality, involve the principle of a contempt for the works and ways of God, For it may be asked, if He, as appears, has chosen to employ inferior organisms as a generative medium for the production of higher ones, even including ourselves, what right have we, his humble creatures, to find fault? There is, also, in this prejudice, an element of unkindliness towards the lower animals, which is utterly out of place. These creatures are all of them part products of the Almighty Conception, as well as ourselves. All of them display wondrous evidences of his wisdom and benevolence. All of them have had assigned to them by their Great Father a part in the drama of the organic world, as well as ourselves. Why should they be held in such contempt? Let us regard them in a proper spirit, as parts of the grand plan, instead of contemplating them in the light of frivolous prejudices, and we shall be altogether at a loss to see how there should be any degradation in the idea of our race having been genealogically connected with them.

FURTHER READING

Bosanquet, S.R. *"Vestiges of the Natural History of Creation:" Its Argument Examined and Exposed.* London: John Hatchard and Son, 1845.

Bowen, Francis. *A Theory of Creation: A Review of "Vestiges of the Natural History of Creation."* Boston: Otis, Broaders, and Company, 1845.

Chambers, Robert. *Vestiges of the Natural History of Creation.* Introduced by Gavin de Beer. 1844. Reprint, Leicester: Leicester UP, 1969.

—. *Vestiges of the Natural History of Creation and Other Evolutionary Writings.* Edited and Introduced by James A. Secord. 1844. Reprint, Chicago: University of Chicago Press, 1994.

Chambers, William. *Memoir of Robert Chambers With Autobiographical Reminiscences of William Chambers.* Edinburgh: W. and R. Chambers, 1872.

Hodge, M.J.S. "The Universal Gestation of Nature: Chambers' *Vestiges* and *Explanations*." *Journal of the History of Biology* 5, no. 1 (1972): 127–51.

Layman, C.H., ed. *Man of Letters: The Early Life and Love Letters of Robert Chambers.* Edinburgh: Edinburgh UP, 1990.

Leonard, David Charles. "Tennyson, Chambers, and Recapitulation." *Victorian Newsletter* 56 (1979): 7–10.

Miller, Hugh. *The Footprints of the Creator; or, the Asterolepis of Stromness, With a Memoir of the Author by Louis Agassiz.* New York: Hurst, 1850.

Millhauser, Milton. *Just Before Darwin: Robert Chambers and Vestiges.* Middletown, CT: Wesleyan UP, 1959.

Raub, Cymbre Quincy. "Robert Chambers and William Whewell: A Nineteenth-Century Debate Over the Origin of Language." *Journal of the History of Ideas* 49, no. 2 (1988): 287–300.

Schwartz, Joel S. "Darwin, Wallace, and Huxley, and *Vestiges of the Natural History of Creation*." *Journal of the History of Biology* 23, no. 1 (1990): 127–53.

Secord, James A. "Behind the Veil: Robert Chambers and *Vestiges*." In *History, Humanity and Evolution: Essays for John C. Greene.* Edited by

James R. Moore. Cambridge: Cambridge UP, 1989, pp. 165–94.

Yeo, Richard. "Science and Intellectual Authority in Mid-Nineteenth-Century Britain: Robert Chambers and *Vestiges of the Natural History of Creation*." In *Energy and Entropy: Science and Culture in Victorian Britain*. Edited by Patrick Brantlinger. Bloomington: Indiana UP, 1989, pp. 1–27.

15. GEORGE COMBE

(1788–1858)

George Combe was an Edinburgh lawyer who retired from the legal profession in 1837 in order to devote his life to popularizing the new science of phrenology, coined from the Greek φρήν-, "mind." His work *The Constitution of Man Considered in Relation to External Objects* (1828; revised 1847), an essay on human behaviour, psychology, and morality founded upon phrenological principles, was one of the most popular books of the nineteenth century, selling close to 300,000 copies world-wide. Combe also wrote the popular *System of Phrenology* in 1825.

Modern phrenology can be specifically traced to the work of Franz Joseph Gall (1758–1828), a Viennese physician, and his student J.G. Spurzheim (1776–1832), who popularized Gall's theories. Gall and Spurzheim published together their *Anatomie et physiologie du système nerveux en général, et du cerveau en particulier, avec des observations sur la possibilité de reconnoître plusiers dispositions intellectuelles et morales de l'homme et des animaux, par la configuration de leurs têtes* (1810–19). The intellectual roots of phrenology, however, stretch back to the ancient art of physiognomy, eighteenth-century faculty psychology, and earlier medieval divisions of the brain into ventricles containing the powers of *memoria*, *phantasia*, and *cogitatio*. After attending Spurzheim's brain dissections and phrenological lectures in Edinburgh, Combe became a firm adherent of the science. George's brother Andrew Combe (known more for his works on physiology) also wrote on phrenology in his *Observations on Mental Derangement* (1831) and *The Principles of Physiology Applied to the Preservation of Health* (1834).

The basic doctrines of phrenological science are "that the brain is the organ of mind; that size is a measure of power, other things being equal; that the formation of the skull bears a relation to the character of the mind, and that efficient moral and intellectual training will influence the development and action of the brain, as physical exercise affects the power and size of the muscles" (Gibbons xiii–xiv). Phrenologists divided the brain into anywhere from 25–37 faculties such as Secretiveness, Self-Esteem, Amativeness, Philoprogenitiveness (affection for young and tender beings), Veneration, etc., and believed that human character could be determined by the size (size being a function of the degree of activity) of the organs of these faculties as indicated by the shape of the skull. Some nineteenth-century phrenologists also classified human character using a modified form of the four temperaments of the Galenic and Hippocratic humour physiology of antiquity, identifying a lymphatic, sanguine, bilious, and nervous disposition or complexion.

The empirical—hence scientific—nature of phrenology, based on correlating craniological measurements with character, was constantly emphasized by its practitioners. Combe, for example, believed that Gall's theories originated in pure observation: "the order of Dr. Gall's discoveries was the following. He *first* distinguished different mental talents and dispositions in his brothers, sisters, and school-fellows; *secondly*, he observed differences in the forms of their heads; *thirdly*, he ascertained that the forms indicated particular developments of brain; and, *lastly*, he ascertained, by extensive observation, that particular forms and particular talents or dispositions, were concomitant in all sane and healthy individuals" (*Letter From George Combe* 34).

By the 1830s, phrenology had gained a large popular following in England, France, and Germany, spawning lectures, journals, and societies. Phrenology also swept through the U.S. beginning in the 1830s, and Clara Barton, U.S. Grant, Walt Whitman, and James A. Garfield all sat for phrenological readings. In New York, the brothers Lorenzo and O.S. Fowler along with Samuel R. Wells organized the booking agency of Fowlers and Wells for phrenological lecturers in 1843, and also published *The American Phrenological Journal*, which reached a circulation of 20,000 in 1847. But the medical and scientific community who had originally looked favourably on phrenology began to abandon the movement by the 1840s, especially when phrenology, derided as "bumpology," became a lucrative business riddled with fraud and outright quackery.

Although phrenology and George Combe in particular were attacked for promoting material atheism and subverting religion (an official charge against Gall in Austria), Combe persevered and became active in the secular education debate, because he strongly believed that negative character traits could be traced to the underdevelopment or misuse of the corresponding mental organ, which could consequently be strengthened by proper vigorous exercise. Roger Cooter has detailed how phrenology became intertwined with a variety of other liberal reform movements in education, penology, and the treatment of the insane. Most phrenologists believed that criminals and the insane were basically reformable through education, medicine, and therapy since they were ultimately suffering from atrophied or improperly exercised brains. Combe directly influenced U.S. educator Horace Mann and Charles Caldwell applied phrenological principles to prison reform.

The influence of phrenology on later activity in the human sciences cannot be denied: anthropology, psychometrics, criminology, and behavioural psychology took up questions originally posed by the phrenologists. In the field of neuro-anatomy, the French neurologist Paul Broca related a specific region of the left frontal lobe to the production of speech—a portion of the brain now known as Broca's Area. Ironically Broca, an avowed anti-phrenologist, had vindicated Gall's general principle of functional localization. Skull and brain measurement continued unabated into the twentieth century. And even today, the idea that physical and biochemical (material) processes completely govern emotional, cognitive, and behavioural development has its adherents in various research communities.

The unprecedented popularity of Combe's book, however, perhaps rests more on its utopian vision of a scientifically-based system which could predict and therefore regulate human behaviour on the basis of easily obtained measurements of the skull. In *The Constitution of Man*, Combe also flattered his audience by reinforcing many of the prejudices of the day: for example, Europeans are praised for their rationality, and non-Europeans castigated for their lack of mental development. Combe also took a fashionable, although paternalistic, stand against human slavery. Combe provided his age with a comforting moral handbook, and rules for predicting human behaviour, stamped with the approval of an allegedly empirical science.

The Constitution of Man Considered in Relation to External Objects.
8th ed. Edinburgh: MacLachlan, Stewart, and Co., 1847, pp. 443–61.

CONCLUSION

The question has frequently been asked, What is the practical use of Phrenology, even supposing it to be true? A few observations will suffice to answer this inquiry, and, at the same time, to present a brief summary of the doctrine of the preceding work.

Prior to the age of Copernicus, the earth and sun presented to the eye phenomena exactly similar to those which they now exhibit; but their motions appeared in a very different light to the understanding.

Before the age of Newton, the revolutions of the planets were known as matter of fact; but mankind was ignorant of the principle of their motions.

Previously to the dawn of modern chemistry, many of the qualities of physical substances were ascertained by observation: but their ultimate principles and relations were not understood.

Knowledge, as I observed in the Introduction, may be made beneficial in two ways—either by rendering the substance discovered directly subservient to human enjoyment; or, where this is impossible, by enabling man to modify his conduct in harmony with its qualities. While knowledge of any department of nature remains imperfect and empirical, the unknown qualities of the objects comprehended in it, may render our efforts either to apply, or to act in accordance with those which are known, altogether abortive. Hence it is only after ultimate qualities and modes of action of nature have been discovered, their relations ascertained, and this knowledge systematized, that science can attain its full character of utility. The merits of Copernicus and Newton consist in having rendered this service to astronomy.

Before the appearance of Drs Gall and Spurzheim, mankind were practicably acquainted with the feelings and intellectual operations of their own minds, and anatomists knew the appearances of the brain. But the science of mind was very much in the same state as that of the heavenly bodies prior to the times of Copernicus and Newton.

First, No unanimity prevailed among philosophers concerning the elementary feelings and intellectual powers of man. Individuals deficient in Conscientiousness, for instance, denied that the sentiment of justice is a primitive mental quality: others, deficient in Veneration, asserted that man is not naturally prone to worship, and ascribed religion to the invention of priests.

Secondly, The extent to which the primitive faculties differ in strength, in different individuals, was matter of dispute, or of vague conjecture; and, concerning many attainments, there was no agreement among philosophers whether they were the gifts of nature or the results of mere cultivation.

Thirdly, Different modes or states of the same feeling were often mistaken for different feelings; and modes of action of all the intellectual faculties were mistaken for distinct faculties.

Fourthly, The brain, confessedly the most important organ of the body, and that with which the nerves of the senses, of motion, and of feeling communicate, had no ascertained functions. Mankind were ignorant of its uses, and of its influence on the mental faculties. They indeed still dispute that its different parts are the organs of different mental powers, and that the vigour of each faculty bears a proportion, *coeteris paribus*, to the size of its organ.

If, in physics, imperfect and empirical knowledge renders the unknown qualities of bodies liable to frustrate the efforts of man to apply or to accommodate his conduct to their known qualities,—and if science becomes useful only in proportion as it attains to a complete and systematic exhibition of ultimate principles, and their relations,—the same doctrine applies with equal or greater force to the philosophy of mind.

The science of POLITICS embraces forms of government, and the relations between different states. All government is designed to combine the efforts of individuals, and to regulate their conduct when united. To arrive at the best means of accomplishing this end, systematic knowledge of the nature of man seems to be highly important. A despotism, for example, may restrain some abuses of the propensities, but it assuredly impedes the exercise of reflection, and others of the highest and noblest powers. A form of government can be suited to the nature of man only when it is calculated to permit the legitimate use, and to restrain the abuses, of all his mental feelings and capacities: and how can such a government be devised, while these faculties, with their spheres of action and external relations, are imperfectly known? Again, all relations between different states must also be in accordance with the nature of man, to prove permanently beneficial; and the question recurs, How are these to be framed while that nature is a matter of conjecture? Napoleon disbelieved in a sentiment of justice as an innate quality of the mind, and, in his relations with other states, relied on fear and interest as the grand motives of conduct: but that sentiment existed, and, combined with other faculties which he outraged, prompted Europe to hurl him from his throne. If Napoleon had comprehended the principles of human nature, and their relations, as forcibly and clearly as he did the principles of mathematics, in which he excelled, his understanding would have greatly modified his conduct, and Europe would have escaped numerous calamities.

LEGISLATION, civil and criminal, is intended to regulate and direct the human faculties in their efforts at gratification; and laws, to be useful, must accord with the constitution of these faculties. But how can salutary laws be enacted, while the subject to be governed, or human nature, is not accurately understood? The inconsistency and intricacy of the laws, even in enlightened nations, have afforded themes for the satirist in every age;—yet how could the case be otherwise? Legislators provided rules for directing the qualities of human nature, which they conceived themselves to know; but either error in their conceptions, or the effects of other qualities unknown or unattended to, defeated their intentions. The law, for example, punishing heresy with burning, was addressed by our ancestors to Cautiousness and the Love of Life; but, Intellect, Veneration, Conscientiousness, and Firmness, were omitted in their estimate of human principles of action;—and these set the law at defiance. There are many laws still in the statute-book, equally at variance with the nature of man.[1]

EDUCATION is intended to enlighten the intellect, to train it and the moral sentiments to vigour, and to repress the too great activity of the selfish feelings. But how can this be successfully accomplished, when the faculties and sentiments themselves, the laws to which they are subjected, and their relations to external objects are unascertained? Accordingly, the theories and practices observed in education are innumerable and contradictory; which could not happen if men knew the constitution and relations of the object which they were training.

In an "Essai sur la Statistique morale de la France," by Mons. A.M. Guerry, published at Paris in 1833, it is stated that crimes against property and person are most numerous in proportion to the population in those departments of France—the north and east—in which the people are the best educated, the richest, and the most industrious. This must be owing in part to the increased power which education confers of doing either good or evil, and partly to defects in the education afforded.

It is proper to remark, moreover, that M. Guerry's statement, supposing it to be grounded on sufficient data, does not shew that education tends to increase rather than diminish crime; for, as a writer in the Phrenological Journal observes, "until it be proved that education has the same kind of subjects to operate on in every part of France, its effects cannot be judged of from such data as those furnished by M. Guerry." After stating reasons for concluding that the generality of heads are better in some parts of France than in others, the writer adds: "Now, this important fact ought not to be overlooked, as it has hitherto been, in judging of the influence of education; for it can hardly be doubted, that educated but inferior minds will display less morality than minds

1 See an example of Phrenology applied to Legislation, in "Essays on Human Rights and their Political Guaranties," by E. P. Hurlbut, Royal 8vo, price 2s. Simpkin, Marshall, and Co., 1847.

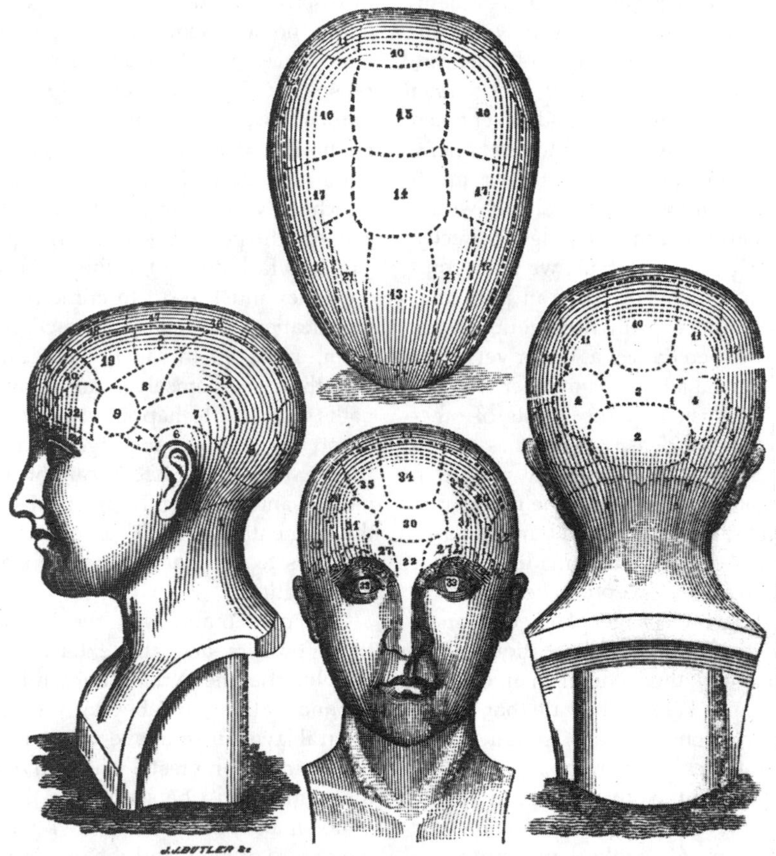

NAMES OF THE PHRENOLOGICAL ORGANS,

REFERRING TO THE FIGURES INDICATING THEIR RELATIVE POSITIONS.

AFFECTIVE.		INTELLECTUAL.	
1. ***Propensities.***	2. ***Sentiments.***	1. ***Perceptive.***	2. ***Reflective.***
1 Amativeness.	10 Self-esteem.	22 Individuality.	34 Comparison.
2 Philoprogenitiveness.	11 Love of Approbation.	23 Form.	35 Causality.
3 Concentrativeness.	12 Cautiousness.	24 Size.	
4 Adhesiveness.	13 Benevolence.	25 Weight.	
5 Combativeness.	14 Veneration.	26 Coloring.	
6 Destructiveness.	15 Firmness.	27 Locality.	
† Alimentiveness.	16 Consciousness.	28 Number.	
7 Secretiveness.	17 Hope.	29 Order.	
8 Acquisitiveness.	18 Wonder.	30 Eventuality.	
9 Constructiveness.	19 Ideality.	31 Time.	
	? Unascertained.	32 Tune.	
	20 Wit or Mirthfulness.	33 Language.	
	21 Imitation.		

Diagram of a phrenological head showing the location of the mental organs, from George Combe, *A System of Phrenology* (New York: William H. Colyer, 1846). See the selection on George Combe in this volume for an explanation of the diagram.

which are uneducated but naturally much superior. What should we say of a man who should call in question the efficacy of medical treatment, because a patient tainted from birth with consumption, and who had been long under the care of a physician, was not so healthy as a person with naturally sound lungs, who had never taken medical advice in his life? But for the treatment, the consumptive man would have been much worse than he actually was, and probably would have died in early youth. To judge correctly, therefore, of the question at issue, we must compare the present amount of crime in particular departments of France, with its amount *in the same departments* when there was either very little instruction or none at all. In this manner we shall also avoid being misled by the effects of other influences; such as the density or thinness of the population,—the employment of the people in agriculture or manufactures, and their residence on the coast, in the interior, or in mountainous or fertile districts. Were such a trial made, I think it would almost without exception be found, in cases where no great change of circumstances had occurred, that in exact proportion to the increase of education, there had been an obvious diminution of crime. I am well aware that, by the system of instruction generally pursued, the moral feelings, which restrain from crime, are wholly neglected: but cultivation even of the intellect appears favourable to morality; *first*, by giving periods of repose to the lower propensities, of whose excessive activity crime is the result; *secondly*, by promoting the formation of habits of regularity, subordination, and obedience; and, *thirdly*, by strengthening and informing the intellect, and thereby enabling it to see more clearly the dangerous consequences of crime. No doubt there are criminals on whom an excellent intellectual education has been bestowed; but instead of thence inferring that education increases the liability of mankind to crime, I think it may with great reason be asked, whether, had the same individuals wanted education altogether, their crimes would not have been ten times more atrocious?"—*Phrenological Journal*, vol. ix., p. 268.

The philosophy of man being unknown, children are not taught any rational views of the order of God's Providence on earth, nor are they trained to venerate and obey it; they are not instructed in the constitution of society, and obtain no sufficient information concerning the real sources of individual enjoyment, and social prosperity. They are not taught any system of morals based on the nature of man and his social relations, but are left each to grope his way to happiness guided by creeds and catechisms, which they see many men neglecting in their actions. The poor observe the rich pursuing pleasure and fashion; and, if they follow such examples, they must resort to crime for the means of gratification. No solid instruction is given to them, sufficient to satisfy their understandings that the rich themselves are straying from the paths that lead to happiness, and that it is to be found only in other and higher occupations.

MORALS and RELIGION, cannot assume a systematic and thoroughly practical character, until the elementary faculties of the mind, and their relations to the external creation be discovered and taught.

It is presumable that the Deity, in creating these powers and the external world, really adapted the one to the other; and that individuals and nations, in obeying the dictates of the natural laws must, in every instance, be promoting their best interests, while, in departing from them, they must be sacrificing these to passion or to illusory notions of advantage. But, until the nature of man, and the relationship between it and the external world, shall be scientifically ascertained, and systematically expounded, it will be impossible to support morality by the powerful demonstration that interest coincides with it, and to render religion practical, by shewing that all nature is in harmony with the sentiment of veneration in the mind. The tendency in most men to view expediency as not always coincident with justice, affords a striking proof of the limited knowledge of the constitution of man and the external world still existing in society.

The PROFESSION, PURSUITS, HOURS OF EXERTION, and AMUSEMENTS of individuals, should also bear reference to their physical and mental constitution; but hitherto no guiding principle has been possessed, to regulate practice in these important particulars—another evidence that the science of man has been unknown.

In consequence of the want of a philosophy of man, there is little harmony between the different departments of human pursuit. God is one; and as He is intelligent, benevolent, and powerful, we may reasonably conclude that creation is one harmonious system, in which the physical is adapted to the moral, the moral to the physical, and every department of these grand divisions to the whole. But at present, many principles clearly revealed by philosophy are impracticable because the institutions of society have not been founded with a due regard to their existence. An educated lady, for example, and a member of one of the learned professions, may possess the clearest conviction that God, by the manner in which he has constituted the body, and connected the mind with the brain, has positively enjoined muscular exertion, as indispensable to the possession of sound health, the enjoyment of life, and the rearing of a healthy offspring; and, nevertheless, they may find themselves so hedged round by routine of employment, the fashions of society, the influence of opinion, and the positive absence of all arrangements suited to the purpose, that they may be rendered nearly as incapable of yielding this obedience to God's law as if they were imprisoned in a dungeon.

By religion we are commanded to set our affections on things above, and not to permit our minds to be engrossed with the cares of this world; we are desired to seek godliness, and eschew selfishness, contention, and the vanities of life. These precepts must have been intended to be practically followed, otherwise it was a mockery of mankind to give them forth: But if they were intended to be practised, God must have arranged the inherent constitution of man, and that of the world, in such a manner as to admit of their being obeyed,—and not only so, but to render men happy in proportion as they should practise, and miserable as they should neglect them. Nevertheless, when we survey human society in the forms in which it has hitherto existed, and in which it now exists, these precepts appear to have been, and to be now, absolutely impracticable to ninety-nine out of every hundred of civilized men. Suppose the most eloquent and irresistibly convincing discourse on the Christian duties to be delivered on Sunday to a congregation of Manchester manufacturers and their operatives, or to London merchants, Essex farmers, or Westminster lawyers, how would they find their respective spheres of life adapted for acting practically on their convictions? They are all commanded to love God with their whole heart and soul, and to resist the world and the flesh, or, in philosophical language, to support their moral affections and intellectual powers in habitual activity—to direct them to noble, elevating, and beneficial objects—and to resist the subjugation of these higher attributes of their minds to animal pleasure, sordid selfishness, and worldly ambition. The moral and intellectual powers assent to the reasonableness of these precepts and rejoice in the prospect of their practical application; but, on Monday morning, the manufacturers, owing to the institutions of society, and the department of life into which their lot has been cast before they had either reason or moral perception to direct their choice, must commence a course of ceaseless toil,—the workmen that they may support life, and the masters that they may avoid ruin, or accumulate wealth. Saturday evening finds them worn out with mental and bodily exertion, continued through all the intermediate days, and directed to pursuits connected with this world alone. Sunday dawns upon them in a state of mind widely at variance with the Christian condition. In like manner, the merchant must devote himself to his bargains, the farmer to his plough, and the lawyer to his briefs, with corresponding assiduity; so that their moral powers have neither objects presented to them, nor vigour left for enjoyments befitting their nature and desires.

It is in vain to say to individuals that they err in acting thus: individuals are carried along in the great stream of social institutions and pursuits. The operative labourer is compelled to follow his routine of toil under pain of absolute starvation. The master-manufacturer, the merchant, the farmer, and the lawyer, are pursued by competitors so active, that if they relax in selfish ardour, they will be speedily plunged into ruin. If God has so constituted the human mind and body, and so arranged external nature, that all this is unavoidably necessary for man, then the Christian precepts are scarcely more suited to human nature and circumstances in this world, than the command to fly would be to the nature of the horse. If, on the other hand, man's nature and circumstances do in themselves admit of the Christian precepts being realized, it is obvious that a great revolution must take place in our notions, principles of action, practices, and social institutions, before this can be accomplished. That many Christian teachers believe this improvement possible, and desire its execution, I cannot doubt; but through want of knowledge of the constituent elements of human nature, and their relations—through want, in short, of a philosophy of mind and of physical nature—they have never been able to perceive what God has rendered man capable of attaining,—how it may be attained,—or on what principles the moral and physical government of the world in regard to man is conducted. Consequently, they have not acted generally on the idea of religion being a branch of an all-comprehending philosophy; they have relied chiefly on inculcating the precepts of their Master, threatening future punishments for disobedience, and promising future rewards for observance,—without proving philosophically to society, not only that its institutions, practices, and principles, must be erected on loftier ground than they are at present before it can become truly Christian,—but that these improvements are actually within the compass of human nature, aided by science and Scripture. Individuals in whom there is a strong aspiration after the realization of the Christian state of society, but whose intellects cannot perceive any natural means by which it can be produced, take refuge in the regions of prophecy, and expect a miraculous reign of saints in the Millennium. How much more profitable would it be to study the philosophy of man's nature, which is obviously the work of God, and endeavour to introduce morality and happiness by the means appointed by him in creation! Supernatural agency has long ceased to interfere with human affairs; and whenever it shall operate again, we may presume that it will be neither assisted nor retarded by human opinions and speculations.

We need only attend to the scenes daily presenting themselves in society, to obtain an irresistible conviction that many evil consequences result from the want of a true theory of human nature, and its relations. Every preceptor in schools—every professor in colleges—every author, editor, and pamphleteer—every member of Parliament, councillor, and judge—has a set of notions of his own, which, in his mind, holds the place of a system of the philosophy of man; and although he may not have methodised his ideas, or even acknowledged them to himself as a theory, yet they constitute a standard to him by which he practically judges of all questions in morals, politics, and religion: he advocates whatever views coincide with them, and condemns all that differ from them, with as unhesitating a dogmatism as the most pertinacious theorist on earth. Each also despises the notions of his fellows, in so far as they differ from his own. In short, the human faculties too generally operate simply as impulses, exhibiting all the conflict and uncertainty of mere feeling, unenlightened by perception of their own nature and objects. Hence, public measures in general, whether relating to education, religion, trade, manufactures, the poor, criminal law, or any other subject linked with the dearest interests of society, instead of being treated as branches of one general system of economy, and adjusted on scientific principles each in harmony with all the rest, are supported, or opposed, on narrow and empirical grounds, and often call forth displays of ignorance, prejudice, selfishness, intolerance, and bigotry, that greatly obstruct the progress of improvement. Indeed, any important approach to unanimity, even among sensible and virtuous men, will be impossible, so long as the order of nature is not acknowledged as an authoritative guide to individual feelings and perceptions.

If, then, the doctrine of the natural laws here expounded be true, it will, when matured, supply the deficiencies now pointed out.

But here another question naturally presents itself—How are the views explained in this work, supposing them to contain some portion of truth, to be rendered practical? Sound views of human nature and of the divine government come home to the feelings and understandings of men; they perceive them to possess a substantive existence and reality, which rivet attention and command respect. If the doctrine unfolded in the present treatise be in any degree true, it is destined to operate on the character of legislation, on practical conduct, and on public instruction,—especially that from the pulpit. Individuals whose minds have embraced the views which it contains, inform me that many sermons appear to them inconsistent in their different propositions, at variance with sound views of human nature, and so vague as to have little relation to practical life and conduct. They partake of the abstractedness of the scholastic philosophy. The first divine of comprehensive intellect and powerful moral feelings who shall take courage and introduce the natural laws into his discourses, and teach the people the works and institutions of the Creator, will reap a great reward in usefulness and pleasure. If this course shall, as heretofore, be neglected, the people, who are daily increasing in knowledge of philosophy and practical science, will, in a few years, look with disrespect on their clerical guides, and probably force them, by "pressure from without," to remodel the entire system of pulpit-instruction.

The institutions and manners of society indicate the state of mind of the influential classes at the time when they prevail. The trial and burning of old women as witches, point out clearly the predominance of Destructiveness and Wonder over Intellect and Benevolence, in those who were guilty of such cruel absurdities. The practices of wager of battle, and ordeal by fire and water, indicate great activity of Combativeness, Destructiveness, and Veneration, in those who permitted them, combined with lamentable ignorance of the natural constitution of the world. In like manner, the enormous sums willingly expended in war, and the small sums grudgingly paid for public improvements,—the intense energy displayed in the pursuit of wealth,—and the general apathy evinced in the search after knowledge and virtue,—unequivocally proclaim activity of Combativeness, Destructiveness, Acquisitiveness, Self-Esteem and Love of Approbation, with comparatively moderate vivacity of Benevolence and Conscientiousness in the present generation. Before, therefore, the practices of mankind can be altered, the state of their minds must be changed. It is an error to impose institutions on a people greatly in advance of their mental condition. The rational method is, first to instruct the intellect, then to interest the sentiments, and, last of all, to form arrangements in harmony with these faculties, and resting on them as their basis.

The views developed in the preceding chapters, if founded in nature, may be expected to lead, ultimately, to considerable changes in many of the customs and pursuits of society; but to accomplish this effect, the principles themselves must first be ascertained to be true; next they must be sedulously taught; and only thereafter can they be practically applied. It appears to me that a long series of years will probably elapse before even nations now regarded as civilized, will model their institutions and manners in harmony with the natural laws.

The first step should be to teach these laws to the young. Their minds, not being occupied by prejudice, will recognise them as congenial to their constitution; the first generation that shall embrace them from infancy will proceed to modify the institutions of society into accordance with their dictates; and in the course of ages, they may at length be found to be practically useful. A perception of the importance of the natural laws will lead to their observance, and this will be attended by an increase of physical prosperity, a higher morality, and in process of time, an improved development of brain, thereby increasing the desire and capacity for farther progress. All true theories have ultimately been adopted and influenced practice; and I see no reason to fear that the present, if true, will prove an exception. The failure of all previous systems is the natural consequence of their having been unfounded; if this resemble them, it will deserve, and assuredly will meet, a similar fate.

The present work may be regarded as, in one sense, an introduction to an essay on education. If the views unfolded in it be in general sound, it will follow that education has scarcely yet commenced. If the Creator has bestowed on the body, on the mind, and on external nature, determinate constitutions, and has arranged them to act on each other, and to produce happiness or misery to man, according to certain definite principles,—and if this action goes on invariably, inflexibly, and irresistibly, whether men attend to it or not,—it is obvious that the very basis of useful knowledge must consist in an acquaintance with these natural arrangements;—and that education will be valuable in the exact degree in which it communicates such information, and trains the faculties to act upon it. Reading, writing, and accounts, which make up the instruction enjoyed by the lower orders, are merely *means of acquiring knowledge*, but do not *constitute* it. Greek, Latin, and Mathematics, which are added in the education of the middle and upper classes, are still only *means* of obtaining information: hence, with the exception of the few who pursue physical science, society dedicates very little attention to the study of the natural laws. And even those who do study science, disconnect it from the moral and religious sentiments, and thus allow more than half of its beneficial influence on human conduct to be lost.

In attempting to give effect to the views now discussed, I respectfully recommend that each individual, according as he becomes acquainted with the natural laws, should obey them, and communicate his experience of their operations to others; avoiding at the same time, the subversion, by violence, of established institutions, and all outrages on public sentiment by intemperate discussions. The doctrines before unfolded, if true, authorise us to predicate that the most successful method of ameliorating the condition of mankind, will be that which appeals most directly to their moral sentiments and intellect; and I may add, from experience and observation, that, in proportion as any individual becomes acquainted with the real constitution of the human mind, will his conviction of the efficacy of this method increase.

Finally, if it be true that the natural laws must be obeyed as a preliminary condition to happiness in this world, and if virtue and happiness be inseparably allied, the religious instructors of mankind may probably discover in the general and prevalent ignorance of these laws, one reason of the limited success which has hitherto attended their efforts to improve the condition of mankind; and they may perhaps perceive it to be not inconsistent with their sacred office, to instruct men in the natural institutions of the Creator, as well as in Scripture doctrines, and to recommend obedience to both. They exercise so vast an influence over the best members of society, that their countenance may hasten, or their opposition retard, by a century, the general adoption of the natural laws as guides to human conduct.

If the excessive toil of the manufacturer be inconsistent with that elevation of the moral and intellectual faculties of man which is commanded by religion, and if the moral and physical welfare of mankind be not at variance with each other (which they cannot be), the institutions of society out of which the necessity for that labour arises, must, philosophically speaking, be pernicious to the interests of the state as a political body, and to the temporal welfare of the individuals who compose it; and whenever we shall be in possession of a correct knowledge of the elements of human nature, and the principles on which God has constituted and governs the world, the *evidence* that these practices are detrimental to *our temporal welfare* will be as clear as that of their inconsistency with our religious duties. Until, however, divines shall become acquainted with this relation between philosophy and religion, they will not possess adequate means of rendering their precepts practical in this world; they will not carry the intellectual perceptions of their hearers fully along with them; they will be incapable of controlling the force of the animal propensities; and they will never lead society to the fulfilment of its highest destinies.

At present, the animal propensities are fortified in the strong entrenchments of social institutions: Acquisitiveness, for example, is protected and fostered by our arrangements for accumulating wealth; a worldly spirit, by our constant struggle to obtain the means of subsistence; pride and vanity, by our artificial distinctions of rank and fashion; and Combativeness and Destructiveness by our warlike professions. The divine assails the vices and inordinate passions of mankind by the denunciations of the gospel; but as long as society shall be animated by different principles, and maintain in vigour institutions whose spirit is diametrically opposite to its doctrines, so long will it be difficult for him to effect the realization of his precepts in practice. Yet it appears to me, that, by teaching mankind the philosophy of their own nature and of the world in which they live—by proving to them the harmony between the order of God's secular providence and Christian morality, and the inconsistency of their own practices with both—they may be induced to modify the latter, and to entrench the moral powers in social institutions; and then the triumph of virtue and religion will be more complete.

Those who advocate the exclusive importance of spiritual religion for the improvement of mankind, appear to me to err in overlooking too much the necessity for complying with the natural conditions on which all improvement depends; and I anticipate, that when schools and colleges shall expound the various branches of science as elucidations of the order of God's providence for the guidance of human conduct on earth,—when the pulpit shall deal with the same principles, shew their practical application to man's duties and enjoyments, and add the sanctions of religion to enforce the observance of the natural laws—and when the busy scenes of life shall be so arranged as to become a field for the practice at once of our philosophy and of our religion—then will man assume his station as a rational being, and religion will achieve her triumph.

FURTHER READING

Bentley, M. "The Psychological Antecedents of Phrenology." *Psychological Monographs* 21 (1966): 102–115.

Capen, Nahum. *Reminiscences of Dr. Spurzheim and George Combe: And a Review of the Science of Phrenology.* New York: Fowler and Wells, 1881.

Colbert, Charles. *A Measure of Perfection: Phrenology and the Fine Arts in America.* Chapel Hill: University of North Carolina Press, 1997.

Combe, George. *Letter from George Combe to Francis Jeffrey, Esq. in Answer to his Criticism on Phrenology.* Edinburgh: John Anderson, 1826.

Cooter, Roger. *The Cultural Meaning of Popular Science: Phrenology and the Organization of Consent in Nineteenth-Century Britain.* Cambridge: Cambridge UP, 1984.

Davies, John D. *Phrenology, Fad and Science: A 19th-Century American Crusade.* New Haven: Yale UP, 1955.

Erikson, Paul A. *Phrenology and Physical Anthropology: The George Combe Connection.* Halifax, NS: Saint Mary's University, 1979.

Gibbon, Charles. *The Life of George Combe.* 2 vols. London: Macmillan and Co., 1878.

de Giustino, David. *Conquest of Mind: Phrenology and Victorian Social Thought.* London: Croom Helm, 1975.

Greenblatt, Samuel H. "Phrenology in the Science and Culture of the 19th Century." *Neurosurgery* 37, no. 4 (1995): 790–805.

Hollander, Bernard. *The Mental Functions of the Brain: An Investigation into Their Localisation and Their Manifestation in Health and Disease.* New York: G.P. Putnam's Sons, 1901.

Kaufman, M.H., and N. Basden. "Marked Phrenological Heads: Their Evolution, With Particular Reference to the Influence of George Combe and the Phrenological Society of Edinburgh." *Journal of the History of Collections* (1997): 139–59.

Seedair, Stephen. *A Tract for All Time. The Christian or True Constitution of Man, Versus the Pernicious Fallacies of Mr. Combe and Other Materialistic Writers.* Edinburgh: Myles MacPhail, 1856.

Shapin, Steven. "The Politics of Observation: Cerebral Anatomy and Social Interests in the Edinburgh Phrenology Disputes." In Roy Wallis, ed., *On the Margins of Science: The Social Construction of Rejected Knowledge.* Sociological Review Monograph 27. Keele: University of Keele Press, 1979, pp. 139–78.

Sizer, Nelson. *Heads and Faces, and How to Study Them; A Manual of Phrenology and Physiognomy for the People.* New York: Fowler and Wells, 1885.

—. *Forty Years in Phrenology.* New York: Fowler and Wells, 1882.

Stern, Madeleine B. *Heads and Headlines: The Phrenological Fowlers.* Norman: University of Oklahoma Press, 1971.

Young, Robert M. *Mind, Brain and Adaptation in the Nineteenth Century.* Oxford: Clarendon Press, 1990.

16. WILLIAM WHEWELL

(1794–1866)

The son of a master carpenter, William Whewell entered Trinity College, Cambridge, in 1812 where he excelled in all subjects, eventually receiving his M.A. (1819) and D.D. (1844) degrees there. He took Priest's orders in 1826. Whewell spent his entire professional career at Cambridge, serving as Vice-Chancellor of the university in 1842 and 1855. In addition to his literary and theological interests, he wrote on meterology, political economy, and mineralogy. He built an improved anemometer for measuring winds, and developed a crystallographic system for measuring crystal cleavage planes. Between the 1830s and 1850s he published his theories of the tides, and won a Royal Society medal for this work. In 1836 he became President of the Geological Society. Whewell was responsible for bringing the word "scientist," among other terms, into general use.

Because Whewell has not been associated with any major scientific discovery, his name has all but been forgotten except to historians and philosophers of science. Very few aspects of nineteenth-century intellectual life, however, were left untouched by Whewell's restless intellect; he corresponded for many years with Herschel and Lyell, disputed with David Brewster over the plurality of worlds, and composed one of the Bridgewater Treatises, *Astronomy and General Physics Considered with Reference to Natural Theology* (1833). His contemporary Sydney Smith once quipped about Whewell that "science is his forte and omniscience is his foible" (Todhunter 1.410). Whewell's numerous books include textbooks, poetry, sermons and notable works on ethics and morality, including *Elements of Morality, Including Polity* (1845), *Lectures on Systematic Morality* (1846), and *Lectures on the History of Moral Philosophy in England* (1852).

Later in life Whewell turned to studies of the history and philosophy of the sciences, composing the two massive works he is best known for, *The History of the Inductive Sciences, From the Earliest to the Present Time* (1837) and *The Philosophy of the Inductive Sciences, Founded Upon Their History* (1840), the most in-depth analyses of scientific method since the *Novum Organum* of Francis Bacon. The *Philosophy* was later revised and republished in three volumes: *The History of Scientific Ideas* (1858), *Novum organum renovatum* (1858), and *On the Philosophy of Discovery* (1860).

In Whewell's historical vision of the growth of the sciences, each science progressed through a "prelude," "inductive epoch," and then a "sequel." The prelude stage involved the sharpening of scientific ideas and the aggregation of precise facts. Then during the inductive epoch some breakthrough in scientific thinking occurred, but one which absorbed and extended previous ideas. In the sequel period, the scientific discovery was confirmed, disseminated, and employed to solve other problems in science. These epochs are not necessarily linear or progressive, since Whewell's theory included stationary periods like the Middle Ages (which he called the "midday slumber" of science), when knowledge did not advance. Whewell clearly conceived of his *History* as an aid to the progress of the modern sciences: "It will be universally expected that a History of Inductive Science should point out to us a philosophical distribution of the existing body of knowledge, and afford us some indication of the most promising mode of directing our future efforts to add to its extent and completeness" (*History*, 1869 ed., 42). More than any other previous work, the *History* provided a justification for the study of the history of science as a serious intellectual discipline and as an aid to the contemporary practice of science.

Two other important works on induction and scientific method, both taking different views from Whewell on various points, appeared in the nineteenth century: John Herschel's popular *A Preliminary Discourse on the Study of Natural Philosophy* (1830), and John Stuart Mill's *A System of Logic* (1843). In "Mr. Mill's Logic" in his *On the Philosophy of Discovery*, Whewell felt compelled to respond at length to Mill's *Logic* on the necessity of ideas in scientific explanation. Herschel, working within the generally accepted tradition of English Baconian empiricism, surveyed Whewell's work in an anonymous review of 1841 ("Whewell on the Inductive Sciences") and like Mill, differed from Whewell in that he grounded all knowledge in experience, not ideas. Bacon and Herschel believed that an induction represented a summary of facts from which we derive a general law, while for Whewell an idea was superinduced on facts by the mind (i.e., demonstrating the priority and importance of the theory or idea). Here Whewell was drawing on Immanuel Kant, who argued that perception could not take place without the imposition of preformed ideas about the world, which he called categories of thought. The Mill-Whewell-Herschel debates were obscured by the three writers' frequent confusion between the logic of scientific discovery (the origin and process of acquiring scientific knowledge) and the logic of justification (the verification and validation of scientific theories).

For Whewell, each science contains its own unique "fundamental ideas" which form the basis of that science. Whewell pointed to the difficulty in separating facts and ideas that together formed the "fundamental antithesis" which, when synthesized, brought about an advance of scientific knowledge. In the first section of his *Philosophy*, Whewell examined the fundamental antithesis of 1) sensations, which form the basis of facts, and 2) ideas in the mind which order and categorize sensation into meaningful knowledge. Whewell's inductive process works as follows: explication of concepts and clarification of ideas first takes place, then decomposition of facts into theoretical and factual components must occur. In the final synthesis, colligation (from Latin *colligare*, to bind together) occurs when the scientist discovers the most suitable idea which binds together and adequately explains the facts which have been drawn together concerning a certain phenomenon. He further believed that science progressed by subsuming specific empirical observations under increasingly generalized theoretical statements: "each induction supplies the materials of fresh inductions," he explained (*Philosophy* 46). If general propositions or laws are mutually supported by "consilience of inductions," then other knowledge can be deduced from these laws. A consilience of induction occurs when an induced theory from one set of data corresponds to the induction from another, mutually re-enforcing the induced theories. Unlike Karl Popper's criteria for scientific hypotheses (the hypothetico-deductive method), which claims truth value for a theory until falsified, Whewell judged the efficacy of theories primarily on their ability to predict other facts and on the scientist's belief that a consilience of induction was not accidental or merely coincidental.

In Whewell's epistemology, no necessary truths can arise directly from experience. The mind grasps a necessary idea by intuition. Readers of Whewell's works will notice a repeated binary structure to Whewell's thought, which Robert Blanché calls the fundamental indecision of his philosophy, "oscillating ceaselessly between idealism and realism, and between epistemology and ontology" (*Rationalisme* 2). Whewell's argments on induction and scientific method are lengthy and minutely worked out, at times stretching to several chapters, and a brief anthology selection unfortunately does not do justice to his thought.

The Philosophy of the Inductive Sciences, Founded Upon Their History.
2d ed. 2 vols. London: John W. Parker, 1847, pp. 16–51.

Chapter II. OF THE FUNDAMENTAL ANTITHESIS OF PHILOSOPHY

Sect. 1.—Thoughts and Things.

In order that we may do something towards determining the nature and conditions of human knowledge, (which I have already stated as the purpose of this work,) I shall have to refer to an antithesis or opposition, which is familiar and generally recognized, and in which the distinction of the things opposed to each other is commonly considered very clear and plain. I shall have to attempt to make this opposition sharper and stronger than it is usually conceived, and yet to shew that the distinction is far from being so clear and definite as it is usually assumed to be: I shall have to point the contrast, yet shew that the things which are contrasted cannot be separated:—I must explain that the antithesis is constant and essential, but yet that there is no fixed and permanent line dividing its members. I may thus appear, in different parts of my discussion, to be proceeding in opposite directions, but I hope that the reader who gives me a patient attention will see that both steps lead to the point of view to which I wish to lead him.

The antithesis or opposition of which I speak is denoted, with various modifications, by various pairs of terms: I shall endeavour to show the connexion of these different modes of expression, and I will begin with that form which is the simplest and most idiomatic.

The simplest and most idiomatic expression of the antithesis to which I refer is that in which we oppose to each other THINGS and THOUGHTS. The opposition is familiar and plain. Our Thoughts are something which belongs to ourselves; something which takes place within us; they are what *we* think; they are actions of our minds. Things, on the contrary, are something different from ourselves and independent of us; something which is without us; they *are*; we see them, touch them, and thus know that they exist; but we do not make them by seeing or touching them, as we make our *Thoughts* by thinking them; we are passive, and *Things* act upon our organs of perception.

Now what I wish especially to remark is this: that in all human KNOWLEDGE both Thoughts and Things are concerned. In every part of my knowledge there must be some *thing* about which I know, and an internal act of *me* who know. Thus, to take simple yet definite parts of our knowledge, if I know that a solar year consists of 365 days, or a lunar month of 30 days, I know something about the sun or the moon; namely, that those objects perform certain revolutions and go through certain changes, in those numbers of days; but I count such numbers and conceive such revolutions and changes by acts of my own thoughts. And both these elements of my knowledge are indispensable. If there were not such external Things as the sun and the moon I could not have any knowledge of the progress of time as marked by them. And however regular were the motions of the sun and moon, if I could not count their appearances and combine their changes into a cycle, or if I could not understand this when done by other men, I could not know anything about a year or a month. In the former case I might be conceived as a human being, possessing the human powers of thinking and reckoning, but kept in a dark world with nothing to mark the progress of existence. The latter is the case of brute animals, which see the sun and moon, but do not know how many days make a month or a year, because they have not human powers of thinking and reckoning.

The two elements which are essential to our knowledge in the above cases, are necessary to human knowledge in all cases. In all cases, Knowledge implies a combination of Thoughts and Things. Without this combination, it would not be Knowledge. Without Thoughts, there could be no connexion; without Things, there could be no reality. Thoughts and Things are so intimately combined in our Knowledge, that we do not look upon them as distinct. One single act of the mind involves them both; and their contrast disappears in their union.

But though Knowledge requires the union of these two elements, Philosophy requires the separation of them, in order that the nature and structure of Knowledge may be seen. Therefore I begin by considering this separation. And I now proceed to speak of another way of looking at the antithesis of which I have spoken; and which I may, for the reasons which I have just mentioned, call the FUNDAMENTAL ANTITHESIS OF PHILOSOPHY.

Sect. 2.—Necessary and Experiential Truths.

Most persons are familiar with the distinction of *necessary* and *contingent* truths. The former kind are Truths which cannot but be true; as that 19 and 11 make 30;—that parallelograms upon the same base and between the same parallels are equal:—that all the angles in the same segment of a circle are equal. The latter are Truths which *it happens* (*contingit*) are true; but which, for any thing which we can see, might have been otherwise; as that a lunar month contains 30 days, or that the stars revolve in circles round the pole. The latter kind of Truths are learnt by experience, and hence we may call them *Truths of Experience*, or, for the sake of convenience, *Experiential* Truths, in contrast with Necessary Truths.

Geometrical propositions are the most manifest examples of Necessary Truths. All persons who have read and understood the elements of geometry, know that the propositions above stated (that parallelograms upon the same base and between the same parallels are equal; that all the angles in the same segment of a circle are equal,) are necessarily true; not only they *are* true, but they *must be* true. The meaning of the terms being understood, and the proof being gone through, the truth of the propositions must be assented to. We learn these propositions to be true by demonstrations deduced from definitions and axioms; and when we have thus learnt them, we see that they could not be otherwise. In the same manner, the truths which concern numbers are necessary truths: 19 and 11 not only *do* make 30, but *must* make that number, and cannot make anything else. In the same manner, it is a necessary truth that half the sum of two numbers added to half their difference is equal to the greater number.

It is easy to find examples of Experiential Truths;—propositions which we know to be true, but know by experience only. We know, in this way, that salt will dissolve in water; that plants cannot live without light;—in short, we know in this way all that we do know in chemistry, physiology, and the material sciences in general. I take the *Sciences* as my examples of human knowledge, rather than the common truths of daily life, or moral or political truths; because, though the latter are more generally interesting, the former are much more definite and certain, and therefore better starting-points for our speculations, as I have already said. And we may take elementary astronomical truths as the most familiar examples of Experiential Truths in the domain of science.

With these examples, the distinction of Necessary and Experiential Truths is, I hope, clear. The former kind, we see to be true by thinking about them, and see that they could not be otherwise. The latter kind, men could never have discovered to be true without looking at them; and having so discovered them, still no one will pretend to say they might not have been otherwise. For aught we can see, the astronomical truths which express the motions and periods of the sun, moon and stars, might have been otherwise. If we had been placed in another part of the solar system, our experiential truths respecting days, years, and the motions of the heavenly bodies, would have been other than they are, as we know from astronomy itself.

It is evident that this distinction of Necessary and Experiential Truths involves the same antithesis which we have already considered;—the antithesis of Thoughts and Things. Necessary Truths are derived from our own Thoughts: Experiential Truths are derived from our observation of Things about us. The opposition of Necessary and Experiential Truths is another aspect of the Fundamental Antithesis of Philosophy.

Sect. 3.—Deduction and Induction.

I have already stated that geometrical truths are established by demonstrations *deduced* from definitions and axioms. The term *Deduction* is specially applied to such a course of demonstration of truths from definitions and axioms. In the case of the parallelograms upon the same base and between the same parallels, we prove certain triangles to be equal, by supposing them placed so that their two bases have the same extremities; and hence, referring to an Axiom respecting straight lines, we infer that the bases coincide. We combine these equal triangles with other equal spaces, and in this way make up both the one and the other of the parallelograms, in such a manner as to shew that they are equal. In this manner, going on step by step, deducing the equality of the triangles from the axiom, and the equality of the parallelograms from that of the triangles, we travel to the conclusion. And this process of successive deduction is the scheme of all geometrical proof. We begin with Definitions of the notions which we reason about, and with Axioms, or self-evident truths, respecting these notions; and we get, by reasoning from these, other truths which are demonstratively evident; and from these truths again, others of the same kind, and so on. We begin with our own Thoughts, which supply us with Axioms to start from; and we reason from these, till we come to propositions which are applicable to the Things about us; as for instance, the propositions respecting circles and spheres are applicable to the motions of the heavenly bodies. This is *Deduction*, or *Deductive Reasoning*.

Experiential truths are acquired in a very different way. In order to obtain such truths, we begin with Things. In order to learn how many days there are in a year, or in a lunar month, we must begin by observing the sun and the moon. We must observe their changes day by day, and try to make the cycle of change fit into some notion of number which we supply from our own Thoughts. We shall find that a cycle of 30 days nearly will fit the changes of phase of the moon;—that a cycle of 365 days nearly will fit the changes of daily motion of the sun. Or, to go on to experiential truths of which the discovery comes within the limits of the history of science—we shall find (as Hipparchus found) that the unequal motion of the sun among the stars, such as observation shews it to be, may be fitly represented by the notion of an *eccentric;*—a circle in which the sun has an equable annual motion, the spectator not being in the center of the circle. Again, in the same manner, at a later period, Kepler started from more exact observations of the sun, and compared them with a supposed motion in a certain ellipse; and was able to shew that, not a circle about an eccentric point, but an ellipse, supplied the mode of conception which truly agreed with the motion of the sun about the earth; or rather, as Copernicus had already shewn, of the earth about the sun. In such cases, in which truths are obtained by beginning from observation of external things and by finding some notion with which the Things, as observed, agree, the truths are said to be obtained by *Induction*. The process is an *Inductive Process*.

The contrast of the Deductive and Inductive process is obvious. In the former, we proceed at each step from general truths to particular applications of them; in the latter, from particular observations to a general truth which includes them. In the former case we may be said to reason *downwards*, in the latter case, *upwards*; for general notions are conceived as standing above particulars. Necessary truths are proved, like arithmetical sums, by adding together the portions of which they consist. An inductive truth is proved, like the guess which answers a riddle, by its agreeing with the facts described. Demonstration is irresistible in its effect on the belief, but does not produce surprize, because all the steps to the conclusion are exhibited, before we arrive at the conclusion. Inductive inference is not demonstrative, but it is often more striking than demonstrative reasoning, because the intermediate links between the particulars and the inference are not shown. Deductive truths are the results of relations among our own Thoughts. Inductive Truths are relations which we discern among existing Things; and thus, this opposition of Deduction and Induction is again an aspect of the Fundamental Antithesis already spoken of.

Sect. 4.—Theories and Facts.

General experiential Truths, such as we have just spoken of, are called *Theories,* and the particular observations from which they are collected, and which they include and explain, are called *Facts.* Thus Hipparchus's doctrine, that the sun moves in an eccentric about the earth, is *his Theory* of the Sun, or the *Eccentric Theory.* The doctrine of Kepler, that the Earth moves in an Ellipse about the Sun, is *Kepler's Theory* of the Earth, the Elliptical Theory. Newton's doctrine that this elliptical motion of the Earth about the Sun is produced and governed by the Sun's attraction upon the Earth, is the *Newtonian* theory, the *Theory of Attraction.* Each of these Theories was accepted, because it included, connected and explained the *Facts;* the Facts being, in the two former cases, the motions of the Sun as observed; and in the other case, the elliptical motion of the Earth as known by Kepler's Theory. This antithesis of *Theory* and *Fact* is included in what has just been said of Inductive Propositions. A Theory is an Inductive Proposition, and the Facts are the particular observations from which, as I have said, such Propositions are inferred by Induction. The Antithesis of Theory and Fact implies the fundamental Antithesis of Thoughts and Things; for a Theory (that is, a true Theory) may be described as a Thought which is contemplated distinct from Things and seen to agree with them; while a Fact is a combination of our Thoughts with Things in so complete agreement that we do not regard them as separate.

Thus the antithesis of Theory and Fact involves the antithesis of Thoughts and Things, but is not identical with it. Facts involve Thoughts, for we know Facts only by thinking about them. The Fact that the year consists of 365 days; the Fact that the month consists of 30 days, cannot be known to us, except we have the Thoughts of Time, Number and Recurrence. But these Thoughts are so familiar, that we have the Fact in our mind as a simple Thing without attending to the Thought which it involves. When we mould our Thoughts into a Theory, we consider the Thought as distinct from the Facts; but yet, though distinct, not independent of them; for it is a true Theory, only by including and agreeing with the Facts.

Sect. 5.—Ideas and Sensations.

We have just seen that the antithesis of Theory and Fact, although it involves the antithesis of Thoughts and Things, is not identical with it. There are other modes of expression also, which involve the same Fundamental Antithesis, more or less modified. Of these, the pair of words which in their relations appear to separate the members of the antithesis most distinctly are *Ideas* and *Sensations.* We see and hear and touch external things, and thus perceive them by our senses; but in perceiving them, we connect the impressions of sense according to relations of space, time, number, likeness, cause, etc. Now some at least of these kinds of connexion, as space, time, number, may be contemplated distinct from the things to which they are applied; and so contemplated, I term them *Ideas.* And the other element, the impressions upon our senses which they connect, are called *Sensations.*

I term space, time, cause, etc., *Ideas,* because they are general relations among our sensations, apprehended by an act of the mind, not by the senses simply. These relations involve something beyond what the senses alone could furnish. By the sense of sight we see various shades and colours and shapes before us, but the *outlines* by which they are separated into distinct objects of definite forms, are the work of the mind itself. And again, when we conceive visible things, not only as surfaces of a certain form, but as *solid bod-*

ies, placed at various distances in space, we again exert an act of the mind upon them. When we see a body move, we see it move in a path or *orbit*, but this orbit is not itself seen; it is constructed by the mind. In like manner when we see the motions of a needle towards a magnet, we do not *see* the attraction or force which produces the effects; but we infer the force, by having in our minds the Idea of Cause. Such acts of thought, such *Ideas*, enter into our perceptions of external things.

But though our perceptions of external things involve some act of the mind, they must involve something else besides an act of the mind. If we must exercise an act of thought in order to see force exerted, or orbits described by bodies in motion, or even in order to see bodies existing in space, and to distinguish one kind of object from another, still the act of thought alone does not make the bodies. There must be something besides, *on which* the thought is exerted. A colour, a form, a sound, are not produced by the mind, however they may be moulded, combined, and interpreted by our mental acts. A philosophical poet has spoken of

> All the world
> Of eye and ear, both what they half create,
> And what perceive.

But it is clear, that though they *half* create, they do not wholly create: there must be an external world of colour and sound to give impressions to the eye and ear, as well as internal powers by which we perceive what is offered to our organs. The mind is in some way passive as well as active: there are objects without as well as faculties within;—Sensations, as well as acts of Thought.

Indeed this is so far generally acknowledged, that according to common apprehension, the mind is passive *rather* than active in acquiring the knowledge which it receives concerning the material world. Its sensations are generally considered more distinct than its operations. The world without is held to be more clearly real than the faculties within. That there is something different from ourselves, something external to us, something independent of us, something which no act of our minds can make or can destroy, is held by all men to be at least as evident, as that our minds can exert any effectual process in modifying and appreciating the impressions made upon them. Most persons are more likely to doubt whether the mind be always actively applying Ideas to the objects which it perceives, than whether it perceive them passively by means of Sensations.

But yet a little consideration will show us that an activity of the mind, and an activity according to certain Ideas, is requisite in all our knowledge of external objects. We see objects, of various solid forms, and at various distances from us. But we do not thus perceive them by sensation alone. Our visual impressions cannot, of themselves, convey to us a knowledge of solid form, or of distance from us. Such knowledge is inferred from what we see:—inferred by conceiving the objects as existing in space, and by applying to them the Idea of Space. Again:—day after day passes, till they make up a year: but we do not know that the days are 365, except we count them; and thus apply to them our Idea of Number. Again:—we see a needle drawn to a magnet: but, in truth, the *drawing* is what we cannot see. We see the needle move, and infer the attraction, by applying to the fact our Idea of Force, as the cause of motion. Again:—we see two trees of different kinds; but we cannot know that they are so, except by applying to them our Idea of the resemblance and difference which makes kinds. And thus Ideas, as well as Sensations, necessarily enter into all our knowledge of objects: and these two words express, perhaps more exactly than any of the pairs before mentioned, that Fundamental Antithesis, in the union of which, as I have said, all knowledge consists.

Sect. 6.—Reflexion and Sensation.

It will hereafter be my business to show what the Ideas are, which thus enter into our knowledge; and how each Idea has been, as a matter of historical fact, introduced into the Science to which it especially belongs. But before I proceed to do this, I will notice some other terms, besides the phrases already noticed, which have a reference, more or less direct, to the Fundamental Antithesis of Ideas and Sensations. I will mention some of these, in order that if they should come under the reader's notice, he may not be perplexed as to their bearing upon the view here presented to him.

The celebrated doctrine of Locke, that all our "Ideas," (that is, in his use of the word, all our objects of thinking,) come from Sensation or Reflexion, will naturally occur to the reader as connected with the antithesis of which I have been speaking. But there is a great difference between Locke's account of Sensation and Reflexion, and our view of Sensation and Ideas. He is speaking of the origin of our knowledge;—we, of its nature and composition. He is content to say that all the knowledge which we do not receive directly by Sensation, we obtain by Reflex Acts of the mind, which make up his Reflexion. But we hold that there is no Sensation without an act of the mind, and that the mind's activity is not only reflexly exerted upon itself, but directly upon objects, so as to perceive in them connexions and relations which are not Sensations. He is content to put together, under the name of Reflexion, everything in our knowledge which is not Sensation: we are to attempt to analyze all that is not Sensation; not only to say it consists of Ideas, but to point out what those Ideas are, and to show the mode in which each of them enters into our knowledge. His purpose was, to prove that there are no Ideas, except the reflex acts of the mind: our endeavour will be to show that the acts of the mind, both direct and reflex, are governed by certain Laws, which may be conveniently termed Ideas. His procedure was, to deny that any knowledge could be derived from the mind alone: our course will be, to show that in every part of our most certain and exact knowledge, those who have added to our knowledge in every age have referred to principles which the mind itself supplies. I do not say that my view is contrary to his: but it is altogether different from his. If I grant that all our knowledge comes from Sensation and Reflexion, still my task then is only begun; for I want further to determine, in each science, what portion comes, not from mere Sensation, but from those Ideas by the aid of which either Sensation or Reflexion can lead to Science.

Locke's use of the word "idea" is, as the reader will perceive, different from ours. He uses the word, as he says, which "serves best to stand for whatsoever is the object of the understanding when a man thinks." "I have used it," he adds, "to express whatever is meant by *phantasm*, *notion*, *species*, or whatever it is to which the mind can be employed about in thinking." It might be shown that this separation of the *mind itself* from the ideal *objects* about which it is employed in thinking, may lead to very erroneous results. But it may suffice to observe that we use the word *Ideas*, in the manner already explained, to express that element, supplied by the mind itself, which must be combined with Sensation in order to produce knowledge. For us, Ideas are not Objects of Thought, but rather Laws of Thought. Ideas are not synonymous with Notions; they are Principles which give to our Notions whatever they contain of truth. But our use of the term *Idea* will be more fully explained hereafter.

Sect. 7.—Subjective and Objective.

The Fundamental Antithesis of Philosophy of which I have to speak has been brought into great prominence in the writings of modern German philosophers, and has conspicuously formed the basis of their systems. They have indicated this antithesis by the terms *subjective* and *objective*. According to the technical language of old writers, a thing and its qualities are described as *subject* and *attributes*; and thus a man's faculties and acts are attributes of which he is the *subject*. The mind is the *subject* in which ideas inhere. Moreover, the man's faculties and acts are employed upon external *objects*; and from objects all his sensations arise. Hence the part of a man's knowledge which belongs to his own mind, is *subjective*: that which flows in upon him from the world external to him, is *objective*. And as in man's contemplation of nature, there is always some act of thought which depends upon himself, and some matter of thought which is independent of him, there is, in every part of his knowledge, a subjective and an objective element. The combination of the two elements, the subjective or ideal, and the objective or observed, is necessary, in order to give us any insight into the laws of nature. But different persons, according to their mental habits and constitution, may be inclined to dwell by preference upon the one or the other of these two elements. It may perhaps interest the reader to see this difference of intellectual character illustrated in two eminent men of genius of modern times, Göthe and Schiller.

Göthe himself gives us the account to which I refer, in his history of the progress of his speculations concerning the Metamorphosis of Plants; a mode of viewing their structure by which he explained, in a very striking and beautiful manner, the relations of the different parts of a plant to each other; as has been narrated in the *History of the Inductive Sciences*. Göthe felt a delight in the passive contemplation of nature, unmingled with the desire of reasoning and theorizing; a delight such as naturally belongs to those poets who merely embody the images which a fertile genius suggests, and do not mix with these pictures, judgments and reflexions of their own. Schiller, on the other hand, both by his own strong feeling of the value of a moral purpose in poetry, and by his adoption of a system of metaphysics in which the subjective element was made very prominent, was well disposed to recognize fully the authority of ideas over external impressions.

Göthe for a time felt a degree of estrangement towards Schiller, arising from this contrariety in their views and characters. But on one occasion they fell into discussion on the study of natural history; and Göthe endeavoured to impress upon his companion his persuasion that nature was to be considered, not as composed of detached and incoherent parts, but as active and alive, and unfolding herself in each portion, in virtue of principles which pervade the whole. Schiller objected that no such view of the objects of natural history had been pointed out by observation, the only guide which the natural historians recommended; and was disposed on this account to think the whole of their study narrow and shallow. "Upon this," says Göthe, "I expounded to him, in as lively a way as I could, the metamorphosis of plants, drawing on paper for him, as I proceeded, a diagram to represent that general form of a plant which shows itself in so many and so various transformations. Schiller attended and understood; and, accepting the explanation, he said, 'This is not observation, but an idea.' I replied," adds Göthe, "with some degree of irritation; for the point which separated us was most luminously marked by this expression: but I smothered my vexation, and merely said, 'I was happy to find that I had got ideas without knowing it; nay, that I saw them before my eyes.'" Göthe then goes on to say, that he had been grieved to the very soul by maxims promulgated by Schiller, that no observed fact ever could correspond with an idea. Since he himself loved best to wander in the domain of external observation, he had been led to look with repugnance and hostility upon anything which professed to depend upon ideas. "Yet," he observes, "it occurred to me that if my Observation was identical with his Idea, there must be some common ground on which we might meet." They went on with their mutual explanations, and became intimate and lasting friends. "And thus," adds the poet, "by means of that mighty and interminable controversy between *object* and *subject*, we two concluded an alliance which remained unbroken, and produced much benefit to ourselves and others."

The general diagram of a plant, of which Göthe here speaks, must have been a combination of lines and marks expressing the relations of position and equivalence among the elements of vegetable forms, by which so many of their resemblances and differences may be explained. Such a symbol is not an Idea in that general sense in which we propose to use the term, but is a particular modification of the general Ideas of symmetry, developement, and the like; and we shall hereafter see, according to the phraseology which we shall explain in the next chapter, how such a diagram might express the *ideal conception* of a plant.

The antithesis of *subjective* and *objective* is very familiar in the philosophical literature of Germany and France; nor is it uncommon in any age of our own literature. But though efforts have recently been made to give currency among us to this phraseology, it has not been cordially received, and has been much complained of as not of obvious meaning. Nor is the complaint without ground: for when we regard the mind as the *subject* in which ideas inhere, it becomes for us an *object*, and the antithesis vanishes. We are not so much accustomed to use *subject* in this sense, as to make it a proper contrast to *object*. The combination "*ideal* and *objective*," would more readily convey to a modern reader the opposition which is intended between the ideas of the mind itself, and the objects which it contemplates around it.

To the antitheses already noticed—Thoughts and Things; Necessary and Experiential Truths; Deduction and Induction; Theory and Fact; Ideas and Sensations; Reflexion and Sensation; Subjective and Objective; we may add others, by which distinctions depending more or less upon the fundamental antithesis have been denoted. Thus we speak of the *internal* and *external* sources of our knowledge; of the world *within* and the world *without* us; of *Man* and *Nature*. Some of the more recent metaphysical writers of Germany have divided the universe into the *Me* and the *Not-me* (Ich and Nicht-ich). Upon such phraseology we may observe, that to have the fundamental antithesis of which we speak really understood, is of the highest consequence to philosophy, but that little appears to be gained by expressing it in any novel manner. The most weighty part of the philosopher's task is to analyze the operations of the mind; and in this task, it can aid us but little to call it, instead of the *mind*, the *subject*, or the *me*.

Sect. 8.—Matter and Form.

There are some other ways of expressing, or rather of illustrating, the fundamental antithesis, which I may briefly notice. The antithesis has been at different times presented by means of various images. One of the most ancient of these, and one which is still very instructive, is that which speaks of Sensations as the *Matter*, and Ideas as the *Form*, of our knowledge; just as ivory is the matter, and a cube the form, of a die. This comparison has the advantage of showing that two elements of an antithesis which cannot be separated in fact, may yet be advantageously separated in our reasonings. For Matter and Form cannot by any means be detached from each other. All matter must have some form; all form must be the form of some material thing. If the ivory be not a cube, it must have a spherical or some other form. And the cube, in order to be a cube, must be of some material;—if not of ivory, of wood, or stone, for instance. A figure without matter is merely a geometrical conception;—a modification of the idea of space. Matter without figure is a mere abstract term;—a supposed union of certain sensible qualities which, so insulated from others, cannot exist. Yet the distinction of Matter and Form is real; and, as a subject of contemplation, clear and plain. Nor is the distinction by any means useless. The speculations which treat of the two subjects, Matter and Figure, are very different. Matter is the subject of the sciences of Mechanics and Chemistry; Figure, of Geometry. These two classes of Sciences have quite different sets of principles. If we refuse to consider the Matter and the Form of bodies separately, because we cannot exhibit Matter and Form separately, we shut the door to all philosophy on such subjects. In like manner, though Sensations and Ideas are necessarily united in all our knowledge, they can be considered as distinct; and this distinction is the basis of all philosophy concerning knowledge.

This illustration of the relation of Ideas and Sensations may enable us to estimate a doctrine which has been put forwards at various times. In a certain school of speculators there has existed a disposition to derive all our Ideas from our Sensations, the term *Idea* being, in this school, used in its wider sense, so as to include all modifications and limitations of our Fundamental Ideas. The doctrines of this school have been summarily expressed by saying that "Every Idea is a transformed Sensation." Now, even supposing this assertion to be exactly true, we easily see, from what has been said, how little we are likely to answer the ends of philosophy by putting forward such a maxim as one of primary importance. For we might say, in like manner, that every statue is but a transformed block of marble, or every edifice but a collection of transformed stones. But what would these assertions avail us, if our object were to trace the rules of art by which beautiful statues were formed, or great works of architecture erected? The question naturally occurs, What is the nature, the principle, the law of this Transformation? In what faculty resides the transforming power? What train of ideas of beauty, and symmetry, and stability, in the mind of the statuary or the architect, has produced those great works which mankind look upon as among their most valuable possessions;—the Apollo of the Belvidere, the Parthenon, the Cathedral of Cologne? When this is what we want to know, how are we helped by learning that the Apollo is of Parian marble, or the Cathedral of basaltic stone? We must know much more than this, in order to acquire any insight into the principles of statuary or of architecture. In like manner, in order that we may make any progress in the philosophy of knowledge, which is our purpose, we must endeavour to learn something further respecting ideas than that they are transformed sensations, even if they were this.

But, in reality, the assertion that our ideas are transformed sensations, is erroneous as well as frivolous. For it conveys, and is intended to convey, the opinion that our sensations have one form which properly belongs to them; and that, in order to become ideas, they are converted into some other form. But the truth is, that our sensations, of themselves, without some act of the mind, such as involves what we have termed an Idea, have no form. We cannot see one object without the idea of space; we cannot see two without the idea of resemblance or difference; and space and difference are not sensations. Thus, if we are to employ the metaphor of Matter and Form, which is implied in the expression to which I have referred, our sensations, from their first reception, have their Form not *changed*, but *given* by our Ideas. Without the relations of thought which we here term *Ideas*, the sensations are matter without form. Matter without form cannot exist: and in like manner sensations cannot become perceptions of objects, without some formative power of the mind. By the very act of being received as perceptions, they have a formative power exercised upon them, the operation of which might be expressed, by speaking of them, not as *transformed*, but simply as *formed*;—as invested with form, instead of being the mere formless material of perception. The word *inform*, according to its Latin etymology, at first implied this process by which matter is invested with form. Thus Virgil[1] speaks of the thunderbolt as *informed* by the hands of Brontes, and Steropes, and Pyracmon. And Dryden introduces the word in another place:—

> Let others better mould the running mass
> Of metals, or *inform* the breathing brass.

Even in this use of the word, the form is something superior to the brute manner, and gives it a new significance and purpose. And hence the term is again used to denote the effect produced by an intelligent principle of a still higher kind:—

> . . . . . . He *informed*
> This ill-shaped body with a daring soul.

And finally even the soul itself, in its original condition, is looked upon as matter, when viewed with reference to education and knowledge, by which it is afterwards moulded; and hence these are, in our language, termed *information*. If we confine ourselves to the first of these three uses of the term, we may correct the erroneous opinion of which we have just been speaking, and retain the metaphor by which it is expressed, by saying, that ideas are not *transformed*, but *informed* sensations.

Sect. 9.—Man the Interpreter of Nature.

There is another image by which writers have represented the acts of thought through which knowledge is obtained from the observation of the external world. Nature is the Book, and Man is the *Interpreter*. The facts of the external world are marks, in which man discovers a meaning, and so reads them. Man is the Interpreter of Nature, and Science is the right Interpretation. And this image also is, in many respects, instruc-

1 Ferrum exercebant vasto Cyclopes in Antro
Brontesque Steropesque et nudus membra Pyracmon;
His informatum manibus, jam parte polita
Fulmen erat.—*Aen.* viii. 424.

tive. It exhibits to us the necessity of both elements;—the marks which man has to look at, and the knowledge of the alphabet and language which he must possess and apply before he can find any meaning in what he sees. Moreover this image presents to us, as the ideal element, an activity of the mind of that very kind which we wish to point out. Indeed the illustration is rather an example than a comparison of the composition of our knowledge. The letters and symbols which are presented to the Interpreter are really objects of sensation: the notion of letters as signs of words, the notion of connexions among words by which they have meaning, really are among our Ideas;—*Signs* and *Meaning* are Ideas, supplied by the mind, and added to all that sensation can disclose in any collection of visible marks. The Sciences are not figuratively, but really, Interpretations of Nature. But this image, whether taken as example or comparison, may serve to show both the opposite character of the two elements of knowledge, and their necessary combination, in order that there may be knowledge.

This illustration may also serve to explain another point in the conditions of human knowledge which we shall have to notice:—namely, the very different degrees in which, in different cases, we are conscious of the mental act by which our sensations are converted into knowledge. For the same difference occurs in reading an inscription. If the inscription were entire and plain, in a language with which we were familiar, we should be unconscious of any mental act in reading it. We should seem to collect its meaning by the sight alone. But if we had to decipher an ancient inscription, of which only imperfect marks remained, with a few entire letters among them, we should probably make several suppositions as to the mode of reading it, before we found any mode which was quite successful; and thus, our guesses, being separate from the observed facts, and at first not fully in agreement with them, we should be clearly aware that the conjectured meaning, on the one hand, and the observed marks on the other, were distinct things, though these two things would become united as elements of one act of knowledge when we had hit upon the right conjecture.

Sect. 10.—The Fundamental Antithesis inseparable.

The illustration just referred to, as well as other ways of considering the subject, may help us to get over a difficulty which at first sight appears perplexing. We have spoken of the common opposition of *Theory* and *Fact* as important, and as involving what we have called the Fundamental Antithesis of Philosophy. But after all, it may be asked, Is this distinction of Theory and Fact really tenable? Is it not often difficult to say whether a special part of our knowledge is a Fact or a Theory? Is it a Fact or a Theory that the stars revolve round the pole? Is it a Fact or a Theory that the earth is a globe revolving on its axis? Is it a Fact or a Theory that the earth travels in an ellipse round the sun? Is it a Fact or a Theory that the sun attracts the earth? Is it a Fact or a Theory that the loadstone attracts the needle? In all these cases, probably some persons would answer one way, and some persons the other. There are many persons by whom the doctrine of the globular form of the earth, the doctrine of the earth's elliptical orbit, the doctrine of the sun's attraction on the earth, would be called *theories*, even if they allowed them to be true theories. But yet if each of these propositions be true, is it not a *fact*? And even with regard to the simpler facts, as the motion of the stars round the pole, although this may be a Fact to one who has watched and measured the motions of the stars, one who has not done this, and who has only carelessly looked at these stars from time to time, may naturally speak of the circles which the astronomer makes them describe as Theories. It would seem, then, that we cannot in such cases expect general assent, if we say, *This is a Fact and not a Theory, or, This is a Theory and not a Fact*. And the same is true in a vast range of cases. It would seem, therefore, that we cannot rest any reasoning upon this distinction of Theory and Fact; and we cannot avoid asking whether there is any real distinction in this antithesis, and if so, what it is.

To this I reply: the distinction between Theory (that is, true Theory) and Fact, is this: that in Theory the Ideas are considered as distinct from the Facts: in Facts, though Ideas may be involved, they are not, in our apprehension, separated from the sensations. In a Fact, the Ideas are applied so readily and familiarly, and incorporated with the sensations so entirely, that we do not see *them*, we see *through them*. A person who carefully notes the motion of a star all night, sees the circle which it describes, as he sees the star, though the circle is, in fact, a result of his own Ideas. A person who has in his mind the measures of different lines and countries on the earth's surface, and who can put them together into one conception, finds that they can make no figure but a globular one: to him, the earth's globular form is a Fact, as much as the square form of his chamber. A person to whom the grounds of believing the earth to travel round the sun are as familiar as the grounds for believing the movements of the mail-coaches in this country, looks upon the former event as a Fact, just as he looks upon the latter events as Facts. And a person who, knowing the Fact of the earth's annual motion, refers it distinctly to its mechanical cause, conceives the sun's attraction as a Fact, just as he conceives as a Fact, the action of the wind which turns the sails of a mill. He cannot *see* the force in either case; he supplies it out of his own Ideas. And thus, a true Theory is a Fact; a Fact is a familiar Theory. That which is a Fact under one aspect, is a Theory under another. The most recondite Theories when firmly established are Facts: the simplest Facts involve something of the nature of Theory. Theory and Fact correspond, in a certain degree, with Ideas and Sensations, as to the nature of their opposition. But the Facts are Facts, so far as the Ideas have been combined with the Sensations and absorbed in them: the Theories are Theories, so far as the Ideas are kept distinct from the Sensations, and so far as it is considered still a question whether those can be made to agree with these.

We may, as I have said, illustrate this matter by considering man as *interpreting* the phenomena which he sees. He often interprets without being aware that he does so. Thus when we see the needle move towards the magnet, we assert that the magnet exercises an attractive force on the needle. But it is only by an interpretative act of our own minds that we ascribe this motion to attraction. That, in this case, a force is exerted—something of the nature of the pull which we could apply by our own volition—is our interpretation of the phenomena; although we may be conscious of the act of interpretation, and may then regard the attraction as a Fact.

Nor is it in such cases only that we interpret phenomena in our own way, without being conscious of what we do. We see a tree at a distance, and judge it to be a chestnut or a lime; yet this is only an inference from the colour or form of the mass according to preconceived classifications of our own. Our lives are full of such unconscious interpretations. The farmer recognizes a good or a bad soil; the artist a picture of a favourite master; the geologist a rock of a known locality, as we recognize the faces and voices of our friends; that is, by judgments formed on what we see and hear; but judgments in which we do not analyze the steps, or distinguish the inference from the appearance. And in these mixtures of observation and inference, we speak of the judgment thus formed, as a Fact directly observed.

Even in the case in which our perceptions appear to be most direct, and least to involve any interpretations of our own,—in the simple process of seeing,—who does not know how much we, by an act of the mind, add to that which our senses receive? Does any one fancy that he sees a solid cube? It is easy to show that the solidity of the figure, the relative position of its faces and edges to each other, are inferences of the spectator; no more conveyed to his conviction by the eye alone, than they would be if he were looking at a painted representation of a cube. The scene of nature is a picture without depth of substance, no less than the scene of art; and in the one case as in the other, it is the mind which, by an act of its own, discovers that colour and shape denote distance and solidity. Most men are unconscious of this perpetual habit of reading the language of the external world, and translating as they read. The draughtsman, indeed, is compelled, for his purposes, to return back in thought from the solid bodies which he has inferred, to the shapes of surface which he really sees. He knows that there is a mask of theory over the whole face of nature, if it be *theory* to infer more than we *see*. But other men, unaware of this masquerade, hold it to be a fact that they see cubes and spheres, spacious apartments and winding avenues. And these things are facts to them, because they are unconscious of the mental operation by which they have penetrated nature's disguise.

And thus, we still have an intelligible distinction of Fact and Theory, if we consider Theory as a conscious, and Fact as an unconscious inference, from the phenomena which are presented to our senses.

But still, Theory and Fact, Inference and Perception, Reasoning and Observation, are antitheses in none of which can we separate the two members by any fixed and definite line.

Even the simplest terms by which the antithesis is expressed cannot be separated. Ideas and Sensations, Thoughts and Things, Subject and Object, cannot in any case be applied absolutely and exclusively. Our Sensations require Ideas to bind them together, namely, Ideas of space, time, number, and the like. If not so bound together, Sensations do not give us any apprehension of Things or Objects. All Things, all Objects, must exist in space and in time—must be one or many. Now space, time, number, are not Sensations or Things. They are something different from, and opposed to Sensations and Things. We have termed them Ideas. It may be said they are *Relations* of Things, or of Sensations. But granting this form of expression, still a *Relation* is not a Thing or a Sensation; and therefore we must still have another and opposite element, along with our Sensations. And yet, though we have thus these two elements in every act of perception, we cannot designate any portion of the act as absolutely and exclusively belonging to one of the elements. Perception involves Sensation, along with Ideas of time, space, and the like; or, if any one prefers the expression, we may say, Perception involves Sensations along with the apprehension of Relations. Perception is Sensation, along with such Ideas as make Sensation into an apprehension of Things or Objects.

And as Perception of Objects implies Ideas,—as Observation implies Reasoning;—so, on the other hand, Ideas cannot exist where Sensation has not been; Reasoning cannot go on when there has not been previous Observation. This is evident from the necessary order of developement of the human faculties. Sensation necessarily exists from the first moments of our existence, and is constantly at work. Observation begins before we can suppose the existence of any Reasoning which is not involved in Observation. Hence, at whatever period we consider our Ideas, we must consider them as having been already engaged in connecting our Sensations, and as having been modified by this employment. By being so employed, our Ideas are unfolded and defined; and such development and definition cannot be separated from the Ideas themselves. We cannot conceive space, without boundaries or forms; now Forms involve Sensations. We cannot conceive time, without events which mark the course of time; but events involve Sensations. We cannot conceive number, without conceiving things which are numbered; and Things imply sensations. And the forms, things, events, which are thus implied in our Ideas, having been the objects of Sensation constantly in every part of our life, have modified, unfolded, and fixed our Ideas, to an extent which we cannot estimate, but which we must suppose to be essential to the processes which at present go on in our minds. We cannot say that Objects create Ideas; for to perceive Objects we must already have Ideas. But we may say, that Objects and the constant Perception of Objects have so far modified our Ideas, that we cannot, even in thought, separate our Ideas from the perception of Objects.

We cannot say of any Ideas, as of the Idea of space, or time, or number, that they are absolutely and exclusively Ideas. We cannot conceive what space, or time, or number, would be in our minds, if we had never perceived any Thing or Things in space or time. We cannot conceive ourselves in such a condition as never to have perceived any Thing or Things in space or time. But, on the other hand, just as little can we conceive ourselves becoming acquainted with space and time or numbers as objects of Sensation. We cannot reason without having the operations of our minds affected by previous Sensations; but we cannot conceive Reasoning to be merely a series of Sensations. In order to be used in Reasoning, Sensation must become Observation; and, as we have seen, Observation already involves Reasoning. In order to be connected by our Ideas, Sensations must be Things or Objects, and Things or Objects already include Ideas. And thus, none of the terms by which the fundamental antithesis is expressed can be absolutely and exclusively applied.

I will make a remark suggested by the views which have thus been presented. Since, as we have just seen, none of the terms which express the fundamental antithesis can be applied absolutely and exclusively, the absolute application of the antithesis in any particular case can never be a conclusive or immoveable principle. This remark is the more necessary to be borne in mind, as the terms of this antithesis are often used in a vehement and peremptory manner. Thus we are often told that such a thing is *a Fact*; A FACT and not a Theory, with all the emphasis which, in speaking or writing, tone or italics or capitals can give. We see from what has been said, that when this is urged, before we can estimate the truth, or the value of the assertion, we must ask to whom is it a Fact? what habits of thought, what previous information, what Ideas does it imply, to conceive the Fact as a Fact? Does not the apprehension of the Fact imply assumptions which may with equal justice be called Theory, and which are perhaps false Theory? in which case, the Fact is no Fact. Did not the ancients assert it as a Fact, that the earth stood still, and the stars moved? and can any Fact have stronger apparent evidence to justify persons in asserting it emphatically than this had?

These remarks are by no means urged in order to shew that no Fact can be certainly known to be true; but only, to shew that no Fact can be certainly shown to be a Fact, merely by calling it a Fact, however emphatically. There is by no means any ground of general skepticism with regard to truth, involved in the doctrine of the necessary combination of two elements in all our knowledge. On the contrary, Ideas are requisite to the essence, and Things to the reality of our knowledge in every case. The proportions of Geometry and Arithmetic are examples of knowledge respecting our Ideas of space and number, with regard to which there is no room for doubt. The doctrines of Astronomy are examples of truths not less certain respecting the Facts of the external world.

Sect. 11.—Successive Generalization.

In the preceding pages we have been led to the doctrine, that though, in the Antithesis of Theory and Fact, there is involved an essential opposition; namely the opposition of the thoughts within us and the phenomena without us; yet that we cannot distinguish and define the members of this antithesis separately. Theories become Facts, by becoming certain and familiar: and thus, as our knowledge becomes more sure and more extensive, we are constantly transferring to the class of facts, opinions which were at first regarded as theories.

Now we have further to remark, that in the progress of human knowledge respecting any branch of speculation, there may be *several* such steps in succession, each depending upon and including the preceding. The theoretical views which one generation of discoverers establishes, become the facts from which the next generation advances to new theories. As men rise from the particular to the general, so, in the same manner, they rise from what is general to what is more general. Each induction supplies the materials of fresh inductions; each generalization, with all that it embraces in its circle, may be found to be but one of many circles, comprehended within the circuit of some wider generalization.

This remark has already been made, and illustrated, in the *History of the Inductive Sciences*;[2] and, in truth, the whole of the history of science is full of suggestions and exemplifications of this course of things. It may be convenient, however, to select a few instances which may further explain and confirm this view of the progress of scientific knowledge.

The most conspicuous instance of this succession is to be found in that science which has been progressive from the beginning of the world to our own times, and which exhibits by far the richest collection of successive discoveries: I mean Astronomy. It is easy to see that each of these successive discoveries depended on those antecedently made, and that in each, the truths which were the highest point of the knowledge of

2 *Hist. Inductive Sciences*, B. VII. c. ii. Sect. 5.

one age were the fundamental basis of the efforts of the age which came next. Thus we find, in the days of Greek discovery, Hipparchus and Ptolemy combining and explaining the particular *facts* of the motion of the sun, moon, and planets, by means of the *theory* of epicycles and eccentrics;—a highly important step, which gave an intelligible connexion and rule to the motions of each of these luminaries. When these cycles and epicycles, thus truly representing the apparent motions of the heavenly bodies, had accumulated to an inconvenient amount, by the discovery of many inequalities in the observed motions, Copernicus showed that their effects might all be more simply included, by making the sun the center of motion of the planets, instead of the earth. But in this new view, he still retained the epicycles and eccentrics which governed the motion of each body. Tycho Brahe's observations, and Kepler's calculations, showed that, besides the vast number of facts which the epicyclical theory could account for, there were some which it would not exactly include, and Kepler was led to the persuasion that the planets move in ellipses. But this view of motion was at first conceived by Kepler as a modification of the conception of epicycles. On one occasion he blames himself for not sooner seeing that such a modification was possible. "What an absurdity on my part!" he cries;[3] "as if libration in the diameter of the epicycle might not come to the same thing as motion in the ellipse." But again; Kepler's *laws* of the elliptical motion of the planets were established; and these laws immediately became the *facts* on which the mathematicians had to found their mechanical theories. From these facts, Newton, as we have related, proved that the central force of the sun retains the planets in their orbits, according to the law of the inverse square of the distance. The same *law* was shown to prevail in the gravitation of the earth. It was shown, too, by induction from the motions of Jupiter and Saturn, that the planets attract each other; by calculations from the figure of the earth, that the parts of the earth attract each other; and, by considering the course of the tides, that the sun and moon attract the waters of the ocean. And all these curious discoveries being established as *facts*, the subject was ready for another step of generalization. By an unparalleled rapidity in the progress of discovery in this case, not only were all the inductions which we have first mentioned made by one individual, but the new advance, the higher flight, the closing victory, fell to the lot of the same extraordinary person.

The attraction of the sun upon the planets, of the moon upon the earth, of the planets on each other, of the parts of the earth on themselves, of the sun and moon upon the ocean;—all these truths, each of itself a great discovery, were included by Newton in the higher *generalization*, of the universal gravitation of matter, by which each particle is drawn to each other according to the law of the inverse square: and thus this long advance from discovery to discovery, from truths to truths, each justly admired when new, and then rightly used as old, was closed in a worthy and consistent manner, by a truth which is the most worthy admiration, because it includes all the researches of preceding ages of Astronomy.

We may take another example of a succession of this kind from the history of a science, which, though it has made wonderful advances, has not yet reached its goal, as physical astronomy appears to have done, but seems to have before it a long prospect of future progress. I now refer to Chemistry, in which I shall try to point out how the preceding discoveries afforded the materials of the succeeding; although this subordination and connexion is, in this case, less familiar to men's minds than in Astronomy, and is, perhaps, more difficult to present in a clear and definite shape. Sylvius saw, in the facts which occur, when an acid and an alkali are brought together, the evidence that they neutralize each other. But cases of neutralization, and acidification, and many other effects of mixture of the ingredients of bodies, being thus viewed as *facts*, had an aspect of unity and law given them by Geoffroy and Bergman,[4] who introduced the *conception* of

3 *Hist. Inductive Sciences*, B. v. c. iv. Sect. 3.

4 *Hist. Inductive Sciences*, B. XIV. c. iii.

the Chemical Affinity or Elective Attraction, by which certain elements select other elements, as if by preference. That combustion, whether a chemical union or a chemical separation of ingredients, is of the same nature with acidification, was the doctrine of Beecher and Stahl, and was soon established as a truth which must form a part of every succeeding physical theory. That the rules of affinity and chemical composition may include gaseous elements, was established by Black and Cavendish. And all these truths, thus brought to light by chemical discoverers,—affinity, the identity of acidification and combustion, the importance of gaseous elements,—along with all the facts respecting the weight of ingredients and compounds which the balance disclosed,—were taken up, connected, and included as *particulars* in the oxygen *theory* of Lavoisier. Again, the results of this theory, and the quantity of the several ingredients which entered into each compound—(such results, for the most part, being now no longer mere theoretical speculations, but recognized facts)—were the *particulars* from which Dalton derived that wide law of chemical combination which we term the Atomic *Theory*. And this law, soon generally accepted among chemists, is already in its turn become one of the *facts* included in Faraday's *Theory* of the identity of Chemical Affinity and Electric Attraction.

It is unnecessary to give further exemplifications of this constant ascent from one step to a higher;—this perpetual conversion of true theories into the materials of other and wider theories. It will hereafter be our business to exhibit, in a more full and formal manner, the mode in which this principle determines the whole scheme and structure of all the most exact sciences. And thus, beginning with the facts of sense, we gradually climb to the highest forms of human knowledge, and obtain from experience and observation a vast collection of the most wide and elevated truths.

There are, however, truths of a very different kind, to which we must turn our attention, in order to pursue our researches respecting the nature and grounds of our knowledge. But before we do this, we must notice one more feature in that progress of science which we have already in part described.

FURTHER READING

Blanché, Robert. *Le Rationalisme de Whewell.* Paris: Librairie Félix Alcan, 1935.

Butts, Robert E., ed. *William Whewell's Theory of Scientific Method.* Pittsburgh: University of Pittsburgh Press, 1968.

Elkana, Yehuda. *William Whewell: Selected Writings on the History of Science.* Chicago: University of Chicago Press, 1984.

Fisch, Menachem. *William Whewell: Philosopher of Science.* Oxford: Clarendon Press, 1991.

Fisch, Menachem, and Simon Schaffer, eds. *William Whewell: A Composite Portrait.* Oxford: Clarendon Press, 1991.

Giere, Ronald N., and Richard S. Westfall, eds. *Foundations of Scientific Method: The Nineteenth Century.* Bloomington: Indiana UP, 1973.

Gower, Barry. *Scientific Method: An Historical and Philosophical Introduction.* London: Routledge, 1997.

Henderson, James P. *Early Mathematical Economics: William Whewell and the British Case.* New York: Rowman and Littlefield Publishers, Inc., 1996.

Herschel, John. *A Preliminary Discourse on the Study of Natural Philosophy.* 1830. Reprint, New York: Johnson Reprint Corporation, 1966.

—. *A Preliminary Discourse on the Study of Natural Philosophy.* Edited by Arthur Fine. 1830. Reprint, Chicago: University of Chicago Press, 1987.

Klee, Robert. *Introduction to the Philosophy of Science: Cutting Nature at its Seams.* Oxford: Oxford UP, 1997.

Laudan, L. "William Whewell on the Consilience of Inductions." *Monist* 55 (1971): 368–91.

Marcucci, Silvestro. *L'"idealismo" scientifico di William Whewell.* Pisa: Istituto di Filosofia, 1963.

Morrell, J., and Arnold Thackray. *Gentlemen of Science: Early Years of the British Association for the Advancement of Science.* Oxford: Clarendon Press, 1981.

Morrison, Margaret. "Whewell on the Ultimate Problem of Philosophy." *Studies in History and Philosophy of Science* 28, no. 3 (1997): 417–37.

Richards, Joan L. "Observing Science in Early

Victorian England: Recent Scholarship on William Whewell." *Perspectives on Science* 4 (1996): 231–47.

Ruse, M. "The Scientific Methodology of William Whewell." *Centaurus* 20 (1976): 227–57.

Snyder, Laura J. "The Mill-Whewell Debate: Much Ado About Induction." *Perspectives on Science* 5, no. 2 (1997): 159–98.

Stair Douglas, Janet Mary. *The Life and Selections from the Correspondence of William Whewell, D.D.* 2d ed. London: Kegan Paul, Trench, and Co., 1882.

Stoll, M.R. *Whewell's Philosophy of Induction.* Lancaster, 1928.

Todhunter, Isaac. *William Whewell, D.D.: An Account of His Writings With Selections from His Literary and Scientific Correspondence.* 2 vols. London: Macmillan and Co., 1876.

Wettersten, John R. *The Roots of Critical Rationalism.* Amsterdam: Rodopi, 1992.

Yeo, Richard. *Defining Science: William Whewell, Natural Knowledge, and Public Debate in Early Victorian Britain.* Cambridge: Cambridge UP, 1993.

17. HERBERT SPENCER

(1820–1903)

Herbert Spencer was recognized in his lifetime as one of the most important English philosophers and synthesizers of scientific thought, although he did almost no original experimental research. Spencer's books can be found in almost every section of a modern library, attesting to the encyclopedic range of his interests. His works were translated and read by admirers in the United States, Germany, France, Russia, and Japan. Born in Derby, Spencer grew up under a critical, authoritarian father who was a school teacher. The fact that Spencer never married, his periodic withdrawal from society, and his social theories of individualism are in part traceable to his upbringing and family circumstances.

His Uncle Thomas Spencer tutored him in mathematics and classical languages for three years beginning in 1833, and Spencer became involved in his Uncle's radical politics. After this private tutoring by his family, Spencer was largely self-taught. Like John Tyndall, Spencer became a railway engineer in 1837 during the boom years, but later began a journalism career in London in 1848 as an editor of the *Economist*. His friends and acquaintances in London included the novelist George Eliot (Marian Evans), whom he once considered marrying, her companion George Henry Lewes, another scientific popularizer and writer on psychology, and members of the X-Club, including John Lubbock, John Tyndall, and T.H. Huxley.

Spencer wrote a ten-volume series of textbooks on various scientific fields which he called his *Synthetic Philosophy*, including *First Principles* (1862), *Principles of Psychology* (1855), *The Principles of Biology* (1864–67), *Principles of Sociology* (1876–96), and *The Principles of Ethics* (1879–93). Spencer's *First Principles* outlined his epistemology and presented his theory of evolution, which unlike Darwin he generalized to all cosmic and natural processes—from the primal forces of consolidation and aggregation, to the condensation of the stars from hot gas, the progression of the earth from molten to solid state, the development of the embryo, and the evolution of both the individual and society. His basic definition of evolution as "a change from a less coherent form to a more coherent form, consequent on the dissipation of motion and integration of matter" (*First Principles* 327) was too general and all-encompassing to be productively applied to any specific case of evolutionary change, and Spencer ignored many crucial distinctions among the mechanisms of very different types of progressive development.

Spencer is credited with developing the idea of Social Darwinism, an application of Darwinian natural selection and evolution to the development of societies, although Spencer pointed out that several years before Darwin's publication of the theory of evolution in 1859, Spencer had published evolutionary ideas in his *Principles of Psychology* (1855), as well as in his first book *Social Statics* (1851), of which the section on "Poor Laws" is reprinted below. Spencer's *Social Statics* reflects the same intellectual sources influential on Darwin's thinking—primarily Comte, the *Vestiges of Natural Creation*, Lamarckianism (in Charles Lyell's summary), Thomas Malthus, and the Benthamites. Although in his essay "A Theory of Population" (1852) Spencer had coined the term "survival of the fittest" which was eventually adopted by Darwin, he later differed with Darwin in the 1886 essay "The Factors of Organic Evolution" by insisting on Lamarckian use-inheritance as the primary mechanism of evolution.

The less subtle of the Social Darwinists, especially in the United States, advocated minimizing state interference in society, and opposed helping the poor because charity interrupted the natural evolutionary mechanisms which eliminated the less fit from society. Although Darwinian thought applied to the social realm could be used strategically by socialists who believed that science had demonstrated the inevitability of change and the possible perfectibility of humankind and society through evolution, or perhaps violent revolution, Spencer's philosophy was generally taken up by middle-class conservatives—who objected to being taxed for social inequalities they believed to be caused by aristocratic landowners—and the wealthy mercantile classes who used social Darwinism to justify the status quo and to support unrestricted economic activity (laissez-faire capitalism) as naturally ordained. For the successful industrialists, their wealth "was simply the wealth of the fittest survivors of the commercial struggle for existence" (Taylor 89). Obviously eugenic ideas—encouraging the best-adapted to breed and discouraging the unfit from producing offspring—were intermingled in Spencer's and the Social Darwinists' theories.

Unsatisfied with the quality of the reviews of *Social Statics* which did not engage its evolutionary consequences, Spencer wrote his own review which he later published in his posthumous *An Autobiography* (1904). He characterized his own work as: "it considers Man as an organized being subject to the laws of life at large, and considers him as forced by increase of numbers into a social state which necessitates certain limitations to the actions by which he carries on his life; and a cardinal doctrine, much emphasized by Mr. Spencer, is that Man has been, and is, undergoing modifications of nature which fit him for the social state, by making conformity to these conditions spontaneous" (*Autobiography* 1.360–1). The primary ethical premise of the work was that "every man is free to do whatsoever he wills provided he does not infringe the equal freedom of any other man" (*Autobiography* 361).

In the passage below, Spencer argues that the imposition of a national system of poor relief laws would circumvent natural human charity and the natural purifying process by which society is cleansed of the incapable, the improvident, and the idle. Although Spencer believed that humanity would evolve from selfish individual egoism into altruism, his statements on the frailer and needier members of society were frequently harsh, perhaps unconsciously reflecting his own self-disgust with his lifelong neuroses and periodic nervous exhaustion. In *The Study of Sociology* (1873), he attacked the idea of providing charity to the unfit and unworthy: "it may be doubted whether the maudlin philanthropy which, looking only at direct mitigations, persistently ignores indirect mischiefs, does not inflict a greater total of misery than the extremest selfishness inflicts. Refusing to consider the remote influences of his incontinent generosity, the thoughtless giver stands but a degree above the drunkard who thinks only of today's pleasure and ignores tomorrow's pain, or the spendthrift who seeks immediate delights at the cost of ultimate poverty" (344).

By the end of the nineteenth century, Spencer's sociology and politics founded on evolutionary ideas were already widely viewed as obsolete for the modern European state. As J.D.Y. Peel has noted: "Spencer was desperately aware of the emergence of phenomena, from as early as the late 1860s, that were contrary to his ideals: the growth of a professional civil service with ever-widening scope, compulsory primary education, the halting beginnings of a welfare state and new conceptions of the positive role of the state, and later the revival of working-class militancy, demands for the re-imposition of tariffs, imperial preference, accelerated acquisition of colonies, rearmament, jingoism" (*Herbert Spencer* 224).

"Poor-Laws." *Social Statics: or, the Conditions Essential to Human Happiness Specified, and the First of Them Developed.* London: John Chapman, 1851, pp. 311–29.

§ 1.

In common with its other assumptions of secondary offices, the assumption by a government of the office of Reliever-general to the poor, is necessarily forbidden by the principle that a government cannot rightly do anything more than protect. In demanding from a citizen contributions for the mitigation of distress—contributions not needed for the due administration of men's rights—the state is, as we have seen, reversing its function, and diminishing that liberty to exercise the faculties which it was instituted to maintain. Possibly, unmindful of the explanations already given, some will assert that by satisfying the wants of the pauper, a government is in reality extending *his* liberty to exercise his faculties, inasmuch as it is giving him something without which the exercise of them is impossible; and that hence, though it decreases the rate-payer's sphere of action, it compensates by increasing that of the rate-receiver. But this statement of the case implies a confounding of two widely-different things. To enforce the fundamental law—to take care that every man has freedom to do all that he wills, provided he infringes not the equal freedom of any other man—this is the special purpose for which the civil power exists. Now insuring to each the right to pursue within the specified limits the objects of his desires without let or hindrance, is quite a separate thing from insuring him satisfaction. Of two individuals, one may use his liberty of action successfully—may achieve the gratifications he seeks after, or accumulate what is equivalent to many of them—property; whilst the other, having like privileges, may fail to do so. But with these results the state has no concern. All that lies within its commission is to see that each man is allowed to use such powers and opportunities as he possesses; and if it takes from him who has prospered to give to him who has not, it violates its duty towards the one to do more than its duty towards the other. Or, repeating the idea elsewhere expressed (p. 278), it breaks down the vital law of society, that it may effect what social vitality does not call for.

§ 2.

The notion popularized by Cobbett, that every one has a right to a maintenance out of the soil, leaves those who adopt it in an awkward predicament. Do but ask them to specify, and they are set fast. Assent to their principle; tell them you will assume their title to be valid; and then, as a needful preliminary to the liquidation of their claim, ask for some precise definition of it—inquire "What is a maintenance?" They are dumb. "Is it," say you, "potatoes and salt, with rags and a mud cabin? or is it bread and bacon, in a two-roomed cottage? Will a joint on Sundays suffice? or does the demand include meat and malt liquor daily? Will tea, coffee, and tobacco be expected? and if so, how many ounces of each? Are bare walls and brick floors all that is needed? or must there be carpets and paper-hangings? Are shoes considered essential? or will the Scotch practice be approved? Shall the clothing be of fustian? if not, of what quality must the broadcloth be? In short, just point out where, between the two extremes of starvation and luxury, this something called a maintenance lies." Again they are dumb. You expostulate. You explain that nothing can be done until the question is satisfactorily answered. You show that the claim must be reduced to a detailed, intelligible shape before a step can be taken towards its settlement. "How else," you ask, " shall we know whether enough has been awarded, or whether too much?" Still they are dumb. And, indeed, there is no possible reply for them. Opinions they may offer in plenty; but not a precise, unanimous answer. One thinks that a bare subsistence is all that can fairly be demanded. Here is another who hints at something beyond mere necessaries. A third maintains that a few of the enjoyments of life should be provided for. And some of the more consistent, pushing the doctrine to its legitimate result, will rest satisfied with nothing short of community of property. Who now shall decide amongst these conflicting notions? Or, rather, how shall their propounders be brought to an agreement? Can any one of them prove that his definition is tenable and the others not? Yet he must do this if he would make out a case. Before he can prosecute his claim against society, in the high court of morality; he must "file his bill of particulars." If he accomplishes this he is entitled to a hearing. If not, he must evidently be non-suited.

The right to labour—that French translation of our poor-law doctrine—may be similarly treated. A criticism parallel to the foregoing would place its advocates in a parallel dilemma. But there is another way in which the fallacy of this theory, either in its English or its continental form, may be made manifest—a way that may here be fitly employed.

And first let us make sure of the meaning wrapped up in this expression—right to labour. Evidently if we would avoid mistakes we must render it literally—right to *the* labour; for the thing demanded is not the liberty of labouring: this, no one disputes; but it is the opportunity of labouring—the having remunerative employment provided, which is contended for. Now, without dwelling upon the fact that the word *right* as here used, bears a signification quite different from its legitimate one—that it does not here imply something inherent in man, but something depending upon external circumstances—not something possessed in virtue of his faculties, but something springing out of his relationship to others—not something true of him as a solitary individual, but something which can be true of him only as one of a community—not something antecedent to society, but something necessarily subsequent to it—not something expressive of a claim to *do*, but of a claim to be *done unto*—without dwelling upon this, let us take the expression as it stands, and see how it looks when reduced to its lowest terms. When the artizan asserts his right to have work provided for him, he presupposes the existence of some power on which devolves the duty of providing such work. What power is this? The government, he says. But the government is not an original power, it is a deputed one—is subject therefore, to the instruction of its employer—must do that only which its employer directs—and can be held responsible for nothing save the performance of its employer's behests. Now who is its employer? Society. Strictly speaking, therefore, the assertion of our artizan is, that it is the duty of society to find work for him. But he is himself a member of society—is consequently a unit of that body who ought, as he says, to find work for every man—has hence a share in the *duty* of finding work for every man. Whilst, therefore, it is the duty of all other men to find work for him, it is his duty to help in finding work for all other men. And hence, if we indicate his fellows alphabetically, his theory is that A, B, C, D, and the rest of the nation, are bound to employ him; that he is bound, in company with B, C, D, and the rest, to employ A; that he is bound, in company with A, C, D, and the rest, to employ B; is bound, with A, B, D, and the rest, to employ C, with A, B, C, and the rest, to employ D; and so on with each individual of the half score or score millions, of whom the society may be composed!

Thus do we see how readily imaginary rights are distinguishable from real ones. They need no disproof: they disprove themselves. The ordeal of a definition breaks the illusion at once. Bubble-like, they will bear a cursory glance; but disappear in the grasp of any one who tries to lay hold of them.

Meanwhile we must not overlook the fact that, erroneous as are these poor-law and communist theories—these assertions of a man's right to a maintenance, and of his right to have work provided for him—they are, nevertheless, nearly related to a truth. They are unsuccessful efforts to express the fact, that whoso is born on to this planet of ours thereby obtains some interest in it—may not be summarily dismissed again—may not have his existence ignored by those in possession. In other words, they are attempts to embody that thought which finds its legitimate utterance in the law—all men have equal rights to the use of the Earth (Chap. IX.). The prevalence of these crude ideas is natural enough. A vague perception that there is something wrong about the relationship in which the great mass of mankind stand to the soil and to life, was sure eventually to grow up. After getting from under the grosser injustice of slavery, men could not help beginning in course of time to feel what a monstrous thing it was that nine people out of ten should live in the world on sufferance, not having even standing room, save by allowance of those who claimed the Earth's surface (p. 114). Could it be right that all these human beings should not only be without claim to the necessaries of life—should not only be denied the use of those elements from which such necessaries are obtainable—but should further be unable to exchange their labour for such necessaries, except by leave of their more fortunate fellows? Could it be that the majority had thus no better title to existence than one based upon the good-will or convenience of the minority? Could it be that these landless men had "been *mis*-sent to this earth, where all the seats were already taken?" Surely not. And if not, how ought matters to stand? To all which questions, now forced upon men's minds in more or less definite shapes, there come, amongst other answers, these theories of a right to a maintenance and a right of labour. Whilst, therefore, they must be rejected as untenable, we may still recognise in them the imperfect utterances of the moral sense in its efforts to express equity.

§ 3.

The wrong done to the people at large by robbing them of their birthright—their heritage in the earth—is, indeed, thought by some a sufficient excuse for a poor-law, which is regarded by such as an instrumentality for distributing compensation. There is much plausibility in this construction of the matter. But as a defence of national organizations for the support of paupers, it will not bear criticism. Even were there no better reason for demurring to the supposed compromise, it might still be objected that to counterbalance one injury by inflicting another, and to perpetuate these mutual injuries without knowing whether they are or are not equivalents, is at best a very questionable policy. Why organize a diseased state? Some time or other this morbid constitution of things, under which the greater part of the body-politic is cut off from direct access to the source of life, must be changed. Difficult, no doubt, men will find it to establish a normal condition. There is no knowing how many generations may pass away before the task is accomplished. But accomplished it will eventually be. All arrangements, however, which disguise the evils entailed by the present inequitable relationship of mankind to the soil, postpone the day of rectification. "A generous poor-law" is openly advocated as the best means of pacifying an irritated people. Workhouses are used to mitigate the more acute symptoms of social unhealthiness. Parish pay is hush-money. Whoever, then, desires the radical cure of national maladies, but especially of this atrophy of one class, and hypertrophy of another, consequent upon unjust land tenure, cannot consistently advocate any kind of compromise.

But a poor-law is *not* the means of distributing compensation. Neither in respect of those from whom it comes, nor in respect of those to whom it goes, does pauper-relief fulfil the assumed purpose. According to the hypothesis poors'-rates should bear wholly upon the land. But they do not. And at least that part of them which bears upon the land should come from the usurpers or their descendants. But it does not. According to the hypothesis the burden should not fall upon the innocent. But it does; for poors'-rates were imposed after landed property had in many cases changed hands by purchase. According to the hypothesis the burden should not fall upon those already defrauded. But it does; for the majority of rate-payers belong to the non-landowning class. According to the hypothesis all men kept out of their inheritance should receive a share of this so-called compensation. But they do not; for only here and there one, gets any of it. In no way, therefore, is the theory carried out. The original depredators are beyond reach. The guiltless are taxed in their place. A large proportion of those already robbed are robbed afresh. And of the rest, only a few receive the proceeds.

§ 4.

The usual reason assigned for supporting a poor-law is, that it is an indispensable means of mitigating popular suffering. Given by a churchman such a reason is natural enough; but coming, as it often does, from a dissenter, it is strangely inconsistent. Most of the objections raised by the dissenter to an established religion will tell with equal force against established charity. He asserts that it is unjust to tax him for the support of a creed he does not believe. May not another as reasonably protest against being taxed for the maintenance of a system of relief he disapproves? He denies the right of any bishop or council to choose for him which doctrines he shall accept and which he shall reject. Why does he not also deny the right of any commissioner or vestry to choose for him who are worthy of his charity and who are not? If he dissents from a national church on the ground that religion will be more general and more sincere when voluntarily sustained, should he not similarly dissent from a poor-law on the ground that spontaneous beneficence will produce results both wider and better? Might not the corruption which he points out as neutralizing the effects of a state-taught creed, be paralleled by those evils of pauperism accompanying a state-provision for the poor? Should not his nonconformity in respect to *faith* be accompanied by nonconformity in respect to *good works*? Certainly his present opinions are incongruous beyond all reconciling. He resists every attempt to interfere with the *choice* of his religion, but submits to despotic dictation as to the *exercise* of that religion. Whilst he denies the right of a legislature to explain the *theory*, he yet argues the necessity of its direction in the *practice*. It is inconceivable that these positions can be harmonized. Whoso believes that spiritual destitution is to be remedied only by a national church, may with some show of reason propose to deal with physical destitution by an analogous instrumentality. But the advocate of voluntaryism is bound to stand by his principle in the one case as much as in the other.

§ 5.

Whether the sufferings of the unfortunate shall be soothed in obedience to the gentle whisperings of benevolence, or whether fear of the harsh threats of law shall be the motive for relieving them, is indeed a question of no small importance. In deciding how misery is best alleviated we have to consider, not only what is done for the afflicted, but what is the reactive effect upon those who do it. The relationship that springs up between benefactor and beneficiary is, for this present state of the world, a refining one. Having power to muzzle awhile those propensities of the savage which yet linger in us—corrective as it is of that cold, hard state of feeling in which the every-day business of life is pursued—and drawing closer as it does those links of mutual dependence which keep society together—charity is in its nature essentially civilizing. The emotion accompanying every generous act adds an atom to the fabric of the ideal man. As no cruel thing can be done without character being thrust a degree back towards barbarism, so no kind thing can be done without character being moved a degree forward towards perfection. Doubly efficacious, therefore, are all assuagings of distress instigated by sympathy; for not only do they remedy the particular evils to be met, but they help to mould humanity into a form by which such evils will one day be precluded.

Far otherwise is it with law-enforced plans of relief. These exercise just the opposite influence. "The quality of mercy (or pity) is not strained," says the poet. But a poor-law tries to make men pitiful by force. "It droppeth as the gentle rain from heaven," continues the poet. By a poor-law it is wrung from the unwilling. "It blesses him that gives, and him that takes," adds the poet. A poor-law makes it curse both; the one with discontent and recklessness, the other with complainings and often-renewed bitterness.

This turning of balm into poison must have been remarked by the most careless. Watch a ratepayer when the collector's name is announced. You shall observe no kindling of the eye at some thought of happiness to be conferred—no relaxing of the mouth as though selfish cares had for the moment been forgotten—no softening of the voice to tell of compassionate emotion: no, none of these; but rather shall you see contracted features, a clouded brow, a sudden disappearance of what habitual kindliness of expression there may be; the tax-paper is glanced over half in fear and half in vexation; there are grumblings about the short time that has elapsed since the last rate; the purse comes slowly from the pocket; every coin is grudgingly parted with; and after the collector (who is treated with bare civility) has made his exit, some little time passes before the usual equanimity is regained. Is there anything in this to remind us of the virtue which is "twice blessed?" Note again how this act-of-parliament charity perpetually supersedes men's better sentiments. Here is a respectable citizen with enough and to spare: a man of some feeling; liberal, if there is need; generous even, if his pity is excited. A beggar knocks at his door; or he is accosted in his walk by some way-worn tramp. What does he do? Does he listen, investigate, and, if proper, assist? No; he commonly cuts short the tale with—"I have nothing for you, my good man; you must go to your parish." And then he shuts the door, or walks on, as the case may be, with evident unconcern. Should it strike him the next moment that there was something very wo-begone in the petitioner's look, this uncomfortable thought is met by the reflection, that so long as there is a poor-law, he cannot starve, and that it will be time enough to consider his claims when he applies for relief. Thus does the consciousness that there exists a legal provision for the indigent, act as an opiate to the yearnings of sympathy. Had there been no ready-made excuse, the behaviour would probably have been different. Commiseration, pleading for at least an inquiry into the case, would most likely have prevailed; and, in place of an application to the board of guardians, ending in a pittance coldly handed across the pay-table to be thanklessly received, might have commenced a relationship good for both parties—a generosity humanizing to the one, and a succour made doubly valuable to the other by a few words of consolation and encouragement, followed, it may be, by a lift into some self-supporting position.

In truth there could hardly be found a more efficient device for estranging men from each other, and decreasing their fellow-feeling, than this system of state-almsgiving. Being kind by proxy!—could anything be more blighting to the finer instincts? Here is an institution through which, for a few shillings periodically paid, the citizen may compound for all kindness owing from him to his poorer brothers. Is he troubled with twinges of conscience? here is an anodyne for him, to be had by subscribing so much in the pound on his rental. Is he indifferent as to the welfare of others? why then in return for punctual payment of rates he shall have absolution for hardness of heart. Look: here is the advertisement. "Gentlemen's benevolence done for them, in the most business-like manner, and on the lowest terms. Charity doled out by a patent apparatus, warranted to save all soiling of fingers and offence to the nose. Good works undertaken by contract. Infallible remedies for self-reproach always on hand. Tender feelings kept easy at per annum."

And thus we have the gentle, softening, elevating intercourse that should be habitually taking place between rich and poor, superseded by a cold, hard, lifeless mechanism, bound together by dry parchment acts and regulations—managed by commissioners, boards, clerks, and collectors, who perform their respective functions as tasks—and kept a-going by money forcibly taken from all classes indiscriminately. In place of the music breathed by feelings attuned to kind deeds, we have the harsh creaking and jarring of a thing that cannot stir without creating discord—a thing whose every act, from the gathering of its funds to their final distribution, is prolific of grumblings, discontent, anger—a thing that breeds squabbles about authority, disputes as to claims, brow-beatings, jealousies, litigations, corruption, trickery, lying, ingratitude—a thing that supplants, and therefore makes dormant, men's nobler feelings, whilst it stimulates their baser ones.

And now mark how we find illustrated in detail the truth elsewhere expressed in the abstract, that whenever a government oversteps its duty—the maintaining of men's rights—it inevitably retards the process of adaptation. For what faculty is it whose work a poor-law so officiously undertakes? Sympathy. The very faculty above all others needing to be exercised. The faculty which distinguishes the social man from the savage. The faculty which originates the idea of justice—which makes men regardful of each other's claims—which renders society possible. The faculty of whose growth civilization is a history—on whose increased strength the future ameliorations of man's state mainly depend—and by whose ultimate supremacy, human morality, freedom, and happiness will be secured. Of this faculty poor-laws partially supply the place. By doing which they diminish the demands made upon it, limit its exercise, check its development, and therefore retard the process of adaptation.

§ 6.

Pervading all nature we may see at work a stern discipline, which is a little cruel that it may be very kind. That state of universal warfare maintained throughout the lower creation, to the great perplexity of many worthy people, is at bottom the most merciful provision which the circumstances admit of. It is much better that the ruminant animal, when deprived by age of the vigour which made its existence a pleasure, should be killed by some beast of prey, than that it should linger out a life made painful by infirmities, and eventually die of starvation. By the destruction of all such, not only is existence ended before it becomes burdensome, but room is made for a younger generation capable of the fullest enjoyment; and, moreover, out of the very act of substitution happiness is derived for a tribe of predatory creatures. Note further, that their carnivorous enemies not only remove from herbivorous herds individuals past their prime, but also weed out the sickly, the malformed, and the least fleet or powerful. By the aid of which purifying process, as well as by the fighting, so universal in the pairing season, all vitiation of the race through the multiplication of its inferior samples is prevented; and the maintenance of a constitution completely adapted to surrounding conditions, and therefore most productive of happiness, is ensured.

The development of the higher creation is a progress towards a form of being capable of a happiness undiminished by these drawbacks. It is in the human race that the consummation is to be accomplished. Civilization is the last stage of its accomplishment. And the ideal man is the man in whom all the conditions of that accomplishment are fulfilled. Meanwhile the well-being of existing humanity, and the unfolding of it into this ultimate perfection, are both secured by that same beneficent, though severe discipline, to which the animate creation at large is subject: a discipline which is pitiless in the working out of good: a felicity-pursuing law which never swerves for the avoidance of partial and temporary suffering. The poverty of the incapable, the distresses that come upon the imprudent, the starvation of the idle, and those shoulderings aside of the weak by the strong, which leave so many "in shallows and in miseries," are the decrees of a large, far-seeing benevolence. It seems hard that an unskilfulness which with all his efforts he cannot overcome, should entail hunger upon the artizan. It seems hard that a labourer incapacitated by sickness from competing with his stronger fellows, should have to bear the resulting privations. It seems hard that widows and orphans should be left to struggle for life or death. Nevertheless, when regarded not separately, but in connection with the interests of universal humanity, these harsh fatalities are seen to be full of the highest beneficence—the same beneficence which brings to early graves the children of diseased parents, and singles out the low-spirited, the intemperate, and the debilitated as the victims of an epidemic.

There are many very amiable people—people over whom in so far as their feelings are concerned we may fitly rejoice—who have not the nerve to look this matter fairly in the face. Disabled as they are by their sympathies with present suffering, from duly regarding ultimate consequences, they pursue a course which is very injudicious, and in the end even cruel. We do not consider it true kindness in a mother to gratify her child with sweetmeats that are certain to make it ill. We should think it a very foolish sort of benevolence which led a surgeon to let his patient's disease progress to a fatal issue, rather than inflict pain by an operation. Similarly, we must call those spurious philanthropists, who, to prevent present misery, would entail greater misery upon future generations. All defenders of a poor-law must, however, be classed amongst such. That rigorous necessity which, when allowed to act on them, becomes so sharp a spur to the lazy, and so strong a bridle to the random, these paupers' friends would repeal, because of the wailings it here and there produces. Blind to the fact, that under the natural order of things society is constantly excreting its unhealthy, imbecile, slow, vacillating, faithless members, these unthinking, though well-meaning, men advocate an interference which not only stops the purifying process, but even increases the vitiation—absolutely encourages the multiplication of the reckless and incompetent by offering them an unfailing provision, and *dis*courages the multiplication of the competent and provident by heightening the prospective difficulty of maintaining a family. And thus, in their eagerness to prevent the really salutary sufferings that surround us, these sigh-wise and groan-foolish people bequeath to posterity a continually increasing curse.

Returning again to the highest point of view, we find that there is a second and still more injurious mode in which law-enforced charity checks the process of adaptation. To become fit for the social state, man has not only to lose his savageness, but he has to acquire the capacities needful for civilized life. Power of application must be developed; such modification of the intellect as shall qualify it for its new tasks must take place; and, above all, there must be gained the ability to sacrifice a small immediate gratification for a future great one. The state of transition will of course be an unhappy state. Misery inevitably results from incongruity between constitution and conditions. All these evils, which afflict us, and seem to the uninitiated the obvious consequences of this or that removable cause, are unavoidable attendants on the adaptation now in progress. Humanity is being pressed against the inexorable necessities of its new position—is being moulded into harmony with them, and has to bear the resulting unhappiness as best it can. The process *must* be undergone, and the sufferings *must* be endured. No power on earth, no cunningly-devised laws of statesmen, no world-rectifying schemes of the humane, no communist panaceas, no reforms that men ever did broach or ever will broach, can diminish them one jot. Intensified they may be, and are; and in preventing their intensification, the philanthropic will find ample scope for exertion. But there is bound up with the change a *normal* amount of suffering, which cannot be lessened without altering the very laws of life. Every attempt at mitigation of this eventuates in exacerbation of it. All that a poor-law, or any kindred institution can do, is to partially suspend the transition—to take off for awhile, from certain members of society, the painful pressure which is effecting their transformation. At best this is merely to postpone what must ultimately be borne. But it is more than this: it is to undo what has already been done. For the circumstances to which adaptation is taking place cannot be superseded without causing a retrogression—a partial loss of the adaptation previously effected; and as the whole process must some time or other be passed through, the lost ground must be gone over again, and the attendant pain borne afresh. Thus, besides retarding adaptation, a poor-law adds to the distresses inevitably attending it.

At first sight these considerations seem conclusive against *all* relief to the poor—voluntary as well as compulsory; and it is no doubt true that they imply a condemnation of whatever private charity enables the recipients to elude the necessities of our social existence. With this condemnation, however, no rational man will quarrel. That careless squandering of pence which has fostered into perfection a system of organized begging—which has made skilful mendicancy more profitable than ordinary manual labour—which induces the simulation of palsy, epilepsy, cholera, and no end of diseases and deformities—which has called into existence warehouses for the sale and hire of impostor's dresses—which has given to pity-inspiring babes a market value of 9*d.* per day—the unthinking benevolence which has generated all this, cannot but be disapproved by every one. Now it is only against this injudicious charity that the foregoing argument tells. To that charity which may be described as helping men to help themselves, it makes no objection—countenances it rather. And in helping men to help themselves, there remains abundant scope for the exercise of a people's sympathies. Accidents will still supply victims on whom generosity may be legitimately expended. Men thrown upon their backs by unforeseen events, men who have failed for want of knowledge inaccessible to them, men ruined by the dishonesty of others, and men in whom hope long delayed has made the heart sick, may, with advantage to all parties, be assisted. Even the prodigal, after severe hardship has branded his memory with the unbending conditions of social life to which he must submit, may properly have another trial afforded him. And, although by these ameliorations the process of adaptation must be remotely interfered with, yet in the majority of cases, it will not be so much retarded in one direction as it will be advanced in another.

§ 7.

Objectionable as we find a poor-law to be, even under the supposition that it does what it is intended to do—diminish present suffering—how shall we regard it on finding that in reality it does no such thing—cannot do any such thing? Yet, paradoxical as the assertion looks, this is absolutely the fact. Let but the observer cease to contemplate so fixedly one side of the phenomenon—pauperism and its relief, and begin to examine the other side—rates and the *ultimate* contributors of them, and he will discover that to suppose the sum-total of distress diminishable by act-of-parliament bounty is a delusion. A statement of the case in terms of labour and produce will quickly make this clear.

Here, at any specified period, is a given quantity of food and things exchangable for food, in the hands or at the command of the middle and upper classes. A certain portion of this food is needed by these classes themselves, and is consumed by them at the same rate, or very near it, be there scarcity or abundance. Whatever variation occurs in the sum-total of food and its equivalents must therefore affect the remaining portion, not used by these classes for personal sustenance. This remaining portion is given by them to the people in return for their labour, which is partly expended in the production of a further supply of necessaries, and partly in the production of luxuries. Hence, by how much this portion is deficient, by so much must the people come short. Manifestly a re-distribution by legislative or other agency cannot make that sufficient for them which was previously insufficient. It can do nothing but change the parties by whom the insufficiency is felt. If it gives enough to some who else would not have enough, it must inevitably reduce certain others to the condition of not having enough. And thus, to the extent that a poor-law mitigates distress in one place, it unavoidably produces distress in another.

Should there be any to whom this abstract reasoning is unsatisfactory, a concrete statement of the case will, perhaps, remove their doubts. A poors'-rate collector takes from the citizen a sum of money equivalent to bread and clothing for one or more paupers. Had not this sum been so taken, it would either have been used to purchase superfluities, which the citizen now does without, or it would have been paid by him into a bank, and lent by the banker to a manufacturer, merchant, or tradesman; that is, it would ultimately have been given in wages either to the producer of the superfluities or to an operative, paid out of the banker's loan. But this sum having been carried off as poors'-rate, whoever would have received it as wages must now to that extent go without wages. The food which it represented having been taken to sustain a pauper, the artizan to whom that food would have been given in return for work done, must now lack food. And thus, as at first said, the transaction is simply a change of the parties by whom the insufficiency of food is felt.

Nay, the case is even worse. Already it has been pointed out, that by suspending the process of adaptation, a poor-law increases the distress to be borne at some future day; and here we shall find that it also increases the distress to be borne now. For be it remembered, that of the sum taken in any year to support paupers, a large portion would otherwise have gone to support labourers employed in new reproductive works—land-drainage, machine-building, etc. An additional stock of commodities would by-and-by have been produced, and the number of those who go short would consequently have been diminished. Thus the astonishment expressed by some that so much misery should exist, notwithstanding the distribution of fifteen millions a year by endowed charities, benevolent societies, and poor-law unions, is quite uncalled for; seeing that the larger the sum gratuitously administered, the more intense will shortly become the suffering. Manifestly, out of a given population, the greater the number living on the bounty of others, the smaller must be the number living by labour; and the smaller the number living by labour, the smaller must be the production of food and other necessaries; and the smaller the production of necessaries, the greater must be the distress.

§ 8.

We find, then, that the verdict given by the law of state-duty against a public provision for the indigent is enforced by sundry independent considerations. A critical analysis of the alleged rights, for upholding which a poor-law is defended, shows them to be fictitious. Nor does the plea that a poor-law is a means of distributing compensation for wrongs done to the disinherited people turn out to be valid. The assumption that only by law-administered relief can physical destitution be met, proves to be quite analogous to the assumption that spiritual destitution necessitates a law-administered religion; and consistency requires those who assert the sufficiency of voluntary effort in the one case to assert it in the other also. The substitution of a mechanical charity for charity prompted by the heart is manifestly unfavourable to the growth of men's sympathies, and therefore adverse to the process of adaptation. Legal bounty further retards adaptation by interposing between the people and the conditions to which they must become adapted, so as partially to suspend those conditions. And, to crown all, we find, not only that a poor-law must necessarily fail to diminish popular suffering, but that it must inevitably increase that suffering, both directly by checking the production of commodities, and indirectly by causing a retrogression of character, which painful discipline must at some future day make good.

FURTHER READING

Burrow, J.W. *Evolution and Society: A Study in Victorian Social Theory.* Cambridge: Cambridge UP, 1966.

Carneiro, Robert L., ed. *The Evolution of Society: Selections from Herbert Spencer's Principles of Sociology.* Chicago: University of Chicago Press, 1967.

Collins, F. Howard. *An Epitome of the Synthetic Philosophy.* New York: D. Appleton and Company, 1889.

Duncan, David. *The Life and Letters of Herbert Spencer.* London: Methuen and Company, 1908.

Elliot, Hugh. *Herbert Spencer.* New York: Henry Holt and Company, 1917.

Ferri, Enrico. *Socialism and Modern Science (Darwin, Spencer, Marx).* Translated by Robert Rives La Monte. 3rd. ed. Chicago: Charles H. Kerr and Co., 1909.

Gaupp, Otto. *Herbert Spencer.* Stuttgart: Fr. Frommanns Verlag, 1903.

George, Henry. *A Perplexed Philosopher.* New York: Doubleday Page and Co., 1906.

Hofstadter, Richard. *Social Darwinism in American Thought.* Revised Edition. New York: George Braziller, 1969.

MacPherson, Hector. *Spencer and Spencerism.* New York: Doubleday, Page and Co., 1900.

Paxton, Nancy L. *George Eliot and Herbert Spencer: Feminism, Evolutionism, and the Reconstruction of Gender.* Princeton: Princeton UP, 1991.

Peel, J.D.Y. *Herbert Spencer: The Evolution of a Sociologist.* London: Heinemann, 1971.

—. ed. *Herbert Spencer On Social Evolution.* Chicago: University of Chicago Press, 1972.

Perrin, Robert G. *Herbert Spencer: A Primary and Secondary Bibliography.* New York: Garland, 1993.

Royce, Josiah. *Herbert Spencer: An Estimate and*

Review. New York: Fox, Duffield and Co., 1904.

Rumney, Jay. *Herbert Spencer's Sociology.* 1937. Reprint, New York: Atherton Press, 1966.

Spencer, Herbert. *Essays: Scientific, Political, and Speculative.* Revised Edition. London: Williams and Norgate, 1890.

—. *An Autobiography.* 2 vols. London: Williams and Norgate, 1904.

—. *The Man Versus the State.* Edited by Truxtun Beale. New York: Mitchell Kennerley, 1916.

—. *On Education.* Edited by F.A. Cavenagh. Cambridge: Cambridge UP, 1932.

Taylor, M.W. *Men Versus the State: Herbert Spencer and Late Victorian Liberalism.* Oxford: Clarendon Press, 1992.

Thomson, J. Arthur. *Herbert Spencer*. London: J.M. Dent and Sons, 1906.

Watson, John. *Comte, Mill, and Spencer: An Outline of Philosophy.* Glasgow: James MacLehose and Sons, 1895.

Wiltshire, David. *The Social and Political Thought of Herbert Spencer.* Oxford: Oxford UP, 1978.

Interior of the Great Exhibition Hall, known as the Crystal Palace, a glass and metal building housing the famous exhibition of Victorian technology and science held in 1851. *The London Illustrated News*, from which this drawing is taken (25 January 1851), commented "to our view, the general effect is cruelly destroyed by the trees at the north end of the Transept, which occupy so much of the space which might be otherwise so much more usefully appropriated" (57).

18. AUGUSTE COMTE

(1798–1857)

Auguste Comte, the founder of the Positive Philosophy, was an original thinker who so profoundly shaped nineteenth-century science, history of science, and culture that there is still no consensus on the complex nature of his influence. His writings won him disciples in England, France, and the Americas—most notably John Stuart Mill, George Henry Lewes, Harriet Martineau, and Émile Littré—although his periodic mental instability, paranoia, and megalomania eventually destroyed many of his personal and professional relationships. The intellectual roots of his philosophy can be traced from the French Revolution and its "cult of reason" to Montesquieu, Condorcet, and Saint-Simon, who hired the young Comte as his secretary. Because of his Republican sympathies, Comte was expelled from the Paris École Polytechnique in 1816, although he later lectured there on mechanics and was appointed an examiner for the school.

Comte's thought went through two distinct phases—one marked by the *Cours de philosophie positive* (1830–42) and the other characterized by the *Calendrier positiviste* (1849), the *Catéchisme positiviste* (1852), and *Le Système de politique positive* (1851–54), in which he attempted to establish a "Religion of Humanity." Comte himself recognized these two careers, and he revealingly compared himself to Aristotle in the first, and to St. Paul in the second.

In the *Cours*, Comte surveyed the current state of the sciences and introduced his most important concept, the law of the three stages or states of society. According to Comte, a society naturally and inevitably passes through three stages: 1) the Theological or "fictitious" stage (further subdivided into the phases of Fetichism, Polytheism, and Monotheism), in which the universe is perceived to be ruled by gods, and matter is animated by living wills or souls; 2) the Metaphysical or "abstract" stage, a transitional period, in which these divine beings are abstracted into impersonal powers, forces, or vital principles; and finally 3) the Positive or "scientific" stage, in which mankind no longer searches for absolute, first, or ultimate causes, but seeks an understanding of the observable relationships governing phenomena, empirically obtained and verified. Each of these stages may coexist with one another at any one given historical period, but one stage always predominates. Comte suggested that the normal intellectual development of children into adulthood imitated the three stages, and perhaps his model of progress was ultimately biological.

Drawing on the law of the three stages, Comte organized the knowledge of his day into a hierarchy of sciences according to their state of positive development. Each science depended on the one directly preceding it and consequently each science grew more complex in the hierarchy, with "Social Physics" or Sociology at the pinnacle. Comte established the following positive order of the sciences: Mathematics, Astronomy, Physics, Chemistry, Physiology (Biology), and Social Physics or Sociology. He viewed Mathematics as the most positive or perfect science and Sociology as the most neglected, complex, and important. In fact, Comte believed the great power of his Positive Philosophy lay in its ability to examine previous historical developments in order to uncover the laws governing human society. Under his leadership, the positive era would be inaugurated: the study of the sciences of the past would create a better organized, happier, and more just society. Comte has rightly been called the father of modern sociology, as he was the first to use the term in its modern sense in 1839.

Comte further divided Sociology into Social Statics (the study of the present order of society) and Social Dynamics (the progress of societies according to the law of the three stages). His ideas on social progress link him to other nineteenth-century progressive and linear philosophies such as Hegelian dialectic and Marxism, which described a succession of social states leading to a utopian withering away of government. It is not surprising that Comte became popular in England at mid-century, since there was a widespread belief in scientific and technical progress and a consensus that "things were getting better." The Crystal Palace exhibition of 1851, which showcased Britain's industrial might and technological advances, was ample evidence to all that the march of science was improving society.

Comte's later works were inspired by his love for Clotilde de Vaux, whom he met and idolized after the dissolution of his unhappy marriage with Caroline Massin in 1842. Comte gradually developed a "Religion of Humanity," outlined in the *Système de politique positive*, in which he was to play the role of "High Priest." The motto of the 'Culte systématique de l'humanité' was "Love as our Principle, Order as our basis, Progress as our end." With its calendar of positive saints, and "positive catechism," his new system took on many aspects of the Catholicism he had rejected in his youth, except that in this new religion, the worship of Humanity would replace the worship of God. Women would represent the "sexe animante," the guiding principle of the Grand Être (Great Being or collective humanity). Comte throughout the *Système de politique* centred the moral well-being of the state around women.

Comte's new state would be divided into a spiritual power headed by a "High Priest," responsible for the intellectual and moral training of the Great Being, and a Temporal Power, made up of a capitalist patriciate, who would guide the financial and material production of society. Comte even suggested a timetable for the inauguration of the Positive State: at the end of seven years, Comte would assume control over public education; five years later the Emperor Napoléon or his successor would resign to a transitional proletarian triumvirate, and during the next twenty-one years, society would prepare itself for the new dispensation. As with all utopian schemes, however, "there was no guarantee that his new elite would have any interest beyond self-aggrandizement" (Pickering 707).

The passage below from the *Cours* was translated by Harriet Martineau, an important intellectual figure in her own right. Martineau wrote extensively on political economy. Her most notable achievements include the novel *Deerbrook* (1839) and *Illustrations of Political Economy* (1832–34), a series of instructive fables each illustrating a principle of economics. In 1851, a Mr. Lombe of Florence advanced her £500 for her translation of Comte's *Cours*, and Martineau proceeded to condense and abridge the six-volume work into two. For Martineau, as for many who took up the positivist banner in the nineteenth century, the Positive Philosophy promised a regeneration of society founded on scientific (rational) principles: she writes in the preface, that in the Positive State "we find ourselves living, not under capricious and arbitrary conditions, unconnected with the constitution and movements of the whole, but under great, general, invariable laws, which operate on us as a part of the whole" ("Preface" 10). When Comte learned of Martineau's translation, he praised her work somewhat paternalistically as an illustration of the feminine spirit which would guide the future Positive State: "common spirits only will be astonished that such a work has issued from one of your sex; but true philosophers will feel on the contrary that it could not have arisen from any other source, since above all it is in your work that one finds today the spirit of accord and generous freedom, almost incompatible with the depressing regime which today prevails among men, most importantly even among the lettered" (Littré 647–48). John Stuart Mill, like Martineau another admirer of Comte's earlier work but who later recoiled from the messianic Comte of the final writings, viewed Comte's religion of humanity as an unwarranted despotism of society over the individual. Mill's *Auguste Comte and Positivism*, however, is still one of the best short introductions to Comte's thought.

The Positive Philosophy of Auguste Comte.
Translated by Harriet Martineau. 2 vols. London: John Chapman, 1853, pp. 1.1–17.

Account of the Aim of this Work—View of the Nature and Importance of the Positive Philosophy.

A general statement of any system of philosophy may be either a sketch of a doctrine to be established, or a summary of a doctrine already established. If greater value belongs to the last, the first is still important, as characterizing from its origin the subject to be treated. In a case like the present, where the proposed study is vast and hitherto indeterminate, it is especially important that the field of research should be marked out with all possible accuracy. For this purpose, I will glance at the considerations which have originated this work, and which will be fully elaborated in the course of it.

In order to understand the true value and character of the Positive Philosophy, we must take a brief general view of the progressive course of the human mind, regarded as a whole; for no conception can be understood otherwise than through its history.

From the study of the development of human intelligence, in all directions, and through all times, the discovery arises of a great fundamental law, to which it is necessarily subject, and which has a solid foundation of proof, both in the facts of our organization and in our historical experience. The law is this:—that each of our leading conceptions,—each branch of our knowledge,—passes successively through three different theoretical conditions: the Theological, or fictitious; the Metaphysical, or abstract; and the Scientific, or positive. In other words, the human mind, by its nature, employs in its progress three methods of philosophizing, the character of which is essentially different, and even radically opposed: viz., the theological method, the metaphysical, and the positive. Hence arise three philosophies, or general systems of conceptions on the aggregate of phenomena, each of which excludes the others. The first is the necessary point of departure of the human understanding; and the third is its fixed and definitive state. The second is merely a state of transition.

In the theological state, the human mind, seeking the essential nature of beings, the first and final causes (the origin and purpose) of all effects,—in short, Absolute knowledge,—supposes all phenomena to be produced by the immediate action of supernatural beings.

In the metaphysical state, which is only a modification of the first, the mind supposes, instead of supernatural beings, abstract forces, veritable entities (that is, personified abstractions) inherent in all beings, and capable of producing all phenomena. What is called the explanation of phenomena is, in this stage, a mere reference of each to its proper entity.

In the final, the positive state, the mind has given over the vain search after Absolute notions, the origin and destination of the universe, and the causes of phenomena, and applies itself to the study of their laws,—that is, their invariable relations of succession and resemblance. Reasoning and observation, duly combined, are the means of this knowledge. What is now understood when we speak of an explanation of facts is simply the establishment of a connection between single phenomena and some general facts, the number of which continually diminishes with the progress of science.

The Theological system arrived at the highest perfection of which it is capable when it substituted the providential action of a single Being for the varied operations of the numerous divinities which had been before imagined. In the same way, in the last stage of the Metaphysical system, men substitute one great entity (Nature) as the cause of all phenomena, instead of the multitude of entities at first supposed. In the same way, again, the ultimate perfection of the Positive system would be (if such perfection could be hoped for) to represent all phenomena as particular aspects of a single general fact;—such as Gravitation, for instance.

The importance of the working of this general law will be established hereafter. At present, it must suffice to point out some of the grounds of it.

There is no science which, having attained the positive stage, does not bear marks of having passed through the others. Some time since it was (whatever it might be) composed, as we can now perceive, of metaphysical abstractions; and, further back in the course of time, it took its form from theological conceptions. We shall have only too much occasion to see, as we proceed, that our most advanced sciences still bear very evident marks of the two earlier periods through which they have passed.

The progress of the individual mind is not only an illustration, but an indirect evidence of that of the general mind. The point of departure of the individual and of the race being the same, the phases of the mind of a man correspond to the epochs of the mind of the race. Now, each of us is aware, if he looks back upon his own history, that he was a theologian in his childhood, a metaphysician in his youth, and a natural philosopher in his manhood. All men who are up to their age can verify this for themselves.

Besides the observation of facts, we have theoretical reasons in support of this law.

The most important of these reasons arises from the necessity that always exists for some theory to which to refer our facts, combined with the clear impossibility that, at the outset of human knowledge, men could have formed theories out of the observation of facts. All good intellects have repeated, since Bacon's time, that there can be no real knowledge but that which is based on observed facts. This is incontestable, in our present advanced stage; but, if we look back to the primitive stage of human knowledge, we shall see that it must have been otherwise then. If it is true that every theory must be based upon observed facts, it is equally true that facts cannot be observed without the guidance of some theory. Without such guidance, our facts would be desultory and fruitless; we could not retain them: for the most part we could not even perceive them.

Thus, between the necessity of observing facts in order to form a theory, and having a theory in order to observe facts, the human mind would have been entangled in a vicious circle, but for the natural opening afforded by Theological conceptions. This is the fundamental reason for the theological character of the primitive philosophy. This necessity is confirmed by the perfect suitability of the theological philosophy to the earliest researches of the human mind. It is remarkable that the most inaccessible questions,—those of the nature of beings, and the origin and purpose of phenomena,—should be the first to occur in a primitive state, while those which are really within our reach are regarded as almost unworthy of serious study. The reason is evident enough:—that experience alone can teach us the measure of our powers; and if men had not begun by an exaggerated estimate of what they can do, they would never have done all that they are capable of. Our organization requires this. At such a period there could have been no reception of a positive philosophy, whose function is to discover the laws of phenomena, and whose leading characteristic it is to regard as interdicted to human reason those sublime mysteries which theology explains, even to their minutest details, with the most attractive facility. It is just so under a practical view of the nature of the researches with which men first occupied themselves. Such inquiries offered the powerful charm of unlimited empire over the external world,—a world destined wholly for our use, and involved in every way with our existence. The theological philosophy, presenting this view, administered exactly the stimulus necessary to incite the human mind to the irksome labour without which it could make no progress. We can now scarcely conceive of such a state of things, our reason having become sufficiently mature to enter upon laborious scientific researches, without needing any such stimulus as wrought upon the imaginations of astrologers and alchemists. We have motive enough in the hope of discovering the laws of phenomena, with a view to the confirmation or rejection of a theory. But it could not be so in the earliest days; and it is to the chimeras of astrology and alchemy that we owe the long series of observations and experiments on which our positive science is based. Kepler felt this on behalf of astronomy, and Berthollet on behalf of chemistry. Thus was a spontaneous philosophy, the theological, the only possible beginning, method, and provisional system, out of which the Positive philosophy could grow. It is easy, after this, to perceive how Metaphysical methods and doctrines must have afforded the means of transition from the one to the other.

The human understanding, slow in its advance, could not step at once from the theological into the positive philosophy. The two are so radically opposed, that an intermediate system of conceptions has been necessary to render the transition possible. It is only in doing this, that Metaphysical conceptions have any utility whatever. In contemplating phenomena, men substitute for supernatural direction a corresponding entity. This entity may have been supposed to be derived from the supernatural action: but it is more easily lost sight of, leaving attention free for the facts themselves, till, at length, metaphysical agents have ceased to be anything more than the abstract names of phenomena. It is not easy to say by what other process than this our minds could have passed from supernatural considerations to natural; from the theological system to the positive.

The Law of human development being thus established, let us consider what is the proper nature of the Positive Philosophy.

As we have seen, the first characteristic of the Positive Philosophy is that it regards all phenomena as subjected to invariable natural *Laws*. Our business is,—seeing how vain is any research into what are called *Causes*, whether first or final,—to pursue an accurate discovery of these Laws, with a view to reducing them to the smallest possible number. By speculating upon causes, we could solve no difficulty about origin and purpose. Our real business is to analyse accurately the circumstances of phenomena, and to connect them by the natural relations of succession and resemblance. The best illustration of this is in the case of the doctrine of Gravitation. We say that the general phenomena of the universe are *explained* by it, because it connects under one head the whole immense variety of astronomical facts; exhibiting the constant tendency of atoms towards each other in direct proportion to their masses, and in inverse proportion to the squares of their distances; whilst the general fact itself is a mere extension of one which is perfectly familiar to us, and which we therefore say that we know;—the weight of bodies on the surface of the earth. As to what weight and attraction are, we have nothing to do with that, for it is not a matter of knowledge at all. Theologians and metaphysicians may imagine and refine about such questions; but positive philosophy rejects them. When any attempt has been made to explain them, it has ended only in saying that attraction is universal weight, and that weight is terrestrial attraction: that is, that the two orders of phenomena are identical; which is the point from which the question set out. Again, M. Fourier, in his fine series of researches on Heat, has given us all the most important and precise laws of the phenomena of heat, and many large and new truths, without once inquiring into its nature, as his predecessors had done when they disputed about calorific matter and the action of an universal ether. In treating his subject in the Positive method, he finds inexhaustible material for all his activity of research, without betaking himself to insoluble questions.

Before ascertaining the stage which the Positive Philosophy has reached, we must bear in mind that the different kinds of our knowledge have passed through the three stages of progress at different rates, and have not therefore arrived at the same time. The rate of advance depends on the nature of the knowledge in question, so distinctly that, as we shall see hereafter, this consideration constitutes an accessary to the fundamental law of progress. Any kind of knowledge reaches the positive stage early in proportion to its generality, simplicity, and independence of other departments. Astronomical science, which is above all made up of facts that are general, simple, and independent of other sciences, arrived first; then terrestrial Physics; then Chemistry; and, at length, Physiology.

It is difficult to assign any precise date to this revolution in science. It may be said, like everything else, to have been always going on, and especially since the labours of Aristotle and the school of Alexandria; and then from the introduction of natural science into the West of Europe by the Arabs. But, if we must fix upon some marked period, to serve as a rallying point, it must be that,—about two centuries ago,—when the human mind was astir under the precepts of Bacon, the conceptions of Descartes, and the discoveries of Galileo. Then it was that the spirit of the Positive philosophy rose up in opposition to that of the superstitious and scholastic systems which had hitherto obscured the true character of all science. Since that date, the progress of the Positive philosophy, and the decline of the other two, have been so marked that no rational mind now doubts that the revolution is destined to go on to its completion,—every branch of knowledge being, sooner or later, brought within the operation of Positive philosophy. This is not yet the case. Some are still lying outside: and not till they are brought in will the Positive philosophy possess that character of universality which is necessary to its definitive constitution.

In mentioning just now the four principal categories of phenomena,—astronomical, physical, chemical, and physiological,—there was an omission which will have been noticed. Nothing was said of Social phenomena. Though involved with the physiological, Social phenomena demand a distinct classification, both on account of their importance and of their difficulty. They are the most individual, the most complicated, the most dependent on all others; and therefore they must be the latest,—even if they had no special obstacle to encounter. This branch of science has not hitherto entered into the domain of Positive philosophy. Theological and metaphysical methods, exploded in other departments, are as yet exclusively applied, both in the way of inquiry and discussion, in all treatment of Social subjects, though the best minds are heartily weary of eternal disputes about divine right and the sovereignty of the people. This is the great, while it is evidently the only gap which has to be filled, to constitute, solid and entire, the Positive Philosophy. Now that the human mind has grasped celestial and terrestrial physics,—mechanical and chemical; organic physics, both vegetable and animal,—there remains one science, to fill up the series of sciences of observation,—Social physics. This is what men have now most need of: and this it is the principal aim of the present work to establish.

It would be absurd to pretend to offer this new science at once in a complete state. Others, less new, are in very unequal conditions of forwardness. But the same character of positivity which is impressed on all the others will be shown to belong to this. This once done, the philosophical system of the moderns will be in fact complete, as there will then be no phenomenon which does not naturally enter into some one of the five great categories. All our fundamental conceptions having become homogeneous, the Positive state will be fully established. It can never again change its character, though it will be for ever in course of development by additions of new knowledge. Having acquired the character of universality which has hitherto been the only advantage resting with the two preceding systems, it will supersede them by its natural superiority, and leave to them only an historical existence.

We have stated the special aim of this work. Its secondary and general aim is this:—to review what has been effected in the Sciences, in order to show that they are not radically separate, but all branches from the same trunk. If we had confined ourselves to the first and special object of the work, we should have produced merely a study of Social physics: whereas, in introducing the second and general, we offer a study of Positive philosophy, passing in review all the positive sciences already formed.

The purpose of this work is not to give an account of the Natural Sciences. Besides that would be endless, and that it would require a scientific preparation such as no one man possesses, it would be apart from our object, which is to go through a course of not Positive Science, but Positive Philosophy. We have only to consider each fundamental science in its relation to the whole positive system, and to the spirit which characterizes it; that is, with regard to its methods and its chief results.

The two aims, though distinct, are inseparable; for, on the one hand, there can be no positive philosophy without a basis of social science, without which it could not be all-comprehensive; and, on the other hand, we could not pursue Social science without having been prepared by the study of phenomena less complicated than those of society, and furnished with a knowledge of laws and anterior facts which have a bearing upon social science. Though the fundamental sciences are not all equally interesting to ordinary minds, there is no one of them that can be neglected in an inquiry like the present; and, in the eye of philosophy, all are of equal value to human welfare. Even those which appear the least interesting have their own value, either on account of the perfection of their methods, or as being the necessary basis of all the others.

Lest it should be supposed that our course will lead us into a wilderness of such special studies as are at present the bane of a true positive philosophy, we will briefly advert to the existing prevalence of such special pursuit. In the primitive state of human knowledge there is no regular division of intellectual labour. Every student cultivates all the sciences. As knowledge accrues, the sciences part off; and students devote themselves each to some one branch. It is owing to this division of employment, and concentration of whole minds upon a single department, that science has made so prodigious an advance in modern times; and the perfection of this division is one of the most important characteristics of the Positive philosophy. But, while admitting all the merits of this change, we cannot be blind to the eminent disadvantages which arise from the limitation of minds to a particular study. It is inevitable that each should be possessed with exclusive notions, and be therefore incapable of the general superiority of ancient students, who actually owed that general superiority to the inferiority of their knowledge. We must consider whether the evil can be avoided without losing the good of the modern arrangement; for the evil is becoming urgent. We all acknowledge that the divisions established for the convenience of scientific pursuit are radically artificial; and yet there are very few who can embrace in idea the whole of any one science: each science moreover being itself only a part of a great whole. Almost every one is busy about his own particular section, without much thought about its relation to the general system of positive knowledge. We must not be blind to the evil, nor slow in seeking a remedy. We must not forget that this is the weak side of the positive philosophy, by which it may yet be attacked, with some hope of success, by the adherents of the theological and metaphysical systems. As to the remedy, it certainly does not lie in a return to the ancient confusion of pursuits, which would be mere retrogression, if it were possible, which it is not. It lies in perfecting the division of employments itself,—in carrying it one degree higher,—in constituting one more specialty from the study of scientific generalities. Let us have a new class of students, suitably prepared, whose business it shall be to take the respective sciences as they are, determine the spirit of each, ascertain their relations and mutual connection, and reduce their respective principles to the smallest number of general principles, in conformity with the fundamental rules of the Positive Method. At the same time, let other students be prepared for their special pursuit by an education which recognizes the whole scope of positive science, so as to profit by the labours of the students of generalities, and so as to correct reciprocally, under that guidance, the results obtained by each. We see some approach already to this arrangement. Once established, there would be nothing to apprehend from any extent of division of employments. When we once have a class of learned men, at the disposal of all others, whose business it shall be to connect each new discovery with the general system, we may dismiss all fear of the great whole being lost sight of in the pursuit of the details of knowledge. The organization of scientific research will then be complete; and it will henceforth have occasion only to extend its development, and not to change its character. After all, the formation of such a new class as is proposed would be merely an extension of the principle which has created all the classes we have. While science was narrow, there was only one class: as it expanded, more were instituted. With a further advance a fresh need arises, and this new class will be the result.

The general spirit of a course of Positive Philosophy having been thus set forth, we must now glance at the chief advantages which may be derived, on behalf of human progression, from the study of it. Of these advantages, four may be especially pointed out.

I. The study of the Positive Philosophy affords the only rational means of exhibiting the logical laws of the human mind, which have hitherto been sought by unfit methods. To explain what is meant by this, we may refer to a saying of M. de Blainville, in his work on Comparative Anatomy, that every active, and especially every living being, may be regarded under two relations—the Statical and the Dynamical; that is, under conditions or in action. It is clear that all considerations range themselves under the one or the other of these heads. Let us apply this classification to the intellectual functions.

If we regard these functions under their Statical aspect—that is, if we consider the conditions under which they exist—we must determine the organic circumstances of the case, which inquiry involves it with anatomy and physiology. If we look at the Dynamic aspect, we have to study simply the exercise and results of the intellectual powers of the human race, which is neither more nor less than the general object of the Positive Philosophy. In short, looking at all scientific theories as so many great logical facts, it is only by the thorough observation of these facts that we can arrive at the knowledge of logical laws. These being the only means of knowledge of intellectual phenomena, the illusory psychology, which is the last phase of theology, is excluded. It pretends to accomplish the discovery of the laws of the human mind by contemplating it in itself; that is, by separating it from causes and effects. Such an attempt, made in defiance of the physiological study of our intellectual organs, and of the observation of rational methods of procedure, cannot succeed at this time of day.

The Positive Philosophy, which has been rising since the time of Bacon, has now secured such a preponderance, that the metaphysicians themselves profess to ground their pretended science on an observation of facts. They talk of external and internal facts, and say that their business is with the latter. This is much like saying that vision is explained by luminous objects painting their images upon the retina. To this the physiologists reply that another eye would be needed to see the image. In the same manner, the mind may observe all phenomena but its own. It may be said that a man's intellect may observe his passions, the seat of the reason being somewhat apart from that of the emotions in the brain; but there can be nothing like scientific observation of the passions, except from without, as the stir of the emotions disturbs the observing faculties more or less. It is yet more out of the question to make an intellectual observation of intellectual processes. The observing and observed organ are here the same, and its action cannot be pure and natural. In order to observe, your intellect must pause from activity; yet it is this very activity that you want to observe. If you cannot effect the pause, you cannot observe: if you do effect it, there is nothing to observe. The results of such a method are in proportion to its absurdity. After two thousand years of psychological pursuit, no one proposition is established to the satisfaction of its followers. They are divided, to this day, into a multitude of schools, still disputing about the very elements of their doctrine. This interior observation gives birth to almost as many theories as there are observers. We ask in vain for any one discovery, great or small, which has been made under this method. The psychologists have done some good in keeping up the activity of our understandings, when there was no better work for our faculties to do; and they may have added something to our stock of knowledge. If they have done so, it is by practising the Positive method—by observing the progress of the human mind in the light of science; that is, by ceasing, for the moment, to be psychologists.

The view just given in relation to logical Science becomes yet more striking when we consider the logical Art.

The Positive Method can be judged of only in action. It cannot be looked at by itself, apart from the work on which it is employed. At all events, such a contemplation would be only a dead study, which could produce nothing in the mind which loses time upon it. We may talk for ever about the method, and state it in terms very wisely, without knowing half so much about it as the man who has once put it in practice upon a single particular of actual research, even without any philosophical intention. Thus it is that psychologists, by dint of reading the precepts of Bacon and the discourses of Descartes, have mistaken their own dreams for science.

Without saying whether it will ever be possible to establish *à priori* a true method of investigation, independent of a philosophical study of the sciences, it is clear that the thing has never been done yet, and that we are not capable of doing it now. We cannot as yet explain the great logical procedures, apart from their applications. If we ever do, it will remain as necessary then as now to form good intellectual habits by studying the regular application of the scientific methods which we shall have attained.

This, then, is the first great result of the Positive Philosophy—the manifestation by experiment of the laws which rule the Intellect in the investigation of truth; and, as a consequence the knowledge of the general rules suitable for that object.

II. The second effect of the Positive Philosophy, an effect not less important and far more urgently wanted, will be to regenerate Education. The best minds are agreed that our European education, still essentially theological, metaphysical, and literary, must be superseded by a Positive training, conformable to our time and needs. Even the governments of our day have shared, where they have not originated, the attempts to establish positive instruction; and this is a striking indication of the prevalent sense of what is wanted. While encouraging such endeavours to the utmost, we must not however conceal from ourselves that everything yet done is inadequate to the object. The present exclusive specialty of our pursuits, and the consequent isolation of the sciences, spoil our teaching. If any student desires to form an idea of natural philosophy as a whole, he is compelled to go through each department as it is now taught, as if he were to be only an astronomer, or only a chemist; so that, be his intellect what it may, his training must remain very imperfect. And yet his object requires that he should obtain general positive conceptions of all the classes of natural phenomena. It is such an aggregate of conceptions, whether on a great or on a small scale, which must henceforth be the permanent basis of all human combinations. It will constitute the mind of future generations. In order to [achieve] this regeneration of our intellectual system, it is necessary that the sciences, considered as branches from one trunk, should yield us, as a whole, their chief methods and their most important results. The specialties of science can be pursued by those whose vocation lies in that direction. They are indispensable; and they are not likely to be neglected; but they can never of themselves renovate our system of Education; and, to be of their full use, they must rest upon the basis of that general instruction which is a direct result of the Positive Philosophy.

III. The same special study of scientific generalities must also aid the progress of the respective positive sciences: and this constitutes our third head of advantages.

The divisions which we establish between the sciences are, though not arbitrary, essentially artificial. The subject of our researches is one: we divide it for our convenience, in order to deal the more easily with its difficulties. But it sometimes happens—and especially with the most important doctrines of each science—that we need what we cannot obtain under the present isolation of the sciences,—a combination of several special points of view; and for want of this, very important problems wait for their solution much longer than they otherwise need do. To go back into the past for an example: Descartes' grand conception with regard to analytical geometry is a discovery which has changed the whole aspect of mathematical science, and yielded the germ of all future progress; and it issued from the union of two sciences which had always before been separately regarded and pursued. The case of pending questions is yet more impressive; as, for instance, in Chemistry, the doctrine of Definite Proportions. Without entering upon the discussion of the fundamental principle of this theory, we may say with assurance that, in order to determine it—in order to determine whether it is a law of nature that atoms should necessarily combine in fixed numbers,—it will be indispensable that the chemical point of view should be united with the physiological. The failure of the theory with regard to organic bodies indicates that the cause of this immense exception must be investigated; and such an inquiry belongs as much to physiology as to chemistry. Again, it is as yet undecided whether azote is a simple or a compound body. It was concluded by almost all chemists that azote is a simple body; the illustrious Berzelius hesitated, on purely chemical considerations; but he was also influenced by the physiological observation that animals which receive no azote in their food have as much of it in their tissues as carnivorous animals. From this we see how physiology must unite with chemistry to inform us whether azote is simple or compound, and to institute a new series of researches upon the relation between the composition of living bodies and their mode of alimentation.

Such is the advantage which, in the third place, we shall owe to Positive philosophy—the elucidation of the respective sciences by their combination. In the fourth place:

IV. The Positive Philosophy offers the only solid basis for that Social Reorganization which must succeed the critical condition in which the most civilized nations are now living.

It cannot be necessary to prove to anybody who reads this work that Ideas govern the world, or throw it into chaos; in other words, that all social mechanism rests upon Opinions. The great political and moral crisis that societies are now undergoing is shown by a rigid analysis to arise out of intellectual anarchy. While stability in fundamental maxims is the first condition of genuine social order, we are suffering under an utter disagreement which may be called universal. Till a certain number of general ideas can be acknowledged as a rallying-point of social doctrine, the nations will remain in a revolutionary state, whatever palliatives may be devised; and their institutions can be only provisional. But whenever the necessary agreement on first principles can be obtained, appropriate institutions will issue from them, without shock or resistance; for the causes of disorder will have been arrested by the mere fact of the agreement. It is in this direction that those must look who desire a natural and regular, a normal state of society.

Now, the existing disorder is abundantly accounted for by the existence, all at once, of three incompatible philosophies,—the theological, the metaphysical, and the positive. Any one of these might alone secure some sort of social order; but while the three co-exist, it is impossible for us to understand one another upon any essential point whatever. If this is true, we have only to ascertain which of the philosophies must, in the nature of things, prevail; and, this ascertained, every man, whatever may have been his former views, cannot but concur in its triumph. The problem once recognized cannot remain long unsolved; for all considerations whatever point to the Positive Philosophy as the one destined to prevail. It alone has been advancing during a course of centuries, throughout which the others have been declining. The fact is incontestable. Some may deplore it, but none can destroy it, nor therefore neglect it but under penalty of being betrayed by illusory speculations. This general revolution of the human mind is nearly accomplished. We have only to complete the Positive Philosophy by bringing Social phenomena within its comprehension, and afterwards consolidating the whole into one body of homogeneous doctrine. The marked preference which almost all minds, from the highest to the commonest, accord to positive knowledge over vague and mystical conceptions, is a pledge of what the reception of this philosophy will be when it has acquired the only quality that it now wants—a character of due generality. When it has become complete, its supremacy will take place spontaneously, and will re-establish order throughout society. There is, at present, no conflict but between the theological and the metaphysical philosophies. They are contending for the task of reorganizing society; but it is a work too mighty for either of them. The positive philosophy has hitherto intervened only to examine both, and both are abundantly discredited by the process. It is time now to be doing something more effective, without wasting our forces in needless controversy. It is time to complete the vast intellectual operation begun by Bacon, Descartes, and Galileo, by constructing the system of general ideas which must henceforth prevail among the human race. This is the way to put an end to the revolutionary crisis which is tormenting the civilized nations of the world.

Leaving these four points of advantage, we must attend to one precautionary reflection.

Because it is proposed to consolidate the whole of our acquired knowledge into one body of homogeneous doctrine, it must not be supposed that we are going to study this vast variety as proceeding from a single principle, and as subjected to a single law. There is something so chimerical in attempts at universal explanation by a single law, that it may be as well to secure this Work at once from any imputation of the kind, though its development will show how undeserved such an imputation would be. Our intellectual resources are too narrow, and the universe is too complex, to leave any hope that it will ever be within our power to carry scientific perfection to its last degree of simplicity. Moreover, it appears as if the value of such an attainment, supposing it possible, were greatly overrated. The only way, for instance, in which we could achieve the business, would be by connecting all natural phenomena with the most general law we know,—which is that of Gravitation, by which astronomical phenomena are already connected with a portion of terrestrial physics. Laplace has indicated that chemical phenomena may be regarded as simple atomic effects of the Newtonian attraction, modified by the form and mutual position of the atoms. But supposing this view proveable (which it cannot be while we are without data about the constitution of bodies), the difficulty of its application would doubtless be found so great that we must still maintain the existing division between astronomy and chemistry, with the difference that we now regard as natural that division which we should then call artificial. Laplace himself presented his idea only as a philosophic device, incapable of exercising any useful influence over the progress of chemical science. Moreover, supposing this insuperable difficulty overcome, we should be no nearer to scientific unity, since we then should still have to connect the whole of physiological phenomena with the same law, which certainly would not be the least difficult part of the enterprise. Yet, all things considered, the hypothesis we have glanced at would be the most favourable to the desired unity.

The consideration of all phenomena as referable to a single origin is by no means necessary to the systematic formation of science, any more than to the realization of the great and happy consequences that we anticipate from the positive philosophy. The only necessary unity is that of Method, which is already in great part established. As for the doctrine, it need not be *one*; it is enough that it should be *homogeneous*. It is, then, under the double aspect of unity of method and homogeneousness of doctrine that we shall consider the different classes of positive theories in this work. While pursing the philosophical aim of all science, the lessening of the number of general laws requisite for the explanation of natural phenomena, we shall regard as presumptuous every attempt, in all future time, to reduce them rigorously to one.

Having thus endeavoured to determine the spirit and influence of the Positive Philosophy, and to mark the goal of our labours, we have now to proceed to the exposition of the system; that is, to the determination of the universal, or encyclopaedic order, which must regulate the different classes of natural phenomena, and consequently the corresponding positive sciences.

FURTHER READING

Alengry, Franck. *Essai historique et critique sur la sociologie chez Auguste Comte.* Paris: Félix Alcan, 1900.

Besant, Annie. *Auguste Comte: His Philosophy, His Religion, and His Sociology.* London: C. Watts, no date.

Caird, Edward. *The Social Philosophy and Religion of Comte.* Glasgow: J. MacLehose and Sons, 1885.

Charlton, D.G. *Positivist Thought in France During the Second Empire 1852–1870.* Oxford: Clarendon Press, 1959.

Comte, Auguste. *The Catechism of Positive Religion.* Translated by Richard Congreve. London: John Chapman, 1858.

—. *Confessions and Testament of Auguste Comte: and his Correspondence with Clotilde de Vaux.* Edited by Albert Crompton. Liverpool: Henry Young and Sons, 1910.

—. *The Crisis of Industrial Civilization: The Early Essays of Auguste Comte.* Introduced by Ronald Fletcher. Translated by Henry Dix Hutton. London: Heinemann Educational Books, Ltd., 1974.

—. *A General View of Positivism.* Translated by J.H. Bridges. London: Trübner and Co., 1865.

—. *Philosophie première: Cours de philosophie positive, leçons 1 à 45.* Edited by Michel Serres, François Dagognet, and Allal Sinaceur. Paris: Hermann, 1975.

—. *Physique sociale: Cours de philosophie positive, leçons 46 à 60.* Edited by Jean-Paul Enthoven. Paris: Hermann, 1975.

—. *System of Positive Polity.* Translated by J.H. Bridges, Frederic Harrison, *et al.* 4 vols. London: Longmans, Green, and Co., 1875–77.

Ducassé, Pierre. *Méthode et intuition chez Auguste Comte.* Paris: Félix Alcan, 1939.

Dupuy, Paul. *Le Positivisme d'Auguste Comte.* Paris: Félix Alcan, 1911.

Gouhier, Henri. *La Jeunesse d'Auguste Comte et la formation du positivisme.* 2 vols. Paris: J. Vrin, 1933–36.

—. *La vie d'Auguste Comte.* 2d ed. Paris: J. Vrin, 1997.

Grange, Juliette. *La philosophie d'Auguste Comte: Science, politique, religion.* Paris: Presses Universitaires de France, 1996.

Hawkins, Richmond Laurin. *Auguste Comte and the United States (1816–1853).* Cambridge: Harvard UP, 1936.

Kofman, Sarah. *Aberrations: Le devenir-femme d'Auguste Comte.* Paris: Aubier Flammarion, 1978.

Kremer-Marietti, Angèle. *Le Concept de science positive.* Paris: Klincksieck, 1983.

—. *Entre le signe et l'histoire: L'Anthropologie positiviste d'Auguste Comte.* Paris: Klincksieck, 1982.

Lenzer, Gertrud, ed. *Auguste Comte and Positivism: The Essential Writings.* Chicago: The University of Chicago Press, 1975.

Lévy-Bruhl, Lucien. *The Philosophy of Auguste Comte.* Translated by Frederic Harrison. London: Swan Sonnenschein and Co., 1903.

Lewes, G.H. *Philosophy of the Sciences: Being an Exposition of the Principles of the "Cours de Philosophie Positive" of Auguste Comte.* London: Henry G. Bohn, 1853.

Littré, Émile. *Auguste Comte et la philosophie positive.* Paris: L. Hachette et Compagnie, 1863.

—. "Preface d'un Disciple." In Auguste Comte, *Principes de philosphie positive.* Paris: J.B. Baillière et Fils, 1868.

Lonchampt, Joseph. *Précis de la vie et des écrits d'Auguste Comte. Extrait de la Revue Occidentale.* Paris: Rue Monsieur-Le-Prince, 1889.

Mill, John Stuart. *Auguste Comte and Positivism.* London: Trübner, 1865.

Muglioni, Jacques. *Auguste Comte: Un philosophe pour notre temps.* Paris: Éditions Kimé, 1995.

Pickering, Mary. *Auguste Comte: An Intellectual Biography.* Vol. 1. Cambridge: Cambridge UP, 1993.

de Roberty, E. *Auguste Comte et Herbert Spencer.* Paris: Félix Alcan, 1894.

de Rouvre, Charles. *Auguste Comte et le Catholicisme.* Paris: Éditions Rieder, 1928.

Scharff, Robert C. *Comte After Positivism.* Cambridge: Cambridge UP, 1995.

Seillière, Ernest. *Auguste Comte.* Paris: Félix Alcan, 1924.

Simon, W.M. *European Positivism in the Nineteenth Century: An Essay in Intellectual History.* Ithaca, NY: Cornell UP, 1963.

Simpson, George. *Auguste Comte: Sire of Sociology.* New York: Thomas Y. Crowell, 1969.

Thompson, Kenneth. *Auguste Comte: The Foundation of Sociology.* New York: John Wiley and Sons, 1975.

Uta, Michel. *La Loi des trois états dans la philosophie d'Auguste Comte.* Paris: Félix Alcan, 1928.

—. *La Théorie du savoir dans la philosophie d'Auguste Comte.* Paris: Félix Alcan, 1928.

19. CHARLES ROBERT DARWIN

(1809–1882)

Charles Darwin was one of the most influential and prolific scientists of the nineteenth century. He has been written about so extensively by twentieth-century historians that a "Darwin industry" of secondary literature has arisen. His grandfather was the celebrated author Erasmus (see the selection above) and his father Robert practised medicine very successfully in Shrewsbury. Darwin also studied medicine at Edinburgh from 1825–27, but finding his studies uncongenial, transferred to Cambridge to train as a clergyman. Adam Sedgwick and John Henslow (who suggested the Beagle Voyage) were early scientific influences on Darwin. To the dismay of his father, many of Darwin's activities at Cambridge clearly fell into the category of extra-curricular, including rat-catching, shooting, and beetle collecting: "I will give a proof of my zeal: one day, on tearing off some old bark, I saw two rare beetles and seized one in each hand; then I saw a third and new kind, which I could not bear to lose, so that I popped the one which I held in my right hand into my mouth. Alas it ejected some intensely acrid fluid, which burnt my tongue so that I was forced to spit the beetle out, which was lost, as well as the third one" (*Autobiography* 62).

Darwin served as a naturalist aboard *H.M.S. Beagle* from 1831–36, visiting South America and the Pacific islands. Darwin collected specimens and observed variations in related species of birds and animals. He read the first edition of Charles Lyell's *Principles of Geology* on the voyage and became convinced of the immensity of geologic time, a time period which would allow natural selection to take place. Upon his return, he married his cousin Emma Wedgwood and published several volumes describing the scientific findings of the Beagle voyage, including *The Zoology of the Voyage of the H.M.S. Beagle* (1839–43), *Journal of Researches into the Geology and Natural History of the Various Countries Visited by H.M.S. Beagle* (1839), *Geological Observations on the Volcanic Islands Visited During the Voyage of H.M.S. Beagle, Together With Some Brief Notices of the Geology of Australia and the Cape of Good Hope* (1844), and *Geological Observations on South America* (1846). Darwin also advanced his original theory of the structure and distribution of *Coral Reefs* (1842), arguing that atolls developed by the deposition of polyp skeletons on gradually subsiding underlying strata, rather than on submerged volcanic craters at a fixed depth, as Lyell had proposed. He then spent eight years classifying the subclass Cirripedia, or the barnacles. At this time, Darwin began to suffer from the mysterious recurrent illness—perhaps Chagas' disease, hypochondria, overwork, or a neurological disorder—which forced him into a reclusive life at Downe in Kent. He eventually sought relief in hydrotherapathy.

Through observation of the similarities between extinct and related living species in South America, Darwin began to question the orthodox position that species had remained unchanged since first placed on earth by God. In 1837, Darwin started the first of a series of *Transmutation Notebooks* on the species question which later evolved into the 1842 and 1844 drafts of an essay which Darwin called "my big book" on species, later to be rewritten as *The*

Origin of Species, with the first two chapters eventually forming *The Variation of Animals and Plants Under Domestication* (1868). Darwin's reading of Thomas Malthus's *Essay on Population* demonstrated to him that population growth would always outstrip food supply, inevitably resulting in competition for limited resources. Those individuals possessing the traits, produced by random variation, which best allowed the organism to survive would pass these traits to their offspring, ensuring the survival of these traits and thereby slowly modifying the species to the extent that intermediate varieties would supplant or exterminate the parent type. For this process, Darwin adopted Herbert Spencer's phrase "survival of the fittest," although the fit should not be seen as qualitatively "better" than other individuals, but simply as those which are naturally selected by the environment to leave more offspring. As Robert Young has shown, Darwin was led to the idea of natural selection by the example of artificial selection of domestic animals in which breeders selected animals for specific desirable traits. In Darwin's scheme, nature simply acts as an unconscious and more perfect selector. Darwin investigated artificial selection in his *The Variation of Animals and Plants Under Domestication* (1868); he also bred pigeons himself, joined breeding societies, and like his cousin Francis Galton, circulated questionnaires to plant and animal breeders.

In 1858, the English naturalist Alfred Russel Wallace sent Darwin an essay from Malaysia entitled "On the Tendency of Varieties to Depart Indefinitely from the Original Type" (see the Wallace selections below). Darwin immediately recognized his own views on species transmutation in Wallace's work. Subsequently, Joseph Hooker and Charles Lyell arranged a meeting of the Linnean society at which Wallace's and Darwin's ideas were jointly presented. A year later, Darwin published his *Origin of Species by Means of Natural Selection, or the Preservation of Favoured Races in the Struggle for Life*. The book sold out instantly and at the 1860 Oxford meeting of the BAAS, Thomas Huxley defended Darwin's views against Bishop Samuel Wilberforce and the theologians who were shocked by the implication (not stated by Darwin) that man and apes shared a common origin and that Paley's natural theology with its purposeful creator was no longer tenable. Darwin eventually applied his evolutionary views to mankind in *The Descent of Man* (1871), stating clearly that man had evolved from lower life forms. It must be pointed out that Darwin had no clear conception of how characters were transmitted from parent to offspring, as knowledge of genetic inheritance had to wait until the twentieth century with the rediscovery of the work of Gregor Mendel. By 1900, Darwin's theories were being disputed from a number of quarters and his pangenesis theory of blending inheritance had to be abandoned in the twentieth century in favour of particulate inheritance by genes.

Darwin's *Descent of Man, and Selection in Relation to Sex*, of which the concluding chapter is reprinted below, introduced the concept of sexual selection—an intra-species mechanism operating between males and females—which like natural selection modified species over time and produced sexual dimorphism, to the extent, as Darwin points out, that sometimes males and females of the same species had been assigned to different genera by various naturalists. The topic of sexual selection has recently been of great interest, and Nancy Etcoff's *Survival of the Prettiest* (1999) provides a bibliography of scientific and popular writings on the evolutionary role of beauty and the factors believed to be involved in mate selection in humans and animals. Carl Jay Bajema has in addition compiled an anthology of pre-1900 essays on *Evolution by Sexual Selection Theory Prior to 1900* (1984) which supplements Bernard Campbell's edition of modern essays entitled *Sexual Selection and the Descent of Man, 1871–1971* (1972).

Darwin's final works include *On the Various Contrivances by which British and Foreign Orchids are Fertilized by Insects* (1862), *The Expression of the Emotions in Man and Animals* (1872), *Insectivorous Plants* (1875), *The Movements and Habits of Climbing Plants* (1875), *The Effects of Cross and Self Fertilization in the Vegetable Kingdom* (1876), *The Different Forms of Flowers on Plants of the Same Species* (1877), *The Power of Movement in Plants* (1880), and *The Formation of the Vegetable Mould Through the Action of Worms* (1881). Darwin's complete works have been edited by Paul H. Barrett and R.B. Freeman for Pickering and Chatto. Darwin's *Autobiography*, with deleted passages restored by Nora Barlow, remains a classic of scientific biography.

On the Origin of Species By Means of Natural Selection, or the Preservation of Favoured Races in the Struggle for Life.

London: John Murray, 1859, pp. 459–490.

Chapter XIV. Recapitulation and Conclusion.

Recapitulation of the difficulties on the theory of Natural Selection—Recapitulation of the general and special circumstances in its favour—Causes of the general belief in the immutability of species—How far the theory of natural selection may be extended—Effects of its adoption on the study of Natural history—Concluding remarks.

As this whole volume is one long argument, it may be convenient to the reader to have the leading facts and inferences briefly recapitulated.

That many and grave objections may be advanced against the theory of descent with modification through natural selection, I do not deny. I have endeavoured to give to them their full force. Nothing at first can appear more difficult to believe than that the more complex organs and instincts should have been perfected, not by means superior to, though analogous with, human reason, but by the accumulation of innumerable slight variations, each good for the individual possessor. Nevertheless, this difficulty, though appearing to our imagination insuperably great, cannot be considered real if we admit the following propositions, namely,—that gradations in the perfection of any organ or instinct, which we may consider, either do now exist or could have existed, each good of its kind,—that all organs and instincts are, in ever so slight a degree, variable,—and, lastly, that there is a struggle for existence leading to the preservation of each profitable deviation of structure or instinct. The truth of these propositions cannot, I think, be disputed.

It is, no doubt, extremely difficult even to conjecture by what gradations many structures have been perfected, more especially amongst broken and failing groups of organic beings; but we see so many strange gradations in nature, as is proclaimed by the canon, "Natura non facit saltum," that we ought to be extremely cautious in saying that any organ or instinct, or any whole being, could not have arrived at its present state by many graduated steps. There are, it must be admitted, cases of special difficulty on the theory of natural selection; and one of the most curious of these is the existence of two or three defined castes of workers or sterile females in the same community of ants; but I have attempted to show how this difficulty can be mastered.

With respect to the almost universal sterility of species when first crossed, which forms so remarkable a contrast with the almost universal fertility of varieties when crossed, I must refer the reader to the recapitulation of the facts given at the end of the eighth chapter, which seem to me conclusively to show that this sterility is no more a special endowment than is the incapacity of two trees to be grafted together; but that it is incidental on constitutional differences in the reproductive systems of the intercrossed species. We see the truth of this conclusion in the vast difference in the result, when the same two species are crossed reciprocally; that is, when one species is first used as the father and then as the mother.

The fertility of varieties when intercrossed and of their mongrel offspring cannot be considered as universal; nor is their very general fertility surprising when we remember that it is not likely that either their constitutions or their reproductive systems should have been profoundly modified. Moreover, most of the varieties which have been experimentised on have been produced under domestication; and as domestication apparently tends to eliminate sterility we ought not to expect it also to produce sterility.

The sterility of hybrids is a very different case from that of first crosses, for their reproductive organs are more or less functionally impotent; whereas in first crosses the organs on both sides are in a perfect condition. As we continually see that organisms of all kinds are rendered in some degree sterile from their constitutions having been disturbed by slightly different and new conditions of life, we need not feel surprise at hybrids being in some degree sterile, for their constitutions can hardly fail to have been disturbed from being compounded of two distinct organisations. This parallelism is supported by another parallel, but directly opposite, class of facts; namely, that the vigour and fertility of all organic beings are increased by slight changes in their conditions of life, and that the offspring of slightly modified forms or varieties acquire from being crossed increased vigour and fertility. So that, on the one hand, considerable changes in the conditions of life and crosses between greatly modified forms, lessen fertility; and on the other hand, lesser changes in the conditions of life and crosses between less modified forms, increase fertility.

Turning to geographical distribution, the difficulties encountered on the theory of descent with modification are grave enough. All the individuals of the same species, and all the species of the same genus, or even higher group, must have descended from common parents; and therefore, in however distant and isolated parts of the world they are now found, they must in the course of successive generations have passed from some one part to the others. We are often wholly unable even to conjecture how this could have been effected. Yet, as we have reason to believe that some species have retained the same specific form for very long periods, enormously long as measured by years, too much stress ought not to be laid on the occasional wide diffusion of the same species; for during very long periods of time there will always be a good chance for wide migration by many means. A broken or interrupted range may often be accounted for by the extinction of the species in the intermediate regions. It cannot be denied that we are as yet very ignorant of the full extent of the various climatal and geographical changes which have affected the earth during modern periods; and such changes will obviously have greatly facilitated migration. As an example, I have attempted to show how potent has been the influence of the Glacial period on the distribution both of the same and of representative species throughout the world. We are as yet profoundly ignorant of the many occasional means of transport. With respect to distinct species of the same genus inhabiting very distant and isolated regions, as the process of modification has necessarily been slow, all the means of migration will have been possible during a very long period; and consequently the difficulty of the wide diffusion of species of the same genus is in some degree lessened.

As on the theory of natural selection an interminable number of intermediate forms must have existed, linking together all the species in each group by gradations as fine as our present varieties, it may be asked, Why do we not see these linking forms all around us? Why are not all organic beings blended together in an inextricable chaos? With respect to existing forms, we should remember that we have no right to expect (excepting in rare cases) to discover *directly* connecting links between them, but only between each and some extinct and supplanted form. Even on a wide area, which has during a long period remained continuous, and of which the climate and other conditions of life change insensibly in going from a district occupied by one species into another district occupied by a closely allied species, we have no just right to expect often to find intermediate varieties in the intermediate zone. For we have reason to believe that only a few species are undergoing change at any one period; and all changes are slowly effected. I have also shown that the intermediate varieties which will at first probably exist in the intermediate zones, will be liable to be supplanted by the allied forms on either hand; and the latter, from existing in greater numbers, will generally be modified and improved at a quicker rate than the intermediate varieties, which exist in lesser numbers; so that the intermediate varieties will, in the long run, be supplanted and exterminated.

On this doctrine of the extermination of an infinitude of connecting links, between the living and extinct inhabitants of the world, and at each successive period between the extinct and still older species, why is not every geological formation charged with such links? Why does not every collection of fossil remains afford plain evidence of the gradation and mutation of the forms of life? We meet with no such evidence, and this is the most obvious and forcible of the many objections which may be urged against my theory. Why, again, do whole groups of allied species appear, though certainly they often falsely appear, to have come in suddenly on the several geological stages? Why do we not find great piles of strata beneath the Silurian system, stored with the remains of the progenitors of the Silurian groups of fossils? For certainly on my theory such strata must somewhere have been deposited at these ancient and utterly unknown epochs in the world's history.

I can answer these questions and grave objections only on the supposition that the geological record is far more imperfect than most geologists believe. It cannot be objected that there has not been time sufficient for any amount of organic change; for the lapse of time has been so great as to be utterly inappreciable by the human intellect. The number of specimens in all our museums is absolutely as nothing compared with the countless generations of countless species which certainly have existed. We should not be able to recognise a species as the parent of any one or more species if we were to examine them ever so closely, unless we likewise possessed many of the intermediate links between their past or parent and present states; and these many links we could hardly ever expect to discover, owing to the imperfection of the geological record. Numerous existing doubtful forms could be named which are probably varieties; but who will pretend that in future ages so many fossil links will be discovered, that naturalists will be able to decide, on the common view, whether or not these doubtful forms are varieties? As long as most of the links between any two species are unknown, if any one link or intermediate variety be discovered, it will simply be classed as another and distinct species. Only a small portion of the world has been geologically explored. Only organic beings of certain classes can be preserved in a fossil condition, at least in any great number. Widely ranging species vary most, and varieties are often at first local,—both causes rendering the discovery of intermediate links less likely. Local varieties will not spread into other and distant regions until they are considerably modified and improved; and when they do spread, if discovered in a geological formation, they will appear as if suddenly created there, and will be simply classed as new species. Most formations have been intermittent in their accumulation; and their duration, I am inclined to believe, has been shorter than the average duration of specific forms. Successive formations are separated from each other by enormous blank intervals of time; for fossiliferous formations, thick enough to resist future degradation, can be accumulated only where much sediment is deposited on the subsiding bed of the sea. During the alternate periods of elevation and of stationary level the record will be blank. During these latter periods there will probably be more variability in the forms of life; during periods of subsidence, more extinction.

With respect to the absence of fossiliferous formations beneath the lowest Silurian strata, I can only recur to the hypothesis given in the ninth chapter. That the geological record is imperfect all will admit; but that it is imperfect to the degree which I require, few will be inclined to admit. If we look to long enough intervals of time, geology plainly declares that all species

London's satiric journal *Punch* enjoyed lampooning Charles Darwin and his theories throughout his career. By 1861, primarily through the efforts of T.H. Huxley, the theory that man had descended from Apes was being discussed in all quarters of English society.

have changed; and they have changed in the manner which my theory requires, for they have changed slowly and in a graduated manner. We clearly see this in the fossil remains from consecutive formations invariably being much more closely related to each other, than are the fossils from formations distant from each other in time.

Such is the sum of the several chief objections and difficulties which may justly be urged against my theory; and I have now briefly recapitulated the answers and explanations which can be given to them. I have felt these difficulties far too heavily during many years to doubt their weight. But it deserves especial notice that the more important objections relate to questions on which we are confessedly ignorant; nor do we know how ignorant we are. We do not know all the possible transitional gradations between the simplest and the most perfect organs; it cannot be pretended that we know all the varied means of Distribution during the long lapse of years, or that we know how imperfect the Geological Record is. Grave as these several difficulties are, in my judgment they do not overthrow the theory of descent with modification.

Now let us turn to the other side of the argument. Under domestication we see much variability. This seems to be mainly due to the reproductive system being eminently susceptible to changes in the conditions of life; so that this system, when not rendered impotent, fails to reproduce offspring exactly like the parent-form. Variability is governed by many complex laws,—by correlation of growth, by use and disuse, and by the direct action of the physical conditions of life. There is much difficulty in ascertaining how much modification our domestic productions have undergone; but we may safely infer that the amount has been large, and that modifications can be inherited for long periods. As long as the conditions of life remain the same, we have reason to believe that a modification, which has already been inherited for many generations, may continue to be inherited for an almost infinite number of generations. On the other hand we have evidence that variability, when it has once come into play, does not wholly cease; for new varieties are still occasionally produced by our most anciently domesticated productions.

Man does not actually produce variability; he only unintentionally exposes organic beings to new conditions of life, and then nature acts on the organisation, and causes variability. But man can and does select the variations given to him by nature, and thus accumulate them in any desired manner. He thus adapts animals and plants for his own benefit or pleasure. He may do this methodically, or he may do it unconsciously by preserving the individuals most useful to him at the time, without any thought of altering the breed. It is certain that he can largely influence the character of a breed by selecting, in each successive generation, individual differences so slight as to be quite inappreciable by an uneducated eye. This process of selection has been the great agency in the production of the most distinct and useful domestic breeds. That many of the breeds produced by man have to a large extent the character of natural species, is shown by the inextricable doubts whether very many of them are varieties or aboriginal species.

There is no obvious reason why the principles which have acted so efficiently under domestication should not have acted under nature. In the preservation of favoured individuals and races, during the constantly-recurrent Struggle for Existence, we see the most powerful and ever-acting means of selection. The struggle for existence inevitably follows from the high geometrical ratio of increase which is common to all organic beings. This high rate of increase is proved by calculation, by the effects of a succession of peculiar seasons, and by the results of naturalisation, as explained in the third chapter. More individuals are born than can possibly survive. A grain in the balance will determine which individual shall live and which shall die,—which variety or species shall increase in number, and which shall decrease, or finally become extinct. As the individuals of the same species come in all respects into the closest competition with each other, the struggle will generally be most severe between them; it will be almost equally severe between the varieties of the same species, and next in severity between the species of the same genus. But the struggle will often be very severe between beings most remote in the scale of nature. The slightest advantage in one being, at any age or during any season, over those with which it comes into competition, or better adaptation in however slight a degree to the surrounding physical conditions, will turn the balance.

With animals having separated sexes there will in most cases be a struggle between the males for possession of the females. The most vigorous individuals, or those which have most successfully struggled with their conditions of life, will generally leave most progeny. But success will often depend on having special weapons or means of defence, or on the charms of the males; and the slightest advantage will lead to victory.

As geology plainly proclaims that each land has undergone great physical changes, we might have expected that organic beings would have varied under nature, in the same way as they generally have varied under the changed conditions of domestication. And if there be any variability under nature, it would be an unaccountable fact if natural selection had not come into play. It has often been asserted, but the assertion is quite incapable of proof, that the amount of variation under nature is a strictly limited quantity. Man, though acting on external characters alone and often capriciously, can produce within a short period a great result by adding up mere individual differences in his domestic productions; and every one admits that there are at least individual differences in species under nature. But, besides such differences, all naturalists have admitted the existence of varieties, which they think sufficiently distinct to be worthy of record in systematic works. No one can draw any clear distinction between individual differences and slight varieties; or between more plainly marked varieties and sub-species, and species. Let it be observed how naturalists differ in the rank which they assign to the many representative forms in Europe and North America.

If then we have under nature variability and a powerful agent always ready to act and select, why should we doubt that variations in any way useful to beings, under their excessively complex relations of life, would be preserved, accumulated, and inherited? Why, if man can by patience select variations most useful to himself, should nature fail in selecting variations useful, under changing conditions of life, to her living products? What limit can be put to this power, acting during long ages and rigidly scrutinising the whole constitution, structure, and habits of each creature,—favouring the good and rejecting the bad? I can see no limit to this power, in slowly and beautifully adapting each form to the most complex relations of life. The theory of natural selection, even if we looked no further than this, seems to me to be in itself probable. I have already recapitulated, as fairly as I could, the opposed difficulties and objections: now let us turn to the special facts and arguments in favour of the theory.

On the view that species are only strongly marked and permanent varieties, and that each species first existed as a variety, we can see why it is that no line of demarcation can be drawn between species, commonly supposed to have been produced by special acts of creation, and varieties which are acknowledged to have been produced by secondary laws. On this same view we can understand how it is that in each region where many species of a genus have been produced, and where they now flourish, these same species should present many varieties; for where the manufactory of species has been active, we might expect, as a general rule, to find it still in action; and this is the case if varieties be incipient species. Moreover, the species of the larger genera, which afford the greater number of varieties or incipient species, retain to a certain degree the character of varieties; for they differ from each other by a less amount of difference than do the species of smaller genera. The closely allied species also of the larger genera apparently have restricted ranges, and they are clustered in little groups round other species—in which respects they resemble varieties. These are strange relations on the view of each species having been independently created, but are intelligible if all species first existed as varieties.

As each species tends by its geometrical ratio of reproduction to increase inordinately in number; and as the modified descendants of each species will be enabled to increase by so much the more as they become more diversified in habits and structure, so as to be enabled to seize on many and widely different places in the economy of nature, there will be a constant tendency in natural selection to preserve the most divergent offspring of any one species. Hence during a long-continued course of modification, the slight differences, characteristic of varieties of the same species, tend to be augmented into the greater differences characteristic of species of the same genus. New and improved varieties will inevitably supplant and exterminate the older, less improved and intermediate varieties; and thus species are rendered to a large extent defined and distinct objects. Dominant species belonging to the larger groups tend to give birth to new and dominant forms; so that each large group tends to become still larger, and at the same time more divergent in character. But as all groups cannot thus succeed in increasing in size, for the world would not hold them, the more dominant groups beat the less dominant. This tendency in the large groups to go on increasing in size and diverging in character, together with the almost inevitable contingency of much extinction, explains the arrangement of all the forms of life, in groups subordinate to groups, all within a few great classes, which we now see everywhere around us, and which has prevailed throughout all time. This grand fact of the grouping of all organic beings seems to me utterly inexplicable on the theory of creation.

As natural selection acts solely by accumulating slight, successive, favourable variations, it can produce no great or sudden modification; it can act only by very short and slow steps. Hence the canon of "Natura non facit saltum," which every fresh addition to our knowledge tends to make more strictly correct, is on this theory simply intelligible. We can plainly see why nature is prodigal in variety, though niggard in innovation. But why this should be a law of nature if each species has been independently created, no man can explain.

Many other facts are, as it seems to me, explicable on this theory. How strange it is that a bird, under the form of woodpecker, should have been created to prey on insects on the ground; that upland geese, which never or rarely swim, should have been created with webbed feet; that a thrush should have been created to dive and feed on sub-aquatic insects; and that a petrel should have been created with habits and structure fitting it for the life of an auk or grebe! and so on in endless other cases. But on the view of each species constantly trying to increase in number, with natural selection always ready to adapt the slowly varying descendants of each to any unoccupied or ill-occupied place in nature, these facts cease to be strange, or perhaps might even have been anticipated.

As natural selection acts by competition, it adapts the inhabitants of each country only in relation to the degree of perfection of their associates; so that we need feel no surprise at the inhabitants of any one country, although on the ordinary view supposed to have been specially created and adapted for that country, being beaten and supplanted by the naturalised productions from another land. Nor ought we to marvel if all the contrivances in nature be not, as far as we can judge, absolutely perfect; and if some of them be abhorrent to our ideas of fitness. We need not marvel at the sting of the bee causing the bee's own death; at drones being produced in such vast numbers for one single act, and being then slaughtered by their sterile sisters; at the astonishing waste of pollen by our fir-trees; at the instinctive hatred of the queen bee for her own fertile daughters; at ichneumonidae feeding within the live bodies of caterpillars; and at other such cases. The wonder indeed is, on the theory of natural selection, that more cases of the want of absolute perfection have not been observed.

The complex and little known laws governing variation are the same, as far as we can see, with the laws which have governed the production of so-called specific forms. In both cases physical conditions seem to have produced but little direct effect; yet when varieties enter any zone, they occasionally assume some of the characters of the species proper to that zone. In both varieties and species, use and disuse seem to have produced some effect; for it is difficult to resist this conclusion when we look, for instance, at the logger-headed duck, which has wings incapable of flight, in nearly the same condition as in the domestic duck; or when we look at the burrowing tucutucu, which is occasionally blind, and then at certain moles, which are habitually blind and have their eyes covered with skin; or when we look at the blind animals inhabiting the dark caves of America and Europe. In both varieties and species correlation of growth seems to have played a most important part, so that when one part has been modified other parts are necessarily modified. In both varieties and species reversions to long-lost characters occur. How inexplicable on the theory of creation is the occasional appearance of stripes on the shoulder and legs of the several species of the horse-genus and in their hybrids! How simply is this fact explained if we believe that these species have descended from a striped progenitor, in the same manner as the several domestic breeds of pigeon have descended from the blue and barred rock-pigeon!

On the ordinary view of each species having been independently created, why should the specific characters, or those by which the species of the same genus differ from each other, be more variable than the generic characters in which they all agree? Why, for instance, should the colour of a flower be more likely to vary in any one species of a genus, if the other species, supposed to have been created independently, have differently coloured flowers? If species are only well-marked varieties, of which the characters have become in a high degree permanent, we can understand this fact; for they have already varied since they branched off from a common progenitor in certain characters, by which they have come to be specifically distinct from each other; and therefore these same characters would be more likely still to be variable than the genetic characters which have been inherited without change for an enormous period. It is inexplicable on the theory of creation why a part developed in a very unusual manner in any one species of a genus, and therefore, as we may naturally infer, of great importance to the species, should be eminently liable to variation; but, on my view, this part has undergone, since the several species branched off from a common progenitor, an unusual amount of variability and modification, and therefore we might expect this part generally to be still variable. But a part may be developed in the most unusual manner, like the wing of a bat, and yet not be more variable than any other structure, if the part be common to many subordinate forms, that is, if it has been inherited for a very long period; for in this case it will have been rendered constant by long-continued natural selection.

Glancing at instincts, marvellous as some are, they offer no greater difficulty than does corporeal structure on the theory of the natural selection of successive, slight, but profitable modifications. We can thus understand why nature moves by graduated steps in endowing different animals of the same class with their several instincts. I have attempted to show how much light the principle of gradation throws on the admirable architectural powers of the hive-bee. Habit no doubt sometimes comes into play in modifying instincts; but it certainly is not indispensable, as we see, in the case of neuter insects, which leave no progeny to inherit the effects of long-continued habit. On the view of all the species of the same genus having descended from a common parent, and having inherited much in common, we can understand how it is that allied species, when placed under considerably different conditions of life, yet should follow nearly the same instincts; why the thrush of South America, for instance, lines her nest with mud like our British species. On the view of instincts having been slowly acquired through natural selection we need not marvel at some instincts being apparently not perfect and liable to mistakes, and at many instincts causing other animals to suffer.

If species be only well-marked and permanent varieties, we can at once see why their crossed offspring should follow the same complex laws in their degrees and kinds of resemblance to their parents,—in being absorbed into each other by successive crosses, and in other such points,—as do the crossed offspring of acknowledged varieties. On the other hand, these would be strange facts if species have been independently created, and varieties have been produced by secondary laws.

If we admit that the geological record is imperfect in an extreme degree, then such facts as the record gives, support the theory of descent with modification. New species have come on the stage slowly and at successive intervals; and the amount of change, after equal intervals of time, is widely different in different groups. The extinction of species and of whole groups of species, which has played so conspicuous a part in the history of the organic world, almost inevitably follows on the principle of natural selection; for old forms will be supplanted by new and improved forms. Neither single species nor groups of species reappear when the chain of ordinary generation has once been broken. The gradual diffusion of dominant forms, with the slow modification of their descendants, causes the forms of life, after long intervals of time, to appear as if they had changed simultaneously throughout the world. The fact of the fossil remains of each formation being in some degree intermediate in character between the fossils in the formations above and below, is simply explained by their intermediate position in the chain of descent. The grand fact that all extinct organic beings belong to the same system with recent beings, falling either into the same or into intermediate groups, follows from the living and the extinct being the offspring of common parents. As the groups which have descended from an ancient progenitor have generally diverged in character, the progenitor with its early descendants will often be intermediate in character in comparison with its later descendants; and thus we can see why the more ancient a fossil is, the oftener it stands in some degree intermediate between existing and allied groups. Recent forms are generally looked at as being, in some vague sense, higher than ancient and extinct forms; and they are in so far higher as the later and more improved forms have conquered the older and less improved organic beings in the struggle for life. Lastly, the law of the long endurance of allied forms on the same continent,—of marsupials in Australia, of edentata in America, and other such cases,—is intelligible, for within a confined country, the recent and the extinct will naturally be allied by descent.

Looking to geographical distribution, if we admit that there has been during the long course of ages much migration from one part of the world to another, owing to former climatal and geographical changes and to the many occasional and unknown means of dispersal, then we can understand, on the theory of descent with modification, most of the great leading facts in Distribution. We can see why there should be so striking a parallelism in the distribution of organic beings throughout space, and in their geological succession throughout time; for in both cases the beings have been connected by the bond of ordinary generation, and the means of modification have been the same. We shall see the full meaning of the wonderful fact, which must have struck every traveller, namely, that on the same continent, under the most diverse conditions, under heat and cold, on mountain and lowland, on deserts and marshes, most of the inhabitants within each great class are plainly related; for they will generally be descendants of the same progenitors and early colonists. On this same principle of former migration, combined in most cases with modification, we can understand, by the aid of the Glacial period, the identity of some few plants, and the close alliance of many others, on the most distant mountains, under the most different climates; and likewise the close alliance of some of the inhabitants of the sea in the northern and southern temperate zones, though separated by the whole intertropical ocean. Although two areas may present the same physical conditions of life, we need feel no surprise at their inhabitants being widely different, if they have been for a long period completely separated from each other; for as the relation of organism to organism is the most important of all relations, and as the two areas will have received colonists from some third source or from each other, at various periods and in different proportions, the course of modification in the two areas will inevitably be different.

On this view of migration, with subsequent modification, we can see why oceanic islands should be inhabited by few species, but of these, that many should be peculiar. We can clearly see why those animals which cannot cross wide spaces of ocean, as frogs and terrestrial mammals, should not inhabit oceanic islands; and why, on the other hand, new and peculiar species of bats, which can traverse the ocean, should so often be found on islands far distant from any continent. Such facts as the presence of peculiar species of bats, and the absence of all other mammals, on oceanic islands, are utterly inexplicable on the theory of independent acts of creation.

The existence of closely allied or representative species in any two areas, implies, on the theory of descent with modification, that the same parents formerly inhabited both areas; and we almost invariably find that wherever many closely allied species inhabit two areas, some identical species common to both still exist. Wherever many closely allied yet distinct species occur, many doubtful forms and varieties of the same species likewise occur. It is a rule of high generality that the inhabitants of each area are related to the inhabitants of the nearest source whence immigrants might have been derived. We see this in nearly all the plants and animals of the Galapagos archipelago, of Juan Fernandez, and of the other American islands being related in the most striking manner to the plants and animals of the neighbouring American mainland; and those of the Cape de Verde archipelago and other African islands to the African mainland. It must be admitted that these facts receive no explanation on the theory of creation.

The fact, as we have seen, that all past and present organic beings constitute one grand natural system, with group subordinate to group, and with extinct groups often falling in between recent groups, is intelligible on the theory of natural selection with its contingencies of extinction and divergence of character. On these same principles we see how it is, that the mutual affinities of the species and genera within each class are so complex and circuitous. We see why certain characters are far more serviceable than others for classification;—why adaptive characters, though of paramount importance to the being, are of hardly any importance in classification; why characters derived from rudimentary parts, though of no service to the being, are often of high classificatory value; and why embryological characters are the most valuable of all. The real affinities of all organic beings are due to inheritance or community of descent. The natural system is a genealogical arrangement, in which we have to discover the lines of descent by the most permanent characters, however slight their vital importance may be.

The framework of bones being the same in the hand of a man, wing of a bat, fin of the porpoise, and leg of the horse,—the same number of vertebrae forming the neck of the giraffe and of the elephant,—and innumerable other such facts, at once explain themselves on the theory of descent with slow and slight successive modifications. The similarity of pattern in the wing and leg of a bat, though used for such different purpose,—in the jaws and legs of a crab,—in the petals, stamens, and pistils of a flower, is likewise intelligible on the view of the gradual modification of parts or organs, which were alike in the early progenitor of each class. On the principle of successive variations not always supervening at an early age, and being inherited at a corresponding not early period of life, we can clearly see why the embryos of mammals, birds, reptiles, and fishes should be so closely alike, and should be so unlike the adult forms. We may cease marvelling at the embryo of an air-breathing mammal or bird having branchial slits and arteries running in loops, like those in a fish which has to breathe the air dissolved in water, by the aid of well-developed branchiæ.

Disuse, aided sometimes by natural selection, will often tend to reduce an organ, when it has become useless by changed habits or under changed conditions of life; and we can clearly understand on this view the meaning of rudimentary organs. But disuse and selection will generally act on each creature, when it has come to maturity and has to play its full part in the struggle for existence, and will thus have little power of acting on an organ during early life; hence the organ will not be much reduced or rendered rudimentary at this early age. The calf, for instance, has inherited teeth, which never cut through the gums of the upper jaw, from an early progenitor having well-developed teeth; and we may believe, that the teeth in the mature animal were reduced, during successive generations, by disuse or by the tongue and palate having been fitted by natural selection to browse without their aid; whereas in the calf, the teeth have been left untouched by selection or disuse, and on the principle of inheritance at corresponding ages have been inherited from a remote period to the present day. On the view of each organic being and each separate organ having been specially created, how utterly inexplicable it is that parts, like the teeth in the embryonic calf or like the

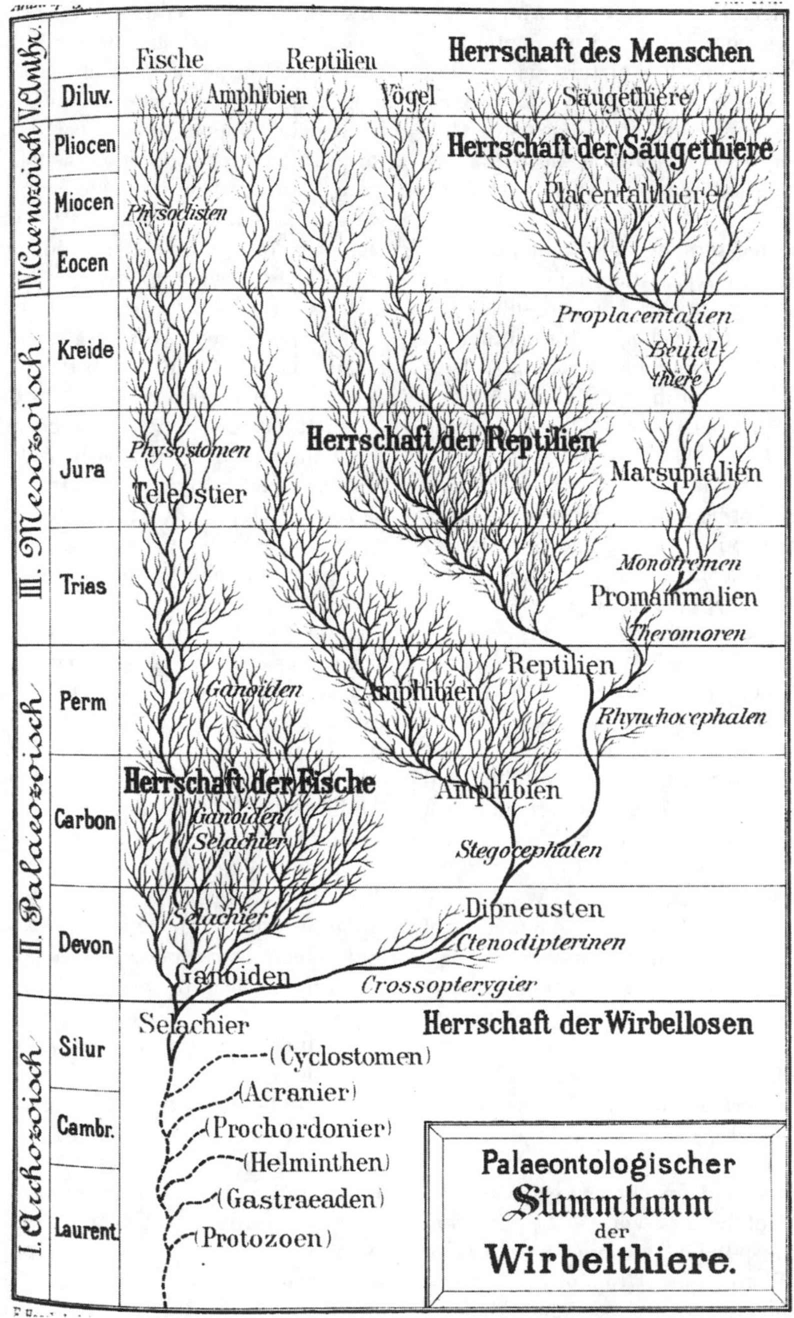

Ernst Haeckel's palaeontological descent tree of the vertebrates from *Anthropogenie*, part 2, *Stammesgeschichte des Menschen: Wissenschaftliche Vorträge über die Grundzüge der Menschlichen Phylogenie* (Leipzig: Wilhelm Engelmann, 1891). See also the similar branching diagrams in Robert Chambers's *Vestiges of Natural Creation* and Charles Darwin's *Origin of Species*.

shrivelled wings under the soldered wing-covers of some beetles, should thus so frequently bear the plain stamp of inutility! Nature may be said to have taken pains to reveal, by rudimentary organs and by homologous structures, her scheme of modification, which it seems that we wilfully will not understand.

I have now recapitulated the chief facts and considerations which have thoroughly convinced me that species have changed, and are still slowly changing by the preservation and accumulation of successive slight favourable variations. Why, it may be asked, have all the most eminent living naturalists and geologists rejected this view of the mutability of species? It cannot be asserted that organic beings in a state of nature are subject to no variation; it cannot be proved that the amount of variation in the course of long ages is a limited quantity; no clear distinction has been, or can be, drawn between species and well-marked varieties. It cannot be maintained that species when intercrossed are invariably sterile, and varieties invariably fertile; or that sterility is a special endowment and sign of creation. The belief that species were immutable productions was almost unavoidable as long as the history of the world was thought to be of short duration; and now that we have acquired some idea of the lapse of time, we are too apt to assume, without proof, that the geological record is so perfect that it would have afforded us plain evidence of the mutation of species, if they had undergone mutation.

But the chief cause of our natural unwillingness to admit that one species has given birth to other and distinct species, is that we are always slow in admitting any great change of which we do not see the intermediate steps. The difficulty is the same as that felt by so many geologists, when Lyell first insisted that long lines of inland cliffs had been formed, and great valleys excavated, by the slow action of the coast-waves. The mind cannot possibly grasp the full meaning of the term of a hundred million years; it cannot add up and perceive the full effects of many slight variations, accumulated during an almost infinite number of generations.

Although I am fully convinced of the truth of the views given in this volume under the form of an abstract, I by no means expect to convince experienced naturalists whose minds are stocked with a multitude of facts all viewed, during a long course of years, from a point of view directly opposite to mine. It is so easy to hide our ignorance under such expressions as the "plan of creation," "unity of design," etc., and to think that we give an explanation when we only restate a fact. Any one whose disposition leads him to attach more weight to unexplained difficulties than to the explanation of a certain number of facts will certainly reject my theory. A few naturalists, endowed with much flexibility of mind, and who have already begun to doubt on the immutability of species, may be influenced by this volume; but I look with confidence to the future, to young and rising naturalists, who will be able to view both sides of the question with impartiality. Whoever is led to believe that species are mutable will do good service by conscientiously expressing his conviction; for only thus can the load of prejudice by which this subject is overwhelmed be removed.

Several eminent naturalists have of late published their belief that a multitude of reputed species in each genus are not real species; but that other species are real, that is, have been independently created. This seems to me a strange conclusion to arrive at. They admit that a multitude of forms, which till lately they themselves thought were special creations, and which are still thus looked at by the majority of naturalists, and which consequently have every external characteristic feature of true species,—they admit that these have been produced by variation, but they refuse to extend the same view to other and very slightly different forms. Nevertheless they do not pretend that they can define, or even conjecture, which are the created forms of life, and which are those produced by secondary laws. They admit variation as a *vera causa* in one case, they arbitrarily reject it in another, without assigning any distinction in the two cases. The day will come when this will be given as a curious illustration of the blindness of preconceived opinion. These authors seem no more startled at a miraculous act of creation than at an ordinary birth. But do they really believe that at innumerable periods in the earth's history certain elemental atoms have been commanded suddenly to flash into living tissues? Do they believe that at each supposed act of creation one individual or many were produced? Were all the infinitely numerous kinds of animals and plants created as eggs or seed, or as full grown? and in the case of mammals, were they created bearing the false marks of nourishment from the mother's womb? Although naturalists very properly demand a full explanation of every difficulty from those who believe in the mutability of species, on their own side they ignore the whole subject of the first appearance of species in what they consider reverent silence.

It may be asked how far I extend the doctrine of the modification of species. The question is difficult to answer, because the more distinct the forms are which we may consider, by so much the arguments fall away in force. But some arguments of the greatest weight extend very far. All the members of whole classes can be connected together by chains of affinities, and all can be classified on the same principle, in groups subordinate to groups. Fossil remains sometimes tend to fill up very wide intervals between existing orders. Organs in a rudimentary condition plainly show that an early progenitor had the organ in a fully developed state; and this in some instances necessarily implies an enormous amount of modification in the descendants. Throughout whole classes various structures are formed on the same pattern, and at an embryonic age the species closely resemble each other. Therefore I cannot doubt that the theory of descent with modification embraces all the members of the same class. I believe that animals have descended from at most only four or five progenitors, and plants from an equal or lesser number.

Analogy would lead me one step further, namely, to the belief that all animals and plants have descended from some one prototype. But analogy may be a deceitful guide. Nevertheless all living things have much in common, in their chemical composition, their germinal vesicles, their cellular structure, and their laws of growth and reproduction. We see this even in so trifling a circumstance as that the same poison often similarly affects plants and animals; or that the poison secreted by the gall-fly produces monstrous growths on the wild rose or oak-tree. Therefore I should infer from analogy that probably all the organic beings which have ever lived on this earth have descended from some one primordial form, into which life was first breathed.

When the views entertained in this volume on the origin of species, or when analogous views are generally admitted, we can dimly foresee that there will be a considerable revolution in natural history. Systematists will be able to pursue their labours as at present; but they will not be incessantly haunted by the shadowy doubt whether this or that form be in essence a species. This I feel sure, and I speak after experience, will be no slight relief. The endless disputes whether or not some fifty species of British brambles are true species will cease. Systematists will have only to decide (not that this will be easy) whether any form be sufficiently constant and distinct from other forms, to be capable of definition; and if definable, whether the differences be sufficiently important to deserve a specific name. This latter point will become a far more essential consideration than it is at present; for differences, however slight, between any two forms, if not blended by intermediate gradations, are looked at by most naturalists as sufficient to raise both forms to the rank of species. Hereafter we shall be compelled to acknowledge that the only distinction between species and well-marked varieties is, that the latter are known, or believed, to be connected at the present day by intermediate gradations, whereas species were formerly thus connected. Hence, without quite rejecting the consideration of the present existence of intermediate gradations between any two forms, we shall be led to weigh more carefully and to value higher the actual amount of difference between them. It is quite possible that forms now generally acknowledged to be merely varieties may hereafter be thought worthy of specific names, as with the primrose and cowslip; and in this case scientific and common language will come into accordance. In short, we shall have to treat species in the same manner as those naturalists treat genera, who admit that genera are merely artificial combinations made for convenience. This may not be a cheering prospect; but we shall at least be freed from the vain search for the undiscovered and undiscoverable essence of the term species.

The other and more general departments of natural history will rise greatly in interest. The terms used by naturalists of affinity, relationship, community of type, paternity, morphology, adaptive characters, rudimentary and aborted organs, etc., will cease to be metaphorical, and will have a plain signification. When we no longer look at an organic being as a savage looks at a ship, as at something wholly beyond his comprehension; when we regard every production of nature as one which has had a history; when we contemplate every complex structure and instinct as the summing up of many contrivances, each useful to the possessor, nearly in the same way as when we look at any great mechanical invention as the summing up of the labour, the experience, the reason, and even the blunders of numerous workmen; when we thus view each organic being, how far more interesting, I speak from experience, will the study of natural history become!

A grand and almost untrodden field of inquiry will be opened, on the causes and laws of variation, on correlation of growth, on the effects of use and disuse, on the direct action of external conditions, and so forth. The study of domestic productions will rise immensely in value. A new variety raised by man will be a far more important and interesting subject for study than one more species added to the infinitude of already recorded species. Our classifications will come to be, as far as they can be so made, genealogies; and will then truly give what may be called the plan of creation. The rules for classifying will no doubt become simpler when we have a definite object in view. We possess no pedigrees or armorial bearings; and we have to discover and trace the many diverging lines of descent in our natural genealogies, by characters of any kind which have long been inherited. Rudimentary organs will speak infallibly with respect to the nature of long-lost structures. Species and groups of species, which are called aberrant, and which may fancifully be called living fossils, will aid us in forming a picture of the ancient forms of life. Embryology will reveal to us the structure, in some degree obscured, of the prototypes of each great class.

When we can feel assured that all the individuals of the same species, and all the closely allied species of most genera, have within a not very remote period descended from one parent, and have migrated from some one birthplace; and when we better know the many means of migration, then, by the light which geology now throws, and will continue to throw, on former changes of climate and of the level of the land, we shall surely be enabled to trace in an admirable manner the former migrations of the inhabitants of the whole world. Even at present, by comparing the differences of the inhabitants of the sea on the opposite sides of a continent, and the nature of the various inhabitants of that continent in relation to their apparent means of immigration, some light can be thrown on ancient geography.

The noble science of Geology loses glory from the extreme imperfection of the record. The crust of the earth with its embedded remains must not be looked at as a well-filled museum, but as a poor collection made at hazard and at rare intervals. The accumulation of each great fossiliferous formation will be recognised as having depended on an unusual concurrence of circumstances, and the blank intervals between the successive stages as having been of vast duration. But we shall be able to gauge with some security the duration of these intervals by a comparison of the preceding and succeeding organic forms. We must be cautious in attempting to correlate as strictly contemporaneous two formations, which include few identical species, by the general succession of their forms of life. As species are produced and exterminated by slowly acting and still existing causes, and not by miraculous acts of creation and by catastrophes; and as the most important of all causes of organic change is one which is almost independent of altered and perhaps suddenly altered physical conditions, namely, the mutual relation of organism to organism,—the improvement of one being entailing the improvement or the extermination of others; it follows, that the amount of organic change in the fossils of consecutive formations probably serves as a fair measure of the lapse of actual time. A number of species, however, keeping in a body might remain for a long period unchanged, whilst within this same period, several of these species, by migrating into new countries and coming into competition with foreign associates, might become modified; so that we must not overrate the accuracy of organic change as a measure of time. During early periods of the earth's history, when the forms of life were probably fewer and simpler, the rate of change was probably slower; and at the first dawn of life, when very few forms of the simplest structure existed, the rate of change may have been slow in an extreme degree. The whole history of the world, as at present known, although of a length quite incomprehensible by us, will hereafter be recognised as a mere fragment of time, compared with the ages which have elapsed since the first creature, the progenitor of innumerable extinct and living descendants, was created.

In the distant future I see open fields for far more important researches. Psychology will be based on a new foundation, that of the necessary acquirement of each mental power and capacity by gradation. Light will be thrown on the origin of man and his history.

Authors of the highest eminence seem to be fully satisfied with the view that each species has been independently created. To my mind it accords better with what we know of the laws impressed on matter by the Creator, that the production and extinction of the past and present inhabitants of the world should have been due to secondary causes, like those determining the birth and death of the individual. When I view all beings not as special creations, but as the lineal descendants of some few beings which lived long before the first bed of the Silurian system was deposited, they seem to me to become ennobled. Judging from the past, we may safely infer that not one living species will transmit its unaltered likeness to a distant futurity. And of the species now living very few will transmit progeny of any kind to a far distant futurity; for the manner in which all organic beings are grouped, shows that the greater number of species of each genus, and all the species of many genera, have left no descendants, but have become utterly extinct. We can so far take a prophetic glance into futurity as to foretel that it will be the common and widely-spread species, belonging to the larger and dominant groups, which will ultimately prevail and procreate new and dominant species. As all the living forms of life are the lineal descendants of those which lived long before the Silurian epoch, we may feel certain that the ordinary succession by generation has never once been broken and that no cataclysm has desolated the whole world. Hence we may look with some confidence to a secure future of equally inappreciable length. And as natural selection works solely by and for the good of each being, all corporeal and mental endowments will tend to progress towards perfection.

It is interesting to contemplate an entangled bank, clothed with many plants of many kinds, with birds singing on the bushes, with various insects flitting about, and with worms crawling through the damp earth, and to reflect that these elaborately constructed forms, so different from each other, and dependent on each other in so complex a manner, have all been produced by laws acting around us. These laws, taken in the largest sense, being Growth with Reproduction; Inheritance which is almost implied by reproduction; Variability from the indirect and direct action of the external conditions of life, and from use and disuse; a Ratio of Increase so high as to lead to a Struggle for Life, and as a consequence to Natural Selection, entailing Divergence of Character and the Extinction of less-improved forms. Thus, from the war of nature, from famine and death, the most exalted object which we are capable of conceiving, namely, the production of the higher animals, directly follows. There is grandeur in this view of life, with its several powers, having been originally breathed into a few forms or into one; and that, whilst this planet has gone cycling on according to the fixed law of gravity, from so simple a beginning endless forms most beautiful and most wonderful have been, and are being, evolved.

Punch's satiric depiction of Darwinian evolution from worms to man on the occasion of the publication of Charles Darwin's *The Formation of the Vegetable Mould Through the Action of Worms* (1881). The bearded figure in the upper left representing Father Time is Darwin.

"General Summary and Conclusion." *The Descent of Man, and Selection in Relation to Sex.* London: John Murray, 1871, pp. 385–405.

Main conclusion that man is descended from some lower form — Manner of development — Genealogy of man — Intellectual and moral faculties — Sexual selection — Concluding remarks.

A brief summary will here be sufficient to recall to the reader's mind the more salient points in this work. Many of the views which have been advanced are highly speculative, and some no doubt will prove erroneous; but I have in every case given the reasons which have led me to one view rather than to another. It seemed worth while to try how far the principle of evolution would throw light on some of the more complex problems in the natural history of man. False facts are highly injurious to the progress of science, for they often long endure; but false views, if supported by some evidence, do little harm, as every one takes a salutary pleasure in proving their falseness; and when this is done, one path towards error is closed and the road to truth is often at the same time opened.

The main conclusion arrived at in this work, and now held by many naturalists who are well competent to form a sound judgment, is that man is descended from some less highly organised form. The grounds upon which this conclusion rests will never be shaken, for the close similarity between man and the lower animals in embryonic development, as well as in innumerable points of structure and constitution, both of high and of the most trifling importance,—the rudiments which he retains, and the abnormal reversions to which he is occasionally liable,—are facts which cannot be disputed. They have long been known, but until recently they told us nothing with respect to the origin of man. Now when viewed by the light of our knowledge of the whole organic world, their meaning is unmistakeable. The great principle of evolution stands up clear and firm, when these groups of facts are considered in connection with others, such as the mutual affinities of the members of the same group, their geographical distribution in past and present times, and their geological succession. It is incredible that all these facts should speak falsely. He who is not content to look, like a savage, at the phenomena of nature as disconnected, cannot any longer believe that man is the work of a separate act of creation. He will be forced to admit that the close resemblance of the embryo of man to that, for instance, of a dog—the construction of his skull, limbs, and whole frame, independently of the uses to which the parts may be put, on the same plan with that of other mammals—the occasional reappearance of various structures, for instance of several distinct muscles, which man does not normally possess, but which are common to the Quadrumana—and a crowd of analogous facts—all point in the plainest manner to the conclusion that man is the co-descendant with other mammals of a common progenitor.

We have seen that man incessantly presents individual differences in all parts of his body and in his mental faculties. These differences or variations seem to be induced by the same general causes, and to obey the same laws as with the lower animals. In both cases similar laws of inheritance prevail. Man tends to increase at a greater rate than his means of subsistence; consequently he is occasionally subjected to a severe struggle for existence, and natural selection will have effected whatever lies within its scope. A succession of strongly-marked variations of a similar nature are by no means requisite; slight fluctuating differences in the individual suffice for the work of natural selection. We may feel assured that the inherited effects of the long-continued use or disuse of parts will have done much in the same direction with natural selection. Modifications formerly of importance, though no longer of any special use, will be long inherited. When one part is modified, other parts will change through the principle of correlation, of which we have instances in many curious cases of correlated monstrosities. Something may be attributed to the direct and definite action of the surrounding conditions of life, such as abundant food, heat, or moisture; and lastly, many characters of slight physiological importance, some indeed of considerable importance, have been gained through sexual selection.

No doubt man, as well as every other animal, presents structures, which as far as we can judge with our little knowledge, are not now of any service to him, nor have been so during any former period of his existence, either in relation to his general conditions of life, or of one sex to the other. Such structures cannot be accounted for by any form of selection, or by the inherited effects of the use and disuse of parts. We know, however, that many strange and strongly-marked peculiarities of structure occasionally appear in our domesticated productions, and if the unknown causes which produce them were to act more uniformly, they would probably become common to all the individuals of the species. We may hope hereafter to understand something about the causes of such occasional modifications, especially through the study of monstrosities: hence the labours of experimentalists, such as those of M. Camille Dareste, are full of promise for the future. In the greater number of cases we can only say that the cause of each slight variation and of each monstrosity lies much more in the nature or constitution of the organism, than in the nature of the surrounding conditions; though new and changed conditions certainly play an important part in exciting organic changes of all kinds.

Through the means just specified, aided perhaps by others as yet undiscovered, man has been raised to his present state. But since he attained to the rank of manhood, he has diverged into distinct races, or as they may be more appropriately called sub-species. Some of these, for instance the Negro and European, are so distinct that, if specimens had been brought to a naturalist without any further information, they would undoubtedly have been considered by him as good and true species. Nevertheless all the races agree in so many unimportant details of structure and in so many mental peculiarities, that these can be accounted for only through inheritance from a common progenitor; and a progenitor thus characterised would probably have deserved to rank as man.

It must not be supposed that the divergence of each race from the other races, and of all the races from a common stock, can be traced back to any one pair of progenitors. On the contrary, at every stage in the process of modification, all the individuals which were in any way best fitted for their conditions of life, though in different degrees, would have survived in greater numbers than the less well fitted. The process would have been like that followed by man, when he does not intentionally select particular individuals, but breeds from all the superior and neglects all the inferior individuals. He thus slowly but surely modifies his stock, and unconsciously forms a new strain. So with respect to modifications, acquired independently of selection, and due to variations arising from the nature of the organism and the action of the surrounding conditions, or from changed habits of life, no single pair will have been modified in a much greater degree than the other pairs which inhabit the same country, for all will have been continually blended through free intercrossing.

By considering the embryological structure of man,—the homologies which he presents with the lower animals,—the rudiments which he retains,—and the reversions to which he is liable, we can partly recall in imagination the former condition of our early progenitors; and can approximately place them in their proper position in the zoological series. We thus learn that man is descended from a hairy quadruped, furnished with a tail and pointed ears, probably arboreal in its habits, and an inhabitant of the Old World. This creature, if its whole structure had been examined by a naturalist, would have been classed amongst the Quadrumana, as surely as would the common and still more ancient progenitor of the Old and New World monkeys. The Quadrumana and all the higher mammals are probably derived from an ancient marsupial animal, and this through a long line of diversified forms, either from some reptile-like or some amphibian-like creature, and this again from some fish-like animal. In the dim obscurity of the past we can see that the early progenitor of all the Vertebrata must have been an aquatic animal, provided with branchiae, with the two sexes united in the same individual, and with the most important organs of the body (such as the brain and heart) imperfectly developed. This animal seems to have been more like the larvae of our existing marine Ascidians than any other known form.

The greatest difficulty which presents itself, when we are driven to the above conclusion on the origin of man, is the high standard of intellectual power and of moral disposition which he has attained. But every one who admits the general principle of evolution, must see that the mental powers of the higher animals, which are the same in kind with those of mankind, though so different in degree, are capable of advancement. Thus the interval between the mental powers of one of the higher apes and of a fish, or between those of an ant and scale-insect, is immense. The development of these powers in animals does not offer any special difficulty; for with our domesticated animals, the mental faculties are certainly variable, and the variations are inherited. No one doubts that these faculties are of the utmost importance to animals in a state of nature. Therefore the conditions are favourable for their development through natural selection. The same conclusion may be extended to man; the intellect must have been all-important to him, even at a very remote period, enabling him to use language, to invent and make weapons, tools, traps, etc.; by which means, in combination with his social habits, he long ago became the most dominant of all living creatures.

A great stride in the development of the intellect will have followed, as soon as, through a previous considerable advance, the half-art and half-instinct of language came into use; for the continued use of language will have reacted on the brain, and produced an inherited effect; and this again will have reacted on the improvement of language. The large size of the brain in man, in comparison with that of the lower animals, relatively to the size of their bodies, may be attributed in chief part, as Mr. Chauncey Wright has well remarked,[1] to the early use of some simple form of language,—that wonderful engine which affixes signs to all sorts of objects and qualities, and excites trains of thought which would never arise from the mere impression of the senses, and if they did arise could not be followed out. The higher intellectual powers of man, such as those of ratiocination, abstraction, self-consciousness, etc., will have followed from the continued improvement of other mental faculties; but without considerable culture of the mind, both in the race and in the individual, it is doubtful whether these high powers would be exercised, and thus fully attained.

The development of the moral qualities is a more interesting and difficult problem. Their foundation lies in the social instincts, including in this term the family ties. These instincts are of a highly complex nature, and in the case of the lower animals give special tendencies towards certain definite actions; but the more important elements for us are love, and the distinct emotion of sympathy. Animals endowed with the social instincts take pleasure in each other's company, warn each other of danger, defend and aid each other in many ways. These instincts are not extended to all the individuals of the species, but only to those of the same community. As they are highly beneficial to the species, they have in all probability been acquired through natural selection.

A moral being is one who is capable of comparing his past and future actions and motives,—of approving of some and disapproving of others; and the fact that man is the one being who with certainty can be thus designated makes the greatest of all distinctions between him and the lower animals. But in our third chapter I have endeavoured to shew that the moral sense follows, firstly, from the enduring and always present nature of the social instincts, in which respect man agrees with the lower animals; and secondly, from his mental faculties being highly active and his impressions of past events extremely vivid, in which respects he differs from the lower animals. Owing to this condition of mind, man cannot avoid looking backwards and comparing the impressions of past events and actions. He also

1 On the "Limits of Natural Selection," in the "North American Review," Oct. 1870, p. 295.

continually looks forward. Hence after some temporary desire or passion has mastered his social instincts, he will reflect and compare the now weakened impression of such past impulses, with the ever present social instinct; and he will then feel that sense of dis-satisfaction which all unsatisfied instincts leave behind them. Consequently he resolves to act differently for the future—and this is conscience. Any instinct which is permanently stronger or more enduring than another, gives rise to a feeling which we express by saying that it ought to be obeyed. A pointer dog, if able to reflect on his past conduct, would say to himself, I ought (as indeed we say of him) to have pointed at that hare and not have yielded to the passing temptation of hunting it.

Social animals are partly impelled by a wish to aid the members of the same community in a general manner, but more commonly to perform certain definite actions. Man is impelled by the same general wish to aid his fellows, but has few or no special instincts. He differs also from the lower animals in being able to express his desires by words, which thus become the guide to the aid required and bestowed. The motive to give aid is likewise somewhat modified in man: it no longer consists solely of a blind instinctive impulse, but is largely influenced by the praise or blame of his fellow men. Both the appreciation and the bestowal of praise and blame rest on sympathy; and this emotion, as we have seen, is one of the most important elements of the social instincts. Sympathy, though gained as an instinct, is also much strengthened by exercise or habit. As all men desire their own happiness, praise or blame is bestowed on actions and motives, according as they lead to this end; and as happiness is an essential part of the general good, the greatest-happiness principle indirectly serves as a nearly safe standard of right and wrong. As the reasoning powers advance and experience is gained, the more remote effects of certain lines of conduct on the character of the individual, and on the general good, are perceived; and then the self-regarding virtues, from coming within the scope of public opinion, receive praise, and their opposites receive blame. But with the less civilised nations reason often errs, and many bad customs and base superstitions come within the same scope, and consequently are esteemed as high virtues, and their breach as heavy crimes.

The moral faculties are generally esteemed, and with justice, as of higher value than the intellectual powers. But we should always bear in mind that the activity of the mind in vividly recalling past impressions is one of the fundamental though secondary bases of conscience. This fact affords the strongest argument for educating and stimulating in all possible ways the intellectual faculties of every human being. No doubt a man with a torpid mind, if his social affections and sympathies are well developed, will be led to good actions, and may have a fairly sensitive conscience. But whatever renders the imagination of men more vivid and strengthens the habit of recalling and comparing past impressions, will make the conscience more sensitive, and may even compensate to a certain extent for weak social affections and sympathies.

The moral nature of man has reached the highest standard as yet attained, partly through the advancement of the reasoning powers and consequently of a just public opinion, but especially through the sympathies being rendered more tender and widely diffused through the effects of habit, example, instruction, and reflection. It is not improbable that virtuous tendencies may through long practice be inherited. With the more civilised races, the conviction of the existence of an all-seeing Deity has had a potent influence on the advancement of morality. Ultimately man no longer accepts the praise or blame of his fellows as his chief guide, though few escape this influence, but his habitual convictions controlled by reason afford him the safest rule. His conscience then becomes his supreme judge and monitor. Nevertheless the first foundation or origin of the moral sense lies in the social instincts, including sympathy; and these instincts no doubt were primarily gained, as in the case of the lower animals, through natural selection.

The belief in God has often been advanced as not only the greatest, but the most complete of all the distinctions between man and the lower animals. It is however impossible, as we have seen, to maintain that this belief is innate or instinctive in man. On the other hand a belief in all-pervading spiritual agencies seems to be universal; and apparently follows from a considerable advance in the reasoning powers of man, and from a still greater advance in his faculties of imagination, curiosity and wonder. I am aware that the assumed instinctive belief in God has been used by many persons as an argument for His existence. But this is a rash argument, as we should thus be compelled to believe in the existence of many cruel and malignant spirits, possessing only a little more power than man; for the belief in them is far more general than of a beneficent Deity. The idea of a universal and beneficent Creator of the universe does not seem to arise in the mind of man, until he has been elevated by long-continued culture.

He who believes in the advancement of man from some lowly-organised form, will naturally ask how does this bear on the belief in the immortality of the soul. The barbarous races of man, as Sir J. Lubbock has shewn, possess no clear belief of this kind; but arguments derived from the primeval beliefs of savages are, as we have just seen, of little or no avail. Few persons feel any anxiety from the impossibility of determining at what precise period in the development of the individual, from the first trace of the minute germinal vesicle to the child either before or after birth, man becomes an immortal being; and there is no greater cause for anxiety because the period in the gradually ascending organic scale cannot possibly be determined.[2]

I am aware that the conclusions arrived at in this work will be denounced by some as highly irreligious; but he who thus denounces them is bound to shew why it is more irreligious to explain the origin of man as a distinct species by descent from some lower form, through the laws of variation and natural selection, than to explain the birth of the individual through the laws of ordinary reproduction. The birth both of the species and of the individual are equally parts of that grand sequence of events, which our minds refuse to accept as the result of blind chance. The understanding revolts at such a conclusion, whether or not we are able to believe that every slight variation of structure,—the union of each pair in marriage,—the dissemination of each seed,—and other such events, have all been ordained for some special purpose.

Sexual selection has been treated at great length in these volumes; for, as I have attempted to shew, it has played an important part in the history of the organic world. As summaries have been given to each chapter, it would be superfluous here to add a detailed summary. I am aware that much remains doubtful, but I have endeavoured to give a fair view of the whole case. In the lower divisions of the animal kingdom, sexual selection seems to have done nothing: such animals are often affixed for life to the same spot, or have the two sexes combined in the same individual, or what is still more important, their perceptive and intellectual faculties are not sufficiently advanced to allow of the feelings of love and jealousy, or of the exertion of choice. When, however, we come to the Arthropoda and Vertebrata, even to the lowest classes in these two great Sub-Kingdoms, sexual selection has effected much; and it deserves notice that we here find the intellectual faculties developed, but in two very distinct lines, to the highest standard, namely in the Hymenoptera (ants, bees, etc.) amongst the Arthropoda, and in the Mammalia, including man, amongst the Vertebrata.

In the most distinct classes of the animal kingdom, with mammals, birds, reptiles, fishes, insects, and even crustaceans, the differences between the sexes follow almost exactly the same rules. The males are almost always the wooers; and they alone are armed with special weapons for fighting with their rivals. They are generally stronger and larger than the females, and are endowed with the requisite qualities of courage and pugnacity. They are provided, either exclusively or in a much higher degree than the females, with organs for producing vocal or instrumental music, and with odoriferous glands. They are ornamented with infinitely diversified appendages, and with the most brilliant or conspicuous colours, often arranged in elegant patterns, whilst the females are left

2 The Rev. J.A. Picton gives a discussion to this effect in his "New Theories and the Old Faith," 1870.

unadorned. When the sexes differ in more important structures, it is the male which is provided with special sense-organs for discovering the female, with locomotive organs for reaching her, and often with prehensile organs for holding her. These various structures for securing or charming the female are often developed in the male during only part of the year, namely the breeding season. They have in many cases been transferred in a greater or less degree to the females; and in the latter case they appear in her as mere rudiments. They are lost by the males after emasculation. Generally they are not developed in the male during early youth, but appear a short time before the age for reproduction. Hence in most cases the young of both sexes resemble each other; and the female resembles her young offspring throughout life. In almost every great class a few anomalous cases occur in which there has been an almost complete transposition of the characters proper to the two sexes; the females assuming characters which properly belong to the males. This surprising uniformity in the laws regulating the differences between the sexes in so many and such widely separated classes, is intelligible if we admit the action throughout all the higher divisions of the animal kingdom of one common cause, namely sexual selection.

Sexual selection depends on the success of certain individuals over others of the same sex in relation to the propagation of the species; whilst natural selection depends on the success of both sexes, at all ages, in relation to the general conditions of life. The sexual struggle is of two kinds; in the one it is between the individuals of the same sex, generally the male sex, in order to drive away or kill their rivals, the females remaining passive; whilst in the other, the struggle is likewise between the individuals of the same sex, in order to excite or charm those of the opposite sex, generally the females, which no longer remain passive, but select the more agreeable partners. This latter kind of selection is closely analogous to that which man unintentionally, yet effectually, brings to bear on his domesticated productions, when he continues for a long time choosing the most pleasing or useful individuals, without any wish to modify the breed.

The laws of inheritance determine whether characters gained through sexual selection by either sex shall be transmitted to the same sex, or to both sexes; as well as the age at which they shall be developed. It appears that variations which arise late in life are commonly transmitted to one and the same sex. Variability is the necessary basis for the action of selection, and is wholly independent of it. It follows from this, that variations of the same general nature have often been taken advantage of and accumulated through sexual selection in relation to the propagation of the species, and through natural selection in relation to the general purposes of life. Hence secondary sexual characters, when equally transmitted to both sexes can be distinguished from ordinary specific characters only by the light of analogy. The modifications acquired through sexual selection are often so strongly pronounced that the two sexes have frequently been ranked as distinct species, or even as distinct genera. Such strongly-marked differences must be in some manner highly important; and we know that they have been acquired in some instances at the cost not only of inconvenience, but of exposure to actual danger.

The belief in the power of sexual selection rests chiefly on the following considerations. The characters which we have the best reason for supposing to have been thus acquired are confined to one sex; and this alone renders it probable that they are in some way connected with the act of reproduction. These characters in innumerable instances are fully developed only at maturity; and often during only a part of the year, which is always the breeding-season. The males (passing over a few exceptional cases) are the most active in courtship; they are the best armed, and are rendered the most attractive in various ways. It is to be especially observed that the males display their attractions with elaborate care in the presence of the females; and that they rarely or never display them excepting during the season of love. It is incredible that all this display should be purposeless. Lastly we have distinct evidence with some quadrupeds and birds that the individuals of the one sex are capable of feeling a strong antipathy or preference for certain individuals of the opposite sex.

Bearing these facts in mind, and not forgetting the marked results of man's unconscious selection, it seems to me almost certain that if the individuals of one sex were during a long series of generations to prefer pairing with certain individuals of the other sex, characterised in some peculiar manner, the offspring would slowly but surely become modified in this same manner. I have not attempted to conceal that, excepting when the males are more numerous than the females, or when polygamy prevails, it is doubtful how the more attractive males succeed in leaving a larger number of offspring to inherit their superiority in ornaments or other charms than the less attractive males; but I have shewn that this would probably follow from the females,—especially the more vigorous females which would be the first to breed, preferring not only the more attractive but at the same time the more vigorous and victorious males.

Although we have some positive evidence that birds appreciate bright and beautiful objects, as with the Bower-birds of Australia, and although they certainly appreciate the power of song, yet I fully admit that it is an astonishing fact that the females of many birds and some mammals should be endowed with sufficient taste for what has apparently been effected through sexual selection; and this is even more astonishing in the case of reptiles, fish, and insects. But we really know very little about the minds of the lower animals. It cannot be supposed that male Birds of Paradise or Peacocks, for instance, should take so much pains in erecting, spreading, and vibrating their beautiful plumes before the females for no purpose. We should remember the fact given on excellent authority in a former chapter, namely that several peahens, when debarred from an admired male, remained widows during a whole season rather than pair with another bird.

Nevertheless I know of no fact in natural history more wonderful than that the female Argus pheasant should be able to appreciate the exquisite shading of the ball-and-socket ornaments and the elegant patterns on the wing-feathers of the male. He who thinks that the male was created as he now exists must admit that the great plumes, which prevent the wings from being used for flight, and which, as well as the primary feathers, are displayed in a manner quite peculiar to this one species during the act of courtship, and at no other time, were given to him as an ornament. If so, he must likewise admit that the female was created and endowed with the capacity of appreciating such ornaments. I differ only in the conviction that the male Argus pheasant acquired his beauty gradually, through the females having preferred during many generations the more highly ornamented males; the aesthetic capacity of the females having been advanced through exercise or habit in the same manner as our own taste is gradually improved. In the male, through the fortunate chance of a few feathers not having been modified, we can distinctly see how simple spots with a little fulvous shading on one side might have been developed by small and graduated steps into the wonderful ball-and-socket ornaments; and it is probable that they were actually thus developed.

Everyone who admits the principle of evolution, and yet feels great difficulty in admitting that female mammals, birds, reptiles, and fish, could have acquired the high standard of taste which is implied by the beauty of the males, and which generally coincides with our own standard, should reflect that in each member of the vertebrate series the nerve-cells of the brain are the direct offshoots of those possessed by the common progenitor of the whole group. It thus becomes intelligible that the brain and mental faculties should be capable under similar conditions of nearly the same course of development, and consequently of performing nearly the same functions.

The reader who has taken the trouble to go through the several chapters devoted to sexual selection, will be able to judge how far the conclusions at which I have arrived are supported by sufficient evidence. If he accepts these conclusions, he may, I think, safely extend them to mankind; but it would be superfluous here to repeat what I have so lately said on the manner in which sexual selection has apparently acted on both the male and female side, causing the two sexes of man to differ in body and mind, and the several races to differ from each other in various characters, as well as from their ancient and lowly-organised progenitors.

He who admits the principle of sexual selection will be led to the remarkable conclusion that the cerebral system not only regulates most of the existing functions of the body, but has indirectly influenced the progressive development of various bodily structures and of certain mental qualities. Courage, pugnacity, perseverance, strength and size of body, weapons of all kinds, musical organs, both vocal and instrumental, bright colours, stripes and marks, and ornamental appendages, have all been indirectly gained by the one sex or the other, through the influence of love and jealousy, through the appreciation of the beautiful in sound, colour or form, and through the exertion of a choice; and these powers of the mind manifestly depend on the development of the cerebral system.

Man scans with scrupulous care the character and pedigree of his horses, cattle, and dogs before he matches them; but when he comes to his own marriage he rarely, or never, takes any such care. He is impelled by nearly the same motives as are the lower animals when left to their own free choice, though he is in so far superior to them that he highly values mental charms and virtues. On the other hand he is strongly attracted by mere wealth or rank. Yet he might by selection do something not only for the bodily constitution and frame of his offspring, but for their intellectual and moral qualities. Both sexes ought to refrain from marriage if in any marked degree inferior in body or mind; but such hopes are Utopian and will never be even partially realised until the laws of inheritance are thoroughly known. All do good service who aid towards this end. When the principles of breeding and of inheritance are better understood, we shall not hear ignorant members of our legislature rejecting with scorn a plan for ascertaining by an easy method whether or not consanguineous marriages are injurious to man.

The advancement of the welfare of mankind is a most intricate problem: all ought to refrain from marriage who cannot avoid abject poverty for their children; for poverty is not only a great evil, but tends to its own increase by leading to recklessness in marriage. On the other hand, as Mr. Galton has remarked, if the prudent avoid marriage, whilst the reckless marry, the inferior members will tend to supplant the better members of society. Man, like every other animal, has no doubt advanced to his present high condition through a struggle for existence consequent on his rapid multiplication; and if he is to advance still higher he must remain subject to a severe struggle. Otherwise he would soon sink into indolence, and the more highly-gifted men would not be more successful in the battle of life than the less gifted. Hence our natural rate of increase, though leading to many and obvious evils, must not be greatly diminished by any means. There should be open competition for all men; and the most able should not be prevented by laws or customs from succeeding best and rearing the largest number of offspring. Important as the struggle for existence has been and even still is, yet as far as the highest part of man's nature is concerned there are other agencies more important. For the moral qualities are advanced, either directly or indirectly, much more through the effects of habit, the reasoning powers, instruction, religion, etc., than through natural selection; though to this latter agency the social instincts, which afforded the basis for the development of the moral sense, may be safely attributed.

The main conclusion arrived at in this work, namely that man is descended from some lowly-organised form, will, I regret to think, be highly distasteful to many persons. But there can hardly be a doubt that we are descended from barbarians. The astonishment which I felt on first seeing a party of Fuegians on a wild and broken shore will never be forgotten by me, for the reflection at once rushed into my mind—such were our ancestors. These men were absolutely naked and bedaubed with paint, their long hair was tangled, their mouths frothed with excitement, and their expression was wild, startled, and distrustful. They possessed hardly any arts, and like wild animals lived on what they could catch; they had no government, and were merciless to every one not of their own small tribe. He who has seen a savage in his native land will not feel much shame, if forced to acknowledge that the blood of some more humble creature flows in his veins. For my own part I would as soon be descended from that heroic little monkey, who braved his dreaded enemy in order to save the life of his keeper; or from that old baboon, who, descending from the mountains, carried away in triumph his young comrade from a crowd of astonished dogs—as from a savage who delights to torture his enemies, offers up bloody sacrifices, practises infanticide without remorse, treats his wives like slaves, knows no decency, and is haunted by the grossest superstitions.

Man may be excused for feeling some pride at having risen, though not through his own exertions, to the very summit of the organic scale; and the fact of his having thus risen, instead of having been aboriginally placed there, may give him hopes for a still higher destiny in the distant future. But we are not here concerned with hopes or fears, only with the truth as far as our reason allows us to discover it. I have given the evidence to the best of my ability; and we must acknowledge, as it seems to me, that man with all his noble qualities, with sympathy which feels for the most debased, with benevolence which extends not only to other men but to the humblest living creature, with his god-like intellect which has penetrated into the movements and constitution of the solar system—with all these exalted powers—Man still bears in his bodily frame the indelible stamp of his lowly origin.

FURTHER READING

Amigioni, David, and Jeff Wallace, eds. *Charles Darwin's The Origin of Species: New Interdisciplinary Essays.* Manchester: Manchester UP, 1995.

Andersson, M.B. *Sexual Selection.* Princeton: Princeton UP, 1994.

Appleman, Philip. *Darwin: A Norton Critical Edition.* New York: W.W. Norton and Co., Inc., 1970.

Bajema, Carl Jay, ed. *Evolution by Sexual Selection Theory Prior to 1900.* New York: Van Nostrand Reinhold Company, 1984.

Barnett, S.A. *A Century of Darwin.* Cambridge, MA: Harvard UP, 1958.

Bateson, Patrick. *Mate Choice.* Cambridge: Cambridge UP, 1983.

Batten, Mary. *Sexual Strategies: How Females Choose Their Mates.* New York: G.P. Putnam's Sons, 1992.

Beer, Gillian. *Darwin's Plots: Evolutionary Narrative in Darwin, George Eliot, and Nineteenth-century Fiction.* London: Routledge & Kegan Paul, 1983.

Bender, Bert. *The Descent of Love: Darwin and the Theory of Sexual Selection in American Fiction, 1871-1926.* Philadelphia: University of Pennsylvania Press, 1996.

Bowlby, John. *Charles Darwin: A New Life.* New York: W.W. Norton and Co., 1991.

Bowler, Peter J. *Evolution: the History of an Idea.* Rev. ed. Berkeley: University of California Press, 1989.

—. *The Non-Darwinian Revolution: Reinterpreting a Historical Myth.* Baltimore: Johns Hopkins UP, 1988.

—. *Charles Darwin: The Man and His Influence.* London: Basil Blackwell, 1990.

—. *The Eclipse of Darwinism: Anti-Darwinian Evolution Theories in the Decades Around 1900.* Baltimore: Johns Hopkins UP, 1983.

Bradbury, J.W. and M.B. Andersson, eds. *Sexual Selection: Testing the Alternatives.* New York: John Wiley and Sons, 1987.

Browne, Janet. *Charles Darwin: Voyaging.* London: Jonathan Cape, 1995.

Campbell, Bernard Grant, ed. *Sexual Selection and the Descent of Man, 1871-1971.* Chicago: Aldine Pub. Co., 1972.

Colp, Ralph, Jr. *To Be an Invalid: The Illness of Charles Darwin.* Chicago: University of Chicago Press, 1977.

Cronin, Helena. *The Ant and the Peacock: Altruism and Sexual Selection from Darwin to Today.* Cambridge: Cambridge UP, 1991.

Darwin, Charles. *The Autobiography of Charles Darwin 1809–1882.* Edited by Nora Barlow. 1958. Reprint, New York: W.W. Norton and Co., 1969.

—. *Charles Darwin's Natural Selection, Being the Second Part of his Big Species Book Written from 1856 to 1858.* Edited by R.H. Stauffer. Cambridge: Cambridge UP, 1975.

—. *Charles Darwin's Notebooks, 1836–1844: Geology, Transmutation of Species, Metaphysical Enquiries.* Transcribed and Edited by Paul

H. Barrett *et al.* Ithaca, NY: Cornell UP, 1987.

—. *The Collected Papers of Charles Darwin.* Edited by Paul H. Barrett. Forward by Theodosius Dobzhansky. 2 vols. Chicago: University of Chicago Press, 1977.

—. *The Life and Letters of Charles Darwin, Including an Autobiographical Chapter.* Edited by Francis Darwin. 3 vols. London: John Murray, 1888.

Darwin, Francis, ed. *The Foundations of the Origin of Species: Two Essays Written in 1842 and 1844 by Charles Darwin.* Cambridge: Cambridge UP, 1909.

Desmond, Adrian J., and James Moore. *Darwin: The Life of A Tormented Evolutionist.* New York: Warner Books, 1991.

Dobzhansky, Theodosius. *Genetics and the Origin of Species.* New York: Columbia UP, 1937.

Eberhard, William G. *Female Control: Sexual Selection by Cryptic Female Choice.* Princeton: Princeton UP, 1996.

Eiseley, Loren. *Darwin's Century: Evolution and the Men Who Discovered It.* Garden City, NY: Doubleday Anchor Books, 1958.

Ellegård, Alvar. *Darwin and the General Reader: The Reception of Darwin's Theory of Evolution in the British Periodical Press, 1859–1872.* Foreword by David L. Hull. 1958. Reprint, Chicago: University of Chicago Press, 1990.

Etcoff, Nancy. *Survival of the Prettiest: The Science of Beauty.* New York: Doubleday, 1999.

Freeman, R.B. *Charles Darwin: A Companion.* London: William Dawson and Sons, Ltd., 1978.

Glass, Bentley, Owsei Temkin, and William L. Straus, Jr., eds. *Forerunners of Darwin: 1745-1859.* Baltimore: Johns Hopkins UP, 1959.

Gould, James L. and Carol Grant Gould. *Sexual Selection.* New York: Scientific American Library, 1989.

Grant, Verne. *Selection Modes Involved in Sexual Selection.* Jena: G. Fischer, 1995.

Greene, John C. *The Death of Adam: Evolution and Its Impact on Western Thought.* Ames: Iowa State UP, 1959.

Gruber, Howard E. *Darwin on Man: A Psychological Study of Scientific Creativity, Together with Darwin's Early and Unpublished Notebooks, Transcribed and Annotated by Paul H. Barrett.* New York: E.P. Dutton, 1974.

Hersey, George L. *The Evolution of Allure: Sexual Selection from the Medici Venus to the Incredible Hulk.* Cambridge: MIT Press, 1996.

Himmelfarb, Gertude. *Darwin and the Darwinian Revolution.* Garden City, NY: Doubleday Anchor Books, 1959.

Hodge, M.J.S. *Origins and Species: A Study of the Historical Sources of Darwinism and the Contexts of Some Other Accounts of Organic Diversity from Plato and Aristotle On.* New York: Garland Publishing, Inc., 1991.

Hull, David. *Darwin and His Critics: The Reception of Darwin's Theory of Evolution by the Scientific Community.* Cambridge, MA: Harvard UP, 1973.

Irvine, William. *Apes, Angels, and Victorians: The Story of Darwin, Huxley, and Evolution.* New York: McGraw-Hill Book Co., Inc., 1955.

Jones, Doug. *Physical Attractiveness and the Theory of Sexual Selection.* Ann Arbor: University of Michigan Museum of Anthropology, 1996.

Kohn, David, ed. *The Darwinian Heritage.* Princeton, NJ: Princeton UP, 1985.

Mayr, Ernst. *The Growth of Biological Thought: Diversity, Evolution, and Inheritance.* Cambridge, MA: Belknap Press, 1982.

—. *One Long Argument: Charles Darwin and the Genesis of Modern Evolutionary Thought.* Cambridge, MA: Harvard UP, 1991.

McKinney, H. Lewis, ed. and trans. *Lamarck to Darwin: Contributions to Evolutionary Biology 1809–1859.* Lawrence, KS: Coronado Press, 1971.

Oldroyd, David R. *Darwinian Impacts: An Introduction to the Darwinian Revolution.* Atlantic Highlands, NJ: Humanities Press, 1980.

—. "How Did Darwin Arrive at His Theory? The Secondary Literature to 1982." *History of Science* 21 (1984): 325–374.

Osborn, Henry Fairfield. *From the Greeks to Darwin; the Development of the Evolution Idea.* London: Macmillan and Co., 1894.

Ospovat, Dov. *The Development of Darwin's Theory: Natural History, Natural Theology, and Natural Selection, 1838-1859.* Cambridge: Cambridge UP, 1981.

Peckham, Morse. *The Origin of Species by Charles Darwin: A Variorum Text.* Philadelphia: University of Philadelphia Press, 1959.

Ridley, Mark, ed. *The Darwin Reader.* New York: Norton, 1996.

—. *Evolution.* Oxford: Oxford UP, 1997.

Robinson, Daniel N. *Darwinism: Critical Reviews from Dublin Review, Edinburgh Review, Quarterly Review.* Significant Contributions to the History of Psychology 1750–1920. Washington: University Publications of America, 1977.

Ruse, Michael. *The Darwinian Revolution.* Chicago: University of Chicago Press, 1979.

Vorzimmer, Peter J. *Charles Darwin: the Years of Controversy. The Origin of Species and its Critics, 1859-1882.* Philadelphia: Temple UP, 1970.

Young, Robert M. *Darwin's Metaphor: Nature's Place in Victorian Culture.* Cambridge: Cambridge UP, 1985.

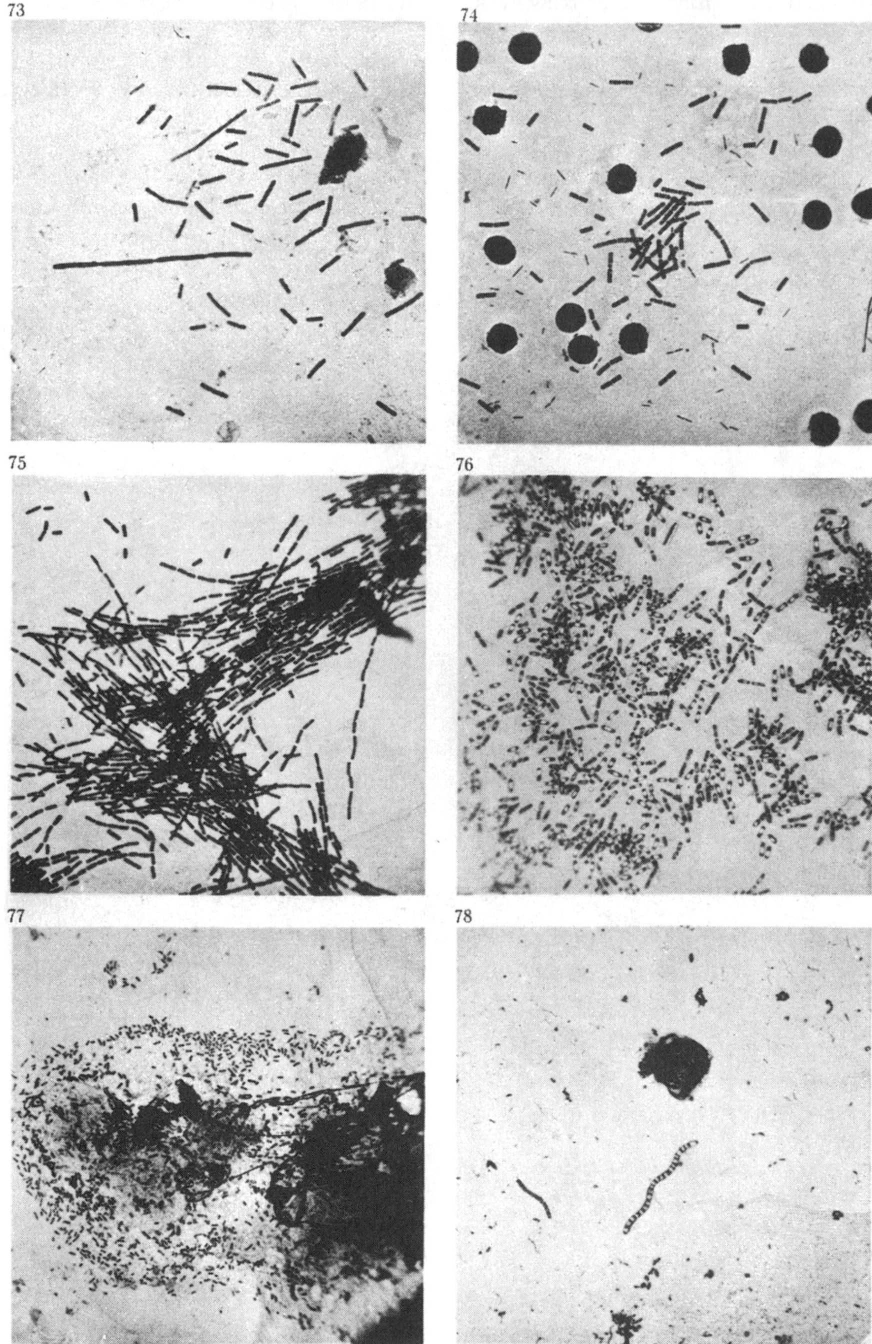

Photographs of *Bacillus* and *Spirillum* (fig. 78) bacteria from the bacteriological work of Robert Koch *Zur Untersuchung von pathogenen Organismen* (1881). Koch isolated a variety of microscopic pathogens, including *Mycobacterium tuberculosis*, the cause of tuberculosis, in 1882 and the anthrax bacillus in 1876.

20. LOUIS PASTEUR

(1822–1895)

Louis Pasteur was descended from a family of millers and tanners in the Franche-Comté region. His great-grandfather bought himself out of serfdom from the Comte d'Udressier in 1763. His father Jean-Joseph served in the Napoléonic wars, and Pasteur probably inherited his sense of order and discipline as well as his sometimes chauvinistic patriotism from the elder Pasteur. For his work on infectious diseases, Pasteur was proclaimed a national hero, a model of the disinterested scientist who dedicated all *pour l'humanité*, but personally he could be bitterly contentious when his theories were attacked by scientific opponents. His elevation to near sainthood in French culture is partially a result of Pasteur's own self-promotion and his obscuring of the scientific contributions of his predecessors and contemporaries; as Gerald Geison has noted, "in a way that few scientists have the opportunity or talent to bring off, Pasteur laid the foundation of his own legend" (267). At the same time, Pasteur repeatedly demonstrated the coherence and practical benefits of his theories of disease and fermentation through careful experimentation and he developed successful vaccines for a number of viral and bacterial diseases.

The young Pasteur excelled in painting and drawing, but also showed great aptitude for the sciences in school. In 1847 he obtained his doctorate at the École Normale and enlisted in the National Guard when the 1848 Republic was proclaimed. Pasteur laid the foundations of stereochemistry in his recognition of chemical isomers—substances with the same chemical formula but possessing different molecular geometries—by attributing the different polarizing properties of the tartrates and paratartrates of sodium and ammonium to differences in crystalline structure. In recognition of his work in crystallography and chemistry, he was appointed professor of chemistry at the University of Strasbourg in 1848. In 1854 he served as dean and professor of chemistry at the University of Lille and was named Administrator of the École Normale Supérieure. From the 1850s onwards, he worked on the related issues of fermentation and spontaneous generation, eventually establishing micro-organisms as the causal agent of fermentation. In the mid-eighteenth century, John Needham and Lazzaro Spallanzani had conducted a vigorous debate over spontaneous generation and fermentation by heating substances in sealed flasks, and watching for the subsequent growth of organisms, although the results of these experiments could be interpreted in a number of ways. It was Pasteur who finally made sense of these previous studies under a coherent theory. In Pasteur's time alcoholic fermentation and putrefaction were widely viewed as analogous processes. Later in his germ theory Pasteur would include infectious disease as another process involving living microbes. In England, Joseph Lister developed Pasteur's germ theory of disease to the greatest extent by introducing Lister's antiseptic method for surgery.

In 1857, Pasteur published his *Mémoire sur la fermentation appelée lactique*, attributing lactic fermentation to an organism similar to yeast. The majority of eighteenth- and nineteenth-century scientists—Lavoisier, Gay-Lussac, Berzelius, Liebig, Helmholtz, and H. Charlton

Bastian—, on the other hand, all believed in a chemical origin for fermentation, denying the primary role of yeast or other biological sources, and the group of scientists before Pasteur, including Schwann, Persoon, Kützing, and Cagniard de la Tour, who attributed fermentation to microbial action were never given a fair hearing. Chemical ferments (enzymes) had been isolated early in the century, such as pepsin by Schwann and diastase by Payen and Persoz in 1833. In 1897, Edouard Buchner isolated zymase from yeast juice, an enzyme which is capable, in conjunction with other phosphates and co-enzymes, of chemically producing alcoholic fermentation without living yeast: In a sense, then, both Pasteur and his opponents were right in the debate over the chemical or biological origin of fermentation. After discovering an "animal ferment" growing in the absence of oxygen, Pasteur introduced the terms "aerobic" and "anaerobic" (with or without oxygen) into the fermentation debate. This discovery led to his later erroneous view that all fermentative processes were anaerobic, resulting from the action of "life without air."

In related work, Pasteur began studying the diseases of wine in the 1860s at the request of Napoléon III and published *Études sur le vin, ses maladies, causes qui les provoquent* (1866). In the 1860s he also determined that the air is filled with living organisms and that spontaneous generation could in fact be attributed to these air-borne germs. In 1865 he patented the process of heating wines to 55 degrees C (Pasteurization) to kill the organisms which cause spoilage; Pasteurization has been found equally effective on milk, beer, orange juice, and cider. Also in 1865, Pasteur began studies of the silkworm diseases *pébrine* and *flacherie* which were decimating the French and Asian silk industry. Although he did not come to a complete understanding of these complex diseases, which also involved viruses undetectable by Pasteur's light microscopes, Pasteur isolated the microbes partially responsible for the sick worms, and his *Études sur la maladie des vers à soie* (1870) can still be read as a model of precise experimental method in the tradition of Claude Bernard.

Pasteur was removed from his administrative post at the École Normale in 1867 after his authoritarian handling of a student protest, but remained affiliated with the school through a new laboratory specially built for him. The following year, he suffered a stroke, followed by two more attacks in 1887 and 1894 which left him increasingly paralyzed. His *Études sur le vinaigre* appeared in 1868 and presented the results of his studies on acetic fermentation, the process which changes wine into vinegar, also previously believed, especially by Liebig, to be a strictly chemical process. Partly in revenge for France's humiliating defeat in the Franco-Prussian War of 1870, Pasteur began research into the spoilage of beer in an effort to challenge the German beer industry. His resulting *Études sur la bière* (1876) recommended Pasteurization of beer to prevent spoilage. In 1876, he ran unsuccessfully for a seat in the French Senate. Pasteur's and Robert Koch's final confirmation in the 1870s that anthrax was caused by a specific bacterium and its highly resistant spores strengthened his germ theory of disease, and by the end of the nineteenth century the causal agents of a great number of bacterial and viral diseases were established.

In a dramatic public experiment in 1881, Pasteur demonstrated the effectiveness of his newly developed anthrax vaccine on sheep, goats, and cows at Pouilly-le-Fort, although he was somewhat deceptive about the actual vaccine used. That same year he was elected to the Académie Française. Although his rabies research was not far advanced, Pasteur reluctantly but successfully administered a rabies vaccine to a nine-year-old boy in 1885 using attenuated virus from spinal marrow.

The Institut Pasteur for microbiological research, built from private donations, opened in 1888. At his death in 1895, Pasteur's body was interred in the Institut after a national outpouring of grief matched only by the funeral of Victor Hugo. Pasteur was fortunate in that three of his descendants have produced detailed accounts of his life and work: his son-in-law René Vallery-Radot's *M. Pasteur, histoire d'un savant par un ignorant* (1883) and *La vie de Pasteur* (1900); Maurice Vallery-Radot's *Pasteur: Un génie au service de l'homme* (1985); and Pasteur Vallery-Radot's series of books. Pasteur Vallery-Radot also edited Pasteur's *Correspondence* (1940–51) and the complete *Oeuvres de Pasteur* (1922–39).

In the first article below, Pasteur describes his discovery of anaerobic microbes (infusoria) producing butyric acid—$CH_3(CH_2)_2COOH$—which gives butter its rancid smell. Infusoria were an ill-defined group of organisms including protozoa, moulds, and bacteria, and Pasteur's use of the term would be closer to the more general "microbe" or "germ." Pasteur was not a zoologist and was criticized for some of his microbial identifications; other researchers identified his butyric acid ferment as a member of the genus *Vibrionidae* (now classified as bacteria). In the second article reprinted below, Pasteur seeks to demonstrate that life in liquid media arises from air-borne germs rather than from spontaneous generation or from the chemical action of the air.

"Infusorian Animalcules Living Without Free Oxygen and Determining Fermentation."

Comptes rendus de l'Académie des sciences 52 (February 25, 1861): 344–47; Reprinted in *Oeuvres de Pasteur*, vol. 2, pp. 136–38. Translation by A.S. Weber.

The variety of products formed by the so-called *lactic* fermentation are well known. Lactic acid, a gum, mannite, butyric acid, alcohol, carbonic acid and hydrogen appear simultaneously or successively in extremely variable and quite unexpected proportions. I have slowly realized that the vegetable ferment which transforms sugar into lactic acid is different from the one or ones (because there are two of them) which determine the production of the gummy material, and that these ferments do not produce lactic acid. Moreover, I have also recognized that these various vegetable ferments, if they are perfectly pure, can in no way give rise to butyric acid.

Therefore there must be a specific butyric ferment. I have focussed my attention on this point for a long time. The communication which I have the honor of addressing to the Academy today precisely concerns the origin of butyric acid in the so-called lactic acid fermentation. I will not enter here into all the details of this research. I will first limit myself to announcing one of the conclusions of my work: that is, *the butyric ferment is an infusorian.*

I had been prepared not to expect this result, to such a degree that for a long time I felt compelled to prevent the appearance of these little animals, for fear that they were not drawing nourishment from the vegetable ferment which I had assumed to be the butyric ferment, the same vegetable ferment which I was searching to discover in the liquid media that I was using. But unsuccessful in uncovering the cause of the origin of the butyric acid, I was in the end struck by the coincidence, that my analyses showed me to be inevitable, between the acid and the infusoria, and conversely between the infusoria and the production of the acid, a fact that I had previously attributed to the favorable and suitable environment that the butyric acid provided to these animalcules.

Since then, a great number of experiments have convinced me that the transformation of sugar, mannite, and lactic acid into butyric acid is due exclusively to these infusoria, and that it is necessary to consider them as the true butyric ferment.

The infusoria can be described as follows: they are little cylindrical rods, rounded at the ends, ordinarily straight, and either isolated or joined in chains of two, three, four or sometimes even more divisions. On average, their size is 0.002 mm. The length of one of the isolated divisions varies from 0.002 mm to 0.015 mm to 0.02 mm. These infusoria propel themselves by gliding. During this movement, their bodies stay rigid, or undergo light undulations. They pirouette, balancing themselves where the anterior or posterior part of their bodies vibrates rapidly. The undulations of their movements become very evident as soon as their length reaches 0.015 mm. Often they are bent back towards one of their extremities, sometimes to both. This peculiarity is rare at the beginning of their lives.

They reproduce by fissiparity [division]. The chains of divisions which some of their bodies join into are evidently a result of this mode of reproduction. A single infusorian which drags other bodies after itself sometimes rapidly shakes as if to detach itself.

Although the bodies of these vibrios have a cylindrical appearance, one might say that they are often formed from a series of particles or very short, scarcely begun divisions which are without doubt the rudimentary beginnings of these little animals.

One can sow these infusoria just like one would sow beer yeast. They multiply if the environment is adapted to their nutrition. But it is essential to remark here that one can sow them in a liquid containing only sugar, ammonia, and phosphates; that is to say, crystalline and completely mineral substances, and they reproduce themselves in correlation with the butyric fermentation, which appears very evident. The weight formed from this process is notable, although always small when compared to the total quantity of butyric acid produced, just as in all ferments.

The existence of infusoria possessing the character of ferments is already a fact which certainly seems worthy of attention; but one unique peculiarity accompanies this fact—these infusorian animalcules live and multiply without limit without the necessity of providing them with the least quantity of air or free oxygen.

It would be tedious to recount here how I absolutely excluded oxygen from the interiors and surfaces of the liquid media where these infusoria live and swarm by the millions, since I have carefully established this elsewhere. I will only add that I did not want to present my findings to the Academy without calling several of its members to witness, who appeared to acknowledge the rigor of the experimental proofs which I placed before their eyes.

Not only do these infusoria live without air, but the air kills them. If one passes a stream of pure carbonic acid for an unspecified amount of time through the liquid in which they live, their life and reproduction are in no way affected. If, on the contrary, a stream of atmospheric air is substituted for the carbonic acid under exactly the same conditions, in only one or two hours the infusoria all die, and the butyric fermentation connected to their existence is soon stopped.

Thus we arrive at this double proposition:

1. *The butyric ferment is an infusorian.*
2. *This infusorian lives without free oxygen gas.*

This is, I believe, the first known example of animal ferments, and also of animals living without free oxygen gas.

The comparison of the way of life and properties of these animalcules with the way of life and properties of the vegetable ferments who live equally without the aid of free oxygen, is self-evident, along with the consequences that may be deduced from it relating to the cause of fermentations. However, I would like to reserve the ideas which these new facts suggest until I have submitted them to the light of experiment.

"Experiments Related to Spontaneous Generation."

Comptes rendus de l'Académie des sciences
50 (February 6, 1860): 303–7; Reprinted in *Oeuvres de Pasteur*, vol. 2, pp. 187–91.
Translation by A.S. Weber.

The research which I have the honor to communicate to the Academy only applies to a single liquid, but one of the most changeable kind. This research has appeared so persuasive to competent judges who wished to examine it, that I took the liberty to fix a date to submit it henceforth to the judgement of the Academy.

In the first part of my work, I apply myself to the microscopical study of the air. With the means of an aspirator with uninterrupted water flow, I passed outside air into a tube in which there was a small wad of cotton, modified so that the cotton would be soluble in a mixture of alcohol and ether. The cotton stops a portion of the solid corpuscles which the air contains. By dissolving the cotton in a small tube with the mixture of ethered alcohol and letting it sit for 24 hours, all of the dust collects at the bottom of the tube where it is easy to concentrate it by decantation, without any loss, if one is careful to separate each decantation by an undisturbed period of 12 to 24 hours. The dust is poured into a watch glass where the rest of the liquid evaporates quickly. It is easy to examine with a microscope the dust collected in this manner, and to add various reagents. This method allows us to isolate the dust in the air during each day, at all times of the year. I propose to apply this method to the examination of air-borne dust of several localities, and comparatively to various elevations.

One discovers from this method that there are corpuscles constantly in the common air in variable quantities whose form and structure reveal that they are organized bodies. These corpuscles are analogous to the ones discovered by microscopists in the dust found on the surface of exterior objects. It is true that M. Pouchet has recognized starch granules among the corpuscles in ordinary dust, but there are comparatively few of these. It is very easy to prove this by dissolving in a drop of concentrated sulfuric acid some air-borne dust collected in the manner I have just indicated. The starch granules dissolve in several seconds, and the majority of the other corpuscles are in no way altered in their form or volume. Many of them even resist the action of the concentrated sulfuric acid for several days. These corpuscles are probably the spores of mucedinea, because I have noticed the same resistance in the spores which have developed in ordinary conditions.

Thus there are organized corpuscles in the air at all times of the year. Are these corpuscles the fertile germs of vegetable organisms or infusoria? This is the question that must be resolved.

I tried three distinct methods. The first, which requires a vat of mercury, leaves doubts in the mind. Blind experiments succeed sometimes. However this method is rather instructive and accounts for many poorly interpreted experiments up to our own times. I will reveal this method in my Mémoire with all the requisite details, and will not pause here.

The second method seems unassailable and completely explanatory. In a round flask of approximately 300 cc, I introduced 100 to 150 cc of albuminous sugar water with the following proportions:

Water 100
Sugar 10
Albuminous matter and minerals
drawn from beer yeast........................0.2 to 0.7

The tapered neck of the flask was attached to a tube of platinum heated red-hot. I boiled the liquid for two or three minutes, and then let it cool completely. It filled up with heated air at normal pressure. Then I sealed the neck of the flask with a torch.

The flask, placed in an oven with a constant temperature of 28 to 32 degrees C can remain there indefinitely without the liquid undergoing the least alteration. After a period of four to six weeks in the stove, I attached the flask by means of a rubber tube (its opening constantly blocked) to a device constructed with the following features: 1) a large glass tube in which I placed a piece of small diameter tubing, open at both ends, free to glide in the larger tube and containing a piece of one of the little wads of cotton saturated with air-born dust; 2) a T-shaped tube provided with three stopcocks; one of the stopcocks was connected to an air pump, another stopcock was connected to a platinum tube heated red hot, and the third stopcock was connected to the tube which I just talked about.

Then after having closed the stopcock connected to the platinum tube, I evacuated the flask. The stopcock was reopened in a manner allowing the calcined (heated) air to gradually enter the apparatus. The two processes of evacuation and introduction of calcined air were alternately repeated six to twelve times. The small tube containing the cotton was thus permeated throughout its smallest interstices with calcined air, but the cotton still held the dust. That done, I broke the end of the flask through the rubber tubing, without disturbing the connections, and let the small tube with the cotton fall into the flask. Finally I resealed the neck of the flask with a torch and placed it back into the oven. Growths never failed to appear in the flask. Here are some of the details of the experiment which deserve the closest scrutiny:

1) Organized growths always begin to appear at the end of 24 to 36 hours. This is precisely the same time necessary for these growths to appear in this kind of liquid when it is exposed to outside air.

2) Moulds arise in the cotton in the most ordinary manner, and they soon fill up the extremities of the cotton wad.

3) The same growths found in ordinary air are formed. Representing the infusoria is the *bacterium*. Representing the mucedinea are the *penicillium*, the *ascophora*, the *aspergillus*, and many other types.

4) Just as in ordinary air, the liquid sometimes gives rise to a type of mucedinea, sometimes another, just as in the experiment various moulds develop.

To sum up, we see on one hand that there are always organized corpuscles in the suspended dust in common air, and on the other hand that if this atmospheric dust is placed into an appropriate liquid, in an atmosphere which is itself completely inactive, then various growths arise, such as the *bacterium termo* and several types of mucedinea, of the same type which would have naturally arisen if the liquid had been freely exposed to ordinary air for the same amount of time.

However, does the cotton, as an organic substance, affect the experiment? And what would happen besides by repeating the same procedure on a flask prepared as I have just described, but excluding the air-borne dust?

I therefore replaced the cotton by a mineral substance, asbestos. The wads of asbestos, after exposure for several hours to the current of air in the aspirator, were introduced into the flasks as I have previously explained, and they gave the same results as the wads of cotton; but with a wad of asbestos previously heated and not charged with air-borne dust, nothing—no disturbance, bacteria, nor mucedinea whatever—was produced. The liquid maintained a perfect clarity.

The following method confirms and enlarges these preliminary results.

I took a certain number of flasks in which I introduced the same fermentable liquid, in the same quantity. I drew out the necks of the flasks with a torch and bent them in various ways, while leaving in all of them an opening of 1–2 mm^2 or more. I boiled the liquid of the great majority of the flasks for several minutes. I only left three or four unboiled. Then I left all of the flasks in a place where the air was calm.

After 24 or 48 hours, according to the temperature, the liquid in the flasks which had not been boiled (but which had attained a temperature of 100 degrees C at the moment of its preparation) were disturbed and became covered little by little with various mucors. The liquid of the other flasks remained clear, not only for several days, but for entire months. However, the flasks were open; undoubtedly it was the sinuosities and angles of their necks which prevented the germs from falling into the liquid. It is true that ordinary air entered abruptly into the flask at the beginning, but during the whole time of its abrupt entry into the flask, the liquid, which was very hot and slow to cool, killed the germs carried by the air; then, when the liquid returned to a temperature low enough to allow the development of these germs, the returning air very slowly dropped its dust at the opening of the neck, or else deposited it on the interior walls of the neck en route to the liquid. Also, if one breaks off the neck of the flask by drawing a file across it, and placing the remaining portion of the flask in a vertical position, mucedinea and bacteria arise in the liquid after a day or two.

M. Chevreul has already previously performed analogous experiments in his courses.

This method, so easily carried out and which explains so well the preceding method, will convince even the most biassed minds. The method, in my opinion, is also especially interesting, because it provides proof that there is nothing in the air, outside of dust, which is necessary for the production of life. Oxygen only intervenes to maintain the life arising from the germs. Gas, fluid, electricity, magnetism, known or hidden things—there is nothing whatsoever in the air, outside of the germs which it carries, which represents a necessary condition of life.

I am going to study other liquids and the growth of other plants and infusoria. I would further like to follow directly the relations of the vegetable seed, from the egg to the adult, in several particular circumstances. I would be eager to communicate to the Academy all the results which appear to me worthy of its attention.

FURTHER READING

Balibar, Françoise, and Marie-Laure Prévost. *Pasteur: Cahiers d'un savant.* Paris: CNRS Éditions, 1995.

Bulloch, William. *The History of Bacteriology.* London: Oxford UP, 1938.

Conant, J.B. *Pasteur's Study of Fermentation.* Harvard Case Histories in Experimental Science, 6. Cambridge, MA: Harvard UP, 1952.

Cuny, Hilaire. *Louis Pasteur: The Man and his Theories.* Translated by Patrick Evans. London: Souvenir Press, 1965.

Dagognet, François. *Pasteur sans la légende.* Paris: Les Empêcheurs de Penser en Rond, 1994.

Darmon, Pierre. *Pasteur.* Paris: Fayard, 1995.

Debré, Patrice. *Louis Pasteur.* Paris: Flammarion, 1994.

Dubos, René J. *Louis Pasteur, Freelance of Science.* Boston: Little, Brown and Company, 1950.

Duclaux, Émile. *Pasteur: The History of a Mind.* Translated by Erwin F. Smith and Florence Hedges. Philadelphia: W.B. Saunders Company, 1920.

Gascar, Pierre. *Du côté de chez Monsieur Pasteur.* Paris: Éditions Odile Jacob, 1986.

Geison, Gerald L. *The Private Science of Louis Pasteur.* Princeton, NJ: Princeton UP, 1995.

Koprowski, Hilary, and Stanley A. Plotkin, eds. *World's Debt to Pasteur.* Wistar Symposium Series, vol. 3. New York: Alan R. Liss, Inc., 1985.

de Kruif, Paul. *Microbe Hunters.* New York: Harcourt, Brace and Co., 1926.

Latour, Bruno. *The Pasteurization of France.* Translated by Alan Sheridan and John Law. Cambridge, MA: Harvard UP, 1988.

Nicolle, Jacques. *Louis Pasteur: A Master of Scientific Enquiry.* London: Hutchinson and Co., 1961.

The Pasteur Fermentation Centennial 1857–1957: A Scientific Symposium. New York: Charles Pfizer and Co., Inc., 1958.

Pasteur, Louis. *Louis Pasteur: Écrits scientifiques et médicaux.* Edited by André Pichot. Paris: Flammarion, 1994.

—. *Pasteur: Extraits de ses oeuvres.* Edited by R. Dujarric de la Rivière. Paris: Gauthier-Villars, 1967.

—. *Researches on the Molecular Asymmetry of Natural Organic Products.* Alembic Club Reprints, 14. Edinburgh: The Alembic Club, 1905.

Salomon-Bayet, Claire, *et al. Pasteur et la révolution pastorienne.* Paris: Payot, 1986.

Temple, Dennis. "Pasteur's Theory of Fermentation: A 'Virtual Tautology.'" *Studies in History and Philosophy of Science* 17, no. 4 (1986): 487–503.

Vallery-Radot, Maurice. *Pasteur: Un génie au service de l'homme.* Lausanne: Éditions Pierre-Marcel Favre, 1985.

Vallery-Radot, Pasteur. *Images de la vie et de l'oeuvre de Pasteur: Documents photographiques, la plupart inédits provenant de la collection Pasteur Vallery-Radot.* Paris: Flammarion, 1956.

—. *Louis Pasteur: A Great Life in Brief.* Translated

by Alfred Joseph. New York: Alfred A. Knopf, 1958.
—. *Pasteur inconnue.* Paris: Flammarion, 1954.
Vallery-Radot, René. *The Life of Pasteur.* Translated by Mrs. R.L. Devonshire. 2 vols. London: Constable, 1911.
—. *Louis Pasteur: His Life and Labours.* Translated by Lady Claud Hamilton. London: Longmans, Green, and Co., 1885.
—. *La Vie de Pasteur.* 2 vols. Paris: Flammarion, 1900.

21. MICHAEL FARADAY

(1791–1867)

Born to a poor blacksmith near London, Michael Faraday educated himself by reading the books he rebound as an apprentice in a book bindery. He attended a series of public lectures by Sir Humphry Davy, the most important chemist of the day, and afterwards sent his own carefully bound notes from the lecture to Davy. Davy subsequently took Faraday on as a laboratory assistant at the Royal Institution in 1813. Although Faraday and Davy made a scientific tour of Europe in 1813–15, their friendship cooled after Davy accused him of plagiarizing research from Wollaston and after Davy attempted to block Faraday's election to the Royal Society.

Faraday was a member of the Protestant religious sect the Sandemanians; the strong work ethic, simplicity, and fierce attachment to the truth of this fundamentalist religion also characterized Faraday's experimental work in chemistry and physics. The influence of Faraday's religion on his scientific work cannot be underestimated, particularly the Sandemanian idea of the unity and harmony of nature: one of Faraday's primary research goals was to develop a unified theory of natural forces. John Tyndall wrote about Faraday that "his religious feeling and his philosophy could not be kept apart; there was an habitual overflow of the one into the other" (181). Tyndall further believed that Faraday drew his week-day strength from his Sunday worship; Faraday's temporary exclusion from the Sandemanians in 1844 over a religious dispute may have contributed to the recurring mental instability, particularly severe in 1839, which plagued his later years.

The growth of Faraday's scientific thought can be traced in the laboratory notebooks he kept from 1820–62, published as *Faraday's Diary* by the Royal Institution. Faraday's original discoveries are numerous and far-ranging: he discovered benzene in 1825, succeeded in liquefying gases such as chlorine, and carried out research on steel alloys. He also instituted the popular Friday evening lecture series at the Royal Institution. In 1819, Hans Oersted had announced the discovery of the relationship between electricity and magnetism—electromagnetism—when he demonstrated that an electric current deflected a magnetized needle. André-Marie Ampère, with his two-fluid theory of electromagnetism which reduced magnetism to an effect of electricity, further demonstrated that an electric current was for all intents and purposes a magnet. Faraday and Davy rushed to replicate the experiments of Oersted, Ampère, and Arago; in 1821 Faraday succeeded in forcing a magnet to rotate around a wire carrying an electric current. This was a basic electric motor. In 1831, by thrusting a bar magnet into a metal cylinder, he produced electricity, creating the opposite of Oersted's effect. He also converted the mechanical energy of a rotating copper disk into electrical energy by spinning the copper disk between the north and south poles of a magnet. This was a simple dynamo or electrical generator, the machine used today to convert the mechanical energy of steam or falling water into electricity. Faraday also discovered that an electrical current changing over time in an insulated coil induced another current in a nearby coil, demonstrating the property of electrical induction, which is a central principle in electromagnetic theory and the basis of electrical transformers. With these electrical experiments, Faraday directly laid the foundation of the electrical power industry of the twentieth century. Despite these achievements and true to the Sandemanian sect, he declined the presidency of the Royal Society. Faraday seldom expressed his scientific thought in complicated mathematical terms, and it was James Clerk Maxwell's mathematical genius which developed Faraday's experimental and theoretical work on electromagnetic fields into classical field theory.

Faraday collected his electrical researches into 3 volumes in *Experimental Researches in Electricity* (1839–55) and his *Experimental Researches in Chemistry and Physics* (1859). He also wrote two books, *Chemical Manipulation, Being Instructions to Students in Chemistry* (1827–29) and a series of lectures on the non-metallic elements (1853). Besides *The Chemical History of a Candle*, reprinted below, William Crookes also edited another series of Faraday's lectures entitled *A Course of Six Lectures on the Various Forces of Matter and Their Relation to Each Other* (1860) for young people.

Faraday believed that there was no better way to enter into the study of natural philosophy than by considering the physical phenomena of a burning candle. Faraday wished to demonstrate to his audience the basic properties of gasses, the rules of chemical combination, and the chemical processes underlying everyday events. Readers will recognize in Faraday's lectures on the candle many of the stock crowd pleasers still used today by introductory chemistry instructors, such as the ignition of hydrogen formed by electrolysis (pop!) or the crushing of a steam-filled can by rapid cooling (crunch!).

Chemical History of a Candle to Which is Added A Lecture on Platinum.
Edited by William Crookes. London: Griffin, Bohn, and Company, 1861, pp. 142–71.[1]

LECTURE VI

Carbon or Charcoal—Coal Gas—Respiration and its Analogy to the Burning of a Candle—Conclusion.

A Lady who honours me by her presence at these lectures, has conferred a still further obligation by sending me these two candles, which are from Japan, and, I presume, are made of that substance to which I referred in a former lecture. You see that they are even far more highly ornamented than the French candles, and, I suppose, are candles of luxury, judging from their appearance. They have a remarkable peculiarity about them; namely, a hollow wick,—that beautiful peculiarity which Argand introduced into the lamp and made so valuable. To those who receive such presents from the East, I may just say that this and such like materials, gradually undergo a change which gives them on the surface a dull and dead appearance; but they may easily be restored to their original beauty if the surface be rubbed with a clean cloth or silk handkerchief, so as to polish the little rugosity or roughness: this will restore the beauty of the colours. I have so rubbed one of these candles, and you see the difference between

1 Due to damage in the 1861 base text consulted by the editor, the illustrations, identical to the first edition, have been drawn from the nineteenth-century reprint edition by Chatto and Windus.

it and the other which has not been polished, but which may be restored by the same process. Observe, also, that these moulded candles from Japan are made more conical than the moulded candles in this part of the world.

I told you, when we last met, a good deal about carbonic acid. We found by the lime-water test, that when the vapour from the top of the candle or lamp was received into bottles and tested by this solution of lime-water (the composition of which I explained to you, and which you can make for yourselves), we had that white opacity which was in fact calcareous matter, like shells and corals, and many of the rocks and minerals in the earth. But I have not yet told you fully and clearly the chemical history of this substance, carbonic acid, as we have it from the candle, and I must now resume that subject. We have seen the products, and the nature of them, as they issue from the candle. We have traced the water to its elements, and now we have to see where are the elements of the carbonic acid supplied by the candle: a few experiments will show this. You remember that when a candle burns badly it produces smoke; but if it is burning well, there is no smoke. And you know that the brightness of the candle is due to this smoke, which becomes ignited. Here is an experiment to prove this: so long as the smoke remains in the flame of the candle and becomes ignited it gives a beautiful light, and never appears to us in the form of black particles. I will light some fuel, which is extravagant in its burning; this will serve our purpose—a little turpentine on a sponge. You see the smoke rising from it, and floating into the air in large quantities; and remember now, the carbonic acid that we have from the candle is from such smoke as that. To make that evident to you, I will introduce this turpentine burning on the sponge into a flask where I have plenty of oxygen, the rich part of the atmosphere, and you now see that the smoke is all consumed. This is the first part of our experiment, and now what follows? The carbon which you saw flying off from the turpentine flame in the air is now entirely burned in this oxygen, and we shall find that it will, by this rough and temporary experiment, give us exactly the same conclusion and result as we had from the combustion of the candle. The reason why I make the experiment in this manner is solely that I may cause the steps of our demonstration to be so simple that you can never for a moment lose the train of reasoning, if you only pay attention. All the carbon which is burned in oxygen, or air, comes out as carbonic acid, whilst those particles which are not so burned show you the second substance in the carbonic acid; namely, the carbon—that body which made the flame so bright whilst there was plenty of air, but which was thrown off in excess when there was not oxygen enough to burn it.

I have also to show you a little more distinctly, the history of carbon and oxygen in their union to make carbonic acid. You are now better able to understand this than before, and I have prepared three or four experiments by way of illustration. This jar is filled with oxygen, and here is some carbon which has been placed in a crucible, for the purpose of being made red-hot. I keep my jar dry, and venture to give you a result imperfect in some degree, in order that I may make the experiment brighter. I am about to put the oxygen and the carbon together. That this is carbon (common charcoal pulverized) you will see by the way in which it burns in the air [letting some of the red-hot charcoal fall out of the crucible]. I am now about to burn it in oxygen gas, and look at the difference. It may appear to you at a distance as if it were burning with a flame; but it is not so. Every little piece of charcoal is burning as a spark, and whilst it so burns it is producing carbonic acid. I specially want these two or three experiments to point out what I shall dwell upon more distinctly by and by—that carbon burns in this way, and not as a flame.

Instead of taking many particles of carbon to burn I will take a rather large piece, which will enable you to see the form and size, and to trace the effects very decidedly. Here is the jar of oxygen, and here is the piece of charcoal, to which I have fastened a little piece of wood, which I can set fire to, and so commence the combustion, which I could not conveniently do without. You now see the charcoal burning, but not as a flame (or if there be a flame it is the smallest possible one, which I know the cause of; namely, the formation of a little carbonic oxide close upon the surface of the carbon). It goes on burning, you

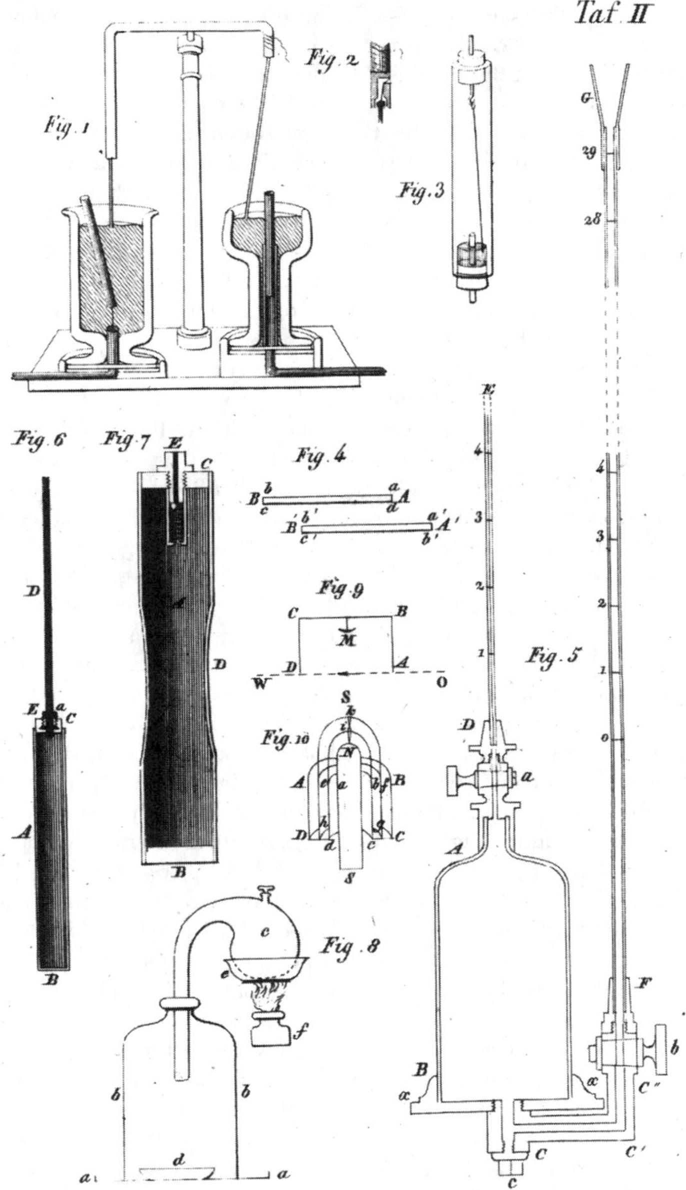

Apparatus from Michael Faraday's electrical experiments, including in the upper left hand corner the famous instrument containing fixed and free bar magnets immersed in mercury-filled containers through which an electric current was run. With this device, Faraday demonstrated electromagnetic rotation, which led eventually to the development of electrical dynamos and motors. The apparatus is described in detail in Faraday's *Experimental Researches in Electricity* (London, 1844), pp. 147–51. Reproduced in L.W. Gilbert, "Ueber neue electrisch-magnetische Bewegungen," *Annalen der Physik* 42 (1822): tafel II.

see, slowly producing carbonic acid by the union of this carbon or charcoal (they are equivalent terms) with the oxygen. I have here another piece of charcoal, a piece of bark, which has the quality of being blown to pieces—exploding—as it burns. By the effect of the heat we shall reduce the lump of carbon into particles that will fly off; still every particle, equally with the whole mass, burns in this peculiar way—it burns as a coal and not like a flame. You observe a multitude of little combustions going on, but no flame. I do not know a finer experiment than this to show that carbon burns with a spark.

Here, then, is carbonic acid formed from its elements. It is produced at once; and if we examined it by lime-water, you will see that we have the same substance which I have previously described to you. By putting together 6 parts of carbon by weight (whether it comes from the flame of a candle or from powdered charcoal) and 16 parts of oxygen by weight, we have 22 parts of carbonic acid; and, as we saw last time, the 22 parts of carbonic acid combined with 28 parts of lime, produced common carbonate of lime. If you were to examine an oyster-shell and weigh the component parts, you would find that every 50 parts would give 6 of carbon and 16 of oxygen combined with 28 of lime. However, I do not want to trouble you with these minutiae; it is only the general philosophy of the matter that we can now go into. See how finely the carbon is dissolving away [pointing to the lump of charcoal burning quietly in the jar of oxygen]. You may say that the charcoal is actually dissolving in the air round about; and if that were perfectly pure charcoal, which we can easily prepare, there would be no residue whatever. When we have a perfectly cleansed and purified piece of carbon, there is no ash left. The carbon burns as a solid dense body, that heat alone cannot change as to its solidity, and yet it passes away into vapour that never condenses into solid or liquid under ordinary circumstances; and what is more curious still is the fact that the oxygen does not change in its bulk by the solution of the carbon in it. Just as the bulk is at first, so it is at last, only it has become carbonic acid.

There is another experiment which I must give you before you are fully acquainted with the general nature of carbonic acid. Being a compound body, consisting of carbon and oxygen, carbonic acid is a body that we ought to be able to take asunder. And so we can. As we did with water, so we can with carbonic acid,—take the two parts asunder. The simplest and quickest way is to act upon the carbonic acid by a substance that can attract the oxygen from it, and leave the carbon behind. You recollect that I took potassium and put it upon water or ice, and you saw that it could take the oxygen from the hydrogen. Now, suppose we do something of the same kind here with this carbonic acid. You know carbonic acid to be a heavy gas: I will not test it with lime-water, as that will interfere with our subsequent experiments, but I think the heaviness of the gas and the power of extinguishing flame will be sufficient for our purpose. I introduce a flame into the gas, and you will see whether it will be put out. You see the light is extinguished. Indeed, the gas may, perhaps, put out phosphorus, which you know has a pretty strong combustion. Here is a piece of phosphorus heated to a high degree. I introduce it into gas, and you observe the light is put out, but it will take fire again in the air, because there it re-enters into combustion. Now let me take a piece of potassium, a substance which even at common temperatures can act upon carbonic acid, though not sufficiently for our present purpose, because it soon gets covered with a protecting coat; but if we warm it up to the burning point in air, as we have a fair right to do, and as we have done with phosphorus, you will see that it can burn in carbonic acid; and if it burns, it will burn by taking oxygen, so that you will see what is left behind. I am going, then, to burn this potassium in the carbonic acid, as a proof of the existence of oxygen in the carbonic acid. [In the preliminary process of heating the potassium exploded.] Sometimes we get an awkward piece of potassium that explodes, or something like it, when it burns. I will take another piece, and now that it is heated I introduce it into the jar, and you perceive that it burns in the carbonic acid—not so well as, in the air, because the carbonic acid contains the oxygen combined, but it does burn, and takes away the oxygen. If I now put this potassium into water, I find that besides the potash formed (which you need not trouble about) there is a quantity of carbon produced. I have here made the experiment in a very rough way, but I assure you that if I were to make it carefully, devoting a day to it, instead of five minutes, we should get all the proper amount of charcoal left in the spoon, or in the place where the potassium was burned, so that there could be no doubt as to the result. Here, then, is the carbon obtained from the carbonic acid, as a common black substance; so that you have the entire proof of the nature of carbonic acid as consisting of carbon and oxygen. And now, I may tell you, that *whenever* carbon burns under common circumstances, it produces carbonic acid.

Suppose I take this piece of wood, and put it into a bottle with lime-water. I might shake that lime-water up with wood and the atmosphere as long as I pleased, it would still remain clear as you see it; but suppose I burn the piece of wood in the air of that bottle. You, of course, know I get water. Do I get carbonic acid ? [The experiment was performed.] There it is, you see—that is to say, the carbonate lime, which results from carbonic acid, and that carbonic acid must be formed from the carbon which comes from the wood, from the candle, or any other thing. Indeed, you have yourselves frequently tried a very pretty experiment, by which you may see the carbon in wood. If you take a piece of wood, and partly burn it, and then blow it out, you have carbon left. There are things that do not show carbon in this way. A candle does not so show it, but it contains carbon. Here also is a jar of coal-gas, which produces carbonic acid abundantly,—you do not see the carbon, but we can soon show it to you. I will light it, and as long as there is any gas in this cylinder it will go on burning. You see no carbon, but you see a flame, and because that is bright it will lead you to guess that there is carbon in the flame. But I will show it to you by another process. I have some of the same gas in another vessel, mixed with a body that will burn the hydrogen of the gas, but will not burn the carbon. I will light them with a burning taper, and you perceive the hydrogen is consumed, but not the carbon, which is left behind as a dense black smoke. I hope that by these three or four experiments you will learn to see when carbon is present, and understand what are the products of combustion, when gas or other bodies are thoroughly burned in the air.

Before we leave the subject of carbon, let us make a few experiments and remarks upon its wonderful condition, as respects ordinary combustion. I have shown you that the carbon in burning burns only as a solid body, and yet you perceive that, after it is burned, it ceases to be a solid. There are very few fuels that act like this. It is in fact only that great source of fuel, the carbonaceous series, the coals, charcoals, and woods, that can do it. I do not know that there is any other elementary substance besides carbon that burns with these conditions; and if it had not been so, what would happen to us? Suppose all fuel had been like iron which, when it burns, burns into a solid substance. We could not then have such a combustion as you have in this fire-place. Here also is another kind of fuel which burns very well—as well as, if not better, than carbon—so well, indeed, as to take fire of itself when it is in the air, as you see. [Breaking a tube full of lead pyrophorus.] This substance is lead, and you see how wonderfully combustible it is. It is very much divided, and is like a heap of coals in the fire-place: the air can get to its surface and inside, and so it burns. But why does it not burn in that way now when it is lying in a mass? [Emptying the contents of the tube in a heap on to a plate of iron.] Simply because the air cannot get to it. Though it can produce a great heat, the great heat which we want in our furnaces and under our boilers, still that which is produced cannot get away from the portion which remains unburned underneath, and that portion, therefore, is prevented from coming in contact with the atmosphere, and cannot be consumed. How different is that from carbon! Carbon burns just in the same way as this lead does, and so gives an intense fire in the furnace, or wherever you choose to burn it; but then the body produced by its combustion passes away, and the remaining carbon is left clear. I showed you how carbon went on dissolving in the oxygen, leaving no ash; whereas, here [pointing to the heap of pyrophorus] we have actually more ash than fuel, for it is heavier by the amount of the oxygen which has united with it. Thus you see the difference between carbon and lead or iron: if we chose iron,

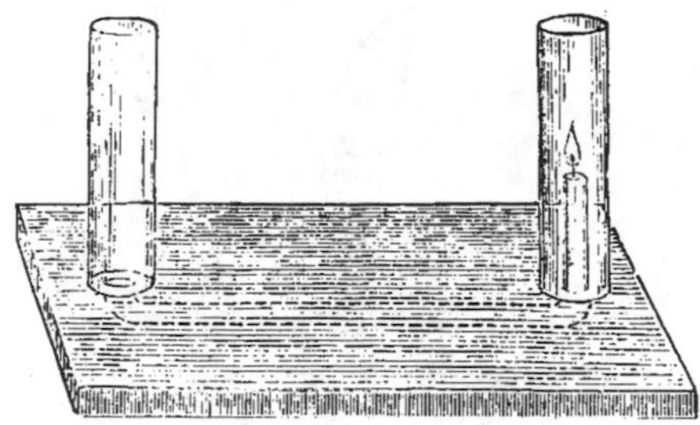

Fig. 32.

which gives so wonderful a result in our applications of this fuel, either as light or heat. If, when the carbon burnt, the product went off as a solid body, you would have had the room filled with an opaque substance, as in the case of the phosphorus; but when carbon burns, everything passes up into the atmosphere. It is in a fixed, almost unchangeable condition before the combustion; but afterwards it is in the form of gas, which it is very difficult (though we have succeeded) to produce in a solid or liquid state.

Now I must take you to a very interesting part of our subject—to the relation between the combustion of a candle and that living kind of combustion which goes on within us. In every one of us there is a living process of combustion going on very similar to that of a candle, and I must try to make that plain to you. For it is not merely true in a poetical sense—the relation of the life of man to a taper; and if you will follow, I think I can make this clear. In order to make the relation very plain, I have devised a little apparatus which we can soon build up before you. Here is a board and a groove cut in it [Fig. 32], and I can close the groove at the top part by a little cover; I can then continue the groove as a channel by a glass tube at each end, there being a free passage through the whole. Suppose I take a taper or candle (we can now be liberal in our use of the word "candle," since we understand what it means), and place it in one of the tubes; it will go on, you see, burning very well. You observe that the air which feeds the flame passes down the tube at one end, then goes along the horizontal tube, and ascends the tube at the other end in which the taper is placed. If I stop the aperture through which the air enters, I stop combustion, as you perceive. I stop the supply of air, and consequently the candle goes out. But now what will you think of this fact? In a former experiment I showed you the air going from one burning candle to a second candle. If I took the air proceeding from another candle, and sent it down by a complicated arrangement into this tube, I should put this burning candle out. But what will you say when I tell you that my breath will put out that candle? I do not mean by blowing at all, but simply that the nature of my breath is such that a candle cannot burn in it. I will now hold my mouth over the aperture, and without blowing the flame in any way, let no air enter the tube but what comes from my mouth. You see the result. I did not blow the candle out. I merely let the air which I expired pass into the aperture, and the result was that the light went out for want of oxygen, and for no other reason. Something or other—namely, my lungs—had taken away the oxygen from the air, and there was no more to supply the combustion of the candle. It is, I think, very pretty to see the time it takes before the bad air which I throw into this part of the apparatus has reached the candle. The candle at first goes on burning, but so soon as the air has had time to reach it it goes out. And now I will show you another experiment, because this is an important part of our philosophy. Here is a jar which contains fresh air, as you can see by the circumstance of a candle or gas-light burning it [Fig. 33]. I make it close for a little time, and by means of a pipe I get my mouth over it so that I can inhale the air. By putting it over water, in the way that you see, I am able to draw up this air (supposing the cork to be quite tight), take it into my lungs, and throw it back into the jar: we can then examine it, and see the result. You observe, I first take up the air, and then throw it back, as is evident from the ascent and descent of the water, and now, by putting a taper into the air, you will see the state in which it is by the light being extinguished. Even one inspiration, you see, has completely spoiled this air, so that it is no use my trying to breathe it a second time. Now you understand the ground of the impropriety of many of the arrangements among the houses of the poorer classes, by which the air is breathed over and over again, for the want of a supply, by means of proper ventilation, sufficient to produce a good result. You see how bad the air becomes by a single breathing; so that you can easily understand how essential fresh air is to us.

To pursue this a little further, let us see what will happen with lime-water. Here is a globe which contains a little lime-water, and it is so arranged as regards the pipes, as to give access to the air within, so that we can ascertain the effect of respired, or unrespired air upon it. [Fig. 34] Of course I can either draw in air (through A), and so make the air that feeds my lungs go through the lime-water, or I can force the air out of my lungs through the tube (B), which goes to the bottom, and so show its effect upon the lime-water. You will observe that however long I draw the external air into the lime-water, and then through it to my lungs, I shall produce no effect upon the water—it will not make the lime-water turbid; but if I throw the air *from* my lungs through the lime-water, several times in succession, you see how white and milky the water is getting, showing the effect which expired air has had upon it; and now you begin to know that the atmosphere which we have spoiled by respiration is spoiled by carbonic acid, for you see it here in contact with the lime-water.

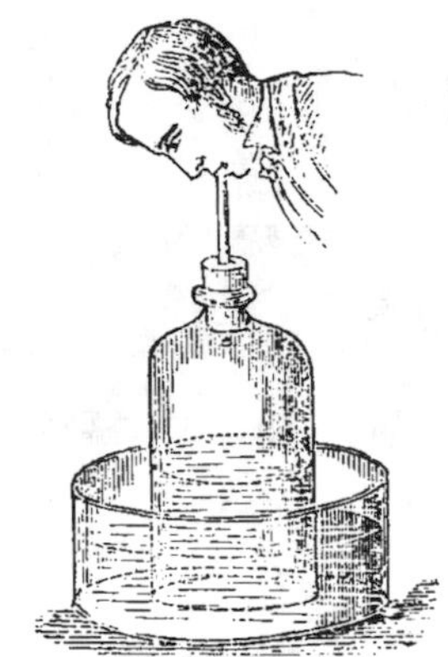

Fig. 33.

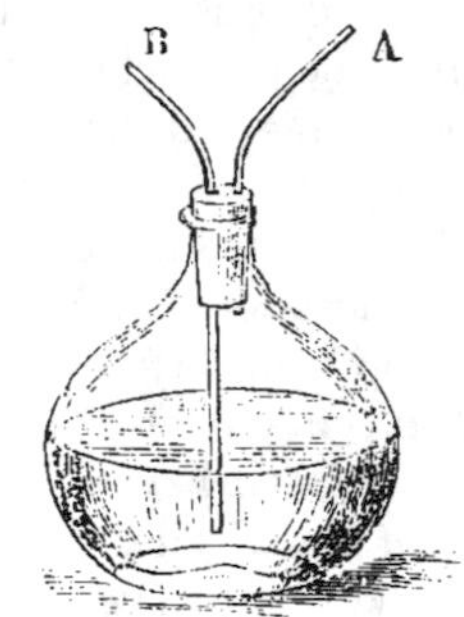

Fig. 34.

I have here two bottles, one containing lime-water and the other common water, and tubes which pass into the bottles and connect them [Fig. 35]. The apparatus is very rough, but it is useful notwithstanding. If I take these two bottles, inhaling here and exhaling there, the arrangement of the tubes will prevent the air going backwards. The air coming in will go to my mouth and lungs, and in going out, will pass through the lime-water, so that I can go on breathing and making an experiment, very refined in its nature, and very good in its results. You will observe that the good air has done nothing to the lime-water; in the other case nothing has come to the lime-water but my respiration, and you see the difference in the two cases.

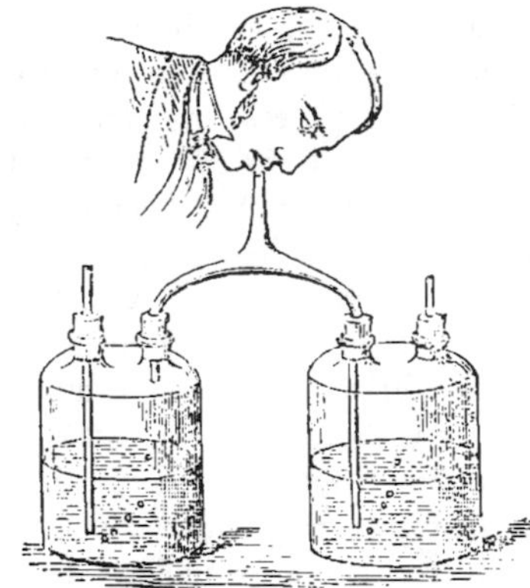

Fig. 35.

Let us now go a little further. What is all this process going on within us which we cannot do without, either day or night, which is so provided for by the Author of all things that He has arranged that it shall be independent of all will? If we restrain our respiration, as we can to a certain extent, we should destroy ourselves. When we are asleep, the organs of respiration and the parts that are associated with them, still go on with their action, so necessary is this process of respiration to us, this contact of the air with the lungs. I must tell you, in the briefest possible manner, what this process is. We consume food: the food goes through that strange set of vessels and organs within us, and is brought into various parts of the system, into the digestive parts especially; and alternately the portion which is so changed is carried through our lungs by one set of vessels, while the air that we inhale and exhale is drawn into and thrown out of the lungs by another set of vessels, so that the air and the food come close together, separated only by an exceedingly thin surface: the air can thus act upon the blood by this process, producing precisely the same results in kind as we have seen in the case of the candle. The candle combines with parts of the air, forming carbonic acid, and evolves heat; so in the lungs there is this curious, wonderful change taking place. The air entering, combines with the carbon (not carbon in a free state, but, as in this case, placed ready for action at the moment), and makes carbonic acid, and is so thrown out into the atmosphere, and thus this singular result takes place; we may thus look upon the food as fuel. Let me take that piece of sugar, which will serve my purpose. It is a compound of carbon, hydrogen, and oxygen, similar to a candle, as containing the same elements, though not in the same proportion; the proportions being as shown in this table:—

Sugar.		
Carbon	72	
Hydrogen	11	} 99
Oxygen	88	

This is, indeed a very curious thing, which you can well remember, for the oxygen and hydrogen are in exactly the proportions which form water, so that sugar may be said to be compounded of 72 parts of carbon and 99 parts of water; and it is the carbon in the sugar that combines with the oxygen carried in by the air in the process of respiration, so making us like candles; producing these actions, warmth, and far more wonderful results besides, for the sustenance of the system, by a most beautiful and simple process. To make this still more striking, I will take a little sugar; or to hasten the experiment I will use some syrup, which contains about three-fourths of sugar and a little water. If I put a little oil of vitriol on it, it takes away the water, and leaves the carbon in a black mass. [The Lecturer mixed the two together.] You see how the carbon is coming out, and before long we shall have a solid mass of charcoal, all of which has come out of sugar. Sugar, as you know, is food, and here we have absolutely a solid lump of carbon where you would not have expected it. And if I make arrangements so as to oxidize the carbon of sugar, we shall have a much more striking result. Here is sugar, and I have here an oxidizer—a quicker one than the atmosphere; and so we shall oxidize this fuel by a process different from respiration in its form, though not different in its kind. It is the combustion of the carbon by the contact of oxygen which the body has supplied to it. If I set this into action at once, you will see combustion produced. Just what occurs in my lungs—taking in oxygen from another source, namely, the atmosphere, takes place here by a more rapid process.

You will be astonished when I tell you what this curious play of carbon amounts to. A candle will burn some four, five, six, or seven hours. What then must be the daily amount of carbon going up into the air in the way of carbonic acid! What a quantity of carbon must go from each of us in respiration! What a wonderful change of carbon must take place under these circumstances of combustion or respiration! A man in twenty-four hours converts as much as seven ounces of carbon into carbonic acid; a milch cow will convert seventy ounces, and a horse seventy-nine ounces, solely by the act of respiration. That is, the horse in twenty-four hours burns seventy-nine ounces of charcoal, or carbon, in his organs of respiration to supply his natural warmth in that time. All the warm-blooded animals get their warmth in this way, by the conversion of carbon, not in a free state, but in a state of combination. And what an extraordinary notion this gives us of the alterations going on in our atmosphere. As much as 5,000,000 pounds, or 548 tons, of carbonic acid is formed by respiration in London alone in twenty-four hours. And where does all this go? Up into the air. If the carbon had been like the lead which I showed you, or the iron which, in burning, produces a solid substance, what would happen? Combustion could not go on. As charcoal burns it becomes a vapour and passes off into the atmosphere, which is the great vehicle, the great carrier for conveying it away to other places. Then what becomes of it? Wonderful is it to find that the change produced by respiration, which seems so injurious to us (for we cannot breathe air twice over), is the very life and support of plants and vegetables that grow upon the surface of the earth. It is the same also under the surface, in the great bodies of water; for fishes and other animals respire upon the same principle, though not exactly by contact with the open air.

Such fish as I have here [pointing to a globe of gold-fish] respire by the oxygen which is dissolved from the air by the water, and form carbonic acid, and they all move about to produce the one great work of making the animal and vegetable kingdoms subservient to each other. And all the plants growing upon the surface of the earth, like that which I have brought here to serve as an illustration, absorb carbon; these leaves are taking up their carbon from the atmosphere to which we have given it in the form of carbonic acid, and they are growing and prospering. Give them a pure air like ours, and they could not live in it; give them carbon with other matters, and they live and rejoice. This piece of wood gets all its carbon, as the trees and plants get theirs, from the atmosphere, which, as we have seen, carries away what is bad for us and at the same time good for them,—what is disease to the one being health to the other. So are we made dependent not merely upon our fellow-creatures, but upon our fellow-existers, all Nature being tied together by the laws that make one part conduce to the good of another.

There is another little point which I must mention before we draw to a close—a point which concerns the whole of these operations, and most curious and beautiful it is to see it clustering upon and associated with the bodies that concern us—oxygen, hydrogen, and carbon, in different states of their existence. I showed you just now some powdered lead, which I set burning;[2] and you saw that the moment the fuel was brought to the air it acted, even before it got out of the bottle,—the moment the air crept in it acted. Now, there is a case of chemical affinity by which all our operations proceed. When we breathe, the same operation is going on within us. When we burn a candle, the attraction of the different parts one to the other is going on. Here it is going on in this case of the lead, and it is a beautiful instance of chemical affinity. If the products of combustion rose off from the surface, the lead would take fire, and go on burning to the end; but you remember that we have this difference between charcoal and lead—that, while the lead can start into action at once if there be access of air to it, the carbon will remain days, weeks, months, or years. The manuscripts of Herculaneum were written with carbonaceous ink, and there they have been for 1800 years or more, not having been at all changed by the atmosphere, though coming in contact with it under various circumstances. Now, what is the circumstance which makes the lead and carbon differ in this respect? It is a striking thing to see that the matter which is appointed to serve the purpose of fuel *waits* in its action; it does not start off burning, like the lead and many other things that I could show you, but which I have not encumbered the table with; but it waits for action. This waiting is a curious and wonderful thing. Candles—those Japanese candles, for instance—do not start into action at once like the lead or iron (for iron finely divided does the same thing as lead), but there they wait for years, perhaps for ages, without undergoing any alteration. I have here a supply of coal-gas. The jet is giving forth the gas, but you see it does not take fire—it comes out into the air, but it waits till it is hot enough before it burns. If I make it hot enough, it takes fire. If I blow it out, the gas that is issuing forth waits till the light is applied to it again. It is curious to see how different substances wait—how some will wait till the temperature is raised a little, and others till it is raised a good deal. I have here a little gunpowder and some gun-cotton; even these things differ in the conditions under which they will burn. The gunpowder is com-

2 *Lead pyrophorus* is made by heating dry tartrate of lead in a glass tube (closed at one end, and drawn out to a fine point at the other) until no more vapours are evolved. The open end of the tube is then to be sealed before the blowpipe. When the tube is broken and the contents shaken out into the air, they burn with a red flash.

posed of carbon and other substances, making it highly combustible; and the gun-cotton is another combustible preparation. They are both waiting, but they will start into activity at different degrees of heat, or under different conditions. By applying a heated wire to them, we shall see which will start first [touching the gun-cotton with the hot iron]. You see the gun-cotton has gone off, but not even the hottest part of the wire is now hot enough to fire the gunpowder. How beautifully that shows you the difference in the degree in which bodies act in this way! In the one case the substance will wait any time until the associated bodies are made active by heat; but, in the other, as in the process of respiration, it waits no time. In the lungs, as soon as the air enters, it unites with the carbon; even in the lowest temperature which the body can bear short of being frozen, the action begins at once, producing the carbonic acid of respiration; and so all things go on fitly and properly. Thus you see the analogy between respiration and combustion is rendered still more beautiful and striking. Indeed, all I can say to you at the end of these lectures (for we must come to an end at one time or other) is to express a wish that you may, in your generation, be fit to compare to a candle; that you may, like it, shine as lights to those about you; that, in all your actions, you may justify the beauty of the taper by making your deeds honourable and effectual in the discharge of your duty to your fellow-men.

FURTHER READING

Agassi, Joseph. *Faraday as a Natural Philosopher.* Chicago: University of Chicago Press, 1971.

Bence Jones, Henry. *The Life and Letters of Faraday.* 2 vols. London: Longmans, Green, and Co., 1870.

Berman, Morris. *Social Change and Scientific Organization: The Royal Institution, 1799–1844.* Ithaca: Cornell UP, 1978.

Cantor, Geoffrey. *Michael Faraday Sandemanian and Scientist: A Study of Science and Religion in the Nineteenth Century.* New York: St. Martin's Press, 1991.

—, D. Gooding, and F.A.J.L. James. *Faraday.* Rev. ed. Atlantic Highlands, NJ: Humanities Press, 1996.

Faraday, Michael. *The Forces of Matter.* Buffalo: Prometheus Books, 1993.

—. *The Chemical History of a Candle.* Edited by William Crookes. Preface by L. Pearce Williams. 1861. Reprint, New York: Collier Books, 1962.

—. *The Correspondence of Michael Faraday.* Edited by Frank A.J.L. James. 3 vols. London: The Institution of Electrical Engineers, 1991–96.

—. *Faraday's Diary: Being the Various Philosophical Notes of Experimental Investigation Made by Michael Faraday, D.C.L., F.R.S.* Edited by Thomas Martin. 7 vols. London: G. Bell and Sons, Ltd., 1932–36.

Gladstone, J.H. *Michael Faraday.* London: Macmillan, 1872.

Gooding, David, and Frank A.J.L. James. *Faraday Rediscovered: Essays on the Life and Work of Michael Faraday, 1791–1867.* London: Macmillan, 1985.

Jeffreys, Alan E. *Michael Faraday: A List of his Lectures and Published Writings.* London: Royal Institution of Great Britain, 1960.

Lemmerich, Jost. *Michael Faraday 1791–1867: Erforscher der Elektrizität.* München: C.H. Beck, 1991.

Martin, Thomas. *Faraday's Discovery of Electromagnetic Induction.* London: Edward Arnold and Co., 1949.

Meyer, Herbert W. *A History of Electricity and Magnetism.* Cambridge, MA: M.I.T. Press, 1971.

Thomas, J.M. *Michael Faraday and the Royal Institution.* New York: Adam Hilger, 1991.

Thompson, S.P. *Michael Faraday. His Life and Work.* London: Cassell and Company, Ltd., 1898.

Tricker, R.A.R. *The Contributions of Faraday and Maxwell to Electrical Science.* Oxford: Pergamon Press, 1966.

Tyndall, John. *Faraday as a Discoverer.* 1868. Reprint, New York: Thomas Y. Crowell Company, 1961.

Williams, L. Pearce. *The Selected Correspondence of Michael Faraday.* 2 vols. Cambridge: Cambridge UP, 1971.

—. *Michael Faraday: A Biography.* New York: Basic Books, 1965.

—. *The Origins of Field Theory.* New York: Random House, 1966.

22. FRIEDRICH MAX MÜLLER

(1823–1900)

Friedrich Max Müller aided in the development of comparative philology (or linguistics), an intellectual movement which attempted to direct the study of languages and culture along the lines of the natural sciences. This attempt demonstrates the desire of many areas of knowledge in the nineteenth as well as the twentieth centuries to legitimize themselves by borrowing the aura of the natural sciences, even when the application of experimental scientific method, mathematical models, and statistical analysis could not solve many of the questions posed by these traditionally non-scientific disciplines.

Müller was born in Dessau, Germany, the son of the poet Wilhelm Müller. He studied classics and oriental languages at the University of Leipzig. His work on the *Rig-Veda* brought him to England in 1846 to consult Vedic manuscripts owned by the East India Company. In 1854, he became Taylorian Professor of Modern European Languages at Oxford, and in 1868 the university created the first chair of comparative philology especially for him.

Friedrich Schlegel had first signalled the interest of philologists in scientific methods with his *Über die Sprache und Weisheit der Indier* (1808) by suggesting that a comparison of grammars "will give us totally new insights into the genealogy of languages, much in the way comparative anatomy has shed light on higher natural history" (Christmann 203). Franz Bopp, who trained Müller and Wilhelm von Humboldt in Sanskrit and who wrote the first comparative Indo-European grammar, similarly spoke of languages as "organic natural bodies." In 1786, Sir William Jones—who was also influenced by the anatomy/language analogy—had clearly demonstrated that Sanskrit was related to Greek, Latin, and the modern European languages. By mid-century, philologists spoke of the "laws" of language and language change, which they modelled on the immutable laws of physics or chemistry; examples include Grimm's Law, Grassmann's Law, and Verner's Law, all describing consonant shifts in the Indo-European languages.

Müller delivered two series of lectures on this "science of language" before the Royal Institution in 1861 and 1863. As the excerpt below demonstrates, he argued for placing language study among the natural sciences. He employed similar arguments for the scientific study of both religion and mental processes in *Introduction to the Science of Religion* (1873) and *The Science of Thought* (1887).

The origin of language and linguistic arguments for and against evolution came to the fore with Robert Chambers's *Vestiges of the Natural History of Creation* (1844), which hypothesized that human language was simply a more developed version of animal communication. These pronouncements shocked those religious thinkers who deified language as the one feature that separated man from the animals, an illustration of the divine gift of reason in man. Darwin later argued similarly in *The Descent of Man* (1871) that there was no fundamental difference in the mental faculties of man and the higher animals (35): "it does not appear altogether incredible," Darwin wrote, "that some unusually wise ape-like animal should have thought of imitating the growl of a beast of prey, so as to indicate to his fellow monkeys the nature of the expected danger. And this would have been a first step in the formation of a language" (57). The idea of attributing rudimentary language skills to primates was extremely offensive to Müller, for whom language was the Rubicon separating man from the animals. For Müller, language was synonymous with human reason.

Müller criticized Darwin's theories of language in three lectures at the Royal Institution published as "Lectures on Mr. Darwin's Philosophy of Language" in *Fraser's Magazine* in 1873. Müller believed that Darwin's Lockean empirical attitude towards language, the position that language arises first in sensation and imprints itself on the blank slate of the mind—*in intellectu nihil est quod non prius fuit in sensu*—had been refuted by Kant's theory of prior categories in the *Critique of Pure Reason*, which Müller had translated. Müller also engaged in a decades-long, often personal, debate with the Yale linguist William Dwight Whitney. Whitney sided with the Darwinians concerning the evolution of language, and in his polemical *Max Müller and the Science of Language: A Criticism* (1892) accused Müller of failing to demonstrate that "there is in the mode of communication possessed by men [anything] that is distinctive, that is different in principle from the methods of other animals" (32). Another important debate involving science and language revolved around the "Aryan Question," discussed in chapter 23 of Charles Lyell's *Antiquity of Man*. Müller endorsed Schlegel's theory that European languages were based on the language of a northern Indian culture which had migrated westward. The controversy drew in anthropologists, evolutionists, and ethnographers who were debating either single or multiple origins of the human species.

Müller's collected essays on language appeared from 1867–75 as *Chips from a German Workshop* and his collected works were issued by Longmans in 1898. Müller originally opposed the onomatopoeic and interjectionist theories of languages, which argued that language originated from the imitation of natural sounds or instinctive utterances, two hypotheses he labelled derisively the "bow-wow" and "pooh-pooh" theories. Müller himself, however, espoused variations of these theories, such as the "ding-dong" theory, that speech sounds arose as a response to environment, and the "Yo-he-ho" theory of Ludwig Noiré that language originated from communal games and work noises (yo-heave-ho). Müller displayed extraordinary energy in translating eastern texts, training linguists, and popularizing the science of language, but he said of himself with his characteristic self-deprecating humour: "I have never done anything; I have never been a doer, a canvasser, a wire-puller, a manager, in the ordinary sense of these words.... I have been a scholar, a *Stubengelehrter*, and *voilà tout*!" (*Autobiography* 308).

The Science of Language.
London: Longmans and Co., 1861, pp. 1–29.

Lecture 1

The Science of Language One of the Physical Sciences.

When I was asked some time ago to deliver a course of lectures on Comparative Philology in this Institution, I at once expressed my readiness to do so. I had lived long enough in England to know that the peculiar difficulties arising from my imperfect knowledge of the language would be more than balanced by the forbearance of an English audience, and I had such perfect faith in my subject that I thought it might be trusted even in the hands of a less skilful expositor. I felt convinced that the researches into the history of languages and into the nature of human speech, which have been carried on during the last fifty years in England, France, and Germany, deserved a larger share of public sympathy than they had hitherto received; nay, it seemed to me, as far as I could judge, that the discoveries in this newly-opened mine of scientific inquiry were not inferior, whether in novelty or importance, to the most brilliant discoveries of our age.

It was not till I began to write my lectures that I became aware of the difficulties of the task I had undertaken. The dimensions of the science of language are so vast that it is impossible in a course of nine lectures to give more than a very general survey of it; and as one of the greatest charms of this science consists in the minuteness of the analysis by which each language, each dialect, each word, each grammatical form is tested, I felt that it was almost impossible to do full justice to my subject, or to place the achievements of those who founded and fostered the science of language in their true light. Another difficulty arises from the dryness of many of the problems which I shall have to discuss. Declensions and conjugations cannot be made amusing, nor can I avail myself of the advantages possessed by most lecturers, who enliven their discussions by experiments and diagrams. If, with all these difficulties and drawbacks, I do not shrink from opening to-day this course of lectures on mere words, on nouns and verbs and particles—if I venture to address an audience accustomed to listen, in this place, to the wonderful tales of the natural historian, the chemist, and geologist, and wont to see the novel results of inductive reasoning invested by native eloquence with all the charms of poetry and romance—it is because, though mistrusting myself, I cannot mistrust my subject. The study of words may be tedious to the school-boy, as breaking of stones is to the wayside labourer, but to the thoughtful eye of the geologist these stones are full of interest—he sees miracles on the high road, and reads chronicles in every ditch. Language, too, has marvels of her own, which she unveils to the inquiring glance of the patient student. There are chronicles below her surface, there are sermons in every word. Language has been called sacred ground, because it is the deposit of thought. We cannot tell as yet what language is. It may be a production of nature, a work of human art, or a divine gift. But to whatever sphere it belongs, it would seem to stand unsurpassed—nay, unequalled in it—by anything else. If it be a production of nature, it is her last and crowning production, which she reserved for man alone. If it be a work of human art, it would seem to lift the human artist almost to the level of a divine creator. If it be the gift of God, it is God's greatest gift; for through it God spake to man and man speaks to God in worship, prayer, and meditation.

Although the way which is before us may be long and tedious, the point to which it tends will be full of interest; and I believe I may promise that the view opened before our eyes from the summit of our science, will fully repay the patient travellers, and perhaps secure a free pardon to their venturous guide.

The Science of Language is a science of very modern date. We cannot trace its lineage much beyond the beginning of our century, and it is scarcely received as yet on a footing of equality by the elder branches of learning. Its very name is still unsettled, and the various titles that have been given to it in England, France, and Germany are so vague and varying that they have led to the most confused ideas among the public at large as to the real objects of this new science. We hear it spoken of as Comparative Philology, Scientific Etymology, Phonology, and Glossology. In France it has received the convenient, but somewhat barbarous, name of *Linguistique*. If we must have a Greek title for our science, we might derive it either from *mythos*, word, or from *logos*, speech. But the title of *Mythology* is already occupied, and *Logology* would jar too much on classical ears. We need not waste our time in criticising these names, as none of them has as yet received that universal sanction which belongs to the titles of other modern sciences, such as Geology or Comparative Anatomy; nor will there be much difficulty in christening our young science after we have once ascertained its birth, its parentage, and its character. I myself prefer the simple designation of the Science of Language, though in these days of high-sounding titles, this plain name will hardly meet with general acceptance.

From the name we now turn to the meaning of our science. But before we enter upon a definition of its subject-matter, and determine the method which ought to be followed in our researches, it will be useful to cast a glance at the history of the other sciences, among which the science of language now, for the first time, claims her place; and examine their origin, their gradual progress, and definite settlement. The history of a science is, as it were, its biography; and as we buy experience cheapest in studying the lives of others, we may, perhaps, guard our young science from some of the follies and extravagances inherent in youth by learning a lesson for which other branches of human knowledge have had to pay more dearly.

There is a certain uniformity in the history of most sciences. If we read such works as Whewell's *History of the Inductive Sciences* or Humboldt's *Kosmos*, we find that the origin, the progress, the causes of failure and success have been the same for almost every branch of human knowledge. There are three marked periods or stages in the history of every one of them, which we may call the *Empirical*, the *Classificatory*, and the *Theoretical*. However humiliating it may sound, every one of our sciences, however grand their present titles, can be traced back to the most humble and homely occupations of half-savage tribes. It was not the true, the good, and the beautiful which spurred the early philosophers to deep researches and bold discoveries. The foundation-stone of the most glorious structures of human ingenuity in ages to come was supplied by the pressing wants of a patriarchal and semi-barbarous society. The names of some of the most ancient departments of human knowledge tell their own tale. Geometry, which at present declares itself free from all sensuous impressions, and treats of its points and lines and planes as purely ideal conceptions, not to be confounded with the coarse and imperfect representations as they appear on paper to the human eye, geometry, as its very name declares, began with measuring a garden or a field. It is derived from the Greek *gē*, land, ground, earth, and *metron*, measure. Botany, the science of plants, was originally the science of *botanē*, which in Greek does not mean a plant in general, but fodder, from *boskein*, to feed. The science of plants would have been called Phytology, from the Greek *phyton*, a plant.[1] The founders of Astronomy were not the poet or the philosopher, but the sailor and the farmer. The early poet may have admired the 'mazy dance of planets,' and the philosopher may have speculated on the heavenly harmonies; but it was to the sailor alone that a knowledge of the glittering guides of heaven became a question of life and death. It was he who calculated their risings and settings with the accuracy of a merchant and the shrewdness of an adventurer; and the names that were given to single stars or constellations clearly show that they were invented by the ploughers of the sea and of the land. The moon, for instance, the golden hand on the dark dial of heaven, was called by them the Measurer—the measurer of time; for time was measured by nights, and moons, and winters, long before it was reckoned by days, and suns, and years.

---

1 See Jessen, *Was heisst Botanik?* 1861.

Moon[2] is a very old word. It was *môna* in Anglo-Saxon, and was used there, not as a feminine, but as a masculine; for the moon was originally a masculine, and the sun a feminine, in all Teutonic languages; and it is only through the influence of classical models that in English moon has been changed into a feminine, and sun into a masculine. It was a most unlucky assertion which Mr. Harris made in his *Hermes*, that all nations ascribe to the sun a masculine, and to the moon a feminine gender.[3] In the mythology of the Edda *Mâni*, the moon, is the son, *Sôl*, the sun, the daughter of *Mundilföri*. In Gothic *mena*, the moon, is masculine; *sunnô*, the sun, feminine. In Anglo-Saxon, too, *môna*, the moon, continues to be used as a masculine; *sunne*, the sun, as a feminine. In Swedish *måne*, the moon, is masculine; *sol*, the sun, feminine. The Lithuanians also give the masculine gender to the moon, *menů*; the feminine gender to the sun, *saule*: and in Sanskrit, though the sun is ordinarily looked upon as a male power, the most current names for the moon, such as *Chandra, Soma, Indu, Vidhu*, are masculine. The names of the moon are frequently used in the sense of month, and these and other names for month retain the same gender. Thus *menoth* in Gothic, *mônâdh* in Anglo-Saxon are both masculine. In Greek we find *mēn*, and the Ionic *meis*, for month, always used in the masculine gender. In Latin we have the derivative *mensis*, month; and in Sanskrit we find *mâs* for moon, and *mâsa* for month, both masculine.[4]

Now, this *mâs* in Sanskrit is clearly derived from a root *mâ*, to measure, to mete. In Sanskrit, I measure is *mâ-mi*; thou measurest, *mâ-si*; he measures, *mâ-ti* (or *mimî-te*). An instrument of measuring is called in Sanskrit *mâ-tram*, the Greek *metron*, our metre. Now, if the moon was originally called by the farmer the measurer, the ruler of days and weeks and seasons, the regulator of the tides, the lord of their festivals, and the herald of their public assemblies, it is but natural that he should have been conceived as a man, and not as the love-sick maiden which our modern sentimental poetry has put in his place.

It was the sailor who, before entrusting his life and goods to the winds and the waves of the ocean, watched for the rising of those stars which he called the Sailing-stars or *Pleiades*, from *plein*, to sail. Navigation in the Greek waters was considered safe after the return of the Pleiades; and it closed when they disappeared. The Latin name for the *Pleiades* is *Vergiliae*, from *virga*, a sprout or twig. This name was given to them by the Italian husbandmen, because in Italy, where they became visible about May, they marked the return of summer.[5] Another constellation, the seven stars in the head of Taurus, received the name of *Hyades* or *Pluviae* in Latin, because at the time when they rose with the sun they were supposed to announce rain. The astronomer retains these and many other names; he still speaks of the pole of heaven, of wandering and fixed stars,[6] yet he is apt to forget that these terms were not the result of scientific observation and classification, but were borrowed from the language of those who were themselves wanderers on the sea or in the desert, and to whom the fixed stars were in full reality what their name implies,

2 Kuhn's *Zeitschrift für vergleichende Sprachforschung*, b. ix. s. 104. In the Edda the moon is called *ârtali*, year-teller; a Bask name for moon is *argi-izari*, light-measure. See *Dissertation Critique et Apologétique sur la langue Basque*, p. 28.

3 Horne Tooke, p. 27, *note*. Pott, *Studien zur griechischen Mythologie*, 1859, p. 304. Grimm, *Deutsche Grammatik*, iii. p. 349.

4 See Curtius, *Griechische Etymologie*, s. 297.

5 Ideler, *Handbuch der Chronologie*, b. i. s. 241, 242. In the Oscan Inscription of Agnone a Jupiter Virgarius (djoveí verehasioí, dat. sing.) occurs, a name which Professor Aufrecht compares with that of Jupiter Viminius, Jupiter who fosters the growth of twigs (Kuhn's *Zeitschrift*, i. s. 89).—See, however, on Jupiter Viminius and his altars near the Porta Viminalis, Hartung, *Religion der Römer*, ii. 61.

6 As early as the times of Anaximenes of the Ionic, and Alcmaeon of the Pythagorean, schools, the stars had been divided into travelling (ἄστρα πλανώμενα or πλανητά), and non-travelling stars (ἀπλανεῖς ἀστέρες or ἀπλανῆ ἄστρα). Aristotle first used ἄστρα ἐνδεδεμένα, or fixed stars. (See Humboldt, *Kosmos*, vol. iii. p. 28.) Πόλος, the pivot, hinge, or the pole of heaven.

stars driven in and fixed, by which they might hold fast on the deep, as by heavenly anchors.

But although historically we are justified in saying that the first geometrician was a ploughman, the first botanist a gardener, the first mineralogist a miner, it may reasonably be objected that in this early stage a science is hardly a science yet: that measuring a field is not geometry, that growing cabbages is very far from botany, and that a butcher has no claim to the title of comparative anatomist. This is perfectly true, yet it is but right that each science should be reminded of these its more humble beginnings, and of the practical requirements which it was originally intended to answer. A science, as Bacon says, should be a rich storehouse for the glory of God, and the relief of man's estate. Now, although it may seem as if in the present high state of our society students were enabled to devote their time to the investigation of the facts and laws of nature, or to the contemplation of the mysteries of the world of thought, without any side-glance at the practical results of their labours, no science and no art have long prospered and flourished among us, unless they were in some way subservient to the practical interests of society. It is true that a Lyell collects and arranges, a Faraday weighs and analyses, an Owen dissects and compares, a Herschel observes and calculates, without any thought of the immediate marketable results of their labours. But there is a general interest which supports and enlivens their researches, and that interest depends on the practical advantages which society at large derives from these scientific studies. Let it be known that the successive strata of the geologist are a deception to the miner, that the astronomical tables are useless to the navigator, that chemistry is nothing but an expensive amusement, of no use to the manufacturer and the farmer—and astronomy, chemistry, and geology would soon share the fate of alchemy and astrology. As long as the Egyptian science excited the hopes of the invalid by mysterious prescriptions (I may observe by the way that the hieroglyphic signs of our modern prescriptions have been traced back by Champollion to the real hieroglyphics of Egypt[7])—and as long as it instigated the avarice of its patrons by the promise of the discovery o' gold, it enjoyed a liberal support at the courts of princes, and under the roofs of monasteries. Though alchemy did not lead to the discovery of gold, it prepared the way to discoveries more valuable. The same with astrology. Astrology was not such mere imposition as it is generally supposed to have been. It is counted a science by so sound and sober a scholar as Melancthon, and even Bacon allows it a place among the sciences, though admitting that 'it had better intelligence and confederacy with the imagination of man than with his reason.' In spite of the strong condemnation which Luther pronounced against it, astrology continued to sway the destinies of Europe; and a hundred years after Luther, the astrologer was the counsellor of princes and generals, while the founder of modern astronomy died in poverty and despair. In our time the very rudiments of astrology are lost and forgotten.[8] Even real and useful arts, as soon as they cease to be useful, die away, and their secrets are sometimes lost beyond the hope of recovery. When after the Reformation our churches and chapels were divested of their artistic ornaments, in order to restore, in outward appearance also, the simplicity and purity of the Christian church, the colours of the painted windows began to fade away, and have never regained their former depth and harmony. The invention of printing gave the death-blow to the art of ornamental writing and of miniature-painting employed in the illumination of manuscripts; and the best artists of the present day despair of rivalling the minuteness, softness, and brilliancy combined by the humble manufacturer of the mediaeval missal.

7 Bunsen's *Egypt*, vol. iv. p. 108.

8 According to a writer in *Notes and Queries* (2nd Series, vol. x. p. 500), astrology is not so entirely extinct as we suppose. 'One of our principal writers,' he states, 'one of our leading barristers, and several members of the various antiquarian societies, are practised astrologers at this hour. But no one cares to let his studies be known, so great is the prejudice that confounds an art requiring the highest education with the jargon of the gipsy fortune-teller.'

I speak somewhat feelingly on the necessity that every science should answer some practical purpose, because I am aware that the science of language has but little to offer to the utilitarian spirit of our age. It does not profess to help us in learning languages more expeditiously, nor does it hold out any hope of ever realising the dream of one universal language. It simply professes to teach what language is, and this would hardly seem sufficient to secure for a new science the sympathy and support of the public at large. There are problems, however, which, though apparently of an abstruse and merely speculative character, have exercised a powerful influence for good or evil in the history of mankind. Men before now have fought for an idea, and have laid down their lives for a word; and many of the problems which have agitated the world from the earliest to our own times, belong properly to the science of language.

Mythology, which was the bane of the ancient world, is in truth a disease of language. A mythe means a word, but a word which, from being a name or an attribute, has been allowed to assume a more substantial existence. Most of the Greek, the Roman, the Indian and other heathen gods are nothing but poetical names, which were gradually allowed to assume a divine personality never contemplated by their original inventors. *Eos* was a name of the dawn before she became a goddess, the wife of *Tithonos*, or the dying day. *Fatum*, or fate, meant originally what had been spoken; and before Fate became a power, even greater than Jupiter, it meant that which had once been spoken by Jupiter, and could never be changed—not even by Jupiter himself. *Zeus* originally meant the bright heaven, in Sanskrit *Dyaus*; and many of the stories told of him as the supreme god, had a meaning only as told originally of the bright heaven, whose rays, like golden rain, descend on the lap of the earth, the *Danae* of old, kept by her father in the dark prison of winter. No one doubts that *Luna* was simply a name of the moon; but so was likewise *Lucina*, both derived from *lucere*, to shine. *Hecate*, too, was an old name of the moon, the feminine of *Hekatos* and *Hekatebolos*, the far-darting sun; and *Pyrrha*, the Eve of the Greeks, was nothing but a name of the red earth, and in particular of Thessaly. This mythological disease, though less virulent in modern languages, is by no means extinct.[9]

During the middle ages the controversy between Nominalism and Realism, which agitated the church for centuries, and finally prepared the way for the Reformation, was again, as its very name shows, a controversy on names, on the nature of language, and on the relation of words to our conceptions on one side, and to the realities of the outer world on the other. Men were called heretics for believing that words such as *justice* or *truth* expressed only conceptions of our mind, not real things walking about in broad daylight.

In modern times the science of language has been called in to settle some of the most perplexing political and social questions. 'Nations and languages against dynasties and treaties,' this is what has remodelled, and will remodel still more, the map of Europe; and in America comparative philologists have been encouraged to prove the impossibility of a common origin of languages and races, in order to justify, by scientific arguments, the unhallowed theory of slavery. Never do I remember to have seen science more degraded than on the title-page of an American publication in which, among the profiles of the different races of man, the profile of the ape was made to look more human than that of the negro.

Lastly, the problem of the position of man on the threshold between the worlds of matter and spirit has of late assumed a very marked prominence among the problems of the physical and mental sciences. It has absorbed the thoughts of men who, after a long life spent in collecting, observing, and analysing, have brought to its solution qualifications unrivalled in any previous age; and if we may judge from the greater warmth displayed in discussions ordinarily conducted with the calmness of judges and not with the passion of pleaders, it might seem, after all, as if the great problems of our being, of the true nobility of our blood, of our descent from heaven or earth, though unconnected with anything that

9 See *Lectures on the Science of Language*, 2nd Series, 12th lecture.

is commonly called practical, have still retained a charm of their own—a charm that will never lose its power on the mind and on the heart of man. Now, however much the frontiers of the animal kingdom have been pushed forward, so that at one time the line of demarcation between animal and man seemed to depend on a mere fold in the brain, there is *one* barrier which no one has yet ventured to touch—the barrier of language. Even those philosophers with whom *penser c'est sentir*,[10] who reduce all thought to feeling, and maintain that we share the faculties which are the productive causes of thought in common with beasts, are bound to confess that *as yet* no race of animals has produced a language. Lord Monboddo, for instance, admits that *as yet* no animal has been discovered in the possession of language, 'not even the beaver, who of all the animals we know, that are not, like the orang-outangs, of our own species, comes nearest to us in sagacity.'

Locke, who is generally classed together with these materialistic philosophers, and who certainly vindicated a large share of what had been claimed for the intellect as the property of the senses, recognized most fully the barrier which language, as such, placed between man and brutes. 'This I may be positive in,' he writes, 'that the power of abstracting is not at all in brutes, and that the having of general ideas is that which puts a perfect distinction between man and brutes. For it is evident we observe no footsteps in these of making use of general signs for universal ideas; from which we have reason to imagine that they have not the faculty of abstracting or making general ideas, since they have no use of *words* or any other general signs.'

If, therefore, the science of language gives us an insight into that which, by common consent, distinguishes man from all other living beings; if it establishes a frontier between man and the brute, which can never be removed, it would seem to possess at the present moment peculiar claims on the attention of all who, while watching with sincere admiration the progress of comparative physiology, yet consider it their duty to enter their manly protest against a revival of the shallow theories of Lord Monboddo.

But to return to our survey of the history of the physical sciences. We had examined the empirical stage through which every science has to pass. We saw that, for instance, in botany, a man who has travelled through distant countries, who has collected a vast number of plants, who knows their names, their peculiarities, and their medicinal qualities, is not yet a botanist, but only a herbalist, a lover of plants, or what the Italians call a *dilettante*, from *dilettare*, to delight. The real science of plants, like every other science, begins with the work of classification. An empirical acquaintance with facts rises to a scientific knowledge of facts as soon as the mind discovers beneath the multiplicity of single productions the unity of an organic system. This discovery is made by means of comparison and classification. We cease to study each flower for its own sake; and by continually enlarging the sphere of our observation, we try to discover what is common to many and offers those essential points on which groups or natural classes may be established. These classes again, in their more general features, are mutually compared; new points of difference, or of similarity of a more general and higher character, spring to view, and enable us to discover classes of classes, or families. And when the whole kingdom of plants has thus been surveyed, and a simple tissue of names been thrown over the garden of nature; when we can lift it up, as it were, and view it in our mind, as a whole, as a system well defined and complete, we then speak of the science of plants, or botany. We have entered into altogether a new sphere of knowledge where the individual is subject to the general, fact to law; we discover thought, order, and purpose pervading the whole realm of nature, and we perceive the dark chaos of matter lighted up by the reflection of a divine mind. Such views may be right or wrong. Too hasty comparisons, or too narrow distinctions, may have prevented

10 'Man has two faculties, or two passive powers, the existence of which is generally acknowledged: 1, the faculty of receiving the different impressions caused by external objects, physical sensibility; and 2, the faculty of preserving the impressions caused by these objects, called memory, or weakened sensation. These faculties, the productive causes of thought, we have in common with beasts.... Everything is reducible to feeling.'—*Helvetius*.

the eye of the observer from discovering the broad outlines of nature's plan. Yet every system, however insufficient it may prove hereafter, is a step in advance. If the mind of man is once impressed with the conviction that there must be order and law everywhere, it never rests again until all that seems irregular has been eliminated, until the full beauty and harmony of nature has been perceived, and the eye of man has caught the eye of God beaming out from the midst of all His works. The failures of the past prepare the triumphs of the future.

Thus, to recur to our former illustration, the systematic arrangement of plants which bears the name of Linnaeus, and which is founded on the number and character of the reproductive organs, failed to bring out the natural order which pervades all that grows and blossoms. Broad lines of demarcation which unite or divide large tribes and families of plants were invisible from his point of view. But in spite of this, his work was not in vain. The fact that plants in every part of the world belonged to one great system was established once for all; and even in later systems most of his classes and divisions have been preserved, because the conformation of the reproductive organs of plants happened to run parallel with other more characteristic marks of true affinity.[11] It is the same in the history of astronomy. Although the Ptolemaean system was a wrong one, yet even from its eccentric point of view, laws were discovered determining the true movements of the heavenly bodies. The conviction that there remains something unexplained is sure to lead to the discovery of our error. There can be no error in nature; the error must be with us. This conviction lived in the heart of Aristotle when, in spite of his imperfect knowledge of nature, he declared 'that there is in nature nothing interpolated or without connection, as in a bad tragedy;' and from his time forward every new fact and every new system have confirmed his faith.

The object of classification is clear. We understand things if we can comprehend them; that is to say, if we can grasp and hold together single facts, connect isolated impressions, distinguish between what is essential and what is merely accidental, and thus predicate the general of the individual, and class the individual under the general. This is the secret of all scientific knowledge. Many sciences, while passing through this second or classificatory stage, assume the title of comparative. When the anatomist has finished the dissection of numerous bodies, when he has given names to every organ, and discovered the distinctive functions of each, he is led to perceive similarity where at first he saw dissimilarity only. He discovers in the lower animals rudimentary indications of the more perfect organisation of the higher; and he becomes impressed with the conviction that there is in the animal kingdom the same order and purpose which pervades the endless variety of plants or any other realm of nature. He learns, if he did not know it before, that things were not created at random or in a lump, but that there is a scale which leads, by imperceptible degrees, from the lowest infusoria to the crowning work of nature—man; that all is the manifestation of one and the same unbroken chain of creative thought, the work of one and the same all-wise Creator.

In this way the second or classificatory leads us naturally to the third or final stage—the theoretical, or metaphysical. If the work of classification is properly carried out, it teaches us that nothing exists in nature by accident; that each individual belongs to a species, each species to a genus; and that there are laws which underlie the apparent freedom and variety of all created things. These laws indicate to us the presence of a purpose in the mind of the Creator; and whereas the material world was looked upon by ancient philosophers as a mere illusion, as an agglomerate of atoms, or as the work of an evil principle, we now read and interpret its pages as the revelation of a divine power, and wisdom, and love. This has given to the study of nature a new character. After the observer has collected his facts, and after the classifier has placed them in order, the student

11 'The generative organs being those which are most remotely related to the habits and food of an animal, I have always regarded as affording very clear indications of its true affinities.'—Owen, as quoted by Darwin, *Origin of species*, p. 414.

asks what is the origin and what is the meaning of all this? and he tries to soar, by means of induction, or sometimes even of divination, into regions not accessible to the mere collector. In this attempt the mind of man no doubt has frequently met with the fate of Phaeton; but, undismayed by failure, he asks again and again for his father's steeds. It has been said that this so-called philosophy of nature has never achieved anything; that it has done nothing but prove that things must be exactly as they had been found to be by the observer and collector. Physical science, however, would never have been what it is without the impulses which it received from the philosopher, nay, even from the poet. 'At the limits of exact knowledge,' (I quote the words of Humboldt,) 'as from a lofty island-shore, the eye loves to glance towards distant regions. The images which it sees may be illusive; but like the illusive images which people imagined they had seen from the Canaries or the Azores, long before the time of Columbus, they may lead to the discovery of a new world.'

Copernicus, in the dedication of his work to Pope Paul III. (it was commenced in 1517, finished 1530, published 1543), confesses that he was brought to the discovery of the sun's central position, and of the diurnal motion of the earth, not by observation or analysis, but by what he calls the feeling of a want of symmetry in the Ptolemaic system. But who had told him that there *must* be symmetry in all the movements of the celestial bodies, or that complication was not more sublime than simplicity? Symmetry and simplicity, before they were discovered by the observer, were postulated by the philosopher. The first idea of revolutionising the heavens was suggested to Copernicus, as he tells us himself, by an ancient Greek philosopher, by Philolaus, the Pythagorean. No doubt with Philolaus the motion of the earth was only a guess, or, if you like, a happy intuition, not, as it was with Tycho de Brahe and his friend Kepler, the result of wearisome observations of the orbits of the planet Mars. Nevertheless, if we may trust the words of Copernicus, it is quite possible that without that guess we should never have heard of the Copernican system. Truth is not found by addition and multiplication only. When speaking of Kepler, whose method of reasoning has been considered as unsafe and fantastic by his contemporaries as well as by later astronomers, Sir David Brewster remarks very truly, 'that, as an instrument of research, the influence of imagination has been much overlooked by those who have ventured to give laws to philosophy.' The torch of imagination is as necessary to him who looks for truth, as the lamp of study. Kepler held both, and more than that, he had the star of faith to guide him in all things from darkness to light.

In the history of the physical sciences, the three stages which we have just described as the empirical, the classificatory, and the theoretical, appear generally in chronological order. I say, generally, for there have been instances, as in the case just quoted of Philolaus, where the results properly belonging to the third have been anticipated in the first stage. To the quick eye of genius one case may be like a thousand, and one experiment, well chosen, may lead to the discovery of an absolute law. Besides, there are great chasms in the history of science. The tradition of generations is broken by political or ethnic earthquakes, and the work that was nearly finished has frequently had to be done again from the beginning, when a new surface had been formed for the growth of a new civilisation. The succession, however, of these three stages is no doubt the natural one, and it is very properly observed in the study of every science. The student of botany begins as a collector of plants. Taking each plant by itself, he observes its peculiar character, its habitat, its proper season, its popular or unscientific name. He learns to distinguish between the roots, the stem, the leaves, the flower, the calyx, the stamina, and pistils. He learns, so to say, the practical grammar of the plant before he can begin to compare, to arrange, and classify. Again, no one can enter with advantage on the third stage of any physical science without having passed through the second. No one can study *the* plant, no one can understand the bearing of such a work as, for instance, Professor Schleiden's *Life of the Plant*,[12] who has not studied the life of plants in the wonderful variety, and in the still more wonderful order, of nature. These last and

12 *Die Pflanze und ihr Leben*, von M. J. Schleiden, Leipzig, 1858.

highest achievements of inductive philosophy are possible only after the way has been cleared by previous classification. The philosopher must command his classes like regiments which obey the order of their general. Thus alone can the battle be fought and truth be conquered.

After this rapid glance at the history of the other physical sciences, we now return to our own, the science of language, in order to see whether it really is a science, and whether it can be brought back to the standard of the inductive sciences. We want to know whether it has passed, or is still passing, through the three phases of physical research; whether its progress has been systematic or desultory, whether its method has been appropriate or not. But before we do this, we shall, I think, have to do something else. You may have observed that I always took it for granted that the science of language, which is best known in this country by the name of comparative philology, is one of the physical sciences, and that therefore its method ought to be the same as that which has been followed with so much success in botany, geology, anatomy, and other branches of the study of nature. In the history of the physical sciences, however, we look in vain for a place assigned to comparative philology, and its very name would seem to show that it belongs to quite a different sphere of human knowledge. There are two great divisions of human knowledge, which, according to their subject-matter, may be called *physical* and *historical*. Physical science deals with the works of God, historical science with the works of man.[13] Now if we were to judge by its name, comparative philology, like classical philology, would seem to take rank, not as a physical, but as an historical science, and the proper method to be applied to it would be that which is followed in the history of art, of law, of politics, and religion. However, the title of comparative philology must not be allowed to mislead us. It is difficult to say by whom that title was invented; but all that can be said in defence of it is, that the founders of the science of language were chiefly scholars or philologists, and that they based their inquiries into the nature and laws of language on a comparison of as many facts as they could collect within their own special spheres of study. Neither in Germany, which may well be called the birth-place of this science, nor in France, where it has been cultivated with brilliant success, has that title been adopted. It will not be difficult to show that, although the science of language owes much to the classical scholar, and though in return it has proved of great use to him, yet comparative philology has really nothing whatever in common with philology in the usual meaning of the word. Philology, whether classical or oriental, whether treating of ancient or modern, of cultivated or barbarous languages, is an historical science. Language is here treated simply as a means. The classical scholar uses Greek or Latin, the oriental scholar Hebrew or Sanskrit, or any other language, as a key to an understanding of the literary monuments which bygone ages have bequeathed to us, as a spell to raise from the tomb of time the thoughts of great men in different ages and different countries, and as a means ultimately to trace the social, moral, intellectual, and religious progress of the human race. In the same manner, if we study living languages, it is not for their own sake that we acquire grammars and vocabularies. We do so on account of their practical usefulness. We use them as letters of introduction to the best society or to the best literature of the leading nations of Europe. In comparative philology the case is totally different. In the science of language, languages are not treated as a means; language itself becomes the sole object of scientific inquiry. Dialects which have never produced any literature at all, the jargons of savage tribes, the clicks of the Hottentots, and the vocal modulations of the Indo-Chinese are as important, nay, for the solution of some of our problems, more important, than the poetry of Homer, or the prose of Cicero. We do not want to know languages, we want to know language; what language is, how it can form an instrument or an organ of thought; we want to know its ori-

13 'Thus the science of optics, including all the laws of light and colour, is a physical science, whereas the science of painting, with all its laws of manipulation and colouring, being that of a man-created art, is a purely historical science.'—*Intellectual Repository*, June 2, 1862, p. 247.

gin, its nature, its laws; and it is only in order to arrive at that knowledge that we collect, arrange, and classify all the facts of language that are within our reach.

And here I must protest, at the very outset of these lectures, against the supposition that the student of language must necessarily be a great linguist. I shall have to speak to you in the course of these lectures of hundreds of languages, some of which, perhaps, you may never have heard mentioned even by name. Do not suppose that I know these languages as you know Greek or Latin, French or German. In that sense I know indeed very few languages, and I never aspired to the fame of a Mithridates or a Mezzofanti. It is impossible for a student of language to acquire a practical knowledge of all the tongues with which he has to deal. He does not wish to speak the Kachikal language, of which a professorship was lately founded in the University of Guatemala,[14] or to acquire the elegancies of the idiom of the Tcheremissians; nor is it his ambition to explore the literature of the Samoyedes, or the New-Zealanders. It is the grammar and the dictionary which form the subject of his inquiries. These he consults and subjects to a careful analysis, but he does not encumber his memory with paradigms of nouns and verbs, or with long lists of words which have never been used for the purposes of literature. It is true, no doubt, that no language will unveil the whole of its wonderful structure except to the scholar who has studied it thoroughly and critically in a number of literary works representing the various periods of its growth. Nevertheless, short lists of vocables, and imperfect sketches of a grammar, are in many instances all that the student can expect to obtain, or can hope to master and to use for the purposes he has in view. He must learn to make the best of this fragmentary information, like the comparative anatomist, who frequently learns his lessons from the smallest fragments of fossil bones, or the vague pictures of animals brought home by unscientific travellers. If it were necessary for the comparative philologist to acquire a critical or practical acquaintance with all the languages which form the subject of his inquiries, the science of language would simply be an impossibility. But we do not expect the botanist to be an experienced gardener, or the geologist a miner, or the ichthyologist a practical fisherman. Nor would it be reasonable to object in the science of language to the same division of labour which is necessary for the successful cultivation of subjects much less comprehensive. Though much of what we might call the realm of language is lost to us for ever, though whole periods in the history of language are by necessity withdrawn from our observation, yet the mass of human speech that lies before us, whether in the petrified strata of ancient literature or in the countless variety of living languages and dialects, offers a field as large, if not larger, than any other branch of physical research. It is impossible to fix the exact number of known languages, but their number can hardly be less than nine hundred.[15] That this vast field should never have excited the curiosity of the natural philosopher before the beginning of our century may seem surprising, more surprising even than the indifference with which former generations treated the lessons which even the stones seemed to teach of the life still throbbing in the veins and on the very surface of the earth. The saying that 'familiarity breeds contempt' would seem applicable to the subjects of both these sciences. The gravel of our walks hardly seemed to deserve a scientific treatment, and the language which every ploughboy can speak could not be raised without an effort to the dignity of a scientific problem. Man had studied every part of nature, the mineral treasures in the bowels of the earth, the flowers of each season, the animals of every continent, the laws of storms, and the movements of the heavenly bodies; he had analysed every substance, dissected every organism, he knew every bone and muscle, every nerve and fibre of his own body to the ultimate elements which compose his flesh and blood; he had meditated on the nature of his soul, on the laws of his mind, and tried to penetrate into the last causes of all being–and yet language, without the aid of

14 Sir J. Stoddart, *Glossology*, p. 22.

15 Balbi in his *Atlas* counts 860. Cf. Pott, *Rassen*, p. 230; *Etymologische Forschungen*, ii. 83. (Second Edition.)

which not even the first step in this glorious career could have been made, remained unnoticed. Like a veil that hung too close over the eye of the human mind, it was hardly perceived. In an age when the study of antiquity attracted the most energetic minds, when the ashes of Pompeii were sifted for the playthings of Roman life; when parchments were made to disclose, by chemical means, the erased thoughts of Grecian thinkers; when the tombs of Egypt were ransacked for their sacred contents, and the palaces of Babylon and Nineveh forced to surrender the clay diaries of Nebuchadnezzar; when everything, in fact, that seemed to contain a vestige of the early life of man was anxiously searched for and carefully preserved in our libraries and museums—language, which in itself carries us back far beyond the cuneiform literature of Assyria and Babylonia and the hieroglyphic documents of Egypt; which connects ourselves, through an unbroken chain of speech, with the very ancestors of our race, and still draws its life from the first utterances of the human mind—language, the living and speaking witness of the whole history of our race, was never cross-examined by the student of history, was never made to disclose its secrets until questioned, and, so to say, brought back to itself within the last fifty years, by the genius of a Humboldt, Bopp, Grimm, Bunsen, and others. If you consider that, whatever view we take of the origin and dispersion of language, nothing new has ever been added to the substance of language,[16] that all its changes have been changes of form, that no new root or radical has ever been invented by later generations, as little as one single element has ever been added to the material world in which we live; if you bear in mind that in one sense, and in a very just sense, we may be said to handle the very words which issued from the mouth of the son of God, when he gave names to 'all cattle, and to the fowl of the air, and to every beast of the field,' you will see, I believe, that the science of language has claims on your attention, such as few sciences can rival or excel.

Having thus explained the manner in which I intend to treat the science of language, I hope in my next lecture to examine the objections of those philosophers who see in language nothing but a contrivance devised by human skill for the more expeditious communication of our thoughts, and who would wish to see it treated, not as a production of nature, but as a work of human art.

16 Pott, *Etym. Forsch.*, ii. 230.

FURTHER READING

Aarsleff, Hans. *From Locke to Saussure: Essays on the Study of Language and Intellectual History.* Minneapolis: University of Minnesota Press, 1982.

—. *The Study of Language in England, 1780–1860.* Princeton: Princeton UP, 1967.

Beer, Gillian. *Open Fields: Science in Cultural Encounter.* Oxford: Clarendon Press, 1996.

Bharti, Brahm Datt. *Max Muller: A Lifelong Masquerade (the inside story of a* secular *Christian missionary who masqueraded all his lifetime from behind the mask of literature and philology and mortgaged his pen, intellect and scholarship to wreck Hinduism).* New Delhi: Erabooks, 1992.

Chaudhuri, Nirad C. *Scholar Extraordinary: The Life of Professor the Rt. Hon. Friedrich Max Müller, P.C.* New York: Oxford UP, 1974.

Christmann, Hans Helmut. "Linguistics and Modern Philology in Germany 1800–1840 as 'Scientific' Subjects and as University Disciplines." In Stefano Poggi and Maurizio Bossi, eds., *Romanticism in Science: Science in Europe, 1790–1840.* Dordrecht: Kluwer Academic Publishers, 1994, pp. 203–14.

Dowling, Linda. "Victorian Oxford and the Science of Language." *Publications of the Modern Language Association* 97, no. 2 (1982): 160–178.

Knoll, Elizabeth. "The Science of Language and the Evolution of Mind: Max Müller's Quarrel with Darwinism." *Journal of the History of the Behavioral Sciences* 22 (1986): 3–22.

Lyell, Charles. "Origin and Development of Languages and Species Compared." *The Geological Evidences of the Antiquity of Man.* Philadelphia: George W. Childs, 1863, pp. 454–70.

Maher, J. Peter and Konrad Koerner. *Linguistics and Evolutionary Theory: Three Essays by August Schleicher, Ernst Haeckel, and Wilhelm Bleek.*

Amsterdam: John Benjamins Publishing Company, 1983.

Müller, F. Max. *My Autobiography: A Fragment.* New York: Charles Scribner's Sons, 1901.

Neufeldt, Ronald W. *F. Max Müller and the Ṛg-Veda: A Study of its Role in his Work and Thought.* Calcutta: Minerva Associates Pvt. Ltd., 1980.

Noiré, Ludwig. *Max Müller and the Philosophy of Language.* London: Longmans, Green, and Co., 1879.

Olender, Maurice. *The Languages of Paradise: Race, Religion, and Philology in the Nineteenth Century.* Translated by Arthur Goldhammer. Cambridge, MA: Harvard UP, 1992.

Pedersen, Holger. *Linguistic Science in the Nineteenth Century: Methods and Results.* Translated by John Webster Spargo. Cambridge, MA: Harvard UP, 1931.

Rothermund, Dietmar. *The German Intellectual Quest for India.* New Delhi: Manohar, 1986.

Schrempp, Gregory. "The Re-Education of Friedrich Max Müller: Intellectual Appropriation and Epistemological Antinomy in Mid-Victorian Evolutionary Thought." *Journal of the Royal Anthropological Institute* 18 (1983): 90–110.

Trompf, G.W. *Friedrich Max Mueller As a Theorist of Comparative Religion.* Bombay: Shakuntala Publishing House, 1978.

Valone, David A. "Language, Race, and History: The Origin of the Whitney-Müller Debate and the Transformation of the Human Sciences." *Journal of the History of the Behavioral Sciences* 32 (1996): 119–134.

Whitney, William Dwight. *Max Müller and the Science of Language: A Criticism.* New York: D. Appleton and Co., 1892.

23. HERMANN VON HELMHOLTZ

(1821–1894)

Helmholtz began his scientific career as a surgeon, writing his inaugural dissertation on *De Fabrica systematis nervosi evertebratorum*, "On the Structure of the Nervous System in Invertebrates." He served as a military surgeon at Potsdam from 1843 to 1848 and began there his work on animal heat, muscular activity, and metabolism, which culminated in one of his most important contributions to physics, *Ueber die Erhaltung der Kraft* ("On the Conservation of Force") of 1847. He was greatly responsible for the powerful world influence of German science in the mid- to late-nineteenth century.

One of his objectives in investigating the area of force or energy conservation was to disprove the vitalism theories of his contemporaries by demonstrating that the forces (energy)[1] of the body are conserved—to believe otherwise, in Helmholtz's opinion, would mean that vital principles within the body could generate energy without limit, an admission which would also support the idea of perpetual motion. The metaphoric resilience of vitalistic thinking in this area of research, however, is indicated by the term Helmholtz uses to measure the quantity of work, *vis viva* (vital or living power), which he defined as $1/2\ mv^2$, the expression in Newtonian mechanics for kinetic energy. Helmholtz based his arguments for the universal conservation of force on two fundamental axioms: "either on the maxim that it is not possible by any combination whatever of natural bodies to derive an unlimited amount of mechanical force, or on the assumption that all actions in nature can be ultimately referred to attractive or repulsive forces, the intensity of which depends solely upon the distances between the points by which the forces are exerted" ("On the Conservation," trans. Tyndall 114). He then derived the following general equation for a system of natural bodies obeying the principle of the conservation of force:

$$\tfrac{1}{2}mQ^2 - \tfrac{1}{2}mq^2 = -\int_r^R \phi dr$$

where m = a mass particle, ϕ = a central force acting on m, the distance R to r = distance of the action of the force, Q = velocity of m at distance R, and q = velocity of m at distance r. Helmholtz described the behaviour of the system as follows: "in this case our principle requires that the quantity of work gained by the passage of the system from the first position to the second, and the quantity lost by the passage of the system from the second position back again to the first, are always equal, it matters not in what way or at what velocity the change has been effected. For were the quantity of work greater in one way than another, we might use the former for the production of work and the latter to carry the bodies

1 During this period of theoretical flux, the terms "work" (*Arbeit*), "force" (*Kraft*), and later "energy" (*Energie*) did not possess stable meanings in mid-nineteenth-century German physics. Helmholtz employed the terms *lebendige Kraft* ("living force") and *Spannkraft* ("tensional force") in *Ueber die Erhaltung*, and he later lent support to W.J.M. Rankine's English phrase "conservation of energy."

back to their primitive positions, and in this way procure an indefinite amount of mechanical force. We should thus have built a *perpetuum mobile* which could not only impart motion to itself, but also to exterior bodies" ("On the Conservation," trans. Tyndall 118–19). Helmholtz wished to demonstrate this principle mathematically throughout all the branches of physics.

Helmholtz also questioned the theory of heat as a material substance (caloric), arguing instead in "Report on Work Done on the Theory of Animal Heat for 1845" that heat must arise from motion, friction, or electric currents and, most importantly, that chemical, electrical, and mechanical forces in their transformations must remain equivalent.

Helmholtz's theories on energy conservation were not entirely new, as Helmholtz himself admitted, and he gave credit to Julius Robert Mayer and Newton, Bernoulli, Rumford, and Davy for earlier but more limited expressions of the law of conservation of force. In Ernst Mach's history of the idea of the conservation of energy (1872), Helmholtz's role is reduced and Mach attempts to prove that the idea "is by no means so new as one tends to believe; that, indeed, almost all eminent investigators had a more or less confused idea of it, and since the time of Stevinus and Galileo, it has served as the foundation of the most important extensions of the physical sciences" (20). Similarly Thomas Kuhn uncovered other authors who in the 1830s, prior to Helmholtz, had argued for the quantitative interconvertibility of heat and work, a special case of energy conservation (Kuhn 321).

In 1849 Helmholtz was appointed professor of physiology at Königsberg, and then assumed a full professorship of anatomy and physiology at Bonn in 1855. From 1858–1871 he served at Heidelberg as professor of physiology and finally in 1871 succeeded to the prestigious chair of physics at Bonn vacated by Gustav Magnus. Over his lifetime, a steady flow of scientific discoveries issued from his research: the first measurement of the speed of the nerve impulse in frogs (a surprisingly slow 30 m/ sec), the invention of the ophthalmoscope for viewing the retina (1850), and work in electrodynamics and sound and colour perception. In 1888, he served as the first president of the Physikalisch-Technische Reichsanstalt. His *Handbuch der Physiologischen Optik* (1856–66) collected together his experiments on colour vision, and in his *Die Lehre von den Tonempfindungen als physiologische Grundlage für die Theorie der Musik* (*On the Sensations of Tone as a Physiological Basis For the Theory of Music*; 1863, revised 1877), Helmholtz investigated the physics of the perception of partial tones, harmonies, and musical scales.

Helmholtz can be credited with the championing of James Clerk Maxwell's work on electric waves and paving the way for the research of Heinrich Hertz and Ludwig Boltzmann, both of whose theories of electromagnetism and thermodynamics eventually superseded those of Helmholtz.

Below is an excerpt from the popular lecture version of his ideas on the conservation of force delivered at Carlsruhe in 1862–63, not to be confused with the more technical and mathematical paper *Ueber die Erhaltung der Kraft* (Berlin: G.A. Reimer, 1847). John Tyndall translated the *Erhaltung der Kraft* into English in Taylor's *Scientific Memoirs* (1853).

"On the Conservation of Force." *Popular Lectures on Scientific Subjects.* Translated by E. Atkinson *et al.* Introduced by Professor Tyndall. 2 vols. London: Longmans, Green, and Co., 1873, pp. 1.317–62.[2]

As I have undertaken to deliver here a series of lectures, I think the best way in which I can discharge that duty will be to bring before you, by means of a suitable example, some view of the special character of those sciences to the study of which I have devoted myself. The natural sciences, partly in consequence of their practical applications, and partly from their intellectual influence on the last four centuries, have so profoundly, and with such increasing rapidity, transformed all the relations of the life of civilised nations; they have given these nations such increase of riches, of enjoyment of life, of the preservation of health, of means of industrial and of social intercourse, and even such increase of political power, that every educated man who tries to understand the forces at work in the world in which he is living, even if he does not wish to enter upon the study of a special science, must have some interest in that peculiar kind of mental labour which works and acts in the sciences in question.

On a former occasion I have already discussed the characteristic differences which exist between the natural and the mental sciences as regards the kind of scientific work. I then endeavoured to show that it is more especially in the thorough conformity with law which natural phenomena and natural products exhibit, and in the comparative ease with which laws can be stated, that this difference exists. Not that I wish by any means to deny, that the mental life of individuals and peoples is also in conformity with law, as is the object of philosophical, philological, historical, moral, and social sciences to establish. But in mental life, the influences are so interwoven, that any definite sequence can but seldom be demonstrated. In Nature the converse is the case. It has been possible to discover the law of the origin and progress of many enormously extended series of natural phenomena with such accuracy and completeness that we can predict their future occurrence with the greatest certainty; or in cases in which we have power over the conditions under which they occur, we can direct them just according to our will. The greatest of all instances of what the human mind can effect by means of a well-recognized law of natural phenomena is that afforded by modern astronomy. The one simple law of gravitation regulates the motions of the heavenly bodies not only of our own planetary system, but also of the far more distant double stars; from which, even the ray of light, the quickest of all messengers, needs years to reach our eye; and just on account of this simple conformity with law, the motions of the bodies in question, can be accurately predicted and determined both for the past and for future years and centuries to a fraction of a minute.

On this exact conformity with law depends also the certainty with which we know how to tame the impetuous force of steam, and to make it the obedient servant of our wants. On this conformity depends, moreover, the intellectual fascination which chains the physicist to his subjects. It is an interest of quite a different kind to that which mental and moral sciences afford. In the latter it is man in the various phases of his intellectual activity who chains us. Every great deed of which history tells us, every mighty passion which art can represent, every picture of manners, of civic arrangements, of the culture of peoples of distant lands, or of remote times, seizes and interests us, even if there is no exact scientific connection among them. We continually find points of contact and comparison in our own conceptions and feelings; we get to know the hidden capacities and desires of the mind, which in the ordinary peaceful course of civilised life remain unawakened.

2 Illustrations are drawn from the more carefully engraved German edition, *Vorträge und Reden von Hermann von Helmholtz*. Band 1. Braunschweig: Friedrich Vieweg und Sohn, 1884.

It is not to be denied that, in the natural sciences, this kind of interest is wanting. Each individual fact, taken of itself, can indeed arouse our curiosity or our astonishment, or be useful to us in its practical applications. But intellectual satisfaction we obtain only from a connection of the whole, just from its conformity with law. *Reason* we call that faculty innate in us of discovering laws and applying them with thought. For the unfolding of the peculiar forces of pure reason in their entire certainty and in their entire bearing, there is no more suitable arena than inquiry into nature in the wider sense, the mathematics included. And it is not only the pleasure at the successful activity of one of our most essential mental powers; and the victorious subjections to the power of our thought and will of an external world, partly unfamiliar, and partly hostile, which is the reward of this labour; but there is a kind, I might almost say, of artistic satisfaction, when we are able to survey the enormous wealth of Nature as a regularly-ordered whole—a kosmos, an image of the logical thought of our own mind.

The last decades of scientific development have led us to the recognition of a new universal law of all natural phenomena, which, from its extraordinarily extended range, and from the connection which it constitutes between natural phenomena of all kinds, even of the remotest times and the most distant places, is especially fitted to give us an idea of what I have described as the character of the natural sciences, which I have chosen as the subject of this lecture.

This law is the *Law of the Conservation of Force*, a term the meaning of which I must first explain. It is not absolutely new; for individual domains of natural phenomena it was enunciated by Newton and Daniel Bernoulli; and Rumford and Humphry Davy have recognised distinct features of its presence in the laws of heat.

The possibility that it was of universal application was first stated by Dr. Julius Robert Mayer, a Schwabian physician (now living in Heilbronn) in the year 1842, while almost simultaneously with, and independently of him, James Prescot Joule, an English manufacturer, made a series of important and difficult experiments on the relation of heat to mechanical force, which supplied the chief points in which the comparison of the new theory with experience was still wanting.

The law in question asserts, that the *quantity of force which can be brought into action in the whole of Nature is unchangeable*, and can neither be increased nor diminished. My first object will be to explain to you what is understood by *quantity of force*; or as the same idea is more popularly expressed with reference to its technical application, what we call *amount of work* in the mechanical sense of the word.

The idea of work for machines, or natural processes, is taken from comparison with the working power of man; and we can therefore best illustrate from human labour, the most important features of the question with which we are concerned. In speaking of the work of machines, and of natural forces, we must, of course, in this comparison eliminate anything in which activity of intelligence comes into play. The latter is also capable of the hard and intense work of thinking, which tries a man just as muscular exertion does. But whatever of the actions of intelligence is met with in the work of machines, of course is due to the mind of the constructor and cannot be assigned to the instrument at work.

Now, the external work of man is of the most varied kind as regards the force or ease, the form and rapidity, of the motions used on it, and the kind of work produced. But both the arm of the blacksmith who delivers his powerful blows with the heavy hammer, and that of the violinist who produces the most delicate variations in sound, and the hand of the lace-maker who works with threads so fine that they are on the verge of the invisible, all these acquire the force which moves them in the same manner and by the same organs, namely, the muscles of the arm. An arm the muscles of which are lamed is incapable of doing any work; the moving force of the muscle must be at work in it, and these must obey the nerves, which bring to them orders from the brain. That member is then capable of the greatest variety of motions; it can compel the most varied instruments to execute the most diverse tasks.

Just so is it with machines: they are used for the most diversified arrangements. We produce by their agency an infinite variety of movements, with the most various degrees of force and rapidity, from powerful steam-hammers and rolling-mills, where gigantic masses of iron are cut and shaped like butter, to spinning and weaving-frames, the work of which rivals that of the spider. Modern mechanism has the richest choice of means of transferring the motion of one set of rolling wheels to another with greater or less velocity; of changing the rotating motion of wheels into the up-and-down motion of the piston-rod, of the shuttle, of falling hammers and stamps; or, conversely, of changing the latter into the former; or it can, on the other hand, change movements of uniform into those of varying velocity, and so forth. Hence this extraordinarily rich utility of machines for so extremely varied branches of industry. But one thing is common to all these differences; they all need a *moving force,* which sets and keeps them in motion, just as the works of the human hand all need the moving force of the muscles.

Now, the work of the smith requires a far greater and more intense exertion of the muscles than that of the violin-player; and there are in machines corresponding differences in the power and duration of the moving force required. These differences, which correspond to the different degree of exertion of the muscles in human labour, are alone what we have to think of when we speak of the *amount of work* of a machine. We have nothing to do here with the manifold character of the actions and arrangements which the machines produce; we are only concerned with an expenditure of force.

This very expression which we use so fluently, 'expenditure of force,' which indicates that the force applied has been expended and lost, leads us to a further characteristic analogy between the effects of the human arm and those of machines. The greater the exertion, and the longer it lasts, the more is the arm *tired,* and the more *is the store of its moving force for the time exhausted.* We shall see that this peculiarity of becoming exhausted by work is also met with in the moving forces of inorganic nature; indeed, that this capacity of the human arm of being tired is only one of the consequences of the law with which we are now concerned. When fatigue sets in, recovery is needed, and this can only be effected by rest and nourishment. We shall find that also in the inorganic moving forces, when their capacity for work is spent, there is a possibility of reproduction, although in general other means must be used to this end than in the case of the human arm.

From the feeling of exertion and fatigue in our muscles, we can form a general idea of what we understand by amount of work; but we must endeavour, instead of the indefinite estimate afforded by this comparison, to form a clear and precise idea of the standard by which we have to measure the amount of work. This we can do better by the simplest inorganic moving forces than by the actions of our muscles, which are a very complicated apparatus, acting in an extremely intricate manner.

Let us now consider that moving force which we know best, and which is simplest—gravity. It acts, for example as such, in those clocks which are driven by a weight. This weight fastened to a string, which is wound round a pulley connected with the first toothed wheel of the clock, cannot obey the pull of gravity without setting the whole clockwork in motion. Now I must beg you to pay special attention to the following points: the weight cannot put the clock in motion without itself sinking; did the weight not move, it could not move the clock, and its motion can only be such a one as obeys the action of gravity. Hence, if the clock is to go, the weight must continually sink lower and lower, and must at length sink so far that the string which supports it is run out. The clock then stops. The useful effect of its weight is for the present exhausted. Its gravity is not lost or diminished: it is attracted by the earth as before, but the capacity of this gravity to produce the motion of the clockwork is lost. It can only keep the weight at rest in the lowest point of its path, it cannot farther put it in motion.

But we can wind up the clock by the power of the arm, by which the weight is again raised. When this has been done, it has regained its former capacity, and can again set the clock in motion.

We learn from this that a raised weight possesses a *moving force,* but that it must necessarily sink if this force is to act, that by sinking, this moving force is exhausted, but by using another extraneous moving force—that of the arm—its activity can be restored.

The work which the weight has to perform in driving the clock is not indeed great. It has continually to overcome the small resistances which the friction of the axles and teeth, as well as the resistance of the air, oppose to the motion of the wheels, and it has to furnish the force for the small impulses and sounds which the pendulum produces at each oscillation. If the weight is detached from the clock, the pendulum swings for a while before coming to rest, but its motion becomes each moment feebler, and ultimately ceases entirely, being gradually used up by the small hindrances I have mentioned. Hence, to keep the clock going, there must be a moving force, which, though small, must be continually at work. Such a one is the weight.

We get, moreover, from this example, a measure for the amount of work. Let us assume that a clock is driven by a weight of a pound, which falls five feet in twenty-four hours. If we fix ten such clocks, each with a weight of one pound, then ten clocks will be driven twenty-four hours; hence, as each has to overcome the same resistances in the same time as the others, ten times as much work is performed for ten pounds fall through five feet. Hence, we conclude that the height of the fall being the same, the work increases directly as the weight.

Now, if we increase the length of the string so that the weight runs down ten feet, the clock will go two days instead of one; and, with double the height of fall, the weight will overcome on the second day the same resistances as on the first, we will therefore do twice as much work as when it can only run down five feet. The weight being the same, the work increases as the height of fall. Hence, we may take the product of the weight into the height of fall as a measure of work, at any rate, in the present case. The application of this measure is, in fact, not limited to the individual case, but the universal standard adopted in manufactures for measuring magnitude of work is a *foot pound*—that is, the amount of work which a pound raised through a foot can produce.[3]

We may apply this measure of work to all kinds of machines, for we should be able to set them all in motion by means of a weight sufficient to turn a pulley. We could thus always express the magnitude of any driving force, for any given machine, by the magnitude and height of fall of such a weight as would be necessary to keep the machine going with its arrangements until it had performed a certain work. Hence it is that the measurement of work by foot pounds is universally applicable. The use of such a weight as a driving force would not indeed be practically advantageous in those cases in which we were compelled to raise it by the power of our own arm; it would in that case be simpler to work the machine by the direct action of the arm. In the clock we use a weight so that we need not stand the whole day at the clockwork, as we should have to do to move it directly. By winding up the clock we accumulate a store of working capacity in it, which is sufficient for the expenditure of the next twenty-four hours.

The case is somewhat different when Nature herself raises the weight, which then works for us. She does not do this with solid bodies, at least not with such regularity as to be utilised; but she does it abundantly with water, which, being raised to the tops of mountains by meteorological processes, returns in streams from them. The gravity of water we use as moving force, the most direct application being in what are called *overshot* wheels, one of which is represented in Fig. 38. Along the circumference of such a wheel are a series of buckets, which act as receptacles for the water, and, on the side turned to the observer, have the tops uppermost; on the opposite side the tops of the buckets are upside-down. The water flows at M into the buckets of the front of the wheel, and at F, where the mouth begins to incline downwards, it flows out. The buckets on the circumference are filled on the side turned to the observer, and empty on the other side. Thus the former are weighted by the water contained in them, the latter not; the weight of the water acts continuously on only one side of the wheel,

3 This is the *technical* measure of work; to convert it into scientific measure, it must be multiplied by the intensity of gravity.

Fig. 38

draws this down, and thereby turns the wheel; the other side of the wheel offers no resistance, for it contains no water. It is thus the weight of the falling water which turns the wheel, and furnishes the motive power. But you will at once see that the mass of water which turns the wheel must necessarily fall in order to do so, and that though, when it has reached the bottom, it has lost none of its gravity, it is no longer in a position to drive the wheel, if it is not restored to its original position, either by the power of the human arm or by means of some other natural force. If it can flow from the mill-stream to still lower levels, it may be used to work other wheels. But when it has reached its lowest level, the sea, the last remainder of the moving force is used up, which is due to gravity—that is, to the attraction of the earth, and it cannot act by its weight until it has been again raised to a high level. As this is actually effected by meteorological processes, you will at once observe that these are to be considered as sources of moving force.

Water-power was the first inorganic force which man learnt to use instead of his own labour or of that of domestic animals. According to Strabo, it was known to King Mithridates, of Pontus, who was also otherwise celebrated for his knowledge of nature; near his palace there was a water-wheel. Its use was first introduced among the Romans in the time of the first Emperors. Even now we find water-mills in all mountains, valleys, or wherever there are rapidly-flowing, regularly-filled, brooks and streams. We find water-power used for all purposes which can possibly be effected by machines. It drives mills which grind corn, saw-mills, hammers and oil-presses, spinning-frames and looms, and so forth. It is the cheapest of all motive powers, it flows spontaneously from the inexhaustible stores of nature; but it is restricted to a particular place, and only in mountainous countries is it present in any quantity; in level countries extensive reservoirs are necessary for damming the rivers to produce any amount of water-power.

Before passing to the discussion of other motive forces, I must answer an objection which may readily suggest itself. We all know that there are numerous machines, systems of pulleys, levers and cranes, by the aid of which heavy burdens may be lifted by a comparatively small expenditure of force. We have all of us often seen one or two workmen hoist heavy masses of stones to great heights, which they would be quite unable to do directly; in like manner, one or two men, by means of a crane, can transfer the largest and heaviest chests from a ship to the quay. Now it may be asked, If a large, heavy weight had been used for driving a machine, would it not be very easy, by means of a crane or a system of pulleys, to raise it anew, so that it could again be used as a motor, and thus acquire motive power, without being compelled to use a corresponding exertion in raising the weight?

The answer to this is, that all these machines, in that degree in which for the moment they facilitate the exertion, also prolong it, so that by their help no motive power is ultimately gained. Let us assume that four labourers have to raise a load of four hundredweight, by means of a rope passing over a single pulley. Every time the rope is pulled down through four feet, the load is also raised through four feet. But now, for the sake of comparison, let us suppose the same load hung to a block of four pulleys, as represented in Fig. 39. A single labourer would now be able to raise the load by the same exertion of force as each one of

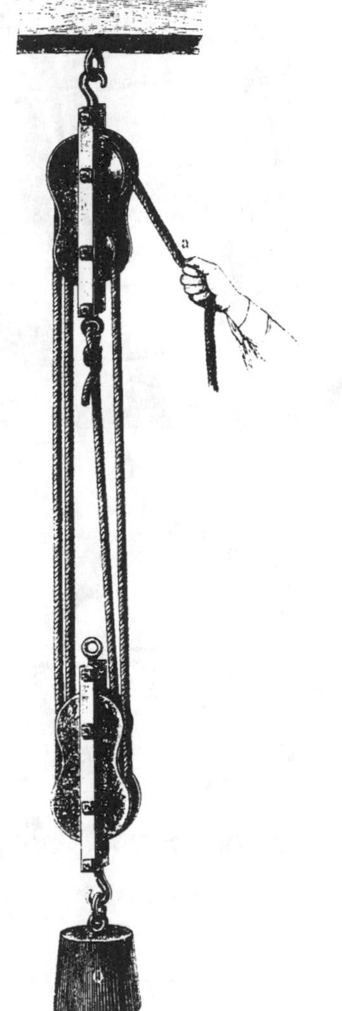

Fig. 39

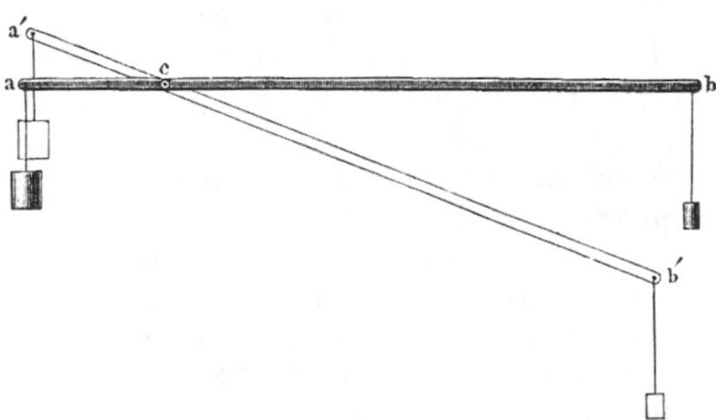

the four put forth. But when he pulls the rope through four feet, the load only rises one foot, for the length through which he pulls the rope, at *a*, is uniformly distributed in the block over four ropes, so that each of these is only shortened by a foot. To raise the load, therefore, to the same height, the one man must necessarily work four times as long as the four together did. But the total expenditure of work is the same, whether four labourers work for a quarter of an hour or one works for an hour.

If, instead of human labour, we introduce the work of a weight, and hang to the block a load of 400, and at *a*, where otherwise the labourer works, a weight of 100 pounds, the block is then in equilibrium, and, without any appreciable exertion of the arm, may be set in motion. The weight of 100 pounds sinks, that of 400 rises. Without any measurable expenditure of force, the heavy weight has been raised by the sinking of the smaller one. But observe that the smaller weight will have sunk through four times the distance that the greater one has risen. But a fall of 100 pounds through four feet is just as much 400 foot pounds as a fall of 400 pounds through one foot.

The action of levers in all their various modifications is precisely similar. Let *ab*, Fig. 40, be a simple lever, supported at *c*, the arm *cb* being four times as long as the other arm *ac*. Let a weight of one pound be hung at *b*, and a weight of four pounds at *a*, the lever is then in equilibrium, and the least pressure of the finger is sufficient, without any appreciable exertion of force, to place it in the position *a′b′*, in which the heavy weight of four pounds has been raised, while the one-pound weight has sunk. But here, also, you will observe no work has been gained, for while the heavy weight has been raised through one inch, the lighter one has fallen through four inches; and four pounds through one inch is, as work, equivalent to the product of one pound through four inches.

Most other fixed parts of machines may be regarded as modified and compound levers; a toothed-wheel, for instance as a series of levers, the ends of which are represented by the individual teeth, and one after the other of which is put in activity, in the degree in which the tooth in question seizes, or is seized by the adjacent pinion. Take, for instance, the crabwinch, represented in Fig. 41. Suppose the pinion on the axis of the barrel of the winch has twelve teeth, and the toothed-wheel, HH, seventy-two teeth, that is six times as many as the former. The winch must now be turned round six times before the toothed-wheel, H, and the barrel, D, have made one turn, and before the rope which raises the load has been lifted by a length equal to the circumference of the barrel. The workman thus requires six times the time, though to be sure only one-sixth of the exertion, which he would have to use if the handle were directly applied to the barrel, D. In all these machines, and parts of machines, we find it confirmed that in proportion as the velocity of the motion increases its power diminishes, and that when the power increases the velocity diminishes, but that the amount of work is never thereby increased.

Fig. 41

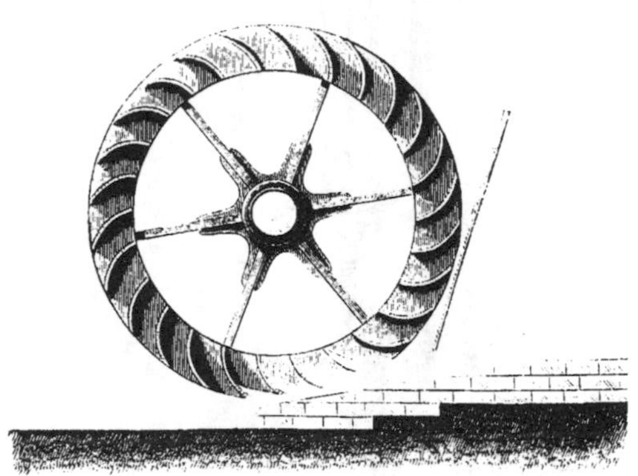
Fig. 42

In the overshot mill-wheel, described above, water acts by its weight. But there is another form of mill-wheels, what is called the *undershot wheel*, in which it only acts by its impact, as represented in Fig. 42. These are used where the height from which the water comes is not great enough to flow on the upper part of the wheel. The lower part of undershot wheels dips in the flowing water which strikes against their float-boards and carries them along. Such wheels are used in swift-flowing streams which have a scarcely perceptible fall, as, for instance, on the Rhine. In the immediate neighbourhood of such a wheel, the water need not necessarily have a great fall if it only strikes with considerable velocity. It is the velocity of the water, exerting an impact against the float-boards, which acts in this case, and which produces the motive power.

Windmills, which are used in the great plains of Holland and North Germany to supply the want of falling water, afford another instance of the action of velocity. The sails are driven by air in motion—by wind. Air at rest could just as little drive a windmill as water at rest a water-wheel. The driving force depends here on the velocity of moving masses.

A bullet resting in the hand is the most harmless thing in the world; by its gravity it can exert no great effect; but when fired and endowed with great velocity it drives all obstacles with the most tremendous force.

If I lay the head of a hammer gently on a nail, neither its small weight nor the pressure of my arm is quite sufficient to drive the nail into wood; but if I swing the hammer and allow it to fall with great velocity, it acquires a new force, which can overcome far greater hindrances.

These examples teach us that the velocity of a moving mass can act as motive force. In mechanics, velocity in so far as it is motive force, and can produce work, is called *vis viva*. The name is not well chosen; it is too apt to suggest to us the force of living beings. Also in this case you will see,

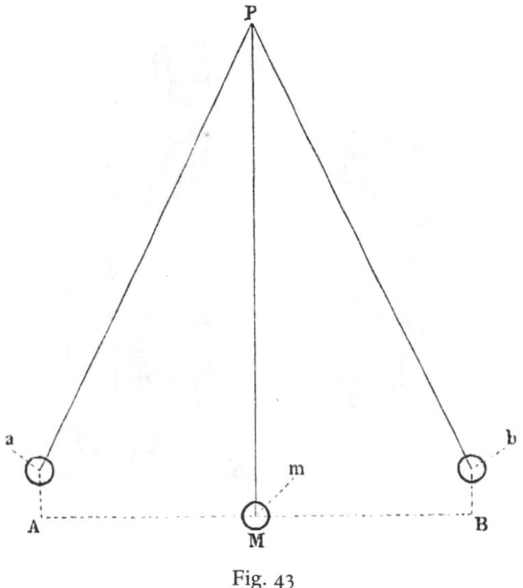

Fig. 43

from the instances of the hammer and of the bullet, that velocity is lost as such, when it produces working power. In the case of the water-mill, or of the windmill, a more careful investigation of the moving masses of water and air is necessary to prove that part of their velocity has been lost by the work which they have performed.

The relation of velocity to working power is most simply and clearly seen in a simple pendulum, such as can be constructed by any weight which we suspend to a cord. Let M, Fig. 43, be such a weight, of a spherical form; AB, a horizontal line drawn through the centre of the sphere; P the point at which the cord is fastened. If now I draw the weight M on one side towards A, it moves in the arc M*a*, the end of which, *a*, is somewhat higher than the point A in the horizontal line. The weight is thereby raised to the height A*a*. Hence my arm must exert a certain force to bring the weight to *a*. Gravity resists this motion and endeavours to bring back the weight to M, the lowest point which it can reach.

Now, if after I have brought the weight to *a* I let it go, it obeys this force of gravity and returns to M, arrives there with a certain velocity, and no longer remains quietly hanging at M as it did before, but swings beyond M towards *b*, where its motion stops as soon as it has traversed on the side of B an arc equal in length to that on the side of A, and after it has risen to a distance B*b* above the horizontal line, which is equal to the height A*a*, to which my arm had previously raised it. In *b* the pendulum returns, swings the same way back through M towards *a*, and so on, until its oscillations are gradually diminished, and ultimately annulled by the resistance of the air and by friction.

You see here that the reason why the weight, when it comes from *a* to M, and does not stop here, but ascends to *b*, in opposition to the action of gravity, is only to be sought in its velocity. The velocity which it has acquired in moving from the height A*a* is capable again raising it to an equal height, B*b*. The velocity of the moving mass, M, is thus capable of raising this mass; that is to say, in the language of mechanics, of performing work. This would also be the case if we had imparted such a velocity to the suspended weight by a blow.

From this we learn further how to measure the working power of velocity—or, what is the same thing, the *vis viva* of the *moving mass*. It is equal to the work, expressed in foot pounds, which the same mass can exert after its velocity has been used to raise it, under the most favourable circumstances, to as great a height as possible.[4] This does not depend on the direction of the velocity; for if we swing a weight attached to a thread in a circle, we can even change a downward motion into an upward one.

The motion of the pendulum shows us very distinctly how the forms of working power hitherto considered—that of a raised weight and that of a moving mass—may merge into one another. In the points *a* and *b*, Fig. 43, the mass has no velocity; at the point M it has fallen as far as possible, but possesses velocity. As the weight goes from *a* to *m* the work of the raised weight is changed into *vis viva*; as the weight goes further from *m* to *b* the *vis viva* is changed into the work

4 The measure of *vis viva* in theoretical mechanics is half the product of the weight into the square of the velocity. To reduce it to the technical measure of the work we must divide it by the intensity of gravity; that is, by the velocity at the end of the first second of a freely falling body.

of a raised weight. Thus the work which the arm originally imparted to the pendulum is not lost in these oscillations, provided we may leave out of consideration the influence of the resistance of the air and of friction. Neither does it increase, but it continually changes the form of its manifestation.

Let us now pass to other mechanical forces, those of elastic bodies. Instead of the weights which drive our clocks, we find in time-pieces and in watches, steel springs which are coiled in winding up the clock, and are uncoiled by the working of the clock. To coil up the spring we consume the force of the arm; this has to overcome the resisting elastic force of the spring as we wind it up, just as in the clock we have to overcome the force of gravity which the weight exerts. The coiled spring can, however, perform work; it gradually expends this acquired capability in driving the clock-work.

If I stretch a crossbow and afterwards let it go, the stretched string moves the arrow; it imparts to it force in the form of velocity. To stretch the cord my arm must work for a few seconds; this work is imparted to the arrow at the moment it is shot off. Thus the crossbow concentrates into an extremely short time the entire work which the arm had communicated in the operation of stretching; the clock, on the contrary, spreads it over one or several days. In both cases no work is produced which my arm did not originally impart to the instrument, it is only expended more conveniently.

The case is somewhat different if by any other natural process I can place an elastic body in a state of tension without having to exert my arm. This is possible and is most easily observed in the case of gases.

If, for instance, I discharge a fire-arm loaded with gunpowder, the greater part of the mass of the powder is converted into gases at a very high temperature, which have a powerful tendency to expand, and can only be retained in the narrow space in which they are formed, by the exercise of the most powerful pressure. In expanding with enormous force they propel the bullet, and impart to it a great velocity, which we have already seen is a form of work.

In this case, then, I have gained work which

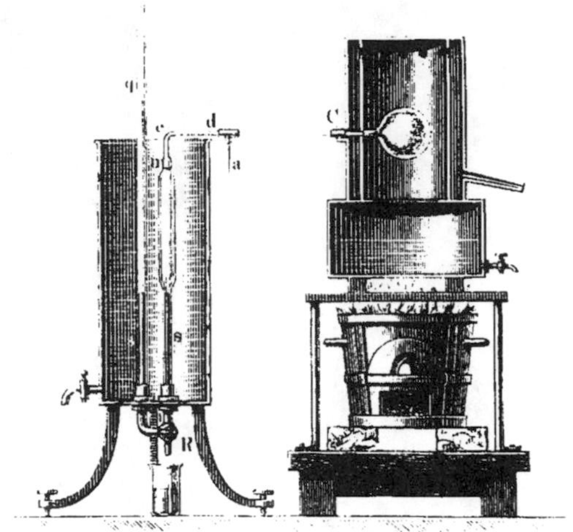

Fig. 44

my arm has not performed. Something, however, has been lost; the gunpowder, that is to say, whose constituents have changed into other chemical compounds, from which they cannot, without further ado, be restored to their original condition. Here, then, a chemical change has taken place, under the influence of which work has been gained.

Elastic forces are produced in gases by the aid of heat, on a far greater scale.

Let us take, as the most simple instance, atmospheric air. In Fig. 44 an apparatus is represented such as Regnault used for measuring the expansive force of heated gases. If no great accuracy is required in the measurement, the apparatus may be arranged more simply. At C is a glass globe filled with dry air, which is placed in a metal vessel, in which it can be heated by steam. It is connected with the U-shaped tube, ss, which contains a liquid, and the limbs of which communicate with each other when the stop-cock R is closed. If the liquid is in equilibrium in the tube ss when the globe is cold, it rises in the leg s, and ultimately overflows when the globe is heated. If, on the contrary, when the globe is heated, equilibrium be restored by allowing some of the liquid to flow out at R, as the globe cools it will be drawn up towards *n*. In both cases liquid is raised, and work thereby produced.

The same experiment is continuously repeated

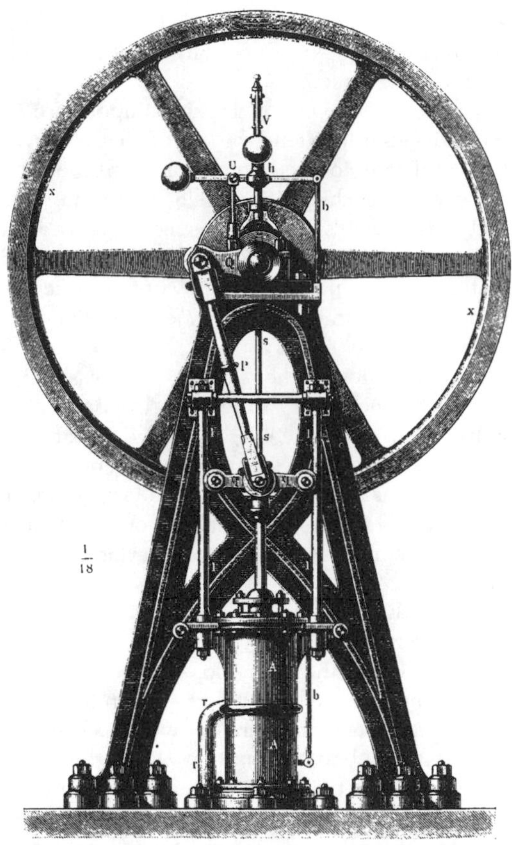

Fig. 45

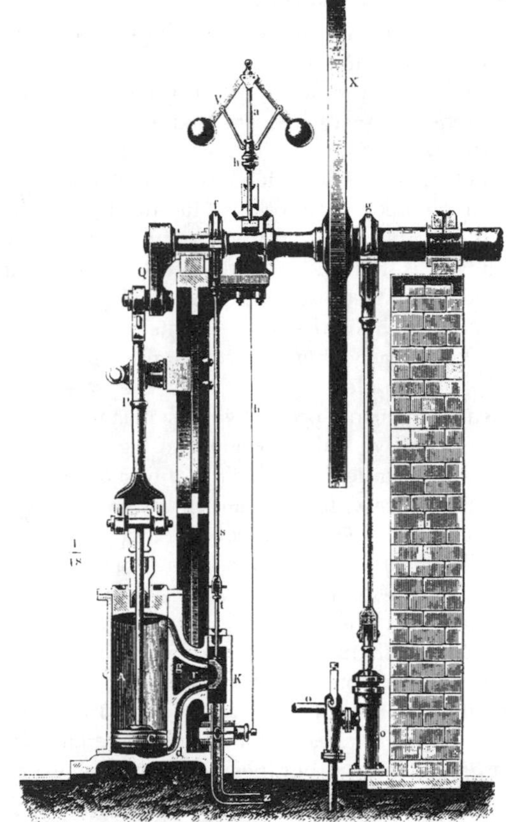

Fig. 46

on the largest scale in steam engines, though in order to keep up a continual disengagement of compressed gases from the boiler, the air in the globe in Fig. 44, which would soon reach the maximum of its expansion, is replaced by water, which is gradually changed into steam by the application of heat. But steam, as long as it remains as such, is an elastic gas which endeavours to expand exactly like atmospheric air. And instead of the column of liquid which was raised in our last experiment, the machine is caused to drive a solid piston which imparts its motion to other parts of the machine. Fig. 45 represents a front view of the working parts of a high pressure engine, and Fig. 46 a section. The boiler in which steam is generated is not represented; the steam passes through the tube zz, Fig. 46, to the cylinder AA in which moves a tightly fitted piston C. The parts between the tube zz and the cylinder AA, that is the slide valve in the valve-chest KK, and the two tubes d and e allow the steam to pass first below and then above the piston, while at the same time the steam has free exit from the other half of the cylinder. When the steam passes under the piston, it forces it upward; when the piston has reached the top of its course the position of the valve in KK changes, and the steam passes above the piston and forces it down again. The piston-rod acts by means of the connecting-rod P, on the crank Q of the fly-wheel X and sets this in motion. By means of the rod s, the motion of the rod regulates the opening and closing of the valve. But we need not here enter into those mechanical arrangements, however ingeniously they have been devised. We are only interested in the manner in which heat produces elastic vapour, and how this vapour, in its endeavour to expand, is compelled to move the solid parts of the machine, and furnish work.

You all know how powerful and varied are the effects of which steam engines are capable; with them has really begun the great development of industry which has characterised our century before all others. Its most essential superiority over motive powers formerly known, is that it is not restricted to a particular place. The store of coal and the small quantity of water which are the sources of its power can be brought everywhere, and steam engines can even be made movable, as is the case with steam-ships and locomotives. By means of these machines we can develop motive power to almost an indefinite extent at any place on the earth's surface, in deep mines and even on the middle of the ocean; while water and wind-mills are bound to special parts of the surface of the land. The locomotive transports travellers and goods over the land in numbers and with a speed which must have seemed an incredible fable to our forefathers, who looked upon the mail-coach with its six passengers in the inside and its ten miles an hour, as an enormous progress. Steam-engines traverse the ocean independently of the direction of the wind, and, successfully resisting storms which would drive sailing-vessels far away, reach their goal at the appointed time. The advantages which the concourse of numerous, and variously skilled workmen in all branches offers in large towns where wind and water power are wanting, can be utilised, for steam-engines find place everywhere, and supply the necessary crude force; thus the more intelligent human force may be spared for better purposes; and, indeed, wherever the nature of the ground or the neighbourhood of suitable lines of communication present a favourable opportunity for the development of industry, the motive power is also present in the form of steam-engines.

We see, then, that heat can produce mechanical power; but in the cases which we have discussed we have seen that the quantity of force which can be produced by a given measure of a physical process is always accurately defined, and that the further capacity for work of the natural forces, is either diminished or exhausted by the work which has been performed. How is it now with *Heat* in this respect?

This question was of decisive importance in the endeavour to extend the law of the Conservation of Force to all natural processes. In the answer lay the chief difference between the older and newer views in these respects. Hence it is that many physicists designate that view of Nature corresponding to the law of the conservation of force with the name of *the Mechanical Theory of Heat*.

The older view of the nature of heat was that it is a substance, very fine and imponderable indeed, but indestructible, and unchangeable in quantity, which is an essential fundamental property of all matter. And, in fact, in a large number of natural processes, the quantity of heat which can be demonstrated by the thermometer is unchangeable.

By conduction and radiation, it can indeed pass from hotter to colder bodies; but the quantity of heat which the former lose can be shown by the thermometer to have reappeared in the latter. Many processes, too, were known, especially in the passage of bodies from the solid to the liquid and gaseous states, in which heat disappeared—at any rate, as regards the thermometer. But when the gaseous body was restored to the liquid, and the liquid to the solid state, exactly the same quantity of heat reappeared which formerly seemed to have been lost. Heat was said *to have become latent*. On this view, liquid water differed from solid ice in containing a certain quantity of heat bound, which, just because it was bound, could not pass to the thermometer, and therefore was not indicated by it. Aqueous vapour contains a far greater quantity of heat thus bound. But if the vapour be precipitated, and the liquid water restored to the state of ice, exactly the same amount of heat is liberated as had become latent in the melting of the ice and in the vaporisation of the water.

Finally, heat is sometimes produced and sometimes disappears in chemical processes. But even here it might be assumed that the various chemical elements and chemical compounds contain certain constant quantities of latent heat, which, when they change their composition, are sometimes liberated and sometimes must be supplied from external sources. Accurate experiments have shown that the quantity of heat which is developed by a chemical process, for instance, in burning a pound of pure carbon into carbonic acid, is perfectly constant, whether the combustion is slow or rapid, whether it takes place all at once or by intermediate stages. This is also agreed very well with the assumption, which was the basis of the theory of heat, that heat is a substance entirely unchangeable in quantity. The natural processes which have here been briefly mentioned, were the subject of extensive experimental and mathematical investigations, especially of the great French physicists in the last decade of the former, and the first decade of the present, century; and a rich and accurately-worked chapter of physics had been developed, in which everything agreed excellently with the hypothesis—that heat is a substance. On the other hand, the invariability in the quantity of heat in all these processes could at that time be explained in no other manner than that heat is a substance.

But one relation of heat—namely, that to mechanical work—had not been accurately investigated. A French engineer, Sadi Carnot, son of the celebrated War Minister of the Revolution, had indeed endeavoured to deduce the work which heat performs, by assuming that the hypothetical caloric endeavoured to expand like a gas; and from this assumption he deduced in fact a remarkable law as to the capacity of heat for work, which even now, though with an essential alteration introduced by Clausius, is among the bases of the modern mechanical theory of heat, and the practical conclusions from which, so far as they could at that time be compared with experiments, have held good.

But it was already known that whenever two bodies in motion rubbed against each other, heat was developed anew, and it could not be said whence it came.

The fact is universally recognised; the axle of a carriage which is badly greased and where the friction is great, becomes hot—so hot, indeed, that it may take fire; machine-wheels with iron axles going at a great rate may become so hot that they weld to their sockets. A powerful degree of friction is not, indeed, necessary to disengage an appreciable degree of heat; thus, a lucifer-match, which by rubbing is so heated that the phosphoric mass ignites, teaches this fact. Nay, it is enough to rub the dry hands together to feel the heat produced by friction, and which is far greater than the heating which takes place when the hands lie gently on each other. Uncivilized people use the friction of two pieces of wood to kindle a fire. With this view, a sharp spindle of hard wood is made to revolve rapidly on a base of soft wood in the manner represented in Fig. 47.

Fig. 47

So long as it was only a question of the friction of solids, in which particles from the surface become detached and compressed, it might be supposed that some changes in structure of the bodies rubbed might here liberate latent heat, which would thus appear as heat of friction.

But heat can also be produced by the friction of liquids, in which there could be no question of changes in structure, or of the liberation of latent heat. The first decisive experiment of this kind was made by Sir Humphry Davy in the commencement of the present century. In a cooled space he made two pieces of ice rub against each other, and thereby caused them to melt. The latent heat which the newly formed water must have here assimilated could not have been conducted to it by the cold ice, or have been produced by a change of structure; it could have come from no other cause than from friction, and must have been created by friction.

Heat can also be produced by the impact of imperfectly elastic bodies as well as by friction. This is the case, for instance, when we produce fire by striking flint against steel, or when an iron bar is worked for some time by powerful blows of the hammer.

If we inquire into the mechanical effects of friction and of inelastic impact, we find at once that these are the processes by which all terrestrial movements are brought to rest. A moving body whose motion was not retarded by any resisting force would continue to move to all eternity. The motions of the planets are an instance of this. This is apparently never the case with the motion of the terrestrial bodies, for they are always in contact with other bodies which are at rest, and rub against them. We can, indeed, very much diminish their friction, but never completely annul it. A wheel which turns about a well-worked axle, once set in motion continues it for a long time; and the longer, the more truly and smoother the axle is made to turn, the better it is greased, and the less the pressure it has to support. Yet the *vis viva* of the motion which we have imparted to such a wheel when we started it, is gradually lost in consequence of friction. It disappears, and if we do not carefully consider the matter, it seems as if the *vis viva* which the wheel had possessed had been simply destroyed without any substitute.

A bullet which is rolled on a smooth horizontal surface continues to roll until its velocity is destroyed by friction on the path, caused by the very minute impacts on its little roughnesses.

A pendulum which has been put in vibration can continue to oscillate for hours if the suspension is good, without being driven by a weight; but by the friction against the surrounding air, and by that at its place of suspension, it ultimately comes to rest.

A stone which has fallen from a height has acquired a certain velocity on reaching the earth; this we know is the equivalent of a mechanical work; so long as this velocity continues as such, we can direct it upwards by means of suitable arrangements, and thus utilise it to raise the stone again. Ultimately the stone strikes against the earth and comes to rest; the impact has destroyed its velocity, and therewith apparently also the mechanical work which this velocity could have effected.

If we review the result of all these instances, which each of you could easily add to from your own daily experience, we shall see that friction and inelastic impact are processes in which mechanical work is destroyed, and heat produced in its place.

The experiments of Joule, which have been already mentioned, lead us a step further. He has measured in foot pounds the amount of work which is destroyed by the friction of solids and by the friction of liquids; and, on the other hand, he has determined the quantity of heat which is thereby produced, and has established a definite relation between the two. His experiments show that when heat is produced by the consumption of work, a definite quantity of work is required to produce that amount of heat which is known to physicists as the *unit of heat*; the heat, that is to say, which is necessary to raise one gramme of water through one degree centigrade. The quantity of work necessary for this is, according to Joule's best experiments, equal to the work which a gramme would perform in falling through a height of 425 metres.

In order to show how closely concordant are his numbers, I will adduce the results of a few series of experiments which he obtained after introducing the latest improvements in his methods.

1. A series of experiments in which water was heated by friction in a brass vessel. In the interior of this vessel a vertical axle provided with sixteen paddles was rotated, the eddies thus produced being broken by a series of projecting barriers, in which parts were cut out large enough for the paddles to pass through. The value of the equivalent was 424.9 metres.

2. Two similar experiments, in which mercury in an iron vessel was substituted for water in a brass one, gave 425 and 426.3 metres.

3. Two series of experiments, in which a conical ring rubbed against another, both surrounded by mercury, gave 426.7 and 425.6 metres.

Exactly the same relations between heat and work were also found in the reverse process—that is, when work was produced by heat. In order to execute this process under physical conditions that could be controlled as perfectly as possible, permanent gases and not vapours were used, although the latter are, in practice, more convenient for producing large quantities of work, as in the case of the steam-engine. A gas which is allowed to expand with moderate velocity becomes cooled. Joule was the first to show the reason of this cooling. For the gas has, in expanding, to overcome the resistance, which the pressure of the atmosphere and the slowly yielding side of the vessel oppose to it; or, if it cannot of itself overcome the resistance, it supports the arm of the observer which does it. Gas thus performs work, and this work is produced at the cost of its heat. Hence the cooling. If, on the contrary, the gas is suddenly allowed to issue into a perfectly exhausted space where it finds no resistance, it does not become cool as Joule has shown; or if individual parts of it become cool, others become warm; and, after the temperature has become equalised, this is exactly as much as before the sudden expansion of the gaseous mass.

How much heat the various gases disengage when they are compressed, and how much work is necessary for their compression; or, conversely, how much heat disappears when they expand under a pressure equal to their own counterpressure, and how much work they thereby effect in overcoming this counterpressure, was partly known from the older physical experiments, and has partly been determined by the recent experiments of Regnault by extremely perfect methods. Calculations with the best data of this kind give us the value of the thermal equivalent from experiments:—

With atmospheric air ....426.0 metres.
" oxygen ....425.7 "
" nitrogen ....431.3 "
" hydrogen ....425.3 "

Comparing these numbers with those which determine the equivalence of heat and mechanical work in friction, as close an agreement is seen as can at all be expected from numbers which have been obtained by such varied investigations of different observers.

Thus then: a certain quantity of heat may be changed into a definite quantity of work; this quantity of work can also be retransformed into heat, and, indeed, into exactly the same quantity of heat as that from which it originated; in a mechanical point of view, they are exactly equivalent. Heat is a new form in which a quantity of work may appear.

These facts no longer permit us to regard heat as a substance, for its quantity is not unchangeable. It can be produced anew from the *vis viva* of motion destroyed; it can be destroyed, and then produces motion. We must rather conclude from this that heat itself is a motion, an internal invisible motion of the smallest elementary particles of bodies. If, therefore, motion seems lost in friction and impact, it is not actually lost, but only passes from the great visible masses to their smallest particles; while in steam-engines the internal motion of the heated gaseous particles is transferred to the piston of the machine, accumulated in it, and combined in a resultant whole.

But what is the nature of this internal motion, can only be asserted with any degree of probability in the case of gases. Their particles probably cross one another in rectilinear paths in all directions, until, striking another particle, or against the side of the vessel, they are reflected in another direction. A gas would thus be analogous to a swarm of gnats, consisting, however, of particles infinitely small and infinitely more closely packed. This hypothesis, which has been developed by Krönig, Clausius, and Maxwell, very well accounts for all the phenomena of gases.

What appeared to the earlier physicists to be the constant quantity of heat is nothing more than the whole motive power of the motion of heat, which remains constant so long as it is not transformed into other forms of work, or results afresh from them.

We turn now to another kind of natural forces which can produce work—I mean the chemical. We have to-day already come across them. They are the ultimate cause of the work which gunpowder and the steam-engine produce; for the heat which is consumed in the latter, for example, originates in the combustion of carbon—that is to say, in a chemical process. The burning of coal is the chemical union of carbon with the oxygen of the air, taking place under the influence of the chemical affinity of the two substances.

We may regard this force as an attractive force between the two, which, however, only acts through them with extraordinary power, if the smallest particles of the two substances are in closest proximity to each other. In combustion this force acts; the carbon and oxygen atoms strike against each other and adhere firmly, inasmuch as they form a new compound—carbonic acid—a gas known to all of you as that which ascends from all fermenting and fermented liquids—from beer and champagne. Now this attraction between the atoms of carbon and of oxygen performs work just as much as that which the earth in the form of gravity exerts upon a raised weight. When the weight falls to the ground, it produces an agitation, which is partly transmitted to the vicinity as sound waves, and partly remains as the motion of heat. The same result we must expect from chemical action. When carbon and oxygen atoms have rushed against each other, the newly-formed particles of carbonic acid must be in the most violent molecular motion—that is, in the motion of heat. And this is so. A pound of carbon burned with oxygen to form carbonic acid, gives as much heat as is necessary to raise 80.9 pounds of water from freezing to the boiling point; and just as the same amount of work is produced when a weight falls, whether it falls slowly or fast, so also the same quantity of heat is produced by the combustion of carbon, whether this is slow or rapid, whether it takes place all at once, or by successive stages.

When the carbon is burned, we obtain in its stead, and in that of the oxygen, the gaseous product of combustion—carbonic acid. Immediately after combustion it is incandescent. When it has afterwards imparted heat to the vicinity, we have in the carbonic acid the entire quantity of carbon and the entire quantity of oxygen, and also the force of affinity quite as strong as before. But the action of the latter is now limited to holding the atoms of carbon and oxygen firmly united; they can no longer produce either heat or work any more than a fallen weight can do work if it has not been again raised by some extraneous force. When the carbon has been burnt we take no further trouble to retain the carbonic acid; it can do no more service, we endeavour to get it out of the chimneys of our houses as fast as we can.

It is possible, then, to tear asunder the particles of carbonic acid, and give to them once more the capacity of work which they had before they were combined, just as we can restore the potentiality of a weight by raising it from the ground? It is indeed possible. We shall afterwards see how it occurs in the life of plants; it can also be effected by inorganic processes, though in roundabout ways, the explanation of which would lead us too far from our present course.

This can, however, be easily and directly shown for another element, hydrogen, which can be burnt just like carbon. Hydrogen with carbon is a constituent of all combustible vegetable substances, among others, it is also an essential constituent of the gas which is used for lighting our streets and rooms; in the free state it is also a gas, the lightest of all, and burns when ignited with a feebly luminous blue flame. In this combustion—that is, in the chemical combination of hydrogen with oxygen, a very considerable quantity of heat is produced; for a given weight of hydrogen, four times as much heat as in the combustion of the same weight of carbon. The product of combustion is water, which, therefore, is not of itself further combustible, for the hydrogen in it is completely saturated with oxygen. The force of affinity, therefore, of hydrogen for oxygen, like that of carbon for oxygen, performs work in combustion, which appears in the form of heat. In the water which has been formed during combustion, the force of affinity is exerted between the elements as before, but its capacity for work is lost. Hence the two elements must be again separated, their atoms torn apart, if new effects are to be produced from them.

This we can do by the aid of currents of electricity. In the apparatus depicted in Fig. 48, we have two glass vessels filled with acidulated water, a and a', which are separated in the middle by a porous plate moistened with water. In both sides are fitted platinum wires, k, which are attached to platinum plates, i and i'. As soon as a galvanic current is transmitted through the water by the platinum wires, k, you see bubbles of gas ascend from the plates i and i'. These bubbles are the two elements of water, hydro-

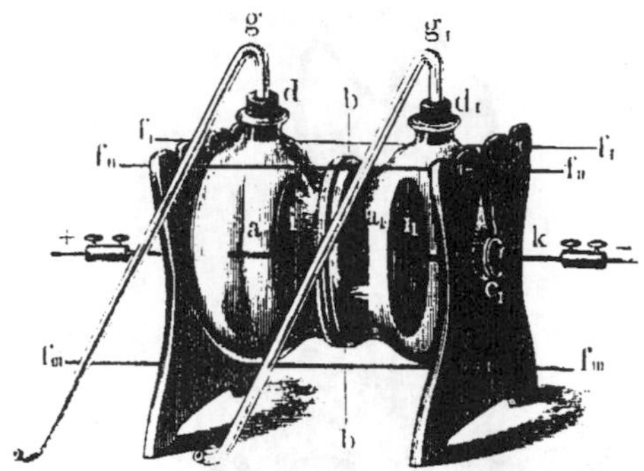

Fig. 48

Fig. 49

gen on the one hand, and oxygen on the other. The gases emerge through the tubes g and g′. If we wait until the upper part of the vessels and the tubes have been filled with it, we can inflame hydrogen at one side; it burns with a blue flame. If I bring a glimmering spill near the mouth of the other tube it bursts into flame, just as happens with oxygen gas, in which the processes of combustion are far more intense than in atmospheric air, where the oxygen mixed with nitrogen is only one-fifth of the whole volume.

If I hold a glass flask filled with water over the hydrogen flame, the water, newly formed in combustion, condenses upon it.

If a platinum wire be held in the almost nonluminous flame, you see how intensely it is ignited; in a plentiful current of a mixture of the gases, hydrogen and oxygen, which have been liberated in the above experiment, the almost infusible platinum might even be melted. The hydrogen which has here been liberated from the water by the electrical current has regained the capacity of producing large quantities of heat by a fresh combination with oxygen; its affinity for oxygen has regained for it its capacity for work.

We here become acquainted with a new source of work, the electric current which decomposes water. This current is itself produced by a galvanic battery, Fig. 49. Each of the four vessels contains nitric acid, in which there is a hollow cylinder of very compact carbon. In the middle of the carbon cylinder is a cylindrical porous vessel of white clay, which contains dilute sulphuric acid; in this dips a zinc cylinder. Each zinc cylinder is connected by a metal ring with the carbon cylinder of the next vessel, the last zinc cylinder n is connected with one platinum plate, and the first carbon cylinder, p, with the other platinum plate of the apparatus for the decomposition of water. If now the conducting circuit of this galvanic apparatus is completed, and the decomposition of water begins, a chemical process takes place simultaneously in the cells of the voltaic battery. Zinc takes oxygen from the surrounding water and undergoes a slow combustion. The product of combustion thereby produced, oxide of zinc, unites further with sulphuric acid, for which it has a powerful affinity, and sulphate of zinc, a saline kind of substance, dissolves in the liquid. The oxygen, moreover, which is withdrawn from it is taken by the water from the nitric acid surrounding the cylinder of carbon, which is very rich in it, and readily gives it up. Thus, in the galvanic battery zinc burns to sulphate of zinc at the cost of the oxygen of nitric acid.

Thus, while one product of combustion, water, is again separated, a new combustion is taking place—that of zinc. While we there reproduce chemical affinity which is capable of work, it is here lost. The electrical current is, as it were, only the carrier which transfers the chemical force of the zinc uniting with oxygen and acid to water in the decomposing cell, and uses it for overcoming the chemical force of hydrogen and oxygen.

In this case, we can restore work which has

been lost, but only by using another force, that of oxidizing zinc.

Here we have overcome chemical forces by chemical forces, through the instrumentality of the electrical current. But we can attain the same object by mechanical forces, if we produce the electrical current by a magneto-electrical machine, Fig. 50. If we turn the handle, the anker RR′, on which is coiled copper wire, rotates in from of the poles of the horse-shoe magnet, and in these coils electrical currents are produced, which can be led from the points a and b. If the ends of these wires are connected with the apparatus for decomposing water we obtain hydrogen and oxygen, though in far smaller quantity than by the aid of the battery which we used before. But this process is interesting, for the mechanical force of the arm which turns the wheel produces the work which is required for separating the combined chemical elements. Just as the steam-engine changes chemical into mechanical force, the magneto-electrical machine transforms mechanical force into chemical.

Fig. 50

The application of electrical currents opens out a large number of relations between the various natural forces. We have decomposed water into its elements by such currents, and should be able to decompose a large number of other chemical compounds. On the other hand, in ordinary galvanic batteries electrical currents are produced by chemical forces.

In all conductors through which electrical currents pass they produce heat; I stretch a thin platinum wire between the ends n and p of the galvanic battery, Fig. 49; it becomes ignited and melts. On the other hand, electrical currents are produced by heat in what are called thermo-electric elements.

Fig. 51

Iron which is brought near a spiral of copper wire, traversed by an electrical current, becomes magnetic, and then attracts other pieces of iron, or a suitably placed steel magnet. We thus obtain mechanical actions which meet with extended applications in the electrical telegraph, for instance. Fig. 51 represents a Morse's telegraph in one-third of the natural size. The essential part is a horse-shoe shaped iron core, which stands in the copper spirals bb. Just over the top of this is a small steel magnet cc, which is attracted the moment an electrical current, arriving by the telegraph wire, traverses the spirals bb. The magnet cc is rigidly fixed in the lever dd, at the other end of which is a style; this makes a mark on a paper band, drawn by a clock-work, as often and as long as cc is attracted by the magnetic action of the electric current. Conversely, by reversing the magnetism in the iron core of the spirals bb, we should obtain in them an electrical current just as we have obtained such currents in the magneto-electrical machine, Fig. 50; in the spirals of that machine there is an iron core which, by being approached to the poles of the large horse-shoe magnet, is sometimes magnetised in one and sometimes in the other direction.

I will not accumulate examples of such relations; in subsequent lectures we shall come across them. Let us review these examples once more, and recognise in them the law which is common to all.

A raised weight can produce work, but in doing so it must necessarily sink from its height, and, when it has fallen as deep as it can fall, its gravity remains as before, but it can no longer do work.

A stretched spring can do work, but in so doing it becomes loose. The velocity of a moving mass can do work, but in doing so it comes to rest. Heat can perform work; it is destroyed in the operation. Chemical forces can perform work, but they exhaust themselves in the effort.

Electrical currents can perform work, but to keep them up we must consume either chemical or mechanical forces, or heat.

We may express this generally. *It is a universal character of all known natural forces that their capacity for work is exhausted in the degree in which they actually perform work.*

We have seen, further, that when a weight fell without performing any work, it *either* acquired velocity or produced heat. We might also drive a magneto-electrical machine by a falling weight; it would then furnish electrical currents.

We have seen that chemical forces, when they come into play, produce either heat or electrical currents or mechanical work.

We have seen that heat may be changed into work; there are apparatus (thermo-electric batteries) in which electrical currents are produced by it. Heat can directly separate chemical compounds; thus, when we burn limestone, it separates carbonic acid from lime.

Thus, whenever the capacity for work of one natural force is destroyed, it is transformed into another kind of activity. Even within the circuit of inorganic natural forces, we can transform each of them into an active condition by the aid of any other natural force which is capable of work. The connections between the various natural forces which modern physics has revealed, are so extraordinarily numerous that several entirely different methods may be discovered for each of these problems.

I have stated how we are accustomed to measure mechanical work, and how the equivalent in work of heat may be found. The equivalent in work of chemical processes is again measured by the heat which they produce. By similar relations, the equivalent in work of the other natural forces may be expressed in terms of mechanical work.

If, now, a certain quantity of mechanical work is lost, there is obtained, as experiments made with the object of determining this point show, an equivalent quantity of heat, or, instead of this, of chemical force; and, conversely, when heat is lost, we gain an equivalent quantity of chemical or mechanical force; and, again, when chemical force disappears, an equivalent of heat or work; so that in all these interchanges between various inorganic natural forces working force may indeed disappear in one form, but then it reappears in exactly equivalent quantity in some other form; it is thus neither increased nor diminished, but always remains in exactly the same quantity. We shall subsequently see that the same law holds good also for processes in organic nature, so far as the facts have been tested.

It follows thence *that the total quantity of all the forces* capable of work *in the whole universe remains eternal and unchanged throughout all their changes.* All change in nature amounts to this, that force can change its form and locality without its quantity being changed. The universe possesses, once for all, a store of force which is not altered by any change of phenomena, can neither be increased nor diminished, and which maintains any change which takes place on it.

You see how, starting from considerations based on the immediate practical interests of technical work, we have been led up to a universal natural law, which, as far as all previous experience extends, rules and embraces all natural processes; which is no longer restricted to the practical objects of human utility, but expresses a perfectly general and particularly characteristic property of all natural forces, and which, as regards generality, is to be placed by the side of the laws of the unalterability of mass, and the unalterability of the chemical elements.

At the same time, it also decides a great practical question which has been much discussed in the last two centuries, to the decision of which an infinity of experiments have been made and an infinity of apparatus constructed—that is, the question of the possibility of a perpetual motion. By this was understood a machine which was to work continuously without the aid of any external driving force. The solution of this problem promised enormous gains. Such a machine would have had all the advantages of steam without requiring the expenditure of fuel. Work is wealth. A machine which could produce work from nothing was as good as one which made gold. This problem had thus for a long time occupied the place of gold making, and had confused many a pondering brain. That a perpetual motion could not be produced by the aid of the then known mechanical forces could be demonstrated in the last century by the aid of the mathematical mechanics which had at that time been developed. But to show also that it is not possible even if heat, chemical forces, electricity, and magnetism were made to co-operate, could not be done without a knowledge of our law in all its generality. The possibility of a perpetual motion was first finally negatived by the law of the conservation of force, and this law might also be expressed in the practical form that no perpetual motion is possible, that force cannot be produced from nothing, something must be consumed.

You will only be ultimately able to estimate the importance and the scope of our law when you have before your eyes a series of its applications to individual processes on nature.

What I have to-day mentioned as to the origin of the moving forces which are at our disposal, directs us to something beyond the narrow confines of our laboratories and our manufactories, to the great operations at work in the life of the earth and of the universe. The force of falling water can only flow down from the hills when rain and snow bring it to them. To furnish these, we must have aqueous vapour in the atmosphere, which can only be effected by the aid of heat, and this heat comes from the sun. The steam-engine needs the fuel which the vegetable life yields, whether it be the still active life of the surrounding vegetation, or the extinct life which has produced the immense coal deposits in the depths of the earth. The forces of man and animals must be restored by nourishment; all nourishment comes ultimately from the vegetable kingdom, and leads us back to the same source.

You see then that when we inquire into the origin of the moving forces which we take into our service, we are thrown back upon the meteorological processes in the earth's atmosphere, on the life of plants in general, and on the sun.

Introduction to a Series of Lectures Delivered at Carlsruhe in the Winter of 1862–1863.

FURTHER READING

Cahan, David, ed. *Hermann von Helmholtz and the Foundations of Nineteenth-Century Science.* Berkeley: University of California Press, 1993.

Caneva, Kenneth L. *Robert Mayer and the Conservation of Energy.* Princeton: Princeton UP, 1993.

Ebert, Hermann. *Hermann von Helmholtz.* Stuttgart: Wissenschaftliche Verlagsgesellschaft m.b.H., 1949.

Eckart, Wolfgang U., and Klaus Volkert, eds. *Hermann von Helmholtz: Vorträge eines Heidelberger Symposiums anläßlich des einhundertsten Todestages.* Pfaffenweiler: Centaurus-Verlagsgesellschaft, 1996.

Hatfield, Gary. *The Natural and the Normative: Theories of Spatial Perception from Kant to Helmholtz.* Cambridge, MA: M.I.T. Press, 1990.

Heimann, Peter M. "Helmholtz and Kant: The Metaphysical Foundations of 'Über die Erhaltung der Kraft.'" *Studies in History and Philosophy of Science* 5, no. 3 (1974): 205–38.

Helmholtz, Hermann von. *Abhandlungen zur Thermodynamik.* Edited by Max Planck. Leipzig: Akademische Verlagsgesellschaft m.b.H., 1921.

—. *Introduction to Physiological Optics.* Translated by James P.C. Southall. London: Oxford UP, 1937.

—. *On the Sensations of Tone as a Physiological Basis for the Theory of Music.* Translated by Alexander J. Ellis. Introduced by Henry Mar-

genau. New York: Dover Publications, Inc., 1954.
—. *Philosophische Vorträge und Aufsätze.* Edited by Herbert Hörz and Siegfried Wollgast. Berlin: Akademie-Verlag, 1971.
—. *Popular Lectures on Scientific Subjects.* Translated by E. Atkinson *et al.* Introduced by Professor Tyndall. 2 vols. London: Longmans, Green, and Co., 1873.
—. *Science and Culture: Popular and Philosophical Essays.* Edited by David Cahan. Chicago: University of Chicago Press, 1995.
—. *Selected Writings of Hermann von Helmholtz.* Edited by Russell Kahl. Middletown, CT: Wesleyan UP, 1971.
—. *Über die Erhaltung der Kraft* [facsimile manuscript]. Weinheim: Physik-Verlag, 1983.
—. *Wissenschaftliche Abhandlungen von Hermann Helmholtz.* 3 vols. Leipzig: Johann Ambrosius Barth, 1882–95.
Koenigsberger, Leo. *Hermann von Helmholtz.* Translated by Frances A. Welby. Oxford: Clarendon Press, 1906.
—. *Hermann von Helmholtz's Untersuchungen über die Grundlagen der Mathematik und Mechanik.* Leipzig: B.G. Teubner, 1896.
Krüger, Lorenz, ed. *Universalgenie Helmholtz: Rückblick nach 100 Jahren.* Berlin: Akademie Verlag, 1994.
Kuhn, Thomas S. "Energy Conservation as an Example of Simultaneous Discovery." In Marshall Clagett, ed. *Critical Problems in the History of Science.* Madison, WI: University of Wisconsin Press, 1969, pp. 321–356.
Mach, Ernst. *History and Root of the Principle of the Conservation of Energy.* Translated by Philip E.B. Jourdain. Chicago: Open Court Publishing Co., 1911.
M'Kendrick, John Gray. *Hermann Ludwig Ferdinand von Helmholtz.* New York: Longmans, Green and Co., 1899.
Ordóñez, Javier. "The Story of a Non-Discovery: Helmholtz and the Conservation of Energy." *Spanish Studies in the Philosophy of Science.* Edited by Gonzalo Munévar. Boston Studies in the Philosophy of Science, vol. 186. Dordrecht: Kluwer Academic Publishers, 1996.
Planck, Max. *Das Prinzip der Erhaltung der Energie.* Leipzig: B.G. Teubner, 1887.
Rechenberg, Helmut. *Hermann von Helmholtz: Bilder seines Lebens und Wirkens.* Weinheim: VCH, 1994.
Reiner, Julius. *Hermann von Helmholtz.* Leipzig: T. Thomas, 1905.
Turner, R. Steven. *In the Eye's Mind: Vision and the Helmholtz–Hering Controversy.* Princeton: Princeton UP, 1994.
Warren, Richard M., and Roslyn P. Warren. *Helmholtz on Perception: Its Physiology and Development.* New York: John Wiley and Sons, Inc., 1968.

24. JAMES CLERK MAXWELL

(1831–1879)

Like Lord Kelvin of Glasgow, James Clerk Maxwell was another mathematically-trained Scottish electrician and physicist who substantially influenced the development of Faraday's field theory of electromagnetism. Born in 1831 in Edinburgh, Maxwell enrolled at Edinburgh University at age 16 and was acquainted with a number of prominent Scottish scientists from an early age, including Kelvin, James David Forbes, and Peter Guthrie Tait. At the age of 14, he developed a new geometrical method of generating perfect oval curves, publishing his results with the help of Forbes in the *Proceedings of the Royal Society of Edinburgh* for 1846. After election to fellowship at Trinity College, Cambridge in 1855 he began teaching hydrostatics and optics, but returned to Scotland to teach at Marischal College in Aberdeen in 1856. In 1860, he was appointed to a professorship at King's College, London.

When the Cambridge Cavendish Laboratory opened in 1871, Maxwell was invited to serve as the first Cavendish Professor of Physics after Kelvin and Helmholtz had declined the position. Maxwell personally designed the equipment and instrumentation and heavily influenced the research program of the laboratory. Maxwell completed his frequently reprinted *Treatise on Heat* in the same year, and published his widely read textbook *Treatise on Electricity and Magnetism* in 1873. He also edited the electrical papers of Henry Cavendish (1879). Maxwell contributed several scientific articles containing considerable technical detail, such as "Constitution of Bodies," "Diffusion," and "Atom," etc., to the ninth edition of the *Encyclopaedia Britannica,* which he co-edited with T.H. Huxley. These articles were reprinted in the 1890 Cambridge edition of Maxwell's *Scientific Papers*. He also published substantial research on thermodynamics, the molecular theory of gases, colour vision (developing quantitative colorimetry by resurrecting Thomas Young's three-receptor model), and theoretical work on the rings of Saturn. Using his colour theory, he produced the first colour photograph in 1861 by projection through coloured filters.

In order to illustrate principles of diffusion and gas thermodynamics, Maxwell invented the celebrated thought experiment "Maxwell's Demon," which Kelvin (who gave the demon its name) described as "a creature of imagination having certain perfectly well-defined powers of action, purely mechanical in their character, invented to help us to understand the 'Dissipation of Energy' in nature" (*Popular Lectures* 1.144). With the demon, Maxwell demonstrated that the Second Law of Thermodynamics, which supports the observation that heat naturally flows from hot to cold bodies, could only be true statistically, since in individual molecular collisions, hot bodies can derive heat from colder ones.

Maxwell's work on electromagnetism can be viewed in many respects as a mathematization and elaboration of ideas advanced by Faraday's experimental work, and illustrates a crucial point in nineteenth-century physics when natural language and the language of mathematics in describing physical phenomena parted company; for example, in his paper "On Faraday's Lines of Force" (1855), Maxwell sought to express mathematically (using an analogy to fluid flow) Faraday's graphical observation of curvilinear magnetic forces operating along lines of iron filings scattered on paper and under the influence of magnets. Upon reading Maxwell's papers, Faraday wrote to Maxwell, "when a mathematician engaged in investigating physical actions and results has arrived at his conclusions, may they not be expressed in common language as fully, clearly and definitely as in mathematical formulae?" (Campbell 290). Maxwell later wrote in the *Treatise on Electricity and Magnetism,* "[Faraday] did not feel called upon either to force his results into a shape acceptable to the mathematical taste of the time, or to express them in a form which mathematicians might attack ... it is mainly in the hope of making these ideas the basis of a mathematical method that I have undertaken this treatise" (176).

"Maxwell's equations," as they have been called after refinement by Oliver Heaviside and Heinrich Hertz, were developed to understand the relationship between electrical and magnetic phenomena in magnetic and electrical fields. The equations grew out of Maxwell's continuing attempts to describe Faraday's fields mathematically. Drawing on a similar model of gases proposed by William Rankine, Maxwell in his paper "On Physical Lines of Force" (1861–62), constructed an elaborate mechanical model of spinning molecular vortices within a fluid medium—the ether—which he likened to mechanical cogs, wheels, and idle wheels. The mechanical properties of the rotating vortices—such as their angular velocity and deformation—could explain by analogy many magnetostatic, electrostatic, electrodynamic, and magnetodynamic effects. Maxwell's model eventually accounted for all of the then-known experimentally derived laws of electricity and magnetism, and predicted others.

Faraday had demonstrated that a magnetic field could shift the plane of polarization of light (the Faraday Effect) suggesting a link between light and magnetism. Maxwell discovered wave properties in his mechanical field model and further found that he could describe light as the elastic displacement of the medium (ether) describing transverse shear waves. In his 1864 paper entitled "A Dynamic Theory of the Electromagnetic Field," of which Part VI "Electromagnetic Theory of Light" is reproduced below, Maxwell working from his model hypothesized that light was an electromagnetic wave disturbance propagating through a field. He found that his calculation for the velocity of propagation of these transverse waves in the ether closely matched the experimentally measured velocity of light. In 1888, Henrich Hertz experimentally verified the existence of electric waves (radio and microwaves) possessing all of the normal optical properties of light such as reflection, refraction, diffraction, etc.

Thus, Maxwell's equations for the spatial and interactive properties of electric and magnetic fields in space, serve to define light and other forms of radiation as electromagnetic waves. Electromagnetic waves exist in a continuous spectrum ranging from the long wavelength and low-frequency radio waves at one end of the scale, to microwaves, infrared radiation, visible light, ultraviolet light, X rays, and finally at the other end of the spectrum to high frequency, short wave-length gamma rays. Although there were many researchers working in the field of electromagnetism at the same time as Maxwell, such as Hermann Helmholtz, Michael Faraday, Hendrik Lorentz, Ludwig Boltzmann, and Wilhelm Eduard Weber, Maxwell's work still stands out as highly original: it is difficult to conceive of modern electronic technology such as radio, television, microwave transmission, or even the atomic bomb without Maxwell's insights. In order to reduce transcription errors, the selection below has been reproduced in photographic facsimile from the 1890 edition of Maxwell's collected papers.

"A Dynamical Theory of the Electromagnetic Field."
Royal Society Transactions, vol. CLV, *1864*.
Reprinted in W.D. Niven, ed., *The Scientific Papers of James Clerk Maxwell*.
2 vols. Cambridge: Cambridge University Press, 1890, 1.577–88.

PART VI.

ELECTROMAGNETIC THEORY OF LIGHT.

(91) At the commencement of this paper we made use of the optical hypothesis of an elastic medium through which the vibrations of light are propagated, in order to shew that we have warrantable grounds for seeking, in the same medium, the cause of other phenomena as well as those of light. We then examined electromagnetic phenomena, seeking for their explanation in the properties of the field which surrounds the electrified or magnetic bodies. In this way we arrived at certain equations expressing certain properties of the electromagnetic field. We now proceed to investigate whether these properties of that which constitutes the electromagnetic field, deduced from electromagnetic phenomena alone, are sufficient to explain the propagation of light through the same substance.

(92) Let us suppose that a plane wave whose direction cosines are l, m, n is propagated through the field with a velocity V. Then all the electromagnetic functions will be functions of

$$w = lx + my + nz - Vt.$$

The equations of Magnetic Force (B), p. 556, will become

$$\mu\alpha = m\frac{dH}{dw} - n\frac{dG}{dw},$$

$$\mu\beta = n\frac{dF}{dw} - l\frac{dH}{dw},$$

$$\mu\gamma = l\frac{dG}{dw} - m\frac{dF}{dw}.$$

If we multiply these equations respectively by l, m, n, and add, we find

$$l\mu\alpha + m\mu\beta + n\mu\gamma = 0 \quad\text{.............................. (62)},$$

which shews that the direction of the magnetization must be in the plane of the wave.

(93) If we combine the equations of Magnetic Force (B) with those of Electric Currents (C), and put for brevity

$$\frac{dF}{dx}+\frac{dG}{dy}+\frac{dH}{dz}=J, \text{ and } \frac{d^2}{dx^2}+\frac{d^2}{dy^2}+\frac{d^2}{dz^2}=\nabla^2 \quad\text{............ (63)},$$

$$\left.\begin{aligned} 4\pi\mu p' &= \frac{dJ}{dx}-\nabla^2 F \\ 4\pi\mu q' &= \frac{dJ}{dy}-\nabla^2 G \\ 4\pi\mu r' &= \frac{dJ}{dz}-\nabla^2 H \end{aligned}\right\} \quad\text{......................... (64).}$$

If the medium in the field is a perfect dielectric there is no true conduction, and the currents p', q', r' are only variations in the electric displacement, or, by the equations of Total Currents (A),

$$p'=\frac{df}{dt}, \qquad q'=\frac{dg}{dt}, \qquad r'=\frac{dh}{dt} \quad\text{...................... (65).}$$

But these electric displacements are caused by electromotive forces, and by the equations of Electric Elasticity (E),

$$P=kf, \qquad Q=kg, \qquad R=kh \quad\text{........................ (66).}$$

These electromotive forces are due to the variations either of the electromagnetic or the electrostatic functions, as there is no motion of conductors in the field; so that the equations of electromotive force (D) are

$$\left.\begin{aligned} P &= -\frac{dF}{dt}-\frac{d\Psi}{dx} \\ Q &= -\frac{dG}{dt}-\frac{d\Psi}{dy} \\ R &= -\frac{dH}{dt}-\frac{d\Psi}{dz} \end{aligned}\right\} \quad\text{.............................. (67).}$$

(94) Combining these equations, we obtain the following:—

$$\left.\begin{aligned} k\left(\frac{dJ}{dx}-\nabla^2 F\right)+4\pi\mu\left(\frac{d^2F}{dt^2}+\frac{d^2\Psi}{dxdt}\right)&=0 \\ k\left(\frac{dJ}{dy}-\nabla^2 G\right)+4\pi\mu\left(\frac{d^2G}{dt^2}+\frac{d^2\Psi}{dydt}\right)&=0 \\ k\left(\frac{dJ}{dz}-\nabla^2 H\right)+4\pi\mu\left(\frac{d^2H}{dt^2}+\frac{d^2\Psi}{dzdt}\right)&=0 \end{aligned}\right\} \quad\text{.................(68).}$$

If we differentiate the third of these equations with respect to y, and the second with respect to z, and subtract, J and Ψ disappear, and by remembering the equations (B) of magnetic force, the results may be written

$$\left.\begin{aligned} k\nabla^2\mu\alpha &= 4\pi\mu\frac{d^2}{dt^2}\mu\alpha \\ k\nabla^2\mu\beta &= 4\pi\mu\frac{d^2}{dt^2}\mu\beta \\ k\nabla^2\mu\gamma &= 4\pi\mu\frac{d^2}{dt^2}\mu\gamma \end{aligned}\right\} \qquad (69).$$

(95) If we assume that α, β, γ are functions of $lx+my+nz-Vt=w$, the first equation becomes

$$k\mu\frac{d^2\alpha}{dw^2} = 4\pi\mu^2V^2\frac{d^2\alpha}{dw^2} \qquad (70),$$

or

$$V = \pm\sqrt{\frac{k}{4\pi\mu}} \qquad (71).$$

The other equations give the same value for V, so that the wave is propagated in either direction with a velocity V.

This wave consists entirely of magnetic disturbances, the direction of magnetization being in the plane of the wave. No magnetic disturbance whose direction of magnetization is not in the plane of the wave can be propagated as a plane wave at all.

Hence magnetic disturbances propagated through the electromagnetic field agree with light in this, that the disturbance at any point is transverse to the direction of propagation, and such waves may have all the properties of polarized light.

(96) The only medium in which experiments have been made to determine the value of k is air, in which $\mu=1$, and therefore, by equation (46),

$$V = v \qquad (72).$$

By the electromagnetic experiments of MM. Weber and Kohlrausch *,

$$v = 310{,}740{,}000 \text{ metres per second}$$

* *Leipzig Transactions*, Vol. v. (1857), p. 260, or Poggendorff's *Annalen*, Aug. 1856, p. 10.

is the number of electrostatic units in one electromagnetic unit of electricity, and this, according to our result, should be equal to the velocity of light in air or vacuum.

The velocity of light in air, by M. Fizeau's * experiments, is

$$V = 314,858,000\,;$$

according to the more accurate experiments of M. Foucault †,

$$V = 298,000,000.$$

The velocity of light in the space surrounding the earth, deduced from the coefficient of aberration and the received value of the radius of the earth's orbit, is

$$V = 308,000,000.$$

(97) Hence the velocity of light deduced from experiment agrees sufficiently well with the value of v deduced from the only set of experiments we as yet possess. The value of v was determined by measuring the electromotive force with which a condenser of known capacity was charged, and then discharging the condenser through a galvanometer, so as to measure the quantity of electricity in it in electromagnetic measure. The only use made of light in the experiment was to see the instruments. The value of V found by M. Foucault was obtained by determining the angle through which a revolving mirror turned, while the light reflected from it went and returned along a measured course. No use whatever was made of electricity or magnetism.

The agreement of the results seems to shew that light and magnetism are affections of the same substance, and that light is an electromagnetic disturbance propagated through the field according to electromagnetic laws.

(98) Let us now go back upon the equations in (94), in which the quantities J and Ψ occur, to see whether any other kind of disturbance can be propagated through the medium depending on these quantities which disappeared from the final equations.

* *Comptes Rendus*, Vol. XXIX. (1849), p. 90.

† Ibid. Vol. LV. (1862), pp. 501, 792.

If we determine χ from the equation

$$\nabla^2\chi = \frac{d^2\chi}{dx^2} + \frac{d^2\chi}{dy^2} + \frac{d^2\chi}{dz^2} = J \qquad (73),$$

and F', G', H' from the equations

$$F' = F - \frac{d\chi}{dx}, \quad G' = G - \frac{d\chi}{dy}, \quad H' = H - \frac{d\chi}{dz} \qquad (74),$$

then

$$\frac{dF'}{dx} + \frac{dG'}{dy} + \frac{dH'}{dz} = 0 \qquad (75),$$

and the equations in (94) become of the form

$$k\nabla^2 F' = 4\pi\mu\left\{\frac{d^2F'}{dt^2} + \frac{d}{dxdt}\left(\Psi + \frac{d\chi}{dt}\right)\right\} \qquad (76).$$

Differentiating the three equations with respect to x, y, and z, and adding, we find that

$$\Psi = -\frac{d\chi}{dt} + \phi(x, y, z) \qquad (77),$$

and that

$$\left.\begin{aligned} k\nabla^2 F' &= 4\pi\mu\frac{d^2F'}{dt^2} \\ k\nabla^2 G' &= 4\pi\mu\frac{d^2G'}{dt^2} \\ k\nabla^2 H' &= 4\pi\mu\frac{d^2H'}{dt^2} \end{aligned}\right\} \qquad (78).$$

Hence the disturbances indicated by F', G', H' are propagated with the velocity $V = \sqrt{\frac{k}{4\pi\mu}}$ through the field; and since

$$\frac{dF'}{dx} + \frac{dG'}{dy} + \frac{dH'}{dz} = 0,$$

the resultant of these disturbances is in the plane of the wave.

(99) The remaining part of the total disturbances F, G, H being the part depending on χ, is subject to no condition except that expressed in the equation

$$\frac{d\Psi}{dt} + \frac{d^2\chi}{dt^2} = 0.$$

If we perform the operation ∇^2 on this equation, it becomes

$$ke = \frac{dJ}{dt} - k\nabla^2\phi\,(x,\ y,\ z)\ldots\ldots\ldots\ldots\ldots\ldots\ldots\ldots(79).$$

Since the medium is a perfect insulator, e, the free electricity, is immoveable, and therefore $\frac{dJ}{dt}$ is a function of x, y, z, and the value of J is either constant or zero, or uniformly increasing or diminishing with the time; so that no disturbance depending on J can be propagated as a wave.

(100) The equations of the electromagnetic field, deduced from purely experimental evidence, shew that transversal vibrations only can be propagated. If we were to go beyond our experimental knowledge and to assign a definite density to a substance which we should call the electric fluid, and select either vitreous or resinous electricity as the representative of that fluid, then we might have normal vibrations propagated with a velocity depending on this density. We have, however, no evidence as to the density of electricity, as we do not even know whether to consider vitreous electricity as a substance or as the absence of a substance.

Hence electromagnetic science leads to exactly the same conclusions as optical science with respect to the direction of the disturbances which can be propagated through the field; both affirm the propagation of transverse vibrations, and both give the same velocity of propagation. On the other hand, both sciences are at a loss when called on to affirm or deny the existence of normal vibrations.

Relation between the Index of Refraction and the Electromagnetic Character of the substance.

(101) The velocity of light in a medium, according to the Undulatory Theory, is

$$\frac{1}{i}V_0,$$

where i is the index of refraction and V_0 is the velocity in vacuum. The velocity, according to the Electromagnetic Theory, is

$$\sqrt{\frac{k}{4\pi\mu}},$$

where, by equations (49) and (71), $k = \frac{1}{D}k_0$, and $k_0 = 4\pi V_0^2$.

Hence $$D=\frac{i^2}{\mu} \qquad \text{(80)},$$

or the Specific Inductive Capacity is equal to the square of the index of refraction divided by the coefficient of magnetic induction.

Propagation of Electromagnetic Disturbances in a Crystallized Medium.

(102) Let us now calculate the conditions of propagation of a plane wave in a medium for which the values of k and μ are different in different directions. As we do not propose to give a complete investigation of the question in the present imperfect state of the theory as extended to disturbances of short period, we shall assume that the axes of magnetic induction coincide in direction with those of electric elasticity.

(103) Let the values of the magnetic coefficient for the three axes be λ, μ, ν, then the equations of magnetic force (B) become

$$\left.\begin{aligned} \lambda\alpha &= \frac{dH}{dy}-\frac{dG}{dz} \\ \mu\beta &= \frac{dF}{dz}-\frac{dH}{dx} \\ \nu\gamma &= \frac{dG}{dx}-\frac{dF}{dy} \end{aligned}\right\} \qquad \text{(81)}.$$

The equations of electric currents (C) remain as before.

The equations of electric elasticity (E) will be

$$\left.\begin{aligned} P &= 4\pi a^2 f \\ Q &= 4\pi b^2 g \\ R &= 4\pi c^2 h \end{aligned}\right\} \qquad \text{(82)},$$

where $4\pi a^2$, $4\pi b^2$, and $4\pi c^2$ are the values of k for the axes of x, y, z.

Combining these equations with (A) and (D), we get equations of the form

$$\frac{1}{\mu\nu}\left(\lambda\frac{d^2F}{dx^2}+\mu\frac{d^2F}{dy^2}+\nu\frac{d^2F}{dz^2}\right)-\frac{1}{\mu\nu}\frac{d}{dx}\left(\lambda\frac{dF}{dx}+\mu\frac{dG}{dy}+\nu\frac{dH}{dz}\right)=\frac{1}{a^2}\left(\frac{d^2F}{dt^2}+\frac{d^2\Psi}{dx\,dt}\right) \ldots \text{(83)}.$$

(104) If l, m, n are the direction-cosines of the wave, and V its velocity, and if

$$lx+my+nz-Vt=w \qquad \text{(84)},$$

then F, G, H, and Ψ will be functions of w; and if we put F', G', H', Ψ' for the second differentials of these quantities with respect to w, the equations will be

$$\left.\begin{aligned}&\left\{V^2 - a^2\left(\frac{m^2}{\nu} + \frac{n^2}{\mu}\right)\right\} F' + \frac{a^2 lm}{\nu} G' + \frac{a^2 ln}{\mu} H' - lV\Psi' = 0\\ &\left\{V^2 - b^2\left(\frac{n^2}{\lambda} + \frac{l^2}{\nu}\right)\right\} G' + \frac{b^2 mn}{\lambda} H' + \frac{b^2 ml}{\nu} F' - mV\Psi' = 0\\ &\left\{V^2 - c^2\left(\frac{l^2}{\mu} + \frac{m^2}{\lambda}\right)\right\} H' + \frac{c^2 nl}{\mu} F' + \frac{c^2 nm}{\lambda} G' - nV\Psi' = 0\end{aligned}\right\} \text{.........} (85).$$

If we now put

$$\left.\begin{aligned}&V^4 - V^2\frac{1}{\lambda\mu\nu}\left\{l^2\lambda(b^2\mu + c^2\nu) + m^2\mu(c^2\nu + a^2\lambda) + n^2\nu(a^2\lambda + b^2\mu)\right\}\\ &\qquad + \frac{a^2b^2c^2}{\lambda\mu\nu}\left(\frac{l^2}{a^2} + \frac{m^2}{b^2} + \frac{n^2}{c^2}\right)(l^2\lambda + m^2\mu + n^2\nu) = U\end{aligned}\right\} \text{......} (86),$$

we shall find

$$F'V^2U - l\Psi'VU = 0 \text{ ....} (87),$$

with two similar equations for G' and H'. Hence either

$$V = 0 \text{ ....} (88),$$

$$U = 0 \text{ ....} (89),$$

or

$$VF' = l\Psi', \quad VG' = m\Psi' \text{ and } VH' = n\Psi' \text{..........} (90).$$

The third supposition indicates that the resultant of F', G', H' is in the direction normal to the plane of the wave; but the equations do not indicate that such a disturbance, if possible, could be propagated, as we have no other relation between Ψ' and F', G', H'.

The solution $V=0$ refers to a case in which there is no propagation.

* The solution $U=0$ gives two values for V^2 corresponding to values of F', G', H', which are given by the equations

$$\frac{l}{a^2}F' + \frac{m}{b^2}G' + \frac{n}{c^2}H' = 0 \text{ ....} (91),$$

$$\frac{a^2 l\lambda}{F'}(b^2\mu - c^2\nu) + \frac{b^2 m\mu}{G'}(c^2\nu - a^2\lambda) + \frac{c^2 n\nu}{H'}(a^2\lambda - b^2\mu) = 0 \text{ ....} (92).$$

* [Although it is not expressly stated in the text it should be noticed that in finding equations (91) and (92) the quantity Ψ' is put equal to zero. See § 98 and also the corresponding treatment of this subject in the Electricity and Magnetism ii. § 796. It may be observed that the

(105) The velocities along the axes are as follows:—

Direction of propagation		x	y	z
Direction of the electric displacements	x		$\frac{a^2}{\nu}$	$\frac{a^2}{\mu}$
	y	$\frac{b^2}{\nu}$		$\frac{b^2}{\lambda}$
	z	$\frac{c^2}{\mu}$	$\frac{c^2}{\lambda}$	

Now we know that in each principal plane of a crystal the ray polarized in that plane obeys the ordinary law of refraction, and therefore its velocity is the same in whatever direction in that plane it is propagated.

If polarized light consists of electromagnetic disturbances in which the electric displacement is in the plane of polarization, then

$$a^2 = b^2 = c^2 \qquad (93).$$

If, on the contrary, the electric displacements are perpendicular to the plane of polarization,

$$\lambda = \mu = \nu \qquad (94).$$

We know, from the magnetic experiments of Faraday, Plücker, &c., that in many crystals λ, μ, ν are unequal.

equations referred to and the table given in § 105 may perhaps be more readily understood from a different mode of elimination. If we write

$$\lambda l^2 + \mu m^2 + \nu n^2 = P\lambda\mu\nu \text{ and } \lambda l F' + \mu m G' + \nu n H' = Q\lambda\mu\nu,$$

it is readily seen that

$$F' = l\frac{V\Psi' - a^2\lambda Q}{V^2 - a^2\lambda P},$$

with similar expressions for G', H'. From these we readily obtain by reasoning similar to that in § 104, the equation corresponding to (86), viz.:

$$\frac{l^2\lambda}{V^2 - a^2\lambda P} + \frac{m^2\mu}{V^2 - b^2\mu P} + \frac{n^2\nu}{V^2 - c^2\nu P} = 0.$$

This form of the equation agrees with that given in the Electricity and Magnetism ii. § 797.

By means of this equation the equations (91) and (92) readily follow when $\Psi' = 0$. The ratios of F' : G' : H' for any direction of propagation may also be determined.]

The experiments of Knoblauch* on electric induction through crystals seem to shew that a, b and c may be different.

The inequality, however, of λ, μ, ν is so small that great magnetic forces are required to indicate their difference, and the differences do not seem of sufficient magnitude to account for the double refraction of the crystals.

On the other hand, experiments on electric induction are liable to error on account of minute flaws, or portions of conducting matter in the crystal.

Further experiments on the magnetic and dielectric properties of crystals are required before we can decide whether the relation of these bodies to magnetic and electric forces is the same, when these forces are permanent as when they are alternating with the rapidity of the vibrations of light.

Relation between Electric Resistance and Transparency.

(106) If the medium, instead of being a perfect insulator, is a conductor whose resistance per unit of volume is ρ, then there will be not only electric displacements, but true currents of conduction in which electrical energy is transformed into heat, and the undulation is thereby weakened. To determine the coefficient of absorption, let us investigate the propagation along the axis of x of the transverse disturbance G.

By the former equations

$$\frac{d^2G}{dx^2} = -4\pi\mu(q')$$

$$= -4\pi\mu\left(\frac{df}{dt} + q\right) \text{ by (A),}$$

$$\frac{d^2G}{dx^2} = +4\pi\mu\left(\frac{1}{k}\frac{d^2G}{dt^2} - \frac{1}{\rho}\frac{dG}{dt}\right) \text{ by (E) and (F)} \quad\ldots\ldots\ldots\ldots\ldots (95).$$

If G is of the form

$$G = e^{-px}\cos(qx + nt) \quad\ldots\ldots\ldots\ldots\ldots (96),$$

we find that

$$p = \frac{2\pi\mu}{\rho}\frac{n}{q} = \frac{2\pi\mu}{\rho}\frac{V}{i} \quad\ldots\ldots\ldots\ldots\ldots (97),$$

where V is the velocity of light in air, and i is the index of refraction. The proportion of incident light transmitted through the thickness x is

$$e^{-2px} \quad\ldots\ldots\ldots\ldots\ldots (98).$$

* *Philosophical Magazine*, 1852.

Let R be the resistance in electromagnetic measure of a plate of the substance whose thickness is x, breadth b, and length l, then

$$R = \frac{l\rho}{bx},$$

$$2px = 4\pi\mu \frac{V}{i} \frac{l}{bR} \quad \text{.............................. (99).}$$

(107) Most transparent solid bodies are good insulators, whereas all good conductors are very opaque.

Electrolytes allow a current to pass easily and yet are often very transparent. We may suppose, however, that in the rapidly alternating vibrations of light, the electromotive forces act for so short a time that they are unable to effect a complete separation between the particles in combination, so that when the force is reversed the particles oscillate into their former position without loss of energy.

Gold, silver, and platinum are good conductors, and yet when reduced to sufficiently thin plates they allow light to pass through them. If the resistance of gold is the same for electromotive forces of short period as for those with which we make experiments, the amount of light which passes through a piece of gold-leaf, of which the resistance was determined by Mr C. Hockin, would be only 10^{-50} of the incident light, a totally imperceptible quantity. I find that between $\frac{1}{500}$ and $\frac{1}{1000}$ of green light gets through such gold-leaf. Much of this is transmitted through holes and cracks; there is enough, however, transmitted through the gold itself to give a strong green hue to the transmitted light. This result cannot be reconciled with the electromagnetic theory of light, unless we suppose that there is less loss of energy when the electromotive forces are reversed with the rapidity of the vibrations of light than when they act for sensible times, as in our experiments.

Absolute Values of the Electromotive and Magnetic Forces called into play in the Propagation of Light.

(108) If the equation of propagation of light is

$$F = A \cos \frac{2\pi}{\lambda}(z - Vt),$$

the electromotive force will be

$$P = -A\frac{2\pi}{\lambda}V\sin\frac{2\pi}{\lambda}(z - Vt);$$

and the energy per unit of volume will be

$$\frac{P^2}{8\pi\mu V^2},$$

where P represents the greatest value of the electromotive force. Half of this consists of magnetic and half of electric energy.

The energy passing through a unit of area is

$$W = \frac{P^2}{8\pi\mu V};$$

so that

$$P = \sqrt{8\pi\mu V W},$$

where V is the velocity of light, and W is the energy communicated to unit of area by the light in a second.

According to Pouillet's data, as calculated by Professor W. Thomson *, the mechanical value of direct sunlight at the Earth is

83·4 foot-pounds per second per square foot.

This gives the maximum value of P in direct sunlight at the Earth's distance from the Sun,

$$P = 60{,}000{,}000,$$

or about 600 Daniell's cells per metre.

At the Sun's surface the value of P would be about

13,000 Daniell's cells per metre.

At the Earth the maximum magnetic force would be ·193 †.

At the Sun it would be 4·13.

These electromotive and magnetic forces must be conceived to be reversed twice in every vibration of light; that is, more than a thousand million million times in a second.

* *Transactions of the Royal Society of Edinburgh*, 1854 ("Mechanical Energies of the Solar System").

† The horizontal magnetic force at Kew is about 1·76 in metrical units.

FURTHER READING

Berger, M.S., ed. *J.C. Maxwell, the Sesquicentennial Symposium: New Vistas in Mathematics, Science, and Technology.* Amsterdam: Elsevier Science Pub. Co., 1984.

Blumtritt, Oskar. *Zur Genese der elektromagnetischen Feldtheorie.* Hildesheim: Gerstenberg Verlag, 1986.

Brush, Stephen G., C.W.F. Everitt, and Elizabeth Garber, eds. *Maxwell on Saturn's Rings.* Cambridge: MIT Press, 1983.

Buchwald, Jed Z. *The Creation of Scientific Effects: Heinrich Hertz and Electric Waves.* Chicago: University of Chicago Press, 1994.

—. *From Maxwell to Microphysics: Aspects of Electromagnetic Theory in the Last Quarter of the Nineteenth Century.* Chicago: University of Chicago Press, 1985.

Campbell, Lewis and William Garnett. *The Life of James Clerk Maxwell.* 1882. Reprint, New York: Johnson Reprint Corporation, 1969.

Crowther, J.G. *Men of Science: Humphry Davy, Michael Faraday, James Prescott Joule, William Thomson, James Clerk Maxwell.* New York: W.W. Norton and Co., 1936.

Duhem, Pierre. *Les Théories électriques de J. Clerk Maxwell.* Paris: A. Hermann, 1902.

Everitt, C.W.F. *James Clerk Maxwell: Physicist and Natural Philosopher.* New York: Scribner, 1975.

Garber, Elizabeth, Stephen G. Brush, and C.W.F. Everitt, eds. *Maxwell on Heat and Statistical Mechanics: On "Avoiding All Personal Enquiries" of Molecules.* Bethlehem: Lehigh UP, 1995.

—. *Maxwell on Molecules and Gases.* Cambridge, MA: MIT Press, 1986.

Glazebrook, R.T. *James Clerk Maxwell and Modern Physics.* London: Cassell and Co., Ltd., 1901.

Goldman, Martin. *The Demon in the Aether: The Story of James Clerk Maxwell.* Edinburgh: Paul Harris Publishing, 1983.

Harman, P.M. *Energy, Force, and Matter: The Conceptual Development of Nineteenth-century Physics.* Cambridge: Cambridge UP, 1982.

—. *The Natural Philosophy of James Clerk Maxwell.* Cambridge: Cambridge UP, 1998.

—. ed. *Wranglers and Physicists: Studies on Cambridge Physics in the Nineteenth Century.* Manchester: Manchester UP, 1985.

Hendry, John. *James Clerk Maxwell and the Theory of the Electromagnetic Field.* Bristol: A. Hilger, 1986.

Hunt, Bruce J. *The Maxwellians.* Ithaca: Cornell UP, 1991.

James Clerk Maxwell: A Commemoration Volume, 1831–1931. Essays by Sir J. J. Thomson, Max Planck, Albert Einstein et alia. Cambridge: Cambridge UP, 1931.

MacDonald, D.K.C. *Faraday, Maxwell, and Kelvin.* Garden City: Doubleday Anchor, 1964.

Macfarlane, Alexander. *Lectures on Ten British Physicists of the Nineteenth Century.* New York: John Wiley, 1919.

Maxwell, James Clerk. *A Dynamical Theory of the Electromagnetic Field. With An Appreciation by Albert Einstein.* Edited by Thomas F. Torrance. Edinburgh: Scottish Academic Press, 1982.

—. *The Scientific Letters and Papers of James Clerk Maxwell.* 2 vols. Edited by P.M. Harman. Cambridge: Cambridge UP, 1990.

—. *Theory of Heat.* 3rd ed. 1872. Reprint, Westport: Greenwood Press, 1970.

—. *A Treatise on Electricity and Magnetism.* 3rd ed. 1891. Reprint, New York: Dover Publications, Inc., 1954.

Schaefer, Clemens. *Einführung in die maxwellsche Theorie der Elektrizität und des Magnetismus.* Leipzig: B.G. Teubner, 1922.

Siegel, Daniel M. *Innovation in Maxwell's Electromagnetic Theory: Molecular Vortices, Displacement Current, and Light.* Cambridge: Cambridge UP, 1991.

Tolstoy, Ivan. *James Clerk Maxwell: A Biography.* Edinburgh: Canongate, 1981.

Tricker, R.A.R. *The Contributions of Faraday and Maxwell to Electrical Science.* Oxford: Pergamon Press, 1966.

Whittaker, Edmund T. *A History of the Theories of Aether and Electricity.* Rev. ed. 2 vols. London: T. Nelson, 1951–53.

25. CLAUDE BERNARD

(1813–1878)

Claude Bernard placed experimental medicine and physiology on a firm basis in the nineteenth century. He discovered several fundamental physiological processes and began a textbook for this area of enquiry entitled *Principes de médecine experimentale*, of which only *Introduction à l'étude de la médecine expérimentale* (*The Introduction to the Study of Experimental Medicine*) of 1865, excerpted below, was ever published. The *Principes* was reconstructed and published posthumously in 1947 by Léon Delhoume.

Born in Saint-Julien in the wine-growing region of Beaujolais, Bernard was apprenticed to an apothecary at age 19. The young man had serious literary aspirations and wrote two plays, *La Rose du Rhône* and *Arthur de Bretagne*, but was discouraged by the Paris literary critic Saint-Marc Girardin, who counselled him instead to enter medicine. He subsequently trained as an assistant to the physiologist François Magendie, lecturer at the Collège de France, eventually succeeding him as professor of medicine in 1855.

Bernard's primary discoveries include: 1) the discovery of liver glycogen and its functions, 2) the role of the pancreas in digestion and absorption of fats, 3) vasomotor innervation, 4) neuromuscular effects of curare and other toxins, 5) the principle of physiological determinism, 6) the concept of internal secretion, and 7) the concept of the *milieu intérieur* (internal environment). Bernard also established the idea that the experimental physiologist must artificially and actively intervene in nature (through experiment) in order to achieve scientific knowledge, in contrast to the earlier medical tradition, which he unfairly characterized as passive, observational, and based on *a priori* reasoning. Modern experimental physiology in his view, on the other hand, possessed "the conquering character of the modern sciences" (*Principes* 7).

Bernard's discovery, described in *Recherches sur une nouvelle fonction du foie* (1853), that the liver stored a water-insoluble starchy substance (glycogen) which converted into sugar (glucose), broke down the barriers between animal and plant physiology, leading to an interest in general physiology as well as a better understanding of diabetes mellitus, an illness involving abnormalities in sugar metabolism. It was formerly believed that only plants stored starches or produced sugars.

One of the most important early contributions to the science of physiology was Bernard's elucidation of the concept of the *milieu intérieur* or "internal environment." Formulated in the 1850s in part as a refutation of Xavier Bichat's vitalism, Bernard's concept was not without contradictions and should be understood as a related complex of ideas. Bernard's final formulation of the idea appeared in *Lectures on the Phenomena of Life Common to Animals and Plants* (1878):

> I believe I was the first to insist upon this idea that there are really two environments for the animal, an *external environment* in which the organism is placed, and an *internal environment* [*milieu intérieur*]

> in which the elements of the tissues live. Life does not run its course within the external environment, atmospheric air for the air breathing creatures, fresh or salt water for the aquatic animals, but within the *fluid internal environment* formed by the circulating organic liquid that surrounds and bathes all of the anatomical elements of the tissues; this is the lymph or plasma, the liquid portion of the blood which in the higher animals perfuses the tissues and constitutes the ensemble of all the interstitial fluids, is an expression of all the local nutritions, and is the source and confluence of all the elementary exchanges. A complex organism must be considered as an association of *simple beings*, which are the anatomical elements, and which live in the fluid internal environment. *The constancy of the internal environment is the condition for free and independent life....* (83–84)

Thus, the internal environment in higher animals (Bernard described a higher animal as "an organism that has placed itself in a hot-house") provided physiological independence for the organism, but at the same time allowed the passage of the nutritive elements oxygen and water. Bernard therefore sometimes spoke of the *milieu intérieur* as a buffer or an intermediary between the internal and external environments. Bernard's *milieu intérieur* influenced the later work of Walter B. Cannon, Joseph Barcroft, and J.B.S. Haldane on homeostasis, hormonal systems, and regulatory mechanisms.

The Introduction to the Study of Experimental Medicine discusses Bernard's experimental methodology using examples from his own experiments. In his essay on Bernard's scientific method, Henri Bergson joined Bernard in criticizing the naive view of induction which proposed that facts are disinterestedly and randomly collected, then subjected to a power of synthesis or generalized into laws. Bergson and Bernard both saw invention, intuition, and human creativity (not objectivity) at every level of scientific endeavour: "invention must be everywhere, both in the most humble discovery of fact and in the simplest experiment. Wherever there is no personal or original effort, not even the beginnings of science can be found. This is the great pedagogical maxim which Claude Bernard's work teaches us" (Bergson, "La Philosophie de Claude Bernard" 232).

Although a determinist (i.e., a believer that physico–chemical laws underlie physiology), Bernard's great achievement was that he realized the complexity of living processes and did not merely reduce them to simplistic mechanical and chemical models. Bernard also demonstrated great humility in the face of what he did not know: "Science," he wrote, "is like a rope that we hold by one end, which we see. The other end is in the water and it is attached to the unknown. Whenever one pretends to present a completed work where nothing remains obscure, one can say that it is false" (*Cahier rouge*, in Grande 89). Not the pure mechanist, determinist or materialist that he sometimes proclaimed himself to be, Bernard continually struggled with the problem of vital force in biology (the idea that an immaterial, self-organizing, irreducible force governs living beings). Bernard partially escaped from these difficulties by proclaiming along with Auguste Comte that only proximate and secondary causes, not ultimate or final ones, were available to the experimenter.

Bernard's *Introduction* inspired Émile Zola's essay "The Experimental Novel," in which Zola argued that the novelist is essentially an experimental observer in that he sets characters in motion and then scrupulously records the ensuing events. Bernard became a symbol and archetype of both the scientist and experimental science in nineteenth-century literature and thought—Bernard's unfortunate marriage originally formed the basis of Zola's *Dr. Pascal*, and in Dostoyevsky's *Brothers Karamazov*, Dmitri curses scientists as "blackguards" and

"Bernards." In the passages below, Bernard discusses induction, his opposition to vitalism, the concept of the *milieu intérieur*, and some of his specific experimental discoveries.

Introduction à l'étude de la médecine expérimentale.

Paris: J.B. Baillière et Fils, 1865, pp. 85–92, 101–4, 107–112, 265–301. Translated by A.S. Weber.

On Doubt in Experimental Reasoning.

I will summarize the preceding paragraph in saying that there seems to me to be only one form of reasoning: *deduction* by syllogism. Our mind could not reason otherwise even if it so desired, and if this were the proper place, I could attempt to support my views with physiological arguments. But in order to discover scientific truth, it matters little in the end how our mind reasons; it suffices to let it reason naturally, and in this case the mind will always start from a principle in order to arrive at a conclusion. The one thing we need to do here is to insist on a precept which will constantly arm the mind against the innumerable causes of error which one encounters in the application of the experimental method.

This general precept, which is one of the foundations of the experimental method, is doubt; and this precept is expressed by saying that the conclusion of our reasoning must always remain doubtful when the point of departure or the principle is not an absolute truth. We have already seen that there is no absolute truth except in mathematical principles; for all natural phenomena, the principles from which we start, just like the conclusions which we reach, only represent relative truths. The stumbling block of the experimenter consists in believing he knows what he doesn't know, and in understanding as absolute those truths which are only relative. Thus the unique and fundamental rule of scientific investigation boils down to doubt, just as great philosophers, moreover, have already stated.

Experimental reasoning is precisely the inverse of scholastic reasoning. Scholasticism always needs a fixed and confirmed point of departure, and not being able to find it either in exterior things, or in reason, Scholasticism borrows this point of departure from whatever *irrational* source it can find: either a revelation, a tradition, or a conventional or arbitrary authority. Once the starting point is in place, the scholastic or systematic thinker logically deduces all the consequences from this point, even invoking observation and the testing[1] of facts as arguments when they are in his favor; the only condition is that the starting point must remain fixed and will not vary according to experiment and observation, but on the contrary, the facts will be interpreted in order to adapt them to the starting point. The experimenter, on the other hand, never admits a fixed starting point; his principle is a postulate from which he deduces logically all the consequences, without ever considering that postulate as absolute or beyond the realm of experiment. The elements of the chemists are only elements until there is proof to the contrary. All the theories which serve as starting points for the physicist, the chemist, and even moreso for the physiologist, are only true until facts are discovered which the theories cannot encompass, or which contradict the theory. When these contradictory facts are shown to be firmly established, the experimenter hastens to modify his theory because he knows that is the only way to advance and make progress in the sciences, unlike the scholastic and systematic thinker, who becomes

---

1 [Bernard's 'expérience' has been translated variously as "experiment," "experience," "knowledge," or "test", as Bernard implies all of these meanings in his usage. At times, 'expérience' is almost synonymous with "knowledge." Editor's note.]

inflexible in the face of this knowledge in order to protect his starting point. Thus the experimenter always doubts even his starting point; by necessity he possesses a modest and flexible mind, and accepts contradiction, on the one condition that it may be proved to him. The scholastic or systematic thinker—they are one and the same—never doubts his starting point, to which he tries to refer everything; he possesses a proud and intolerant mind and does not accept contradiction, since he refuses to admit that his starting point can change. What also separates the systematic thinker from the experimental thinker is that the first one imposes his idea, while the second one only advances it for what it is worth. Finally, another essential characteristic which distinguishes experimental reasoning from scholastic reasoning is the fecundity of the one, and the sterility of the other. The scholastic who believes himself in possession of absolute certainty amounts to nothing; this is understandable, since through his absolute principle, he places himself outside of nature in which everything is relative. This is in contrast to the experimenter, who always doubts and who never believes to possess absolute certainty about anything, and who succeeds in mastering the phenomena which surround him and in extending his power over nature. Thus man *can do more than he knows*, and the true experimental science only gives him enough power to show him his ignorance. Possessing the absolute truth matters little to the scientist, as long as he understands the certainty of the inter-relations of phenomena among themselves. Our mind is so limited, in fact, that we can know neither the beginning nor the ending of things; but we can grasp the middle, that is to say, that which immediately surrounds us.

Systematic and scholastic reasoning is natural to the inexperienced and proud mind; it is only through the thorough experimental study of nature that one succeeds in acquiring the doubting spirit of the experimenter. It takes a long time to acquire this doubt; and among those who believe they are following the experimental way in physiology and medicine, there are still many scholastics, as we will see later on. As for me, I am convinced that only the study of nature can give the scientist a true understanding of science. Philosophy, which I consider an excellent exercise for the mind, has systematic and scholastic tendencies in spite of itself, which would be harmful to the scientist proper. After all, no method can replace that study of nature which makes the true scientist; without that study, everything that the philosophers were able to say, and everything that I was able to repeat after them in this introduction, would remain inapplicable and sterile.

I do not therefore believe, as I have previously mentioned, that there is any great profit for the scientist to debate the definitions of induction and deduction, nor to discuss the question of whether one proceeds by one or the other of these so-called processes of the mind. However, Baconian induction has become famous and has been made the foundation of all scientific philosophy. Bacon was an extraordinary genius and his idea of the great restoration of the sciences was sublime; one is seduced and carried along in spite of oneself in reading the *Novum organum* and the *Augmentum scientiarum*. One remains in a sort of trance before that amalgam of scientific illumination, clothed in the most elevated poetical forms. Bacon sensed the sterility of Scholasticism; he understood well and completely foresaw the importance of experiment for the future of the sciences. Bacon, however, was not a scientist, and he had no understanding of the mechanism of the experimental method. To prove this, it would suffice to cite the unsuccessful attempts he made in this line. Bacon recommends that we flee from hypotheses and theories;[2] we have seen, however, that these are the auxiliaries of the experimental method, indispensable just as scaffolding is necessary in constructing a house. Bacon had, as always happens, extravagant admirers and detractors. Without placing myself on either side of the question, I will say that, while recognizing the genius of Bacon, I do not believe any more than J. de Maistre,[3] that he endowed human intellect with a new instrument, and it seems to

2 Bacon, *Works*, edition of Fr. Riaux, *Introduction*, p. 30.

3 J. de Maistre, *Examen de la philosophie de Bacon*.

me, along with M. de Rémusat,[4] that induction does not differ from the syllogism. Besides, I believe that great experimenters appeared before the precepts of experimentation, in the same way that great orators preceded treatises on rhetoric. Consequently, it does not appear permissible for me to say, even in speaking of Bacon, that he invented the experimental method; a method which Galileo and Torricelli practiced so admirably, and which Bacon could never use.

When Descartes[5] starts from universal doubt and repudiates authority, he provides much more practical precepts for the experimenter than those which Bacon gave for induction. We have seen, in fact, that doubt alone calls forth experiment; it is doubt in the end which determines the form of experimental reasoning.

Yet, in medicine and the physiological sciences, it is important to determine properly to what point doubt should be extended, so as to distinguish it from skepticism, and to show how scientific doubt becomes an element of the greatest certitude. The skeptic is someone who does not believe in science, but who believes in himself; he believes enough in himself to dare to deny science and to affirm that science is not bound by fixed and determined laws. The doubter is the true scientist; he only doubts himself and his interpretations, but he believes in science; he even admits a criterion or an absolute scientific principle in the experimental sciences. This principle is the *determinism* of phenomena, which is as absolute in the phenomena of living bodies as in inorganic bodies,[6] as we will discuss later (p. 114).

Finally, in concluding this section, we can say that, in all experimental reasoning, there are two possible cases: the hypothesis of the experimenter will either be invalidated or confirmed by experiment. When experiment invalidates a preconceived idea, the experimenter must reject or modify his idea. But even when experiment fully confirms the preconceived idea, the experimenter must still doubt; because it concerns a truth not fully known, his reason still demands a counterproof.

The Spontaneity of Living Bodies is not an Obstacle to the Use of Experimentation.

The spontaneity enjoyed by beings endowed with life has been one of the principal objections that has been raised against the use of experimentation in biological studies. Each living being, in effect, appears to us as if provided with a kind of interior force which presides over vital manifestations, which become more and more independent of the general influence of the cosmos, the more the being in question rises in the scale of organization. In the higher animals and in man, for example, this vital force appears to result in the withdrawal of the living body from general physico-chemical influences, and thus renders experimental access to it very difficult.

Inorganic bodies offer nothing similar, and whatever their nature, they are all lacking in spontaneity. Since the manifestation of their properties is linked absolutely to the physico-chemical conditions which surround them and which serve as their environment, it follows that the experimenter can easily access them and modify them at will.

On the other hand, all the phenomena of a living body exist in a reciprocal harmony, such that it seems to be impossible to separate one part from the organism without immediately disturbing the entire system. In the higher animals in particular, their extreme sensitivity leads to even greater reactions and disturbances.

Many doctors and speculative physiologists, along with some anatomists and naturalists, have exploited these various arguments to protest against experimentation on living beings. They have supposed that the vital force was in opposition to the physico-chemical forces, and that this vital force dominated all the phenomena of life, subjecting them to altogether special laws, and making the organism an organized whole which the experimenter could not touch without destroying the very character of life itself. They have even gone so far as to say that inorganic bodies and living bodies differed radically from

4 De Rémusat, *Bacon, sa vie, son temps et sa philosophie*, 1857.

5 Descartes, *Discours sur la méthode.*

6 ["Corps bruts." Editor.]

this point of view, in such a manner that experimentation was applicable to the one kind of body and not to the other. Cuvier, who shares this opinion, and who thinks that physiology should be a science of observation and of anatomical deduction, expresses himself thus: "All parts of a living body are connected; they can only act in so far as they all act together. Separating one part from the whole means sending it back to the realm of dead substances; it would change the essence of the body entirely."[7]

If the preceding objections had some foundation, we would have to acknowledge that either determinism is impossible in the phenomena of life—which would simply deny biological science—or that the vital force must be studied by particular procedures and that the science of life must rest on other principles than the science of inert bodies. These ideas, which were current in other time periods, are undoubtedly fading away now more and more; however, it is important to wipe out the final traces of them, because whatever remains of these so-called vitalist ideas in certain minds represents a real obstacle to the progress of experimental medicine.

I propose therefore to establish that the science of the phenomena of life must have the same basis as that of the science of inorganic bodies, and that there is no difference in this respect between the principles of the biological sciences and those of the physico-chemical sciences. In fact, just as we have said previously, the goal which the experimental method proposes is the same everywhere; it consists in reconnecting by experiment natural phenomena with their conditions of existence or to their immediate causes. In biology, these causes being known, the physiologist will be able to direct the manifestations of the phenomena of life just as the physicist and the chemist direct the natural phenomena whose laws they have discovered; but in doing this, the experimenter does not act on life.

Yet there is an absolute determinism in all of the sciences because, each phenomenon being linked necessarily to physico-chemical conditions, the scientist can modify these conditions to master the phenomenon, that is to say, to hinder or favor its manifestation. In the case of inorganic bodies, there is no debate on this subject. I would like to prove that it is the same for living bodies, and that, for them also, determinism exists.

The Physiological Phenomena of Higher Organisms Occur in Organic Interior Environments Perfected by and Endowed with Constant Physico-Chemical Properties.

It is very important, in order to understand completely the application of experimentation to living beings, to be perfectly clear about the ideas which we are developing at this point. When we examine a higher living organism, that is to say, a complex being, and when we observe it carrying out its different functions in the general cosmic environment common to all phenomena of nature, the organism seems independent of that environment up to a certain point. But this appearance results from our illusions concerning the simplicity of the phenomena of life. The external phenomena which we perceive in that living being are fundamentally very complex, and they represent the result of a host of intimate properties of organic elements whose manifestations are linked to the physico-chemical conditions of the internal environments in which these elements are immersed. In our explanations, we do away with this internal environment and only see the external environment which is before our eyes. But the real explanation of the phenomena of life rests on the study and knowledge of the most tenuous and subtle particles which form the organic elements of the body. This idea, set down in biology long ago by the great physiologists, appears more and more valid as the study of the organization of living beings makes greater progress. We must learn in addition that these *intimate particles* of the organization only manifest their vital activity by a necessary physico-chemical relation with *intimate environments* which we should equally study and know. Otherwise, if we limit ourselves to the examination of the total phenomena visible from the outside, we might falsely believe that there is a unique force

7 *Letter to J.C. Mertrud*, p. 5. Year VIII.

in living beings which violates the physico-chemical laws of the general cosmic environment, in the same way an ignorant person could believe that there is a special force which violates the laws of gravity in a machine which mounts into the air or runs along the ground. Now a living organism is nothing but an amazing machine endowed with the most marvelous properties and activated by the aid of the most complex and delicate mechanisms. There are no opposing forces struggling with one another; nature only knows order and disorder, harmony and discord.

In experimentation on inorganic bodies, we only need to take into account a single environment, the external cosmic environment; while in higher living beings, there are at least two environments to consider: the *external environment* or extra-organic, and the internal environment [*milieu intérieur*] or intra-organic. Each year in my course on physiology in the Faculty of Sciences, I develop these new ideas on organic environments, ideas which I consider the basis of general physiology; they also necessarily form the basis of general pathology, and these same concepts will guide us in the application of experimentation to living beings. As I have said elsewhere, the complexity due to the existence of an organic internal environment is the only reason for the great difficulties which we encounter in the experimental determination of the phenomena of life and in the application of the means capable of modifying these phenomena.[8]

With the aid of the thermometer, the barometer, and all the instruments which record and measure the properties of the external environment, the physicist and the chemist who experiment on inert bodies, having only to consider the external environment, can always set up identical conditions. For the physiologist, these instruments no longer suffice, and besides, it is in the *milieu intérieur* which he must employ them. In effect it is the *milieu intérieur* of living beings which is always in immediate relation with the normal or pathological vital manifestations of the organic elements. The higher one travels up the scale of living beings, the more the organization becomes more complicated, and the more the organic elements become more delicate and require a more perfect *milieu intérieur*. All the circulating fluids, the blood serum and the intra-organic fluids, in reality constitute this *milieu intérieur*.

In all living beings, the *milieu intérieur*, which is a true *product of the organism*, preserves the necessary relations of exchange and equilibrium with the external cosmic environment; but, as the organism becomes more perfect, the organic environment becomes more specialized and in some manner isolates itself more and more from the ambient environment. In plants and cold-blooded animals, as we have said before, this isolation is less complete than in hot-blooded animals; in hot-blooded animals, the blood maintains an almost fixed and constant temperature and composition. But these different conditions do not create differences in nature among different living beings; they only represent improvements in the environmental mechanisms of isolation and protection. The vital manifestations of animals only vary because the physico-chemical conditions of their internal environments vary; thus a mammal whose blood has been cooled, either by natural hibernation, or by certain lesions of the nervous system, completely resembles, in the properties of its tissues, a true cold-blooded animal.

In sum, one can, according to what has been said, construct an idea of the enormous complexity of the phenomena of life and the almost insurmountable difficulties in exactly determining them which confront the physiologist when he is forced to carry out experimentation in these interior or organic environments. Nevertheless, these obstacles will not frighten us if we are convinced that we are travelling along the right path. In effect, there is an absolute determinism in all vital phenomena; hence there exists a biological science, and consequently, all the studies to which we devote ourselves will not be in vain. General physiology is the fundamental biological science towards which all the other biological sciences converge. Its main concern is to determine the

8 Claude Bernard, *Leçons sur la physiologie et la pathologie du système nerveux*. Leçon d'ouverture, December 17, 1856. Paris, 1858, t. 1—Cours de pathologie expérimentale, *The Medical Times*, 1860.

elementary conditions of the phenomena of life. Pathology and therapeutics also rest on this common foundation. It is through the normal activity of the organic elements that life maintains a state of health; it is the abnormal manifestation of these same elements which characterizes disease, and in the end through the intermediary of the organic environment modified by the means of certain toxic or medicinal substances, therapeutics can act on the organic elements. To arrive at a resolution of these various problems, it is necessary to break down the organism successively, as one takes apart a machine in order to understand it and to study all the workings; this means, that before experimenting on the elements, it is first necessary to experiment on the apparatus and the organs. We must have recourse to a successive analytical study of the phenomena of life by using the same experimental method which the physicist and the chemist use to analyze the phenomena of inorganic bodies. The difficulties which result from the complexity of the phenomena of living bodies, arise solely in the application of experimentation, because in the end the goal and the principles of the method always remain exactly the same.

Third Part

Applications of the Experimental Method to the Study of Vital Phenomena

Examples of Experimental Physiological Investigation

The ideas which we developed in the first two parts of this introduction will be better understood if we can apply them to investigations of physiology and experimental medicine and show that these ideas can serve as easily remembered precepts for experimenters. That is why I have brought together in what follows a certain number of examples which appeared most suitable to me in order to make my point. In all these examples, I have, as much as possible, cited my own work, for the sole reason that in the case of reasoning and intellectual processes, I will be much more certain about what I am advancing in describing what happened to me than what may have taken place in the minds of others. Besides, I do not presume to offer these examples as models to follow; I only employ them to express my ideas better and to make my thought easier to grasp.

Many diverse circumstances can serve as starting points for scientific research; I will summarize, however, all these varieties under two principal cases:

1. Where the starting point of experimental research is an observation.
2. Where the starting point of experimental research is an hypothesis or a theory.

I.—Where the Starting Point for Experimental Research is an Observation.

Experimental ideas are often born by accident or on the occasion of a fortuitous observation. Nothing is more common, and this is the simplest way to begin a scientific endeavour. We take a walk, so to speak, in the realm of science, and we pursue whatever happens to present itself before our eyes by accident. Bacon compares scientific investigation with hunting; the observations that present themselves are the game. To continue the same comparison, one can add that if the game appears even when we are looking for it, it also happens that it appears when we are not looking for it or when we are looking for another kind of game. I will cite an example in which these two cases ocurred in succession. I will be careful at the same time to analyze every circumstance in this physiological investigation, in order to demonstrate the application of the principles which we developed in the first part of this introduction, principally in chapters I and II.

First example.—Rabbits were brought into my laboratory one day from the market. They were placed on a table where they urinated, and I happened to notice that their urine was clear and acid. This fact struck me, because normally rabbits, because they are herbivores, have cloudy and alkaline urine, while carnivores on the other hand, as is well known, have clear and acid urine. This observation of the acidity of the rabbits' urine made me think that these animals must have been in the nutritional condition of carnivores. I assumed that they had probably not eaten for a long time, and thus they found themselves transformed because of this starvation into true carnivorous animals living off their own blood. Nothing was easier than to verify this preconceived idea or hypothesis by experiment. I gave the rabbits grass to eat, and several hours later, their urine became cloudy and alkaline. These same rabbits were then starved, and after 24 or 36 hours at most their urine became clear and strongly acid again; and then it became alkaline again by giving them grass, etc. I repeated this very simple experiment a great number of times on the rabbits, and always with the same result. I then repeated it on a horse, an herbivorous animal which also has cloudy and alkaline urine. I found that, just as with rabbits, starvation produced a prompt acidification of the urine with a very considerable increase of urea, to the point that it sometimes crystallizes spontaneously in the cooled urine. The result of my experiments was that I thus arrived at this general proposition which was unknown at the time, that is, namely, *all fasting animals feed on meat*, such that herbivores have urine similar to that of carnivores.

Here we are dealing with a very simple, particular fact which allows us to follow easily the evolution of experimental reasoning. When we see a phenomena which we are not in the habit of seeing, we must always ask ourselves what it is related to, or to put it another way, what is its proximate cause; then a response or an idea arises in the mind which must be submitted to experiment. When I saw the acid urine of the rabbits, I instinctively asked myself what could be the cause. The experimental *idea* consisted of the spontaneous connection my mind made between the acidity of the rabbits' urine and the state of starvation which I considered as the true diet of flesh-eaters. The inductive *reasoning* which I implicitly made consisted of the following syllogism: the urine of carnivores is acid; now, the rabbits which I observed had acid urine; therefore they are carnivores, that is to say, in a state of fasting. This is what remained to be established by *experiment*.

But in order to prove that my fasting rabbits were really carnivores, a counterproof was needed. It was necessary to create a carnivorous rabbit experimentally by feeding it meat, in order to see if its urine would then be clear, acidic and filled with urea, just as it would be during a period of starvation. That is why I had the rabbits fed on cold boiled beef (a food they readily eat when they are given nothing else). My expectation was again verified, and during the entire time of their animal diet, the rabbits maintained clear and acid urine.

To complete my experiment, I wanted in addition to see by means of autopsy if the digestion of meat in a rabbit was carried out in the same manner as in a carnivore. I found, in fact, all the phenomena of an excellent digestion in their intestinal reactions, and I established that all the chyliferous vessels were gorged with a very abundant, white and milky chyle just like in carnivores. But in connection with these autopsies, which offered a confirmation of my ideas on the digestion of meat in rabbits, a fact presented itself here which I had never considered and which became for me, as we shall see, the starting point for a new endeavour.

Second example (sequel to the preceding).—I happened to notice in sacrificing the rabbits which I had forced to eat meat, that the white and milky lymphatic vessels were first visible in the small intestine at the lower part of the duodenum, about 30 centimetres below the pylorus. This fact attracted my attention, because in dogs, the lymphatic vessels are first visible much higher in the duodenum and immediately below the pylorus. In examining the situation more closely, I noticed that this particularity in the rabbit coincided with the insertion of the pancreatic canal, which is situated in a very low point and precisely in the neighborhood where the lymphatic vessels begin to contain chyle made white and milky by the emulsion of fatty nutritive elements.

The fortuitous *observation* of this fact awakened in me an idea and generated in my mind the thought that the pancreatic juice could well be the cause of the emulsion of the fatty matter and consequently the cause of its absorption by the lymphatic vessels. Again I instinctively made the following syllogism: the white chyle is due to the emulsion of the fat; now, in rabbits, white chyle forms at the level where the pancreatic juice flows into the intestine; thus it is the pancreatic

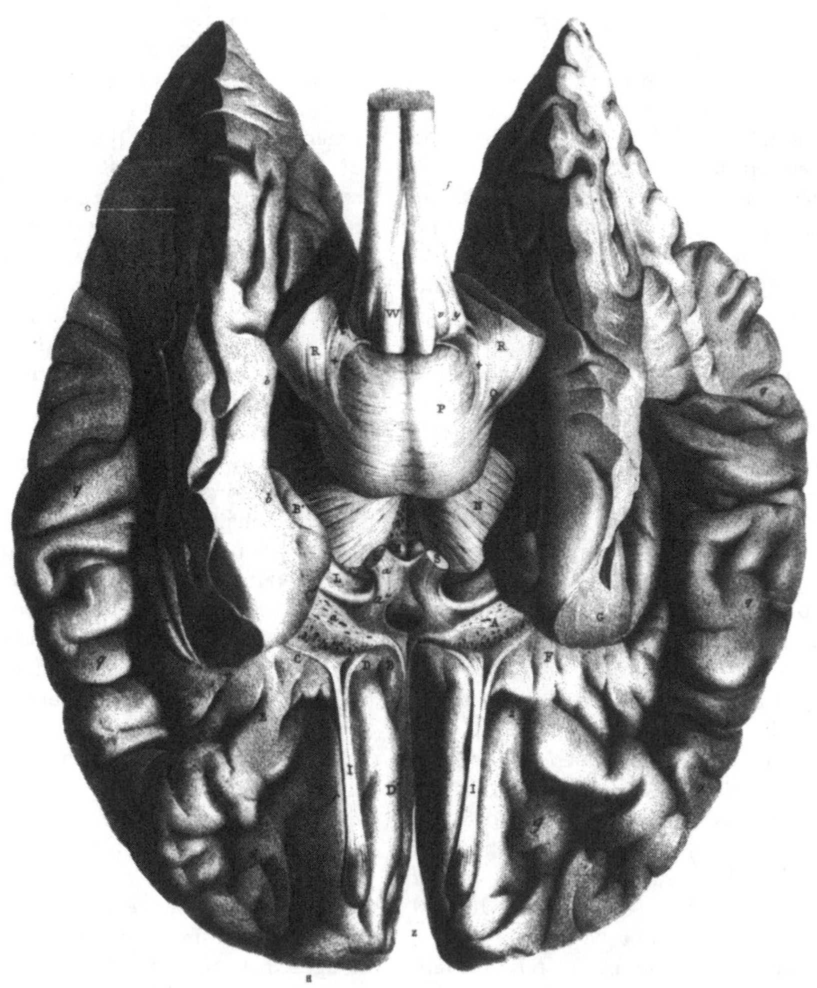

Illustration of cerebral anatomy from M. Foville, *Traité complet de l'anatomie, de la physiologie et de la pathologie du système nerveux cérébro-spinal* (Paris, 1844). The accompanying Atlas, from which this plate was taken, was prepared by Émile Beau and F. Bion.

juice which emulsifies the fat and forms white chyle. This is what needed to be decided by experiment.

In view of this preconceived idea, I imagined and immediately carried out an *experiment* suitable for verifying the truth or falseness of my supposition. This experiment consisted of testing the properties of the pancreatic juice directly on neutral or alimentary fats. But pancreatic juice does not flow naturally outside of the body like saliva or urine, for example; its secreting organ is, on the contrary, deeply buried in the abdominal cavity. I was therefore forced to use the methods of *experimentation* in order to procure pancreatic fluid from a living animal in suitable physiological conditions and in a sufficient quantity. Only then could I carry out my experiment, that is to say, to control my preconceived idea, and the experiment proved that my idea was correct. In fact, pancreatic juice obtained in suitable conditions from dogs, rabbits, and many other kinds of animals, mixed with oil or melted fat, always emulsified instantaneously, and later acidified these fatty bodies in decomposing them, with the aid of a particular ferment, into fatty acids, cerin, etc., etc.

I will not elaborate on these experiments any further, since I have developed them at length in a specialized study.[9] I have only wished to show how an initial observation made by accident on the acidity of rabbit urine gave me the idea to carry out experiments on their carnivorous nutrition, and how subsequently by following up on these experiments, I brought to light, without searching for it, another observation related to the special arrangement of the juncture of the pancreatic duct in rabbits. This unexpected second observation generated by the experiment, gave me in turn the idea to carry out experiments on the action of the pancreatic juice.

We see from the preceding examples how the *observation* of a fact or phenomenon, arriving unexpectedly by accident, engenders by anticipation a preconceived *idea* or hypothesis on the probable cause of the observed phenomenon; how the preconceived idea gives birth to reasoning which deduces the proper experiment to verify it; and how, in one case, it was necessary to have recourse to experimentation in order to work out that verification, that is to say, by using more or less complicated operative processes, etc. In the last example, experiment played a double role; it first judged and confirmed the predictions of the reasoning which had engendered it, but moreso it provoked a new observation. One can thus call this kind of observation, an *observation provoked or engendered by experiment*. This proves that it is necessary, as we have said, to observe every result of an experiment, both those which are related to the preconceived idea and even those which have no relation to it. If we get into the habit of only seeing the facts related to our preconceived idea, we would often deprive ourselves of making discoveries; because it frequently happens that an unsuccessful experiment can provoke a very good observation, as the example which follows proves.

Third example.—In 1857, I undertook a series of experiments on the elimination of substances in the urine, and this time the results of the experiment did not confirm, as they did in the previous examples, my predictions or my preconceived ideas on the mechanism of the elimination of substances in the urine. I thus made what is normally called an unsuccessful experiment or rather experiments. But we have previously advanced the principle that there are no unsuccessful experiments, because when they do not serve the investigation for which they were devised, we must still profit from the observations that they can furnish which give occasion to other experiments.

While investigating how the substances which I had injected into animals were eliminated from the blood leaving the kidney, I observed by accident that the blood of the renal vein was crimson, while the blood of the neighboring veins was black like ordinary venous blood. This unexpected peculiarity struck me, and I thus made the *observation* of a new fact which the experiment had engendered and which was alien to the experimental goal which I was following in that

9 Claude Bernard, *Mémoire sur le pancréas et sur le rôle du suc pancréatique dans les phénomènes digestifs*. Paris, 1856.

same experiment. I thus gave up my original idea, which had not been verified, and I directed all my attention to that singular coloration of the venous renal blood, and when I had well established and assured myself that there was no cause for error in the observation of the fact, I naturally asked myself what could be the cause. Upon examining the urine which passed through the urethra, and reflecting upon it, the idea occurred to me that the red coloration of the venous blood might be related to the secretory or functional state of the kidney. In this hypothesis, by stopping the renal secretion, the venous blood should become dark: that is what happened. By reestablishing the renal secretion, the venous blood should become crimson again: this is what I was able to verify each time that I excited the secretion of urine. I thus obtained experimental proof that there is a connection between the secretion of urine and the coloration of the blood of the renal vein.

But that is not all. In the normal state, the venous blood of the kidney is almost constantly crimson, because the urinary organ continuously secretes, although alternately for each kidney. Now, I wanted to know if the crimson color of the venous blood constituted a general fact characteristic of other glands, and I desired to obtain in this way a succinct counter-proof which would demonstrate that it was the secretory phenomenon by itself which led to the modification in the coloration of the venous blood. Here is how I reasoned: if, I told myself, it is the secretion, which causes, as it appears to be, the reddening of the venous glandular blood, then in the glandular organs like salivary glands which secrete fluid intermittently, the venous blood will change color intermittently and will become dark during the dormancy of the gland, and red during secretion. I therefore exposed the sub-maxillary gland of a dog, along with its ducts, nerves, and vessels. In its normal operation, this gland supplies an intermittent secretion which one can excite or stop as desired. Now I clearly established that during the dormancy of the gland, when nothing was flowing from the salivary duct, the venous blood in fact was dark, while as soon as the secretion appeared, the blood became crimson and reverted to a dark color when the secretion stopped, then remained dark during the entire time of the intermission, etc.[10]

These last observations later became the starting point for new ideas which guided me to make investigations concerning the chemical cause of the change of color of the glandular blood during secretion. I will not further describe these experiments here as I have published the details elsewhere.[11] It will suffice for me to have proved that scientific investigation or experimental ideas can give birth to fortuitous, and in some sense *involuntary*, observations which present themselves to us, either spontaneously or on the occasion of an experiment made for a different purpose.

But there is still another case, in which the experimenter provokes and *voluntarily* gives birth to an observation. This case returns us, so to speak, to the preceding example; the only difference is that, instead of waiting for an observation to accidently present itself to us under fortuitous circumstances, we provoke it by an experiment. To take up the comparison of Bacon again, we could say that the experimenter resembles in this case a hunter who, instead of tranquilly waiting for the game, beats the brush in the areas where he supposes it to be. This is what we have called the *experiment to see* (p. 37 and 38). We use this method every time that we do not have a preconceived idea to undertake investigations on a subject in which we have no previous observations. Then one experiments to give birth to observations which can in turn give birth to ideas. This is what normally happens in medicine when one wants to investigate the action of a poison or of some medicinal substance on the animal economy; we make these experiments in order to see, and subsequently we are guided by what we have seen.

Fourth example.—In 1845, M. Pelouze sent me a

---

10 Claude Bernard, *Leçons sur les propriétés physiologiques et les altérations pathologiques des liquides de l'organisme*. Paris, 1859, vol. 2.

11 Claude Bernard, *Sur la quantité d'oxygène que contient le sang veineux des organes glandulaires* (*Compt. rend. de l'Acad. des sciences*, vol. 47, September 6, 1858).

toxic substance called *curare* which had been brought to him from America. At the time, no one knew anything about the physiological mode of action of this substance. We only knew, according to previous observations and from the interesting reports of Alex. de Humboldt, and MM. Boussingault and Roulin, that this substance was difficult to prepare and determine, and killed an animal very rapidly when introduced under the skin. But I was unable through previous observations to develop a preconceived idea concerning the mechanism of death by curare; to achieve this, I had to obtain new observations relating to the organic disturbances which this substance might produce. I provoked the appearance of these observations, that is to say, I made experiments *to see* things about which I had absolutely no preconceived ideas. First I placed curare under the skin of a frog, and it died after several minutes; I opened it immediately and with a physiological autopsy, I examined in succession what had happened to the known physiological properties of the various tissues. I say *physiological autopsy* on purpose, because it is the only really instructive kind. The disappearance of the physiological properties explains death and not anatomical changes. In fact, in the present state of science, we see physiological properties disappear in a number of cases without being able to show by our present means of investigation, any corresponding anatomical alteration; such is the case of curare, for example. Meanwhile, we will discover cases, on the contrary, in which the physiological properties persist despite the very marked anatomical alterations with which the functions are in no way incompatible. Now in the case of my frog poisoned by curare, the heart continued its movements, the blood corpuscles were not altered in their appearance with respect to their physiological properties no more than the muscles which had preserved their normal contractility. But while the nervous system had preserved its normal anatomical appearance, the properties of the nerves had however completely disappeared. There were no longer any voluntary or reflexive movements, and when the motor nerves were directly excited, they no longer caused the muscles to contract. In order to know whether there was anything accidental or erroneous in this first observation, I repeated it several times and verified it in several ways; because the most indispensable thing when one wants to reason experimentally is to be a good observer and to assure oneself that there is no error in the observation which serves as the starting point of the reasoning. Now, I found the same phenomena in mammals and birds as in frogs, and the disappearance of the physiological properties of the motor nervous system became the constant fact. Starting from this well established fact, I was able to advance my analysis of the phenomena and to determine the mechanism of death by curare. I always proceeded by reasonings analogous to those described in the preceding example, and from idea to idea and from experience to experience, I progressed to more and more definite facts. I finally arrived at this general proposition: *curare causes death by the destruction of all the motor nerves without affecting the sensory nerves*.[12]

In the cases in which we make an experiment *to see*, the preconceived idea and the reasoning, as we have said, seems to be completely missing, yet we necessarily reason unknowingly by syllogism. In the case of curare, I instinctively reasoned in the following manner: there is no phenomenon without a cause, and consequently no poisoning without a necessary and determinable physiological lesion, peculiar or specific to the poison employed; now, I thought curare must produce death by a unique action operating on certain definite organic parts. Thus in poisoning the animal by curare and in immediately examining the properties of its various tissues after death, I would perhaps be able find and study a lesion peculiar to this poison.

The mind here is thus still *active* and the *experiment to see*, which seems made for the occasion, however returns us to our general definition of experiment (p. 20). In fact, in every enterprise, the mind always reasons, and even when one seems to act without motivation, an instinctive logic guides our mind. Only we don't realize it, for the

12 Cf. Claude Bernard, *Leçons sur les effets des substances toxiques*. Paris, 1857; *Du curare* (*Revue des deux mondes*, September 1, 1864).

simple reason that we begin to reason before we know and say that we are reasoning, in the same way that we begin to speak before noticing that we are speaking, and in the same way begin to see and hear before knowing what we are seeing and what we are hearing.

Fifth example.—Towards 1846, I wanted to make experiments on the cause of poisoning by carbon monoxide. I knew that this gas had been described as toxic, but I knew absolutely nothing about the mechanism of this poisoning; I therefore could not possess any preconceived opinion. What was to be done, then? I had to give birth to an idea by making a fact appear, that is to say, make an experiment *to see*. In fact, I poisoned a dog by forcing him to breathe carbon monoxide, and immediately after his death, I opened his body. I looked at the state of the organs and the fluids. What instantly attracted my attention was that the blood was crimson in all the vessels; in the veins as well as in the arteries, and in the right and left chambers of the heart. I repeated this experiment on rabbits, birds, and frogs; everywhere I found the same general crimson coloration of the blood. But I was distracted from following this investigation and I kept this *observation* for a long time without using it except to cite it in my courses in regards to the coloration of the blood.

In 1856, no one had investigated this experimental question further, and in my course at the Collège de France on *toxic and medicinal substances*, I took up again the study of carbon monoxide poisoning which I had begun in 1846. I found myself then in an ambiguous situation, because, at that time, I already knew that carbon monoxide poisoning renders the blood crimson throughout the entire circulatory system. It was necessary to make hypotheses and to establish a preconceived idea based on this first observation before proceeding further. Now, on reflecting on the fact of the reddening of the blood, I attempted to interpret it within the framework of all the preceding knowledge that I possessed on the cause of the color of the blood, and the following reflections presented themselves to my mind. The crimson color of the blood, I told myself, is peculiar to the arterial blood and is related to the presence of oxygen in great proportions, while the dark coloration is caused by the disappearance of oxygen and to the presence of a greater proportion of carbonic acid, so the idea came to me that the carbon monoxide, in preserving the crimson color in the venous blood, may have perhaps hindered the oxygen from changing into carbonic acid in the capillaries. However, it seemed difficult to understand how all this could be the cause of death. But still continuing my internal and preconceived reasoning, I added: if all of this were true, blood taken from the veins of animals poisoned by carbon monoxide should contain oxygen like arterial blood; that was what was necessary to see.

Following this reasoning founded on the interpretation of my observation, I carried out an *experiment* to verify my *hypothesis* relating to the persistence of oxygen in the venous blood. To do this, I passed a stream of hydrogen through the crimson venous blood taken from an animal poisoned by carbon monoxide, but I could not, as usual, displace the oxygen. I attempted the same on the arterial blood with no further success. My preconceived idea was therefore false. But this impossibility of obtaining oxygen from the blood of a dog poisoned by carbon monoxide provided a second *observation* for me which suggested new ideas according to which I formed a new hypothesis. What could have happened to the oxygen in the blood? It had not changed into carbonic acid, because I had not displaced any great quantity of this gas in forcing a current of hydrogen through the blood of the poisoned animal. Besides this supposition was in opposition to the color of the blood. I exhausted myself in conjecture concerning the manner in which the carbon monoxide could have made the oxygen disappear and, since gases displace one another, I naturally thought that the carbon monoxide could have displaced the oxygen and driven it from the blood. In order to confirm this, I resolved to change the experiment and to place the blood in artificial conditions which would permit me to recover the displaced oxygen. I then studied the action of carbon monoxide on the blood by *artificial poisoning*. To do this, I took a certain quantity of arterial blood from a healthy animal, and I placed this blood under mercury in a test tube containing carbon monoxide, and I consequently agitated the entire set up in order to poison the blood while protecting it from contact with the outside air. Then after a certain time, I looked to see if the air in the test tube, in contact with the poisoned blood, had been modified, and I determined that the air in contact with the blood was notably enriched with oxygen, at the same time that the proportion of carbon monoxide was diminished. It appeared to me after repeating these experiments under the same conditions that there had been a simple exchange volume for volume between the carbon monoxide and the oxygen in the blood. But the carbon monoxide in displacing the oxygen which it had driven from the blood, remained fixed in the blood corpuscles and could no longer be displaced by oxygen or by any other gas, such that death occurred by the death of the blood corpuscles, or to put it another way, by the cessation of the exercise of their physiological property which is essential to life.

This last example, which I have just related in a very succinct manner, is complete, and it shows from one end to the other how the experimental method proceeds and succeeds in coming to an understanding of the proximate cause of phenomena. First I knew absolutely nothing about the mechanism of the phenomenon of poisoning by carbon monoxide. I made an experiment *to see*, that is to say, for observation. I gathered a first observation on a special modification of the color of the blood. I interpreted that observation, and I made an *hypothesis* which experiment proved to be false, but this experiment furnished me with a second *observation*, upon which I reasoned anew using it as a starting point to make a new hypothesis on the mechanism of the removal of oxygen from the blood. In constructing these hypotheses successively on the facts as I observed them I finally arrived at demonstrating that carbon monoxide substitutes itself in the blood corpuscle in the place of oxygen, as a result of a combination with the substance of the blood corpuscle.

Here the experimental analysis has achieved its goal. It is one of the rare examples in physiology that I am happy to be able to cite. Here the *proximate cause* of the phenomenon of poisoning has been discovered and it is translated into a theoretical expression which takes into account all the facts and which includes at the same time all the observations and experiments. The theory thus formulated produces the principal fact from which all the others are deduced: *carbon monoxide combines more strongly than oxygen with hemoglobin in a blood corpuscle*. It has been proved quite recently that carbon monoxide forms a definite combination with hemoglobin,[13] such that the blood corpuscle, as if petrified by the stability of that combination, loses its vital properties. From then on, everything is deduced logically: carbon monoxide, because of its *property* of stronger combination, expels the oxygen from the blood which is essential for life; the blood corpuscles become inert and we watch the animal die with the symptoms of hemorrhage, from a true paralysis of the corpuscles.

But when a theory is good and offers the real and determined *physico-chemical* cause of phenomena, it not only covers the observed facts, but also can predict others and lead to reasoned applications, which will represent the logical consequences of the theory. Here we again encounter this criterion. In fact, if carbon monoxide has the property of removing oxygen by combining with the blood corpuscle in its place, one could use this gas to analyze the gases in the blood and in particular to determine the existence of oxygen. I deduced this application from my experiments, which today has generally been adopted today.[14] Applications to legal medicine have also been made of this principle of carbon monoxide for uncovering the coloring matter of blood, and from the physiological facts pointed out above we can also already derive consequences relating to hygiene, to experimental pathology, and notably to the mechanisms of certain kinds of anemia.

Without doubt, in all the deductions from the theory, experimental verification will always be

13 Hoppe-Seyler, *Handbuch der physiologisch und pathologisch chemischen Analyse*. Berlin, 1865.

14 Claude Bernard, *De l'emploi de l'oxyde de carbone pour la détermination de l'oxygène au sang* (*Compt. rend. de l'Acad. des sciences*, meeting of September 6, 1858, vol. 47).

necessary as usual, and logic is not sufficient; but that is because the conditions of the action of carbon monoxide on the blood can present other complex circumstances and a host of details that the theory cannot yet predict. Otherwise, as we have often said (Cf. p. 52), we could reach conclusions only through logic, and without the need of experimental verification. It is thus because of new and unforeseen variables, which can introduce themselves into the conditions of a phenomenon, that logic alone is never sufficient in the experimental sciences. Even when we have a theory which appears good, it is only relatively good and it always encompasses a certain proportion of the unknown.

II.—Where the Starting Point of Experimental Research is an Hypothesis or a Theory.

We have already said (p. 46) and we shall see later that in noting an observation, we must never go beyond the facts. But it is not the same in carrying out an experiment. I want to demonstrate therefore that hypotheses are indispensable and that their utility lies precisely in leading us outside of the fact and carrying science forward. Hypotheses have as their object to make us not only carry out new experiments, but also often discover new facts which we would not have noticed without them. In the preceding examples, we have seen that one can begin from a particular fact to gradually rise towards more general ideas, that is to say, towards a theory. But it also happens, as we have just seen, that one can begin from an hypothesis deduced from a theory. In this case, although it concerns a reasoning logically deduced from a theory, it is nevertheless still an hypothesis that we must verify by experiment. Here in fact the theories only represent to us an assemblage of previous facts on which the hypothesis is supported, but cannot be used to demonstrate it experimentally. We have said that in this case we must not submit to the yoke of theory, and that the best condition for finding truth and for making progress in science was to retain the independence of the mind (Cf. p. 80). The following examples will prove this.

First example.—In 1843, in one of my first works, I undertook a study to determine what happened to different alimentary substances during nutrition. I began, as I have already said, with sugar, a definite substance which is easier to recognize and to follow throughout the economy of an organism than any other substance. To this end, I injected solutions of cane sugar into the blood of some animals and I established that this sugar, even when injected into the blood in small doses, passed into the urine. I later recognized that the gastric juice in modifying or transforming this cane sugar, rendered it assimilable, that is to say, destructible in the blood.[15]

Then I wanted to know in which organ this alimentary sugar disappeared and I hypothesized that the sugar which nutrition introduced into the blood might be destroyed in either the lungs or the general capillaries. In fact, the dominant theory at that time, which naturally was my starting point, claimed that the sugar which existed in animals originated exclusively in food and that it was destroyed in the animal organism by the phenomenon of combustion, that is to say, respiration. This is why sugar was given the name *respiratory nutriment*. But I was immediately led to see that the theory of the origin of sugar in animals, which served as my starting point, was false. In fact, by a succession of experiments which I will describe later, I did not find the organ which destroyed sugar, but on the contrary I discovered an organ which created that substance, and I also found that the blood of all animals contained sugar, even when they did not consume it. I therefore established a new fact, unexpected from the theory and which no one had ever noticed, undoubtedly, because they were enthralled by contrary theoretical ideas which they had accepted with too much confidence. I then abandoned right away all my hypotheses on the destruction of sugar, to follow up on this unexpected result which has since become the fertile origin of a new path of investigations as well as a mine of discoveries which is still far from being exhausted.

In these investigations I proceeded according to the principles of the experimental method

15 Claude Bernard, doctoral thesis in medicine. Paris, 1843.

which we have established; that is to say, in the presence of a new well-established fact in contradiction to a theory, instead of preserving the theory and abandoning the fact, I kept the fact which I had studied, and hastened to drop the theory, in conformance with that principle which we discussed in the second chapter: *When a fact which we encounter is in opposition to a prevailing theory, we must accept the fact and abandon the theory, even when that theory, upheld by great names, has been generally adopted.*

We must distinguish, as we have said, between principles and theories and never to believe in theories in an absolute manner. Here we had a theory according to which it was believed that only the vegetable kingdom possessed the power to create the primary substances which the animal kingdom was supposed to break down. According to this established theory upheld by the most illustrious contemporary chemists, animals were incapable of producing sugar in their organism. If I had believed in that theory absolutely, I would have had to conclude that my experiment must have been contaminated with error, and perhaps experimenters less distrustful than myself would have immediately condemned my experiment and would have not lingered for long over an observation which could be the source of error according to the theory, since the observation showed sugar in the blood of animals subjected to a diet lacking starchy or sugary materials. But instead of preoccupying myself with the validity of the theory, I only concentrated on the fact and attempted to establish its reality. Thus through new experiments and by the method of suitable counter-proofs, I was led to confirm my original observation and to discover that the liver was an organ where in certain given situations animal sugar is formed which consequently flows throughout the bloodstream and throughout the tissues and organic fluids.

This animal glycogenesis, that is to say the faculty possessed by both animals and plants to produce sugar, is now an established fact in science, although we have not yet determined a plausible theory to account for the phenomenon. The new facts which I have brought to light, have been the source of a great number of studies and many diverse theories in apparent contradiction among themselves as well as with my own. When we enter into new ground, we should not be afraid to express even risky views in order to excite investigations in all directions. We should not, following the expression of Priestly, remain inactive through false modesty or the fear of being mistaken. I therefore advanced some more or less hypothetical theories on glycogenesis: since mine were offered, others have been made: my theories, just as those of others, will only live as necessarily partial and provisional theories must live, when a new series of investigations begins; they will soon be replaced by others which represent a more advanced state of science, and so forth. Theories represent successive degrees by which science ascends in enlarging its horizon more and more, because the further they are advanced, the more facts theories encompass and explain. True progress consists of exchanging our theory for new ones which go further than the original, until we find one that rests on a greater number of facts. In the case we are discussing, the point is not to condemn the older theory to the profit of the newer. What is important is to open a new road, because what will never perish are well observed facts which ephemeral theories have brought to life; these are the only materials upon which the edifice of science will be built on the day when she possesses a sufficient number of facts and when she has penetrated far into the analysis of phenomena to discover the law or the exact causation.

In summation, theories are only hypotheses verified by a lesser or greater number of facts; those which are verified by the greatest number of facts are the best; but still they are never definitive and one should never believe in them absolutely. We have seen from the preceding examples, that if we had placed our entire confidence in the prevailing theory of the destruction of sugar in animals, and if we had only sought its confirmation, we would not have followed up on the new facts which we encountered. The hypothesis founded on a theory provoked the experiment, but as soon as the results of the experiment appeared, the theory and the hypothesis had to disappear, because the experimental fact was no more than an observation made without any preconceived idea (Cf. p. 40).

Thus the main principle in such complex and little developed sciences as physiology is to pay little attention to hypotheses or theories and to always keep an attentive eye for observing everything that happens in an experiment. A circumstance which appears accidental and inexplicable can become the occasion of the discovery of a new important fact, as we shall see in the continuation of the example previously cited.

Second example, sequel to the last.—After having discovered, as I have mentioned above, that animal liver contains sugar in its normal state and with every kind of diet, I wanted to know the proportion of this substance and its variations in certain physiological and pathological states. I thus began determining the amount of sugar in the livers of animals placed in various defined physiological conditions. I always made two simultaneous determinations of the sugary material in the same liver tissue. But one day pressed for time, it happened that I was not able to make my two analyses at the same time; I immediately made a rapid determination after the death of the animal, and left the other for the next day. But I found this time much greater quantities of sugar than those which I had obtained the previous day in the same hepatic tissue, and I noticed in addition that the proportion of sugar which I had found in the liver, immediately examined after the death of the animal, was much lower than what I had encountered in other experiments which I had made known as giving the normal proportion of hepatic sugar. I did not know to what to attribute the peculiar variation obtained from the same liver and with the same analytical method. What was to be done? Should I have considered the two discordant amounts as the result of an unsuccessful experiment and to ignore them? Should I have averaged the results of these two experiments? That is an expedient which many experimenters would have chosen to deliver themselves from this awkward situation. But I do not approve of this manner of operating for reasons which I have given elsewhere. I have said in effect that nothing must be overlooked in the observation of facts, and I regard it as an indispensable rule of experimental criticism (p. 299) to never admit without proof the existence of a cause of error in an experiment, and to always search for a reason for all the abnormal circumstances which we observe. Nothing occurs by accident, and that which seems accidental is nothing more than an unknown fact which could become, if it were explained, the occasion of a more or less important discovery. This is what happened to me in this case.

I wanted to know in fact what the reason was for the two very different amounts of sugar in the liver of my rabbit. After convincing myself that there was no error in the method of measuring the sugar; and after establishing that the different parts of the liver are equally rich in sugar, the only thing left to examine was the influence of time which had elapsed since the death of the animal, up until the time of my second measurement. Up until then, I had always made my experiments, without attaching any importance to the fact, several hours after the death of the animal, and for the first time, I found myself in the situation of immediately taking a measurement several minutes after the death of the animal, and leaving the other until the next day, that is to say 24 hours afterwards. In physiology, questions of time always have great importance, because the organic matter undergoes numerous and incessant changes. Thus some chemical modification could have been produced in the hepatic tissue. To reassure myself, I made a series of new experiments which dissipated all the obscurities by showing me that the liver tissue constantly enriches itself with sugar during a certain time after death, such that one can find very variable quantities of sugar, based on the moment in which one makes the examination. I was thus led to rectify my former measurements and to discover this new fact; namely, that considerable quantities of sugar are produced in the livers of animals after death. I demonstrated, for example, by passing a stream of cold water injected by force through the hepatic vessels into a still warm animal liver soon after death, that one can completely remove the sugar contained in the hepatic tissue; but the next day, or several hours afterward, when I brought the liver to a mild temperature, I found the liver charged again with a great quantity of sugar produced since the washing.[16]

After I had made this first discovery that sugar is formed in animals after death just as during life, I wanted to extend my examination of this singular phenomenon, and it was thus that I was lead to discover that sugar is produced in the liver with the aid of an enzymatic material acting on a starchy substance which I isolated and gave the name of *glycogenic substance*, such that I was able to demonstrate in a brief manner that sugar is formed in animals by a similar mechanism to that found in plants.

This second series of facts is today widely accepted in science and has contributed to great progress in the understanding of glycogenesis in animals. I have just succinctly described how these facts were discovered and how their starting point proceeded from an apparently useless experimental circumstance. I have described these events in order to prove that we should never neglect anything in experimental investigations; because all accidents have their necessary cause. We should never become too absorbed therefore by the thought one is following, nor fool ourselves about the value of our scientific ideas; we must always keep our eyes open to every event, and maintain a doubting and independent mind (p. 138) disposed to examine everything that happens and not to let anything pass without searching for the reason why. We must remain, in a word, in an intellectual disposition which seems paradoxical, but which, in my opinion, represents the true mind of the investigator. We must *have a robust faith and not believe*; I explain this by saying that we must believe firmly in principles and to doubt formulas; that we must remain unshakable with respect to the principles of experimental medicine (determinism) and not to believe absolutely in theories. The aphorism which I have just expressed can be supported by what we have developed elsewhere (Cf. p. 116), namely, that the principles of the experimental sciences are in our mind, while formulas exist in exterior things. In practice we are forced to believe that the truth (or the provisional truth, at least) is represented by the theory or by the formula, but in scientific philosophy those who place their faith in formulas or theories are wrong. All human knowledge is directed towards finding the correct formula or the correct theory of truth in some realm. We always approach truth, but will we ever completely find it? This is not the place to develop these philosophical ideas; let us return to our subject and look at a new experimental example.

Third example.—Around 1852, I was led by my studies to make experiments on the influence of the nervous system on the phenomena of nutrition and heat. It had been observed before me that in a great many cases, complex paralyses situated in the mixed nerves are followed at one time by a heating, and another time by a cooling of the paralyzed parts. Now here is how I reasoned to explain the fact by relying on observations, on one hand, and on the other hand, on the prevailing theories relative to the phenomena of nutrition and heat. The paralysis of the nerves, I told myself, must lead to the cooling of the nerves by slowing down the phenomena of combustion in the blood, since these phenomena are considered the cause of animal heat. Now, on the other hand, anatomists have noticed for a long time that the sympathetic nerves in particular accompany the arteries. Thus, I thought by induction, it must be the sympathetic nerves which, in the lesion of a mixed trunk of nerves, act to slow down the chemical phenomena in the capillary vessels, and it is their paralysis which must bring about the cooling of the parts. If my hypothesis were true, I added, it should be verifiable by cutting only the sympathetic vascular nerves leading to a certain part and carefully leaving the other nerves. I should then obtain a cooling by paralysis of the vascular nerves without the movement or sensibility disappearing, since I would have left the motor and ordinary sensory nerves intact. In order to carry out my experiment I sought a suitable method of experimentation which would allow me to cut the vascular nerves alone while leaving the other nerves. Here the choice of animal became important rela-

16 Claude Bernard, *Sur le mécanisme de la formation du sucre dans le foie* (*Comptes rendus de l'Acad. des sciences*, September 24, 1855).—Sequel (*Comptes rend. de l'Acad. des sciences*, March 23, 1857).

tive to the solution of the question (p. 213); but the anatomical arrangement which isolates the great cervical sympathetic nerve in certain animals such as the rabbit and the horse, made this solution possible.

According to this reasoning, I therefore cut the great sympathetic nerve in the neck of a rabbit to control my hypothesis and to see what would happen to the change in heat in the side of the head where that nerve was branched out. I had been led, as we have just seen, by a reliance on the prevailing theory and by previous observations, to make the hypothesis that the temperature should be lowered by the cutting of the sympathetic nerve. Now precisely the opposite happened. Soon after severing the great sympathetic nerve in the middle of the neck of the rabbit, I noticed in the entire corresponding side of the head a considerable increase in activity in the circulation accompanied by a rise in heat. The result was therefore exactly opposite to what my hypothesis deduced from the theory had made me predict; but I proceeded as usual, that is to say, I abandoned immediately the theories and the hypotheses in order to observe and study the fact in itself in order to determine as exactly as possible the experimental conditions. Today these experiments on the vascular and thermoregulatory nerves have opened up a new road of research and have become the subject of a great number of works which, I hope, will one day furnish results of great importance to physiology and pathology.[17]

This example proves, like the preceding, that in experiments one can encounter results different from those which we expect from the theories and hypotheses. But if I am calling particular attention to this third example, it is because it offers us an important lesson, that is, without this directing hypothesis of the mind, the experimental fact which contradicted it would never have been noticed. In fact, I was not the first experimenter who had cut the cervical portion of the great sympathetic nerve in living animals. Pourfour du Petit carried out this experiment at the beginning of the last century and he discovered the effects of this nerve on the pupil by reference to an anatomical hypothesis according to which this nerve was supposed to carry the animal spirits into the eyes.[18] Since then many physiologists have repeated the same operation in order to verify or to explain the modifications of the eye which Pourfour du Petit was the first to notice. But none of these physiologists had noticed the phenomenon of the rise in temperature of the parts which I am speaking of and never linked it to the severing of the great sympathetic nerve, even though this phenomenon must have necessarily been produced under the eyes of everyone who, before me, had cut that part of the sympathetic nerve. The hypothesis, as we see, had prepared my mind to see these things from a certain point of view provided by the hypothesis itself, and what that proves is that even I, like the other experimenters, had often cut the great sympathetic nerve to repeat the experiment of Pourfour du Petit, without seeing the fact of heating which I discovered later when an hypothesis led me to make investigations in this direction. The influence of the hypothesis is thus very evident here; we had the fact under our eyes and we did not see anything because it said nothing to the mind. It was extremely simple to observe however, and ever since I pointed it out, every physiologist without exception has noticed and verified it with the greatest facility.

In sum, hypotheses and theories, even when unsuccessful, are useful to lead us to discoveries. This remark is true for all the sciences. The alchemists founded chemistry by following chimerical problems and theories known to be false today. In the physical sciences, which are more advanced than biology, we can still cite researchers who make great discoveries by relying on false theories. This appears to be, in fact, a necessary feebleness of our mind that we are

17 Claude Bernard, *Recherches expérimentales sur le grand sympathique, etc.* (*Mémoires de la Société de biologie*, vol. 5, 1853).—*Sur les nerfs vasculaires et calorifiques du grand sympathique* (*Comptes rendus de l'Acad. des sciences*, 1852, vol. 34, 1862, vol. 55).

18 Pourfour du Petit, *Mémoire dans lequel il est démontré que les nerfs intercostaux fournissent des rameaux qui portent des esprits dans les yeux (Historie de l'Académie pour l'année* 1727).

unable to arrive at truth except through a multitude of errors and obstacles.

What general conclusion can the physiologist draw from the preceding examples? He must conclude that in the present state of biological science the ideas and the theories which prevail only represent very circumscribed and precarious truths which are destined to perish. He should consequently hold little confidence in the real value of these theories, but use them as intellectual instruments necessary to the evolution of the science and appropriate for stimulating the discovery of new facts. Today the art of discovering new phenomena and establishing them exactly must be the special object of all biologists. We must at the same time ground experimental criticism by creating rigorous methods of investigation and experimentation which will permit us to establish observations in an indisputable manner and consequently to banish errors of facts which are the source of unsuccessful theories. Any one who attempted now to support a generalization for the entire field of biology would demonstrate that he had no exact idea of the present state of that science. Today the problem of biology has hardly been posed, and even as it necessary to assemble and cut the stones before we dream about constructing a building, so it is also necessary first to assemble and prepare the facts which must constitute the science of living bodies. This role falls to experimentation—its method is fixed—but the phenomena which it must examine are so complex that the real mover of the science for the moment will be the person who provides several principles of simplification in analytical methods or who perfects the instruments of research. When enough clearly established facts exist, generalizations never slow us down. I am convinced that in the evolving experimental sciences, and particularly in those which are as complex as biology, the discovery of a new instrument for observation or experimentation renders greater theoretical service than all the theoretical and philosophical dissertations. In fact, a new process, a new method of investigation increases our power and allows the possibility of discoveries and investigations which would have been impossible without them. Thus investigations into the formation of sugar in animals could not be carried out before chemistry had given us reagents to recognize sugar which were more sensitive than those we had before.

FURTHER READING

Bergson, Henri. "La Philosophie de Claude Bernard." In *La Pensée et le mouvant: Essais et conférences*. Paris: Presses Universitaires de France, 1960.

Bernard, Claude. *An Introduction to the Study of Experimental Medicine.* Translated by Henry Copley Greene. London: Macmillan and Co., Ltd., 1927.

—. *Lectures on the Phenomena of Life Common to Animals and Plants.* Translated by H. Hoff, R. Guillemin, and L. Guillemin. Springfield, IL: Charles C. Thomas, 1974.

—. *Cahier de notes 1850–1860.* Edited by M.D. Grmek. Paris: Gallimard, 1965.

Binet, Léon, ed. *Esquisses et notes de travails inédites de Claude Bernard.* Paris: Masson et Compagnie, 1952.

Buchholz, Gerhard. *Die Medizintheorie Claude Bernards.* Herzogenrath: Murken-Altrogge, 1985.

Coleman, William, and Frederic L. Holmes, eds. *The Investigative Enterprise: Experimental Physiology in Nineteenth-Century Medicine.* Berkeley: University of California Press, 1988.

Cotard, Henri. *Pour connaître la pensée de Claude Bernard.* Grenoble: Éditions Françaises Nouvelles, 1898.

Cranefield, Paul F. *Claude Bernard's Revised Edition of His Introduction à l'étude de la médecine expérimentale.* New York: Science History Publications, 1976.

Delhoume, Léon. *De Claude Bernard à d'Arsonval.* Paris: J.B. Baillière et Fils, 1939.

—. *Principes de médecine experimentale.* Edited by Léon Delhoume. Paris: Presses Universitaires de France, 1947.

Duval, Mathias *et al. L'Oeuvre de Claude Bernard.* Paris: J.B. Baillière et Fils, 1881.

Faure, Jean-Louis. *Claude Bernard, avec un portrait hors texte.* Paris: Les Éditions G. Crès et Compagnie, 1925.

Foster, Michael. *Claude Bernard.* New York: Longmans, Green, and Co., 1899.

Grande, Francisco, and Maurice B. Visscher, eds. *Claude Bernard and Experimental Medicine and the First English Translation of Claude Bernard's* Cahier Rouge. The *Cahier rouge* translated by H. Hoff, L. Guillemin, and R. Guillemin. Cambridge, MA: Schenkman Publishing Co., Inc., 1967.

Grmek, M.D. *Raisonnement expérimental et recherches toxicologiques chez Claude Bernard.* Genève: Librairie Droz, 1973.

Heim, Roger, *et al. Les Concepts de Claude Bernard sur le milieu intérieur.* Paris: Masson et Compagnie, 1967.

Holmes, F.L. *Claude Bernard and Animal Chemistry: The Emergence of a Scientist.* Cambridge, MA: Harvard UP, 1974.

Lesch, John E. *Science and Medicine in France: The Emergence of Experimental Physiology, 1790–1855.* Cambridge, MA: Harvard UP, 1984.

Mauriac, Pierre. *Claude Bernard.* Paris: Éditions Bernard Grasset, 1954.

Michel, Jacques. *La Nécessité de Claude Bernard.* Paris: Méridiens Klincksieck, 1991.

Olmsted, J.M.D. *Claude Bernard, Physiologist.* New York: Harper and Brothers, 1938.

Olmsted, J.M.D., and E. Harris Olmsted. *Claude Bernard and the Experimental Method in Medicine.* New York: Henry Schuman, 1952.

Parvez, H., and S. Parvez. *Advances in Experimental Medicine: A Centenary Tribute to Claude Bernard.* Amsterdam: Elsevier / North-Holland Biomedical Press, 1980.

Prochiantz, Alain. *Claude Bernard: La Révolution physiologique.* Paris: Presses Universitaires de France, 1990.

Robin, Eugene Debs, ed. *Claude Bernard and the Internal Environment: A Memorial Symposium.* New York: Marcel Dekker, Inc., 1979.

Schiller, Joseph. *Claude Bernard et les problèmes scientifiques de son temps.* Paris: Éditions du Cèdre, 1967.

Sertillanges, A.-D. *La philosophie de Claude Bernard.* Paris: Aubier, 1943.

Virtanen, Reino. *Claude Bernard and his Place in the History of Ideas.* Lincoln: University of Nebraska Press, 1960.

Wolff, Etienne, *et al. Philosophie et méthodologie scientifiques de Claude Bernard.* Paris: Masson et Compagnie, 1967.

26. JOSEPH LISTER

(1827–1912)

Joseph Lister was born in Upton, Essex. His father, the prominent Quaker Joseph Jackson Lister, perfected the achromatic microscope lens and participated in the Royal Society. In 1852, Lister obtained his M.B. degree from University College, London and later studied with James Syme in Edinburgh. He eventually married Syme's daughter Agnes. His career was marked by a steady rise in influence and power: 1860—appointment to the Regius Professorship of Surgery at Glasgow University; 1877—appointment to Professor of Clinical Surgery at King's College, London; 1880—recipient of honorary degrees from Oxford and Cambridge; 1895—election to the presidency of the Royal Society; and 1897—elevation to the title of Baron by Queen Victoria.

Lister's first article on Antisepsis (the prevention of post-operation infection), entitled "On a New Method of Treating Compound Fracture, Abscess, Etc.," appeared in the *Lancet* in 1867. At this time, amputation was the common procedure for treating compound fractures of bones since because of hospital overcrowding and general uncleanliness the little understood post-operative bacterial diseases of septicemia (blood poisoning), erysipelas (skin infection), and pyemia (gangrene) frequently followed in patients suffering from open wounds. Sir James Simpson aptly called these three diseases by the collective name "hospitalism." Unsanitary conditions and the re-use of unclean bedding and surgical instruments from patient to patient, especially in institutions attending to the poor, were common practices in early nineteenth-century hospitals.

It was generally believed that post-operative diseases were the result of the chemical oxidation of flesh in contact with the air. Schwann, Cagniard de la Tour, Tyndall, and Pasteur all provided evidence that putrefaction, and by extension disease, was due to microorganisms floating in the air—as opposed to chemical oxidation—, and Lister reasoned that killing these "floating corpuscles" before they entered a wound from the air would prevent suppuration. He decided to try carbolic acid (phenol) as an antiseptic wash for killing germs, after learning that it had been used to control odours from rotting sewage.

Lister also developed a carbolic acid spray to directly sterilize the air, but later in his career admitted its ineffectiveness, after the greater importance of bacterial transmission by surgical instruments and the surgeon's hands became known. He experimented with a variety of methods involving carbolic acid, including wiping the wound directly with the acid and dressing the wound with acid-soaked lint. He also employed a bandage soaked in carbolic acid and held in place by tin sheeting and tape. Antiseptic methods were not entirely unknown in medicine, however: antiseptic soaps and washes, for example, appear in the works of Galen (circa 150 C.E.), and the general relationship between uncleanliness and infection was widely known in many cultures. Unfortunately, carbolic acid was found to be too toxic and in the 1880s bacteriologists became skeptical about its bactericidal properties, so other alternatives such as mercuric chloride were introduced in Germany and later embraced by Lister.

Lister's original antiseptic program was eventually modified in favour of the related modern regime of aseptic surgery (autoclaving and sterilization of instruments and garments by steam and heat, use of disinfectant soap, alcohol, bleach, etc.), a change perhaps in methodology but not in the general principle of preventative hygiene. Although increases in post-operation survival rates in the late nineteenth century may have been related to a variety of factors besides Lister's innovations, including better nutrition, improvements in overall hospital cleanliness (also advocated by Lister), and the general acceptance of the germ theory of disease, Lister is widely viewed as a crusader in advocating modern hospital procedures of infection prevention. His eulogist and biographer W.W. Cheyne, for example, described him "as one of the knights of olden times sallying out single-handed to find and destroy a formidable enemy, armed with a great but imperfect generalisation, or rather, working hypothesis, and a bottle of crude carbolic acid so impure that he tells us in his first paper that carbolic acid was insoluble in water" (Cheyne 13). In his second paper on Antisepsis, excerpted below, Lister provides a history and explanation of the Antiseptic Principle.

"On the Antiseptic Principle in the Practice of Surgery."

The Collected Papers of Joseph, Baron Lister.
2 vols. Oxford: Clarendon Press, 1909, pp. 2.37–45.
Originally published in the *British Medical Journal* 2 (1867): 246ff.

In the course of an extended investigation into the nature of inflammation, and the healthy and morbid conditions of the blood in relation to it, I arrived, several years ago, at the conclusion that the essential cause of suppuration in wounds is decomposition, brought about by the influence of the atmosphere upon blood or serum retained within them, and, in the case of contused wounds, upon portions of tissue destroyed by the violence of the injury.

To prevent the occurrence of suppuration, with all its attendant risks, was an object manifestly desirable; but till lately apparently unattainable, since it seemed hopeless to attempt to exclude the oxygen, which was universally regarded as the agent by which putrefaction was effected. But when it had been shown by researches of Pasteur that the septic property of the atmosphere depended, not on the oxygen or any gaseous constituent, but on minute organisms suspended in it, which owed their energy to their vitality, it occurred to me that decomposition in the injured part might be avoided without excluding the air, by applying as a dressing some material capable of destroying the life of the floating particles.

Upon this principle I have based a practice of which I will now attempt to give a short account.

The material which I have employed is carbolic or phenic acid, a volatile organic compound which appears to exercise a peculiarly destructive influence upon low forms of life, and hence is the most powerful antiseptic with which we are at present acquainted.

The first class of cases to which I applied it was that of compound fractures, in which the effects of decomposition in the injured part were especially striking and pernicious. The results have been such as to establish conclusively the great principle, that *all the local inflammatory mischief and general febrile disturbance which follow severe injuries are due to the irritating and poisoning influence of decomposing blood or sloughs*. For these evils are entirely avoided by the antiseptic treatment, so that limbs which otherwise would be unhesitatingly condemned to amputation may be retained with confidence of the best results.

In conducting the treatment, the first object must be the destruction of any septic germs which may have been introduced into the wound, either at the moment of the accident or during the time which has since elapsed. This is done by introducing the acid of full strength into all accessible recesses of the wound by means of a piece of rag held in dressing-forceps and dipped in the liquid.[1] This I did not venture to do in the earlier cases; but experience has shown that the compound which carbolic acid forms with the blood, and also any portions of tissue killed by its caustic action, including even parts of the bone, are disposed of by absorption and organization, provided they are afterwards kept from decomposing. We are thus enabled to employ the antiseptic treatment efficiently at a period after the occurrence of the injury at which it would otherwise probably fail. Thus I have now under my care in the Glasgow Infirmary a boy who was admitted with compound fracture of the leg as late as eight and a half hours after the accident, in whom nevertheless all local and constitutional disturbance was avoided by means of carbolic acid, and the bones were firmly united five weeks after his admission.

The next object to be kept in view is to guard effectually against the spreading of decomposition into the wound along the stream of blood and serum which oozes out during the first few days after the accident, when the acid originally applied has been washed out, or dissipated by absorption and evaporation. This part of the treatment has been greatly improved during the last few weeks. The method which I have hitherto published[2] consisted in the application of a piece of lint dipped in the acid, overlapping the sound skin to some extent, and covered with a tin cap, which was daily raised in order to touch the surface of the lint with the antiseptic. This method certainly succeeded well with wounds of moderate size; and, indeed, I may say that in all the many cases of this kind which have been so treated by myself or my house surgeons, not a single failure has occurred. When, however, the wound is very large, the flow of blood and serum is so profuse, especially during the first twenty-four hours, that the antiseptic application cannot prevent the spread of decomposition into the interior unless it overlaps the sound skin for a very considerable distance, and this was inadmissible by the method described above, on account of the extensive sloughing of the surface of the cutis which it would involve. This difficulty has, however, been overcome by employing a paste composed of common whitening (carbonate of lime) mixed with a solution of one part of carbolic acid in four parts of boiled linseed oil, so as to form a firm putty. This application contains the acid in too dilute a form to excoriate the skin, which it may be made to cover to any extent that may be thought desirable, while its substance serves as a reservoir of the antiseptic material. So long as any discharge continues, the paste should be changed daily; and, in order to prevent the chance of mischief occurring during the process, a piece of rag dipped in the solution of carbolic acid in oil is put on next the skin, and maintained there permanently, care being taken to avoid raising it along with the putty. This rag is always kept in an antiseptic condition from contact with the paste above it, and destroys any germs that may fall upon it during the short time that should alone be allowed to pass in the changing of the dressing. The putty should be in a layer about a quarter of an inch thick, and may be advantageously applied rolled out between two pieces of thin calico, which maintain it in the form of a continuous sheet, that may be wrapped in a moment round the whole circumference of a limb, if this be thought desirable, while the putty is prevented by the calico from sticking to the rag which is

1 The addition of a few drops of water to a considerable quantity of the crystallized acid induces it to assume permanently the liquid form.

2 See the preceding paper in this volume.

next to the skin.[3] When all discharge has ceased, the use of the paste is discontinued, but the original rag is left adhering to the skin till healing by scabbing is supposed to be complete. I have at present in the hospital a man with severe compound fracture of both bones of the left leg, caused by direct violence, who, after the cessation of the sanious discharge under the use of the paste, without a drop of pus appearing, has been treated for the last two weeks exactly as if the fracture were a simple one. During this time the rag, adhering by means of a crust of inspissated blood collected beneath it, has continued perfectly dry, and it will be left untouched till the usual period for removing the splints in a simple fracture, when we may fairly expect to find a sound cicatrix beneath it.

We cannot, however, always calculate on so perfect a result as this. More or less pus may appear after the lapse of the first week; and the larger the wound the more likely is this to happen. And here I would desire earnestly to enforce the necessity of persevering with the antiseptic application, in spite of the appearance of suppuration, so long as other symptoms are favourable. The surgeon is extremely apt to suppose tha[t] any suppuration is an indication that the antiseptic treatment has failed, and that poulticing or water dressing should be resorted to. But such a course would in many cases sacrifice a limb or a life. I cannot, however, expect my professional brethren to follow my advice blindly in such a matter, and therefore I feel it necessary to place before them, as shortly as I can, some pathological principles, intimately connected not only with the point we are immediately considering, but with the whole subject of this paper.

If a perfectly healthy granulating sore be well washed and covered with a plate of clean metal, such as block-tin, fitting its surface pretty accurately, and overlapping the surrounding skin an inch or so in every direction, and retained in position by adhesive plaster and a bandage, it will be found, on removing it after twenty-four or forty-eight hours, that little or nothing that can be called pus is present, merely a little transparent fluid, while at the same time there is an entire absence of the unpleasant odour invariably perceived when water dressing is changed. Here the clean metallic surface presenting no recesses, like those of porous lint, for the septic germs to develop in, the fluid exuding from the surface of the granulations has flowed away undecomposed, and the result is absence of suppuration. This simple experiment illustrates the important fact, that granulations have no inherent tendency to form pus, but do so only when subjected to a preternatural stimulus. Further, it shows that the mere contact of a foreign body does not of itself stimulate granulations to suppurate; whereas the presence of decomposing organic matter does. These truths are even more strikingly exemplified by the fact, which I have elsewhere recorded,[4] that a piece of dead bone, free from decomposition, may not only fail to induce the granulations around it to suppurate, but may actually be absorbed by them; whereas a bit of dead bone soaked with putrid pus infallibly induces suppuration in its vicinity.

Another instructive experiment is to dress a granulating sore with some of the putty above described, overlapping the sound skin extensively, when we find in the course of twenty-four hours that pus has been produced by the sore, although the application has been perfectly antiseptic; and, indeed, the larger the amount of carbolic acid in the paste the greater is the quantity of pus formed, provided we avoid such a proportion as would act as a caustic. The carbolic acid, though it prevents decomposition, induces suppuration—obviously by acting as a chemical stimulus; and we may safely infer that putrescent organic materials (which we know to be chemically acrid) operate in the same way.

In so far, then, carbolic acid and decomposing substances are alike—namely, that they induce

3 In order to prevent evaporation of the acid, which passes readily through any organic tissue, such as oiled silk or gutta percha, it is well to cover the paste with a sheet of block-tin, or tinfoil strengthened with adhesive plaster. The thin sheet-lead for lining tea-chests will also answer the purpose, and may be obtained from any wholesale grocer.

4 See p. 16 of this volume.

suppuration by chemical stimulation, as distinguished from what may be termed simple inflammatory suppuration, such as that in which ordinary abscesses originate, where the pus appears to be formed in consequence of an excited action of the nerves, independently of any other stimulus.

There is, however, this enormous difference between the effects of carbolic acid and those of decomposition—viz. that carbolic acid stimulates only the surface to which it is first applied, and every drop of discharge that forms weakens the stimulant by diluting it. But decomposition is a self-propagating and self-aggravating poison; and if it occurs at the surface of a severely injured limb, it will spread into all its recesses so far as any extravasated blood or shreds of dead tissue may extend, and, lying in these recesses, it will become from hour to hour more acrid till it acquires the energy of a caustic, sufficient to destroy the vitality of any tissues naturally weak from inferior vascular supply, or weakened by the injury they sustained in the accident.

Hence it is easy to understand how, when a wound is very large, the crust beneath the rag may prove here and there insufficient to protect the raw surface from the stimulating influence of the carbolic acid in the putty, and the result will be, first, the conversion of the tissues so acted on into granulations, and subsequently the formation of more or less pus. This, however, will be merely superficial, and will not interfere with the absorption and organization of extravasated blood or dead tissues in the interior; but, on the other hand, should decomposition set in before the internal parts have become securely consolidated, the most disastrous results may ensue.

I left behind me in Glasgow a boy, thirteen years of age, who between three and four weeks previously met with a most severe injury to the left arm, which he got entangled in a machine at a fair. There was a wound six inches long and three inches broad, and the skin was very extensively undermined beyond its limits, while the soft parts generally were so much lacerated that a pair of dressing-forceps introduced at the wound, and pushed directly inwards, appeared beneath the skin at the opposite aspect of the limb. From this wound several tags of muscle were hanging, and among them there was one consisting of about three inches of the triceps in almost its entire thickness; while the lower fragment of the bone, which was broken high up, was protruding four and a half inches, stripped of muscle, the skin being tucked in under it. Without the assistance of the antiseptic treatment, I should certainly have thought of nothing else but amputation at the shoulder-joint; but as the radial pulse could be felt, and the fingers had sensation, I did not hesitate to try to save the limb, and adopted the plan of treatment above described, wrapping the arm from the shoulder to below the elbow in the antiseptic application, the whole interior of the wound, together with the protruding bone, having previously been freely treated with strong carbolic acid. About the tenth day the discharge, which up to that time had been only sanious and serous, showed a slight admixture of slimy pus, and this increased till, a few days before I left, it amounted to about three drachms in twenty-four hours. But the boy continued, as he had been after the second day, free from unfavourable symptoms, with pulse, tongue, appetite, and sleep natural, and strength increasing, while the limb remained, as it had been from the first, free from swelling, redness, or pain. I therefore persevered with the antiseptic dressing, and before I left, the discharge was already somewhat less, while the bone was becoming firm. I think it likely that in that boy's case I should have found merely a superficial sore had I taken off all the dressings at the end of three weeks, though, considering the extent of the injury, I thought it prudent to let the month expire before disturbing the rag next the skin. But I feel sure that if I had resorted to ordinary dressing when the pus first appeared, the progress of the case would have been exceedingly different.

The next class of cases to which I have applied the antiseptic treatment is that of abscesses. Here, also, the results have been extremely satisfactory, and in beautiful harmony with the pathological principles indicated above. The pyogenic membrane, like the granulations of a sore, which it resembles in nature, forms pus, not from any inherent disposition to do so, but only because it is subjected to some preternatural stimulation. In an ordinary abscess, whether acute or chronic, before it is opened, the stimulus which maintains the suppuration is derived from the presence of the pus pent up within the cavity. When a free opening is made in the ordinary way, this stimulus is got rid of; but the atmosphere gaining access to the contents, the potent stimulus of decomposition comes into operation, and pus is generated in greater abundance than before. But when the evacuation is effected on the antiseptic principle, the pyogenic membrane, freed from the influence of the former stimulus without the substitution of a new one, ceases to suppurate (like the granulations of a sore under metallic dressing), furnishing merely a trifling amount of clear serum, and, whether the opening be dependent or not, rapidly contracts and coalesces. At the same time any constitutional symptoms previously occasioned by the accumulation of the matter are got rid of without the slightest risk of the irritative fever or hectic hitherto so justly dreaded in dealing with large abscesses.

In order that the treatment may be satisfactory, the abscess must be seen before it has opened. Then, except in very rare and peculiar cases,[5] there are no septic organisms in the contents, so that it is needless to introduce carbolic acid into the interior. Indeed, such a proceeding would be objectionable, as it would stimulate the pyogenic membrane to unnecessary suppuration. All that is necessary is to guard against the introduction of living atmospheric germs from without, at the same time that free opportunity is afforded for the escape of discharge from within.

I have so lately given elsewhere[6] a detailed account of the method by which this is effected, that it is needless for me to enter into it at present, further than to say that the means employed are the same as those described above for the superficial dressing of compound fractures—namely, a piece of rag dipped in the solution of carbolic acid in oil, to serve as an antiseptic curtain, under cover of which the abscess is evacuated by free incision; and the antiseptic paste, to guard against decomposition occurring in the stream of pus that flows out beneath it: the dressing being changed daily till the sinus has closed.

The most remarkable results of this practice in a pathological point of view have been afforded by cases where the formation of pus depended upon disease of bone. Here the abscesses, instead of forming exceptions to the general class in the obstinacy of the suppuration, have resembled the rest in yielding in a few days only a trifling discharge; and frequently the production of pus has ceased from the moment of the evacuation of the original contents. Hence it appears that caries, when no longer labouring, as heretofore, under the irritation of decomposing matter, ceases to be an opprobrium of surgery, and recovers like other inflammatory affections. In the publication before alluded to[7] I have mentioned the case of a middle-aged man with psoas abscess depending on diseased bone, in whom the sinus finally closed after months of patient perseverance with the antiseptic treatment. Since that article was written I have had another instance of success, equally gratifying, but differing in the circumstance that the disease and the recovery were both more rapid in their course. The patient was a blacksmith who had suffered four and a half months before I saw him from symptoms of ulceration of cartilage in the left elbow. These had latterly increased in severity, so as to deprive him entirely of his night's rest and of appetite. I found the region of the elbow greatly swollen, and on careful examination discovered a fluctuating point at the outer aspect of the articulation. I opened it on the antiseptic principle, the incision evidently penetrating to the joint, giving exit to a few drachms of pus. The medical gentleman under whose care he was (Dr. Macgregor of Glasgow) supervised the daily dressing with the carbolic-acid paste till the patient went to spend two or three weeks at the coast, when his wife was entrusted with it. Just two months after I opened the abscess he called to show me the limb, stating

5 As an instance of one of these exceptional cases, I may mention that of an abscess in the vicinity of the colon, and afterwards proved by post mortem examination to have once communicated with it. Here the pus was extremely offensive when evacuated, and exhibited vibrios under the microscope.

6 See p. 32 of this volume.

7 See p. 36 of this volume.

that the discharge had for at least two weeks been as little as it then was—a trifling moisture upon the paste, such as might be accounted for by the little sore caused by the incision. On applying a probe guarded with an antiseptic rag, I found that the sinus was soundly closed, while the limb was free from swelling or tenderness; and, although he had not attempted to exercise it much, the joint could already be moved through a considerable angle. Here the antiseptic principle had effected the restoration of a joint which on any other known system of treatment must have been excised.

Ordinary contused wounds are of course amenable to the same treatment as compound fractures, which are a complicated variety of them. I will content myself with mentioning a single instance of this class of cases. In April last a volunteer was discharging a rifle, when it burst, and blew back the thumb with its metacarpal bone, so that it could be bent back as on a hinge at the trapezial joint, which had evidently been opened, while all the soft parts between the metacarpal bones of the thumb and fore-finger were torn through. I need not insist before my present audience on the ugly character of such an injury. My house surgeon, Mr. Hector Cameron, applied carbolic acid to the whole raw surface, and completed the dressing as if for compound fracture. The hand remained free from pain, redness, or swelling, and, with the exception of a shallow groove, all the wound consolidated without a drop of matter, so that if it had been a clean cut, it would have been regarded as a good example of primary union. The small granulating surface soon healed, and at present a linear cicatrix alone tells of the injury he had sustained, while his thumb has all its movements and his hand a firm grasp.

If the severest forms of contused and lacerated wounds heal thus kindly under the antiseptic treatment, it is obvious that its application to simple incised wounds must be merely a matter of detail. I have devoted a good deal of attention to this class, but I have not as yet pleased myself altogether with any of the methods I have employed. I am, however, prepared to go so far as to say that a solution of carbolic acid in twenty parts of water, while a mild and cleanly application, may be relied on for destroying any septic germs that may fall upon the wound during the performance of an operation; and also that for preventing the subsequent introduction of others, the paste above described, applied as for compound fractures, gives excellent results. Thus I have had a case of strangulated inguinal hernia, in which it was necessary to take away half a pound of thickened omentum, heal without any deep-seated suppuration or any tenderness of the sac or any fever; and amputations, including one immediately below the knee, have remained absolutely free from constitutional symptoms.

Further, I have found that when the antiseptic treatment is efficiently conducted, ligatures may be safely cut short and left to be disposed of by absorption or otherwise. Should this particular branch of the subject yield all that it promises, should it turn out on further trial that when the knot is applied on the antiseptic principle, we may calculate as securely as if it were absent on the occurrence of healing without any deep-seated suppuration; the deligation of main arteries in their continuity will be deprived of the two dangers that now attend it—namely, those of secondary haemorrhage and an unhealthy state of the wound. Further, it seems not unlikely that the present objection to tying an artery in the immediate vicinity of a large branch may be done away with; and that even the innominate, which has lately been the subject of an ingenious experiment by one of the Dublin surgeons on account of its well-known fatality under the ligature from secondary haemorrhage, may cease to have this unhappy character, when the tissues in the vicinity of the thread, instead of becoming softened through the influence of an irritating decomposing substance, are left at liberty to consolidate firmly near an unoffending though foreign body.

It would carry me far beyond the limited time which, by the rules of the Association, is alone at my disposal, were I to enter into the various applications of the antiseptic principle in the several special departments of surgery.

There is, however, one point more that I cannot but advert to—namely, the influence of this mode of treatment upon the general healthiness of a hospital. Previously to its introduction, the two large wards in which most of my cases of accident and of operation are treated were amongst the unhealthiest in the whole surgical division of the Glasgow Royal Infirmary, in consequence, apparently, of those wards being unfavourably placed with reference to the supply of fresh air; and I have felt ashamed, when recording the results of my practice, to have so often to allude to hospital gangrene or pyaemia. It was interesting, though melancholy, to observe that, whenever all, or nearly all, the beds contained cases with open sores, these grievous complications were pretty sure to show themselves; so that I came to welcome simple fractures, though in themselves of little interest either for myself or the students, because their presence diminished the proportion of open sores among the patients. But since the antiseptic treatment has been brought into full operation, and wounds and abscesses no longer poison the atmosphere with putrid exhalations, my wards, though in other respects under precisely the same circumstances as before, have completely changed their character; so that during the last nine months not a single instance of pyaemia, hospital gangrene, or erysipelas has occurred in them.

As there appears to be no doubt regarding the cause of this change, the importance of the fact can hardly be exaggerated.

FURTHER READING

Cameron, Hector Charles. *Joseph Lister the Friend of Man.* London: William Heinemann Medical Books Ltd., 1949.

Cameron, Sir Hector Clare. *Reminiscences of Lister.* Glasgow: Jackson, Wylie and Co., 1927.

Cartwright, Frederick F. *Joseph Lister, the Man Who Made Surgery Safe.* London: Weidenfeld and Nicolson, 1963.

Cheyne, W.W. *Lister and His Achievement.* London: Longmans, Green, and Co., 1925.

Fisher, Richard B. *Joseph Lister 1827–1912.* New York: Stein and Day, 1977.

Fox, Nicholas J. "Scientific Theory Choice and Social Structure: The Case of Joseph Lister's Antisepsis, Humoral Theory and Asepsis." *History of Science* 26 (1988): 367–97.

Godlee, Sir Rickman John. *Lord Lister.* London: Macmillan and Co., 1917.

Guthrie, Douglas. *Lord Lister: His Life and Doctrine.* Edinburgh: E. and S. Livingstone Ltd., 1949.

Hamilton, David. "The Nineteenth-Century Surgical Revolution—Antisepsis or Better Nutrition?" *Bulletin of the History of Medicine* 56 (1982): 30–40.

Leeson, John Rudd. *Lister as I Knew Him.* London: Baillière, Tindall and Cox, 1927.

Lister and the Ligature: A Landmark in the History of Modern Surgery. New Brunswick, NJ: Johnson and Johnson, 1925.

Lister, Joseph. *The Collected Papers of Joseph, Baron Lister.* 2 vols. Oxford: Clarendon Press, 1909.

Pennington, T.H. "Listerism, its Decline and its Persistence: the Introduction of Aseptic Surgical Techniques in Three British Teaching Hospitals, 1890–99." *Medical History* 39, no. 1 (1995): 35–60.

Saleeby, C.W. *Surgery and Society: A Tribute to Listerism.* New York: Moffat, Yard and Company, 1912.

Truax, Rhoda. *Joseph Lister: Father of Modern Surgery.* London: George G. Harrap and Co. Ltd., 1947.

Turner, A. Logan. *Joseph, Baron Lister: Centenary Volume 1827–1927.* Edinburgh: Oliver and Boyd, 1927.

Walker, Kenneth. *Joseph Lister.* London: Hutchinson, 1956.

Wrench, G.T. *Lord Lister: His Life and Work.* New York: Frederick A. Stokes Company, 1913.

27. SIR FRANCIS GALTON

(1822–1911)

Sir Francis Galton reportedly demonstrated early signs of genius, although his educational career was not distinguished. Like his cousin Charles Darwin, he never held an academic scientific position, and his financial independence inherited from his Quaker father in 1844 allowed him the leisure to study and make contributions to the fields of heredity, psychology, statistics, and eugenics. In addition to trips to Eastern Europe, Egypt, and Syria, he travelled in remote parts of southern Africa from 1850–52 and later wrote *Tropical South Africa* (1853). His connection to the influential Darwin and Galton families undoubtedly stimulated his life-long interest in mental capacity and hereditary traits. His study *Finger Prints* (1893), for example, which systematized the print patterns used by criminologists, was originally undertaken in an attempt to establish correlations in racial and individual heredity.

Although heredity and mental capacity had been discussed since classical times, Galton can truly be called the founder of modern eugenics (a term he coined), anthropometrics, and psychometrics. In 1904, Galton defined eugenics as "the science which deals with all influences that improve and develop the inborn qualities of a race. ... The aim of eugenics is to represent each class or sect by its best specimens, causing them to contribute *more* than their proportion to the next generation; that done, to leave them to work out their common civilisation in their own way" ("Eugenics" 82). Underlying the rhetoric of the movement was a desire to artificially select the fittest human beings for future societies. By the early twentieth century, even as post-Mendelian geneticists began to discredit some of Galton's theories of heredity, eugenics was hailed by its promoters as a "racial medicine" which endeavoured "to cure and prevent the diseases of the race, bodily and mental, as ordinary medicine seeks to cure and prevent the diseases of the individual" (Saleeby 134).

In addition to *Hereditary Genius* (1869), Galton wrote three other books concerning heredity: *English Men of Science: Their Nature and Nurture* (1874), *Inquiries into Human Faculty* (1883), and *Natural Inheritance* (1889). Galton contributed a number of articles to the proceedings of the Royal Society on heredity and the Darwinian theory of pangenesis, including "Experiments in Pangenesis" (1871) and "Family Likeness in Stature" (1886). Galton originally explained inheritance using Erasmus and Charles Darwin's gemmule theory of pangenesis, the idea that germs or seeds of cells circulate throughout the bodies of the parents and are blended together in the offspring, but in the 1870s he demonstrated that repeated transfusions of blood between differently coloured rabbits did not change the markings of the rabbits' offspring. He proposed a modified theory of pangenesis, suggesting the existence of "latent" and "patent" gemmules, which anticipated later concepts of genotype (genetic makeup) and phenotype (expressed genetic characters). It must be remembered that genes and DNA were unknown until the twentieth century and the process of heredity was very poorly understood throughout the nineteenth century.

The idea of inherited genius significantly first occurred to Galton during an examination of the mental peculiarities of different races. His method in *Hereditary Genius* was to examine family records, published lists, and the nineteenth-century equivalent of our "Who's Who" books and calculate statistically the correlations among men of genius and their immediate family members (grouped as "Statesmen," "Judges," "Divines," etc.). Galton believed that this procedure would determine if mental talents could be inherited. He was probably the first, as he claims, to examine the question using the statistical methods of the "law of deviation from an average" or the Laplace-Gaussian or normal curve of deviations employed by the Belgian statistician Quetelet. Also known as the bell-curve, this graph forms a bell-shaped curve symmetrical around the point of the average measurement.

As the chart below illustrates (Fig. 1), Galton graded men into two equally spaced groups, according to their natural gifts: 1) below average (designated by lower case letters from a–g), with x = all grades below g; and 2) above average (designated by capital letters from A–G), with X = all grades above G. For example, someone of grade G (of exceptionally high gifts) would represent 1 in 79,000 of the English population, while someone of grade b (of slightly below average gifts) would represent 1 in 6 of the population. Galton found a high correlation between nearness in blood and genius. Galton never rigorously defined his criteria for "genius," "natural powers," and "eminence," although his litmus test seems to have been national acclaim and public consensus concerning greatness. Galton made some attempt to study the inheritance of genius through the female line, but since most nineteenth-century professions were closed to women, preventing them from exercising any native genius, women fell outside of Galton's definitions of eminence. The subjectivity of some of Galton's categorizations, as well as his enthusiasm for quantification, is revealed in his proposed "Beauty Map" of England; while walking, Galton would prick a piece of paper in his pocket in one of three places to record the number of "attractive, repellent, or indifferent" women he met in the street. London women ranked the highest in beauty and Aberdeen the lowest in Galton's scheme.

Of course, Galton's method in *Hereditary Genius* is open to the question of the effect of environment and the effects of the obvious social, educational, and economic benefits that eminent and financially successful men bestow upon their children—this question is popularly known as the "nature-nurture" debate and it still rages in a variety of scientific communities. Arguing against Galton, Alphonse de Candolle in *Histoire des sciences et des savants depuis deux siècles* (1872), while acknowledging the importance of heredity, believed that environment primarily determined genius. In another lengthy response to Galton, F.C. Constable in *Poverty and Hereditary Genius* (1905) also attributed individual success to environmental influence: "only in change of environment can there be any hope for raising the level of the average ability of the race" (19). Galton, however, had dismissed environmental influence on heredity outright in *Hereditary Genius*: "I have no patience with the hypothesis occasionally expressed, and often implied, especially in tales written to teach children to be good, that babies are born pretty much alike, and that the sole agencies in creating differences between boy and boy, and man and man, are steady application and moral effort" (14).

Galton recommends explicitly in *Hereditary Genius* that weak races should be encouraged to marry late to reduce their offspring and that "it would be quite practicable to produce a highly-gifted race of men by judicious marriages during several consecutive generations" (1). He believed that in unrestricted breeding in animals and man, there is always a "regression to the mean" or an averaging out of desirable traits. Galton's work contains many *a priori* assumptions about race, nationality, and social class. Studying his work reminds us that any supposedly "purely scientific" search for objective measurements of genius, of intellectual ability and social structure, then as today, have always involved overt political, economic, and social motives.

Galton's ideas were extended by Karl Pearson (1857–1936), who edited *The Life, Letters and Labours of Francis Galton* (1914–1930). Pearson also became the first Galton Professor of Eugenics at the University of London under a bequest from Galton. Together they founded the Francis Galton Laboratory for National Eugenics in 1909 and the journal *Biometrika* (1901). Galton also established the Anthropometric Laboratory in 1884, "for the use of those who desire to be accurately measured in many ways, either to obtain timely warning of remediable faults in development, or to learn their powers." Galton's and Pearson's ideas on eugenics and the inheritance of intelligence, never dormant, found fertile ground in Europe and America before and during the World Wars and have been debated again recently in two books, Stephen J. Gould's *The Mismeasure of Man* (1981) and Richard Herrnstein and Charles Murray's *The Bell Curve* (1994). At the time of his death in 1911, Galton was writing a eugenic utopia entitled *The Eugenic College of Kantsaywhere.*

Hereditary Genius: An Inquiry into Its Laws and Consequences.
London: Macmillan and Co., 1869, pp. 336–350.

THE COMPARATIVE WORTH OF DIFFERENT RACES

I have now completed what I have to say concerning the kinships of individuals, and proceed, in this chapter, to attempt a wider treatment of my subject, through a consideration of nations and races.

Every long-established race has necessarily its peculiar fitness for the conditions under which it has lived, owing to the sure operation of Darwin's law of natural selection. However, I am not much concerned, for the present, with the greater part of those aptitudes, but only with such as are available in some form or other of high civilization. We may reckon upon the advent of a time, when civilization, which is now sparse and feeble and far more superficial than it is vaunted to be, shall overspread the globe. Ultimately it is sure to do so, because civilization is the necessary fruit of high intelligence when found in a social animal, and there is no plainer lesson to be read off the face of Nature than that the result of the operation of her laws is to evoke intelligence in connexion with sociability. Intelligence is as much an advantage to an animal as physical strength or any other natural gift, and therefore, out of two varieties of any race of animal who are equally endowed in other respects, the most intelligent variety is sure to prevail in the battle of life. Similarly, among animals as intelligent as man, the most social race is sure to prevail, other qualities being equal.

Under even a very moderate form of material civilization, a vast number of aptitudes acquired through the "survivorship of the fittest" and the unsparing destruction of the unfit, for hundreds of generations, have become as obsolete as the old mail-coach habits and customs, since the establishment of railroads, and there is not the slightest use in attempting to preserve them; they are hindrances, and not gains, to civilization. I shall refer to some of these a little further on, but I will first speak of the qualities needed in civi-

Grades of natural ability, separated by equal intervals.		Numbers of men comprised in the several grades of natural ability, whether in respect to their general powers, or to special aptitudes.							
		Proportionate, viz. one in	In each million of the same age.	In total male population of the United Kingdom, viz. 15 millions, of the undermentioned ages :—					
Below average.	Above average.			20—30	30—40	40—50	50—60	60—70	70—80
a	A	4	256,791	651,000	495,000	391,000	268,000	171,000	77,000
b	B	6	162,279	409,000	312,000	246,000	168,000	107,000	48,000
c	C	16	63,563	161,000	123,000	97,000	66,000	42,000	19,000
d	D	64	15,696	39,800	30,300	23,900	16,400	10,400	4,700
e	E	413	2,423	6,100	4,700	3,700	2,520	1,600	729
f	F	4,300	233	590	450	355	243	155	70
g	G	79,000	14	35	27	21	15	9	4
x all grades below g	X all grades above G	1,000,000	1	3	2	2	2	—	—
On either side of average . .			500,000	1,268,000	964,000	761,000	521,000	332,000	149,000
Total, both sides			1,000,000	2,536,000	1,928,000	1,522,000	1,042,000	664,000	298,000

The proportions of men living at different ages are calculated from the proportions that are true for England and Wales. (Census 1861, Appendix, p. 107.)

Example.—The class F contains 1 in every 4,300 men. In other words, there are 233 of that class in each million of men. The same is true of class f. In the whole United Kingdom there are 590 men of class F (and the same number of f) between the ages of 20 and 30; 450 between the ages of 30 and 40; and so on.

Fig. 1 Galton's chart of natural ability from *Hereditary Genius* (1869), p. 34.

lized society. They are, speaking generally, such as will enable a race to supply a large contingent to the various groups of eminent men, of whom I have treated in my several chapters. Without going so far as to say that this very convenient test is perfectly fair, we are at all events justified in making considerable use of it, as I will do, in the estimates I am about to give.

In comparing the worth of different races, I shall make frequent use of the law of deviation from an average, to which I have already been much beholden; and, to save the reader's time and patience, I propose to act upon an assumption that would require a good deal of discussion to limit, and to which the reader may at first demur, but which cannot lead to any error of importance in a rough provisional inquiry. I shall assume that the *intervals* between the grades of ability are the *same* in all the races—that is, if the ability of class A of one race be equal to the ability of class C in another, then the ability of class B of the former shall be supposed equal to that of class D of the latter, and so on. I know this cannot be strictly true, for it would be in defiance of analogy if the variability of all races were precisely the same; but, on the other hand, there is good reason to expect that the error introduced by the assumption cannot sensibly affect the offhand results for which alone I propose to employ it; moreover, the rough data I shall adduce, will go far to show the justice of this expectation.

Let us, then, compare the negro race with the Anglo-Saxon, with respect to those qualities alone which are capable of producing judges, statesmen, commanders, men of literature and science, poets, artists, and divines. If the negro race in America had been affected by no social disabilities, a comparison of their achievements with those of the whites in their several branches of intellectual effort, having regard to the total number of their respective populations, would give the necessary information. As matters stand, we must be content with much rougher data.

First, the negro race has occasionally, but very rarely, produced such men as Toussaint l'Ouverture, who are of our class F; that is to say, its X, or its total classes above G, appear to correspond with our F, showing a difference of not less than two grades between the black and white races, and it may be more.

Secondly, the negro race is by no means wholly deficient in men capable of becoming good factors, thriving merchants, and otherwise considerably raised above the average of whites—that is to say, it can not unfrequently supply men corresponding to our class C, or even D. It will be recollected that C implies a selection of 1 in 16, or somewhat more than the natural abilities possessed by average foremen of common juries, and that D is as 1 in 64—a degree of ability that is sure to make a man successful in life. In short, classes E and F of the negro may roughly be considered as the equivalent of our C and D—a result which again points to the conclusion, that

European scientific expeditions often simultaneously served colonial interests by sounding shipping lanes, resupplying existing colonies, and mapping possible habitable lands for settlement. Pictured are three Fuegians from Cape Horn (Tierra del Fuego) in native and European dress, photographed on an 1882 French scientific expedition. Three other European-educated Fuegians — York Minster, Jemmy Button, and Fuegia Basket — sailed with Charles Darwin on the *Beagle* back to their native land. When the ship returned one year later, the three had abandoned their European clothes and habits. Darwin wrote to his sister: "I feel quite a disgust at the very sound of the voices of these miserable savages." The numbers of the original inhabitants of Tierra del Fuego have declined dramatically since the nineteenth century.

the average intellectual standard of the negro race is some two grades below our own.

Thirdly, we may compare, but with much caution, the relative position of negroes in their native country with that of the travellers who visit them. The latter, no doubt, bring with them the knowledge current in civilized lands, but that is an advantage of less importance than we are apt to suppose. A native chief has as good an education in the art of ruling men, as can be desired; he is continually exercised in personal government, and usually maintains his place by the ascendency of his character, shown every day over his subjects and rivals. A traveller in wild countries also fills, to a certain degree, the position of a commander, and has to confront native chiefs at every inhabited place. The result is familiar enough—the white traveller almost invariably holds his own in their presence. It is seldom that we hear of a white traveller meeting with a black chief whom he feels to be the better man. I have often discussed this subject with competent persons, and can only recall a few cases of the inferiority of the white man,—certainly not more than might be ascribed to an average actual difference of three grades, of which one may be due to the relative demerits of native education, and the remaining two to a difference in natural gifts.

Fourthly, the number among the negroes of those whom we should call half-witted men, is very large. Every book alluding to negro servants in America is full of instances. I was myself much impressed by this fact during my travels in Africa. The mistakes the negroes made in their own matters, were so childish, stupid, and simpleton-like, as frequently to make me ashamed of my own species. I do not think it any exaggeration to say, that their c is as low as our e, which would be a difference of two grades, as before. I have no information as to actual idiocy among the negroes—I mean, of course, of that class of idiocy which is not due to disease.

The Australian type is at least one grade below the African negro. I possess a few serviceable data about the natural capacity of the Australian, but not sufficient to induce me to invite the reader to consider them.

The average standard of the Lowland Scotch and the English North-country men is decidedly a fraction of a grade superior to that of the ordinary English, because the number of the former who attain to eminence is far greater than the proportionate number of their race would have led us to expect. The same superiority is distinctly shown by a comparison of the well-being of the masses of the population; for the Scotch labourer is much less of a drudge than the Englishman of the Midland counties—he does his work better, and "lives his life" besides. The peasant women of Northumberland work all day in the fields, and are not broken down by the work; on the contrary, they take a pride in their effective labour as girls, and, when married, they attend well to the comfort of their homes. It is perfectly distressing to me to witness the draggled, drudged, mean look of the mass of individuals, especially of the women, that one meets in the streets of London and other purely English towns. The conditions of their life seem too hard for their constitutions, and to be crushing them into degeneracy.

The ablest race of whom history bears record is unquestionably the ancient Greek, partly because their master-pieces in the principal departments of intellectual activity are still unsurpassed, and in many respects unequalled, and partly because the population that gave birth to the creators of those master-pieces was very small. Of the various Greek sub-races, that of Attica was the ablest, and she was no doubt largely indebted to the following cause, for her superiority. Athens opened her arms to immigrants, but not indiscriminately, for her social life was such that none but very able men could take any pleasure in it; on the other hand, she offered attractions such as men of the highest ability and culture could find in no other city. Thus, by a system of partly unconscious selection, she built up a magnificent breed of human animals, which, in the space of one century—viz. between 530 and 430 B.C.—produced the following illustrious persons, fourteen in number:—

Statesmen and Commanders.—Themistocles (mother an alien), Miltiades, Aristeides, Cimon (son of Miltiades), Pericles (son of Xanthippus, the victor at Mycale).

Literary and Scientific Men.—Thucydides, Socrates, Xenophon, Plato.

Poets.—Aeschylus, Sophocles, Euripides, Aristophanes.

Sculptor.—Phidias.

We are able to make a closely-approximate estimate of the population that produced these men, because the number of the inhabitants of Attica has been a matter of frequent inquiry, and critics appear at length to be quite agreed in the general results. It seems that the little district of Attica contained, during its most flourishing period (Smith's Class. Geog. Dict.), less than 90,000 native free-born persons, 40,000 resident aliens, and a labouring and artisan population of 400,000 slaves. The first item is the only one that concerns us here, namely, the 90,000 free-born persons. Again, the common estimate that population renews itself three times in a century is very close to the truth, and may be accepted in the present case. Consequently, we have to deal with a total population of 270,000 free-born persons, or 135,000 males, born in the century I have named. Of these, about one-half, or 67,500, would survive the age of 26, and one-third, or 45,000 would survive that of 50. As 14 Athenians became illustrious, the selection is only as 1 to 4,822 in respect to the former limitation, and as 1 to 3,214 in respect to the latter. Referring to the table in page 34, it will be seen that this degree of selection corresponds very fairly to the classes F (1 in 4,300) and above, of the Athenian race. Again, as G is one-sixteenth or one-seventeenth as numerous as F, it would be reasonable to expect to find one of class G among the fourteen; we might, however, by accident, meet with two, three, or even four of that class—say Pericles, Socrates, Plato, and Phidias.

Now let us attempt to compare the Athenian standard of ability with that of our own race and time. We have no men to put by the side of Socrates and Phidias, because the millions of all Europe, breeding as they have done for the subsequent 2,000 years, have never produced their equals. They are, therefore, two or three grades above our G—they might rank as I or J. But, supposing we do not count them at all, saying that some freak of nature acting at that time, may have produced them, what must we say about the rest? Pericles and Plato would rank, I suppose, the one among the greatest of philosophical statesmen, and the other as at least the equal of Lord Bacon. They would, therefore, stand somewhere among our unclassed X, one or two grades above G—let us call them between H and I. All the remainder—the F of the Athenian race—would rank above our G, and equal to or close upon our H. It follows from all this, that the average ability of the Athenian race is, on the lowest possible estimate, very nearly two grades higher than our own—that is, about as much as our race is above that of the African negro. This estimate, which may seem prodigious to some, is confirmed by the quick intelligence and high culture of the Athenian commonalty, before whom literary works were recited, and works of art exhibited, of a far more severe character than could possibly be appreciated by the average of our race, the calibre of whose intellect is easily gauged by a glance at the contents of a railway book-stall.

We know, and may guess something more, of the reason why this marvellously-gifted race declined. Social morality grew exceedingly lax; marriage became unfashionable, and was avoided; many of the more ambitious and accomplished women were avowed courtesans, and consequently infertile, and the mothers of the incoming population were of a heterogeneous class. In a small sea-bordered country, where emigration and immigration are constantly going on, and where the manners are as dissolute as were those of Greece in the period of which I speak, the purity of a race would necessarily fail. It can be, therefore, no surprise to us, though it has been a severe misfortune to humanity, that the high Athenian breed decayed and disappeared; for if it had maintained its excellence, and had multiplied and spread over large countries, displacing inferior populations (which it well might have done, for it was exceedingly prolific), it would assuredly have accomplished results advantageous to human civilization, to a degree that transcends our powers of imagination.

If we could raise the average standard of our race only one grade, what vast changes would be produced! The number of men of natural gifts equal to those of the eminent men of the present day, would be necessarily increased more than tenfold, as will be seen by the fourth column of the table p. 34 [Fig. 1], because there would be 2,423 of them in each million instead of only 233; but far more important to the progress of civilization would be the increase in the yet higher orders of intellect. We know how intimately the course of events is dependent on the thoughts of a few illustrious men. If the first-rate men in the different groups had never been born, even if those among them who have a place in my appendices on account of their hereditary gifts, had never existed, the world would be very different to what it is. Now the table shows that the numbers in these, the loftiest grades of intellect, would be increased in a still higher proportion than that of which I have been speaking; thus the men that now rank under class G would be increased seventeenfold, by raising the average ability of the whole nation a single grade. We see by the table that all England contains (on the average, of course, of several years) only six men between the ages of thirty and eighty, whose natural gifts exceed class G; but in a country of the same population as ours, whose average was one grade higher, there would be eighty-two of such men; and in another whose average was two grades higher (such as I believe the Athenian to have been, in the interval 530–430 B.C.) no less than 1,355 of them would be found. There is no improbability in so gifted a breed being able to maintain itself, as Athenian experience, rightly understood, has sufficiently proved; and as has also been proved by what I have written about the Judges, whose fertility is undoubted, although their average natural ability is F, or 5½ degrees above the average of our own, and 3½ above that of the average Athenians.

It seems to me most essential to the well-being of future generations, that the average standard of ability of the present time should be raised. Civilization is a new condition imposed upon man by the course of events, just as in the history of geological changes new conditions have continually been imposed on different races of animals. They have had the effect either of modifying the nature of the races through the process of natural selection, whenever the changes were sufficiently slow and the race sufficiently pliant, or of destroying them altogether, when the changes were too abrupt or the race unyielding. The number of the races of mankind that have been entirely destroyed under the pressure of the requirements of an incoming civilization, reads us a terrible lesson. Probably in no former period of the world has the destruction of the races of any animal whatever, been effected over such wide areas and with such startling rapidity as in the case of savage man. In the North American Continent, in the West Indian Islands, in the Cape of Good Hope, in Australia, New Zealand, and Van Diemen's Land, the human denizens of vast regions have been entirely swept away in the short space of three centuries, less by the pressure of a stronger race than through the influence of a civilization they were incapable of supporting. And we too, the foremost labourers in creating this civilization, are beginning to show ourselves incapable of keeping pace with our own work. The needs of centralization, communication, and culture, call for more brains and mental stamina than the average of our race possess. We are in crying want for a greater fund of ability in all stations of life; for neither the classes of statesmen, philosophers, artisans, nor labourers are up to the modern complexity of their several professions. An extended civilization like ours comprises more interests than the ordinary statesmen or philosophers of our present race are capable of dealing with, and it exacts more intelligent work than our ordinary artisans and labourers are capable of performing. Our race is overweighted, and appears likely to be drudged into degeneracy by demands that exceed its powers. If its average ability were raised a grade or two, our new classes F and G would conduct the complex affairs of the state at home and abroad as easily as our present F and G, when in the position of country squires, are able to manage the affairs of their establishments and tenantry. All other classes of the community would be similarly promoted to the level of the work required by the nineteenth century, if the average standard of the race were raised.

When the severity of the struggle for existence is not too great for the powers of the race, its action is healthy and conservative, otherwise it is deadly, just as we may see exemplified in the scanty, wretched vegetation that leads a precarious existence near the summer snow line of the Alps, and disappears altogether a little higher up. We want as much backbone as we can get, to bear the racket to which we are henceforth to be exposed, and as good brains as possible to contrive machinery, for modern life to work more smoothly than at present. We can, in some degree, raise the nature of man to a level with the new conditions imposed upon his existence, and we can also, in some degree, modify the conditions to suit his nature. It is clearly right that both these powers should be exerted, with the view of bringing his nature and the conditions of his existence into as close harmony as possible.

In proportion as the world becomes filled with mankind, the relations of society necessarily increase in complexity, and the nomadic disposition found in most barbarians becomes unsuitable to the novel conditions. There is a most unusual unanimity in respect to the causes of incapacity of savages for civilization, among writers on those hunting and migratory nations who are brought into contact with advancing colonization, and perish, as they invariably do, by the contact. They tell us that the labour of such men is neither constant nor steady; that the love of a wandering, independent life prevents their settling anywhere to work, except for a short time, when urged by want and encouraged by kind treatment. Meadows says that the Chinese call the barbarous races on their borders by a phrase which means "hither and thither, not fixed." And any amount of evidence might be adduced to show how deeply Bohemian habits of one kind or another, were ingrained in the nature of the men who inhabited most parts of the earth now overspread by the Anglo-Saxon and other civilized races. Luckily there is still room for adventure, and a man who feels the cravings of a roving, adventurous spirit to be too strong for resistance, may yet find a legitimate outlet for it in the colonies, in the army, or on board ship. But such a spirit is, on the whole, an heirloom that brings more impatient restlessness and beating of the wings against cage-bars, than persons of more civilized characters can readily comprehend, and it is directly at war with the more modern portion of our moral natures. If a man be purely a nomad, he has only to be nomadic, and his instinct is satisfied; but no Englishmen of the nineteenth century are purely nomadic. The most so among them have also inherited many civilized cravings that are necessarily starved when they become wanderers, in the same way as the wandering instincts are starved when they are settled at home. Consequently their nature has opposite wants, which can never be satisfied except by chance, through some very exceptional turn of circumstances. This is a serious calamity, and as the Bohemianism in the nature of our race is destined to perish, the sooner it goes, the happier for mankind. The social requirements of English life are steadily destroying it. No man who only works by fits and starts is able to obtain his living nowadays; for he has not a chance of thriving in competition with steady workmen. If his nature revolts against the monotony of daily labour, he is tempted to the public-house, to intemperance, and, it may be, to poaching, and to much more serious crime: otherwise he banishes himself from our shores. In the first case, he is unlikely to leave as many children as men of more domestic and marrying habits, and, in the second case, his breed is wholly lost to England. By this steady riddance of the Bohemian spirit of our race, the artisan part of our population is slowly becoming bred to its duties, and the primary qualities of the typical modern British workman are already the very opposite of those of the nomad. What they are now, was well described by Mr. Chadwick, as consisting of "great bodily strength, applied under the command of a steady, persevering will, mental self-contentedness, impassibility to external irrelevant impressions, which carries them through the continued repetition of toilsome labour, 'steady as time.' "

It is curious to remark how unimportant to modern civilization has become the once famous and thoroughbred looking Norman. The type of his features, which is, probably, in some degree correlated with his peculiar form of adventurous disposition, is no longer characteristic of our rulers, and is rarely found among celebrities of the present day; it is more often met with among the undistinguished members of highly-born families, and especially among the less conspicuous officers of the army. Modern leading men in all paths of eminence, as may easily be seen in a collection of photographs, are of a coarser and more robust breed; less excitable and dashing, but endowed with far more ruggedness and real vigour. Such also is the case, as regards the German portion of the Austrian nation; they are far more high-caste in appearance than the Prussians, who are so plain that it is disagreeable to travel northwards from Vienna, and watch the change; yet the Prussians appear possessed of the greater moral and physical stamina.

Much more alien to the genius of an enlightened civilization than the nomadic habit, is the impulsive and uncontrolled nature of the savage. A civilized man must bear and forbear, he must keep before his mind the claims of the morrow as clearly as those of the passing minute; of the absent, as well as of the present. This is the most trying of the new conditions imposed on man by civilization, and the one that makes it hopeless for any but exceptional natures among savages, to live under them. The instinct of a savage is admirably consonant with the needs of savage life; every day he is in danger through transient causes; he lives from hand to mouth, in the hour and for the hour, without care for the past or forethought for the future: but such an instinct is utterly at fault in civilized life. The half-reclaimed savage, being unable to deal with more subjects of consideration than are directly before him, is continually doing acts through mere maladroitness and incapacity, at which he is afterwards deeply grieved and annoyed. The nearer inducements always seem to him, through his uncorrected sense of moral perspective, to be incomparably larger than others of the same actual size, but more remote; consequently, when the temptation of the moment has been yielded to and passed away, and its bitter result comes in its turn before the man, he is amazed and remorseful at his past weakness. It seems incredible that he should have done that yesterday which to-day seems so silly, so unjust, and so unkindly. The newly-reclaimed barbarian, with the impulsive, unstable nature of the savage, when he also chances to be gifted with a peculiarly generous and affectionate disposition, is of all others the man most oppressed with the sense of sin.

Now it is a just assertion, and a common theme of moralists of many creeds, that man, such as we find him, is born with an imperfect nature. He has lofty aspirations, but there is a weakness in his disposition, which incapacitates him from carrying his nobler purposes into effect. He sees that some particular course of action is his duty, and should be his delight; but his inclinations are fickle and base, and do not conform to his better judgment. The whole moral nature of man is tainted with sin, which prevents him from doing the things he knows to be right.

The explanation I offer of this apparent anomaly, seems perfectly satisfactory from a scientific point of view. It is neither more nor less than that the development of our nature, whether under Darwin's law of natural selection, or through the effects of changed ancestral habits, has not yet overtaken the development of our moral civilization. Man was barbarous but yesterday, and therefore it is not to be expected that the natural aptitudes of his race should already have become moulded into accordance with his very recent advance. We, men of the present centuries, are like animals suddenly transplanted among new conditions of climate and of food: our instincts fail us under the altered circumstances.

My theory is confirmed by the fact that the members of old civilizations are far less sensible than recent converts from barbarism, of their nature being inadequate to their moral needs. The conscience of a negro is aghast at his own wild, impulsive nature, and is easily stirred by a preacher, but it is scarcely possible to ruffle the self-complacency of a steady-going Chinaman.

The sense of original sin would show, according to my theory, not that man was fallen from a high estate, but that he was rising in moral culture with more rapidity than the nature of his race could follow. My view is corroborated by the conclusion reached at the end of each of the many independent lines of ethnological research—that the human race were utter savages in the beginning; and that, after myriads of years of barbarism, man has but very recently found his way into the paths of morality and civilization.

FURTHER READING

Alvarez Pelaez, Raquel. *Sir Francis Galton: Padre de la Eugenesia.* Madrid: C.S.I.C., 1985.

Blacker, C.P. *Eugenics: Galton and After.* Cambridge, MA: Harvard UP, 1952.

Chesterton, G.K. *Eugenics and Other Evils.* London: Cassell and Co., Ltd., 1922.

Constable, F.C. *Poverty and Hereditary Genius: A Criticism of Mr. Francis Galton's Theory of Hereditary Genius.* London: Arthur C. Fifield, 1905.

Cowan, Ruth Schwartz. *Sir Francis Galton and the Study of Heredity in the Nineteenth Century.* The History of Hereditarian Thought, ed. Charles Rosenberg. New York: Garland Publishing, Inc., 1985.

Fancher, Raymond E. "Francis Galton's African Ethnology and Its Role in the Development of His Psychology." *British Journal for the History of Science* 16, part 1, no. 52 (1983): 67–79.

Farrall, Lyndsay A. *The Origins and Growth of the English Eugenics Movement: 1865–1925.* The History of Hereditarian Thought, ed. Charles Rosenberg. New York: Garland Publishing, Inc., 1985.

Forrest, D.W. *Francis Galton: The Life and Work of a Victorian Genius.* New York: Taplinger Publishing Company, Inc., 1974.

Galton, Francis. "Eugenics; Its Definition, Scope and Aims." *Nature* 70, no. 1804 (1904): 82.

—. *Memories of My Life.* London: Methuen, 1908.

Garrett, Henry E. *Great Experiments in Psychology.* 3d ed. New York: Appleton-Century-Crofts, Inc., 1951.

"Genetics, Eugenics, and Evolution." Special issue of *British Journal for the History of Science,* 22, part 3, no. 74 (1988).

Gould, Stephen J. *The Mismeasure of Man.* Rev. ed. New York: W.W. Norton, 1996.

Haller, Mark. *Eugenics, Hereditarian Attitudes in American Thought.* New Brunswick, NJ: Rutgers UP, 1963.

Herrnstein, Richard J., and Charles Murray. *The Bell Curve: Intelligence and Class Structure in American Life.* New York: The Free Press, 1994.

Hilts, Victor L. "A Guide to Francis Galton's English Men of Science." *Transactions of the American Philosophical Society* n.s. 65, part 5 (1975): 1–85.

Jacoby, Russell, and Naomi Glauberman. *The Bell Curve Debate: History, Documents, Opinions.* New York: Random House, 1995.

Jennings, H.S. *Prometheus, or Biology and the Advancement of Man.* New York: E.P. Dutton, 1925.

Kevles, Daniel J. *In the Name of Eugenics: Genetics and the Uses of Human Heredity.* New York: Alfred A. Knopf, 1985.

Keynes, Milo, ed. *Sir Francis Galton, FRS: The Legacy of His Ideas.* London: Macmillan Press, Ltd., 1993.

Mackenzie, D. "Eugenics in Britain." *Social Studies in Science* 6 (1976): 499–532.

Osborn, Frederick. *Preface to Eugenics.* New York: Harper and Brothers, 1940.

Paul, Diane B. *Controlling Human Heredity: 1865 to the Present.* Atlantic Highlands, NJ: Humanities Press, 1995.

Pearson, Karl. *The Life, Letters, and Labours of Francis Galton.* 3 vols. Cambridge: Cambridge UP, 1914–30.

Porter, Theodore. *The Rise of Statistical Thinking 1820–1900.* Princeton: Princeton UP, 1986.

Robbins, Emily F., ed. *Proceedings of the First National Conference on Race Betterment.* Battle Creek, MI: Race Betterment Foundation, 1914.

Robinson, Daniel N, ed. *Psychometrics and Educational Psychology.* Significant Contributions to the History of Psychology 1750–1920. Vol. 4, Series B. Washington, DC: University Publications of America, Inc., 1977.

Saleeby, Caleb Williams. *The Progress of Eugenics.* New York: Funk and Wagnalls Co., 1914.

Schiller, F.C.S. *Eugenics and Politics.* Boston: Houghton Mifflin Co., 1926.

Searle, G. *Eugenics and Politics in Britain: 1900–1914.* Leiden: Noordhoff, 1976.

Stigler, Stephen M. *The History of Statistics: The Measurement of Uncertainty before 1900.* Cambridge, MA: Harvard UP, 1986.

Stocking, G.W., Jr. *Victorian Anthropology.* London: Collier Macmillan Publishing, 1987.

Thomson, Mathew. *The Problem of Mental Deficiency: Eugenics, Democracy, and Social Policy in Britain c. 1870–1959.* Oxford: Clarendon Press, 1998.

Walter, Wolfgang. *Der Geist der Eugenik: Francis Galtons Wissenschaftsreligion in kultursoziologischer Perspektive.* Bielefeld: Universität Bielefeld, 1983.

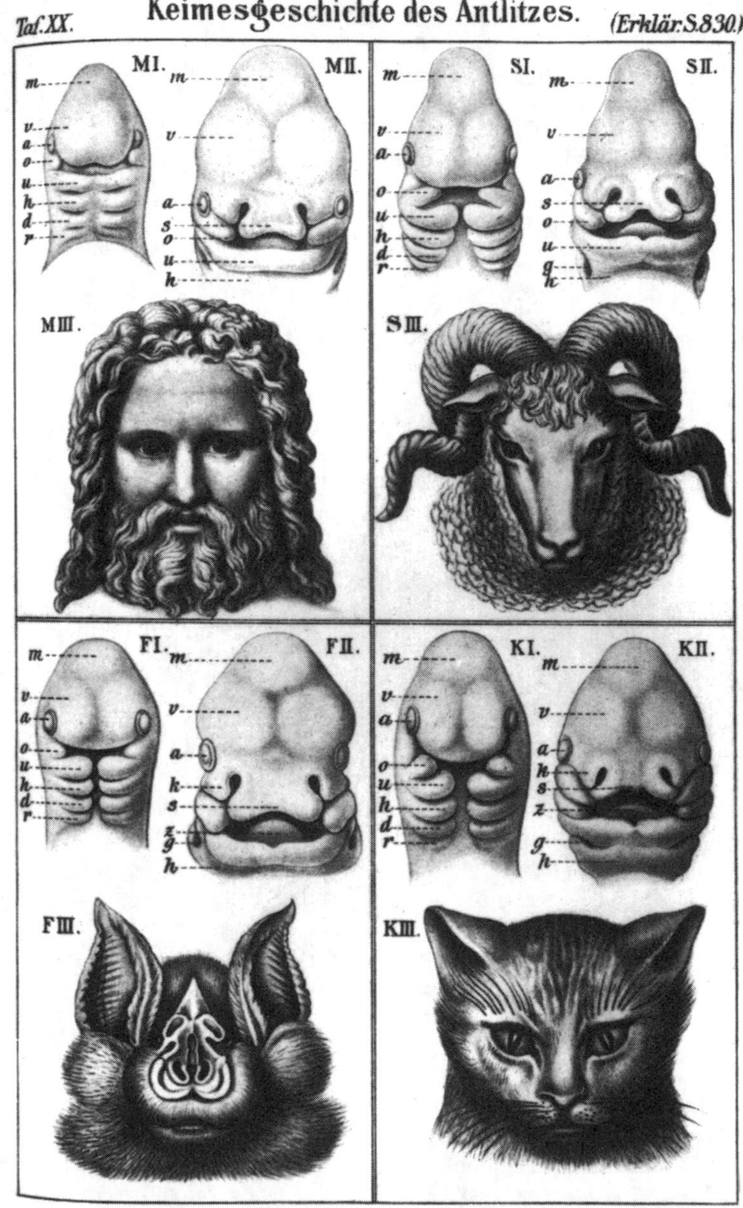

Comparative embryonic development of the face in man, sheep, bats, and cats. From E. Haeckel, *Anthropogenie*, part 2, *Stammesgeschichte des Menschen: Wissenschaftliche Vorträge über die Grundzüge der Menschlichen Phylogenie* (Leipzig: Wilhelm Engelmann, 1891).

28. JOHN TYNDALL

(1820–1893)

John Tyndall, a descendent of the Reformation heretic and Bible translator William Tyndale, was born in 1820 at Leighlin Bridge, County Carlow, Ireland, to an Anglo-Irish Protestant family. After studying mathematics, he joined the Ordnance Survey Office in Ireland as a surveyor. He also worked briefly as a railway engineer for a Manchester firm, and was subsequently offered a teaching post in 1847 at Queenwood College, founded by Richard Owen. With meagre savings, he left Queenwood and proceeded to the University of Marburg in Germany in 1848, where he studied with Karl Knoblauch and R.W. Bunsen (inventor of the spectroscope), obtaining his Ph.D. in 1849. He was elected to the Royal Society in 1852 and was invited by Faraday to accept a professorship of natural philosophy at the Royal Institution, after a successful lecture criticizing Faraday's theories of magnetism. There was no enmity between the two men in spite of their scientific disagreements, however, and after Faraday's death in 1867 Tyndall succeeded him as superintendent of the Royal Institution. Tyndall later summarized Faraday's achievements in his book *Faraday As a Discoverer* (1868).

Tyndall's primary research involved diamagnetism, magne-crystallic action, sound, and thermal (infrared) and solar radiation. His numerous books and collections of lectures in these fields include *Researches on Diamagnetism and Magne-Crystallic Action* (1870), *Heat Considered as a Mode of Motion* (1863), *Notes of a Course of Nine Lectures on Light* (1869), *Notes on Electricity* (1881), and *Sound* (1867). Tyndall collected and annotated some of his papers on physics in *Contributions to Molecular Physics in the Domain of Radiant Heat* (1872).

Tyndall combined science, daring, and pleasure in his frequent mountaineering trips to the Alps where he studied the structure, movement, and "regelation" of glacial ice. He made the first recorded ascent of the Weisshorn in the central Alps. His works on mountaineering and glaciers include *The Glaciers of the Alps* (1860), *Hours of Exercise in the Alps* (1871), and *Forms of Water in Clouds, Rivers, Ice and Glaciers* (1872).

He promoted in England the bacteriological work of Theodor Schwann, Joseph Lister, Robert Koch, and Louis Pasteur and substantiated Pasteur's germ theory of disease by replicating Pasteur's experiments demonstrating that floating "germs" caused putrefaction in sterilized nutrient infusions. Along with Pasteur, he finally put to rest the belief in the spontaneous generation of life, despite the continued opposition of Dr. H.C. Bastian in England. His experiments on putrefaction, infection, and fermentation appear in *The Floating Matter of the Air in Relation to Putrefaction and Disease* (1881).

Tyndall was also known as a popular public lecturer. He toured the United States, lecturing on light from 1872–73, and compiled two popular collections of essays entitled *Fragments of Science for Unscientific People* (2 volumes, 1871) and *New Fragments* (1892). The topics in these collected essays ranged from notes on Pasteur, magnetism, Thomas Carlyle, mountaineering, and Goethe's *Farbenlehre,* to reflections on the reactions to his "Belfast Address."

In 1874 Tyndall delivered the Presidential Address of the British Association for the Advancement of Science, which was published immediately thereafter by Longmans as "The Belfast Address." He further explained and amplified the views of the "Belfast Address" in three essays published in *Fragments of Science*: "Apology for the Belfast Address," "Scientific Materialism," and "The Reverend James Martineau and the Belfast Address." Tyndall provided a history of atomism, summarizing the views of the classical Greek atomists Democritus and Epicurus, as well as Epicurus's Roman popularizer Lucretius, who held that the world could be completely explained by material atoms acting in a void. The name of Epicurus had long been (unjustly) synonymous with atheism in English thought and Tyndall, by expounding his views, was himself branded a "material atheist" and accused of assaulting religion by enquiring into the origins of the universe, an area that scientists had previously left to the domain of theology. As Frank Harris noted, "His praise of matter was like a red flag to the theologic bull" (Harris 77).

There was a widespread popular interest in Lucretius's *De rerum natura* at this time, evidenced by Tennyson's poem "Lucretius" quoted by Tyndall, since Lucretius provided the philosophical underpinnings of the atomic philosophy which dominated nineteenth-century physics and chemistry. The passage in the address that provoked so much consternation was: "By an intellectual necessity I cross the boundary of the experimental evidence, and discern in that Matter which we, in our ignorance of its latent powers, and notwithstanding our professed reverence for its Creator, have hitherto covered with opprobrium, the promise and potency of all terrestrial life" (55). Theologians accused Tyndall, like Lucretius, of suggesting that all activity, including the divine and metaphysical action of Providence, angels, and miracles as well as human consciousness, might be attributable to natural laws and the action of matter. According to St. Augustine's Platonic dualism and the medieval Christian church's *contemptus mundi* philosophy, however, body and soul (or Mind and Matter) were antithetical entities, with matter as inferior, dirty, and unworthy of reflection. Tyndall, on the other hand, was advocating the investigation of matter as the primary intellectual endeavour of thinking men and women of the age—already implicit in most nineteenth-century scientific work, but before Tyndall, seldom placed in such direct opposition to theological authority in a public forum. He also wished to dismiss the book of Genesis as a means of explicating natural philosophy: "it is a poem, not a scientific treatise," he argued.

Tyndall became involved in the "Prayer Gauge Debate," initiated by Henry Thompson's anonymous article in *Contemporary Review* 20 (1872) entitled "The 'Prayer for the Sick'—Hints Towards a Serious Attempt to Estimate Its Value." Thompson proposed to measure scientifically the effect of prayer on the mortality rates of hospital patients over a 3–5 year period (Turner 46). Tyndall contributed his own letter to the *Review*, reprinted among his essays as "On Prayer as a Form of Physical Energy"; while noting the positive moral aspects of prayer, he also claimed that "no good can come of giving [prayer] a delusive value, by claiming for it a power in physical nature" (*New Fragments* 45). A biography of Tyndall started by his wife was completed in 1945 by A. S. Eve and C.H. Creasey under the title *Life and Work of John Tyndall*.

Address Delivered Before the British Association Assembled at Belfast, With Additions.

London: Longmans, Green, and Co., 1874.

An impulse inherent in primeval man turned his thoughts and questionings betimes towards the sources of natural phenomena. The same impulse, inherited and intensified, is the spur of scientific action to-day. Determined by it, by a process of abstraction from experience we form physical theories which lie beyond the pale of experience, but which satisfy the desire of the mind to see every natural occurrence resting upon a cause. In forming their notions of the origin of things, our earliest historic (and doubtless, we might add, our prehistoric) ancestors pursued, as far as their intelligence permitted, the same course. They also fell back upon experience, but with this difference—that the particular experiences which furnished the weft and woof of their theories were drawn, not from the study of nature, but from what lay much closer to them, the observation of men. Their theories accordingly took an anthropomorphic form. To supersensual beings, which, 'however potent and invisible, were nothing but a species of human creatures, perhaps raised from among mankind, and retaining all human passions and appetites,'[1] were handed over the rule and governance of natural phenomena.

Tested by observation and reflection, these early notions failed in the long run to satisfy the more penetrating intellects of our race. Far in the depths of history we find men of exceptional power differentiating themselves from the crowd, rejecting these anthropomorphic notions, and seeking to connect natural phenomena with their physical principles. But long prior to these purer efforts of the understanding the merchant had been abroad, and rendered the philosopher possible; commerce had been developed, wealth amassed, leisure for travel and speculation secured, while races educated under different conditions, and therefore differently informed and endowed, had been stimulated and sharpened by mutual contact. In those regions where the commercial aristocracy of ancient Greece mingled with its eastern neighbours the sciences were born, being nurtured and developed by free-thinking and courageous men. The state of things to be displaced may be gathered from a passage of Euripides quoted by Hume. 'There is nothing in the world; no glory, no prosperity. The gods toss all into confusion; mix everything with its reverse, that all of us, from our ignorance and uncertainty may pay them the more worship and reverence.' Now, as science demands the radical extirpation of caprice and the absolute reliance upon law in nature, there grew with the growth of scientific notions a desire and determination to sweep from the field of theory this mob of gods and demons, and to place natural phenomena on a basis more congruent with themselves.

The problem which had been previously approached from above was now attacked from below; theoretic effort passed from the super- to the sub-sensible. It was felt that to construct the universe in idea it was necessary to have some notion of its constituent parts—of what Lucretius subsequently called the 'First Beginnings.' Abstracting again from experience, the leaders of scientific speculation reached at length the pregnant doctrine of atoms and molecules, the latest developments of which were set forth with such power and clearness at the last meeting of the British Association. Thought, no doubt, had long hovered about this doctrine before it attained the precision and completeness which it assumed in the mind of Democritus,[2] a philosopher who may well for a moment arrest our attention. 'Few great men,' says Lange, a non-materialist, in his excellent 'History of Materialism,' to the spirit and to the letter of which I am equally indebted, 'have been so despitefully used by history as Democritus. In the distorted images sent down to us through unscientific traditions there remains of him almost nothing but the name of "the laughing philosopher," while figures of immeasurably smaller significance spread themselves out at full length before us.' Lange speaks of Bacon's high appreciation of Democritus—for ample illustrations of which I am indebted to my excellent friend Mr. Spedding, the learned editor and biographer of Bacon. It is evident, indeed, that Bacon considered Democritus to be a man of weightier metal than either Plato or Aristotle, though their philosophy 'was noised and celebrated in the schools, amid the din and pomp of professors.' It was not they, but Genseric and Attila and the barbarians, who destroyed the atomic philosophy. 'For at a time when all human learning had suffered shipwreck these planks of Aristotelian and Platonic philosophy, as being of a lighter and more inflated substance, were preserved and came down to us, while things more solid sank and almost passed into oblivion.'

The son of a wealthy farmer, Democritus devoted the whole of his inherited fortune to the culture of his mind. He travelled everywhere; visited Athens when Socrates and Plato were there, but quitted the city without making himself known. Indeed, the dialectic strife in which Socrates so much delighted had no charms for Democritus, who held that 'the man who readily contradicts and uses many words is unfit to learn anything truly right.' He is said to have discovered and educated Protagoras the sophist, being struck as much by the manner in which he, being a hewer of wood, tied up his faggots as by the sagacity of his conversation. Democritus returned poor from his travels, was supported by his brother, and at length wrote his great work entitled 'Diakosmos,' which he read publicly before the people of his native town. He was honoured by his countrymen in various ways, and died serenely at a great age.

The principles enunciated by Democritus reveal his uncompromising antagonism to those who deduced the phenomena of nature from the caprices of the gods. They are briefly these:—1. From nothing comes nothing. Nothing that exists can be destroyed. All changes are due to the combination and separation of molecules. 2. Nothing happens by chance. Every occurrence has its cause from which it follows by necessity. 3. The only existing things are the atoms and empty space; all else is mere opinion. 4. The atoms are infinite in number and infinitely various in form; they strike together, and the lateral motions and whirlings which thus arise are the beginnings of worlds. 5. The varieties of all things depend upon the varieties of their atoms, in number, size, and aggregation. 6. The soul consists of fine, smooth, round atoms, like those of fire. These are the most mobile of all. They interpenetrate the whole body, and in their motions the phenomena of life arise. The first five propositions are a fair general statement of the atomic philosophy, as now held. As regards the sixth, Democritus made his fine smooth atoms do duty for the nervous system, whose functions were then unknown. The atoms of Democritus are individually without sensation; they combine in obedience to mechanical laws; and not only organic forms, but the phenomena of sensation and thought are the result of their combination.

That great enigma, 'the exquisite adaptation of one part of an organism to another part, and to the conditions of life,' more especially the construction of the human body, Democritus made no attempt to solve. Empedocles, a man of more fiery and poetic nature, introduced the notion of love and hate among the atoms to account for their combination and separation. Noticing this gap in the doctrine of Democritus, he struck in with the penetrating thought, linked, however, with some wild speculation, that it lay in the very nature of those combinations which were suited to their ends (in other words, in harmony with their environment) to maintain themselves, while unfit combinations, having no proper habitat, must rapidly disappear. Thus more than 2,000 years ago the doctrine of the 'survival of the fittest,' which in our day, not on the basis of vague conjecture, but of positive knowledge, has been raised to such extraordinary significance, had received at all events partial enunciation.[3]

Epicurus,[4] said to be the son of a poor schoolmaster at Samos, is the next dominant figure in the history of the atomic philosophy. He mastered the writings of Democritus, heard lectures in Athens, went back to Samos, and subsequently wandered through various countries. He finally returned to Athens, where he bought a garden, and surrounded himself by pupils, in the midst of whom he lived a pure and serene life, and died a peaceful death. Democritus looked to the soul as the ennobling part of man; even beauty without understanding partook of animalism. Epicurus also rated the spirit above the body; the pleasure of the body was that of the moment, while the spirit could draw upon the future and the past. His philosophy was almost identical with that of Democritus; be he never quoted either friend or foe. One main object of Epicurus was to free the world from superstition and the fear of death. Death he treated with indifference. It merely robs us of sensation. As long as we are, death is not; and when death is, we are not. Life has no more evil for him who has made up his mind that it is no evil not to live. He adored the gods, but not in the ordinary fashion. The idea of divine power, properly purified, he thought an elevating one. Still he taught, 'Not he is godless who rejects the gods of the crowd, but rather he who accepts them.' The gods were to him eternal and immortal beings, whose blessedness excluded every thought of care or occupation of any kind. Nature pursues her course in accordance with everlasting laws, the gods never interfering. They haunt

> The lucid interspace of world and world
> Where never creeps a cloud or moves a wind,
> Nor ever falls the least white star of snow,
> Nor ever lowest roll of thunder moans,
> Nor sound of human sorrow mounts to mar
> Their sacred everlasting calm.[5]

Lange considers the relation of Epicurus to the gods subjective; the indication probably of an ethical requirement of his own nature. We cannot read history with open eyes, or study human nature to its depths, and fail to discern such a requirement. Man never has been, and he never will be, satisfied with the operations and products of the Understanding alone; hence physical science cannot cover all the demands of his nature. But the history of the efforts made to satisfy these demands might be broadly described as a history of errors—the error, in great part, consisting in ascribing fixity to that which is fluent, which varies as we vary, being gross when we are gross, and becoming, as our capacities widen, more abstract and sublime. On one great point the mind of Epicurus was at peace. He neither sought nor expected, here or hereafter, any personal profit from his relation to the gods. And it is assuredly a fact that loftiness and serenity of thought may be promoted by conceptions which involve no idea of profit of this kind. 'Did I not believe,' said a great man to me once, 'that an Intelligence is at the heart of things, my life on earth would be intolerable.' The utterer of these words is not, in my opinion, rendered less noble but more noble by the fact that it was the need of ethical harmony here, and not the thought of personal profit hereafter, that prompted his observation.

There are persons, not belonging to the highest intellectual zone, nor yet to the lowest, to whom perfect clearness of exposition suggests want of depth. They find comfort and edification in an abstract and learned phraseology. To some such people Epicurus, who spared no pains to rid his style of every trace of haze and turbidity, appeared, on this very account, superficial. He had, however, a disciple who thought it no unworthy occupation to spend his days and nights in the effort to reach the clearness of his master, and to whom the Greek philosopher is mainly indebted for the extension and perpetuation of his fame. A century and a half after the death of Epicurus, Lucretius[6] wrote his great poem, 'On the Nature of Things,' in which he, a Roman, developed with extraordinary ardour the philosophy of his Greek predecessor. He wishes to win over his friend Memnius to the school of Epicurus; and although he has no rewards in a future life to offer, although his object appears to be a purely negative one, he addresses his friend with the heat of an apostle. His object, like that of his great forerunner, is the destruction of superstition; and considering that men trembled before every natural event as a direct monition from the gods, and that everlasting torture was also in prospect, the freedom aimed at by Lucretius might perhaps be deemed a positive good. 'This terror,' he says, 'and darkness of mind must be dispelled, not by the rays of the sun and glittering shafts of day, but by the aspect and the law of nature.' He refutes the notion that anything can come out of nothing, or that that which is once begotten can be recalled to nothing. The first beginnings, the atoms, are indestructible, and into them all things can be resolved at last. Bodies are partly atoms, and partly combinations of atoms; but the atoms nothing can quench. They are strong in solid singleness, and by their denser combination all things can be closely packed and exhibit enduring strength. He denies that matter is infinitely divisible. We come at length to the atoms, without which, as an imperishable substratum, all order in the generation and development of things would be destroyed.

The mechanical shock of the atoms being in his view the all-sufficient cause of things, he combats the notion that the constitution of nature has been in any way determined by intelligent design. The inter-action of the atoms throughout infinite time rendered all manner of combinations possible. Of these the fit ones persisted, while the unfit ones disappeared. Not after sage deliberation did the atoms station themselves in their right places, nor did they bargain what motions they should assume. From all eternity they have been driven together, and after trying motions and unions of every kind, they fell at length into the arrangements out of which this system of things has been formed. 'If you will apprehend and keep in mind these things, nature, free at once, and rid of her haughty lords, is seen to do all things spontaneously of herself, without the meddling of the gods.'[7]

To meet the objection that his atoms cannot be seen, Lucretius describes a violent storm, and shows that the invisible particles of air act in the same way as the visible particles of water. We perceive, moreover, the different smells of things, yet never see them coming to our nostrils. Again, clothes hung up on a shore which waves break upon become moist, and then get dry if spread out in the sun, though no eye can see either the approach or the escape of the water particles. A ring, worn long on the finger, becomes thinner; a water-drop hollows out a stone; the ploughshare is rubbed away in the field; the street pavement is worn by the feet; but the particles that disappear at any moment we cannot see. Nature acts through invisible particles. That Lucretius had a strong scientific imagination the foregoing references prove. A fine illustration of his power in this respect is his explanation of the apparent rest of bodies whose atoms are in motion. He employs the image of a flock of sheep with skipping lambs, which, seen from a distance, presents simply a white patch upon the green hill, the jumping of the individual lambs being quite invisible.

His vaguely-grand conception of the atoms falling eternally through space suggested the nebular hypothesis to Kant, its first propounder. Far beyond the limits of our visible world are to be found atoms innumerable, which have never been united to form bodies, or which, if once united, have been again dispersed, falling silently through immeasurable intervals of time and space. As everywhere throughout the All the same conditions are repeated, so must the phenomena be repeated also. Above us, below us, beside us, therefore, are worlds without end; and this, when considered, must dissipate every thought of a deflection of the universe by the gods. The worlds come and go, attracting new atoms out of limitless space, or dispersing their own particles. The reputed death of Lucretius, which forms the basis of Mr. Tennyson's noble poem, is in strict accordance with his philosophy, which was severe and pure.

During the centuries lying between the first of these three philosophers and the last, the human intellect was active in other fields than theirs. The sophists had run through their career. At Athens had appeared Socrates, Plato, and Aristotle, who ruined the sophists, and whose yoke remains to some extent unbroken to the present hour. Within this period also the School of Alexandria was founded, Euclid wrote his 'Elements,' and made some advance in optics. Archimedes had propounded the theory of the lever and the principles of hydrostatics. Pythagoras had made his experiments on the harmonic intervals, while astronomy was immensely enriched by the discoveries of Hipparchus, who was followed by the historically more celebrated Ptolemy. Anatomy had been made the basis of Scientific medicine; and it is said by Draper[8] that vivisection then began. In fact, the science of ancient Greece had already cleared the world of the fantastic images of divinities operating capriciously through natural phenomena. It had shaken itself free from that fruitless scrutiny 'by the internal light of the mind alone,' which had vainly sought to transcend experience and reach a knowledge of ultimate causes. Instead of accidental observation, it had introduced observation with a purpose;

instruments were employed to aid the senses; and scientific method was rendered in a great measure complete by the union of Induction and Experiment.

What, then, stopped its victorious advance? Why was the scientific intellect compelled, like an exhausted soil, to lie fallow for nearly two millenniums before it could regather the elements necessary to its fertility and strength? Bacon has already let us know one cause; Whewell ascribes this stationary period to four causes—obscurity of thought, servility, intolerance of disposition, enthusiasm of temper—and he gives striking examples of each.[9] But these characteristics must have had their antecedants in the circumstances of the time. Rome and the other cities of the Empire had fallen into moral putrefaction. Christianity had appeared, offering the gospel to the poor, and, by moderation if not asceticism of life, practically protesting against the profligacy of the age. The sufferings of the early Christians, and the extraordinary exaltation of mind which enabled them to triumph over the diabolical tortures to which they were subjected,[10] must have left traces not easily effaced. They scorned the earth, in view of that 'building of God, that house not made with hands, eternal in the heavens.' The Scriptures which ministered to their spiritual needs were also the measure of their Science. When, for example, the celebrated question of antipodes came to be discussed, the Bible was with many the ultimate court of appeal. Augustine, who flourished A.D. 400, would not deny the rotundity of the earth; but he would deny the possible existence of inhabitants at the other side, 'because no such race is recorded in Scripture among the descendants of Adam.' Archbishop Boniface was shocked at the assumption of a 'world of human beings out of the reach of the means of salvation.' Thus reined in, Science was not likely to make much progress. Later on the political and theological strife between the Church and civil governments, so powerfully depicted by Draper, must have done much to stifle investigation.

Whewell makes many wise and brave remarks regarding the spirit of the Middle Ages. It was a menial spirit. The seekers after natural knowledge had forsaken that fountain of living waters, the direct appeal to nature by observation and experiment, and had given themselves up to the remanipulation of the notions of their predecessors. It was a time when thought had become abject, and when the acceptance of mere authority led, as it always does in science, to intellectual death. Natural events, instead of being traced to physical, were referred to moral causes; while an exercise of the phantasy, almost as degrading as the spiritualism of the present day, took the place of scientific speculation. Then came the mysticism of the Middle Ages, Magic, Alchemy, the Neo-platonic philosophy, with its visionary though sublime abstractions, which caused men to look with shame upon their own bodies as hindrances to the absorption of the creature in the blessedness of the Creator. Finally came the Scholastic philosophy, a fusion, according to Lange, of the least-mature notions of Aristotle with the Christianity of the West. Intellectual immobility was the result. As a traveller without a compass in a fog may wander long, imagining he is making way, and find himself after hours of toil at his starting point, so the schoolmen, having 'tied and untied the same knots and formed and dissipated the same clouds,' found themselves at the end of centuries in their old position.

With regard to the influence wielded by Aristotle in the Middle Ages, and which, though to a less extent, he still wields, I would ask permission to make one remark. When the human mind has achieved greatness and given evidence of extraordinary power in any domain, there is a tendency to credit it with similar power in all other domains. Thus theologians have found comfort and assurance in the thought that Newton dealt with the question of revelation, forgetful of the fact that the very devotion of his powers, through all the best years of his life, to a totally different class of ideas, not to speak of any natural disqualification, tended to render him less instead of more competent to deal with theological and historic questions. Goethe, starting from his established greatness as a poet, and indeed from his positive discoveries in Natural History, produced a profound impression among the painters of Germany when he published his 'Farbenlehre,' in which he endeavoured to overthrow Newton's theory of colours. This theory he deemed so obviously absurd that he considered its author a charlatan, and attacked him with a corresponding vehemence of language. In the domain of Natural History Goethe had made really considerable discoveries; and we have high authority for assuming that, had he devoted himself wholly to that side of science, he might have reached in it an eminence comparable with that which he attained as a poet. In sharpness of observation, in the detection of analogies, however apparently remote, in the classification and organization of facts according to the analogies discerned, Goethe possessed extraordinary powers. These elements of scientific inquiry fall in with the discipline of the poet. But, on the other hand, a mind thus richly endowed in the direction of natural history may be almost shorn of endowment as regards the more strictly called physical and mechanical sciences. Goethe was in this condition. He could not formulate distinct mechanical conceptions; he could not see the force of mechanical reasoning; and in regions where such reasoning reigns supreme he became a mere *ignis fatuus* to those who followed him.

I have sometimes permitted myself to compare Aristotle with Goethe, to credit the Stagirite with an almost superhuman power of amassing and systematizing facts, but to consider him fatally defective on that side of the mind in respect to which incompleteness has been just ascribed to Goethe. Whewell refers the errors of Aristotle, not to a neglect of facts, but to 'a neglect of the idea appropriate to the facts; the idea of Mechanical cause, which is Force, and the substitution of vague or inapplicable notions, involving only relations of space or emotions of wonder.' This is doubtless true; but the word 'neglect' implies mere intellectual misdirection, whereas in Aristotle, as in Goethe, it was not, I believe, misdirection, but sheer natural incapacity which lay at the root of his mistakes. As a physicist, Aristotle displayed what we should consider some of the worst attributes of a modern physical investigator—indistinctness of ideas, confusion of mind, and a confident use of language, which led to the delusive notion that he had really mastered his subject, while he had as yet failed to grasp even the elements of it. He put words in the place of things, subject in the place of object. He preached Induction without practising it, inverting the true order of inquiry by passing from the general to the particular, instead of from the particular to the general. He made of the universe a closed sphere, in the centre of which he fixed the earth, proving from general principles, to his own satisfaction and to that of the world for near 2,000 years, that no other universe was possible. His notions of motion were entirely unphysical. It was natural or unnatural, better or worse, calm or violent—no real mechanical conception regarding it lying at the bottom of his mind. He affirmed that a vacuum could not exist, and proved that if it did exist motion in it would be impossible. He determined *à priori* how many species of animals must exist, and shows on general principles why animals must have such and such parts. When an eminent contemporary philosopher, who is far removed from errors of this kind, remembers these abuses of the *à priori* method, he will be able to make allowance for the jealousy of physicists as to the acceptance of so-called *à priori* truths. Aristotle's errors of detail, as shown by Eucken and Lange, were grave and numerous. He affirmed that only in man we had the beating of the heart, that the left side of the body was colder than the right, that men have more teeth than women, and that there is an empty space at the back of every man's head.

There is one essential quality in physical conceptions which was entirely wanting in those of Aristotle and his followers. I wish it could be expressed by a word untainted by its associations; it signifies a capability of being placed as a coherent picture before the mind. The Germans express the act of picturing by the word *vorstellen*, and the picture they call a *Vorstellung*. We have no word in English which comes nearer to our requirements than *Imagination*, and, taken with its proper limitations, the word answers very well; but, as just intimated, it is tainted by its associations, and therefore objectionable to some minds. Compare, with reference to this capacity of mental presentation, the case of the Aristotelian who refers the ascent of water in a pump to Nature's abhorrence of a vacuum, with that of Pascal when he proposed to solve the question of atmospheric pressure by the ascent of the Puy de Dome. In the one case the terms of the explanation refuse to fall into place as a physical image; in the other the image is distinct, the fall and rise of the barometer being clearly figured as the balancing of two varying and opposing pressures.

During the drought of the Middle Ages in Christendom, the Arabian intellect, as forcibly shown by Draper, was active. With the intrusion of the Moors into Spain, he says, order, learning, and refinement took the place of their opposites. When smitten with disease, the Christian peasant resorted to a shrine, the Moorish one to an instructed physician. The Arabs encouraged translations from the Greek philosophers, but not from the Greek poets. They turned in disgust 'from the lewdness of our classical mythology, and denounced as an unpardonable blasphemy all connexion between the impure Olympian Jove and the Most High God.' Draper traces still further than Whewell the Arab elements in our scientific terms, and points out that the under garment of ladies retains to this hour its Arab name. He gives examples of what Arabian men of science accomplished, dwelling particularly on Alhazen, who was the first to correct the Platonic notion that rays of light are emitted by the eye. He discovered atmospheric refraction, and points out that we see the sun and the moon after they have set. He explains the enlargement of the sun and moon, and the shortening of the vertical diameters of both these bodies, when near the horizon. He is aware that the atmosphere decreases in density with increase of elevation, and actually fixes its height at 58½ miles. In the Book of the Balance of Wisdom, he sets forth the connexion between the weight of the atmosphere and its increasing density. He shows that a body will weigh differently in a rare and dense atmosphere: he considers the force with which plunged bodies rise through heavier media. He understands the doctrine of the centre of gravity, and applies it to the investigation of balances and steelyards. He recognises gravity as a force, though he falls into the error of making it diminish simply as the distance increased, and of making it purely terrestrial. He knows the relation between the velocities, spaces, and times of falling bodies, and has distinct ideas of capillary attraction. He improved the hydrometer. The determination of the densities of bodies as given by Alhazen approach very closely to our own. 'I join,' says Draper, in the pious prayer of Alhazen, 'that in the day of judgment the All-Merciful will take pity on the soul of Abur-Raihân, because he was the first of the race of men to construct a table of specific gravities.' If all this be historic truth (and I have entire confidence in Dr. Draper), well may he 'deplore the systematic manner in which the literature of Europe has contrived to put out of sight our scientific obligations to the Mahommedans.'[11]

The strain upon the mind during the stationary period towards ultra-terrestrial things, to the neglect of problems close at hand, was sure to provoke reaction. But the reaction was gradual; for the ground was dangerous, a power being at hand competent to crush the critic who went too far. To elude this power and still allow opportunity for the expression of opinion, the doctrine of 'two-fold truth' was invented, according to which an opinion might be held 'theologically' and the opposite opinion 'philosophically.'[12] Thus in the thirteenth century the creation of the world in six days, and the unchangeableness of the individual soul which had been so distinctly affirmed by St. Thomas Aquinas, were both denied philosophically, but admitted to be true as articles of the Catholic faith. When Protagoras uttered the maxim which brought upon him so much vituperation, that 'opposite assertions are equally true,' he simply meant that human beings differed so much from each other that what was subjectively true to the one might be subjectively untrue to the other. The great Sophist never meant to play fast and loose with the truth by saying that one of two opposite assertions, made by the same individual, could possibly escape being a lie. It was not 'sophistry,' but the dread of theologic vengeance that generated this double dealing with conviction; and it is astonishing to notice what lengths were possible to men who were adroit in the use of artifices of this kind.

Towards the close of the stationary period a word-weariness, if I may so express it, took more and more possession of men's minds. Christendom had become sick of the School philosophy and its verbal wastes, which led to no issue, but left the intellect in everlasting haze. Here and there was heard the voice of one impatiently crying in the wilderness, 'Not unto Aristotle, not unto subtle hypothesis, not unto church, Bible, or blind tradition, must we turn for a knowledge of the universe, but to the direct investigation of Nature by observation and experiment.' In 1543 the epoch-making work of Copernicus on the paths of the heavenly bodies appeared. The total crash of Aristotle's closed universe with the earth at its centre followed as a consequence, and 'the earth moves!' became a kind of watchword among intellectual freemen. Copernicus was Canon of the Church of Frauenburg, in the diocese of Ermeland. For three-and-thirty years he had withdrawn himself from the world and devoted himself to the consolidation of his great scheme of the solar system. He made its blocks eternal; and even to those who feared it and desired its overthrow it was so obviously strong that they refrained for a time from meddling with it. In the last year of the life of Copernicus his book appeared: it is said that the old man received a copy of it a few days before his death, and then departed in peace.

The Italian philosopher Giordano Bruno was one of the earliest converts to the new astronomy. Taking Lucretius as his exemplar, he revived the notion of the infinity of worlds; and, combining with it the doctrine of Copernicus, reached the sublime generalization that the fixed stars are suns, scattered numberless through space and accompanied by satellites, which bear the same relation to them that our earth does to our sun, or our moon to our earth. This was an expansion of transcendent import; but Bruno came closer than this to our present line of thought. Struck with the problem of the generation and maintenance of organisms, and duly pondering it, he came to the conclusion that Nature in her productions does not imitate the technic of man. Her process is one of unravelling and unfolding. The infinity of forms under which matter appears were not imposed upon it by an external artificer; by its own intrinsic force and virtue it brings these forms forth. Matter is not the mere naked, empty *capacity* which philosophers have pictured her to be, but the universal mother who brings forth all things as the fruit of her own womb.

This outspoken man was originally a Dominican monk. He was accused of heresy and had to fly, seeking refuge in Geneva, Paris, England, and Germany. In 1592 he fell into the hands of the Inquisition at Venice. He was imprisoned for many years, tried, degraded, excommunicated, and handed over to the civil power, with the request that he should be treated gently and 'without the shedding of blood.' This meant that he was to be burnt; and burnt accordingly he was, on the 16th of February, 1600. To escape a similar fate Galileo, thirty-three years afterwards, abjured, upon his knees, and with his hand upon the holy gospels, the heliocentric doctrine which he knew to be true. After Galileo came Kepler, who from his German home defied the power beyond the Alps. He traced out from pre-existing observations the laws of planetary motion. Materials were thus prepared for Newton, who bound those empirical laws together by the principle of gravitation.

In the seventeenth century Bacon and Descartes, the restorers of philosophy, appeared in succession. Differently educated and endowed, their philosophic tendencies were different. Bacon held fast to Induction, believing firmly in the existence of an external world, and making collected experiences the basis of all knowledge. The mathematical studies of Descartes gave him a bias towards Deduction; and his fundamental principle was much the same as that of Protagoras, who made the individual man the measure of all things. 'I think, therefore I am,' said Descartes. Only his own identity was sure to him; and the development of this system would have led to an idealism in which the outer world would be resolved into a mere phenomenon of consciousness. Gassendi, one of Descartes's contemporaries, of whom we shall hear more presently, quickly pointed out that the fact of personal existence would be proved as well by reference to any other act as to the act of thinking. I eat, therefore I am; or I love, therefore I am, would be quite as conclusive. Lichtenberg showed that the very thing to be proved was inevitably postulated in the first two words, 'I think;' and that no inference from the postulate could by any possibility be stronger than the postulate itself.

But Descartes deviated strangely from the idealism implied in his fundamental principle. He was the first to reduce, in a manner eminently capable of bearing the test of mental presentation, vital phenomena to purely mechanical principles. Through fear or love, Descartes was a good churchman; he accordingly rejects the notion of an atom, because it was absurd to suppose that God, if he so pleased, could not divide an atom; he puts in the place of the atoms small round particles and light splinters, out of which he builds the organism. He sketches with marvellous physical insight a machine, with water for its motive power, which shall illustrate vital actions. He has made clear to his mind that such a machine would be competent to carry on the processes of digestion, nutrition, growth, respiration, and the beating of the heart. It would be competent to accept impressions from the external sense, to store them up in imagination and memory, to go through the internal movements of the appetites and passions, the external movement of limbs. He deduces these functions of his machine from the mere arrangement of its organs, as the movement of a clock or other automaton is deduced from its weights and wheels. 'As far as these functions are concerned,' he says, 'it is not necessary to conceive any other vegetative or sensitive soul, nor any other principle of motion or of life, than the blood and the spirits agitated by the fire which burns continually in the heart, and which is in no wise different from the fires which exist in inanimate bodies.' Had Descartes been acquainted with the steam-engine, he would have taken it, instead of a fall of water, as his motive power, and shown the perfect analogy which exists between the oxidation of the food in the body and that of the coal in the furnace. He would assuredly have anticipated Mayer in calling the blood which the heart diffuses 'the oil of the lamp of life;' deducing all animal motions from the combustion of this oil, as the motions of a steam-engine are deduced from the combustion of its coal. As the matter stands, however, and considering the circumstances of the time, the boldness, clearness, and precision with which he grasped the problem of vital dynamics constitute a marvellous illustration of intellectual power.[13]

During the Middle Ages the doctrine of atoms had to all appearance vanished from discussion. In all probability it held its ground among sober-minded and thoughtful men, though neither the church nor the world was prepared to hear of it with tolerance. Once, in the year 1348, it received distinct expression. But retraction by compulsion immediately followed, and, thus discouraged, it slumbered till the seventeenth century, when it was revived by a contemporary and friend of Hobbes and Malmesbury, the orthodox Catholic provost of Digne, Gassendi. But before stating his relation to the Epicurean doctrine, it will be well to say a few words on the effect, as regards science, of the general introduction of monotheism among European nations.

'Were men,' says Hume, 'led into the apprehension of invisible intelligent power by contemplation of the works of Nature, they could never possibly entertain any conception but of one single being, who bestowed existence and order on this vast machine, and adjusted all its parts to one regular system.' Referring to the condition of the heathen, who sees a god behind every natural event, thus peopling the world with thousands of beings whose caprices are incalculable, Lange shows the impossibility of any compromise between such notions and those of science, which proceeds on the assumption of never-changing law and causality. 'But,' he continues, with characteristic penetration, 'when the great thought of one God, acting as a unit upon the universe, has been seized, the connexion of things in accordance with the law of cause and effect is not only thinkable, but it is a necessary consequence of the assumption. For when I see ten thousand wheels in motion, and know, or believe, that they are all driven by one, then I know that I have before me a mechanism the action of every part of which is determined by the plan of the whole. So much being assumed, it follows that I may investigate the structure of that machine, and the various motions of its parts. For the time being, therefore, this conception renders scientific action free.' In other words, were a capricious God at the circumference of every wheel and at the end of every lever, the action of the machine would be incalculable by the methods of science. But the action of all its parts being rigidly determined by their connexions and relations, and these being brought into play by a single self-acting driving wheel, then, though this last prime mover may elude me, I am still able to comprehend the machinery which it sets in motion. We have here a conception of the relation of Nature to its Author which seems perfectly acceptable to some minds, but perfectly intolerable to others. Newton and Boyle lived and worked happily under the influence of this conception; Goethe rejected it with vehemence, and the same repugnance to accepting it is manifest in Carlyle.[14]

The analytic and synthetic tendencies of the human mind exhibit themselves throughout history, great writers ranging themselves sometimes on the one side, sometimes on the other. Men of warm feelings and minds open to the elevating impressions produced by Nature as a whole, whose satisfaction, therefore, is rather ethical than logical, lean to the synthetic side, while the analytic harmonizes best with the more precise and more mechanical bias which seeks the satisfaction of the understanding. Some form of pantheism was usually adopted by the one, while a detached Creator, working more or less after the manner of men, was often assumed by the other. Gassendi is hardly to be ranked with either. Having formally acknowledged God as the great first cause, he immediately dropped the idea, applied the known laws of mechanics to the atoms, deducing thence all vital phenomena. He defended Epicurus, and dwelt upon his purity, both of doctrine and of life. True he was a heathen, but so was Aristotle. He assailed superstition and religion, and rightly, because he did not know the true religion. He thought that the gods neither rewarded nor punished, and adored them purely in consequence of their completeness; here we see, says Gassendi, the reverence of the child instead of the fear of the slave. The errors of Epicurus shall be corrected, the body of his truth retained; and then Gassendi proceeds, as any heathen might do, to build up the world, and all that therein is, of atoms and molecules. God, who created earth and water, plants and animals, produced in the first place a definite number of atoms, which constituted the seed of all things. Then began that series of combinations and decompositions which goes on at present, and which will continue in future. The principle of every change resides in matter. In artificial productions the moving principle is different from the material worked upon; but in Nature the agent works within, being the most active and mobile part of the material itself. Thus, this bold ecclesiastic, without incurring the censure of the church or the world, contrives to outstrip Mr. Darwin. The same cast of mind which caused him to detach the Creator from his universe led him also to detach the soul from the body, though to the body he ascribes an influence so large as to render the soul almost unnecessary. The aberrations of reason were in his view an affair of the material brain. Mental disease is brain disease; but then the immortal reason sits apart, and cannot be touched by the disease. The errors of madness are errors of the instrument, not of the performer.

It may be more than a mere result of education, connecting itself probably with the deeper mental structure of the two men, that the idea of Gassendi above enunciated is substantially the same as that expressed by Professor Clerk Maxwell at the close of the very able lecture delivered by him at Bradford last year. According to both philosophers, the atoms, if I might understand aright, are the *prepared materials* which, formed by the skill of the highest, produce by their subsequent inter-action all the phenomena of the material world. There seems to be this difference, however, between Gassendi and Maxwell. The one *postulates*, the other *infers* his first cause. In his 'manufactured articles,' as he calls the atoms, Professor Maxwell finds the basis of an induction which enables him to scale philosophic heights considered inaccessible by Kant, and to take the logical step from the atoms to their Maker.

Accepting here the leadership of Kant, I doubt the legitimacy of Maxwell's logic; but it is impossible not to feel the ethic glow with which his lecture concludes. There is, moreover, a very noble strain of eloquence in his description of the steadfastness of the atoms:—'Natural causes, as we know, are at work, which tend to modify, if they do not at length destroy, all the arrangements and dimensions of the earth and the whole solar system. But though in the course of ages catastrophes have occurred and may yet occur in the heavens, though ancient systems may be dissolved and new systems evolved out of their ruins, the molecules out of which these systems are built—the foundation stones of the material universe—remain unbroken and unworn.'

The atomic doctrine, in whole or in part, was entertained by Bacon, Descartes, Hobbes, Locke, Newton, Boyle, and their successors, until the chemical law of multiple proportions enabled Dalton to confer upon it an entirely new significance. In our day there are secessions from the theory, but it still stands firm. Loschmidt, Stoney, and Sir William Thomson have sought to determine the sizes of the atoms, or rather to fix the limits between which their sizes lie; while only last year the discourses of Williamson and Maxwell illustrate the present hold of the doctrine upon the foremost scientific minds. In fact, it may be doubted whether, wanting this fundamental conception, a theory of the material universe is capable of scientific statement.

Ninety years subsequent to Gassendi the doctrine of bodily instruments, as it may be called, assumed immense importance in the hands of Bishop Butler, who, in his famous 'Analogy of Religion,' developed, from his own point of view, and with consummate sagacity, a similar idea. The Bishop still influences superior minds; and it will repay us to dwell for a moment on his views. He draws the sharpest distinction between our real selves and our bodily instruments. He does not, as far as I remember, use the word soul, possibly because the term was so hackneyed in his day as it had been for many generations previously. But he speaks of 'living powers,' 'perceiving' or 'percipient powers,' 'moving agents,' 'ourselves,' in the same sense as we should employ the term soul. He dwells upon the fact that limbs may be removed, and mortal diseases assail the body, the mind, almost up to the moment of death, remaining clear. He refers to sleep and to swoon, where the 'living powers' are suspended but not destroyed. He considers it quite as easy to conceive of existence out of our bodies as in them: that we may animate a succession of bodies, the dissolution of all of them having no more tendency to dissolve our real selves, or 'deprive us of living faculties—the faculties of perception and action—than the dissolution of any foreign matter which we are capable of receiving impressions from, or making use of for the common occasions of life.' This is the key of the Bishop's position; 'our organized bodies are no more a part of ourselves than any other matter around us.' In proof of this he calls attention to the use of glasses, which 'prepare objects' for the 'percipient power' exactly as the eye does. The eye itself is no more percipient than the glass; is quite as much the instrument of the true self, and also as foreign to the true self, as the glass is. 'And if we see with our eyes only in the same manner as we do with glasses, the like may justly be concluded from analogy of all our senses.'

Lucretius, as you are aware, reached a precisely opposite conclusion; and it certainly would be interesting, if not profitable, to us all, to hear what he would or could urge in opposition to the reasoning of the Bishop. As a brief discussion of the point will enable us to see the bearings of an important question, I will here permit a disciple of Lucretius to try the strength of the Bishop's position, and then allow the Bishop to retaliate, with the view of rolling back, if he can, the difficulty upon Lucretius.

The argument might proceed in this fashion:—

'Subjected to the test of mental presentation (*Vorstellung*), your views, most honoured prelate, would present to many minds a great, if not an insuperable difficulty. You speak of "living powers," "percipient or perceiving powers," and "ourselves;" but can you form a mental picture of any one of these apart from the organism through which it is supposed to act? Test yourself honestly, and see whether you possess any faculty that would enable you to form such a conception. The true self has a local habitation in each of us; thus localized, must it not possess a form? If so, what form? Have you ever for a moment realized it? When a leg is amputated the body is divided into two parts; is the true self in both of them or in one? Thomas Aquinas might say in both; but not you, for you appeal to the consciousness associated with one of the two parts to prove that the other is foreign matter. Is consciousness, then, a necessary element of the true self? If so, what do you say to the case of the whole body being deprived of consciousness? If not, then on what grounds do you deny any portion of the true self to the severed limb? It seems very singular that, from the beginning to the end of your admirable book (and no one admires its sober strength more than I do), you never once mention the brain or nervous system. You begin at one end of the body, and show that its parts may be removed without prejudice to the perceiving power. What if you begin at the other end, and remove, instead of the leg, the brain? The body, as before, is divided into two parts; but both are now in the same predicament, and neither can be appealed to to prove that the other is foreign matter. Or, instead of going so far as to remove the brain itself, let a certain portion of its bony covering be removed, and let a rhythmic series of pressures and relaxations of pressure be applied to the soft substance. At every pressure, "the faculties of perception and of action" vanish; at every relaxation of pressure they are restored. Where, during the intervals of pressure, is the perceiving power? I once had the discharge of a large Leyden battery passed unexpectedly through me: I felt nothing, but was simply blotted out of conscious existence for a sensible interval. Where was my true self during that interval? Men who have recovered from lightning-stroke have been much longer in the same state; and indeed in cases of ordinary concussion of the brain, days may elapse during which no experience is registered in consciousness. Where is the man himself during the period of insensibility? You may say that I beg the question when I assume the man to have been unconscious, that he was really conscious all the time, and has simply forgotten what had occurred to him. In reply to this, I can only say that no one need shrink from the worst tortures that superstition ever invented if only so felt and so remembered. I do not think your theory of instruments goes at all to the bottom of the matter. A telegraph-operator has his instruments, by means of which he converses with the world; our bodies possess a nervous system, which plays a similar part between the perceiving power and external things. Cut the wires of the operator, break his battery, demagnetize his needle: by this means you certainly sever his connexion with the world; but inasmuch as these are real instruments, their destruction does not touch the man who uses them. The operator survives, *and he knows that he survives*. What is it, I would ask, in the human system that answers to this conscious survival of the operator when the battery of the brain is so disturbed as to produce insensibility, or when it is destroyed altogether?

'Another consideration, which you may consider slight, presses upon me with some force. The brain may change from health to disease, and through such a change the most exemplary man may be converted into a debauchee or a murderer. My very noble and approved good master had, as you know, threatenings of lewdness introduced into his brain by his jealous wife's philter; and sooner than permit himself to run even the risk of yielding to these base promptings he slew himself. How could the hand of Lucretius have been thus turned against himself if the real Lucretius remained as before? Can the brain or can it not act in this distempered way without the intervention of the immortal reason? If it can, then it is a prime mover which requires only healthy regulation to render it reasonably self-acting, and there is no apparent need of your immortal reason at all. If it cannot, then the immortal reason, by its mischievous activity in operating upon a broken instrument, must have the credit of committing every imaginable extravagance and crime. I think, if you will allow me to say so, that the gravest consequences are likely to flow from your estimate of the body. To regard the brain as you would a staff or an eyeglass—to shut your eyes to all its mystery, to the perfect correlation of its condition and our consciousness, to the fact that a slight excess or defect of blood in it produces the very swoon to which you refer, and that in relation to it our meat and drink and air and exercise have a perfectly transcendental value and significance—to forget all this does, I think, open a way to innumerable errors in our habits of life, and may possibly in some cases initiate and foster that very disease, and consequent mental ruin, which a wiser appreciation of this mysterious organ would have avoided.'

I can imagine the Bishop thoughtful after hearing this argument. He was not the man to allow anger to mingle with the consideration of a point of this kind. After due reflection, and having strengthened himself by that honest contemplation of the facts which was habitual with him, and which includes the desire to give even adverse facts their due weight, I can suppose the Bishop to proceed thus:—'You will remember that in the "Analogy of Religion," of which you have so kindly spoken, I did not profess to prove anything absolutely, and that I over and over again acknowledged and insisted on the smallness of our knowledge, or rather the depth of our ignorance, as regards the whole system of the universe. My object was to show my deistical friends, who set forth so eloquently the beauty and beneficence of Nature and the Ruler thereof, while they had nothing but scorn for the so-called absurdities of the Christian scheme, that they were in no better condition than we were, and that, for every difficulty found upon our side, quite as great a difficulty was to be found upon theirs. I will now, with your permission, adopt a similar line of argument. You are a Lucretian, and from the combination and separation of insensate atoms deduce all terrestrial things, including organic forms and their phenomena. Let me tell you, in the first instance, how far I am prepared to go with you. I admit that you can build crystalline forms out of this play of molecular force; that the diamond, amethyst, and snow-star are truly wonderful structures which are thus produced. I will go further and acknowledge that even a tree or flower might in this way be organized. Nay, if you can show me an animal without sensation, I will concede to you that it also might be put together by the suitable play of molecular force.

'Thus far our way is clear; but now comes my difficulty. Your atoms are individually without sensation, much more are they without intelligence. May I ask you, then, to try your hand upon this problem? Take your dead hydrogen atoms, your dead oxygen atoms, your dead carbon atoms, your dead nitrogen atoms, your dead phosphorus atoms, and all the other atoms, dead as grains of shot, of which the brain is formed. Imagine them separate and sensationless, observe them running together and forming all imaginable combinations. This, as a purely mechanical process, is *seeable* by the mind. But can you see, or dream, or in any way imagine, how out of that mechanical act, and from those individually dead atoms, sensation, thought, and emotion are to arise? Are you likely to extract Homer out of the rattling of dice, or the Differential Calculus out of the clash of billiard-balls? I am not all bereft of this *Vorstellungs-Kraft* of which you speak, nor am I, like so many of my brethren, a mere vacuum as regards scientific knowledge. I can follow a particle of musk until it reaches the olfactory nerve; I can follow the waves of sound until their tremors reach the water of the labyrinth and set the otoliths and Corti's fibres in motion; I can also visualize the waves of ether as they cross the eye and hit the retina. Nay more, I am able to pursue to the central organ the motion thus imparted at the periphery, and to see in idea the very molecules of the brain thrown into tremors. My insight is not baffled by these physical processes. What baffles and bewilders me, is the notion that from those physical tremors things so utterly incongruous with them as sensation, thought, and emotion can be derived. You may say, or think, that this issue of consciousness from the clash of atoms is not more incongruous than the flash of light from the union of oxygen and hydrogen. But I beg to say that it is. For such incongruity as the flash possesses is that which I now force upon your attention. The flash is an affair of consciousness, the objective counterpart of which is a vibration. It is a flash only by your interpretation. *You* are the cause of the apparent incongruity, and *you* are the thing that puzzles me. I need not remind you that the great Leibnitz felt the difficulty which I feel, and that to get rid of this monstrous deduction of life from death he displaced your atoms by his monads, which were more or less perfect mirrors of the universe, and out of the summation and integration of which he supposed all the phenomena of life—sentient, intellectual, and emotional—to arise.

'Your difficulty, then, as I see you are ready to admit, is quite as great as mine. You cannot satisfy the human understanding in its demand for logical continuity between molecular processes and the phenomena of consciousness. This is a rock on which materialism must inevitably split whenever it pretends to be a complete philosophy of life. What is the moral, my Lucretian? You and I are not likely to indulge in ill-temper in the discussion of these great topics, where we see so much room for honest differences of opinion. But there are people of less wit or more bigotry (I say it with humility) on both sides, who are ever ready to mingle anger and vituperation with such discussions. There are, for example, writers of note and influence at the present day who are not ashamed to assume the "deep personal sin" of a great logician to be the cause of his unbelief in a theologic dogma. And there are others who hold that we, who cherish our noble Bible, wrought as it has been into the constitution of our forefathers, and by inheritance into us, must necessarily be hypocritical and insincere. Let us disavow and discountenance such people, cherishing the unswerving faith that what is good and true in both our arguments will be preserved for the benefit of humanity, while all that is bad or false will disappear.'

I hold the Bishop's reasoning to be unanswerable, and his liberality to be worthy of imitation.

It is worth remarking that in one respect the Bishop was a product of his age. Long previous to his day the nature of the soul had been so favourite and general a topic of discussion, that, when the students of the University of Paris wished to know the leanings of a new Professor, they at once requested him to lecture upon the soul. About the time of Bishop Butler the question was not only agitated but extended. It was seen by the clear-witted men who entered this arena that many of their best arguments applied equally to brutes and men. The Bishop's arguments were of this character. He saw it, admitted it, accepted the consequences, and boldly embraced the whole animal world in his scheme of immortality.

Bishop Butler accepted with unwavering trust the chronology of the Old Testament, describing it as 'confirmed by the natural and civil history of the world, collected from common historians, from the state of the earth, and from the late inventions of arts and sciences.' These words mark progress; and they must seem somewhat hoary to the Bishop's successors of to-day.[15] It is hardly necessary to inform you that since his time the domain of the naturalist has been immensely extended—the whole science of geology, with its astounding revelations regarding the life of the ancient earth, having been created. The rigidity of old conceptions has been relaxed, the public mind being rendered gradually tolerant of the idea that not for six thousand, nor for sixty thousand, nor for six thousand thousand thousand, but for æons embracing untold millions of years, this earth has been the theatre of life and death. The riddle of the rocks has been read by the geologist and palæontologist, from subcambrian depths to the deposits thickening over the sea-bottoms of to-day. And upon the leaves of that stone book are, as you know, stamped the characters, plainer and surer than those formed by the ink of history, which carry the mind back into abysses of past time compared with which the periods which satisfied Bishop Butler cease to have a visual angle.

The lode of discovery once struck, those petrified forms in which life was at one time active increased to multitudes and demanded classification. They were grouped in genera, species, and varieties, according to the degree of similarity subsisting between them. Thus confusion was avoided, each object being found in the pigeon-hole appropriated to it and to its fellows of similar morphological or physiological character. The general fact soon became evident that none but the simplest forms of life lie lowest down, that as we climb higher among the super-imposed strata more perfect forms appear. The change, however, from form to form was not continuous, but by steps—some small, some great. 'A section,' says Mr. Huxley, 'a hundred feet thick will exhibit at different heights a dozen species of Ammonite, none of which passes beyond its particular zone of limestone, or clay, into the zone below it, or into that above it.' In the presence of such facts it was not possible to avoid the question:—Have these forms, showing, though in broken stages and with many irregularities, this unmistakable general advance, been subjected to no continuous law of growth or variation? Had our education been purely scientific, or had it been sufficiently detached from influences which, however ennobling in another domain, have always proved hindrances and delusions when introduced as factors into the domain of physics, the scientific mind never could have swerved from the search for a law of growth, or allowed itself to accept the anthropomorphism which regarded each successive stratum as a kind of mechanic's bench for the manufacture of new species out of all relation to the old.

Biassed, however, by their previous education, the great majority of naturalists invoked a special creative act to account for the appearance of each new group of organisms. Doubtless there were numbers who were clear-headed enough to see that this was no explanation at all, that in point of fact it was an attempt, by the introduction of a greater difficulty, to account for a less. But having nothing to offer in the way of explanation, they for the most part held their peace. Still the thoughts of reflecting men naturally and necessarily simmered round the question. De Maillet, a contemporary of Newton, has been brought into notice by Professor Huxley as one who 'had a notion of the modifiability of living forms.' In my frequent conversations with him, the late Sir Benjamin Brodie, a man of highly philosophic mind, often drew my attention to the fact that, as early as 1794, Charles Darwin's grandfather was the pioneer of Charles Darwin.[16] In 1801, and in subsequent years, the celebrated Lamarck, who produced so profound an impression on the public mind through the vigorous exposition of his views by the author of the 'Vestiges of Creation,' endeavoured to show the development of species out of changes of habit and external condition. In 1813 Dr. Wells, the founder of our present theory of Dew, read before the Royal Society a paper in which, to use the words of Mr. Darwin, 'he distinctly recognises the principle of natural selection; and this is the first recognition that has been indicated.' The thoroughness and skill with which Wells pursued his work, and the obvious independence of his character, rendered him long ago a favourite with me; and it gave me the liveliest pleasure to alight upon this additional testimony to his penetration. Professor Grant, Mr. Patrick Matthew, Von Buch, the author of the 'Vestiges,' D'Halloy, and others,[17] by the enunciation of opinions more or less clear and correct, showed that the question had been fermenting long prior to the year 1858, when Mr. Darwin and Mr. Wallace simultaneously but independently placed their closely concurrent views upon the subject before the Linnean Society.

These papers were followed in 1859 by the publication of the first edition of 'The Origin of Species.' All great things come slowly to the birth. Copernicus, as I informed you, pondered his great work for thirty-three years. Newton for nearly twenty years kept the idea of Gravitation before his mind; for twenty years also he dwelt upon his discovery of Fluxions, and doubtless would have continued to make it the object of his private thought had he not found that Leibnitz was upon his track. Darwin for two and twenty years pondered the problem of the origin of species, and doubtless he would have continued to do so had he not found Wallace upon his track.[18] A concentrated but full and powerful epitome of his labours was the consequence. The book was by no means an easy one; and probably not one in every score of those who then attacked it had read its pages through, or were competent to grasp their significance if they had. I do not say this merely to discredit them; for there were in those days some really eminent scientific men, entirely raised above the heat of popular prejudice, willing to accept any conclusion that science had to offer, provided it was duly backed by fact and argument, and who entirely mistook Mr. Darwin's views. In fact, the work needed an expounder; and it found one in Mr. Huxley. I know nothing more admirable in the way of scientific exposition than those early articles of his on the origin of species. He swept the curve of discussion through the really significant points of the subject, enriched his exposition with profound original remarks and reflections, often summing up in a single pithy sentence an argument which a less compact mind would have spread over pages. But there is one impression made by the book itself which no exposition of it, however luminous, can convey; and that is the impression of the vast amount of labour, both of observation and of thought, implied in its production. Let us glance at its principles.

It is conceded on all hands that what are called varieties are continually produced. The rule is probably without exception. No chick and no child is in all respects and particulars the counterpart of its brother and sister; and in such differences we have 'variety' incipient. No naturalist could tell how far this variation could be carried; but the great mass of them held that never by any amount of internal or external change, nor by the mixture of both, could the offspring of the same progenitor so far deviate from each other as to constitute different species. The function of the experimental philosopher is to combine the conditions of nature and to produce her results; and this was the method of Darwin.[19] He made himself acquainted with what could, without any manner of doubt, be done in the way of producing variation. He associated himself with pigeon-fanciers—bought, begged, kept, and observed every breed that he could obtain. Though derived from a common stock, the diversities of these pigeons were such that 'a score of them might be chosen which, if shown to an ornithologist, and he were told that they were wild birds, would certainly be ranked by him as well-defined species.' The simple principle which guides the pigeon-fancier, as it does the cattle-breeder, is the selection of some variety that strikes his fancy, and the propagation of this variety by inheritance. With his eye still directed to the particular appearance which he wishes to exaggerate, he selects it as it reappears in successive broods, and thus adds increment to increment until an astonishing amount of divergence from the parent type is effected. The breeder in this case does not produce the *elements* of the variation. He simply observes them, and by selection adds them together until the required result has been obtained. 'No man,' says Mr. Darwin, 'would ever try to make a fantail till he saw a pigeon with a tail developed in some slight degree in an unusual manner, or a pouter until he saw a pigeon with a crop of unusual size.' Thus nature gives the hint, man acts upon it, and by the law of inheritance exaggerates the deviation.

Having thus satisfied himself by indubitable facts that the organization of an animal or of a plant (for precisely the same treatment applies to plants) is to some extent plastic, he passes from variation under domestication to variation under nature. Hitherto we have dealt with the adding together of small changes by the conscious selection of man. Can Nature thus select? Mr. Darwin's answer is, 'Assuredly she can.' The number of living things produced is far in excess of the number that can be supported; hence at some period or other of their lives there must be a struggle for existence; and what is the infallible result? If one organism were a perfect copy of the other in regard to strength, skill, and agility, external conditions would decide. But this is not the case. Here we have the fact of variety offering itself to nature, as in the former instance it offered itself to man; and those varieties which are least competent to cope with surrounding conditions will infallibly give way to those that are most competent. To use a familiar proverb, the weakest comes to the wall. But the triumphant fraction again breeds to overproduction, transmitting the qualities which secured its maintenance, but transmitting them in different degrees. The struggle for food again supervenes, and those to whom the favourable quality has been transmitted in excess will assuredly triumph. It is easy to see that we have here the addition of increments favourable to the individual still more rigorously carried out than in the case of domestication; for not only are unfavourable specimens not selected by nature, but they are destroyed. This is what Mr. Darwin calls 'Natural Selection,' which 'acts by the preservation and accumulation of small inherited modifications, each profitable to the preserved being.' With this idea he interpenetrates and leavens the vast store of facts that he and others have collected. We cannot, without shutting our eyes through fear or prejudice, fail to see that Darwin is here dealing, not with imaginary, but with true causes; nor can we fail to discern what vast modifications may be produced by natural selection in periods sufficiently long. Each individual increment may resemble what mathematicians call a 'differential' (a quantity indefinitely small); but definite and great changes may obviously be produced by the integration of these infinitesimal quantities through practically infinite time.

If Darwin, like Bruno, rejects the notion of creative power acting after human fashion, it certainly is not because he is unacquainted with the numberless exquisite adaptations on which this notion of a supernatural artificer has been founded. His book is a repository of the most startling facts of this description. Take the marvellous observation which he cites from Dr. Crüger, where a bucket with an aperture, serving as a spout, is formed in an orchid. Bees visit the flower: in eager search of material for their combs they push each other into the bucket, the drenched ones escaping from their involuntary bath by the spout. Here they rub their backs against the viscid stigma of the flower and obtain glue; then against the pollen-masses, which are thus stuck to the back of the bee and carried away. 'When the bee, so provided, flies to another flower, or to the same flower a second time, and is pushed by its comrades into the bucket, and then crawls out by the passage, the pollen-mass upon its back necessarily comes first into contact with the viscid stigma,' which takes up the pollen; and this is how that orchid is fertilized. Or take this other case of the *Catasetum*. 'Bees visit these flowers in order to gnaw the labellum; in doing this they inevitably touch a long, tapering, sensitive projection. This, when touched, transmits a sensation or vibration to a certain membrane, which is instantly ruptured, setting free a spring, by which the pollen-mass is shot forth like an arrow in the right direction, and adheres by its viscid extremity to the back of the bee.' In this way the fertilising pollen is spread abroad.

It is the mind thus stored with the choicest materials of the teleologist that rejects teleology, seeking to refer these wonders to natural cases. They illustrate, according to him, the method of nature, not the 'technic' of a man-like Artificer. The beauty of flowers is due to natural selection. Those that distinguish themselves by vividly contrasting colours from the surrounding green leaves are most readily seen, most frequently visited by insects, most often fertilized, and hence most favoured by natural selection. Coloured berries also readily attract the attention of birds and beasts, which feed upon them, spread their manured seeds abroad, thus giving trees and shrubs possessing such berries a greater chance in the struggle for existence.

With profound analytic and synthetic skill, Mr. Darwin investigates the cell-making instinct of the hive-bee. His method of dealing with it is representative. He falls back from the more perfectly to the less perfectly developed instinct--from the hive-bee to the humble bee, which uses its own cocoon as a comb, and to classes of bees of intermediate skill, endeavouring to show how the passage might be gradually made from the lowest to the highest. The saving of wax is the most important point in the economy of bees. Twelve to fifteen pounds of dry sugar are said to be needed for the secretion of a single pound of wax. The quantities of nectar necessary for the wax must therefore be vast; and every improvement of constructive instinct which results in the saving of wax is a direct profit to the insect's life. The time that would otherwise be devoted to the making of wax is now devoted to the gathering and storing of honey for winter food. He passes from the humble bee with its rude cells, through the Melipona with its more artistic cells, to the hive-bee with its astonishing architecture. The bees place themselves at equal distances apart upon the wax, sweep and excavate equal spheres round the selected points. The spheres intersect, and the planes of intersection are built up with thin laminae. Hexagonal cells are thus formed. This mode of treating such questions is, as I have said, representative. He habitually retires from the more perfect and complex to the less perfect and simple, and carries you with him through stages of *perfecting*, adds increment to increment of infinitesimal change, and in this way gradually breaks down your reluctance to admit that the exquisite climax of the whole could be a result of natural selection.

Mr. Darwin shirks no difficulty; and, saturated as the subject was with his own thought, he must have known better than his critics the weakness as well as the strength of his theory. This of course would be of little avail were his object a temporary dialectic victory instead of the establishment of a truth which he means to be everlasting. But he takes no pains to disguise the weakness he has discerned; nay, he takes every pains to bring it into the strongest light. His vast resources enable him to cope with objections started by himself and others, so as to leave the final impression upon the reader's mind that, if they be not completely answered, they certainly are not fatal. Their negative force being thus destroyed, you are free to be influenced by the vast positive mass of evidence he is able to bring before you. This largeness of knowledge and readiness of resource render Mr. Darwin the most terrible of antagonists. Accomplished naturalists have levelled heavy and sustained criticisms against him—not always with the view of fairly weighing his theory, but with the express intention of exposing its weak points only. This does not irritate him. He treats every objection with a soberness and thoroughness which even Bishop Butler might be proud to imitate, surrounding each fact with its appropriate detail, placing it in its proper relations, and usually giving it a significance which, as long as it was kept isolated, failed to appear. This is done without a trace of ill-temper. He moves over the subject with the passionless strength of a glacier; and the grinding of the rocks is not always without a counterpart in the logical pulverization of the objector.

But though in handling this mighty theme all passion has been stilled, there is an emotion of the intellect incident to the discernment of new truth which often colours and warms the pages of Mr. Darwin. His success has been great; and this implies not only the solidity of his work, but the preparedness of the public mind for such a revelation. On this head a remark of Agassiz impressed me more than anything else. Sprung from a race of theologians, this celebrated man combated to the last the theory of natural selection. One of the many times I had the pleasure of meeting him in the United States was at Mr. Winthrop's beautiful residence at Brookline, near Boston. Rising from luncheon, we all halted as if by a common impulse in front of a window, and continued there a discussion which had been started at table. The maple was in its autumn glory; and the exquisite beauty of the scene outside seemed, in my case, to interpenetrate without disturbance the intellectual action. Earnestly, almost sadly, Agassiz turned, and said to the gentleman standing round, 'I confess that I was not prepared to see this theory received as it has been by the best intellects of our time. Its success is greater than I could have thought possible.'

In our day grand generalizations have been reached. The theory of the origin of species is but one of them. Another, of still wider grasp and more radical significance, is the doctrine of the Conservation of Energy, the ultimate philosophical issues of which are as yet but dimly seen—that doctrine which 'binds nature fast in fate' to an extent not hitherto recognized, exacting from every antecedent its equivalent consequent, from every consequent its equivalent antecedent, and bringing vital as well as physical phenomena under the dominion of that law of causal connexion which, so far as the human understanding has yet pierced, asserts itself everywhere in nature. Long in advance of all definite experiment upon the subject, the constancy and indestructibility of matter had been affirmed; and all subsequent experience justified the affirmation. Later researches extended the attribute of indestructibility to force. This idea, applied in the first instance to inorganic, rapidly embraced organic nature. The vegetable world, though drawing almost all its nutriment from invisible sources, was proved incompetent to generate anew either matter or force. Its matter is for the most part transmuted gas; its force transformed solar force. The animal world was proved to be equally uncreative, all its motive energies being referred to the combustion of its food. The activity of each animal as a whole was proved to be the transferred activity of its molecules. The muscles were shown to be stores of mechanical force, potential until unlocked by the nerves, and then resulting in muscular contractions. The speed at which messages fly to and fro along the nerves was determined, and found to be, not as had been previously supposed, equal to that of light or electricity, but less than the speed of a flying eagle.

This was the work of the physicist: then came the conquests of the comparative anatomist and physiologist, revealing the structure of every animal, and the function of every organ in the whole biological series, from the lowest zoophyte up to man. The nervous system had been made the object of profound and continued study, the wonderful and, at bottom, entirely mysterious, controlling power which it exercises over the whole organism, physical and mental, being recognized more and more. Thought could not be kept back from a subject so profoundly suggestive. Besides the physical life dealt with by Mr. Darwin, there is a psychical life presenting similar gradations, and asking equally for a solution. How are the different grades and order of Mind to be accounted for? What is the principle of growth of that mysterious power which on our planet culminates in Reason? These are questions which, though not thrusting themselves so forcibly upon the attention of the general public, had not only occupied many reflecting minds, but had been formally broached by one of them before the 'Origin of Species' appeared.

With the mass of materials furnished by the physicist and physiologist in his hands, Mr. Herbert Spencer, twenty years ago, sought to graft upon this basis a system of psychology; and two years ago a second and greatly amplified edition of his work appeared. Those who have occupied themselves with the beautiful experiments of Plateau will remember that when two spherules of olive-oil, suspended in a mixture of alcohol and water of the same density as the oil, are brought together, they do not immediately unite. Something like a pellicle appears to be formed around the drops, the rupture of which is immediately followed by the coalescence of the globules into one. There are organisms whose vital actions are almost as purely physical as that of these drops of oil. They come into contact and fuse themselves thus together. From such organisms to others a shade higher, and from these to others a shade higher still, and on through an ever-ascending series, Mr. Spencer conducts his argument. There are two obvious factors to be here taken into account—the creature and the medium in which it lives, or, as it is often expressed, the organism and its environment. Mr. Spencer's fundamental principle is that between these two factors there is incessant interaction. The organism is played upon by the environment, and is modified to meet the requirements of the environment. Life he defines to be 'a continuous adjustment of internal relations to external relations.'

In the lowest organisms we have a kind of tactual sense diffused over the entire body; then, through impressions from without and their corresponding adjustments, special portions of the surface become more responsive to stimuli than others. The senses are nascent, the basis of all of them being that simple tactual sense which the sage Democritus recognised 2,300 years ago as their common progenitor. The action of light, in the first instance, appears to be a mere disturbance of the chemical processes in the animal organism, similar to that which occurs in the leaves of plants. By degrees the action becomes localized in a few pigment-cells, more sensitive to light than the surrounding tissue. The eye is here incipient. At first it is merely capable of revealing differences of light and shade produced by bodies close at hand. Followed as the interception of the light is in almost all cases by the contact of the closely adjacent opaque body, sight in this condition becomes a kind of 'anticipatory touch.' The adjustment continues; a slight bulging out of the epidermis over the pigment-granules supervenes. A lens is incipient, and, through the operation of infinite adjustments, at length reaches the perfection that it displays in the hawk and eagle. So of the other senses; they are special differentiations of a tissue which was originally vaguely sensitive all over.

With the development of the senses the adjustments between the organism and its environment gradually extend in *space*, a multiplication of experiences and a corresponding modification of conduct being the result. The adjustments also extend in *time*, covering continually greater intervals. Along with this extension in space and time the adjustments also increase in specialty and complexity, passing through the various grades of brute life, and prolonging themselves into the domain of reason. Very striking are Mr. Spencer's remarks regarding the influence of the sense of touch upon the development of intelligence. This is, so to say, the mother-tongue of all the senses, into which they must be translated to be of service to the organism. Hence its importance. The parrot is the most intelligent of birds, and its tactual power is also greatest. From this sense it gets knowledge unattainable by birds which cannot employ their feet as hands. The elephant is the most sagacious of quadrupeds—its tactual range and skill, and the consequent multiplication of experiences, which it owes to its wonderfully adaptable trunk, being the basis of its sagacity. Feline animals, for a similar cause, are more sagacious than hoofed animals—atonement being to some extent made, in the case of the horse, by the possession of sensitive prehensile lips. In the *Primates* the evolution of intellect and the evolution of tactual appendages go hand in hand. In the most intelligent anthropoid apes we find the tactual range and delicacy greatly augmented, new avenues of knowledge being thus open to the animal. Man crowns the edifice here, not only in virtue of his own manipulatory power, but through the enormous extension of his range of experience, by the invention of instruments of precision, which serve as supplemental senses and supplemental limbs. The reciprocal action of these is finely described and illustrated. That chastened intellectual emotion to which I have referred in connexion with Mr. Darwin is not absent in Mr. Spencer. His illustrations possess at times exceeding vividness and force; and from his style on such occasions it is to be inferred that the ganglia of this Apostle of the Understanding are sometimes the seat of a nascent poetic thrill.

It is a fact of supreme importance that actions the performance of which at first requires even painful effort and deliberation may by habit be rendered automatic. Witness the slow learning of its letters by a child, and the subsequent facility of reading in a man, when each group of letters which forms a word is instantly, and without effort, fused to a single perception. Instance the billiard-player, whose muscles of hand and eye, when he reaches the perfection of his art, are unconsciously coördinated. Instance the musician, who, by practice, is enabled to fuse a multitude of arrangements, auditory, tactual, and muscular, into a process of automatic manipulation. Combining such facts with the doctrine of hereditary transmission, we reach a theory of Instinct. A chick, after coming out of the egg, balances itself correctly, runs about, picks up food, thus showing that it possesses a power of directing its movements to definite ends. How did the chick learn this very complex coördination of eye, muscles, and beak? It has not been individually taught; its personal experience is *nil*; but it has the benefit of ancestral experience. In its inherited organization are registered all the powers which it displays at birth. So also as regards the instinct of the hive-bee, already referred to. The distance at which the insects stand apart when they sweep their hemispheres and build their cells is 'organically remembered.'

Man also carries with him the physical texture of his ancestry, as well as the inherited intellect bound up with it. The defects of intelligence during infancy and youth are probably less due to a lack of individual experience than to the fact that in early life the cerebral organization is still incomplete. The period necessary for completion varies with the race and with the individual. As a round shot outstrips a rifled one on quitting the muzzle of the gun, so the lower race in childhood may outstrip the higher. But the higher eventually overtakes the lower, and surpasses it in range. As regards individuals, we do not always find the precocity of youth prolonged to mental power in maturity; while the dulness of boyhood is sometimes strikingly contrasted with the intellectual energy of after years. Newton, when a boy, was weakly, and he showed no particular aptitude at school; but in his eighteenth year he went to Cambridge, and soon afterwards astonished his teachers by his power of dealing with geometrical problems. During his quiet youth his brain was slowly preparing itself to be the organ of those energies which he subsequently displayed.

By myriad blows (to use a Lucretian phrase) the image and superscription of the external world are stamped as states of consciousness upon the organism, the depth of the impression depending upon the number of the blows. When two or more phenomena occur in the environment invariably together, they are stamped to the same depth or to the same relief, and indissolubly connected. And here we come to the threshold of a great question. Seeing that he could in no way rid himself of the consciousness of Space and Time, Kant assumed them to be necessary 'forms of intuition,' the moulds and shapes into which our intuitions are thrown, belonging to ourselves solely and without objective existence. With unexpected power and success Mr. Spencer brings the hereditary experience theory, as he holds it, to bear upon this question. 'If there exist certain external relations which are experienced by all organisms at all instants of their waking lives—relations which are absolutely constant and universal—there will be established answering internal relations that are absolutely constant and universal. Such relations we have in those of Space and Time. As the substratum of all other relations of the Non-Ego, they must be responded to by conceptions that are the substrata of all other relations in the Ego. Being the constant and infinitely repeated elements of thought, they must become the automatic elements of thought—the elements of thought which it is impossible to get rid of—the "forms of intuition."'

Throughout this application and extension of the 'Law of Inseparable Association,' Mr. Spencer stands upon his own ground, invoking, instead of the experiences of the individual, the registered experiences of the race. His overthrow of the restriction of experience to the individual is, I think, complete. That restriction ignores the power of organizing experience furnished at the outset to each individual; it ignores the different degrees of this power possessed by different races and by different individuals of the same race. Were there not in the human brain a potency antecedent to all experience, a dog or cat ought to be as capable of education as a man. These predetermined internal relations are independent of the experiences of the individual. The human brain is the 'organised register of infinitely numerous experiences received during the evolution of life, or rather during the evolution of that series of organisms through which the human organism has been reached. The effects of the most uniform and frequent of these experiences have been successively bequeathed, principal and interest, and have slowly mounted to that high intelligence which lies latent in the brain of the infant. Thus it happens that the European inherits from twenty to thirty cubic inches more of brain than the Papuan. Thus it happens that faculties, as of music, which scarcely exist in some inferior races, become congenital in superior ones. Thus it happens that out of savages unable to count up to the number of their fingers, and speaking a language containing only nouns and verbs, arise at length our Newtons and Shakespeares.'

At the outset of this Address it was stated that physical theories which lie beyond experience are derived by a process of abstraction from experience. It is instructive to note from this point of view the successive introduction of new conceptions. The idea of the attraction of gravitation was preceded by the observation of the attraction of iron by a magnet, and of light bodies by rubbed amber. The polarity of magnetism and electricity appealed to the senses; and thus became the substratum of the conception that atoms and molecules are endowed with definite, attractive, and repellent poles, by the play of which definite forms of crystalline architecture are produced. Thus molecular force becomes *structural*. It required no great boldness of thought to extend its play into organic nature, and to recognize in molecular force the agency by which both plants and animals are built up. In this way out of experience arise conceptions which are wholly ultra-experiential. None of the atomists of antiquity had any notion of this play of molecular polar force, but they had experience of gravity as manifested by falling bodies. Abstracting from this, they permitted their atoms to fall eternally through empty space. Democritus assumed that the larger atoms moved more rapidly than the smaller ones, which they therefore could overtake, and with which they could combine. Epicurus, holding that empty space could offer no resistance to motion, ascribed to all the atoms the same velocity; but he seems to have overlooked the consequence that under such circumstances the atoms could never combine. Lucretius cut the knot by quitting the domain of physics altogether, and causing the atoms to move together by a kind of volition.

Was the instinct utterly at fault which caused Lucretius thus to swerve from his own principles? Diminishing gradually the number of progenitors, Mr. Darwin comes at length to one 'primordial form;' but he does not say, as far as I remember, how he supposes this form to have been introduced. He quotes with satisfaction the words of a celebrated author and divine who had 'gradually learnt to see that it is just as noble a conception of the Deity to believe He created a few original forms, capable of self-development into other and needful forms, as to believe that He required a fresh act of creation to supply the voids caused by the action of His laws.' What Mr. Darwin thinks of this view of the introduction of life I do not know. But the anthropomorphism, which it seemed his object to set aside, is as firmly associated with the creation of a few forms as with the creation of a multitude. We need clearness and thoroughness here. Two courses and two only, are possible. Either let us open our doors freely to the conception of creative acts, or, abandoning them, let us radically change our notions of Matter. If we look at matter as pictured by Democritus, and as defined for generations in our scientific text-books, the notion of any form of life whatever coming out of it is utterly unimaginable. The argument placed in the mouth of Bishop Butler suffices, in my opinion, to crush all such materialism as this. But those who framed these definitions of matter were not biologists, but mathematicians, whose labours referred only to such accidents and properties of matter as could be expressed in their formulae. The very intentness with which they pursued mechanical science turned their thoughts aside from the science of life. May not their imperfect definitions be the real cause of our present dread? Let us reverently, but honestly, look the question in the face. Divorced from matter, where is life to be found? Whatever our *faith* may say, our *knowledge* shows them to be indissolubly joined. Every meal we eat, and every cup we drink, illustrates the mysterious control of Mind by Matter.

Trace the line of life backwards, and see it approaching more and more to what we call the purely physical condition. We come at length to those organisms which I have compared to drops of oil suspended in a mixture of alcohol and water. We reach the *protogenes* of Haeckel, in which we have 'a type distinguishable from a fragment of albumen only by its finely granular character.' Can we pause here? We break a magnet and find two poles in each of its fragments. We continue the process of breaking, but, however small the parts, each carries with it, though enfeebled, the polarity of the whole. And when we can break no longer, we prolong the intellectual vision to the polar molecules. Are we not urged to do *something* similar in the case of life? Is there not a temptation to close to some extent with Lucretius, when he affirms that 'nature is seen to do all things spontaneously of herself without the meddling of the gods?' or with Bruno, when he declares that Matter is not 'that mere empty capacity which philosophers have pictured her to be, but the universal mother who brings forth all things as the fruit of her own womb?' Believing as I do in the continuity of Nature, I cannot stop abruptly where our microscopes cease to be of use. Here the vision of the mind authoritatively supplements the vision of the eye. By an intellectual necessity I cross the boundary of the experimental evidence, and discern in that Matter which we, in our ignorance of its latent powers, and notwithstanding our professed reverence for its Creator, have hitherto covered with opprobrium, the promise and potency of all terrestrial Life.

If you ask me whether there exists the least evidence to prove that any form of life can be developed out of matter, without demonstrable antecedent life, my reply is that evidence considered perfectly conclusive by many has been adduced; and that were some of us who have pondered this question to follow a very common example, and accept testimony because it falls in with our belief, we also should eagerly close with the evidence referred to. But there is in the true man of science a wish stronger than the wish to have his beliefs upheld; namely, the wish to have them true. And this stronger wish causes him to reject the most plausible support if he has reason to suspect that it is vitiated by error. Those to whom I refer as having studied this question, believing the evidence offered in favour of 'spontaneous generation' to be thus vitiated, cannot accept it. They know full well that the chemist now prepares from inorganic matter a vast array of substances which were some time ago regarded as the sole products of vitality. They are intimately acquainted with the structural power of matter as evidenced in the phenomena of crystallization. They can justify scientifically their *belief* in its potency, under the proper conditions, to produce organisms. But in reply to your question they will frankly admit their inability to point to any satisfactory experimental proof that life can be developed save from demonstrable antecedent life. As already indicated, they draw the line from the highest organisms through lower ones down to the lowest, and it is the prolongation of this line by the intellect beyond the range of the senses that leads them to the conclusion which Bruno so boldly enunciated.[20]

The 'materialism' here professed may be vastly different from what you suppose, and I therefore crave your gracious patience to the end. 'The question of an external world,' says Mr. J.S. Mill, 'is the great battleground of metaphysics.'[21] Mr.

Mill himself reduces external phenomena to 'possibilities of sensation.' Kant, as we have seen, made time and space 'forms' of our own intuitions. Fichte, having first by the inexorable logic of his understanding proved himself to be a mere link in that chain of eternal causation which holds so rigidly in Nature, violently broke the chain by making Nature, and all that it inherits, an apparition of his own mind.[22] And it is by no means easy to combat such notions. For when I say I see you, and that I have not the least doubt about it, the reply is, that what I am really conscious of is an affection of my own retina. And if I urge that I can check my sight of you by touching you, the retort would be that I am equally transgressing the limits of fact; for what I am really conscious of is, not that you are there, but that the nerves of my hand have undergone a change. All we hear, and see, and touch, and taste, and smell, are, it would be urged, mere variations of our own condition, beyond which, even to the extent of a hair's breadth, we cannot go. That anything answering to our impressions exists outside of ourselves is not a *fact*, but an *inference*, to which all validity would be denied by an idealist like Berkeley, or by a sceptic like Hume. Mr. Spencer takes another line. With him, as with the uneducated man, there is no doubt or question as to the existence of an external world. But he differs from the uneducated, who think that the world really *is* what consciousness represents it to be. Our states of consciousness are mere *symbols* of an outside entity which produces them and determines the order of their succession, but the real nature of which we can never know.[23] In fact, the whole process of evolution is the manifestation of a Power absolutely inscrutable to the intellect of man. As little in our day as in the days of Job can man by searching find this Power out. Considered fundamentally, then, it is by the operation of an insoluble mystery that life on earth is evolved, species differentiated, and mind unfolded from their prepotent elements in the immeasurable past. There is, you will observe, no very rank materialism here.

The strength of the doctrine of evolution consists, not in an experimental demonstration (for the subject is hardly accessible to this mode of proof), but in its general harmony with scientific thought. From contrast, moreover, it derives enormous relative strength. On the one side we have a theory (if it could with any propriety be so called) derived, as were the theories referred to at the beginning of this Address, not from the study of Nature, but from the observation of men—a theory which converts the Power whose garment is seen in the visible universe into an Artificer, fashioned after the human model, and acting by broken efforts, as man is seen to act. On the other side, we have the conception that all we see around us, and all we feel within us—the phenomena of physical nature as well as those of the human mind—have their unsearchable roots in a cosmical life, if I dare apply the term, an infinitesimal span of which is offered to the investigation of man. And even this span is only knowable in part. We can trace the development of a nervous system, and correlate with it the parallel phenomena of sensation and thought. We see with undoubting certainty that they go hand in hand. But we try to soar in a vacuum the moment we seek to comprehend the connexion between them. An Archimedean fulcrum is here required which the human mind cannot command; and the effort to solve the problem, to borrow a comparison from an illustrious friend of mine, is like the effort of a man trying to lift himself by his own waistband. All that has been here said is to be taken in connexion with this fundamental truth. When 'nascent senses' are spoken of, when 'the differentiation of a tissue at first vaguely sensitive all over' is spoken of, and when these processes are associated with 'the modification of an organism by its environment,' the same parallelism, without contact, or even approach to contact, is implied. Man the *object* is separated by an impassible gulf from man the *subject*. There is no motor energy in intellect to carry it without logical rupture from the one to the other.

Further, the doctrine of evolution derives man in his totality from the inter-action of organism and environment through countless ages past. The Human Understanding, for example—that faculty which Mr. Spencer has turned so skilfully round upon its own antecedents—is itself a result of the play between organism and environment through cosmic ranges of time. Never surely did prescription plead so irresistible a claim.

But then it comes to pass that, over and above his understanding, there are many other things appertaining to man whose perspective rights are quite as strong as those of the understanding itself. It is a result, for example, of the play of organism and environment that sugar is sweet and that aloes are bitter, that the smell of henbane differs from the perfume of a rose. Such facts of consciousness (for which, by the way, no adequate reason has yet been rendered) are quite as old as the understanding; and many other things can boast an equally ancient origin. Mr. Spencer at one place refers to that most powerful of passions—the amatory passion—as one which, when it first occurs, is antecedent to all relative experience whatever; and we may pass its claim as being at least as ancient and valid as that of the understanding. Then there are such things woven into the texture of man as the feeling of Awe, Reverence, Wonder—and not alone the sexual love just referred to, but the love of the beautiful, physical, and moral, in Nature, Poetry, and Art. There is also that deep-set feeling which, since the earliest dawn of history, and probably for ages prior to all history, incorporated itself in the Religions of the world. You who have escaped from these religions into the high-and-dry light of the intellect may deride them; but in so doing you deride accidents of form merely, and fail to touch the immovable basis of the religious sentiment in the nature of man. To yield this sentiment reasonable satisfaction is the problem of problems at the present hour. And grotesque in relation to scientific culture as many of the religions of the world have been and are—dangerous, nay destructive, to the dearest privileges of freemen as some of them undoubtedly have been, and would, if they could, be again—it will be wise to recognize them as the forms of a force, mischievous, if permitted to intrude on the region of *knowledge*, over which it holds no command, but capable of being guided to noble issues in the region of *emotion*, which is its proper and elevated sphere.

All religious theories, schemes and systems, which embrace notions of cosmogony, or which otherwise reach into the domain of science, must, *in so far as they do this*, submit to the control of science, and relinquish all thought of controlling it. Acting otherwise proved disastrous in the past, and it is simply fatuous to-day. Every system which would escape the fate of an organism too rigid to adjust itself to its environment must be plastic to the extent that the growth of knowledge demands. When this truth has been thoroughly taken in, rigidity will be relaxed, exclusiveness diminished, things now deemed essential will be dropped, and elements now rejected will be assimilated. The lifting of the life is the essential point; and as long as dogmatism, fanaticism, and intolerance are kept out, various modes of leverage may be employed to raise life to a higher level. Science itself not unfrequently derives motive power from an ultra-scientific source. Whewell speaks of enthusiasm of temper as a hindrance to science; but he means the enthusiasm of weak heads. There is a strong and resolute enthusiasm in which science finds an ally; and it is to the lowering of this fire, rather than to the diminution of intellectual insight, that the lessening productiveness of men of science in their mature years is to be ascribed. Mr. Buckle sought to detach intellectual achievement from moral force. He gravely erred; for without moral force to whip it into action, the achievements of the intellect would be poor indeed.

It has been said that science divorces itself from literature; but the statement, like so many others, arises from lack of knowledge. A glance at the less technical writings of its leaders—of its Helmholtz, its Huxley, and its Du Bois-Reymond—would show what breadth of literary culture they command. Where among modern writers can you find their superiors in clearness and vigour of literary style? Science desires not isolation, but freely combines with every effort towards the bettering of man's estate. Single-handed, and supported not by outward sympathy, but by inward force, it has built at least one great wing of the many-mansioned home which man in his totality demands. And if rough walls and protruding rafter-ends indicate that on one side the edifice is still incomplete, it is only by wise combination of the parts required with those already irrevocably built that we can hope for completeness. There is no necessary incongruity between what has been accomplished and what remains to be done. The moral glow of

Socrates, which we all feel by ignition, has in it nothing incompatible with the physics of Anaxagoras which he so much scorned, but which he would hardly scorn to-day.

And here I am reminded of one amongst us, hoary, but still strong, whose prophet-voice some thirty years ago, far more than any other of this age, unlocked whatever of life and nobleness lay latent in its most gifted minds—one fit to stand beside Socrates or the Maccabean Eleazar, and to dare and suffer all that they suffered and dared—fit, as he once said of Fichte, 'to have been the teacher of the Stoa, and to have discoursed of Beauty and Virtue in the groves of Academe.' With a capacity to grasp physical principles which his friend Goethe did not possess, and which even total lack of exercise has not been able to reduce to atrophy, it is the world's loss that he, in the vigour of his years, did not open his mind and sympathies to science, and make its conclusions a portion of his message to mankind. Marvellously endowed as he was—equally equipped on the side of the Heart and of the Understanding—he might have done much towards teaching us how to reconcile the claims of both, and to enable them in coming times to dwell together in unity of spirit and in the bond of peace.

And now the end is come. With more time, or greater strength and knowledge, what has been here said might have been better said, while worthy matters here omitted might have received fit expression. But there would have been no material deviation from the views set forth. As regards myself, they are not the growth of a day; and as regards you, I thought you ought to know the environment which, with or without your consent, is rapidly surrounding you, and in relation to which some adjustment on your part may be necessary. A hint of Hamlet's, however, teaches us all how the troubles of common life may be ended; and it is perfectly possible for you and me to purchase intellectual peace at the price of intellectual death. The world is not without refuges of this description; nor is it wanting in persons who seek their shelter and try to persuade others to do the same. The unstable and the weak will yield to this persuasion, and they to whom repose is sweeter than the truth. But I would exhort you to refuse the offered shelter and to scorn the base repose—to accept, if the choice be forced upon you, commotion before stagnation, the leap of the torrent before the stillness of the swamp.

In the course of this Address I have touched on debatable questions and led you over what will be deemed dangerous ground—and this partly with the view of telling you that as regards these questions science claims unrestricted right of search. It is not to the point to say that the views of Lucretius and Bruno, of Darwin and Spencer, may be wrong. Here I should agree with you, deeming it indeed certain that these views will undergo modification. But the point is, that, whether right or wrong, we ask the freedom to discuss them. For science, however, no exclusive claim is here made; you are not urged to erect it into an idol. The inexorable advance of man's understanding in the path of knowledge, and those unquenchable claims of his moral and emotional nature which the understanding can never satisfy, are here equally set forth. The world embraces not only a Newton, but a Shakespeare—not only a Boyle, but a Raphael—not only a Kant, but a Beethoven—not only a Darwin, but a Carlyle. Not in each of these, but in all, is human nature whole. They are not opposed, but supplementary—not mutually exclusive, but reconcilable. And if, unsatisfied with them all, the human mind, with the yearning of a pilgrim for his distant home, will turn to the Mystery from which it has emerged, seeking so to fashion it as to give unity to thought and faith; so long as this is done, not only without intolerance or bigotry of any kind, but with the enlightened recognition that ultimate fixity of conception is here unattainable, and that each succeeding age must be held free to fashion the Mystery in accordance with its own needs—then, casting aside all the restrictions of Materialism, I would affirm this to be a field for the noblest exercise of what, in contrast with the *knowing* faculties, may be called the *creative* faculties of man.

'Fill thy heart with it,' said Goethe, 'and then name it as thou wilt.' Goethe himself did this in untranslateable language.[24] Wordsworth did it in words known to all Englishmen, and which may be regarded as a forecast and religious vitalization of the latest and deepest scientific truth,—

'For I have learned
To look on nature; not as in the hour
Of thoughtless youth; but hearing oftentimes
The still, sad music of humanity,
Nor harsh nor grating, though of ample power
To chasten and subdue. *And I have felt*
A presence that disturbs me with the joy
Of elevated thoughts; a sense sublime
Of something far more deeply interfused,
Whose dwelling is the light of setting suns,
And the round ocean, and the living air,
And the blue sky, and in the mind of man:
A motion and a spirit, that impels
All thinking things, all objects of all thought,
And rolls through all things.'[25]

Notes

1 Hume, *Natural History of Religion.*

2 Born 460 B.C.

3 Lange, 2nd edit., p. 23.

4 Born 342 B.C.

5 Tennyson's *Lucretius.*

6 Born 99 B.C.

7 Monro's translation. In his criticism of this work, *Contemporary Review*, 1867, Dr. Hayman does not appear to be aware of the really sound and subtile observations on which the reasoning of Lucretius, though erroneous, sometimes rests.

8 *History of the Intellectual Development of Europe*, p. 295.

9 *History of the Inductive Sciences*, vol. i.

10 Depicted with terrible vividness in Renan's *Antichrist.*

11 *Intellectual Development of Europe*, p. 359.

12 Lange, 2nd edit. pp. 181, 182.

13 See Huxley's admirable Essay on Descartes. *Lay Sermons*, pp. 364, 365.

14 Boyle's model of the universe was the Strasburg clock with an outside Artificer. Goethe, on the other hand, sang 'Ihm ziemt's die Welt im Innern zu bewegen,/ Natur in sich, sich in Natur zu hegen.' See also Carlyle, *Past and Present*, Chap. V.

15 Only to some; for there are dignitaries who even now speak of the earth's rocky crust as so much building material prepared for man at the Creation. Surely it is time that this loose language should cease.

16 *Zoonomia*, vol. i. pp. 500–510.

17 In 1855 Mr. Herbert Spencer (*Principles of Psychology*, 2nd edit. vol. i. p. 465) expressed 'the belief that life under all its forms has arisen by an unbroken evolution, and through the instrumentality of what are called natural causes.'

18 The behaviour of Mr. Wallace in relation to this subject has been dignified in the highest degree.

19 The first step only towards experimental demonstration has been taken. Experiments now begun might, a couple of centuries hence, furnish data of incalculable value, which ought to be supplied to the science of the future.

20 Bruno was a 'Pantheist,' not an 'Atheist' or a 'Materialist.'

21 *Examination of Hamilton*, p. 154.

22 *Bestimmung des Menschen.*

23 In a paper, at once popular and profound, entitled *Recent Progress in the Theory of Vision*, contained in the volume of Lectures by Helmholtz, published by Longmans, this symbolism of our states of consciousness is also dwelt upon. The impressions of sense are the mere *signs* of external things. In this paper Helmholtz contends strongly against the view that the consciousness of space is inborn; and he evidently doubts the power of the chick to pick up grains of corn without preliminary lessons. On this point, he says, further experiments are needed. Such experiments have been since made by Mr. Spalding, aided, I believe, in some of his observations by the accomplished and deeply lamented Lady Amberly; and they seem to prove conclusively that the chick does not need a single moment's tuition to enable it to stand, run, govern the muscles of its eyes, and to peck. Helmholtz, however, is contending against the notion of pre-established harmony; and I am not aware of his views as to the organisation of experiences of race or breed.

24 Procemium to 'Gott und Welt.'

25 *Tintern Abbey.*

FURTHER READING

Barton, Ruth. "John Tyndall, Pantheist: A Re-Reading of the Belfast Address." *Osiris*, 2d ser., 3 (1987): 111–34.

Basalla, George, William Coleman, and Robert H. Kargon. *Victorian Science: A Self-Portrait from the Presidential Addresses of the British Association for the Advancement of Science.* Garden City, NY: Doubleday and Co., 1970.

Brock, W.H., N.D. MacMillan, and R.C. Mollan, eds. *John Tyndall: Essays on a Natural Philosopher.* Dublin: Royal Dublin Society, 1981.

Conant, James Bryant. *Pasteur's and Tyndall's Study of Spontaneous Generation.* Harvard Case Histories in Experimental Science, no. 7. Cambridge, MA: Harvard UP, 1953.

Corsi, Pietro. *Science and Religion: Baden Powell and the Anglican Debate, 1800-1860.* Cambridge: Cambridge UP, 1988.

Cosslett, Anna Therese. "Science and Value: The Writings of John Tyndall." *Prose Studies* 2 (1978): 41–57.

Draper, John W. *History of the Conflict Between Religion and Science.* London: Henry S. King and Co., 1875.

Eisen, Sydney and Bernard V. Lightman. *Victorian Science and Religion: A Bibliography with Emphasis on Evolution, Belief, and Unbelief, Comprised of Works Published from c. 1900-1975.* Hamden, CT: Archon Books, 1984.

Eve, A.S., and C.H. Creasey. *Life and Work of John Tyndall.* London: Macmillan and Co., Ltd., 1945.

Friday, J.R., R.M. MacLeod, and P. Shepherd, eds. *John Tyndall, Natural Philosopher, 1820-1893: Catalogue of Correspondence, Journals and Collected Papers.* London: Mansell, 1974.

Godkin, Edwin Lawrence. "Tyndall and the Theologians." *Reflections and Comments: 1865–1895.* Westminster: Archibald Constable and Co., 1896, pp. 129–37.

Griffin, John Nash. *Atoms: A Lecture in Reply to Professor Tyndall's Inaugural Address.* Dublin: Hodges, Foster and Co., 1875.

Harris, Frank. "John Tyndall." *Contemporary Portraits.* 4th Series. New York: Brentano's, Inc., 1923, pp. 77–85.

Helmstadter, Richard J. and Bernard Lightman. *Victorian Faith in Crisis: Essays on Continuity and Change in Nineteenth-Century Religious Belief.* London: Macmillan, 1990.

Howard, John Eliot. *An Examination of the Belfast Address of the British Association, 1874 from a Scientific Point of View.* London: R. Hardwicke, 1875.

Jeans, William T. *Lives of the Electricians: Professors Tyndall, Wheatstone, and Morse.* London: Whittaker and Co., 1887.

Lange, F.A. *History of Materialism and Criticism of Its Present Importance.* Trans. E.C. Thomas. London: K. Paul, Trench, Trübner, 1890–92.

Lightman, Bernard. *The Origins of Agnosticism: Victorian Unbelief and the Limits of Knowledge.* Baltimore: Johns Hopkins UP, 1987.

McCosh, James. *Ideas in Nature Overlooked by Dr. Tyndall.* New York: R. Carter and Brothers, 1875.

Turner, Frank M. "Rainfall, Plagues, and the Prince of Wales: A Chapter in the Conflict of Religion and Science." *The Journal of British Studies* 13, no. 2 (1974): 45–65.

Tyndall, John. *New Fragments of Science.* London: Longmans, Green, and Co., 1892.

Young, H. *A Record of the Scientific Work of John Tyndall, D.C.L., LL.D., F.R.S. (1850–1888).* London: Chiswick Press, 1935.

White, Andrew Dickson. *The Warfare of Science with Theology in Christendom.* Preface by John Tyndall. London: H.S. King, 1876.

29. WILLIAM THOMSON, LORD KELVIN

(1824–1907)

Thomson's father James Thomson, upon appointment as Professor of Mathematics at Glasgow University, moved his family to Scotland from Belfast, Ireland where William had been born in 1824. Kelvin remained connected to the Universty for all of his professional life beginning with his election to the Chair of Natural Philosophy (Physics) in 1846 at the age of 22, an appointment he held until his retirement in 1899. His brother James (not to be confused with the unrelated J.J. Thomson, discoverer of the electron) served as a Professor of Engineering also at Glasgow. Thomson's honorary title Baron Kelvin of Netherhall, Largs, bestowed by Queen Victoria in 1892 for his work on the Atlantic telegraph cables, comes from the River Kelvin which flows through Glasgow.

Before permanently installing himself at Glasgow, Kelvin studied at Cambridge University and also worked briefly in the laboratory of Victor Regnault in Paris. He received almost every scientific medal and honour available to scientists in the United Kingdom, served as President of the Royal Society from 1890–95, and was buried in Westminster Abbey, an honour reserved only for the highest dignitaries. He made substantial contributions to hydrostatics, hydrodynamics, electromagnetic theory, electrolysis, thermodynamics, and geophysics. With Peter Guthrie Tait, Kelvin wrote the widely used advanced textbooks *Treatise on Natural Philosophy* (1867) and *Elements of Natural Philosophy* (1873). Kelvin expressed some skepticism about James Clerk Maxwell's electromagnetic theory and specifically Maxwell's 1864 electromagnetic theory of light (hypothesizing that light propagates in waves in a universal ether by alternating electric and magnetic fields), despite Heinrich Hertz's later description of electric waves in the 1880s. Maxwell's equations now stand as the fundamental description of electromagnetic phenomena, having eclipsed Kelvin's pioneering theoretical insights in electromagnetism.

Kelvin gained international attention from his work on the series of trans-Atlantic telegraph cables laid between Newfoundland and Ireland beginning in 1858. Kelvin's research on electrostatics and electrodynamics—which he translated into practical benefits for the Atlantic cables—stemmed directly from the relationship he had discovered between Fourier's theory of heat conduction described by Fourier analysis (which likens heat flow to an incompressible fluid), and electrostatic potential or inductance (electricity likewise having been conceived as a fluid since the eighteenth century). In one of his first papers on the heat and electricity analogy, "[Kelvin] did not prove the results mathematically at all, but employed the physical process of heat conduction to justify mathematical conclusions which he transferred to electricity by mathematical analogy. The physical reasoning made certain theorems in electricity obvious, theorems which, considered analytically, were anything but palpable" (Smith and Wise 205).

Kelvin differed on several crucial points with the chief engineer of the Atlantic cable project E.O.W. Whitehouse, and based on his theory of electricity and his discovery of a law of retardation of the electrical signal, Kelvin suggested to the directors of the telegraph project a thicker cable and the use of his sensitive mirror galvanometer for detecting weak currents. These two ideas were eventually incorporated into the successful 1865 and 1866 cables after Whitehouse's dismissal. Kelvin also invented an improved mariner's compass and a depth-sounder. He was knighted Sir William Thomson as a result of his success with the Atlantic cables and became a public symbol of the ability of applied science to overcome technological problems.

Kelvin developed the still used Kelvin absolute scale of temperature, defining the coldest possible temperature as 0 Kelvin or –273.15° Celsius, the point at which all molecular motion (a source of heat) stops. His interest in temperature and heat conduction led him naturally to geological questions concerning the age of the earth and the physics of its interior, the subject of his 1876 address to the British Association reprinted below in which he summarizes the arguments involved in the determination of the earth's physical state and age. According to his interpretation of the Second Law of Thermodynamics, Kelvin believed that the sun, earth and planets were slowly losing energy and cooling. He wrote two essays in 1862 on the subject, "On the Age of the Sun's Heat" in *Macmillan's Magazine*, and "On the Secular Cooling of the Earth," in *Transactions of the Royal Society of Edinburgh*, both reprinted in the *Treatise on Natural Philosophy*. He calculated a mean temperature of 14,000° C for the sun, examined the hypothesis that meteoric material colliding with the sun had been fuelling the sun's thermal output, and suggested that the light and heat of the sun might die out in the next several million years. Using Fourier's equations again for the linear conduction of heat, Kelvin calculated the age of the earth as 98,000,000 years using the factors of the earth's initial temperature at formation (which he estimated at 3,870° C), the average thermal conductivity of terrestrial rock, and the current geothermal gradient of the earth's outer crust, estimated from data obtained from mines and wells. Kelvin thus proposed upper and lower limits on the earth's age between 20 and 400 million years to allow for unknown variables in his calculations, such as how the intense heat and pressure inside the earth might alter the thermal conductivity and heat of fusion of rock. Kelvin's age of the earth of about 100 million years—which became widely accepted because of its presumed mathematical precision and elegance—immediately challenged Charles Lyell's uniformitarianism by suggesting that the rate and intensity of geological processes could not be viewed as constant over time. Additionally, Kelvin's calculations in the hands of Fleeming Jenkin posed a serious threat to Charles Darwin's theory of natural selection, which Jenkin argued required a geological time span longer than 100 million years in order for natural selection to modify species.

Kelvin's original assumptions were speculative, and recognized as such in a rebuttal by T.H. Huxley who complained that "pages of formulae will not get a definite result out of loose data" (Burchfield 84). After the discovery by the Curies of the enormous amounts of heat generated by radium salts, Ernest Rutherford (1871–1937) of McGill University proposed in 1904 that Kelvin's hypothesis of the age of the earth would have to be altered since naturally occurring radioactive materials in the earth's crust could provide the necessary energy to slow the cooling of a molten planet, increasing the possible age of the earth. Geophysicists today estimate the earth's age to be 4.5–4.6 Ga (giga-annums = 1×10^9 years) based on data from lead isotopes (Dalrymple 403).

"Review of Evidence Regarding the Physical Condition of the Earth. [Being extract from Address to the Mathematical and Physical Section of the British Association, Glasgow, September 7th, 1876.]"

Popular Lectures and Addresses. 3 vols. London: Macmillan and Co., 1894, pp. 2.38–72.

The evidence of a high internal temperature is too well known to need any quotation of particulars at present. Suffice it to say that below the uppermost ten metres stratum of rock or soil sensibly affected by diurnal and annual variations of temperature there is generally found a gradual increase of temperature downwards, approximating roughly in ordinary localities to an average rate of 1° centigrade per thirty metres of descent, but much greater in the neighbourhood of active volcanoes and certain other special localities, of comparatively small area, where hot springs and perhaps also sulphurous vapours prove an intimate relationship to volcanic quality. It is worthy of remark in passing that, so far as we know at present, there are no localities of exceptionally *small* rate of augmentation of underground temperature, and none where temperature diminishes at any time through any considerable depth downwards below the stratum sensibly influenced by summer heat and winter cold. Any considerable area of the earth of, say, not less than a kilometre in any horizontal diameter, which for several thousand years had been covered by snow or ice, and from which the ice had melted away and left an average surface temperature of 13° cent., would, during 900 years, show a decreasing temperature for some depth down from the surface; and 3600 years after the clearing away of the ice would still show residual effect of the ancient cold, in a half rate of augmentation of temperature downwards in the upper strata, gradually increasing to the whole normal rate, which would be sensibly reached at a depth of 600 metres.

By a simple effort of geological calculus it has been estimated that 1° per 30 metres gives 1000° per 30,000 metres, and 3,333 per 100 kilometres. The arithmetical result is irrefragable; but what of the physical conclusion drawn from it with marvellous frequency and pertinacity, that at depths of from 30 to 100 kilometres the temperatures are so high as to melt all substances composing the earth's upper crust? It has been remarked, indeed, that if observation showed any diminution or augmentation of the rate of increase of underground temperature in great depths, it would not be right to reckon on the uniform rate of 1° per 30 metres or thereabouts down to 30 or 60 or 100 kilometres. "But observation has shown nothing of the kind; and therefore surely it is most consonant with inductive philosophy to admit no great deviation in any part of the earth's solid crust from the rate of increase proved by observation as far as the greatest depths to which we have reached!" Now I have to remark upon this argument that the greatest depth to which we have reached in observations of underground temperature is scarcely one kilometre; and that if a ten per cent. diminution of the rate of augmentation of underground temperature downwards were found at a depth of one kilometre, this would demonstrate[1] that within the last 100,000 years the upper surface of the earth must have been at a higher temperature than that now found at the depth of one kilometre. Such a result is no doubt to be found by observation in places which have been overflown by lava in the memory of man or a few thousand years further back; but if, without going deeper than a kilometre, a ten per cent. diminution of the rate of increase of temperature downwards were found for the whole earth, it would limit the whole of geological history to within 100,000 years, or, at all events, would interpose an absolute barrier against the continuous descent of life on the earth from earlier periods than 100,000 years ago. Therefore, although search in particular localities for a diminution of the rate of augmentation of underground temperature in depths of less than a kilometre may be of intense interest, as helping us to fix the dates of extinct volcanic actions which have taken place within 100,000 years or so, we know enough from thoroughly sure geological evidence not to expect to find it, except in particular localities, and to feel quite sure that we shall not find it under any considerable portion of the earth's surface. If we admit as possible any such discontinuity within 900,000 years, we might be prepared to find a sensible diminution of the rate at three kilometres depth; but not at anything less than 30 kilometres if geologists validly claim as much as 90,000,000 of years for the length of the time with which their science is concerned. Now this implies a temperature of 1,000° cent. at the depth of 30 kilometres, allows something less than 2,000° for the temperature at 60 kilometres, and does not require much more than 4,000° cent. at any depth however great, but does require at the great depths a temperature of, at all events, not less than about 4,000° cent. It would not take much "hurrying up" of the actions with which they are concerned to satisfy geologists with the more moderate estimate of 50,000,000 of years. This would imply at least about 3,000° cent. for the limiting temperature at great depths. If the actual substance of the earth, whatever it may be, rocky or metallic, at depths of from 60 to 100 kilometres, under the pressure actually there experienced by it, can be solid at temperatures of from 3,000° to 4,000°, then we may hold the former estimate (90,000,000) to be as probable as the latter (50,000,000), so far as evidence from underground temperature can guide us. If 4,000° would melt the earth's substance at a depth of 100 kilometres, we must reject the former estimate though we might still admit the latter; if 3,000° would melt the substance at a depth of 60 kilometres, we should be compelled to conclude that 50,000,000 of years is an overestimate. Whatever may be its age, we may be quite sure the earth is solid in its interior; not, I admit, throughout its whole volume, for there certainly are spaces in volcanic regions occupied by liquid lava; but whatever portion of the whole mass is liquid, whether the waters of the ocean, or melted matter in the interior, these portions are small in comparison with the whole; and we must utterly reject any geological hypothesis which, whether for explaining underground heat or ancient upheavals and subsidences of the solid crust, or earthquakes, or existing volcanoes, assumes the solid earth to be a shell of 30, or 100, or 500, or 1,000 kilometres thickness, resting on an interior liquid mass.

If the inner boundary of the imagined rigid shell of the earth were rigorously spherical, the interior liquid could experience no precessional or nutational influence from the pressure on its bounding surface, and therefore if homogeneous could have no precession or nutation at all, or if heterogeneous only as much precession and nutation as would be produced by attraction from without in virtue of non-sphericity of its surfaces of equal density, and therefore the shell would have enormously more rapid precession and nutation than it actually has—forty times as much, for instance, if the thickness of the shell is 60 kilometres. A very slight deviation of the inner surface of the shell from perfect sphericity would suffice, in virtue of the quasi-rigidity due to vortex motion, to hold back the shell from taking sensibly more precession than it would give to the liquid, and to cause the liquid (homogeneous or heterogeneous) and the shell to have sensibly the same precessional motion as if the whole constituted one rigid body. But it is only because of the very long period (26,000 years) of precession, in comparison with the period of rotation (one day), that a very slight deviation from sphericity would suffice to cause the whole to move as if it were a rigid body. A little further consideration showed me:—

(1) That an ellipticity of inner surface equal to $\frac{1}{26000 \times 365}$ would be too small, but that an ellipticity of one or two hundred times this amount would not be too small to compel approximate equality of precession throughout liquid and shell.

(2) That with an ellipticity of interior surface equal to $\frac{1}{300}$, if the precessional motion were 26,000 times as great as it is, the motion of the liquid would be very different from that of a rigid mass rigidly connected with the shell.

(3) That with the actual forces and the supposed interior ellipticity of $\frac{1}{300}$, the lunar nineteen-yearly nutation might be affected to about

five per cent. of its amount by interior liquidity.

(4) Lastly, that the lunar semiannual nutation must be largely, and the lunar fortnightly nutation enormously affected by interior liquidity.

But although so much could be foreseen readily enough, I found it impossible to discover without thorough mathematical investigation what might be the characters and amounts of the deviations from a rigid body's motion which the several cases of precession and nutation contemplated would present. The investigation, limited to the case of a homogeneous liquid inclosed in an ellipsoidal shell, has brought out results which I confess have greatly surprised me. When the interior ellipticity of the shell is just too small, or the periodic speed of the disturbance just too great to allow the motion of the whole to be sensibly that of a rigid body, the deviation first sensible renders the precessional or nutational motion of the shell smaller than if the whole were rigid, instead of greater, as I expected. The amount of this difference bears the same proportion to the actual precession or nutation as the fraction measuring the periodic speed of the disturbance (in terms of the period of rotation as unity) bears to the fraction measuring the interior ellipticity of the shell; and it is remarkable that this result is independent of the thickness of the shell, assumed, however, to be small in proportion to the earth's radius. Thus in the case of precession the effect of interior liquidity would be to diminish the precessional motion, that is to say the periodic speed of the precession, in the proportion stated; in other words, it would add to the precessional period a number of days equal to the number whose reciprocal measures the ellipticity. Thus, in the actual case of the earth, if we still take $\frac{1}{300}$ as the ellipticity of the inner boundary of the supposed rigid shell, the effect would be to augment by 300 days the precessional period of 2,600 years, or to diminish by about $\frac{1}{60}$″ the annual precession of about 51″, an effect which I need not say would be wholly insensible. But on the lunar nutation of 18.6 years period, the effect of interior liquidity would be quite sensible: 18.6 years being twenty-three times 300 days, the effect would be to diminish the axes of the ellipse which the earth's pole describes in this period each by $\frac{1}{23}$ of its own amount. The semiaxes of this ellipse, calculated on the theory of perfect rigidity from the very accurately known amount of precession, and the fairly accurate knowledge which we have of the ratio of the lunar to the solar part of the precessional motion, are 9″.22 and 6″.86, with an uncertainty not amounting to one-half per cent. on account of want of perfect accuracy in the latter part of data. If the true values were less each by $\frac{1}{23}$ of its own amount, the discrepance might have escaped detection, or might *not* have escaped detection; but certainly it could be found if looked for. So far nothing can be considered as absolutely proved with reference to the interior solidity of the earth from precession and nutation; but now think of the solar semiannual and the lunar fortnightly nutations. The period of each of these is less than 300 days. Now the hydrodynamical theory shows that, irrespectively of the thickness of the shell, the nutation of the crust would be zero if the period of the nutational disturbance were 300 times the period of rotation (the ellipticity being $\frac{1}{300}$); if the nutational period were anything between this and a certain smaller critical value depending on the thickness of the crust, the nutation would be negative; if the period were equal to this second critical value, the nutation would be infinite; and if the period were still less, the nutation would be again positive. Further, the 183 days period of the solar nutation falls so little short of the critical 300 days that the amount of the nutation is not sensibly influenced by the thickness of the crust, and is negative and equal in absolute value to $\frac{61}{39}$ (being the reciprocal of $\frac{300}{183} - 1$) times what the amount would be were the earth solid throughout. Now this amount, as calculated in the *Nautical Almanac*, makes 0″.55 and 0″.51 the semiaxes of the ellipse traced by the earth's axis round its mean position; and if the true nutation placed the earth's axis on the opposite side of an ellipse having 0.″86 and 0.″81 for its semiaxes, the discrepance could not possibly have escaped detection. But, lastly, think of the lunar fortnightly nutation. Its period is $\frac{1}{20}$ of 300 days, and its amount, calculated in the *Nautical Almanac* on the theory of complete solidity, is such that the greater semiaxis of the approximately circular ellipse described by the pole is 0″.0325. Were the crust infinitely thin this nutation would be negative, but its amount nineteen times that corresponding to solidity. This would make the greater semiaxis of the approximately circular ellipse described by the pole amount to 19 x 0″.0885, which is 1″.7. It would be negative and of some amount between 1″.7 and infinity, if the thickness of the crust were anything from zero to 120 kilometres. This conclusion is absolutely decisive against the geological hypothesis of a thin rigid shell full of liquid.

But interesting in a dynamical point of view as Hopkins's problem is, it cannot afford a decisive argument against the earth's interior liquidity. It assumes the crust to be perfectly stiff and unyielding in its figure. This, of course, it cannot be because no material is infinitely rigid; but, composed of rock and possibly of continuous metal in the great depths, may the crust not, as a whole, be stiff enough to practically fulfil the condition of unyieldingness? No, decidedly it could not; on the contrary, were it of continuous steel and 500 kilometres thick, it would yield very nearly as much as if it were india-rubber to the deforming influences of centrifugal force and of the sun's and moon's attractions. Now although the full problem of precession and nutation, and, what is now necessarily included in it, tides, in a continuous revolving liquid spheroid, whether homogeneous or heterogeneous, has not yet been completely worked out, I think I see far enough towards a complete solution to say that precession and nutations will be practically the same in it as in a solid globe, and that the tides will be practically the same as those of the equilibrium theory. From this it follows that precession and nutations of the solid crust, with the practically perfect flexibility which it would have even though it were 100 kilometres thick and as stiff as steel, would be sensibly the same as if the whole earth from surface to centre were solid and perfectly stiff. Hence precession and nutations yield nothing to be said against such hypotheses as that of Darwin,[2] that the earth as a whole takes approximately the figure due to gravity and centrifugal force, because of the fluidity of the interior and the flexibility of the crust. But, alas for this "attractive sensational idea that a molten interior to the globe underlies a superficial crust, its surface agitated by tidal waves, and flowing freely towards any issue that may here and there be opened for its outward escape" (as Poulett Scrope called it)! the solid crust would yield so freely to the deforming influence of sun and moon that it would simply carry the waters of the ocean up and down with it, and there would be no sensible tidal rise and fall of water relatively to land.

The state of the case is shortly this:—The hypothesis of a perfectly rigid crust containing liquid violates physics by assuming preternaturally rigid matter, and violates dynamical astronomy in the solar semiannual and lunar fortnightly nutations; but tidal theory has nothing to say against it. On the other hand, the tides decide against any crust flexible enough to perform the nutations correctly with a liquid interior, or as flexible as the crust must be unless of preternaturally rigid matter.

But now thrice to slay the slain: suppose the earth this moment to be a thin crust of rock or metal resting on liquid matter; its equilibrium would be unstable! And what of the upheavals and subsidences? They would be strikingly analogous to those of a ship which has been rammed—one portion of crust up and another down, and then all down. I may say, with some degree of confidence, that whatever may be the relative densities of rock, solid and melted, at or about the temperature of liquefaction, it is, I think, probable that cold solid rock is denser than hot melted rock; and no possible degree of rigidity in the crust could prevent it from breaking in pieces and sinking wholly below the liquid lava. Something like this may have gone on, and probably did go on, for thousands of years after solidification commenced—surface-portions of the melted material losing heat, freezing, sinking immediately, or growing to thicknesses of a few metres, when the surface would be cool and the whole solid dense enough to sink. "This process must go on until the sunk portions of crust build up from the bottom a sufficiently close-ribbed skeleton or frame to allow fresh incrustations to remain, bridging across the now small areas of lava pools or lakes.

"In the honeycombed solid and liquid mass thus formed there must be a continual tendency for the liquid, in consequence of its less specific gravity, to work its way up; whether by masses of solid falling from the roofs of vesicles or tunnels and causing earthquake-shocks, or by the roof breaking quite through when very thin, so as to cause two such hollows to unite, or the liquid of any of them to flow out freely over the outer surface of the earth, or by gradual subsidence of the solid owing to the thermodynamic melting which portions of it under intense stress must experience, according to my brother's theory. The results which must follow from this tendency seem sufficiently great and various to account for all that we learn from geological evidence of earthquakes, of upheavals and subsidences of solid, and of eruptions of melted rock."[3]

Leaving altogether now the hypothesis of a hollow shell filled with liquid, we must still face the question, How much does the earth, solid throughout, except small cavities or vesicles filled with liquid, yield to the deforming (or tide-generating) influences of sun and moon? This question can only be answered by observation. A single infinitely accurate spirit-level or plummet far enough away from the sea to be not sensibly affected by the attraction of the rising and falling water would enable us to find the answer. Observe by level or plummet the changes of direction of apparent gravity relatively to an object rigidly connected with the earth, and compare these changes with what they would be were the earth perfectly rigid, according to the known masses and distances of sun and moon. The discrepance, if any is found, would show distortion of the earth and would afford data for determining the dimensions of the elliptic spheroid into which a non-rotating globular mass of the same dimensions and elasticity as the earth would be distorted by centrifugal force if set in rotation, or by tide-generating influences of sun or moon. The effect on the plumb-line of the lunar tide-generating influence is to deflect it towards or from the point of the horizon nearest to the moon, according as the moon is above or below the horizon. The effect is zero when the moon is on the horizon or overhead, and is greatest in either direction when the moon is 45° above or below the horizon. When this greatest value is reached, the plummet is drawn from its mean position through a space equal to $\frac{1}{12000000}$ of the length of the thread. No ordinary plummet or spirit-level could give any perceptible indication whatever of this effect; and to measure its amount it would be necessary to be able to observe angles as small as $\frac{1}{120000000}$ of the radian, or about $\frac{1}{600}''$. At present no apparatus exists within small compass by which it could be done. A submerged water-pipe of considerable length, say 12 kilometres, with its two ends turned up and open, might answer. Suppose, for example, the tube to lie north and south, and its two ends to open into two small cisterns, one of them, the southern for example, of half a decimetre diameter (to escape disturbance from capillary attraction), and the other of two or three decimetres diameter (so as to throw nearly the whole rise and fall into the smaller cistern). For simplicity, suppose the time of observation to be when the moon's declination is zero. The water in the smaller or southern cistern will rise from its lowest position to its highest position while the moon is rising to maximum altitude, and fall again after the moon crosses the meridian till she sets; and it will rise and fall again through the same range from moonset to moonrise. If the earth were perfectly rigid, and if the locality is in latitude 45°, the rise and fall would be half a millimetre on each side of the mean level, or a little short of half a millimetre if the place is within 10° north or south of latitude 45°. If the air were so absolutely quiescent during the observations as to give no varying differential pressure on the two water-surfaces to the amount of $\frac{1}{100}$ millimetre of water or $\frac{1}{1400}$ of mercury, the observation would be satisfactorily practicable, as it would not be difficult by aid of a microscope to observe the rise and fall of the water in the smaller cistern to $\frac{1}{100}$ of a millimetre; but no such quiescence of the atmosphere could be expected at any time; and it is probable that the variations of the water-level due to difference of the barometric pressure at the two ends would, in all ordinary weather, quite overpower the small effect of the lunar tide-generating motive. If, however, the two cisterns, instead of being open to the atmosphere, were connected air-tightly by a return-pipe with no water in it, it is probable that the observation might be successfully made: but Siemens's level or some other apparatus on a similarly small scale would probably be preferable to any elaborate method of obtaining the result by aid of very long pipes laid in the ground; and I have only called your attention to such an ideal method as leading up to the natural phenomenon of tides.

Tides in an open canal or lake of 12 kilometres length would be of just the amount which we have estimated for the cisterns connected by submerged pipe; but would be enormously more disturbed by wind and variations of atmospheric pressure. A canal or lake of 240 kilometres length in a proper direction and in a suitable locality would give but 10 millimetres rise and fall at each end, an effect which might probably be analysed out of the much greater disturbance produced by wind and differences of barometric pressure; but no open liquid level short of the *ingens aequor*, the ocean, will probably be found so well adapted as it for measuring the absolute value of the disturbance produced on terrestrial gravity by the lunar and solar tide-generating motive. But observations of the diurnal and semi-diurnal tides in the ocean do not (as they would on smaller and quicker levels) suffice for this purpose, because their amounts differ enormously from the equilibrium-values on account of the smallness of their periods in comparison with the periods of any of the grave enough modes of free vibration of the ocean as a whole. On the other hand, the lunar fortnightly declinational and the lunar monthly elliptic and the solar semiannual and annual elliptic tides have their periods so long that their amounts must certainly be very approximately equal to the equilibrium values. But there are large annual and semiannual changes of sea-level, probably both differential, on account of wind and differences of barometric pressure and differences of temperature of the water, and absolutely depending on rainfall and the melting away of snow and return evaporation, which altogether swamp the small semiannual and annual tides due to the sun's attraction. Happily, however, for our object, there is no meteorological or other disturbing cause which produces periodic changes of sea-level in either the fortnightly declinational or the monthly elliptic period; and the lunar gravitational tides in these periods are therefore to be carefully investigated in order that we may obtain the answer to the interesting question, How much does the earth as an elastic spheroid yield to the tide-generating influence of sun or moon? Hitherto in the British Association Committee's reductions of Tidal Observations we have not succeeded in obtaining any trustworthy indications of either of these tides. The St. George's Pier landing-stage pontoon, unhappily chosen for the Liverpool tide-gauge, cannot be trusted for such a delicate investigation: the available funds for calculation were expended before the long-period tides for Hilbre Island could be attacked, and three years of Kurrachee gave our only approach to a result. Comparisons of this with an indication of a result of calculations on West Hartlepool tides, conducted with the assistance of a grant from the Royal Society seem to show possibly no sensible yielding, or perhaps more probably some degree of yielding, of the earth's figure. The absence from all the results of any indication of a 18.6 yearly tide (according to the same law as the other long-period tides) is not easily explained without assuming or admitting a considerable degree of yielding.

Closely connected with the question of the earth's rigidity, and of as great scientific interest and of even greater practical moment, is the question, How nearly accurate is the earth as a timekeeper? and another of, at all events, equal scientific interest, How about the permanence of the earth's axis of rotation?

Peters and Maxwell, about 35 and 25 years ago, separately raised the question, How much does the earth's axis of rotation deviate from being a principal axis of inertia? and pointed out that an answer to this question is to be obtained by looking for a variation in latitude of any or every place on the earth's surface in a period of 306 days. The model before you illustrates the travelling round of the instantaneous axis relatively to the earth in an approximately circular cone whose axis is the principal axis of inertia, and relatively to space in a cone round a fixed axis. In the model the former of these cones, fixed relatively to the earth, rolls internally on the latter, supposed to be fixed in space. Peters gave a minute investigation of observations at Pulkova in the years 1841–42, which seem to indicate at that time a deviation amounting to about $\frac{3}{40}''$ of the axis of rotation from the principal axis. Maxwell, from Greenwich observations of the years 1851–54, found seeming indications of a very slight deviation, something less than half a second, but differing altogether in phase from that which the deviation indicated by Peters, if real and permanent, would have produced at Maxwell's later time. On my begging Professor Newcomb to take up the subject, he kindly did so at once, and undertook to analyse a series of observations suitable for the purpose which had been made in the United States Naval Observatory, Washington. A few weeks later I received from him a letter referring me to a paper by Dr. Nysen, of Pulkova Observatory, in which a similar negative conclusion as to constancy of magnitude or direction in the deviation sought for is arrived at from several series of the Pulkova observations between the years 1842 and 1872, and containing the following statement of his conclusions:—[4]

"The investigation of the ten-month period of latitude from the Washington prime vertical observations from 1862 to 1867 is completed, indicating a coefficient too small to be measured with certainty. The declinations with this instrument are subject to an annual period which made it necessary to discuss those of each month separately. As the series extended through a full five years, each month thus fell on five nearly equidistant points of the period. If x and y represent the coordinates of the axis of instantaneous rotation on June 30, 1864, then the observations of the separate months give the following values of x and y:—

	x.	Weight.	*y.*	Weight.
	″		″	
January	−0.35	10	+0.32	
February	−0.03	14	+0.09	
March	+0.17	10	+0.16	
April	+0.44	5	+0.05	
May	+0.08	16	+0.02	
June	−0.01	14	−0.01	
July	−0.05	14	0.00	
August	−0.24	14	+0.29	
September	+0.18	14	+0.21	
October	+0.13	14	−0.01	
November	+0.08	17	−0.20	
December	−0.08	16	−0.08	
Mean	0.01±0.03		+0.05±0.03	

"Accepting these results as real, they would indicate a radius of rotation of the instantaneous axis amounting, at the earth's surface, to 5 feet and a longitude of the point in which this axis intersects the earth's surface near the North Pole, such that on July 11, 1864, it was 180° from Washington, or 103° east of Greenwich. The excess of the coefficient over its probable error is so slight that this result cannot be accepted as anything more than a consequence of the unavoidable errors of observation."

From the discordant character of these results we must not, however, infer that the deviations indicated by Peters, Maxwell, and Newcomb are unreal. On the contrary, any that fall within the limits of probable error of the observations ought properly to be regarded as real. There is, in fact, a *vera causa* in the temporary changes of sea-level due to meteorological causes, chiefly winds, and to meltings of ice in the polar regions and return evaporations, which seems amply sufficient to account for irregular deviations of from $\frac{1}{2}''$ to $\frac{1}{20}''$ of the earth's instantaneous axis from the axis of maximum inertia, or, as I ought rather to say, of the axis of maximum inertia from the instantaneous axis.

As for geological upheavals and subsidences, if on a very large scale of area, they must produce, on the period and axis of the earth's rotation, effects comparable with those produced by changes of sea-level equal to them in vertical amount. For simplicity, calculating as if the earth were of equal density throughout, I find that an upheaval of all the earth's surface in north latitude and east longitude and south latitude and west longitude with equal depression in the other two quarters, amounting at greatest to 10 centimetres, and graduating regularly from the points of maximum elevation to the points of maximum depression in the middles of the four quarters, would shift the earth's axis of maximum moment of inertia through 1″ on the north side towards the meridian of 90° W. longitude, and on the south side towards the meridian of ″90° E. longitude. If such a change were to take place suddenly, the earth's instantaneous axis would experience a sudden shifting of but $\frac{1}{300}''$ (which we may neglect), and then, relatively to the earth, would commence travelling, in a period of 306 days, round the fresh axis of maximum moment of inertia. The sea would be set into vibration, one ocean up and another down through a few centimetres, like water in a bath set aswing. The period of these vibrations would be from 12 to 24 hours, or at most a day or two; their subsidence would probably be so rapid that after at most a few months they would become insensible. Then a regular 306-days period tide of 11 centimetres from lowest to highest would be to be observed, with gradually diminishing amount from century to century, as through the dissipation of energy produced by this tide the instantaneous axis of the earth is gradually brought into coincidence with the fresh axis of maximum moment of inertia. If we multiply these figures by 3600, we find what would be the result of a similar sudden upheaval and subsidence of the earth to the extent of 360 metres above and below previous levels. It is not impossible that in the very early ages of geological history such an action as this, and the consequent 400-metres tide producing a succession of deluges every 306 days for many years, may have taken place; but it seems more probable that even in the most ancient times of geological history the great world-wide changes, such as the upheavals of the continents and subsidences of the ocean-beds from the general level of their supposed molten origin, took place gradually through the thermodynamic melting of solids and the squeezing out of liquid lava from the interior, to which I have already referred. A slow distortion of the earth as a whole would never produce any great angular separation between the instantaneous axis and the axis of maximum moment of inertia for the time being. Considering, then, the great facts of the Himalayas and Andes, and Africa and the depths of the Atlantic, and America and the depths of the Pacific, and Australia, and considering further the ellipticity of the equatorial section of the sea-level estimated by Capt. Clarke at about $^1/_{10}$ of the mean ellipticity of meridional sections of the sea-level, we need no brush from the comet's tail (a wholly chimerical cause which can never have been put forward seriously except in ignorance of elementary dynamical principles) to account for a change in the earth's axis; we need no violent convulsion producing a sudden distortion on a great scale, with change of the axis of maximum moment of inertia followed by gigantic deluges; and we may not merely admit, but assert as highly probable, that the axis of maximum inertia and axis of rotation, always very near one another, may have been in ancient times very far from their present geographical position and may have gradually shifted through 10, 20, 30, 40, or more degrees without at any time any perceptible sudden disturbance of either land or water.

Lastly, as to variations in the earth's rotational period. You all no doubt know how, in 1853, Adams discovered a correction to be needed in the theoretical calculation with which Laplace followed up his brilliant discovery of the dynamical explanation of an apparent acceleration of the moon's mean motion shown by records of ancient eclipses, and how he found that when his correction was applied the dynamical theory of the moon's motion accounted for only about half of the observed apparent acceleration, and how Delaunay in 1866 verified Adams's result and suggested that the explanation may be a retardation of the earth's rotation by tidal friction. The conclusion is that, since the 19th of March, 721 B.C., a day on which an eclipse of the moon was seen in Babylon, commencing "when one hour after her rising was fully passed," the earth has lost rather more than $\frac{1}{3{,}000{,}000}$ of her rotational velocity, or, as a timekeeper, is going slower by 11 1/2 seconds per annum now than then. According to this rate of retardation, if uniform, the earth at the end of a century would, as a timekeeper, be found 22 seconds behind a perfect clock, rated and set to agree with her at the beginning of the century. Newcomb's subsequent investigations in the lunar theory have on the whole tended to confirm this result; but they have also brought to light some remarkable apparent irregularities in the moon's motion, which, if real, refuse to be accounted for by the gravitational theory without the influence of some unseen body or bodies passing near enough to the moon to influence her mean motion. This hypothesis Newcomb considers not so probable as that the apparent irregularities of the moon are not real, and are to be accounted for by irregularities in the earth's rotational velocity. If this is the true explanation, it seems that the earth was going slow from 1850 to 1862, so much as to have got behind by 7 seconds in these 12 years, and then to have begun going faster again so as to gain 8 seconds from 1862 to 1872. So great an irregularity as this would require somewhat greater changes of sea-level, but not many times greater than the British Association Committee's reductions of tidal observations for several places in different parts of the world allow us to admit to have possibly taken place. The assumption of a fluid interior, which Newcomb suggests, and the flow of a large mass of the fluid "from equatorial regions to a position nearer the axis," is not, from what I have said to you, admissible as a probable explanation of the remarkable acceleration of rotational velocity which seems to have taken place about 1862; but happily it is not necessary. A settlement of 14 centimetres in the equatorial regions, with corresponding rise of 28 centimetres at the poles (which is so slight as to be absolutely undiscoverable in astronomical observatories, and which would involve no change of sea-level absolutely disproved by reductions of tidal observations hitherto made), would suffice. Such settlements must occur from time to time; and a settlement of the amount suggested might result from the diminution of centrifugal force due to 150 or 200 centuries tidal retardation of the earth's rotational speed.

NOTES

1 For proof of this and following statements regarding underground heat, I refer to "Secular Cooling of the Earth," *Transactions of the Royal Society of Edinburgh*, 1862; and Thomson and Tait's *Natural Philosophy*, Appendix D.

2 "Observations on the Parallel Roads of Glen Roy and other parts of Lochaber in Scotland, with an attempt to prove that they are of marine origin." *Transactions of the Royal Society* for February 1839, p. 81.

3 "Secular Cooling of the Earth," *Transactions of the Royal Society of Edinburgh*, 1862 (W. Thomson), and Thomson and Tait's *Natural Philosophy*, §§ (*ee*), (*ff*).

4 [For later investigations by Newcomb and others on this subject see extracts from my Presidential Addresses of 1891 and 1892 to the Royal Society.—K., March 22, 1893.]

FURTHER READING

Burchfield, Joe D. *Lord Kelvin and the Age of the Earth.* New York: Science History Publications, 1975.

Crowther, J.G. *Men of Science: Humphry Davy, Michael Faraday, James Prescott Joule, William Thomson, James Clerk Maxwell.* New York: W.W. Norton and Co., 1936.

Dalrymple, G. Brent. *The Age of the Earth.* Stanford: Stanford UP, 1991.

Gray, Andrew. *Lord Kelvin: An Account of His Scientific Life and Work.* 1908. Reprint, New York: Chelsea Publishing Company, 1973.

Harman, P.M., ed. *Wranglers and Physicists: Studies on Cambridge Physics in the Nineteenth Century.* Manchester: Manchester UP, 1985.

Kargon, Robert and Peter Achinstein. *Kelvin's Baltimore Lectures and Modern Theoretical Physics.* Cambridge: MIT Press, 1987.

King, Agnes Gardner. *Kelvin the Man: A Biographical Sketch by His Niece.* London: Hodder and Stoughton, 1925.

Lord Kelvin Professor of Natural Philosophy in the University of Glasgow 1846–1899. Glasgow: James MacLehose and Sons, 1899.

MacDonald, D.K.C. *Faraday, Maxwell, and Kelvin.* Garden City: Doubleday Anchor, 1964.

Macfarlane, Alexander. *Lectures on Ten British*

Physicists of the Nineteenth Century. New York: John Wiley, 1919.

Magie, William Francis. *The Second Law of Thermodynamics: Memoirs by Carnot, Clausius, and Thomson.* New York: Harper & Brothers, 1899.

Munro, John. *Lord Kelvin.* London: H.J. Drane, 1902.

Russell, Alexander. *Lord Kelvin.* London: Blackie & Son, 1938.

—. *Lord Kelvin: His Life and Work.* London: Dodge Publishing Co., 1924.

Sharlin, Harold Issadore, and Tiby Sharlin. *Lord Kelvin the Dynamic Victorian.* University Park: Pennsylvania State UP, 1979.

Smith, Crosbie, and M. Norton Wise. *Energy and Empire: A Biographical Study of Lord Kelvin.* Cambridge: Cambridge UP, 1989.

Thompson, Silvanus P. *The Life of Lord Kelvin.* 2 vols. 1910. Reprint, New York: Chelsea Publishing Company, 1976.

Thomson, William [Lord Kelvin], and Peter Guthrie Tait. *Principles of Mechanics and Dynamics* [*Treatise on Natural Philosophy*]. 1912. Reprint, New York: Dover Publications, 1962.

Thomson, William Sir [Lord Kelvin]. *Notes of Lectures on Molecular Dynamics and the Wave Theory of Light.* Ed. A.S. Hathaway. Baltimore: Johns Hopkins UP, 1884.

—. *Reprint of Papers on Electrostatics and Magnetism.* 2nd ed. London: Macmillan and Co., 1884.

Wilson, David B. *Kelvin and Stokes: A Comparative Study in Victorian Physics.* Bristol: Adam Hilger, 1987.

Young, A.P. *Lord Kelvin.* London: For the British Council by Longmans, Green, 1948.

30. DMITRII IVANOVICH MENDELEEV (also Mendeleeff, Mendeleyev)

(1834–1907)

Mendeleev's father Ivan directed the gymnasium in Tobolsk, Siberia. When Ivan Mendeleev lost his sight, his wife Maria Dmitrievna supported the family by running a glass factory. Dmitrii was sent east to study at St. Petersburg in the Pedagogical Institute. As a student, he suffered from tuberculosis, which claimed his sister. After teaching appointments in the Crimea and Odessa, he became private docent at the University of St. Petersburg. In 1859–60, Mendeleev established a chemical laboratory in Heidelberg after collaboration with Robert Bunsen. He participated in the chemical congress at Karlsruhe in 1860, which met to establish international standards in terminology, chemical formulae, and measurement. He was appointed professor of chemistry at St. Petersburg in 1867. Mendeleev's areas of research included thermal expansion of liquids, specific gravities of aqueous solutions of alcohol, and gas elasticity, but his primary contribution to chemistry was the construction of the periodic chart of the elements. He also advanced his "hydrate" theory of chemical solutions, arguing that solutions do not simply consist of static mechanical mixtures, but rather constitute combinations of solutes and solvents in constant dynamic equilibrium.

After visiting the Pennsylvania oilfields in 1876, Mendeleev developed a new theory of the origin of petroleum based on the chemical action of water on metal carbides rather than the now more widely accepted hypothesis involving compression and decomposition of organic matter. He acted as an industrial consultant in the 1860s to the growing Russian oil industry and published *The Petroleum Industry in Pennsylvania and in the Caucasus* (1877). Mendeleev also became involved with educational issues, publishing *Remarks on Public Education in Russia* (1901). He insisted on more natural science training in Russian schools and opposed the traditional classics-based Russian curriculum: "commenting on Lomonosov's famous pronouncement that the time was approaching when Russia would produce its own Platos and Newtons, [Mendeleev] stated that Russia would do much better if it forgot all about the Platos and doubled its search for Newtons" (Vucinich 160). Mendeleev resigned from his professorship at St. Petersburg in 1890 after troubles over his sponsorship of a student petition presented to the conservative Minister of Education. He had been turned down for a post at the Imperial Academy of Sciences in 1880 for similar political reasons. After his university career, he proceeded to the Bureau of Weights and Measures in 1893.

Mendeleev's *Osnovy Khimii* or *Principles of Chemistry* (1868–71) laid the foundation of the "Periodic Law" which forms the basis of the modern periodic table. On March 6, 1869, Mendeleev stated in a lecture at the Russian Chemical Society that if the elements are arranged according to atomic weights, they exhibit a repetition or periodicity of chemical properties and that extrapolating from gaps in this arrangement we must expect the discovery of unknown elements. He also presented his first periodic table at the meeting. In the same year, Julius Lothar Meyer in Germany published a similar representation of the elements arranged by periodicity.

Through his periodic law, Mendeleev predicted the existence and weights of gallium (discovered 1875), scandium (discovered 1879), and germanium (discovered 1886), which he had prophetically designated with empty spaces in his chart as eka-aluminium, eka-boron, and eka-silicon respectively. On the basis of the periodic table, chemists were also able to correct some chemical formulae as well as erroneous atomic weights such as beryllium.

Klaus Danzer observes that "the time was simply ripe for a revolutionary generalization such as the periodic system of the elements represented" (11). Mendeleev was foreshadowed in his great generalization by De Chancourtois's helix of elements of 1863, J.A.R. Newlands's "law of octaves" (1864–5)—which uncovered periodicity in the 8th elements of his chemical groupings—and W. Odling's work, which suggested that recurrent chemical properties in elements arranged according to atomic weight could not be accidental. In 1882, Mendeleev won the Davy medal from the Royal Society; in 1889 he received the Faraday Medal from the Chemical Society, and in 1905 he was awarded the Copley Medal by the Royal Society. The modern periodic chart still in use today follows Mendeleev's model, except that elements are arranged not by their atomic weight, but by their atomic *number* (reflecting the number of protons in the nucleus), a more reliable indicator of periodicity, as it corresponds to the number of each element's electrons, which primarily determine chemical activity.

The selection below, a transcript of Mendeleev's 1889 Faraday Lecture, was included as an appendix to later editions of the *Principles of Chemistry* and was also reprinted in *Faraday Lectures* 1869–1928 (London, 1928).

"The Periodic Law of the Chemical Elements."

The Principles of Chemistry. Vol. 2.
Translated by George Kamensky from the 6th Russian Edition.
Edited by T.A. Lawson. London: Longmans, Green, and Co., 1897, pp. 471–90.

THE PERIODIC LAW OF THE CHEMICAL ELEMENTS

Faraday Lecture Delivered Before the Fellows of the Chemical Society in the Theatre of the Royal Institution, on Tuesday, June 4, 1889

The high honour bestowed by the Chemical Society in inviting me to pay a tribute to the world-famed name of Faraday by delivering this lecture has induced me to take for its subject the Periodic Law of the Elements—this being a generalisation in chemistry which has of late attracted much attention.

While science is pursuing a steady onward movement, it is convenient from time to time to cast a glance back on the route already traversed, and especially to consider the new conceptions which aim at discovering the general meaning of the stock of facts accumulated from day to day in our laboratories. Owing to the possession of laboratories, modern science now bears a new character, quite unknown, not only to antiquity, but even to the preceding century.

Bacon's and Descartes' idea of submitting the mechanism of science simultaneously to experiment and reasoning has been fully realised in the case of chemistry, it having become not only possible but always customary to experiment. Under the all-penetrating control of experiment, a new theory, even if crude, is quickly strengthened, provided it be founded on a sufficient basis; the asperities are removed, it is amended by degrees, and soon loses the phantom light of a shadowy form or of one founded on mere prejudice; it is able to lead to logical conclusions, and to submit to experimental proof. Willingly or not, in science we all must submit not to what seems to us attractive from one point of view or from another, but to what represents an agreement between theory and experiment; in other words, to demonstrated generalisation and to the approved experiment. Is it long since many refused to accept the generalisations involved in the law of Avogadro and Ampère, so widely extended by Gerhardt? We still may hear the voices of its opponents; they enjoy perfect freedom, but vainly will their voices rise so long as they do not use the language of demonstrated facts. The striking observations with the spectroscope which have permitted us to analyse the chemical constitution of distant worlds, seemed, at first, applicable to the task of determining the nature of the atoms themselves; but the working out of the idea in the laboratory soon demonstrated that the characters of spectra are determined, not directly by the atoms, but by the molecules into which the atoms are packed; and so it became evident that more verified facts must be collected before it will be possible to formulate new generalisations capable of taking their place beside those ordinary ones based upon the conception of simple substances and atoms. But as the shade of the leaves and roots of living plants, together with the relics of a decayed vegetation, favour the growth of the seedling and serve to promote its luxurious development, in like manner sound generalisations—together with the relics of those which have proved to be untenable—promote scientific productivity, and ensure the luxurious growth of science under the influence of rays emanating from the centres of scientific energy. Such centres are scientific associations and societies. Before one of the oldest and most powerful of these I am about to take the liberty of passing in review the twenty years' life of a generalisation which is known under the name of the Periodic Law. It was in March 1869 that I ventured to lay before the then youthful Russian Chemical Society the ideas upon the same subject which I had expressed in my just written 'Principles of Chemistry.'

Without entering into details, I will give the conclusions I then arrived at in the very words I used:—

'1. The elements, if arranged according to their atomic weights, exhibit an evident *periodicity* of properties.

'2. Elements which are similar as regards their chemical properties have atomic weights which are either of nearly the same value (*e.g.* platinum, iridium, osmium) or which increase regularly (*e.g.* potassium, rubidium, caesium).

'3. The arrangement of the elements, or of groups of elements, in the order of their atomic weights, corresponds to their so-called *valencies* as well as, to some extent, to their distinctive chemical properties—as is apparent, among other series, in that of lithium, beryllium, barium, carbon, nitrogen, oxygen, and iron.

'4. The elements which are the most widely diffused have *small* atomic weights.

'5. The *magnitude* of the atomic weight determines the character of the element, just as the magnitude of the molecule determines the character of a compound.

'6. We must expect the discovery of many yet *unknown* elements—for example, elements analogous to aluminium and silicon, whose atomic weight would be between 65 and 75.

'7. The atomic weight of an element may sometimes be amended by a knowledge of those of the contiguous elements. Thus, the atomic weight of tellurium must lie between 123 and 126, and cannot be 128.

'8. Certain characteristic properties of the elements can be foretold from their atomic weights.

'The aim of this communication will be fully attained if I succeed in drawing the attention of investigators to those relations which exist between the atomic weights of dissimilar elements, which, so far as I know, have hitherto been almost completely neglected. I believe that the solution of some of the most important problems of our science lies in researches of this kind.'

To-day, twenty years after the above conclusions here formulated, they may still be considered as expressing the essence of the now well-known periodic law.

Reverting to the epoch terminating with the sixties, it is proper to indicate three series of data without the knowledge of which the periodic law could not have been discovered, and which rendered its appearance natural and intelligible.

In the first place, it was at that time that the numerical value of atomic weights became definitely known. Ten years earlier such knowledge did not exist, as may be gathered from the fact that in 1860 chemists from all parts of the world met at Karlsruhe in order to come to some agreement, if not with respect to views relating to atoms, at any rate as regards their definite representation. Many of those present probably remember how vain were the hopes of coming to an understanding, and how much ground was gained at that Congress by the followers of the unitary theory so brilliantly represented by Cannizzaro. I vividly remember the impression produced by his speeches, which admitted of no compromise, and seemed to advocate truth itself, based on the conceptions of Avogadro, Gerhardt, and Regnault, which at that time were far from being generally recognised. And though no understanding could be arrived at, yet the objects of the meeting were attained, for the ideas of Cannizzaro proved, after a few years, to be the only ones which could stand criticism, and which represented an atom as—'the smallest portion of an element which enters into a molecule of its compound.' Only such real atomic weights—not conventional ones—could afford a basis for generalisation. It is sufficient, by way of example, to indicate the following cases in which the relation is seen at once and is perfectly clear:—

K = 39	Rb = 85	Cs = 133
Ca = 40	Sr = 87	Ba = 137

whereas with the equivalents then in use—

K = 39	Rb = 85	Cs = 133
Ca = 20	Sr = 43.5	Ba = 68.5

the consecutiveness of change in atomic weight, which with the true values is so evident, completely disappears.

Secondly, it had become evident during the period 1860–70, and even during the preceding decade, that the relations between the atomic weights of analogous elements were governed by some general and simple laws. Cooke, Cremers, Gladstone, Gmelin, Lenssen, Pettenkofer, and especially Dumas, had already established many facts bearing on that view. Thus Dumas compared the following groups of analogous elements with organic radicles:—

	Diff.		Diff.
		Mg = 12	
			8
Li = 7		Ca = 20	
	16		3 × 8
Na = 23		Sr = 44	
	16		3 × 8
K = 39		Ba = 68	

	Diff.		Diff.
P = 31		O = 8	
	44		8
As = 75		S = 16	
	44		3 × 8
Sb = 119		Se = 40	
	2 × 44		3 × 8
Bi = 207		Te = 64	

and pointed out some really striking relationships, such as the following:—

F = 19.
Cl = 35.5 = 19 + 16.5
Br = 80 = 19 + 2 x 16.5 + 28.
I = 127 = 2 x 19 + 2 x 16.5 + 2 x 28.

A. Strecker, in his work 'Theorien und Experimente zur Bestimmung der Atomgewichte der

Elemente' (Braunschweig, 1859), after summarising the data relating to the subject, and pointing out the remarkable series of equivalents—

Cr = 26.2 Mn = 27.6 Fe = 28 Ni = 29
Co = 30 Cu = 31.7 Zn = 32.5

remarks that: 'It is hardly probable that all the above-mentioned relations between the atomic weights (or equivalents) of chemically analogous elements are merely accidental. We must, however, leave to the future the discovery of the *law* of the relations which appears in these figures.'[1]

In such attempts at arrangement and in such views are to be recognised the real forerunners of the periodic law; the ground was prepared for it between 1860 and 1870, and that it was not expressed in a determinate form before the end of the decade may, I suppose, be ascribed to the fact that only analogous elements had been compared. The idea of seeking for a relation between the atomic weights of all the elements was foreign to the ideas then current, so that neither the *vis tellurique* of De Chancourtois, nor the *law of octaves* of Newlands, could secure anybody's attention. And yet both De Chancourtois and Newlands like Dumas and Strecker, more than Lenssen and Pettenkofer, had made an approach to the periodic law and had discovered its germs. The solution of the problem advanced but slowly, because the facts, but not the law, stood foremost in all attempts; and the law could not awaken a general interest so long as elements, having no apparent connection with each other, were included in the same octave, as for example:—

1st octave of Newlands . .	H	F	Cl	Co & Ni	Br	Pd	I	Pt & Ir
7th Ditto	O	S	Fe	Se	Rh & Ru	Te	Au	Os or Th

Analogies of the above order seemed quite accidental, and the more so as the octave contained occasionally ten elements instead of eight, and when two such elements as Ba and V, Co and Ni, or Rh and Ru, occupied one place in the octave.[2] Nevertheless, the fruit was ripening, and I now see clearly that Strecker, De Chancourtois, and Newlands stood foremost in the way towards the discovery of the periodic law, and that they merely wanted the boldness necessary to place the whole question at such a height that its reflection on the facts could be clearly seen.

A third circumstance which revealed the periodicity of chemical elements was the accumulation, by the end of the sixties, of new information respecting the rare elements, disclosing their many-sided relations to the other elements and to each other. The researches of Marignac on niobium, and those of Roscoe on vanadium, were of special moment. The striking analogies between vanadium and phosphorus on the one hand, and between vanadium and chromium on the other, which became so apparent in the investigations connected with that element, naturally induced the comparison of V = 51 with Cr = 52, Nb = 94 with Mo = 96, and Ta = 192 with W = 194; while, on the other hand, P = 31 could be compared with S = 32, As = 75 with Se = 79, and Sb = 120 with Te = 125. From such approximations there remained but one step to the discovery of the law of periodicity.

The law of periodicity was thus a direct outcome of the stock of generalisations and established facts which had accumulated by the end of the decade 1860–1870: it is an embodiment of those data in a more or less systematic expression. Where, then, lies the secret of the special importance which has since been attached to the periodic law, and has raised it to the position of a generalisation which has already given to chemistry unexpected aid, and which promises to be

1 'Es ist wohl kaum anzunehmen, dass alle im Vorhergehenden hervorgehobenen Beziehungen zwischen den Atomgewichten (oder Aequivalenten) in chemischen Verhältnissen einander ähnliche Elemente bloss zufällig sind. Die Auffindung der in diesen Zahlen *gesetzlichen* Beziehungen müssen wir jedoch der Zukunft überlassen.'

2 To judge from J.A.R. Newlands's work, *On the Discovery of the Periodic Law*, London, 1884, p. 149; 'On the Law of Octaves' (from the *Chemical News*, 12, 83, August 18, 1865).

far more fruitful in the future and to impress upon several branches of chemical research a peculiar and original stamp? The remaining part of my communication will be an attempt to answer this question.

In the first place we have the circumstance that, as soon as the law made its appearance, it demanded a revision of many facts which were considered by chemists as fully established by existing experience. I shall return, later on, briefly to this subject, but I wish now to remind you that the periodic law, by insisting on the necessity for a revision of supposed facts, exposed itself at once to destruction in its very origin. Its first requirements, however, have been almost entirely satisfied during the last 20 years; the supposed facts have yielded to the law, thus proving that the law itself was a legitimate induction from the verified facts. But our inductions from data have often to do with such details of a science so rich in facts, that only generalisations which cover a wide range of important phenomena can attract general attention. What were the regions touched on by the periodic law? This is what we shall now consider.

The most important point to notice is, that periodic functions, used for the purpose of expressing changes which are dependent on variations of time and space, have been long known. They are familiar to the mind when we have to deal with motion in closed cycles, or with any kind of deviation from a stable position, such as occurs in pendulum-oscillations. A like periodic function became evident in the case of the elements, depending on the mass of the atom. The primary conception of the masses of bodies, or of the masses of atoms, belongs to a category which the present state of science forbids us to discuss, because as yet we have no means of dissecting or analysing the conception. All that was known of functions dependent on masses derived its origin from Galileo and Newton, and indicated that such functions either decrease or increase with the increase of mass, like the attraction of celestial bodies. The numerical expression of the phenomena was always found to be proportional to the mass, and in no case was an increase of mass followed by a recurrence of properties such as is disclosed by the periodic law of the elements. This constituted such a novelty in the study of the phenomena of nature that, although it did not lift the veil which conceals the true conception of mass, it nevertheless indicated that the explanation of that conception must be searched for in the masses of the atoms; the more so, as all masses are nothing but aggregations, or additions, of chemical atoms which would be best described as chemical individuals. Let me remark, by the way, that though the Latin word 'individual' is merely a translation of the Greek word 'atom,' nevertheless history and custom have drawn a sharp distinction between the two words, and the present chemical conception of atoms is nearer to that defined by the Latin word than by the Greek, although this latter also has acquired a special meaning which was unknown to the classics. The periodic law has shown that our chemical individuals display a harmonic periodicity of properties dependent on their masses. Now natural science has long been accustomed to deal with periodicities observed in nature, to seize them with the vice of mathematical analysis, to submit them to the rasp of experiment. And these instruments of scientific thought would surely, long since, have mastered the problem connected with the chemical elements, were it not for a new feature which was brought to light by the periodic law, and which gave a peculiar and original character to the periodic function.

If we mark on an axis of abscissae a series of lengths proportional to angles, and trace ordinates which are proportional to sines or other trigonometrical functions, we get periodic curves of a harmonic character. So it might seem, at first sight, that with the increase of atomic weights the function of the properties of the elements should also vary in the same harmonious way. But in this case there is no such continuous change as in the curves just referred to, because the periods do not contain the infinite number of points constituting a curve, but a *finite* number only of such points. An example will better illustrate this view. The atomic weights—

Ag = 108 Cd = 112 In = 113 Sn = 118
Sb = 120 Te = 125 I = 127

steadily increase, and their increase is accompanied by a modification of many properties which constitutes the essence of the periodic law. Thus, for example, the densities of the above elements decrease steadily, being respectively—

10.5 8.6 7.4 7.2 6.7 6.4 4.9

while their oxides contain an increasing quantity of oxygen—

Ag_2O Cd_2O_2 In_2O_3 Sn_2O_4
Sb_2O_5 Te_2O_6 I_2O_7

But to connect by a curve the summits of the ordinates expressing any of these properties would involve the rejection of Dalton's law of multiple proportions. Not only are there no intermediate elements between silver, which gives AgCl, and cadmium, which gives $CdCl_2$, but, according to the very essence of the periodic law, there can be none; in fact a uniform curve would be inapplicable in such a case, as it would lead us to expect elements possessed of special properties at any point of the curve. The periods of the elements have thus a character very different from those which are so simply represented by geometers. They correspond to points, to numbers, to sudden changes of the masses, and not to a continuous evolution. In these sudden changes destitute of intermediate steps or positions, in the absence of elements intermediate between, say, silver and cadmium, or aluminium and silicon, we must recognise a problem to which no direct application of the analysis of the infinitely small can be made. Therefore, neither the trigonometrical functions proposed by Ridberg and Flavitzky, nor the pendulum-oscillations suggested by Crookes, nor the cubical curves of the Rev. Mr. Haughton, which have been proposed for expressing the periodic law, from the nature of the case, can represent the periods of the chemical elements. If geometrical analysis is to be applied to this subject, it will require to be modified in a special manner. It must find the means of representing in a special way, not only such long periods as that comprising

K Ca Sc Ti V Cr Mn Fe Co Ni Cu Zn Ga Ge As Se Br,

but short periods like the following:—

Na Mg Al Si P S Cl.

In the theory of numbers only do we find problems analogous to ours, and two attempts at expressing the atomic weights of the elements by algebraic formulae seem to be deserving of attention, although neither of them can be considered as a complete theory, nor as promising finally to solve the problem of the periodic law. The attempt of E. J. Mills (1886) does not even aspire to attain this end. He considers that all atomic weights can be expressed by a logarithmic function,

$$15(n - 0.9375^t),$$

in which the variables n and t are *whole numbers*. Thus, for oxygen, $n = 2$, and $t = 1$, whence its atomic weight is = 15.94; in the case of chlorine, bromine, and iodine, n has respective values of 3, 6, and 9, whilst $t = 7$, 6, and 9; in the case of potassium, rubidium, and caesium, $n = 4$, 6, and 9, and $t = 14$, 18, and 20.

Another attempt was made in 1888 by B. N. Tchitchérin. Its author places the problem of the periodic law in the first rank, but as yet he has investigated the alkali metals only. Tchitchérin first noticed the simple relations existing between the atomic volumes of all alkali metals; they can be expressed, according to his views, by the formula

$$A(2 - 0.00535An),$$

where A is the atomic weight, and n is equal to 8 for lithium and sodium, to 4 for potassium, to 3 for rubidium, and to 2 for caesium. If n remained equal to 8 during the increase of A, the volume would become zero at A = 46 2/3, and it would reach its maximum at A = 23 1/3. The close approximation of the number 46 2/3 to the differences between the atomic weights of analogous elements (such as Cs — Rb, I — Br, and so on); the close correspondence of the number 23 1/3 to the atomic weight of sodium; the fact of n being necessarily a whole number, and several other aspects of the question, induce Tchitchérin to believe that they afford a clue to the understanding of the nature of the elements; we must, however, await the full development of his theory before pronouncing judgment on it. What we can at present only be certain of is this: that attempts like the two above named must be repeated and multiplied, because the periodic law has clearly shown that the masses of the atoms increase abruptly, by steps, which are clearly connected in some way with Dalton's law of multiple proportions; and because the periodicity of the elements finds expression in the transition from RX to RX_2, RX_3, RX_4, and so on till RX_8, at which point, the energy of the combining forces being exhausted, the series begins anew from RX to RX_2, and so on.

While connecting by new bonds the theory of the chemical elements with Dalton's theory of multiple proportions, or atomic structure of bodies, the periodic law opened for natural philosophy a new and wide field for speculation. Kant said that there are in the world 'two things which never cease to call for the admiration and reverence of man: the moral law within ourselves, and the stellar sky above us.' But when we turn our thoughts towards the nature of the elements and the periodic law, we must add a third subject, namely, 'the nature of the elementary individuals which we discover everywhere around us.' Without them the stellar sky itself is inconceivable; and in the atoms we see at once their peculiar individualities, the infinite multiplicity of the individuals, and the submission of their seeming freedom to the general harmony of Nature.

Having thus indicated a new mystery of Nature, which does not yet yield to rational conception, the periodic law, together with the revelations of spectrum analysis, have contributed to again revive an old but remarkably long-lived hope—that of discovering, if not by experiment, at least by a mental effort, the *primary matter*—which had its genesis in the minds of the Grecian philosophers, and has been transmitted, together with many other ideas of the classic period, to the heirs of their civilisation. Having grown, during the times of the alchemists up to the period when experimental proof was required, the idea has rendered good service; it induced those careful observations and experiments which later on called into being the works of Scheele, Lavoisier, Priestley, and Cavendish. It then slumbered awhile, but was soon awakened by the attempts either to confirm or to refute the ideas of Prout as to the multiple proportion relationship of the atomic weights of all the elements. And once again the inductive or experimental method of studying Nature gained a direct advantage from the old Pythagorean idea: because atomic weights were determined with an accuracy formerly unknown. But again the idea could not stand the ordeal of experimental test, yet the prejudice remains and has not been uprooted, even by Stas; nay, it has gained a new vigour, for we see that all which is imperfectly worked out, new and unexplained, from the still scarcely studied rare metals to the hardly perceptible nebulae, have been used to justify it. As soon as spectrum analysis appears as a new and powerful weapon of chemistry, the idea of a primary matter is immediately attached to it. From all sides we see attempts to constitute the imaginary substance *helium*[3] the so much longed for primary matter. No attention is paid to the circumstance that the helium line is only seen in the spectrum of the solar protuberances, so that its universality in Nature remains as problematic as the primary matter itself; nor to the fact that the helium line is wanting amongst the Fraunhofer lines of the solar spectrum, and thus does not answer to the brilliant fundamental conception which gives its real force to spectrum analysis.

And finally, no notice is even taken of the indubitable fact that the brilliancies of the spectral lines of the simple substances vary under different temperatures and pressures; so that all probabilities are in favour of the helium line simply belonging to some long since known element placed under such conditions of temperature, pressure, and gravity as have not yet been realised in our experiments. Again, the idea that the excellent investigations of Lockyer of the spectrum of iron can be interpreted in favour of the compound nature of that element, evidently must have arisen from some misunderstanding.

3 That is, a substance having a wave-length equal to 0.0005875 millimetre.

The spectrum of a compound certainly does not appear as a sum of the spectra of its components; and therefore the observations of Lockyer can be considered precisely as a proof that iron undergoes no other changes at the temperature of the sun than those which it experiences in the voltaic arc—provided the spectrum of iron is preserved. As to the shifting of some of the lines of the spectrum of iron while the other lines maintain their positions, it can be explained, as shown by M. Kleiber ('Journal of the Russian Chemical and Physical Society,' 1885, 147), by the relative motion of the various strata of the sun's atmosphere, and by Zöllner's laws of the relative brilliancies of different lines of the spectrum. Moreover, it ought not to be forgotten that if iron were really proved to consist of two or more unknown elements, we should simply have an increase in the number of our elements—not a reduction, and still less a reduction of all of them to one single primary matter.

Feeling that spectrum analysis will not yield a support to the Pythagorean conception, its modern promoters are so bent upon its being confirmed by the periodic law, that the illustrious Berthelot, in his work 'Les origines de l'Alchimie,' 1885, 313, has simply mixed up the fundamental idea of the law of periodicity with the ideas of Prout, the alchemists, and Democritus about primary matter.[4] But the periodic law, based as it is on the solid and wholesome ground of experimental research, has been evolved independently of any conception as to the nature of the elements; it does not in the least originate in the idea of a unique matter; and it has no historical connection with that relic of the torments of classical thought, and therefore it affords no more indication of the unity of matter or of the compound character of our elements, than the law of Avogadro, or the law of specific heats, or even the conclusions of spectrum analysis. None of the advocates of a unique matter have ever tried to explain the law from the standpoint of ideas taken from a remote antiquity when it was found convenient to admit the existence of many gods—and of a unique matter.

When we try to explain the origin of the idea of a unique primary matter, we easily trace that in the absence of inductions from experiment it derives its origin from the scientifically philosophical attempt at discovering some kind of unity in the immense diversity of individualities which we see around. In classical times such a tendency could only be satisfied by conceptions about the immaterial world. As to the material world, our ancestors were compelled to resort to some hypothesis, and they adopted the idea of unity in the formative material, because they were not able to evolve the conception of any other possible unity in order to connect the multifarious relations of matter. Responding to the same legitimate scientific tendency, natural science has discovered throughout the universe a unity of plan, a unity of forces, and a unity of matter, and the convincing conclusions of modern science compel every one to admit these kinds of unity. But while we admit unity in many things, we none the less must also explain the individuality and the apparent diversity which we cannot fail to trace everywhere. It has been said of old, 'Give us a fulcrum, and it will become easy to displace the earth.' So also we must say, 'Give us something that is individualised, and the apparent diversity will be easily understood.' Otherwise, how could unity result in a multitude?

After a long and painstaking research, natural science has discovered the individualities of the chemical elements, and therefore it is now capable not only of analysing, but also of synthesising; it can understand and grasp generality and unity, as well as the individualised and the multifarious. The general and universal, like time and space, like force and motion, vary uniformly; the uniform admit of interpolations, revealing every intermediate phase. But the multitudinous, the individualised—such as ourselves, or the chemical elements, or the members of a peculiar periodic function of the elements, or Dalton's multiple proportions—is characterised in another way: we see in it, side by side with a connecting general principle, leaps, breaks of continuity, points which escape from the analysis of the infi-

4 He maintains (on p. 309) that the periodic law requires two new analogous elements, having atomic weights of 48 and 64, occupying positions between sulphur and selenium, although nothing of the kind results from any of the different readings of the law.

nitely small—an absence of complete intermediate links. Chemistry has found an answer to the question as to the causes of multitudes; and while retaining the conception of many elements, all submitted to the discipline of a general law, it offers an escape from the Indian Nirvana—the absorption in the universal, replacing it by the individualised. However, the place for individuality is so limited by the all-grasping, all-powerful universal, that it is merely a point of support for the understanding of multitude in unity.

Having touched upon the metaphysical bases of the conception of a unique matter which is supposed to enter into the composition of all bodies I think it necessary to dwell upon another theory, akin to the above conception—the theory of the compound character of the elements now admitted by some—and especially upon one particular circumstance which, being related to the periodic law, is considered to be an argument in favour of that hypothesis.

Dr. Pelopidas, in 1883, made a communication to the Russian Chemical and Physical Society on the periodicity of the hydrocarbon radicles, pointing out the remarkable parallelism which was to be noticed in the change of properties of hydrocarbon radicles and elements when classed in groups. Professor Carnelley, in 1886, developed a similar parallelism. The idea of M. Pelopidas will be easily understood if we consider the series of hydrocarbon radicles which contain, say, 6 atoms of carbon:—

I.	II.	III.	IV.	V.	VI.	VII.	VIII.
C_6H_{13}	C_6H_{12}	C_6H_{11}	C_6H_{10}	C_6H_9	C_6H_8	C_6H_7	C_6H_6

The first of these radicles, like the elements of the 1st group, combines with Cl, OH, and so on, and gives the derivatives of hexyl alcohol, $C_6H_{13}(OH)$; but, in proportion as the number of hydrogen atoms decreases, the capacity of the radicles of combining with, say, the halogens increases. C_6H_{12} already combines with 2 atoms of chlorine; C_6H_{11} with 3 atoms, and so on. The last members of the series comprise the radicles of acids: thus C_6H_8, which belongs to the 6th group, gives, like sulphur, a bibasic acid, $C_6H_8O_2(OH)_2$, which is homologous with oxalic acid. The parallelism can be traced still further, because C_6H_5 appears as a monovalent radicle of benzene, and with it begins a new series of aromatic derivatives, so analogous to the derivatives of the aliphatic series. Let me also mention another example from among those which have been given by M. Pelopidas. Starting from the alkaline radicle of monomethylammonium, $N(CH_3)H_3$, or NCH_6, which presents many analogies with the alkaline metals of the 1st group, he arrives, by successively diminishing the number of the atoms of hydrogen, at a 7th group which contains cyanogen, CN, which has long since been compared to the halogens of the 7th group.

The most important consequence which, in my opinion, can be drawn from the above comparison is that the periodic law, so apparent in the elements, has a wider application than might appear at first sight; it opens up a new vista of chemical evolutions. But, while admitting the fullest parallelism between the periodicity of the elements and that of the compound radicles, we must not forget that in the periods of the hydrocarbon radicles we have a *decrease* of mass as we pass from the representatives of the first group to the next, while in the periods of the elements the mass *increases* during the progression. It thus becomes evident that we cannot speak of an identity of periodicity in both cases, unless we put aside the ideas of mass and attraction, which are the real corner-stones of the whole of natural science, and even enter into those very conceptions of simple substances which came to light a full hundred years later than the immortal principles of Newton.[5]

5 It is noteworthy that the year in which Lavoisier was born (1743)—the author of the idea of elements and of the indestructibility of matter—is later by exactly one century than the year in which the author of the theory of gravitation and mass was born (1643 N.S.). The affiliation of the ideas of Lavoisier and those of Newton is beyond doubt.

From the foregoing, as well as from the failures of so many attempts at finding in experiment and speculation a proof of the compound character of the elements and of the existence of primordial matter, it is evident, in my opinion, that this theory must be classed among mere utopias. But utopias can only be combated by freedom of opinion, by experiment, and by new utopias. In the republic of scientific theories freedom of opinions is guaranteed. It is precisely that freedom which permits me to criticise openly the widely-diffused idea as to the unity of matter in the elements. Experiments and attempts at confirming that idea have been so numerous that it really would be instructive to have them all collected together, if only to serve as a warning against the repetition of old failures. And now as to new utopias which may be helpful in the struggle against the old ones, I do not think it quite useless to mention a *phantasy* of one of my students who imagined that the weight of bodies does not depend upon their mass, but upon the character of the motion of their atoms. The atoms, according to this new utopian, may all be homogeneous or heterogeneous, we know not which; we know them in motion only, and that motion they maintain with the same persistence as the stellar bodies maintain theirs. The weights of atoms differ only in consequence of their various modes and quantity of motion; the heaviest atoms may be much simpler than the lighter ones: thus an atom of mercury may be simpler than an atom of hydrogen—the manner in which it moves causes it to be heavier. My interlocutor even suggested that the view which attributes the greater complexity to the lighter elements finds confirmation in the fact that the hydrocarbon radicles mentioned by Pelopidas, while becoming lighter as they lose hydrogen, change their properties periodically in the same manner as the elements change theirs, according as the atoms grow heavier.

The French proverb, *La critique est facile, mais l'art est difficile,* however, may well be reversed in the case of all such ideal views, as it is much easier to formulate than to criticise them. Arising from the virgin soil of newly-established facts, the knowledge relating to the elements, to their masses, and to the periodic changes of their properties has given a motive for the formation of utopian hypotheses, probably because they could not be foreseen by the aid of any of the various metaphysical systems, and exist, like the idea of gravitation, as an independent outcome of natural science, requiring the acknowledgment of general laws, when these have been established with the same degree of persistency as is indispensable for the acceptance of a thoroughly established fact. Two centuries have elapsed since the theory of gravitation was enunciated, and although we do not understand its cause, we still must regard gravitation as a fundamental conception of natural philosophy, a conception which has enabled us to perceive much more than the metaphysicians did or could with their seeming omniscience. A hundred years later the conception of the elements arose; it made chemistry what it now is; and yet we have advanced as little in our comprehension of simple substances since the times of Lavoisier and Dalton as we have in our understanding of gravitation. The periodic law of the elements is only twenty years old; it is not surprising, therefore, that, knowing nothing about the causes of gravitation and mass, or about the nature of the elements, we do not comprehend the *rationale* of the periodic law. It is only by collecting established laws—that is, by working at the acquirement of truth—that we can hope gradually to lift the veil which conceals from us the causes of the mysteries of Nature and to discover their mutual dependency. Like the telescope and the microscope, laws founded on the basis of experiment are the instruments and means of enlarging our mental horizon.

In the remaining part of my communication I shall endeavour to show, and as briefly as possible, in how far the periodic law contributes to enlarge our range of vision. Before the promulgation of this law the chemical elements were mere fragmentary, incidental facts in Nature; there was no special reason to expect the discovery of new elements, and the new ones which were discovered from time to time appeared to be possessed of quite novel properties. The law of periodicity first enabled us to perceive undiscovered elements at a distance which formerly was inaccessible to chemical vision; and long ere they were discovered new elements appeared before our eyes possessed of a number of well-defined properties. We now know three cases of elements whose existence and properties were foreseen by the instrumentality of the periodic law. I need but mention the brilliant discovery of *gallium*, which proved to correspond to eka-aluminium of the periodic law, by Lecoq de Boisbaudran; of *scandium*, corresponding to ekaboron, by Nilson; and of *germanium*, which proved to correspond in all respects to ekasilicon, by Winkler. When, in 1871, I described to the Russian Chemical Society the properties, clearly defined by the periodic law, which such elements ought to possess, I never hoped that I should live to mention their discovery to the Chemical Society of Great Britain as a confirmation of the exactitude and the generality of the periodic law. Now that I have had the happiness of doing so, I unhesitatingly say that, although greatly enlarging our vision, even now the periodic law needs further improvements in order that it may become a trustworthy instrument in further discoveries.[6]

I will venture to allude to some other matters which chemistry has discerned by means of its new instrument, and which it could not have made out without a knowledge of the law of periodicity, and I will confine myself to simple substances and to oxides.

Before the periodic law was formulated the atomic weights of the elements were purely empirical numbers, so that the magnitude of the equivalent, and the atomicity, or the value in substitution possessed by an atom, could only be tested by critically examining the methods of determination, but never directly by considering the numerical values themselves; in short, we were compelled to move in the dark, to submit to the facts, instead of being masters of them. I need not recount the methods which permitted the periodic law at last to master the facts relating to atomic weights, and I would merely call to mind that it compelled us to modify the valencies of *indium* and *cerium*, and to assign to their compounds a different molecular composition. Determinations of the specific heats of these two metals fully confirmed the change. The trivalency of *yttrium*, which makes us now represent its oxide as Y_2O_3 instead of as YO, was also foreseen (in 1870) by the periodic law, and it has now become so probable that Clève, and all other subsequent investigators of the rare metals, have not only adopted it, but have also applied it without any new demonstration to substances so imperfectly known as those of the cerite and gadolinite group, especially since Hillebrand determined the specific heats of lanthanum and didymium and confirmed the expectations suggested by the periodic law. But here, especially in the case of didymium, we meet with a series of difficulties long since foreseen through the periodic law, but only now becoming evident, and chiefly arising from the relative rarity and insufficient knowledge of the elements which usually accompany didymium.

Passing to the results obtained in the case of the rare elements *beryllium*, *scandium*, and *thorium*, it is found that these have many points of contact with the periodic law. Although Avdéeff long since proposed the magnesia formula to represent beryllium oxide, yet there was so much to be said in favour of the alumina formula, on account of the specific heat of the metals and the isomorphism of the two oxides, that it became generally adopted and seemed to be well established. The periodic law, however, as Brauner repeatedly insisted ('Berichte,' 1878, 872; 1881, 53), was against the formula Be_2O_3; it required the magnesia formula BeO—that is, an atomic

6 I foresee some more new elements, but not with the same certitude as before. I shall give one example, and yet I do not see it quite distinctly. In the series which contains Hg=204, Pb=206, and Bi=208, we can imagine the existence (at the place VI—11) of an element analogous to tellurium, which we can describe as dvi-tellurium, Dt, having an atomic weight of 212, and the property of forming the oxide DtO_3. If this element really exists, it ought in the free state to be an easily fusible, crystalline, non-volatile metal of a grey colour, having a density of about 9.3, capable of giving a dioxide, DtO_2, equally endowed with feeble acid and basic properties. This dioxide must give an active oxidation an unstable higher oxide, DtO_3, which should resemble in its properties PbO_2 and Bi_2O_5. Dvi-tellurium hydride, if it be found to exist, will be a less stable compound than even H_2Te. The compounds of dvi-tellurium will be easily reduced, and it will form characteristic definite alloys with other metals.

weight of 9—because there was no place in the system for an element like beryllium having an atomic weight of 13.5. This divergence of opinion lasted for years, and I often heard that the question as to the atomic weight of beryllium threatened to disturb the generality of the periodic law, or, at any rate, to require some important modifications of it. Many forces were operating in the controversy regarding beryllium, evidently because a much more important question was at issue than merely that involved in the discussion of the atomic weight of a relatively rare element: and during the controversy the periodic law became better understood, and the mutual relations of the elements became more apparent than ever before. It is most remarkable that the victory of the periodic law was won by the researches of the very observers who previously had discovered a number of facts in support of the trivalency of beryllium. Applying the higher law of Avogadro, Nilson and Petterson have finally shown that the density of the vapour of the beryllium chloride, $BeCl_2$, obliges us to regard beryllium as bivalent in conformity with the periodic law.[7] I consider the confirmation of Avdéeff's and Brauner's view as important in the history of the periodic law as the discovery of scandium, which, in Nilson's hands, confirmed the existence of ekaboron.

The circumstance that *thorium* proved to be quadrivalent, and Th=232, in accordance with the views of Chydenius and the requirements of the periodic law, passed almost unnoticed, and was accepted without opposition, and yet both thorium and uranium are of great importance in the periodic system, as they are its last members, and have the highest atomic weights of all the elements.

The alteration of the atomic weight of *uranium* from U=120 into U=240 attracted more attention, the change having been made on account of the periodic law, and for no other reason. Now that Roscoe, Rammelsberg, Zimmermann, and several others have admitted the various claims of the periodic law in the case of uranium, its high atomic weight is received without objection, and it endows that element with a special interest.

While thus demonstrating the necessity for modifying the atomic weights of several insufficiently known elements, the periodic law enabled us also to detect errors in the determination of the atomic weights of several elements whose valencies and true position among other elements were already well known. Three such cases are especially noteworthy: those of tellurium, titanium and platinum. Berzelius had determined the atomic weight of *tellurium* to be 128, while the periodic law claimed for it an atomic weight below that of iodine, which had been fixed by Stas at 126.5, and which was certainly not higher than 127. Brauner then undertook the investigation, and he has shown that the true atomic weight of tellurium is lower than that of iodine, being near to 125. For *titanium* the extensive researches of Thorpe have confirmed the atomic weight of Ti=48, indicated by the law, and already foreseen by Rose, but contradicted by the analyses of Pierre and several other chemists. An equally brilliant confirmation of the expectations based on the periodic law has been given in the case of the series osmium, iridium, platinum, and gold. At the time of the promulgation of the periodic law, the determinations of Berzelius, Rose, and many others gave the following figures:—

7 Let me mention another proof of the bivalency of beryllium which may have passed unnoticed, as it was only published in the Russian chemical literature. Having remarked (in 1884) that the density of such solutions of chlorides of metals, MCl_n, as contain 200 mols. of water (or a large and constant amount of water) regularly increases as the molecular weight of the dissolved salt increases, I proposed to one of our young chemists, M. Burdakoff, that he should investigate beryllium chloride. If its molecule be $BeCl_2$ its weight must be =80; and in such a case it must be heavier than the molecule of KCl=74.5, and lighter than that of $MgCl_2$=93. On the contrary, if beryllium chloride is a trichloride, $BeCl_3$=120, its molecule must be heavier than that of $CaCl_2$=111, and lighter than that of $MnCl_2$=126. Experiment has shown the correctness of the former formula, the solution $BeCl_2 + 200H_2O$ having (at 15°/ 4°) a density of 1.0138, this being a higher density than that of the solution $KCl + 200H_2O$ (=1.0121), and lower than that of $MgCl_2 + 200H_2O$ (=1.0203). The bivalency of beryllium was thus confirmed in the case both of the dissolved and the vaporised chloride.

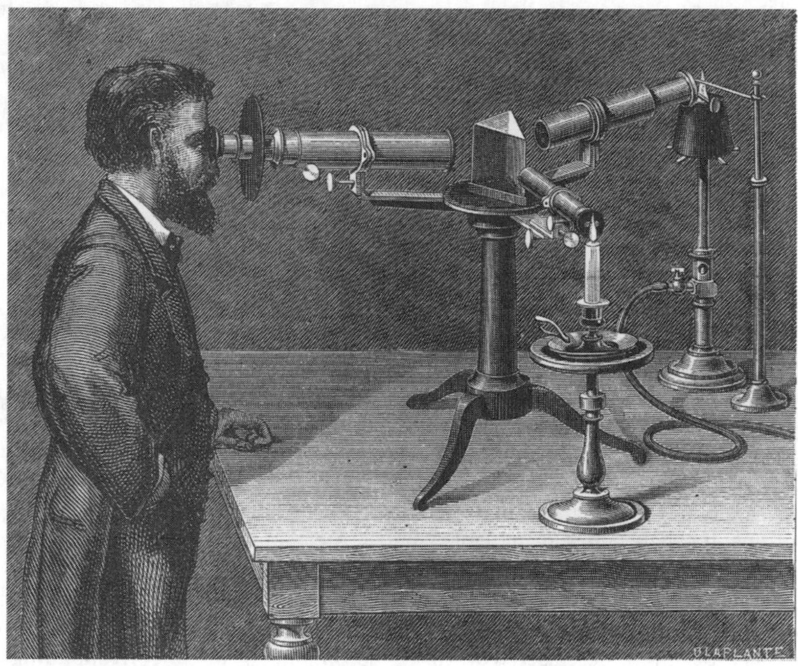

Man using a simple single-prism spectroscope perfected by Robert Bunsen in 1859 from the work of Wollaston earlier in the century. Different elements and compounds produce a characteristic spectrum of light and dark coloured bands which can be used to identify the chemical composition of unknown materials. From an article on spectroscopes by J. Norman Lockyer in *Nature,* 23 January 1873, p. 227.

Huggins' star spectroscope illustrated in *Nature,* 23 January 1873, p. 227.

Os = 200; Ir = 197; Pt = 198; Au = 196.

The expectations of the periodic law[8] have been confirmed, first, by new determinations of the atomic weight of *platinum* (by Seubert, Dittmar, and M'Arthur), which proved to be near to 196 (taking O=16, as proposed by Marignac, Brauner, and others); secondly, by Seubert having proved that the atomic weight of *osmium* is really lower than that of platinum, being near to 191; and thirdly, by the investigations of Krüss, Thorpe and Laurie, proving that the atomic weight of *gold* exceeds that of platinum, and approximates to 197. The atomic weights which were thus found to require correction were precisely those which the periodic law had indicated as affected with errors; and it has been proved, therefore, that the periodic law affords a means of testing experimental results. If we succeed in discovering the exact character of the periodic relationships between the increments in atomic weights of allied elements discussed by Ridberg in 1885, and again by Bazaroff in 1887, we may expect that our instrument will give us the means of still more closely controlling the experimental data relating to atomic weights.

Let me next call to mind that, while disclosing the variation of chemical properties,[9] the periodic law has also enabled us to systematically discuss many of the physical properties of elementary bodies, and to show that these properties are also subject to the law of periodicity. At the Moscow Congress of Russian Naturalists in August, 1869, I dwelt upon the relations which existed between density and the atomic weight of the elements. The following year Professor Lothar Meyer, in his well-known paper,[10] studied the same subject in more detail, and thus contributed to spread information about the periodic law. Later on, Carnelley, Laurie, L. Meyer, Roberts-Austen, and several others applied the periodic system to represent the order in the changes of the magnetic properties of the elements, their melting points, the heats of formation of their haloid compounds, and even of such mechanical properties as the co-efficient of elasticity, the breaking stress, etc., etc. These deductions, which have received further support in the discovery of new elements endowed not only with chemical but even with physical properties, which were foreseen by the law of periodicity, are well known; so I need not dwell upon the subject, and may pass to the consideration of oxides.[11]

In indicating that the gradual increase of the power of elements of combining with oxygen is accompanied by a corresponding decrease in their power of combining with hydrogen, the periodic law has shown that there is a limit of oxidation, just as there is a well-known limit to the capacity of elements for combining with hydrogen. A single atom of an element combines with at most four atoms of either hydrogen or oxygen; and while CH_4 and SiH_4 represent the highest hydrides, so RuO_4 and OsO_4 are the high-

8 I pointed them out in the *Liebig's Annalen*, Supplement Band., viii. 1871, p. 211.

9 Thus, in the typical small period of

Li, Be, B, C, N, O, F,

we see at once the progression from the alkali metals to the acid non-metals, such as are the halogens.

10 *Liebig's Annalen*, Supplement Band., vii. 1870.

11 A distinct periodicity can also be discovered in the spectra of the elements. Thus the researches of Hartley, Ciamician, and others have disclosed, first, the homology of the spectra of analogous elements: secondly, that the alkali metals have simpler spectra than the metals of the following groups; and thirdly, that there is a certain likeness between the complicated spectra of manganese and iron on the one hand, and the no less complicated spectra of chlorine and bromine on the other hand, and their likeness corresponds to the degree of analogy between those elements which is indicated by the periodic law.

est oxides. We are thus led to recognise types of oxides, just as we have had to recognise types of hydrides.[12]

The periodic law has demonstrated that the maximum extent to which different non-metals enter into combination with oxygen is determined by the extent to which they combine with hydrogen, and that the sum of the number of equivalents of both must be equal to 8. Thus chlorine, which combines with 1 atom or 1 equivalent of hydrogen, cannot fix more than 7 equivalents of oxygen, giving Cl_2O_7; while sulphur, which fixes 2 equivalents of hydrogen, cannot combine with more than 6 equivalents or 3 atoms of oxygen. It thus becomes evident that we cannot recognise as a fundamental property of the elements the atomic valencies deduced from their hydrides; and that we must modify, to a certain extent, the theory of atomicity if we desire to raise it to the dignity of a general principle capable of affording an insight into the constitution of all compound molecules. In other words, it is only to carbon, which is quadrivalent with regard both to oxygen and hydrogen, that we can apply the theory of constant valency and of bond, by means of which so many still endeavour to explain the structure of compound molecules. But I should go too far if I ventured to explain in detail the conclusions which can be drawn from the above considerations. Still, I think it necessary to dwell upon one particular fact which must be explained from the point of view of the periodic law in order to clear the way to its extension in that particular direction.

The higher oxides yielding salts the formation of which was foreseen by the periodic system—for instance, in the short series beginning with sodium—

Na_2O, MgO, Al_2O_3, SiO_2, P_2O_5, SO_3, Cl_2O_7,

must be clearly distinguished from the higher degrees of oxidation which correspond to hydrogen peroxide and bear the true character of peroxides. Peroxides such as Na_2O_2, BaO_2, and the like have long been known. Similar peroxides have also recently become known in the case of chromium, sulphur, titanium, and many other elements, and I have sometimes heard it said that discoveries of this kind weaken the conclusions of the periodic law in so far as it concerns the oxides. I do not think so in the least, and I may remark, in the first place, that all these peroxides are endowed with certain properties obviously common to all of them, which distinguish them from the actual, higher, salt-forming oxides, especially their easy decomposition by means of simple contact agencies; their incapability of forming salts of the common type; and their capability of combining with other peroxides (like the faculty which hydrogen peroxide possesses of combining with barium peroxide, discovered by Schoene). Again, we remark that some groups are especially characterised by their capacity of generating peroxides. Such is, for instance, the case in the sixth group, where we find the well-known peroxides of sulphur, chromium, and uranium; so that further investigation of peroxides will

12 Formerly it was supposed that, being a bivalent element, oxygen can enter into any grouping of the atoms, and there was no limit foreseen as to the extent to which it could further enter into combination. We could not explain why bivalent sulphur, which forms compounds such as

$$\mathrm{S}{<}\substack{\mathrm{O}\\ \mathrm{O}}{>} \text{ and } \mathrm{S}{<}\substack{\mathrm{O}\\ \mathrm{O}}{>}\mathrm{O},$$

could not also form oxides such as—

$$\mathrm{S}{<}\substack{\mathrm{O-O}\\ \mathrm{O-O}}{>} \text{ or } \mathrm{S}{<}\substack{\mathrm{O-O}\\ \mathrm{O-O}}{>}\mathrm{O},$$

while other elements, as, for instance, chlorine, form compounds such as—

Cl–O–O–O–O–K

probably establish a new periodic function, foreshadowing that molybdenum and tungsten will assume peroxide forms with comparative readiness. To appreciate the constitution of such peroxides, it is enough to notice that the peroxide form of sulphur (so-called persulphuric acid) stands in the same relation to sulphuric acid as hydrogen peroxide stands to water:—

$H(OH)$, or H_2O, responds to $(OH)(OH)$, or H_2O_2,

and so also—

$H(HSO_4)$, or H_2SO_4, responds to
$(HSO_4)(HSO_4)$, or $H_2S_2O_8$.

Similar relations are seen everywhere, and they correspond to the principle of substitutions which I long since endeavoured to represent as one of the chemical generalisations called into life by the periodic law. So also sulphuric acid, if considered with reference to hydroxyl, and represented as follows—

$$HO(SO_2OH),$$

has its corresponding compound in dithionic acid—

$$(SO_2OH)(SO_2OH), \text{ or } H_2S_2O_6.$$

Therefore, also, phosphoric acid, $HO(POH_2O_2)$, has, in the same sense, its corresponding compound in the subphosphoric acid of Saltzer:—

$$(POH_2O_2)(POH_2O_2), \text{ or } H_4P_2O_6;$$

and we must suppose that the peroxide compound corresponding to phosphoric acid, if it be discovered, will have the following structure:–

$$(H_2PO_4)_2 \text{ or } H_4P_2O_8 = 2H_2O + 2PO_3.$$ [13]

So far as is known at present, the highest form of peroxides is met with in the peroxide of uranium, UO_4, prepared by Fairley;[14] while OsO_4 is the highest oxide giving salts. The line of argument which is inspired by the periodic law, so far from being weakened by the discovery of peroxides, is thus actually strengthened, and we must hope that a further exploration of the region under consideration will confirm the applicability to chemistry generally of the principles deduced from the periodic law.

Permit me now to conclude my rapid sketch of the oxygen compounds by the observation that the periodic law is especially brought into evidence in the case of the oxides which constitute the immense majority of bodies at our disposal on the surface of the earth.

The oxides are evidently subject to the law, both as regards their chemical and their physical properties, especially if we take into account the cases of polymerism which are so obvious when comparing CO_2 with Si_nO_{2n}. In order to prove this I give the densities s and the specific volumes v of the higher oxides of two short periods. To render comparison easier, the oxides are all represented as of the form R_2O_n. In the column headed Δ the differences are given between the volume of the oxygen com-

13 In this sense, oxalic acid, $(COOH)_2$, also corresponds to carbonic acid, $OH(COOH)$, in the same way that dithionic acid corresponds to sulphuric acid, and subphosphoric acid to phosphoric; hence, if a peroxide corresponding to carbonic acid be obtained, it will have the structure of $(HCO_3)_2$, or $H_2C_2O_6 = H_2O + C_2O_5$. So also lead must have a real peroxide, Pb_2O_5.

14 The compounds of uranium prepared by Fairley seem to me especially instructive in understanding the peroxides. By the action of hydrogen peroxide on uranium oxide, UO_3, a peroxide of uranium, $UO_4,4H_2O$, is obtained (U=240) if the solution be acid; but if hydrogen peroxide act on uranium oxide in the presence of caustic soda, a crystalline deposit is obtained which has the composition $Na_4UO_8,4H_2O$, and evidently is a combination of sodium peroxide, Na_2O_2, with uranium peroxide, UO_4. It is possible that the former peroxide, $UO_4,4H_2O$, contains the elements of hydrogen peroxide and uranium peroxide, U_2O_7, or even $U(OH)_6,H_2O_2$, like the peroxide of tin recently discovered by Spring, which has the constitution Sn_2O_5,H_2O_2.

pound and that of the parent element, divided by *n*—that is, by the number of atoms of oxygen in the compound:—[15]

	s.	*v.*	Δ
Na_2O	2.6	24	–22
Mg_2O_2	3.6	22	–3
Al_2O_3	4.0	26	+1.3
Si_2O_4	2.65	45	5.2
P_2O_5	2.39	59	6.2
S_2O_6	1.96	82	8.7
K_2O	2.7	35	–55
Ca_2O	3.15	36	–7
Sc_2O_3	3.86	35	0
Li_2O_4	4.2	38	+5
V_2O_5	3.49	52	6.7
Cr_2O_6	2.74	73	9.5

I have nothing to add to these figures, except that like relations appear in other periods as well. The above relations were precisely those which made it possible for me to be certain that the relative density of ekasilicon oxide would be about 4.7; germanium oxide, actually obtained by Winkler, proved, in fact, to have the relative density 4.703.

The foregoing account is far from being an exhaustive one of all that has already been discovered by means of the periodic law telescope in the boundless realms of chemical evolution. Still less is it an exhaustive account of all that may yet be seen, but I trust that the little which I have said will account for the philosophical interest attached in chemistry to this law. Although but a recent scientific generalisation, it has already stood the test of laboratory verification, and appears as an instrument of thought which has not yet been compelled to undergo modification; but it needs not only new applications, but also improvements, further development, and plenty of fresh energy. All this will surely come, seeing that such an assembly of men of science as the Chemical Society of Great Britain has expressed the desire to have the history of the periodic law described in a lecture dedicated to the glorious name of Faraday.

FURTHER READING

Atkins, P.W. *The Periodic Kingdom: A Journey into the Land of the Chemical Elements.* New York: Basic Books, 1995.

Bensaude-Vincent, Bernadette. "Mendeleev's Periodic System of the Chemical Elements." *British Journal for the History of Science* 19 (1986): 3–17.

Brush, Stephen G. "The Reception of Mendeleev's Periodic Law in America and Britain." *Isis* 87 (1996): 595–628.

Danzer, Klaus. *Dmitri I. Mendelejew und Lothar Meyer: Die Schöpfer des Periodensystems der chemischen Elemente.* Leipzig: B.G. Teubner, 1974.

D.I. Mendeleev (1834–1907) und sein wissenschaftliches Werk. Rostocker Wissenschaftshistorische Manuskripte, Heft 11. Rostock: Wilhelm-Pieck-Universität Rostock, 1985.

Farber, Eduard, ed. *Great Chemists.* New York: Interscience Publishers, 1961.

Kolodkine, Paul. *Mendeléïèv.* Paris: Éditions Seghers, 1963.

Mendeleev, D.I. *An Attempt Towards a Chemical Conception of the Ether.* Translated by George Kamensky. London: Longmans, Green, and Co., 1904.

Pisarzhevsky, O.N. *Dmitry Ivanovich Mendeleyev: His Life and Work.* Moscow: Foreign Languages Publishing House, 1954.

Posin, Daniel Q. *Mendeleyev: The Story of a Great Scientist.* New York: McGraw-Hill Book Company, Inc., 1948.

Rabinowitsch, Eugen, and Erich Thilo. *Periodisches System: Geschichte und Theorie.* Stuttgart: Ferdinand Enke, 1930.

Seshadri, T.R., ed. *Mendeleyev's Periodic Classification of Elements and its Applications.* Delhi: Hindustan Publishing Corporation, 1973.

van Spronsen, Johannes Willem. *The Periodic System of Chemical Elements: A History of the First Hundred Years.* Amsterdam: Elsevier Publishing Co., 1969.

Thorpe, Sir Edward. *Essays in Historical*

15 Δ thus represents the average increase of volume for each atom of oxygen contained in the higher salt-forming oxide. The acid oxides give, as a rule, a higher value of Δ, while in the case of the strongly alkaline oxides its value is usually negative.

Chemistry. London: Macmillan and Co., Ltd., 1894.

Tilden, Sir William A. *Famous Chemists: The Men and Their Work.* London: George Routledge and Sons, Ltd., 1921.

Vucinich, Alexander. *Science in Russian Culture 1861–1917.* Stanford: Stanford UP, 1970.

Weeks, Mary Elvira. *Discovery of the Elements.* Revised by Henry M. Leicester. Easton, PA: Journal of Chemical Education, 1968.

31. WILLIAM JAMES

(1842–1910)

William James's grandfather, William James of Albany (1771–1832), amassed a considerable fortune as a merchant and land speculator in New York, which allowed his son Henry James, Sr. the leisure to study philosophy and Swedenborgian theology. Thus Henry's son William James the younger grew up in an atmosphere of independent wealth in which art, literature, and philosophy were liberally cultivated. These early influences, along with his own struggle to define himself, led to James's deep interest in human behaviour, the nature of the self, and psychological motivation, interests he shared with his expatriate brother, the novelist Henry James, Jr.

James agonized over a future career before beginning the study of chemistry at Harvard in 1861, later switching to medicine in 1864. While at Harvard, he accompanied Louis Agassiz on his 1865 Amazon journey, but his obvious impatience with the natural sciences is revealed in this passage from a letter back home: "although several bushels of different things have already been collected, *nothing* has been done which could not have been done just as well from Boston" (quoted in Feinstein 172). After his return from Brazil, he often joined Chauncey Wright, Oliver Wendell Holmes, and Charles S. Pierce in Cambridge for philosophical discussions. James obtained his M.D. in 1869 from Harvard after a series of illnesses, with periods of depression and self-doubt which led to a serious spiritual crisis and near suicide in the late 1860s and early 1870s. In 1873, he began teaching physiology at Harvard.

James's seemingly erratic career at Harvard from professor of physiology in 1876 to professor of psychology in 1880, and then finally to professor of philosophy in 1885 parallels the restless movement of his most important book, *The Principles of Psychology* (1890). James consciously set out to define the problems of mental functioning from both biological and philosophical perspectives in the *Principles*, although he admitted to the impossibility of experimentally testing some psychological phenomena. As James wrote in 1892 in "A Plea for Psychology as a 'Natural Science'" in the *Philosophical Review*: "I wished, by treating Psychology *like* a natural science, to help her to become one" (*Essays in Psychology* 146). Although psychology—literally 'knowledge of the soul'—only first arose as a separate discipline from philosophy, medicine, or theology in the nineteenth century, the questions which James analyzes in the *Principles* had been examined in detail since classical times in treatises entitled *De anima* and later by theologians and philosophers such as Locke, Hume, Kant, Stewart, and Hartley.

The *Principles* is unique from earlier attempts at codifying the field of psychology such as Thomas Upham's *Elements of Mental Philosophy* (1831), G.H. Lewes's *Problems of Life and Mind* (1874–79), or Herbert Spencer's *Principles of Psychology* (1855)—which James himself used as a textbook—in that James could speak from both his medical training as a materialist physiologist and from his reading in philosophy and literature as a metaphysical philosopher on both sides of the mind/body dualism. Numerous critics have called attention to the paradoxes, contradictions, and equivocations of the unsystematic *Principles*, but many feel that these aspects of James's prose enrich rather than detract from his thought. James's *Principles* does provide the best summary and critique of previous work in the field written up to his time. James examined three prevalent theories of psychology in the *Principles*: 1) the mind-stuff theory—or complete mechanical materialism, stating that ideas act like material atoms; 2) associationism, defined below; and 3) transcendentalism, which believed in a knowing "Ego" or the traditional immortal soul as the source of mental states. In his critique of psychology, James dissented from the prevailing British associationist psychology of John Locke, David Hume, David Hartley, and John Stuart Mill which viewed mental functioning as a series of individual ideas, originating in sensation and strung together in the mind usually in some sort of logical or even mechanical pattern. In associationism, simple ideas "built up" or added together formed complex ideas. Without abandoning associationism, James argued that these connections among thoughts were not necessarily uniform, simple, or even rational.

James on the other hand emphasized the continuity and totality of mental experience and impressions, and employed the terms "stream of thought" and "stream of consciousness" to characterize the state of the working mind. Not surprisingly at this time, stream of consciousness literary techniques were being employed in literature, such as Édouard Dujardin's novel *Les Lauriers sont coupés* (1888) and later in works by James Joyce, William Faulkner, Dorothy Richardson, and Virginia Woolf.

James's *Principles of Psychology* begins with a survey of brain physiology, demonstrating the influence of Claude Bernard and Wilhelm Wundt's *Principles of Physiological Psychology* (1874). He was uncomfortable with pure experimental psychology, however, and argued against the epiphenomenalism of Huxley, the view that mental states are simply an effect of the physiological functioning of the brain, not free and self-motivating acts.

In *Pragmatism: A New Name for Some Old Ways of Thinking* (1907), James, borrowing from the terminology of C.S. Pierce, brought forth a theory of reality grounded in actual processes and experience, although not necessarily in opposition to metaphysics in the way empiricism traditionally has been. The central idea of pragmatism, implicit in many of James's earlier works, is that concepts and truths must be evaluated according to how effectively they operate in the world. James examined religion closely in his 1902 book *The Varieties of Religious Experience*, concluding similarly that religious faith must be judged according to its efficacious effects.

James's later writings blurred his original belief in the dualism (mental states/physical objects; mind/ body; Subject/ Object) which he believed essential for philosophy and psychology, and in the posthumously published *Essays in Radical Empiricism* (1912) he discussed the philosophy of radical empiricism, suggesting that consciousness could be reduced to "pure experience." James summarized his final philosophy as: "the generalized conclusion is that therefore *the parts of experience hold together from next to next by relations that are themselves parts of experience. The directly apprehended universe needs, in short, no extraneous trans-empirical connective support, but possesses in its own right a concatenated or continuous structure*" (*Radical Empiricism* xii). The monism of radical empiricism suggests that for James mind and body both formed part of a more primary unitary essence, "pure experience." The complete works of James have appeared from Harvard University Press under the direction of Frederick H. Burkhardt, Fredson Bowers, and Ignas K. Skrupskelis (1975–88).

The Principles of Psychology.
New York: Henry Holt and Co., 1890, pp. 183–198.

THE METHODS AND SNARES OF PSYCHOLOGY

We have now finished the physiological preliminaries of our subject and must in the remaining chapters study the mental states themselves whose cerebral conditions and concomitants we have been considering hitherto. Beyond the brain, however, there is an outer world to which the brain-states themselves 'correspond.' And it will be well, ere we advance farther, to say a word about the relation of the mind to this larger sphere of physical fact.

PSYCHOLOGY IS A NATURAL SCIENCE.

That is, the mind which the psychologist studies is the mind of distinct individuals inhabiting definite portions of a real space and of a real time. With any other sort of mind, absolute Intelligence, Mind unattached to a particular body, or Mind not subject to the course of time, the psychologist as such has nothing to do. 'Mind,' in his mouth, is only a class name for *minds*. Fortunate will it be if his more modest inquiry result in any generalizations which the philosopher devoted to absolute Intelligence as such can use.

To the psychologist, then, the minds he studies are *objects*, in a world of other objects. Even when he introspectively analyzes his own mind, and tells what he finds there, he talks about it in an objective way. He says, for instance, that under certain circumstances the color gray appears to him green, and calls the appearance an illusion. This implies that he compares two objects, a real color seen under certain conditions, and a mental perception which he believes to represent it, and that he declares the relation between them to be of a certain kind. In making this critical judgment, the psychologist stands as much outside of the perception which he criticises as he does of the color. Both are his objects. And if this is true of him when he reflects on his own conscious states, how much truer is it when he treats of those of others! In German philosophy since Kant the word *Erkenntnisstheorie*, criticism of the faculty of knowledge, plays a great part. Now the psychologist necessarily becomes such an *Erkenntnisstheoretiker*. But the knowledge he theorizes about is not the bare function of knowledge which Kant criticises—he does not inquire into the possibility of knowledge *überhaupt*. He assumes it to be possible, he does not doubt its presence in himself at the moment he speaks. The knowledge he criticises is the knowledge of particular men about the particular things that surround them. This he may, upon occasion, in the light of his *own* unquestioned knowledge, pronounce true or false, and trace the reasons by which it has become one or the other.

It is highly important that this natural-science point of view should be understood at the outset. Otherwise more may be demanded of the psychologist than he ought to be expected to perform.

A diagram will exhibit more emphatically what the assumptions of Psychology must be:

1 The Psychologist	2 The Thought Studied	3 The Thought's Object	4 The Psychologist's Reality

These four squares contain the irreducible data of psychology. No. 1, the psychologist, believes Nos. 2, 3, and 4, which together form *his* total object, to be realities, and reports them and their mutual relations as truly as he can without troubling himself with the puzzle of how he can report them at all. About such *ultimate* puzzles he in the main need trouble himself no more than the geometer, the chemist, or the botanist do, who make precisely the same assumptions as he.[1]

Of certain fallacies to which the psychologist is

1 On the relation between Psychology and General Philosophy, see G. C. Robertson, 'Mind,' vol. VIII. p. 1, and J. Ward, *ibid.* p. 153; J. Dewey, *ibid.* vol. IX. p. 1.

exposed by reason of his peculiar point of view—that of being a reporter of subjective as well as of objective facts, we must presently speak. But not until we have considered the methods he uses for ascertaining what the facts in question are.

THE METHODS OF INVESTIGATION.

Introspective Observation is what we have to rely on first and foremost and always. The word introspection need hardly be defined—it means, of course, the looking into our own minds and reporting what we there discover. *Every one agrees that we there discover states of consciousness.* So far as I know, the existence of such states has never been doubted by any critic, however sceptical in other respects he may have been. That we have *cogitations* of some sort is the *inconcussum* in a world most of whose other facts have at some time tottered in the breath of philosophic doubt. All people unhesitatingly believe that they feel themselves thinking, and that they distinguish the mental state as an inward activity or passion, from all the objects with which it may cognitively deal. *I regard this belief as the most fundamental of all the postulates of Psychology*, and shall discard all curious inquiries about its certainty as too metaphysical for the scope of this book.

A Question of Nomenclature. We ought to have some general term by which to designate all states of consciousness merely as such, and apart from their particular quality or cognitive function. Unfortunately most of the terms in use have grave objections. 'Mental state,' 'state of consciousness,' 'conscious modification,' are cumbrous and have no kindred verbs. The same is true of 'subjective condition.' 'Feeling' has the verb 'to feel,' both active and neuter, and such derivatives as 'feelingly,' 'felt,' 'feltness,' etc., which make it extremely convenient. But on the other hand it has specific meanings as well as its generic one, sometimes standing for pleasure and pain, and being sometimes a synonym of '*sensation*' as opposed to *thought*; whereas we wish a term to cover sensation and thought indifferently. Moreover, 'feeling' has acquired in the hearts of platonizing thinkers a very opprobrious set of implications; and since one of the great obstacles to mutual understanding in philosophy is the use of words eulogistically and disparagingly, impartial terms ought always, if possible, to be preferred. The word *psychosis* has been proposed by Mr. Huxley. It has the advantage of being correlative to *neurosis* (the name applied by the same author to the corresponding nerve-process), and is moreover technical and devoid of partial implications. But it has no verb or other grammatical form allied to it. The expressions 'affection of the soul,' 'modification of the ego,' are clumsy, like 'state of consciousness,' and they implicitly assert theories which it is not well to embody in terminology before they have been openly discussed and approved. 'Idea' is a good vague neutral word, and was by Locke employed in the broadest generic way; but notwithstanding his authority it has not domesticated itself in the language so as to cover bodily sensations, and it moreover has no verb. 'Thought' would be by far the best word to use if it could be made to cover sensations. It has no opprobrious connotation such as 'feeling' has, and it immediately suggests the omnipresence of cognition (or reference to an object other than the mental state itself), which we shall soon see to be of the mental life's essence. But can the expression 'thought of a toothache' ever suggest to the reader the actual present pain itself? It is hardly possible; and we thus seem about to be forced back on some *pair* of terms like Hume's 'impression and idea,' or Hamilton's 'presentation and representation,' or the ordinary 'feeling and thought,' if we wish to cover the whole ground.

In this quandary we can make no definitive choice, but must, according to the convenience of the context, use sometimes one, sometimes another of the synonyms that have been mentioned. *My own partiality is for either* FEELING *or* THOUGHT. I shall probably often use both words in a wider sense than usual, and alternately startle two classes of readers by their unusual sound; but if the connection makes it clear that mental states at large, irrespective of their kind, are meant, this will do no harm, and may even do some good.[2]

---

2 Compare some remarks in Mill's Logic, bk. I. chap. III. §§ 2, 3.

The inaccuracy of introspective observation has been made a subject of debate. It is important to gain some fixed ideas on this point before we proceed.

The commonest spiritualistic opinion is that the Soul or *Subject* of the mental life is a metaphysical entity, inaccessible to direct knowledge, and that the various mental states and operations of which we reflectively become aware are objects of an inner sense which does not lay hold of the real agent in itself, any more than sight or hearing gives us direct knowledge of matter in itself. From this point of view introspection is, of course, incompetent to lay hold of anything more than the Soul's *phenomena*. But even then the question remains, How well can it know the phenomena themselves?

Some authors take high ground here and claim for it a sort of infallibility. Thus Ueberweg:

"When a mental image, as such, is the object of my apprehension, there is no meaning in seeking to distinguish its existence in my consciousness (in me) from its existence out of my consciousness (in itself); for the object apprehended is, in this case, one which does not even exist, as the objects of external perception do, in itself outside of my consciousness. It exists only with me."[3]

And Brentano:

"The phenomena inwardly apprehended are true in themselves. As they appear—of this the evidence with which they are apprehended is a warrant—so they are in reality. Who, then, can deny that in this a great superiority of Psychology over the physical sciences comes to light?"

And again:

"No one can doubt whether the psychic condition he apprehends in himself *be*, and be *so*, as he apprehends it. Whoever should doubt this would have reached that *finished* doubt which destroys itself in destroying every fixed point from which to make an attack upon knowledge."[4]

Others have gone to the opposite extreme, and maintained that we can have no introspective cognition of our own minds at all. A deliverance of Auguste Comte to this effect has been so often quoted as to be almost classical; and some reference to it seems therefore indispensable here.

Philosophers, says Comte,[5] have

"in these latter days imagined themselves able to distinguish, by a very singular subtlety, two sorts of observation of equal importance, one external, the other internal, the latter being solely destined for the study of intellectual phenomena. ... I limit myself to pointing out the principal consideration which proves clearly that this pretended direct contemplation of the mind by itself is a pure illusion. ... It is in fact evident that, by an invincible necessity, the human mind can observe directly all phenomena except its own proper states. For by whom shall the observation of these be made? It is conceivable that a man might observe himself with respect to the *passions* that animate him, for the anatomical organs of passion are distinct from those whose function is observation. Though we have all made such observations on ourselves, they can never have much scientific value, and the best mode of knowing the passions will always be that of observing them from without; for every strong state of passion ... is necessarily incompatible with the state of observation. But, as for observing in the same way *intellectual* phenomena at the time of their actual presence, that is a manifest impossibility. The thinker cannot divide himself into two, of whom one reasons whilst the other observes him reason. The organ observed and the organ observing being, in this case, identical, how could observation take place? This pretended psychological method is then radically null and void. On the one hand, they advise you to isolate yourself, as far as possible, from every external sensation, especially every intellectual work,—for if you were to busy yourself even with the simplest calculation, what would become of *internal* observation?—on the other

3 Logic, § 40.

4 Psychologie, bk. II. chap. III. §§ 1, 2.

5 Cours de Philosophie Positive, I. 34–8.

hand, after having with the utmost care attained this state of intellectual slumber, you must begin to contemplate the operations going on in your mind, when nothing there takes place! Our descendants will doubtless see such pretensions some day ridiculed upon the stage. The results of so strange a procedure harmonize entirely with its principle. For all the two thousand years during which metaphysicians have thus cultivated psychology, they are not agreed about one intelligible and established proposition. '*Internal observation*' gives almost as many divergent results as there are individuals who think they practise it."

Comte hardly could have known anything of the English, and nothing of the German, empirical psychology. The 'results' which he had in mind when writing were probably scholastic ones, such as principles of internal activity, the faculties, the ego, the *liberum arbitrium indifferentiae*, etc. John Mill, in replying to him,[6] says:

"It might have occurred to M. Comte that a fact may be studied through the medium of memory, not at the very moment of our perceiving it, but the moment after: and this is really the mode in which our best knowledge of our intellectual acts is generally acquired. We reflect on what we have been doing when the act is past, but when its impression in the memory is still fresh. Unless in one of these ways, we could not have acquired the knowledge which nobody denies us to have, of what passes in our minds. M. Comte would scarcely have affirmed that we are not aware of our own intellectual operations. We know of our observings and our reasonings, either at the very time, or by memory the moment after; in either case, by direct knowledge, and not (like things done by us in a state of somnambulism) merely by their results. This simple fact destroys the whole of M. Comte's argument. Whatever we are directly aware of, we can directly observe."

Where now does the truth lie? Our quotation from Mill is obviously the one which expresses the most of *practical* truth about the matter. Even the writers who insist upon the absolute veracity of our immediate inner apprehension of a conscious state have to contrast with this the fallibility of our *memory* or *observation* of it, a moment later. No one has emphasized more sharply than Brentano himself the difference between the immediate *feltness* of a feeling, and its perception by a subsequent reflective act. But which mode of consciousness of it is that which the psychologist must depend on? If to *have* feelings or thoughts in their immediacy were enough, babies in the cradle would be psychologists, and infallible ones. But the psychologist must not only *have* his mental states in their absolute veritableness, he must report them and write about them, name them, classify and compare them and trace their relations to other things. Whilst alive they are their own property; it is only *post-mortem* that they become his prey.[7] And as in the naming, classing, and knowing of things in general we are notoriously fallible, why not also here? Comte is quite right in laying stress on the fact that a feeling, to be named, judged, or perceived, must be already past. No subjective state, whilst present, is its own object; its object is always something else. There are, it is true, cases in which we appear to be naming our present feeling, and so to be experiencing and observing the same inner fact at a single stroke, as when we say 'I feel tired,' 'I am angry,' etc. But these are illusory, and a little attention unmasks the illusion. The present conscious state, when I say 'I feel tired,' is not the direct state of tire; when I say 'I feel angry,' it is

6 Auguste Comte and Positivism, 3d edition (1882), p. 64.

7 Wundt says: "The first rule for utilizing inward observation consists in taking, as far as possible, experiences that are accidental, unexpected, and not intentionally brought about ... *First* it is best as far as possible to rely on *Memory* and not on immediate Apprehension . ... *Second*, internal observation is better fitted to grasp clearly conscious states, especially voluntary mental acts: such inner processes as are obscurely conscious and involuntary will almost entirely elude it, because the effort to observe interferes with them, and because they seldom abide in memory." (Logik, II. 432.)

not the direct state of anger. It is the state of *saying-I-feel-tired*, of *saying-I-feel-angry*,—entirely different matters, so different that the fatigue and anger apparently included in them are considerable modifications of the fatigue and anger directly felt the previous instant. The act of naming them has momentarily detracted from their force.[8]

The only sound grounds on which the infallible veracity of the introspective judgment might be maintained are empirical. If we had reason to think it has never yet deceived us, we might continue to trust it. This is the ground actually maintained by Herr Mohr.

"The illusions of our senses," says this author, "have undermined our belief in the reality of the outer world; but in the sphere of inner observation our confidence is intact, for we have never found ourselves to be in error about the reality of an act of thought or feeling. We have never been misled into thinking we were *not* in doubt or in anger when these conditions were really states of our consciousness."[9]

But sound as the reasoning here would be, were the premises correct, I fear the latter cannot pass. However it may be with such strong feelings as doubt or anger, about weaker feelings, and about the *relations to each other* of all feelings, we find ourselves in continual error and uncertainty so soon as we are called on to name and class, and not merely to feel. Who can be sure of the exact *order* of his feelings when they are excessively rapid? Who can be sure, in his sensible perception of a chair, how much comes from the eye and how much is supplied out of the previous knowledge of the mind? Who can compare with precision the *quantities* of disparate feelings even where the feelings are very much alike? For instance, where an object is felt now against the back and now against the cheek, which feeling is most extensive? Who can be sure that two given feelings are or are not exactly the same? Who can tell which is briefer or longer than the other when both occupy but an instant of time? Who knows, of many actions, for what motive they were done, or if for any motive at all? Who can enumerate all the distinct ingredients of such a complicated feeling as *anger*? and who can tell offhand whether or no a perception of *distance* be a compound or a simple state of mind? The whole mind-stuff controversy would stop if we could decide conclusively by introspection that what seem to us elementary feelings are really elementary and not compound.

Mr. Sully, in his work on Illusions, has a chapter on those of Introspection from which we might now quote. But, since the rest of this volume will be little more than a collection of illustrations of the difficulty of discovering by direct introspection exactly what our feelings and their relations are, we need not anticipate our own future details, but just state our general conclusion that *introspection is difficult and fallible; and that the difficulty is simply that of all observation of whatever kind*. Something is before us; we do our best to tell what it is, but in spite of our good will we may go astray, and give a description more applicable to some other sort of thing. The only safeguard is in the final *consensus* of our farther knowledge about the thing in question, later views correcting earlier ones, until at last the harmony of a consistent system is reached. Such a system, gradually worked out, is the best guarantee the psychologist can give for the soundness of any particular psychologic observation which he may report. Such a system we ourselves must strive, as far as may be, to attain.

The English writers on psychology, and the

8 In cases like this, where the state outlasts the act of naming it, exists before it, and recurs when it is past, we probably run little practical risk of error when we talk as if the state knew itself. The state of feeling and the state of naming the feeling are continuous, and the infallibility of such prompt introspective judgments is probably great. But even here the certainty of our knowledge ought not to be argued on the *a priori* ground that *percipi* and *esse* are in psychology the same. The states are really two; the naming state and the named state are apart; '*percipi* is *esse*' is not the principle that applies.

9 J. Mohr: Grundlage der Empirischen Psychologie (Leipzig, 1882), p. 47.

school of Herbart in Germany, have in the main contented themselves with such results as the immediate introspection of single individuals gave, and shown what a body of doctrine they may make. The works of Locke, Hume, Reid, Hartley, Stewart, Brown, the Mills, will always be classics in this line; and in Professor Bain's Treatises we have probably the last word of what this method taken mainly by itself can do—the last monument of the youth of our science, still untechnical and generally intelligible, like the Chemistry of Lavoisier, or Anatomy before the microscope was used.

The Experimental Method. But psychology is passing into a less simple phase. Within a few years what one may call a microscopic psychology has arisen in Germany, carried on by experimental methods, asking of course every moment for introspective data, but eliminating their uncertainty by operating on a large scale and taking statistical means. This method taxes patience to the utmost, and could hardly have arisen in a country whose natives could be *bored*. Such Germans as Weber, Fechner, Vierordt, and Wundt obviously cannot; and their success has brought into the field an array of younger experimental psychologists, bent on studying the *elements* of the mental life, dissecting them out from the gross results in which they are embedded, and as far as possible reducing them to quantitative scales. The simple and open method of attack having done what it can, the method of patience, starving out, and harassing to death is tried; the Mind must submit to a regular *siege*, in which minute advantages gained night and day by the forces that hem her in must sum themselves up at last into her overthrow. There is little of the grand style about these new prism, pendulum, and chronograph-philosophers. They mean business, not chivalry. What generous divination, and that superiority in virtue which was thought by Cicero to give a man the best insight into nature, have failed to do, their spying and scraping, their deadly tenacity and almost diabolic cunning, will doubtless some day bring about.

No general description of the methods of experimental psychology would be instructive to one unfamiliar with the instances of their application, so we will waste no words upon the attempt. *The principal fields of experimentation* so far have been: 1) the connection of conscious states with their physical conditions, including the whole of brain-physiology, and the recent minutely cultivated physiology of the sense-organs, together with what is technically known as 'psycho-physics,' or the laws of correlation between sensations and the outward stimuli by which they are aroused; 2) the analysis of space-perception into its sensational elements; 3) the measurement of the *duration* of the simplest mental processes; 4) that of the *accuracy of reproduction* in the memory of sensible experiences and of intervals of space and time; 5) that of the manner in which simple mental states *influence each other*, call each other up, or inhibit each other's reproduction; 6) that of the *number of facts* which consciousness can simultaneously discern; finally, 7) that of the elementary laws of oblivescence and retention. It must be said that in some of these fields the results have as yet borne little theoretic fruit commensurate with the great labor expended in their acquisition. But facts are facts, and if we only get enough of them they are sure to combine. New ground will from year to year be broken, and theoretic results will grow. Meanwhile the experimental method has quite changed the face of the science so far as the latter is a record of mere work done.

The *comparative method,* finally, supplements the introspective and experimental methods. This method presupposes a normal psychology of introspection to be established in its main features. But where the origin of these features, or their dependence upon one another, is in question, it is of the utmost importance to trace the phenomenon considered through all its possible variations of type and combination. So it has come to pass that instincts of animals are ransacked to throw light on our own; and that the reasoning faculties of bees and ants, the minds of savages, infants, madmen, idiots, the deaf and blind, criminals, and eccentrics, are all invoked in support of this or that special theory about some part of our own mental life. The history of sciences, moral and political institutions, and languages, as types of mental product, are pressed into the same service. Messrs. Darwin and Galton have set the example of circulars of questions sent out by the hundred to those supposed able to reply. The custom has spread, and it will be well for us in the next generation if such circulars be not ranked among the common pests of life. Meanwhile information grows, and results emerge. There are great sources of error in the comparative method. The interpretation of the 'psychoses' of animals, savages, and infants is necessarily wild work, in which the personal equation of the investigator has things very much its own way. A savage will be reported to have no moral or religious feeling if his actions shock the observer unduly. A child will be assumed without self-consciousness because he talks of himself in the third person, etc., etc. No rules can be laid down in advance. Comparative observations, to be definite, must usually be made to test some pre-existing hypothesis; and the only thing then is to use as much sagacity as you possess, and to be as candid as you can.

THE SOURCES OF ERROR IN PSYCHOLOGY.

The first of them arises from the Misleading Influence of Speech. Language was originally made by men who were not psychologists, and most men to-day employ almost exclusively the vocabulary of outward things. The cardinal passions of our life, anger, love, fear, hate, hope, and the most comprehensive divisions of our intellectual activity, to remember, expect, think, know, dream, with the broadest genera of aesthetic feeling, joy, sorrow, pleasure, pain, are the only facts of a subjective order which this vocabulary deigns to note by special words. The elementary qualities of sensation, bright, loud, red, blue, hot, cold, are, it is true, susceptible of being used in both an objective and a subjective sense. They stand for outer qualities and for the feelings which these arouse. But the objective sense is the original sense; and still to-day we have to describe a large number of sensations by the name of the object from which they have most frequently been got. An orange color, an odor of violets, a cheesy taste, a thunderous sound, a fiery smart, etc., will recall what I mean. This absence of a special vocabulary for subjective facts hinders the study of all but the very coarsest of them. Empiricist writers are very fond of emphasizing one great set of delusions which language inflicts on the mind. Whenever we have made a word, they say, to denote a certain group of phenomena, we are prone to suppose a substantive entity existing beyond the phenomena, of which the word shall be the name. But the *lack* of a word quite as often leads to the directly opposite error. We are then prone to suppose that no entity can be there; and so we come to overlook phenomena whose existence would be patent to us all, had we only grown up to hear it familiarly recognized in speech.[10] It is hard to focus our attention on the nameless, and so there results a certain vacuousness in the descriptive parts of most psychologies.

But a worse defect than vacuousness comes from the dependence of psychology on common speech. Naming our thought by its own objects, we almost all of us assume that as the objects are, so the thought must be. The thought of several distinct things can only consist of several distinct bits of thought, or 'ideas;' that of an abstract or universal object can only be an abstract or universal idea. As each object may come and go, be forgotten and then thought of again, it is held that the thought of it has a precisely similar independence, self-identity, and mobility. The thought of the object's recurrent identity is regarded as the identity of its recurrent thought; and the perceptions of multiplicity, of coexistence, of succession, are severally conceived to be brought about only through a multiplicity, a coexistence, a succession, of perceptions. The continuous flow of the mental stream is sacrificed, and in its place an atomism, a brickbat plan of construction, is preached, for the existence of which no good introspective grounds can be brought forward, and out of which presently

10 In English we have not even the generic distinction between the-thing-thought-of and the-thought-thinking-it, which in German is expressed by the opposition between *Gedachtes* and *Gedanke*, in Latin by that between *cogitatum* and *cogitatio*.

grow all sorts of paradoxes and contradictions, the heritage of woe of students of the mind.

These words are meant to impeach the entire English psychology derived from Locke and Hume, and the entire German psychology derived from Herbart, so far as they both treat 'ideas' as separate subjective entities that come and go. Examples will soon make the matter clearer. Meanwhile our psychologic insight is vitiated by still other snares.

'The Psychologist's Fallacy.' The *great* snare of the psychologist is the *confusion of his own standpoint with that of the mental fact* about which he is making his report. I shall hereafter call this the 'psychologist's fallacy' *par excellence*. For some of the mischief, here too, language is to blame. The psychologist, as we remarked above (p. 183), stands outside of the mental state he speaks of. Both itself and its object are objects for him. Now when it is a *cognitive* state (percept, thought, concept, etc.), he ordinarily has no other way of naming it than as the thought, percept, etc., *of that object*. He himself, meanwhile, knowing the self-same object in *his* way, gets easily led to suppose that the thought, which is *of* it, knows it in the same way in which he knows it, although this is often very far from being the case.[11] The most fictitious puzzles have been introduced into our science by this means. The so-called question of presentative or representative perception, of whether an object is present to the thought that thinks it by a counterfeit image of itself, or directly and without any intervening image at all; the question of nominalism and conceptualism, of the shape in which things are present when only a general notion of them is before the mind; are comparatively easy questions when once the psychologist's fallacy is eliminated from their treatment,—as we shall ere long see (in Chapter XII).

Another variety of the psychologist's fallacy is the assumption that the mental state studied must be conscious of itself as the psychologist is conscious of it. The mental state is aware of itself only from within; it grasps what we call its own content, and nothing more. The psychologist, on the contrary, is aware of it from without, and knows its relations with all sorts of other things. What the thought sees is only its own object; what the psychologist sees is the thought's object, plus the thought itself, plus possibly all the rest of the world. We must be very careful therefore, in discussing a state of mind from the psychologist's point of view, to avoid foisting into its own ken matters that are only there for ours. We must avoid substituting what we know the consciousness *is*, for what it is a consciousness *of*, and counting its outward, and so to speak physical, relations with other facts of the world, in among the objects of which we set it down as aware. Crude as such a confusion of standpoints seems to be when abstractly stated, it is nevertheless a snare into which no psychologist has kept himself at all times from falling, and which forms almost the entire stock-in-trade of certain schools. We cannot be too watchful against its subtly corrupting influence.

Summary. To sum up the chapter, Psychology assumes that thoughts successively occur, and that they know objects in a world which the psychologist also knows. *These thoughts are the subjective data of which he treats, and their relations to their objects, to the brain, and to the rest of the world constitute the subject-matter of psychologic science.* Its methods are introspection, experimentation, and comparison. But introspection is no sure guide to truths *about* our mental states; and in particular the poverty of the psychological vocabulary leads us to drop out certain states from our consideration, and to treat others as if they knew themselves and their objects as the psychologist knows both, which is a disastrous fallacy in the science.

FURTHER READING

Allen, G.W. *William James: A Biography.* New York: Viking Press, 1967.

Ayer, A.J. *The Origins of Pragmatism: Studies in the Philosophy of Charles Sanders Pierce, and William James.* San Francisco: Freeman, Cooper, 1968.

Bird, Graham. *William James.* London: Routledge and Kegan Paul, 1986.

11 Compare B. P. Bowne's Metaphysics (1882), p. 408.

Bjork, Daniel W. *The Compromised Scientist: William James in the Development of American Psychology.* New York: Columbia UP, 1983.

—. *William James: The Center of his Vision.* New York: Columbia UP, 1988.

Brožek, Josef, ed. *Explorations in the History of Psychology in the United States.* Lewisburg: Bucknell UP, 1984.

Cotkin, George. *William James, Public Philosopher.* Baltimore: Johns Hopkins UP, 1990.

Donnelly, Margaret E., ed. *Reinterpreting the Legacy of William James.* Washington, DC: American Psychological Association, 1992.

Feinstein, Howard M. *Becoming William James.* Ithaca: Cornell UP, 1984.

Ford, Marcus Peter. *William James's Philosophy: A New Perspective.* Amherst: University of Massachusetts Press, 1982.

James, William. *Essays in Psychology.* Edited by Frederick H. Burkhardt, Fredson Bowers, and Ignas K. Skrupskelis. Cambridge, MA: Harvard UP, 1983.

—. *Essays in Radical Empiricism.* Edited by Ralph Barton Perry. London: Longmans, Green, and Co., 1912.

Johnson, Michael G., and Tracy B. Henley, eds. *Reflections on* The Principles of Psychology: *William James After A Century.* Hillsdale, NJ: Lawrence Erlbaum Associates, 1990.

Linschoten, Hans. *On the Way Toward a Phenomenological Psychology: The Psychology of William James.* Edited by Amedeo Giorgi. Pittsburgh: Duquesne UP, 1968.

Myers, G. *William James: His Life and Thought.* New Haven, CT: Yale UP, 1986.

Perry, Ralph Barton. *The Thought and Character of William James.* 2 vols. Boston: Little, Brown, 1935.

Putnam, Ruth. *The Cambridge Companion to William James.* Cambridge: Cambridge UP, 1997.

Reck, Andrew J. *Introduction to William James: An Essay and Selected Texts.* Bloomington: Indiana UP, 1967.

Reed, Edward S. *From Soul to Mind: The Emergence of Psychology from Erasmus Darwin to William James.* New Haven: Yale UP, 1997.

Rieber, R.W., and Kurt Salzinger, eds. *Psychology: Theoretical-Historical Perspectives.* New York: Academic Press, 1980.

Robinson, Daniel N. *Toward a Science of Human Nature: Essays on the Psychologies of Mill, Hegel, Wundt, and James.* New York: Columbia UP, 1982.

Seigfried, Charlene Haddock. *William James's Radical Reconstruction of Philosophy.* Albany: State University of New York Press, 1990.

Simon, Linda. *William James Remembered.* Lincoln: University of Nebraska Press, 1996.

Taylor, Eugene I., and Robert H. Wozniak, eds. *Pure Experience: The Response to William James.* Bristol: Thoemmes Press, 1996.

—. *William James on Consciousness beyond the Margin.* Princeton: Princeton UP, 1996.

Wild, John. *The Radical Empiricism of William James.* Garden City, NY: Doubleday and Co., Inc., 1969.

Woodward, William R., and Mitchell G. Ash, eds. *The Problematic Science: Psychology in Nineteenth-Century Thought.* New York: Praeger, 1982.

32. THOMAS HENRY HUXLEY

(1825–1895)

T.H. Huxley not only made substantial contributions as a naturalist, anatomist, and ethnographer, but also won acclaim as a widely-read essayist on the subjects of science, science education, science and religion, and man's place in the cosmos. For financial reasons, Huxley originally settled on a medical career. He applied for Navy medical service, and first served Nelson's ship H.M.S. *Victory* at Haslar Naval Hospital and then shipped out on H.M.S. *Rattlesnake* from 1846–50 as Assistant Surgeon and naturalist. Julian Huxley edited Huxley's *Rattlesnake* diary in 1935. The *Rattlesnake* sailed throughout the South Pacific to Australia, Melanesia, and the Louisiade Archipelago. He met his future wife Henrietta Heathorn near Sydney on these voyages. Huxley's scientific work resulting from the *Rattlesnake* voyage won him election to the Royal Society in 1851 at the age of twenty-six; he later served as president of the society in 1883.

While first opposed to species transmutation, Huxley supported Charles Darwin's theory of evolution in the *Origin of Species* and gained his reputation as "Darwin's Bulldog" at the contentious 1860 meeting of the British Association for the Advancement of Science at Oxford by attacking Bishop Samuel "Soapy Sam" Wilberforce. Unfortunately no verbatim account of the proceedings has survived, but later accounts generally agree that Wilberforce flippantly asked whether Huxley were descended from a monkey on his grandfather's or grandmother's side and that Huxley turned the tables on him by retorting to the effect that "I should feel it no shame to have risen from such an origin; but I should feel it a shame to have sprung from one who prostituted the gifts of culture and eloquence to the service of prejudice and of falsehood" (*Life and Letters* 1.201). Later in *Evidence as to Man's Place in Nature* (1863), Huxley demonstrated that there was no strong line of structural demarcation between human anatomy and the anatomy of apes, gorillas, and chimpanzees—he even went so far as to suggest that the highest human faculties begin to appear in rudimentary form in lower forms of life. Huxley's theory of human–ape anatomical resemblances were strongly opposed by the leading palaeontologist of the day, Sir Richard Owen, who maintained that there were vast differences between the brains of humans and other primates. Huxley thrived on controversy and he engaged in a number of other intellectual battles during his lifetime—with Frederic Harrison and the English Comteans, with John Tyndall over Governor Eyre's suppression of the Jamaica rebellion, and with William Gladstone in the journal *Nineteenth Century* over creationism and the book of Genesis.

In an autobiographical sketch of 1889, Huxley explained his willingness to engage in public and scientific controversy for the sake of truth: his lifetime scientific goals were "to promote the increase of natural knowledge and to forward the application of scientific methods of investigation to all the problems of life to the best of my ability, in the conviction—which has grown with my growth and strengthened with my strength—that there is no alleviation for the sufferings of mankind except veracity of thought and of action, and the resolute facing of the world as it is, when the garment of makebelieve, by which pious hands have hidden its uglier features, is stripped off" (de Beer 109). Recent scholarship has provided a more rounded portrait of Huxley, however, by emphasizing his ambition, rebelliousness, and constant attacks on established scientific and ecclesiastical authority, which reveal a spirit ill at ease with mid- and late-Victorian class structure and Anglican theology.

Huxley lectured and wrote extensively on education. He wrote textbooks (*Introduction to the Study of Zoology Illustrated by the Crayfish*) and lectured to workingmen's associations. He was also elected to the first London School Board in 1870. Huxley and Matthew Arnold carried on a published debate in the 1880s on the merits of a scientific versus a liberal arts education. In 1879, he published a study of the philosopher David Hume. Huxley coined the term agnosticism (the existence of God is unknowable), which reflects the influence of Hume's skepticism on his thinking. Although Huxley denied that agnosticism could be classified as a formal religion, James R. Moore has not unjustly called Huxley a "priest of the 'Church scientific,' preacher of 'lay sermons,' and protagonist of a 'New Reformation'" (*Post-Darwinian* 348).

The selection excerpted below derives from Huxley's G.J. Romanes lecture of 1893 on "Evolution and Ethics" delivered at Oxford. Romanes had earlier invited Gladstone to deliver the first lecture on the topic. Since he was instructed not to speak on religion or politics, Huxley wrote a "Prolegomena" in 1894 to explain some omissions in his speech. Some of the central questions that grew out of later discussions of Darwinian evolution revolved around the moral status of both man and nature operating under the regime of a seemingly amoral natural selection—was man competitive and self-interested or altruistic and possessed of an ethics or faculty of reason which stood outside of natural processes? As Adrian Desmond summarizes the main argument of *Evolution and Ethics*, Huxley "portrayed ethical man revolting 'against the moral indifference of nature.' Care had replaced the crushing competition, altruism the competitive aggression. The flight was not to the swift, nor the battle to the strong; among humans a mitigating ethics dictated that the weak be rescued from the wall" (*Huxley* 597). The dichotomies suggested here of altruism / competition and ethics / nature, etc. only provide a hint of the complexity of the social issues posed by Darwinism when it was applied to humankind in the late nineteenth century.

Huxley's lecture on evolution and ethics caused a stir, because of his surprising break with Herbert Spencer (Huxley and Spencer were believed to have held similar views on evolution); Huxley, in contrast to Spencer's "Social Darwinism," maintained that ethical systems based on cosmic or Darwinian evolution were not logically sound or valid. Humans stood outside of the natural laws of creation because of their intellect and moral sensibility, illustrated by Huxley's frequently used example of the specially tended garden (the "horticultural process"), which was exempt from the harsh laws of natural selection (the "cosmic process"). Humans in fact, according to Huxley, constantly battled the Evil cosmic forces with the civilizing moral forces of the Good. As Leonard Huxley pointed out, although similar views of Huxley criticizing Spencer's "Social Darwinism" had been in print for ten years, "nevertheless, the doctrine seemed to take everybody by surprise. The drift of the lecture was equally misunderstood by critics of opposite camps.... On the one hand he was branded as a deserter from free thought; on the other, hailed almost as a convert to orthodoxy" (*Life and Letters* 2.374).

"Prolegomena." *Evolution and Ethics.* London: Macmillan and Co., 1894, pp. 1–45.

I

It may be safely assumed that, two thousand years ago, before Caesar set foot in southern Britain, the whole country-side visible from the windows of the room in which I write, was in what is called "the state of nature." Except, it may be, by raising a few sepulchral mounds, such as those which still, here and there, break the flowing contours of the downs, man's hands had made no mark upon it; and the thin veil of vegetation which overspread the broad-backed heights and the shelving sides of the coombs was unaffected by his industry. The native grasses and weeds, the scattered patches of gorse, contended with one another for the possession of the scanty surface soil; they fought against the droughts of summer, the frosts of winter, and the furious gales which swept, with unbroken force, now from the Atlantic, and now from the North Sea, at all times of the year; they filled up, as they best might, the gaps made in their ranks by all sorts of underground and overground animal ravagers. One year with another, an average population, the floating balance of the unceasing struggle for existence among the indigenous plants, maintained itself. It is as little to be doubted, that an essentially similar state of nature prevailed, in this region, for many thousand years before the coming of Caesar; and there is no assignable reason for denying that it might continue to exist through an equally prolonged futurity, except for the intervention of man.

Reckoned by our customary standards of duration, the native vegetation, like the "everlasting hills" which it clothes, seems a type of permanence. The little Amarella Gentians, which abound in some places to-day, are the descendants of those that were trodden underfoot by the prehistoric savages who have left their flint tools about, here and there; and they followed ancestors which, in the climate of the glacial epoch, probably flourished better than they do now. Compared with the long past of this humble plant, all the history of civilized men is but an episode.

Yet nothing is more certain than that, measured by the liberal scale of time-keeping of the universe, this present state of nature, however it may seem to have gone and to go on for ever, is but a fleeting phase of her infinite variety; merely the last of the series of changes which the earth's surface has undergone in the course of the millions of years of its existence. Turn back a square foot of the thin turf, and the solid foundation of the land, exposed in cliffs of chalk five hundred feet high on the adjacent shore, yields full assurance of a time when the sea covered the site of the "everlasting hills"; and when the vegetation of what land lay nearest, was as different from the present Flora of the Sussex downs, as that of Central Africa now is.[1] No less certain is it that, between the time during which the chalk was formed and that at which the original turf came into existence, thousands of centuries elapsed, in the course of which, the state of nature of the ages during which the chalk was deposited, passed into that which now is, by changes so slow that, in the coming and going of the generations of men, had such witnessed them, the contemporary conditions would have seemed to be unchanging and unchangeable.

But it is also certain that, before the deposition of the chalk, a vastly longer period had elapsed, throughout which it is easy to follow the traces of the same process of ceaseless modification and of the internecine struggle for existence of living things; and that even when we can get no further back, it is not because there is any reason to think we have reached the beginning, but because the trail of the most ancient life remains hidden, or has become obliterated.

Thus that state of nature of the world of plants, which we began by considering, is far from possessing the attribute of permanence. Rather its very essence is impermanence. It may have lasted twenty or thirty thousand years, it may last for twenty or thirty thousand years more, without obvious change; but, as surely as it has followed upon a very different state, so it will be followed by an equally different condition. That which endures is not one or another association of living forms, but the process of which the cosmos is the product, and of which these are among the transitory expressions. And in the living world, one of the most characteristic features of this cosmic process is the struggle for existence, the competition of each with all, the result of which is the selection, that is to say, the survival of those forms which, on the whole, are best adapted to the conditions which at any period obtain; and which are, therefore, in that respect, and only in that respect, the fittest.[2] The acme reached by the cosmic process in the vegetation of the downs is seen in the turf, with its weeds and gorse. Under the conditions, they have come out of the struggle victorious; and, by surviving, have proved that they are the fittest to survive.

That the state of nature, at any time, is a temporary phase of a process of incessant change, which has been going on for innumerable ages, appears to me to be a proposition as well established as any in modern history. Paleontology assures us, in addition, that the ancient philosophers who, with less reason, held the same doctrine, erred in supposing that the phases formed a cycle, exactly repeating the past, exactly foreshadowing the future, in their rotations. On the contrary, it furnishes us with conclusive reasons for thinking that, if every link in the ancestry of these humble indigenous plants had been preserved and were accessible to us, the whole would present a converging series of forms of gradually diminishing complexity, until, at some period in the history of the earth, far more remote than any of which organic remains have yet been discovered, they would merge in those low groups among which the boundaries between animal and vegetable life become effaced.[3]

The word "evolution," now generally applied to the cosmic process, has had a singular history, and is used in various senses.[4] Taken in its popular signification it means progressive development, that is, gradual change from a condition of relative uniformity to one of relative complexity; but its connotation has been widened to include the phenomena of retrogressive metamorphosis, that is, of progress from a condition of relative complexity to one of relative uniformity.

As a natural process, of the same character as the development of a tree from its seed, or of a fowl from its egg, evolution excludes creation and all other kinds of supernatural intervention. As the expression of a fixed order, every stage of which is the effect of causes operating according to definite rules, the conception of evolution no less excludes that of chance. It is very desirable to remember that evolution is not an explanation of the cosmic process, but merely a generalized statement of the method and results of that process. And, further, that, if there is proof that the cosmic process was set going by any agent, then that agent will be the creator of it and of all its products, although supernatural intervention may remain strictly excluded from its further course.

So far as that limited revelation of the nature of things, which we call scientific knowledge, has yet gone, it tends, with constantly increasing emphasis, to the belief that, not merely the world of plants, but that of animals; not merely living things, but the whole fabric of the earth; not merely our planet, but the whole solar system; not merely our star and its satellites, but the millions of similar bodies which bear witness to the order which pervades boundless space, and has endured through boundless time; are all working out their predestined courses of evolution.

With none of these have I anything to do, at present, except with that exhibited by the forms of life which tenant the earth. All plants and animals exhibit the tendency to vary, the causes of which have yet to be ascertained; it is the tendency of the conditions of life, at any given time, while favouring the existence of the variations best adapted to them, to oppose that of the rest and thus to exercise selection; and all living things tend to multiply without limit, while the means of support are limited; the obvious cause of which is the production of offspring more numerous than their progenitors, but with equal expectation of life in the actuarial sense. Without the first tendency there could be no evolution. Without the second, there would be no good reason why one variation should disappear and another take its place; that is to say, there would be no selection. Without the third, the struggle for existence, the agent of the selective process in the state of nature, would vanish.[5]

Granting the existence of these tendencies, all the known facts of the history of plants and of animals may be brought into rational correlation. And this is more than can be said for any other hypothesis that I know of. Such hypotheses, for example, as that of the existence of a primitive, orderless chaos; of a passive and sluggish eternal matter moulded, with but partial success, by archetypal ideas; of a brand-new world-stuff suddenly created and swiftly shaped by a supernatural power; receive no encouragement, but the contrary, from our present knowledge. That our earth may once have formed part of a nebulous cosmic magma is certainly possible, indeed seems highly probable; but there is no reason to doubt that order reigned there, as completely as amidst what we regard as the most finished works of nature or of man.[6] The faith which is born of knowledge, finds its object in an eternal order, bringing forth ceaseless change, through endless time, in endless space; the manifestations of the cosmic energy alternating between phases of potentiality and phases of explication. It may be that, as Kant suggests,[7] every cosmic magma predestined to evolve into a new world, has been the no less predestined end of a vanished predecessor.

II

Three or four years have elapsed since the state of nature, to which I have referred, was brought to an end, so far as a small patch of the soil is concerned, by the intervention of man. The patch was cut off from the rest by a wall; within the area thus protected, the native vegetation was, as far as possible, extirpated; while a colony of strange plants was imported and set down in its place. In short, it was made into a garden. At the present time, this artificially treated area presents an aspect extraordinarily different from that of so much of the land as remains in the state of nature, outside the wall. Trees, shrubs, and herbs, many of them appertaining to the state of nature of remote parts of the globe, abound and flourish. Moreover, considerable quantities of vegetables, fruits, and flowers are produced, of kinds which neither now exist, nor have ever existed, except under conditions such as obtain in the garden; and which, therefore, are as much works of the art of man as the frames and glass-houses in which some of them are raised. That the "state of Art," thus created in the state of nature by man, is sustained by and dependent on him, would at once become apparent, if the watchful supervision of the gardener were withdrawn, and the antagonistic influences of the general cosmic process were no longer sedulously warded off, or counteracted. The walls and gates would decay; quadrupedal and bipedal intruders would devour and tread down the useful and beautiful plants; birds, insects, blight, and mildew would work their will; the seeds of the native plants, carried by winds or other agencies, would immigrate, and in virtue of their long-earned special adaptation to the local conditions, these despised native weeds would soon choke their choice exotic rivals. A century or two hence, little beyond the foundations of the wall and of the houses and frames would be left, in evidence of the victory of the cosmic powers at work in the state of nature, over the temporary obstacles to their supremacy, set up by the art of the horticulturalist.

It will be admitted that the garden is as much a work of art,[8] or artifice, as anything that can be mentioned. The energy localised in certain human bodies, directed by similarly localised intellects, has produced a collocation of other material bodies which could not be brought about in the state of nature. The same proposition is true of all the works of man's hands, from a flint implement to a cathedral or a chronometer; and it is because it is true, that we call these things artificial, term them works of art, or artifice, by way of distinguishing them from the products of the cosmic process, working outside man, which we call natural, or works of nature. The distinction thus drawn between the works of nature and those of man, is universally recognised; and it is, as I conceive, both useful and justifiable.

III

No doubt, it may be properly urged that the operation of human energy and intelligence, which has brought into existence and maintains the garden, by what I have called "the horticultural process," is, strictly speaking, part and parcel of the cosmic process. And no one could more readily agree to that proposition than I. In fact, I do not know that any one has taken more pains than I have, during the last thirty years, to insist upon the doctrine, so much reviled in the early part of that period, that man, physical, intellectual, and moral, is as much a part of nature, as purely a product of the cosmic process, as the humblest weed.[9]

But if, following up this admission, it is urged that, such being the case, the cosmic process cannot be in antagonism with that horticultural process which is part of itself—I can only reply, that if the conclusion that the two are antagonistic is logically absurd, I am sorry for logic, because, as we have seen, the fact is so. The garden is in the same position as every other work of man's art; it is a result of the cosmic process working through and by human energy and intelligence; and, as is the case with every other artificial thing set up in the state of nature, the influences of the latter are constantly tending to break it down and destroy it. No doubt, the Forth bridge and an ironclad in the offing, are, in ultimate resort, products of the cosmic process; as much so as the river which flows under the one, or the sea-water on which the other floats. Nevertheless, every breeze strains the bridge a little, every tide does something to weaken its foundations; every change of temperature alters the adjustment of its parts, produces friction and consequent wear and tear. From time to time, the bridge must be repaired, just as the ironclad must go into dock; simply because nature is always tending to reclaim that which her child, man, has borrowed from her and has arranged in combinations which are not those favoured by the general cosmic process.

Thus, it is not only true that the cosmic energy, working through man upon a portion of the plant world, opposes the same energy as it works through the state of nature, but a similar antagonism is everywhere manifest between the artificial and the natural. Even in the state of nature itself, what is the struggle for existence but the antagonism of the results of the cosmic process in the region of life, one to another?[10]

IV

Not only is the state of nature hostile to the state of art of the garden; but the principle of the horticultural process, by which the latter is created and maintained, is antithetic to that of the cosmic process. The characteristic feature of the latter is the intense and unceasing competition of the struggle for existence. The characteristic of the former is the elimination of that struggle, by the removal of the conditions which give rise to it. The tendency of the cosmic process is to bring about the adjustment of the forms of plant life to the current conditions; the tendency of the horticultural process is the adjustment of the conditions to the needs of the forms of plant life which the gardener desires to raise.

The cosmic process uses unrestricted multiplication as the means whereby hundreds compete for the place and nourishment adequate for one; it employs frost and drought to cut off the weak and unfortunate; to survive, there is need not only of strength, but of flexibility and of good fortune.

The gardener, on the other hand, restricts multiplication; provides that each plant shall have sufficient space and nourishment; protects from frost and drought; and, in every other way, attempts to modify the conditions, in such a manner as to bring about the survival of those forms which most nearly approach the standard of the useful, or the beautiful, which he has in his mind.

If the fruits and the tubers, the foliage and the flowers thus obtained, reach, or sufficiently approach, that ideal, there is no reason why the *status quo* attained should not be indefinitely prolonged. So long as the state of nature remains approximately the same, so long will the energy and intelligence which created the garden suffice to maintain it. However, the limits within which this mastery of man over nature can be maintained are narrow. If the conditions of the cretaceous epoch returned, I fear the most skilful of gardeners would have to give up the cultivation of apples and gooseberries; while, if those of the glacial period once again obtained, open asparagus beds would be superfluous, and the training of fruit trees against the most favourable of south walls, a waste of time and trouble.

But it is extremely important to note that, the state of nature remaining the same, if the produce does not satisfy the gardener, it may be made to approach his ideal more closely. Although the struggle for existence may be at end, the possibility of progress remains. In discussions on these topics, it is often strangely forgotten that the essential conditions of the modification, or evolution, of living things are variation and hereditary transmission. Selection is the means by which certain variations are favoured and their progeny preserved. But the struggle for existence is only one of the means by which selection may be effected. The endless varieties of cultivated flowers, fruits, roots, tubers, and bulbs are not products of selection by means of the struggle for existence, but of direct selection, in view of an ideal of utility or beauty. Amidst a multitude of plants, occupying the same station and subjected to the same conditions, in the garden, varieties arise. The varieties tending in a given direction are preserved, and the rest are destroyed. And the same process takes place among the varieties until, for example, the wild kale becomes a cabbage, or the wild *Viola tricolor* a prize pansy.

V

The process of colonization presents analogies to the formation of a garden which are highly instructive. Suppose a shipload of English colonists sent to form a settlement, in such a country as Tasmania was in the middle of the last century. On landing, they find themselves in the midst of a state of nature, widely different from that left behind them in everything but the most general physical conditions. The common plants, the common birds and quadrupeds, are as totally distinct as the men from anything to be seen on the side of the globe from which they come. The colonists proceed to put an end to this state of things over as large an area as they desire to occupy. They clear away the native vegetation, extirpate or drive out the animal population, so far as may be necessary, and take measures to defend themselves from the re-immigration of either. In their place, they introduce English grain and fruit trees; English dogs, sheep, cattle, horses; and English men; in fact, they set up a new Flora and Fauna and a new variety of mankind, within the old state of nature. Their farms and pastures represent a garden on a great scale, and themselves the gardeners who have to keep it up, in watchful antagonism to the old *régime*. Considered as a whole, the colony is a composite unit introduced into the old state of nature; and, thenceforward, a competitor in the struggle for existence, to conquer or be vanquished.

Under the conditions supposed, there is no doubt of the result, if the work of the colonists be carried out energetically and with intelligent combination of all their forces. On the other hand, if they are slothful, stupid, and careless; or if they waste their energies in contests with one another, the chances are that the old state of nature will have the best of it. The native savage will destroy the immigrant civilized man; of the English animals and plants some will be extirpated by their indigenous rivals, others will pass into the feral state and themselves become components of the state of nature. In a few decades, all other traces of the settlement will have vanished.

VI

Let us now imagine that some administrative authority, as far superior in power and intelligence to men, as men are to their cattle, is set over the colony, charged to deal with its human elements in such a manner as to assure the victory of the settlement over the antagonistic influences of the state of nature in which it is set down. He would proceed in the same fashion as that in which the gardener dealt with his garden. In the first place, he would, as far as possible, put a stop to the influence of external competition by thoroughly extirpating and excluding the native rivals, whether men, beasts, or plants. And our administrator would select his human agents, with a view to his ideal of a successful colony, just as the gardener selects his plants with a view to his ideal of useful or beautiful products.

In the second place, in order that no struggle for the means of existence between these human agents should weaken the efficiency of the corporate whole in the battle with the state of nature, he would make arrangements by which each would be provided with those means; and would be relieved from the fear of being deprived of them by his stronger or more cunning fellows. Laws, sanctioned by the combined force of the colony, would restrain the self-assertion of each man within the limits required for the maintenance of peace. In other words, the cosmic struggle for existence, as between man and man, would be rigorously suppressed; and selection, by its means, would be as completely excluded as it is from the garden.

At the same time, the obstacles to the full development of the capacities of the colonists by other conditions of the state of nature than those already mentioned, would be removed by the creation of artificial conditions of existence of a more favourable character. Protection against extremes of head and cold would be afforded by houses and clothing; drainage and irrigation works would antagonise the effects of excessive rain and excessive drought; roads, bridges, canals, carriages, and ships would overcome the natural obstacles to locomotion and transport; mechanical engines would supplement the natural strength of men and of their draught animals; hygienic precautions would check, or remove, the natural causes of disease. With every step of this progress in civilization, the colonists would become more and more independent of the state of nature; more and more, their lives would be conditioned by a state of art. In order to attain his ends, the administrator would have to avail himself of the courage, industry, and co-operative intelligence of the settlers; and it is plain that the interest of the community would be best served by increasing the proportion of persons who possess such qualities, and diminishing that of persons devoid of them. In other words, by selection directed towards an ideal.

Thus the administrator might look to the establishment of an earthly paradise, a true garden of Eden, in which all things should work together towards the well-being of the gardeners: within which the cosmic process, the coarse struggle for existence of the state of nature, should be abolished; in which that state should be replaced by a state of art; where every plant and every lower animal should be adapted to human wants, and would perish if human supervision and protection were withdrawn; where men themselves should have been selected, with a view to their efficiency as organs for the performance of the functions of a perfected society. And this ideal polity would have been brought about, not by gradually adjusting the men to the conditions around them, but by creating artificial conditions for them; not by allowing the free play of the struggle for existence, but by excluding that struggle; and by substituting selection directed towards the administrator's ideal for the selection it exercises.

VII

But the Eden would have its serpent, and a very subtle beast too. Man shares with the rest of the living world the mighty instinct of reproduction and its consequence, the tendency to multiply with great rapidity. The better the measures of the administrator achieved their object, the more completely the destructive agencies of the state of nature were defeated, the less would that multiplication be checked.

On the other hand, within the colony, the enforcement of peace, which deprives every man of the power to take away the means of existence from another, simply because he is the stronger, would have put an end to the struggle for existence between the colonists, and the competition for the commodities of existence, which would alone remain, is no check upon population.

Thus, as soon as the colonists began to multiply, the administrator would have to face the tendency to the reintroduction of the cosmic struggle into his artificial fabric, in consequence of the competition, not merely for the commodities, but for the means of existence. When the colony reached the limit of possible expansion, the surplus population must be disposed of somehow; or the fierce struggle for existence must recommence and destroy that peace, which is the fundamental condition of the maintenance of the state of art against the state of nature.

Supposing the administrator to be guided by purely scientific considerations, he would, like the gardener, meet this most serious difficulty by systematic extirpation, or exclusion, of the superfluous. The hopelessly diseased, the infirm aged, the weak or deformed in body or in mind, the excess of infants born, would be put away, as the gardener pulls up defective and superfluous plants, or the breeder destroys undesirable cattle. Only the strong and the healthy, carefully matched, with a view to the progeny best adapted to the purposes of the administrator, would be permitted to perpetuate their kind.

VIII

Of the more thoroughgoing of the multitudinous attempts to apply the principles of cosmic evolution, or what are supposed to be such, to social and political problems, which have appeared of late years, a considerable proportion appear to me to be based upon the notion that human society is competent to furnish, from its own resources, an administrator of the kind I have imagined. The pigeons, in short, are to be their own Sir John Sebright.[11] A despotic government, whether individual or collective, is to be endowed with the preternatural intelligence, and with what, I am afraid, many will consider the preternatural ruthlessness, required for the purpose of carrying out the principle of improvement by selection, with the somewhat drastic thoroughness upon which the success of the method depends. Experience certainly does not justify us in limiting the ruthlessness of individual "saviours of society"; and, on the well-known grounds of the aphorism which denies both body and soul to corporations, it seems probable (indeed the belief is not without support in history) that a collective despotism, a mob got to believe in its own divine right by demagogic missionaries, would be capable of more thorough work in this direction than any single tyrant, puffed up with the same illusion, has ever achieved. But intelligence is another affair. The fact that "saviours of society" take to that trade is evidence enough that they have none to spare. And such as they possess is generally sold to the capitalists of physical force on whose resources they depend. However, I doubt whether even the keenest judge of character, if he had before him a hundred boys and girls under fourteen, could pick out, with the least chance of success, those who should be kept, as certain to be serviceable members of the polity, and those who should be chloroformed, as equally sure to be stupid, idle, or vicious. The "points" of a good or of a bad citizen are really far harder to discern than those of a puppy or a short-horn calf; many do not show themselves before the practical difficulties of life stimulate manhood to full exertion. And by that time the mischief is done. The evil stock, if it be one, has had time to multiply, and selection is nullified.

IX

I have other reasons for fearing that this logical ideal of evolutionary regimentation—this pigeon-fancier's polity—is unattainable. In the absence of any such a severely scientific administrator as we have been dreaming of, human society is kept together by bonds of such a singular character, that the attempt to perfect society after his fashion would run serious risk of loosening them.

Social organization is not peculiar to men. Other societies, such as those constituted by bees and ants, have also arisen out of the advantage of cooperation in the struggle for existence; and their resemblances to, and their differences from, human society are alike instructive. The society formed by the hive bee fufils the ideal of the communistic aphorism "to each according to his needs, from each according to his capacity." Within it, the struggle for existence is strictly limited. Queen, drones, and workers have each their allotted sufficiency of food; each performs the function assigned to it in the economy of the hive, and all contribute to the success of the whole co-operative society in its competition with rival collectors of nectar and pollen and with other enemies, in the state of nature without. In the same sense as the garden, or the colony, is a work of human art, the bee polity is a work of apiarian art, brought about by the cosmic process, working through the organization of the hymenopterous type.

Now this society is the direct product of an organic necessity, impelling every member of it to a course of action which tends to the good of the whole. Each bee has its duty and none has any rights. Whether bees are susceptible of feeling and capable of thought is a question which cannot be dogmatically answered. As a pious opinion, I am disposed to deny them more than the merest rudiments of consciousness.[12] But it is curious to reflect that a thoughtful drone (workers and queens would have no leisure for speculation) with a turn for ethical philosophy, must needs profess himself an intuitive moralist of the purest water. He would point out, with perfect justice, that the devotion of the workers to a life of ceaseless toil for a mere subsistence wage, cannot be accounted for either by enlightened selfishness, or by any other sort of utilitarian motives; since these bees begin to work, without experience or reflection, as they emerge from the cell in which they are hatched. Plainly, an eternal and immutable principle, innate in each bee, can alone account for the phenomena. On the other hand, the biologist, who traces out all the extant stages of gradation between solitary and hive bees, as clearly sees in the latter, simply the perfection of an automatic mechanism, hammered out by the blows of the struggle for existence upon the progeny of the former, during long ages of constant variation.

X

I see no reason to doubt that, at its origin, human society was as much a product of organic necessity as that of the bees.[13] The human family, to begin with, rested upon exactly the same conditions as those which gave rise to similar associations among animals lower in the scale. Further, it is easy to see that every increase in the duration of the family ties, with the resulting co-operation of a larger and larger number of descendants for protection and defence, would give the families in which such modification took place a distinct advantage over the others. And, as in the hive, the progressive limitation of the struggle for existence between the members of the family would involve increasing efficiency as regards outside competition.

But there is this vast and fundamental difference between bee society and human society. In the former, the members of the society are each organically predestined to the performance of one particular class of functions only. If they are endowed with desires, each could desire to perform none but those offices for which its organization specially fits it; and which, in view of the good of the whole, it is proper it should do. So long as a new queen does not make her appearance, rivalries and competition are absent from the bee polity.

Among mankind, on the contrary, there is no such predestination to a sharply defined place in the social organism. However much men may differ in the quality of their intellects, the intensity of their passions, and the delicacy of their sensations, it cannot be said that one is fitted by his organization to be an agricultural labourer and nothing else, and another to be a landowner and nothing else. Moreover, with all their enormous differences in natural endowment, men agree in one thing, and that is their innate desire to enjoy the pleasures and to escape the pains of life; and, in short, to do nothing but that which it pleases them to do, without the least reference to the welfare of the society into which they are born. That is their inheritance (the reality at the bottom of the doctrine of original sin) from the long series of ancestors, human and semi-human and brutal, in whom the strength of this innate tendency to self-assertion was the condition of victory in the struggle for existence. That is the reason of the *aviditas vitae*[14]—the insatiable hunger for enjoyment—of all mankind, which is one of the essential conditions of success in the war with the state of nature outside; and yet the sure agent of the destruction of society if allowed free play within.

The check upon this free play of self-assertion, or natural liberty, which is the necessary condition for the origin of human society, is the product of organic necessities of a different kind from those upon which the constitution of the hive depends. One of these is the mutual affection of parent and offspring, intensified by the long infancy of the human species. But the most important is the tendency, so strongly developed in man, to reproduce in himself actions and feelings similar to, or correlated with, those of other men. Man is the most consummate of all mimics in the animal world; none but himself can draw or model; none comes near him in the scope, variety, and exactness of vocal imitation; none is such a master of gesture; while he seems to be impelled thus to imitate for the pure pleasure of it. And there is no such another emotional chameleon. By a purely reflex operation of the mind, we take the hue of passion of those who are about us, or, it may be, the complementary colour. It is not by any conscious "putting one's self in the place" of a joyful or a suffering person that the state of mind we call sympathy usually arises;[15] indeed, it is often contrary to one's sense of right, and in spite of one's will, that "fellow-feeling makes us wondrous kind," or the reverse. However complete may be the indifference to public opinion, in a cool, intellectual view, of the traditional sage, it has not yet been my fortune to meet with any actual sage who took its hostile manifestations with entire equanimity. Indeed, I doubt if the philosopher lives, or ever has lived, who could know himself to be heartily despised by a street boy without some irritation. And, though one cannot justify Haman for wishing to hang Mordecai on such a very high gibbet, yet, really, the consciousness of the Vizier of Ahasuerus, as he went in and out of the gate, that this obscure Jew had no respect for him, must have been very annoying.[16]

It is needful only to look around us, to see that the greatest restrainer of the anti-social tendencies of men is fear, not of the law, but of the opinion of their fellows. The conventions of honour bind men who break legal, moral, and religious bonds; and, while people endure the extremity of physical pain rather than part with life, shame drives the weakest to suicide.

Every forward step of social progress brings men into closer relations with their fellows, and increases the importance of the pleasures and pains derived from sympathy. We judge the acts of others by our own sympathies, and we judge our own acts by the sympathies of others, every day and all day long, from childhood upwards, until associations, as indissoluble as those of language, are formed between certain acts and the feelings of approbation or disapprobation. It becomes impossible to imagine some acts without disapprobation, or others without approbation of the actor, whether he be one's self, or any one else. We come to think in the acquired dialect of morals. An artificial personality, the "man within," as Adam Smith[17] calls conscience, is built up beside the natural personality. He is the watchman of society, charged to restrain the anti-social tendencies of the natural man within the limits required by social welfare.

XI

I have termed this evolution of the feelings out of which the primitive bonds of human society are so largely forged, into the organized and personified sympathy we call conscience, the ethical process.[18] So far as it tends to make any human society more efficient in the struggle for existence with the state of nature, or with other societies, it works in harmonious contrast with the cosmic process. But it is none the less true that, since law and morals are restraints upon the struggle for existence between men in society, the ethical process is in opposition to the principle of the cosmic process, and tends to the suppression of the qualities best fitted for success in that struggle.[19]

It is further to be observed that, just as the self-assertion, necessary to the maintenance of society against the state of nature, will destroy that society if it is allowed free operation within; so the self-restraint, the essence of the ethical process, which is no less an essential condition of the existence of every polity, may, by excess, become ruinous to it.

Moralists of all ages and of all faiths, attending only to the relations of men towards one another in an ideal society, have agreed upon the "golden rule," "Do as you would be done by." In other words, let sympathy be your guide; put yourself in the place of the man towards whom your action is directed; and do to him what you would like to have done to yourself under the circumstances. However much one may admire the generosity of such a rule of conduct; however confident one may be that average men may be thoroughly depended upon not to carry it out to its full logical consequences; it is nevertheless desirable to recognise the fact that these consequences are incompatible with the existence of a civil state, under any circumstances of this world which have obtained, or, so far as one can see, are, likely to come to pass.

For I imagine there can be no doubt that the great desire of every wrongdoer is to escape from the painful consequences of his actions. If I put myself in the place of the man who has robbed me, I find that I am possessed by an exceeding desire not to be fined or imprisoned; if in that of the man who has smitten me on one cheek, I contemplate with satisfaction the absence of any worse result than the turning of the other cheek for like treatment. Strictly observed, the "golden rule" involves the negation of law by the refusal to put it in motion against law-breakers; and, as regards the external relations of a polity, it is the refusal to continue the struggle for existence. It can be obeyed, even partially, only under the protection of a society which repudiates it. Without such shelter, the followers of the "golden rule" may indulge in hopes of heaven, but they must reckon with the certainty that other people will be masters of the earth.

What would become of the garden if the gardener treated all the weeds and slugs and birds and trespassers as he would like to be treated, if he were in their place?

XII

Under the preceding heads, I have endeavoured to represent in broad, but I hope faithful, outlines the essential features of the state of nature and of that cosmic process of which it is the outcome, so far as was needful for my argument; I have contrasted with the state of nature the state of art, produced by human intelligence and energy, as it is exemplified by a garden; and I have shown that the state of art, here and elsewhere, can be maintained only by the constant counteraction of the hostile influences of the state of nature. Further, I have pointed out that the "horticultural process" which thus sets itself against the "cosmic process" is opposed to the latter in principle, in so far as it tends to arrest the struggle for existence, by restraining the multiplication which is one of the chief causes of that struggle, and by creating artificial conditions of life, better adapted to the cultivated plants than are the conditions of the state of nature. And I have dwelt upon the fact that, though the progressive modification, which is the consequence of the struggle for existence in the state of nature, is at an end, such modification may still be effected by that selection, in view of an ideal of usefulness, or of pleasantness, to man, of which the state of nature knows nothing.

I have proceeded to show that a colony, set down in a country in the state of nature, presents close analogies with a garden; and I have indicated the course of action which an administrator, able and willing to carry out horticultural principles, would adopt, in order to secure the success of such a newly formed polity, supposing it to be capable of indefinite expansion. In the contrary case, I have shown that difficulties must arise; that the unlimited increase of the population over a limited area must, sooner or later, reintroduce into the colony that struggle for the means of existence between the colonists, which it was the primary object of the administrator to exclude, insomuch as it is fatal to the mutual peace which is the prime condition of the union of men in society.

I have briefly described the nature of the only radical cure, known to me, for the disease which would thus threaten the existence of the colony; and, however regretfully, I have been obliged to admit that this rigorously scientific method of applying the principles of evolution to human society hardly comes within the region of practical politics; not for want of will on the part of a great many people; but because, for one reason, there is no hope that mere human beings will ever possess enough intelligence to select the fittest. And I have adduced other grounds for arriving at the same conclusion.

I have pointed out that human society took its rise in the organic necessities expressed by imitation and by the sympathetic emotions; and that, in the struggle for existence with the state of nature and with other societies, as part of it, those in which men were thus led to close co-operation had a great advantage.[20] But, since each man retained more or less of the faculties common to all the rest, and especially a full share of the desire for unlimited self-gratification, the struggle for existence within society could only be gradually eliminated. So long as any of it remained, society continued to be an imperfect instrument of the struggle for existence and, consequently, was improvable by the selective influence of that struggle. Other things being alike, the tribe of savages in which order was best maintained; in which there was most security within the tribe and the most loyal mutual support outside it, would be the survivors.

I have termed this gradual strengthening of the social bond, which, though it arrests the struggle for existence inside society, up to a certain point improves the chances of society, as a corporate whole, in the cosmic struggle–the ethical process. I have endeavoured to show that, when the ethical process has advanced so far as to secure every member of the society in the possession of the means of existence, the struggle for existence, as between man and man, within that society is, *ipso facto*, at an end. And, as it is undeniable that the most highly civilized societies have substantially reached this position, it follows that, so far as they are concerned, the struggle for existence can play no important part within them.[21] In other words, the kind of evolution which is brought about in the state of nature cannot take place.

I have further shown cause for the belief that direct selection, after the fashion of the horticulturist and the breeder, neither has played, nor can play, any important part in the evolution of society; apart from other reasons, because I do not see how such selection could be practised without a serious weakening, it may be the destruction, of the bonds which hold society together. It strikes me that men who are accustomed to contemplate the active or passive extirpation of the weak, the unfortunate, and the superfluous; who justify that conduct on the ground that it has the sanction of the cosmic process, and is the only way of ensuring the progress of the race; who, if they are consistent, must rank medicine among the black arts and count the physician a mischievous preserver of the unfit; on whose matrimonial undertakings the principles of the stud have the chief influence; whose whole lives, therefore, are an education in the noble art of suppressing natural affection and sympathy, are not likely to have any large stock of these commodities left. But, without them, there is no conscience, nor any restraint on the conduct of men, except the calculation of self-interest, the balancing of certain present gratifications against doubtful future pains; and experience tells us how much that is worth. Every day, we see firm believers in the hell of the theologians commit acts by which, as they believe when cool, they risk eternal punishment; while they hold back from those which are opposed to the sympathies of their associates.

XIII

That progressive modification of civilization which passes by the name of the "evolution of society," is, in fact, a process of an essentially different character, both from that which brings about the evolution of species, in the state of nature, and from that which gives rise to the evolution of varieties, in the state of art.

There can be no doubt that vast changes have taken place in English civilization since the reign of the Tudors. But I am not aware of a particle of evidence in favour of the conclusion that this evolutionary process has been accompanied by any modification of the physical, or the mental, characters of the men who have been the subjects of it. I have not met with any grounds for suspecting that the average Englishmen of to-day are sensibly different from those that Shakspere knew and drew. We look into his magic mirror of the Elizabethan age, and behold, nowise darkly, the presentment of ourselves.

During these three centuries, from the reign of Elizabeth to that of Victoria, the struggle for existence between man and man has been so largely restrained among the great mass of the population (except for one or two short intervals of civil war), that it can have had little, or no, selective operation. As to anything comparable to direct selection, it has been practised on so small a scale that it may also be neglected. The criminal law, in so far as by putting to death, or by subjecting to long periods of imprisonment, those who infringe its provisions, it prevents the propagation of hereditary criminal tendencies; and the poor-law, in so far as it separates married couples, whose destitution arises from hereditary defects of character, are doubtless selective agents operating in favour of the non-criminal and the more effective members of society. But the proportion of the population which they influence is very small; and, generally, the hereditary criminal and the hereditary pauper have propagated their kind before the law affects them. In a large proportion of cases, crime and pauperism have nothing to do with heredity; but are the consequence, partly, of circumstances and, partly, of the possession of qualities, which, under different conditions of life, might have excited esteem and even admiration. It was a shrewd man of the world who, in discussing sewage problems, remarked that dirt is riches in the wrong place; and that sound aphorism has moral applications. The benevolence and open-handed generosity which adorn a rich man, may make a pauper of a poor one; the energy and courage to which the successful soldier owes his rise, the cool and daring subtlety to which the great financier owes his fortune, may very easily, under unfavourable conditions, lead their possessors to the gallows, or to the hulks. Moreover, it is fairly probable that the children of a 'failure' will receive from their other parent just that little modification of character which makes all the difference. I sometimes wonder whether people, who talk so freely about extirpating the unfit, ever dispassionately consider their own history. Surely, one must be very 'fit,' indeed, not to know of an occasion, or perhaps two, in one's life, when it would have been only too easy to qualify for a place among the 'unfit.'

In my belief the innate qualities, physical, intellectual, and moral, of our nation have remained substantially the same for the last four or five centuries. If the struggle for existence has affected us to any serious extent (and I doubt it) it has been, indirectly, through our military and industrial wars with other nations.

XIV

What is often called the struggle for existence in society (I plead guilty to having used the term too loosely myself), is a contest, not for the means of existence, but for the means of enjoyment. Those who occupy the first places in this practical competitive examination are the rich and the influential; those who fail, more or less, occupy the lower places, down to the squalid obscurity of the pauper and the criminal. Upon the most liberal estimate, I suppose the former group will not amount to two per cent. of the population. I doubt if the latter exceeds another two per cent.; but let it be supposed, for the sake of argument, that it is as great as five per cent.[22]

As it is only in the latter group that anything comparable to the struggle for existence in the state of nature can take place; as it is only among this twentieth of the whole people that numerous men, women, and children die of rapid or slow starvation, or of the diseases incidental to permanently bad conditions of life; and as there is nothing to prevent their multiplication before they are killed off, while, in spite of greater infant mortality, they increase faster than the rich; it seems clear that the struggle for existence in this class can have no appreciable selective influence upon the other 95 per cent. of the population.

What sort of a sheep breeder would he be who should content himself with picking out the worst fifty out of a thousand, leaving them on a barren common till the weakest starved, and then letting the survivors go back to mix with the rest? And the parallel is too favourable; since in a large number of cases, the actual poor and the convicted criminals are neither the weakest nor the worst.

In the struggle for the means of enjoyment, the qualities which ensure success are energy, industry, intellectual capacity, tenacity of purpose, and, at least as much sympathy as is necessary to make a man understand the feelings of his fellows. Were there none of those artificial arrangements by which fools and knaves are kept at the top of society instead of sinking to their natural place at the bottom,[23] the struggle for the means of enjoyment would ensure a constant circulation of the human units of the social compound, from the bottom to the top and from the top to the bottom. The survivors of the contest, those who continued to form the great bulk of the polity, would not be those 'fittest' who got to the very top, but the great body of the moderately "fit," whose numbers and superior propagative power, enable them always to swamp the exceptionally endowed minority.

I think it must be obvious to every one, that, whether we consider the internal or the external interests of society, it is desirable they should be in the hands of those who are endowed with the largest share of energy, of industry, of intellectual capacity, of tenacity of purpose, while they are not devoid of sympathetic humanity; and, in so far as the struggle for the means of enjoyment tends to place such men in possession of wealth and influence, it is a process which tends to the good of society. But the process, as we have seen, has no real resemblance to that which adapts living beings to current conditions in the state of nature; nor any to the artificial selection of the horticulturist.

XV

To return, once more, to the parallel of horticulture. In the modern world, the gardening of men by themselves is practically restricted to the performance, not of selection, but of that other function of the gardener, the creation of conditions more favourable than those of the state of nature; to the end of facilitating the free expansion of the innate faculties of the citizen, so far as it is consistent with the general good. And the business of the moral and political philosopher appears to me to be the ascertainment, by the same method of observation, experiment, and ratiocination, as is practised in other kinds of scientific work, of the course of conduct which will best conduce to that end.

But, supposing this course of conduct to be scientifically determined and carefully followed out, it cannot put an end to the struggle for existence in the state of nature; and it will not so much as tend, in any way, to the adaptation of man to that state. Even should the whole human race be absorbed in one vast polity, within which "absolute political justice" reigns, the struggle for existence with the state of nature outside it, and the tendency to the return of the struggle within, in consequence of over-multiplication, will remain; and, unless men's inheritance from the ancestors who fought a good fight in the state of nature, their dose of original sin, is rooted out by some method at present unrevealed, at any rate to disbelievers in supernaturalism, every child born into the world will still bring with him the instinct of unlimited self-assertion. He will have to learn the lesson of self-restraint and renunciation. But the practice of self-restraint and renunciation is not happiness, though it may be something much better.

That man, as a 'political animal,' is susceptible of a vast amount of improvement, by education, by instruction, and by the application of his intelligence to the adaptation of the conditions of life to his higher needs, I entertain not the slightest doubt. But, so long as he remains liable to error, intellectual or moral; so long as he is compelled to be perpetually on guard against the cosmic forces, whose ends are not his ends, without and within himself; so long as he is haunted by inexpugnable memories and hopeless aspirations; so long as the recognition of his intellectual limitations forces him to acknowledge his incapacity to penetrate the mystery of existence; the prospect of attaining untroubled happiness, or of a state which can, even remotely, deserve the title of perfection, appears to me to be as misleading an illusion as ever was dangled before the eyes of poor humanity. And there have been many of them.

That which lies before the human race is a constant struggle to maintain and improve, in opposition to the State of Nature, the State of Art of an organized polity; in which, and by which, man may develop a worthy civilization, capable of maintaining and constantly improving itself, until the evolution of our globe shall have entered so far upon its downward course that the cosmic process resumes its sway; and, once more, the State of Nature prevails over the surface of our planet.

Note (see p. 30).—It seems the fashion nowadays to ignore Hartley; though, a century and a half ago, he not only laid the foundations but built up much of the superstructure of a true theory of the Evolution of the intellectual and moral faculties. He speaks of what I have termed the ethical process as "our Progress from Self-interest to Self-annihilation." *Observations on Man* (1749), vol. ii. p. 281.

Notes

1 See "On a piece of Chalk" in the preceding volume of these Essays (vol. viii. p. 1).

2 That every theory of evolution must be consistent not merely with progressive development, but with indefinite persistence in the same condition and with retrogressive modification, is a point which I have insisted upon repeatedly from the year 1862 till now. See *Collected Essays*, vol. ii. pp.

461–89; vol. iii. p. 33; vol. viii. p. 304. In the address on "Geological Contemporaneity and Persistent Types" (1862), the paleontological proofs of this proposition were, I believe, first set forth.

3 "On the Border Territory between the Animal and the Vegetable Kingdoms," Essays, vol. viii. p. 162.

4 See "Evolution in Biology," Essays, vol. ii. p. 187.

5 *Collected Essays*, vol. ii. *passim*.

6 *Ibid.*, vol. iv. p. 138; vol. v. pp. 71–73.

7 *Ibid.*, vol. viii. p. 321.

8 The sense of the term "Art" is becoming narrowed; "work of Art" to most people means a picture, a statue, or a piece of *bijouterie*; by way of compensation "artist" has included in its wide embrace cooks and ballet girls, no less than painters and sculptors.

9 See "Man's Place in Nature," *Collected Essays*, vol. vii., and "On the Struggle for Existence in Human Society" (1888), below.

10 Or to put the case still more simply. When a man lays hold of the two ends of a piece of string and pulls them, with intent to break it, the right arm is certainly exerted in antagonism to the left arm; yet both arms derive their energy from the same original source.

11 Not that the conception of such a society is necessarily based upon the idea of evolution. The Platonic state testifies to the contrary.

12 *Collected Essays*, vol. i., "Animal Automatism"; vol. v., "Prologue," pp. 45 *et seq*.

13 *Collected Essays*, vol. v., Prologue, pp. 50–54.

14 See below. Romanes' Lecture, note 7.

15 Adam Smith makes the pithy observation that the man who sympathises with a woman in childbed, cannot be said to put himself in her place. ("The Theory of the Moral Sentiments," Part vii. sec. iii. chap. i.) Perhaps there is more humour than force in the example; and, in spite of this and other observations of the same tenor, I think that the one defect of the remarkable work in which it occurs is that it lays too much stress on conscious substitution, too little on purely reflex sympathy.

16 Esther v. 9–13. "... but when Haman saw Mordecai in the king's gate, that he stood not up, nor moved for him, he was full of indignation against Mordecai ... And Haman told them of the glory of his riches ... and all the things wherein the king had promoted him ... Yet all this availeth me nothing, so long as I see Mordecai the Jew sitting at the king's gate." What a shrewd exposure of human weakness it is!

17 "Theory of the Moral Sentiments," Part iii. chap. 3. *On the influence and authority of conscience.*

18 Worked out, in its essential features, chiefly by Hartley and Adam Smith, long before the modern doctrine of evolution was thought of. See *Note* below, p. 45.

19 See the essay "On the Struggle for Existence in Human Society" below; and *Collected Essays*, vol. i. p. 276, for Kant's recognition of these facts.

20 *Collected Essays*, vol. v., Prologue, p. 52.

21 Whether the struggle for existence with the state of nature and with other societies, so far as they stand in the relation of the state of nature with it, exerts a selective influence upon modern society, and in what direction, are questions not easy to answer. The problem of the effect of military and industrial warfare upon those who wage it is very complicated.

22 Those who read the last Essay in this volume will not accuse me of wishing to attenuate the evil of the existence of this group, whether great or small.

23 I have elsewhere lamented the absence from society of a machinery for facilitating the descent of incapacity. "Administrative Nihilism." *Collected Essays*, vol. i. p. 54.

FURTHER READING

Alexander, Richard D. *The Biology of Moral Systems*. New York: Aldine De Gruyter, 1987.

Ayres, Clarence. *Huxley*. New York: W.W. Norton and Co., Inc., 1932.

Barr, Alan P., ed. *Thomas Henry Huxley's Place in Science and Letters: Centenary Essays*. Athens: University of Georgia Press, 1997.

de Beer, Gavin, ed. *Autobiographies: Charles Darwin, Thomas Henry Huxley*. London: Oxford UP, 1974.

Bibby, Cyril. *Scientist Extraordinary: The Life and Scientific Work of Thomas Henry Huxley 1825–1895*. Oxford: Pergamon Press, 1972.

—, ed. *T.H. Huxley on Education*. Cambridge: Cambridge UP, 1971.

—. *T.H. Huxley: Scientist, Humanist and Educator.* London: Watts, 1959.

Clodd, Edward. *Thomas Henry Huxley*. Edinburgh: W. Blackwood, 1902.

Desmond, Adrian. *Huxley: From Devil's Disciple to Evolution's High Priest*. Reading, MA: Addison-Wesley, 1997.

—. *Archetypes and Ancestors: Palaeontology in Victorian London 1850–1875.* Chicago: University of Chicago Press, 1982.

Foster, Sir Michael, and E. Ray Lankester. *The Scientific Memoirs of Thomas Henry Huxley.* 4 vols. London: Macmillan and Co., 1898–1902.

di Gregorio, Mario A. *T.H. Huxley's Place in Natural Science.* New Haven, CT: Yale UP, 1984.

Huxley, Leonard. *Life and Letters of Thomas Henry Huxley.* 2 vols. New York: D. Appleton and Company, 1909.

—. *Thomas Henry Huxley: A Character Sketch.* 1920. Reprint, Freeport, NY: Books for Libraries Press, 1969.

Huxley, T.H. *Collected Essays.* 9 vols. London: Macmillan and Co., 1893–4.

—. *Evolution and Ethics.* Edited by James G. Paradis and George C. Williams. Princeton: Princeton UP, 1989.

—. *T.H. Huxley's Diary of the Voyage of the H.M.S. Rattlesnake.* Edited by Julian Huxley. London: Chatto and Windus, 1935.

—. *The Major Prose Works of Thomas Henry Huxley.* Edited by Alan P. Barr. Athens: University of Georgia Press, 1997.

—. *Science and Christian Tradition.* New York: D. Appleton and Company, 1897.

Irvine, William. *Apes, Angels, and Victorians: The Story of Darwin, Huxley, and Evolution.* New York: McGraw-Hill Book Co., Inc., 1955.

Jensen, J. Vernon. "Return to the Wilberforce-Huxley Debate." *British Journal for the History of Science* 21 (1988): 161–79.

—. *Thomas Henry Huxley: Communicating for Science.* Newark: University of Delaware Press, 1991.

Mitchell, P. Chalmers. *Thomas Henry Huxley: A Sketch of His Life and Work.* New York: G.P. Putnam's Sons, 1900.

Moore, James R. *The Post-Darwinian Controversies: A Study of the Protestant Struggle to Come to Terms with Darwin in Great Britain and America 1870–1900.* Cambridge: Cambridge UP, 1979.

Paradis, James G. *T.H. Huxley: Man's Place in Nature.* Lincoln: University of Nebraska Press, 1978.

Peterson, Houston. *Huxley: Prophet of Science.* London: Longmans, Green and Co., 1932.

Pyle, Andrew. *Agnosticism: Contemporary Responses to Spencer and Huxley.* Bristol: Thoemmes Press, 1995.

Smith, C.U.M. "Worlds in Collision: Owen and Huxley on the Brain." *Science in Context* 10, no. 2 (1997): 343–65.

Ulke, Karl-Dieter. *Agnostisches Denken im Viktorianischen England.* Freiburg: Karl Alber, 1980.

Weber, Bruce H., David J. Depew, and James D. Smith, eds. *Entropy, Information, and Evolution: New Perspectives on Physical and Biological Evolution.* Cambridge, MA: M.I.T. Press, 1988.

Zygon. Special issue on *Evolutionary Biology and the Problem of Evil* 23, no. 4 (1988): 379–480.

33. WILHELM CONRAD RÖNTGEN

(1845–1923)

No other scientific event of the nineteenth century captured the public imagination so completely as the discovery of X-rays by Wilhelm Conrad Röntgen in 1895. The announcement of the discovery of X-rays reached the popular press immediately and provoked disbelief as well as widespread curiosity and wonder. After the first x-ray photographs were published in 1896, men and women in Europe and America rushed to have their hands photographed by the mysterious rays or to see their own body parts on Thomas Edison's portable fluoroscope. The almost god-like power of X-rays to reveal hidden worlds also inspired great fear and apprehension; a London firm, for example, soon began advertising "X-ray proof" undergarments.

Röntgen was born in Lennep, Germany, but moved to Apeldoorn, Holland as a child with his family, who were engaged in the cloth industry. His early schooling was marred by his expulsion from a Utrecht school for a prank. He obtained a mechanical engineering degree from the Zürich Polytechnic School and his Ph.D. from the University of Zurich in 1869 after submitting a thesis on gases. He was appointed as a professor of physics at the University of Giessen from 1879–88. Röntgen served in 1894 as rector of the University of Würzburg, where he discovered X-rays. He also made significant contributions to the understanding of light, the specific heats of gases, crystallography, and piezoelectric materials.

On November 8, 1895, using Hittorf and Crookes vacuum tubes covered with black paper to study cathode rays (following up work by Lenard and Hertz), Röntgen noticed the fluorescence of barium platinocyanide crystals next to his experimental apparatus; he immediately began experimenting with these new kind of rays, which were obviously different from cathode rays (streams of electrons). He determined that X-rays were invisible to the eye, penetrated a variety of objects, including his own hand, and could not be deflected by a magnet. He communicated his results in December to the Physical-Medical Society of Würzburg as "Eine Neue Art von Strahlen" ("A New Kind of Rays"). He followed up this report with two other communications. Unexplained images and the fogging of photographic plates while using vacuum tubes had been independently noted by several researchers prior to Röntgen, but no one had investigated these effects further. By early 1896, the Ruhmkorff Coils, and Lenard, Hittorf, or Crookes vacuum tubes needed to produce X-rays could be found in most major laboratories throughout the world as scientists rushed to replicate Röntgen's experiments. The medical and industrial uses (detection of manufacturing defects) of X-rays were immediately recognized by Röntgen and his contemporaries.

To deal with the wealth of scientific material relating to Röntgen- or X-rays, several new journals appeared in 1896 and 1897, including *Archives of Clinical Skiagraphy* (later *Archives of the Roentgen Ray*), *The American X-ray Journal*, and *Fortschritte auf dem Gebiete der Röntgenstrahlen*. Otto Glasser's bibliography lists over 1,000 books and articles on X-rays which appeared in 1896 alone.

Röntgen's discovery brought him instant international fame, including awards, medals, and honorary degrees. He was the obvious candidate for the first Nobel prize in physics awarded in 1901. He also received the Rumford Gold Medal from the Royal Society and the Barnard Medal from Columbia University in 1900. In 1900, he moved to the Physical Science Institute at Ludwig-Maximilians Universität in Munich.

Although x-ray equipment was used by medical units during World War I, Röntgen did not directly profit financially from his discovery; on the contrary, he suffered great losses during the debilitating post-war inflation in Germany. Röntgen never enjoyed public lecturing or engagements and spent his happiest hours quietly in his laboratory.

The other major scientific discovery of Röntgen, eclipsed by his work on X-rays, was his demonstration in 1898 of the so-called "Röntgen Current," the existence of magnetic fields in dielectric materials, such as glass, when they are moved between electrically charged condenser plates, an effect predicted from the Faraday-Maxwell electromagnetic theory. These experiments and Röntgen's explanation of the effect provided further proof of James Clerk Maxwell's later theories of electromagnetism. Below is the first English translation of Röntgen's first scientific communication on X-rays containing the first published x-ray photographs.

"On a New Kind of Rays."[1]

Nature 53, no. 1369 (1896): 274–6.

(1) A DISCHARGE from a large induction coil is passed through a Hittorf's vacuum tube, or through a well-exhausted Crookes' or Lenard's tube. The tube is surrounded by a fairly close-fitting shield of black paper; it is then possible to see, in a completely darkened room, that paper covered on one side with barium platinocyanide lights up with brilliant fluorescence when brought into the neighbourhood of the tube, whether the painted side or the other be turned towards the tube. The fluorescence is still visible at two metres distance. It is easy to show that the origin of the fluorescence lies within the vacuum tube.

(2) It is seen, therefore, that some agent is capable of penetrating black cardboard which is quite opaque to ultra-violet light, sunlight, or arc-light. It is therefore of interest to investigate how far other bodies can be penetrated by the same agent. It is readily shown that all bodies possess this same transparency, but in very varying degrees. For example, paper is very transparent; the fluorescent screen will light up when placed behind a book of a thousand pages; printer's ink offers no marked resistance. Similarly the fluorescence shows behind two packs of cards; a single card does not visibly diminish the brilliancy of the light. So, again, a single thickness of tinfoil hardly casts a shadow on the screen; several have to be superposed to produce a marked effect. Thick blocks of wood are still transparent. Boards of pine two or three centimetres thick absorb only very little. A piece of sheet aluminium, 15 mm. thick, still allowed the X-rays (as I will call the rays, for the sake of brevity) to pass, but greatly reduced the fluorescence. Glass plates of similar thickness behave similarly; lead glass is, however, much more opaque than glass free from lead. Ebonite several centimetres thick is transparent. If the hand be held before the fluorescent screen,

1 By W.C. Röntgen. Translated by Arthur Stanton from the *Sitzungsberichte der Würzburger Physik-medic. Gesellschaft*, 1895.

the shadow shows the bones darkly, with only faint outlines of the surrounding tissues.

Water and several other fluids are very transparent. Hydrogen is not markedly more permeable than air. Plates of copper, silver, lead, gold, and platinum also allow the rays to pass, but only when the metal is thin. Platinum .2 mm. thick allows some rays to pass; silver and copper are more transparent. Lead 1.5 mm. thick is practically opaque. If a square rod of wood 20 mm. in the side be painted on one face with white lead, it casts little shadow when it is so turned that the painted face is parallel to the X-rays, but a strong shadow if the rays have to pass through the painted side. The salts of the metals, either solid or in solution, behave generally as the metals themselves.

(3) The preceding experiments lead to the conclusion that the density of the bodies is the property whose variation mainly affects their permeability. At least no other property seems so marked in this connection. But that the density alone does not determine the transparency is shown by an experiment wherein plates of similar thickness of Iceland spar, glass, aluminium, and quartz were employed as screens. Then the Iceland spar showed itself much less transparent than the other bodies, though of approximately the same density. I have not remarked any strong fluorescence of Iceland spar compared with glass (see below, No. 4).

(4) Increasing thickness increases the hindrance offered to the rays by all bodies. A picture has been impressed on a photographic plate of a number of superposed layers of tinfoil, like steps, presenting thus a regularly increasing thickness. This is to be submitted to photometric processes when a suitable instrument is available.

(5) Pieces of platinum, lead, zinc, and aluminium foil were so arranged as to produce the same weakening of the effect. The annexed table shows the relative thickness and density of the equivalent sheets of metal.

	Thickness.	Relative thickness.	Density.
Platinum	.018 mm.	1	21.5
Lead	.050 "	3	11.3
Zinc	.100 "	6	7.1
Aluminium	3.500 "	200	2.6

From these values it is clear that in no case can we obtain the transparency of a body from the product of its density and thickness. The transparency increases much more rapidly than the product decreases.

(6) The fluorescence of barium platinocyanide is not the only noticeable action of the X-rays. It is to be observed that other bodies exhibit fluorescence, *e.g.* calcium sulphide, uranium glass, Iceland spar, rock-salt, etc.

Of special interest in this connection is the fact that photographic dry plates are sensitive to the X-rays. It is thus possible to exhibit the phenomena so as to exclude the danger of error. I have thus confirmed many observations originally made by eye observation with the fluorescent screen. Here the power of the X-rays to pass through wood or cardboard becomes useful. The photographic plate can be exposed to the action without removal of the shutter of the dark slide or other protecting case, so that the experiment need not be conducted in darkness. Manifestly, unexposed plates must not be left in their box near the vacuum tube.

It seems now questionable whether the impression on the plate is a direct effect of the X-rays, or a secondary result induced by the fluorescence of the material of the plate. Films can receive the impression as well as ordinary dry plates.

I have not been able to show experimentally that the X-rays give rise to any calorific effects. These, however, may be assumed, for the phenomena of fluorescence show that the X-rays are capable of transformation. It is also certain that all the X-rays falling on a body do not leave it as such.

The retina of the eye is quite insensitive to these rays: the eye placed close to the apparatus sees nothing. It is clear from the experiments that this is not due to want of permeability on the part of the structures of the eye.

(7) After my experiments on the transparency of increasing thicknesses as of different media, I proceeded to investigate whether the X-rays could be deflected by a prism. Investigations with water and carbon bisulphide in mica prisms of 30° showed no deviation either on the photographic or the fluorescent plate. For comparison, light rays were allowed to fall on the prism as the

apparatus was set up for the experiment. They were deviated 10 mm. and 20 mm. respectively in the case of the two prisms.

With prisms of ebonite and aluminium, I have obtained images on the photographic plate, which point to a possible deviation. It is, however, uncertain, and at most would point to a refractive index 1.05. No deviation can be observed by means of the fluorescent screen. Investigations with the heavier metals have not as yet led to any result, because of their small transparency and the consequent enfeebling of the transmitted rays.

On account of the importance of the question it is desirable to try in other ways whether the X-rays are susceptible of refraction. Finely powdered bodies allow in thick layers but little of the incident light to pass through, in consequence of refraction and reflection. In the case of the X-rays, however, such layers of powder are for equal masses of substance equally transparent with the coherent solid itself. Hence we cannot conclude any regular reflection or refraction of the X-rays. The research was conducted by the aid of finely-powdered rock-salt, fine electrolytic silver powder, and zinc dust already many times employed in chemical work. In all these cases the result, whether by the fluorescent screen or the photographic method, indicated no difference in transparency between the powder and the coherent solid.

It is, hence, obvious that lenses cannot be looked upon as capable of concentrating the X-rays; in effect, both an ebonite and a glass lens of large size prove to be without action. The shadow photograph of a round rod is darker in the middle than at the edge; the image of a cylinder filled with a body more transparent than its walls exhibits the middle brighter than the edge.

(8) The preceding experiments, and others which I pass over, point to the rays being incapable of regular reflection. It is, however, well to detail an observation which at first sight seemed to lead to an opposite conclusion.

I exposed a plate, protected by a black paper sheath, to the X-rays so that the glass side lay next to the vacuum tube. The sensitive film was partly covered with star-shaped pieces of platinum, lead, zinc, and aluminium. On the developed negative the star-shaped impression showed dark under platinum, lead, and, more markedly, under zinc; the aluminium gave no image. It seems, therefore, that these three metals can reflect the X-rays; as, however, another explanation is possible, I repeated the experiment with this only difference, that a film of thin aluminium foil was interposed between the sensitive film and the metal stars. Such an aluminium plate is opaque to ultra-violet rays, but transparent to X-rays. In the result the images appeared as before, this pointing still to the existence of reflection at metal surfaces.

If one considers this observation in connection with others, namely, on the transparency of powders, and on the state of the surface not being effective in altering the passage of the X-rays through a body, it leads to the probable conclusion that regular reflection does not exist, but that bodies behave to the X-rays as turbid media to light.

Since I have obtained no evidence of refraction at the surface of different media, it seems probable that the X-rays move with the same velocity in all bodies, and in a medium which penetrates everything, and in which the molecules of bodies are embedded. The molecules obstruct the X-rays, the more effectively as the density of the body concerned is greater.

(9) It seemed possible that the geometrical arrangement of the molecules might affect the action of a body upon the X-rays, so that, for example, Iceland spar might exhibit different phenomena according to the relation of the surface of the plate to the axis of the crystal. Experiments with quartz and Iceland spar on this point lead to a negative result.

(10) It is known that Lenard, in his investigations on kathode rays, has shown that they belong to the ether, and can pass through all bodies. Concerning the X-rays the same may be said.

In his latest work, Lenard has investigated the absorption coefficients of various bodies for the kathode rays, including air at atmospheric pressure, which gives 4.10, 3.40, 3.10 for 1 cm., according to the degree of exhaustion of the gas in [the] discharge tube. To judge from the nature of the discharge, I have worked at about the same pressure, but occasionally at greater or smaller pressures. I find, using a Weber's photometer, that the intensity of the fluorescent light varies nearly as the inverse square of the distance between screen and discharge tube. This result is obtained from three very consistent sets of observations at distances of 100 and 200 mm. Hence air absorbs the X-rays much less than the kathode rays. This result is in complete agreement with the previously described result, that the fluorescence of the screen can be still observed at 2 metres from the vacuum tube. In general, other bodies behave like air; they are more transparent for the X-rays than for the kathode rays.

(11) A further distinction, and a noteworthy one, results from the action of a magnet. I have not succeeded in observing any deviation of the X-rays even in very strong magnetic fields.

The deviation of kathode rays by the magnet is one of their peculiar characteristics; it has been observed by Hertz and Lenard, that several kinds of kathode rays exist, which differ by their power of exciting phosphorescence, their susceptibility of absorption, and their deviation by the magnet; but a notable deviation has been observed in all cases which have yet been investigated, and I think that such deviation affords a characteristic not to be set aside lightly.

(12) As the result of many researches, it appears that the place of most brilliant phosphorescence of the walls of the discharge-tube is the chief seat whence the X-rays originate and spread in all directions; that is, the X-rays proceed from the front where the kathode rays strike the glass. If one deviates the kathode rays within the tube by means of a magnet, it is seen that the X-rays proceed from a new point, *i.e.* again from the end of the kathode rays.

Also for this reason the X-rays, which are not deflected by a magnet, cannot be regarded as kathode rays which have passed through the glass, for that passage cannot, according to Lenard, be the cause of the different deflection of the rays. Hence I conclude that the X-rays are not identical with the kathode rays, but are produced from the kathode rays at the glass surface of the tube.

(13) The rays are generated not only in glass. I have obtained them in an apparatus closed by an aluminium plate 2 mm. thick. I purpose later to investigate the behaviour of other substances.

(14) The justification of the term "rays" applied to the phenomena, lies partly in the regular shadow pictures produced by the interposition of a more or less permeable body between the source and a photographic plate or fluorescent screen.

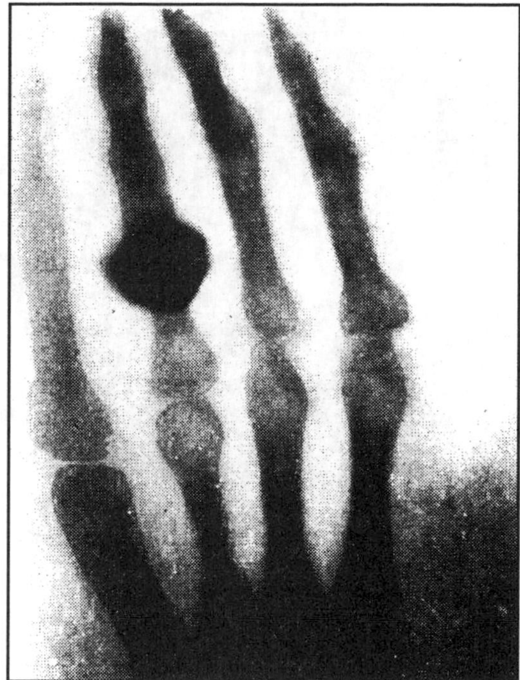

Fig. 1. Photograph of the bones in the fingers of a living human hand. The third finger has a ring upon it.

I have observed and photographed many such shadow pictures. Thus, I have an outline of part of a door covered with lead paint; the image was produced by placing the discharge-tube on one side of the door, and the sensitive plate on the other. I have also a shadow of the bones of the hand (Fig. 1), of a wire wound upon a bobbin, of a set of weights in a box, of a compass card and needle completely enclosed in a metal case (Fig. 2), of a piece of metal where the X-rays show the want of homogeneity, and of other things.

For the rectilinear propagation of the rays, I have a pin-hole photograph of the discharge apparatus covered with black paper. It is faint but unmistakable.

(15) I have sought for interference effects of the X-rays, but possibly, in consequence of their small intensity, without result.

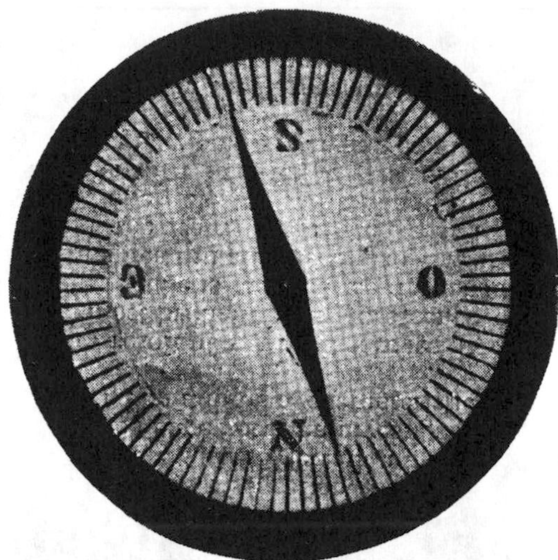

Fig. 2. Photograph of a compass card and needle completely enclosed in a metal case.

(16) Researches to investigate whether electrostatic forces act on the X-rays are begun but not yet concluded.

(17) If one asks, what then are these X-rays; since they are not kathode rays, one might suppose, from their power of exciting fluorescence and chemical action, them to be due to ultra-violet light. In opposition to this view a weighty set of considerations presents itself. If X-rays be indeed ultra-violet light, then that light must possess the following properties.

(*a*) It is not refracted in passing from air into water, carbon bisulphide, aluminium, rock-salt, glass or zinc.

(*b*) It is incapable of regular reflection at the surfaces of the above bodies.

(*c*) It cannot be polarised by any ordinary polarising media.

(*d*) The absorption by various bodies must depend chiefly on their density.

That is to say, these ultra-violet rays must behave quite differently from the visible, infrared, and hitherto known ultra-violet rays.

These things appear so unlikely that I have sought for another hypothesis.

A kind of relationship between the new rays and light rays appears to exist; at least the formation of shadows, fluorescence, and the production of chemical action point in this direction. Now it has been known for a long time, that besides the transverse vibrations which account for the phenomena of light, it is possible that longitudinal vibrations should exist in the ether, and, according to the view of some physicists, must exist. It is granted that their existence has not yet been made clear, and their properties are not experimentally demonstrated. Should not the new rays be ascribed to longitudinal waves in the ether?

I must confess that I have in the course of this research made myself more and more familiar with this thought, and venture to put the opinion forward, while I am quite conscious that the hypothesis advanced still requires a more solid foundation.

FURTHER READING

Beier, Walter. *Wilhelm Conrad Röntgen.* Leipzig: B.G. Teubner, 1985.

Bleich, Alan Ralph. *The Story of X-Rays from Röntgen to Isotopes.* New York: Dover Publications, Inc., 1960.

Claxton, K.T. *Wilhelm Roentgen.* Geneva: Editio–Service, 1970.

Dessauer, Friedrich. *Wilhelm C. Röntgen: Die Offenbarung einer Nacht.* Olten, Switzerland: Verlag Otto Walter, 1945.

Dibner, Bern. *Wilhelm Conrad Röntgen and the Discovery of X rays.* New York: Franklin Watts, Inc., 1968.

Glasser, Otto. *Dr. W.C. Röntgen.* Springfield, IL: Charles C. Thomas, 1945.

—. *Wilhelm Conrad Röntgen and the Early History of the Roentgen Rays.* Springfield, IL: Charles C. Thomas, 1934.

Klickstein, Herbert S. *Wilhelm Conrad Röntgen: On a New Kind of Rays, A Bibliographic Study.* Mallinckrodt Classics of Radiology. Mallinckrodt, 1966.

Leicht, Hans. *Wilhelm Conrad Röntgen: Biographie.* München: Ehrenwirth, 1994.

Michette, Alan, and Slawka Pfauntsch. *X-Rays: The First Hundred Years.* New York: John Wiley and Sons, 1996.

Nicolle, Jacques. *Röntgen.* Paris: Éditions Seghers, 1965.

Nitske, W. Robert. *The Life of Wilhelm Conrad Röntgen.* Tucson: University of Arizona Press, 1971.

Streller, Ernst, Rolf Winau, and Armin Hermann. *Wilhelm Conrad Roentgen 1845–1923.* Bonn: Inter Nationes, 1973.

Thomas, A.M.K. *The Invisible Light: 100 Years of Medical Radiology.* Oxford: Blackwell Science Ltd., 1995.

Unger, Hellmuth. *Wilhelm Conrad Röntgen.* Hamburg: Hoffmann und Campe, 1949.

34. MARIE SKLODOWSKA CURIE

(1867–1934)

Considérons, avec silence,
Celle qui sut donner au jour
Un feu plus beau que l'existence
Et plus dévorant que l'amour.
Maurice Rostand, *Ode à Madame Curie.*

The remarkable story of the discovery of radium, told by Marie Curie's daughter Eve below, began with Henri Becquerel's discovery in 1896 of the spontaneous emission of ionizing radiation (later identified as α-, β-, and γ- rays) from uranium and its compounds. Röntgen in Germany had described X-rays a few months earlier. Curie, who had married the French physicist Pierre Curie in 1895, began an investigation of the properties of these "uranium" or "Becquerel" rays for her doctoral thesis.

Marie Sklodowska's marriage to Pierre Curie initiated one of the greatest scientific collaborations of the nineteenth and early twentieth centuries. As their daughter Irène (who also married and worked with a nuclear physicist, Frédéric Joliot) later wrote: "during the eleven years of their life together, they never left each other's side, neither during their work nor on their vacations" ("Ma Mère" 90). Marie or "Manya" was the daughter of a Polish school teacher. She had endured great hardships, working as a governess, in order to save money to study at the Sorbonne in France. Her perseverance under difficult conditions, both personal and scientific, was characteristic of her entire life and career.

Curie began her study of uranium rays by measuring a number of elements, minerals, and compounds for ionizing activity with a sensitive electrometer designed by Jacques and Pierre Curie, the quartz piezo-electroscope. She studied thorium, which also emitted Becquerel rays, and proposed the term "radioactivity" for the optical and magnetic properties observed by Becquerel, which were unaffected by chemical combination with other elements. Curie examined such minerals as clevite, chalcolite, carnotite, and autunite and determined that those which displayed significant radioactivity commonly contained either uranium or thorium. She also discovered that a uranium-bearing ore, pitchblende, released up to 4 times more radioactivity than uranium oxide and suspected an unknown element. With several tons of ore obtained from the Austrian government, Curie, joined by Pierre, began the laborious process of chemically isolating the source of the extraordinary radiation. They extracted an unstable substance in 1898 and proposed the name "polonium," after Curie's homeland. A second substance was isolated and spectroscopic measurement (each element absorbs and emits a characteristic spectrum of light) by Eugène Demarçay indicated that this was not a known element. They suggested the name "radium" for the new element. In 1902 the Curies isolated a decigram of pure radium chloride by fractional crystallization and received substantial funds from the French Academy of Science for the extraction of radioactive elements. In the meantime, André Debierne had isolated another previously unknown radioactive element, actinium, from pitchblende in 1900. The actual discovery of radium, therefore, consisted of a mixture of brute determination, intuition, and an hypothesis with very few supporting facts, as Curie later explained: "our research method could only be based on the radioactivity of the hypothetical substance, because we were ignorant of any other of its properties" (*Traité* 1.145).

The Curies shared the 1903 Nobel Prize with Becquerel for their work on radium and radioactivity. After Pierre Curie was fatally mangled by a passing carriage in 1906, Curie was offered the chair at the Sorbonne which had been created for her husband. Curie continued her efforts to determine an accurate atomic weight for radium. Radium, which had quickly become the most expensive substance in the world, would soon become medically and commercially valuable with the development of radiotherapy, or curietherapy as it was called in France, for cancer. Like many new pharmaceuticals introduced into western culture, such as opium, tobacco, and cannabis, radium was hailed as a panacea for an astonishing number of diseases, before the devastating effects of large doses of it were illustrated by the deaths of watch dial painters in the 1920s exposed to radium-based paints.

Curie was awarded a second Nobel Prize in 1911 for her continuing work on radium. Several days later, she was denied membership in the Académie des Sciences, demonstrating the considerable professional obstacles encountered by women scientists during the period. French public sentiment concerning this foreign-born and unconventional woman was ambivalent, especially after the publication of letters in 1911 exposing an affair between Curie and Paul Langevin, who was married at the time.

In 1912, construction began on the Institut du Radium in Paris. During World War I, Curie organized a fleet of trucks with portable x-ray units for field use and later wrote *Radiology and War* (1921). Marie Meloney, a U.S. journalist, organized a donations campaign to produce a gram of radium for Curie's laboratory. In 1921, President Warren G. Harding personally presented Curie with the key to a lead-lined casket for housing the gift. Curie raised two daughters, the writer Eve and the physicist Irène Joliot-Curie, who along with her husband Frédéric made important advancements in artificial radioactivity and nuclear fission. In 1934, Curie died from complications from the years of radiation exposure to the element which she had discovered.

Curie, Eve. *Madame Curie: A Biography.*
Translated by Vincent Sheean.
Garden City, NY: Doubleday, Doran and Company, Inc., 1937, pp. 152–164.

THE DISCOVERY OF RADIUM

While a young wife kept house, washed her baby daughter and put pans on the fire, in a wretched laboratory at the School of Physics a woman physicist was making the most important discovery of modern science.

At the end of 1897 the balance sheet of Marie's activity showed two university degrees, a fellowship and a monograph on the magnetization of tempered steel. No sooner had she recovered from childbirth than she was back again at the laboratory.

The next stage in the logical development of

her career was the doctor's degree. Several weeks of indecision came in here. She had to choose a subject of research which would furnish fertile and original material. Like a writer who hesitates and asks himself questions before settling the subject of his next novel, Marie, reviewing the most recent work in physics with Pierre, was in search of a subject for a thesis.

At this critical moment Pierre's advice had an importance which cannot be neglected. With respect to her husband, the young woman regarded herself as an apprentice: he was an older physicist, much more experienced than she. He was even, to put it exactly, her chief, her "boss."

But without a doubt Marie's character, her intimate nature, had a great part in this all-important choice. From childhood the Polish girl had carried the curiosity and daring of an explorer within her. This was the instinct that had driven her to leave Warsaw for Paris and the Sorbonne, and had made her prefer a solitary room in the Latin Quarter to the Dluskis' downy nest. In her walks in the woods she always chose the wild trail or the unfrequented road.

At this moment she was like a traveler musing on a long voyage. Bent over the globe and pointing out, in some far country, a strange name that excites his imagination, the traveler suddenly decides to go there and nowhere else: so Marie, going through the reports of the latest experimental studies, was attracted by the publication of the French scientist Henri Becquerel of the preceding year. She and Pierre already knew this work; she read it over again and studied it with her usual care.

After Roentgen's discovery of X rays, Henri Poincaré conceived the idea of determining whether rays like the X ray were emitted by "fluorescent" bodies under the action of light. Attracted by the same problem, Henri Becquerel examined the salts of a "rare metal," uranium. Instead of finding the phenomenon he had expected, he observed another, altogether different and incomprehensible: he found that uranium salts *spontaneously* emitted, without exposure to light, some rays of unknown nature. A compound of uranium, placed on a photographic plate surrounded by black paper, made an impression on the plate through the paper. And, like the X ray these astonishing "uranic" salts discharged an electroscope by rendering the surrounding air a conductor.

Henri Becquerel made sure that these surprising properties were not caused by a preliminary exposure to the sun and that they persisted when the uranium compound had been maintained in darkness for several months. For the first time, a physicist had observed the phenomenon to which Marie Curie was later to give the name of *radioactivity*. But the nature of the radiation and its origin remained an enigma.

Becquerel's discovery fascinated the Curies. They asked themselves whence came the energy—tiny, to be sure—which uranium compounds constantly disengaged in the form of radiation. And what was the nature of this radiation? Here was an engrossing subject of research, a doctor's thesis! The subject tempted Marie most because it was a virgin field: Becquerel's work was very recent and so far as she knew nobody in the laboratories of Europe had yet attempted to make a fundamental study of uranium rays. As a point of departure, and as the only bibliography, there existed some communications presented by Henri Becquerel at the Academy of Science during the year 1896. It was a leap into great adventure, into an unknown realm.

There remained the question of where she was to make her experiments—and here the difficulties began. Pierre made several approaches to the director of the School of Physics with practically no results: Marie was given the free use of a little glassed-in studio on the ground floor of the school. It was a kind of storeroom, sweating with damp, where unused machines and lumber were put away. Its technical equipment was rudimentary and its comfort nil.

Deprived of an adequate electrical installation and of everything that forms material for the beginning of scientific research, she kept her patience, sought and found a means of making her apparatus work in this hole.

It was not easy. Instruments of precision have sneaking enemies: humidity, changes of temperature. Incidentally the climate of this little workroom, fatal to the sensitive electrometer, was not much better for Marie's health. But this had no importance. When she was cold, the young woman took her revenge by noting the degrees of temperature in centigrade in her notebook. On February 6, 1898, we find, among the formulas and figures: "Temperature here 6° 25." Six degrees...![1] Marie, to show her disapproval, added ten little exclamation points.

The candidate for the doctor's degree set her first task to be the measurement of the "power of ionization" of uranium rays—that is to say, their power to render the air a conductor of electricity and so to discharge an electroscope. The excellent method she used, which was to be the key to the success of her experiments, had been invented for the study of other phenomena by two physicists well known to her: Pierre and Jacques Curie. Her technical installation consisted of an "ionization chamber," a Curie electrometer and a piezoelectric quartz.

At the end of several weeks the first result appeared: Marie acquired the certainty that the intensity of this surprising radiation was proportional to the quantity of uranium contained in the samples under examination, and that this radiation, which could be measured with precision, was not affected either by the chemical state of combination of the uranium or by external factors such as lighting or temperature.

These observations were perhaps not very sensational to the uninitiated, but they were of passionate interest to the scientist. It often happens in physics that an inexplicable phenomenon can be subjected, after some investigation, to laws already known, and by this very fact loses its interest for the research worker. Thus, in a badly constructed detective story, if we are told in the third chapter that the woman of sinister appearance who might have committed the crime is in reality only an honest little housewife who leads a life without secrets, we feel discouraged and cease to read.

Nothing of the kind happened here. The more Marie penetrated into intimacy with uranium rays, the more they seemed without precedent, essentially unknown. They were like nothing else. Nothing affected them. In spite of their very feeble power, they had an extraordinary individuality.

Turning this mystery over and over in her head, and pointing toward the truth, Marie felt and could soon affirm that the incomprehensible radiation was an *atomic* property. She questioned: Even though the phenomenon had only been observed with uranium, nothing proved that uranium was the only chemical element capable of emitting such radiation. Why should not other bodies possess the same power? Perhaps it was only by chance that this radiation had been observed in uranium first, and had remained attached to uranium in the minds of physicists. Now it must be sought for elsewhere....

No sooner said than done. Abandoning the study of uranium, Marie undertook to examine *all known chemical bodies*, either in the pure state or in compounds. And the result was not long in appearing: compounds of another element, thorium, also emitted spontaneous rays like those of uranium and of similar intensity. The physicist had been right: the surprising phenomenon was by no means the property of uranium alone, and it became necessary to give it a distinct name. Mme Curie suggested the name of *radioactivity*. Chemical substances like uranium and thorium, endowed with this particular "radiance," were called *radio elements*.

Radioactivity so fascinated the young scientist that she never tired of examining the most diverse forms of matter, always by the same method. Curiosity, a marvelous feminine curiosity, the first virtue of a scientist, was developed in Marie to the highest degree. Instead of limiting her observation to simple compounds, salts and oxides, she had the desire to assemble samples of minerals from the collection at the School of Physics, and of making them undergo almost at hazard, for her own amusement, a kind of customs inspection which is an electrometer test. Pierre approved, and chose with her the veined fragments, hard or crumbly, oddly shaped, which she wanted to examine.

Marie's idea was simple—simple as the stroke of genius. At the crossroads where Marie now stood, hundreds of research workers might have remained, nonplussed, for months or even years.

1 About 44° Fahrenheit.

After examining all known chemical substances, and discovering—as Marie had done—the radiation of thorium, they would have continued to ask themselves in vain whence came this mysterious radioactivity. Marie, too, questioned and wondered. But her surprise was translated into fruitful acts. She had used up all evident possibilities. Now she turned toward the unplumbed and the unknown.

She knew in advance what she would learn from an examination of the minerals, or rather she thought she knew. The specimens which contained neither uranium nor thorium would be revealed as totally "inactive." The others, containing uranium or thorium, would be radioactive.

Experiment confirmed this prevision. Rejecting the inactive minerals, Marie applied herself to the others and measured their radioactivity. Then came a dramatic revelation: the radioactivity was a *great deal stronger* than could have been normally foreseen by the quantity of uranium or thorium contained in the products examined!

"It must be an error in experiment," the young woman thought; for doubt is the scientist's first response to an unexpected phenomenon.

She started her measurements over again, unmoved, using the same products. She started over again ten times, twenty times. And she was forced to yield to the evidence: the quantities of uranium and of thorium found in these minerals were by no means sufficient to justify the exceptional intensity of the radiation she observed.

Where did this excessive and abnormal radiation come from? Only one explanation was possible: the minerals must contain, in small quantity, a *much more powerfully radioactive substance* than uranium and thorium.

But what substance? In her preceding experiments, Marie had already examined *all known chemical elements*.

The scientist replied to the question with the sure logic and the magnificent audaciousness of a great mind: The minerals certainly contained a radioactive substance, which was at the same time a chemical element unknown until this day: *a new element*.

A new element! It was a fascinating and alluring hypothesis—but still a hypothesis. For the moment this powerfully radioactive substance existed only in the imagination of Marie and of Pierre. But it did exist there. It existed strongly enough to make the young woman go to see Bronya one day and tell her in a restrained, ardent voice:

"You know, Bronya, the radiation that I couldn't explain comes from a new chemical element. The element is there and I've got to find it. We are sure! The physicists we have spoken to believe we have made an error in experiment and advise us to be careful. But I am convinced that I am not mistaken."

These were unique moments in her unique life. The layman forms a theatrical—and wholly false—idea of the research worker and of his discoveries. "The moment of discovery" does not always exist: the scientist's work is too tenuous, too divided, for the certainty of success to crackle out suddenly in the midst of his laborious toil like a stroke of lightning, dazzling him by its fire. Marie, standing in front of her apparatus, perhaps never experienced the sudden intoxication of triumph. This intoxication was spread over several days of decisive labor, made feverish by a magnificent hope. But it must have been an exultant moment when, convinced by the rigorous reasoning of her brain that she was on the trail of new matter, she confided the secret to her elder sister, her ally always....Without exchanging one affectionate word, the two sisters must have lived again, in a dizzying breath of memory, their years of waiting, their mutual sacrifices, their bleak lives as students, full of hope and faith.

It was barely four years before that Marie had written:

> Life is not easy for any of us. But what of that? we must have perseverance and above all confidence in ourselves. We must believe that we are gifted for something, and that this thing, at whatever cost, must be attained.

That "something" was to throw science upon a path hitherto unsuspected.

In a first communication to the Academy, presented by Prof. Lippmann and published in the *Proceedings* on April 12, 1898, "Marie Sklodovska

Curie" announced the probable presence in pitchblende ores of a new element endowed with powerful radioactivity. This was the first stage of the discovery of radium.

By the force of her own intuition the physicist had shown to herself that the wonderful substance must exist. She decreed its existence. But its incognito still had to be broken. Now she would have to verify hypothesis by experiment, isolate this material and see it. She must be able to announce with certainty: "It is there."

Pierre Curie had followed the rapid progress of his wife's experiments with passionate interest. Without directly taking part in Marie's work, he had frequently helped her by his remarks and advice. In view of the stupefying character of her results, he did not hesitate to abandon his study of crystals for the time being in order to join his efforts to hers in the search for the new substance.

Thus, when the immensity of a pressing task suggested and exacted collaboration, a great physicist was at Marie's side—a physicist who was the companion of her life. Three years earlier, love had joined this exceptional man and woman together—love, and perhaps some mysterious foreknowledge, some sublime instinct for the work in common.

The available force was now doubled. Two brains, four hands now sought the unknown element in the damp little workroom in the Rue Lhomond. From this moment onward it is impossible to distinguish each one's part in the work of the Curies. We know that Marie, having chosen to study the radiation of uranium as the subject of her thesis, discovered that other substances were also radioactive. We know that after the examination of minerals she was able to announce the existence of a new chemical element, powerfully radioactive, and that it was the capital importance of this result which decided Pierre Curie to interrupt his very different research in order to try to isolate this element with his wife. At that time—May or June 1898—a collaboration began which was to last for eight years, until it was destroyed by a fatal accident.

We cannot and must not attempt to find out what should be credited to Marie and what to Pierre during these eight years. It would be exactly what the husband and wife did not want. The personal genius of Pierre Curie is known to us by the original work he had accomplished before this collaboration. His wife's genius appears to us in the first intuition of discovery, the brilliant start; and it was to reappear to us again, solitary, when Marie Curie the widow unflinchingly carried the weight of a new science and conducted it, through research, step by step, to its harmonious expansion. We therefore have formal proof that in the fusion of their two efforts, in this superior alliance of man and woman, the exchange was equal.

Let this certainty suffice for our curiosity and admiration. Let us not attempt to separate these creatures full of love, whose handwriting alternates and combines in the working notebooks covered with formulae, these creatures who were to sign nearly all their scientific publications together. They were to write "We found" and "We observed"; and when they were constrained by fact to distinguish between their parts, they were to employ this moving locution:

> Certain minerals containing uranium and thorium (pitchblende, chalcolite, uranite) are very active from the point of view of the emission of Becquerel rays. In a preceding communication, *one of us* showed that their activity was even greater than that of uranium and thorium, and stated the opinion that this effect was due to some other very active substance contained in small quantity in these minerals. (Pierre and Marie Curie: *Proceedings of the Academy of Science*, July 18, 1898.)

Marie and Pierre looked for this "very active" substance in an ore of uranium called pitchblende, which in the crude state had shown itself to be four times more radioactive than the pure oxide of uranium that could be extracted from it. But the composition of this ore had been known for a long time with considerable precision. The new element must therefore be present in very small quantity or it would not have escaped the notice of scientists and their chemical analysis.

According to their calculations—"pessimistic" calculations, like those of true physicists, who always take the less attractive of two probabilities—the collaborators thought the ore should contain the new element to a maximum quantity of one per cent. They decided that this was very little. They would have been in consternation if they had known that the radioactive element they were hunting down did not count for more than a millionth part of pitchblende ore.

They began their prospecting patiently, using a method of chemical research invented by themselves; based on radioactivity they separated all the elements in pitchblende by ordinary chemical analysis and then measured the radioactivity of each of the bodies thus obtained. By successive eliminations they saw the "abnormal" radioactivity take refuge in certain parts of the ore. As they went on, the field of investigation was narrowed. It was exactly the technique used by the police when they search the houses of a neighborhood, one by one, to isolate and arrest a malefactor.

But there was more than one malefactor here: the radioactivity was concentrated principally in two different chemical fractions of the pitchblende. For M. and Mme Curie it indicated the existence of two new elements instead of one. By July 1898 they were able to announce the discovery of one of these substances with certainty.

"You will have to name it," Pierre said to his young wife, in the same tone as if it were a question of choosing a name for little Irène.

The one-time Mlle Sklodovska reflected in silence for a moment. Then, her heart turning toward her own country which had been erased from the map of the world, she wondered vaguely if the scientific event would be published in Russia, Germany and Austria—the oppressor countries—and answered timidly:

"Could we call it 'polonium'?"

In the *Proceedings of the Academy* for July 1898 we read:

> We believe the substance we have extracted from pitchblende contains a metal not yet observed, related to bismuth by its analytical properties. If the existence of this new metal is confirmed we propose to call it *polonium*, from the name of the original country of one of us.

The choice of this name proves that in becoming a Frenchwoman and a physicist Marie had not disowned her former enthusiasms. Another thing proves it for us: even before the note "On a New Radioactive Substance Contained in Pitchblende" had appeared in the *Proceedings of the Academy*, Marie had sent the manuscript to her native country, to that Joseph Boguski who directed the little laboratory at the Museum of Industry and Agriculture where she had made her first experiments. The communication was published in Warsaw in a monthly photographic review called *Swiatlo* almost as soon as in Paris.

Life was unchanged in the little flat in the Rue de la Glacière. Marie and Pierre worked even more than usual; that was all. When the heat of summer came, the young wife found time to buy some baskets of fruit in the markets and, as usual, she cooked and put away preserves for the winter, according to the recipes used in the Curie family. Then she locked the shutters on her windows, which gave on burned leaves; she registered their two bicycles at the Orleans station, and, like thousands of other young women in Paris, went off on holiday with her husband and her child.

This year the couple had rented a peasant's house at Auroux, in Auvergne. Happy to breathe good air after the noxious atmosphere of the Rue Lhomond, the Curies made excursions to Mende, Puy, Clermont, Mont-Dore. They climbed hills, visited grottoes, bathed in rivers. Every day, alone in the country, they spoke of what they called their "new metals," polonium and "the other"—the one that remained to be found. In September they would go back to the damp workroom and the dull minerals; with freshened ardor they would take up their search again. . . .

One grief interfered with Marie's intoxication for work: the Dluskis were on the point of leaving Paris. They had decided to settle in Austrian Poland and to build a sanatorium for tubercular sufferers at Zakopane in the Carpathian Mountains. The day of separation arrived: Marie and Bronya exchanged broken-hearted farewells; Marie was losing her friend and protector, and for the first time she had the feeling of exile.

Marie to Bronya, December 2, 1898:

> You can't imagine what a hole you have made in my life. With you two, I have lost everything

I clung to in Paris except my husband and child. It seems to me that Paris no longer exists, aside from our lodging and the school where we work.

Ask Mme Dluska if the green plant you left behind should be watered, and how many times a day. Does it need a great deal of heat and sun?

We are well, in spite of the bad weather, the rain and the mud. Irène is getting to be a big girl. She is very difficult about her food, and aside from milk tapioca she will eat hardly anything regularly, not even eggs. Write me what would be a suitable menu for persons of her age....

In spite of their prosaic character—or perhaps because of it—some notes written by Mme Curie in that memorable year 1898 seem to us worth quoting. Some are to be found in the margins of a book called *Family Cooking*, with respect to a recipe for gooseberry jelly:

I took eight pounds of fruit and the same weight in crystallized sugar. After an ebullition of ten minutes, I passed the mixture through a rather fine sieve. I obtained fourteen pots of very good jelly, not transparent, which "took" perfectly.

In a school notebook covered with gray linen, in which the young mother had written little Irène's weight day by day, her diet and the appearance of her first teeth, we read under the date of July 20, 1898, some days after the publication of the discovery of polonium:

Irène says "thanks" with her hand. She can walk very well now on all fours. She says "Gogli, gogli, go." She stays in the garden all day at Sceaux on a carpet. She can roll, pick herself up, and sit down.

On August 15, at Auroux:

Irène has cut her seventh tooth, on the lower left. She can stand for half a minute alone. For the past three days we have bathed her in the river. She cries, but today (fourth bath) she stopped crying and played with her hands in the water.

She plays with the cat and chases him with war cries. She is not afraid of strangers any more. She sings a great deal. She gets up on the table when she is in her chair.

Three months later, on October 17, Marie noted with pride:

Irène can walk very well, and no longer goes on all fours.

On January 5, 1899:

Irène has fifteen teeth!

Between these two notes—that of October 17, 1898, in which Irène no longer goes on all fours, and that of January 5 in which Irène has fifteen teeth—and a few months after the note on the gooseberry preserve, we find another note worthy of remark.

It was drawn up by Marie and Pierre Curie and a collaborator called G. Bémont. Intended for the Academy of Science, and published in the *Proceedings* of the session of December 26, 1898, it announced the existence of a second new chemical element in pitchblende.

Some lines of this communication read as follows:

The various reasons we have just enumerated lead us to believe that the new radioactive substance contains a new element to which we propose to give the name of RADIUM.

The new radioactive substance certainly contains a very strong proportion of barium; in spite of that its radioactivity is considerable. The radioactivity of radium therefore must be enormous.

FURTHER READING

Badash, Lawrence. "Radium, Radioactivity, and the Popularity of Scientific Discovery." *Proceedings of the American Philosophical Society* 122, no. 3 (1978): 145–54.

—. *Radioactivity in America: Growth and Decay of a Science.* Baltimore: Johns Hopkins UP, 1979.

Cotton, Eugénie. *Les Curie et la radioactivité.* Paris: Éditions Seghers, 1963.

Curie, Eve. *Madame Curie: A Biography.* Translated by Vincent Sheean. Garden City, NY: Doubleday, Doran and Company, Inc., 1937.

Curie, Marie. *Pierre Curie.* Translated by Charlotte and Vernon Kellogg. New York: Macmillan, 1923.

—. *Radioactive Substances: A Translation from the French of the Classical Thesis Presented to the Faculty of Sciences in Paris by the Distinguished Nobel Prize Winner.* New York: Philosophical Library, 1961.

—. *Traité de radioactivité.* 2 vols. Paris: Gauthier-Villars, 1910.

En L'honneur de Madame Pierre Curie et de la découverte du radium. Paris: Gauthier-Villars et Compagnie, 1922.

Giroud, Françoise. *Marie Curie: A Life.* Translated by Lydia Davis. New York: Holmes and Meier, 1986.

Joliot-Curie, Irène. "Marie Curie, ma mère." *Europe* 108 (1954): 89–121.

—, ed. *Oeuvres de Marie Sklodowska Curie.* Warsaw: Académie Polonaise des Sciences, 1954.

Klickstein, Herbert S. *Marie Sklodowska Curie, Recherches sur les substances radioactives: A Bio-bibliographic Study.* Mallinckrodt Classics of Radiology. Vol. 2. Mallinckrodt, 1966.

Maria Sklodowska-Curie: Centenary Lectures. Vienna: International Atomic Energy Agency, 1968.

Pflaum, Rosalynd. *Grand Obsession: Madam Curie and Her World.* New York: Doubleday, 1989.

Pycior, Helena M., Nancy G. Slack, and Pnina G. Abir-Am, eds. *Creative Couples in the Sciences.* New Brunswick, NJ: Rutgers UP, 1995.

Quinn, Susan. *Marie Curie: A Life.* New York: Simon and Schuster, 1995.

Reid, Robert. *Marie Curie.* New York: E.P. Dutton and Co., Inc., 1974.

Romer, Alfred, ed. *The Discovery of Radioactivity and Transmutation.* New York: Dover Publications, 1964.

Rutherford, Ernest. *Radioactive Substances and Their Radiations.* Cambridge: Cambridge UP, 1913.

—. *Radioactive Transformations.* New Haven: Yale UP, 1906.

Soddy, Frederick. *The Interpretation of Radium and the Structure of the Atom.* 4th ed. New York: G.P. Putnam's Sons, 1920.

Strutt, R.J. *The Becquerel Rays and the Properties of Radium.* London: Edward Arnold, 1904.

35. GEORGE WASHINGTON CARVER

(1865–1943)

George Washington Carver was born in 1865 in Diamond, a frontier town in southwest Missouri, and was raised by Moses and Susan Carver after the disappearance of his mother Mary. Carver wrote later in life, "when just a mere tot in short dresses my very soul thirsted for an education. I literally lived in the woods. I wanted to know every strange stone, flower, insect, bird, or beast. No one could tell me. My only book was an old Webster's Elementary Spelling Book. I would seek the answer here without satisfaction" (Kremer 23). Carver attended school in nearby Neosho, but moved around Kansas frequently, working at odd jobs and seeking what education he could find. In 1886 he homesteaded on 160 acres south of Beeler, Kansas. Carver enrolled in Simpson College in Indianola, Iowa in 1890, intent on studying art, but left in 1891 to pursue studies in botany at the University of Iowa, where he made a number of powerful friends including the later United States Secretary of Agriculture James Wilson. His bachelor's thesis entitled "Plants as Modified by Man," on grafting and cross-breeding, was well received and Carver was assigned to mycologist L.H. Pammel for postgraduate work. Carver later published a list of the Alabama *Cercosporae*, and contributed to *Fungus Diseases of Plants at Ames, Iowa* (Iowa Academy of Science, 1895). Two new species of fungi were named for Carver from his specimens.

In 1896, Booker T. Washington invited Carver to head the newly established Experiment Station at the Tuskegee Normal and Industrial Institute in Alabama. Carver never married and his life from then onwards was completely dedicated to teaching and agricultural and chemical research at the Tuskegee Institute. The Tuskegee Institute, however, was constantly underfunded and Carver often clashed with the superintendent of the school, Washington's brother John H. Washington, over supplies, laboratory space, and staffing problems. Carver also entered a long, bitter dispute with another faculty member George R. Bridgeforth, who challenged Carver's ability to administer the school properly. The dispute led to this blunt assessment of Carver's strengths and weaknesses by Washington:

> when it comes to the organization of classes, the ability required to secure a properly organized and large school or section of a school, you are wanting in ability. When it comes to the matter of practical farm managing which will secure definite, practical, financial results, you are wanting again in ability. You are not to be blamed for this. It is rare that one individual anywhere combines all the elements of success. You are a great teacher, a great lecturer, a great inspirer of young men and old men; that is your forté and we have all been trying as best we could to help you do the work for which you are best fitted and to leave aside that for which you are least fitted. You also have great ability in original research, in making experiments in the soil and elsewhere on untried plants. You have great ability in the direction of showing what can be done in the use of foods and the preserving of foods (McMurray 68).

Carver's championing of "scientific agriculture"—the idea that a chemical knowledge of soils, the introduction of hybrid plants and animals, modern farming methods such as crop rotation, and the controlled testing of plants and fertilizers would increase crop yields and the quality of meat, dairy, and poultry—was essential for the survival of agriculture in Alabama, one of the poorest states in the South, whose soils had been exhausted by decades of the monoculture of "King Cotton" and tobacco. Washington wrote a pamphlet *The Need for Scientific Agriculture in the South* (Tuskegee Institute Farmer's Leaflet No. 7, 1902) describing the distressing situation. Carver thus became part of a U.S. federal government initiative to promote the science of agriculture, beginning with the establishment of the Morrill Land Grant agricultural colleges in the 1860s and 1870s and the disbursement of federal research funds for agriculture under the Hatch Act of 1887. Scientific agriculture can be traced to early eighteenth-century Europe to reformers such as Jethro Tull—who advocated improvement of soils and the use of his mechanical seed drill—as well as to Liebig's studies of agricultural chemistry in the nineteenth century.

Carver demonstrated dried sweet potatoes as a wheat flour substitute for the United States Department of Agriculture in 1918, and gained national prominence in securing tariffs against foreign peanut products by testifying in 1921 before the House Ways and Means Committee, demonstrating the many uses of American peanuts. Carver's international scientific reputation derives primarily from his work on the peanut; he became a frequent spokesman for the peanut industry, and for humanitarian reasons advocated the use of peanuts as an inexpensive, nutritious, multi-use foodstuff to supplement the diet of molasses, meal and fatback of Alabama tenant farmers and sharecroppers. His pamphlet *How to Grow the Peanut and 105 Ways of Preparing it for Human Consumption* (1916) was widely reprinted, and from peanuts, peanut hulls, and peanut oil he synthesized "a dozen beverages, mixed pickles, sauces (Worcestershire and chili), meal, instant and dry coffee, salve, bleach, tan remover, wood filler, washing powder, metal polish, paper, ink, plastics, shaving cream, rubbing oil, linoleum, shampoo, axle grease, synthetic rubber" (Holt, 241). In his later years, he was presented with two honorary science Doctorates from Simpson College and the University of Rochester. At the outbreak of World War II, Carver advised the U.S. government on his fertilizer, soap, and rubber substitutes and he suggested peanuts as a meat substitute. In 1942 his friend industrialist Henry Ford opened a nutrition laboratory named after Carver.

The pamphlet by George Washington Carver reprinted below includes the Alabama state law establishing the Tuskegee Experiment Station and provides Carver's practical advice, including a scientific analysis of their nutritional content, on using inexpensive and plentiful acorns as animal feed instead of, or in conjunction with, corn and cotton seed meal.

Feeding Acorns.
Tuskegee Normal and Industrial Institute Bulletin No. 1.
Tuskegee: Normal School Steam Press, 1898.

Organization and Work.

By the Director.

AGRICULTURAL EXPERIMENT STATION

At the session of the Alabama Legislature of 1896–97, the following bill was passed:

AN ACT

To establish two Branch Agricultural Experiment Stations for the Colored Race and to make appropriations therefor.

Section 1. Be it enacted by the General Assembly of Alabama, That a Branch Agricultural Experiment Station and Agricultural School for the colored race is hereby established and located at Tuskegee, Macon County, Alabama, to be run in connection with the Tuskegee Normal and Industrial Institute and to be known as the Tuskegee Agricultural Experiment Station and Agricultural School.

Section 2. Be it further enacted, That the Board of Control of said Station and School shall be composed of the State Commissioner of Agriculture, the President of the A. and M. College and the Director of the State Experiment Station at Auburn, Alabama, and the members of the Board of Trustees of the Tuskegee Normal and Industrial Institute who reside in the town of Tuskegee, Alabama. The members of said Board shall not receive any compensation, other than expenses actually incurred in visiting the Station and School and while there supervising its affairs.

Section 3. Be it further enacted, That for the equipment and improvement of said Station and School, there is hereby appropriated out of the agricultural fund in the treasury, not otherwise appropriated, the sum of fifteen hundred dollars; one fourth of said sum to be paid quarterly, to-wit: January 1st, April 1st, July 1st, and October 1st; to the Treasurer of said Board of Control, who shall give bond in double the amount of the appropriation, for the safe keeping and faithful application of the sum appropriated, the bond to be approved by the Judge of Probate of Macon County, Alabama, and filed in his office, a certified copy of which shall be forwarded to the Commissioner of Agriculture, to be placed on file in his office.

Section 4. Be it further enacted, That the Trustees of the said Tuskegee Normal and Industrial Institute shall furnish for the use of said Station and School, all the necessary lands and buildings, and that for such use they shall make no charge against the State of Alabama.

Section 5. Be it further enacted, That the Board of Control must cause such experiments to be made at said Station as will advance the interest of scientific agriculture, and to cause such chemical analyses to be made as are deemed necessary, all such analyses, if requested, to be under the supervision of the Commissioner of Agriculture, by the chemist of the agricultural department, without charge.

Section 6. Be it further enacted, That said Board of Control may adopt such rules and regulations as they may deem necessary for the purpose of carrying out the provisions of this act, so that the colored race may have an opportunity of acquiring intelligent practical knowledge of agriculture in all its various branches.

Section 7. Be it further enacted, That it is the purpose of this act to appropriate to the support of the Experiment Station established by this act; the sums appropriated in this act are appropriated only for the purpose of maintaining and operating Experiment Stations, with the view of educating and training colored students, as herein named, in scientific agriculture.

* * * *

Approved February 15th, 1897.

—— * * * * ——

We dare not attempt to state explicitly what this Station will do or even attempt to do, but we take pleasure in saying that as far as we are able, neither time nor expense will be spared to make our work of direct benefit to every farmer.

But few technical terms will be used, and where such are introduced, an explanation will always accompany them.

With reference to soils, our Station is very happily located, as out of 2,200 acres of land, with such a wide difference as to locality and quality, that nearly or quite all of the leading varieties of soil in the State are represented, which permits our work to be of a much broader scope than it otherwise would.

In a State so extremely varied as to soil and agricultural possibilities as Alabama, many problems of interest present themselves to the Experiment Station worker, but we will endeavor to turn our attention first to the problems that are of the deepest interest to the farmers, orchardist and gardeners.

As far as possible and practicable, soils will be studied with reference to the varieties to be found on the same farm and also in different sections of the State, their special adaptability to certain crops and the special treatment they require as to fertilizers, moisture, conservation, etc. It is also our aim to determine, by careful experiments, to what extent drainage, subsoiling and green manuring are beneficial, and what gains should result from a proper rotation of crops. The testing of peas, clovers, vetches, etc., as soil builders upon different kinds of soil.

Special attention will be given to the kinds and varieties of fruit and vegetables best suited for the home and market.

Our best native and introduced forage plants will receive their share of attention, as they bear directly upon our dairy interests, which, we believe, are not excelled by any of the Southern States in butter and cheese possibilities.

Along the lines of live stock, the rational comparison of breeds, the study of foods in their great and increasing variety, and their relation to the production of milk, fat, wool, bone and muscular tissue; and the effect of climate upon soils, plants and animals, will be studied.

We shall feel glad at any time to come in direct personal communication with the farmers and planters, gardeners and fruit growers of the State, and all others to whom we can be of any service, and all questions within the scope of our Station work will be answered as promptly as possible.

And we not only invite, but urge, every farmer to send samples of soils, grains, fertilizers, fodders, grasses, insects, feeding stuffs, etc. All such samples may be sent by mail or express. We further invite every farmer within reach, to visit our Station frequently and come in more direct touch with us.

We hesitate to make a definite statement as to the quarterly issue of the Bulletins, as the matter under consideration may require them more or less frequent. On an average, we think it safe to say that one will appear every quarter.

Neither do we think it wisdom to give much attention to the analysis of the waters from wells and springs, commercial fertilizers and old feed stuffs, that have been analyzed many times in this country as well as Europe.

The excellent work done in Auburn, Ala., at the Agricultural Experiment Station along these lines are quite sufficient, and we will take pleasure in quoting from them from time to time as the opportunity presents itself.

We will also be glad to quote from other Stations when their work bears directly upon our interests.

As far as possible, Bulletins will be sent free of charge to all who ask for them.

In conclusion, we desire to say that every effort will be put forth to carry out the two-fold object of the Station, viz: That of thoroughly equipping the student along the lines of practical and scientific agriculture; also the solving of many vexing problems that are too complex for the average farmer to work out for himself.

Along all lines our motto shall be to meet the full requirements of the act that made this Station possible.

Geo. W. Carver, Director.

Note.—The object and aims or our Station are essentially the same as those of nearly or quite all of the other Stations. I have freely used the ideas and, in some cases, without especial reference their language, for which I wish to acknowledge assistance. G.W.C.

Feeding Acorns.

In this beautiful Southland of ours, with so many natural resources, and the repeated failures of the North, East and West to supply the ever-increasing demand for pork, dairy products, etc., has led us to turn much of our attention in this direction.

The great quantity of acorns produced in our oak forests, which have been hitherto practically a waste product, forms the subject of this Bulletin.

I presume that acorns have been used as an article of food ever since their production and that of the animals to eat them, and man has used them in a half-unconscious way, from the time America's first settlers recognized the food value of the wild hog.

It did not take these sturdy colonists long to learn that the meat of the wild hog was fattest and sweetest when the mast (or acorns) were plentiful, as they were then and are now frequently called.

It is not at all strange that so little attention was paid to this food product, as forests were plenty and the supply far in excess of the demand. The people's custom was to turn their hogs out and let them run at large in the fall, and when wanted for meat, all that was necessary was to take the dogs, wagon and gun; recognize the "split," half-crop" or some such disfigurement of the ear which served as a mark; shoot them down and return home, to repeat the same process when more were wanted. In some sections of the South this system still exists to some extent.

But as the population increased, the forests gave way to cultivated fields, which, of course, decreased the supply of acorns. Yet our soils were rich and the production of other food stuffs comparatively easy, so that the feeding value of this natural product was, in a great measure, lost sight of, and we think it quite time now to consider more closely their value.

I append here the composition of corn and acorns, in order that the reader may compare them. The figures as given are taken from Dr. Armsby's splendid book on feeding, and the figures are essentially those of Julius Kuhn (Mentzel and v. Lengerke's, Landw. Kalender, 1880.)

Before giving these figures, there are a few terms that need explanation before one can make intelligent comparisons. They are as follows:

Total dry matter.—Refers to the weight of any substance dried in the air.

Protein.—By this we mean the class of substances in feeding stuffs that supplies the material for the growth of tissue.

Fat and Carbohydrates.—Belonging to that class of substances which produce fat and heat in the animal body.

Crude Fibre.—Meaning the woody part, and is more or less indigestible; foods with much woody fibre possess a low feeding value.

Ash.—This represents the mineral portion, or what is left after the substance has been burned.

This table represents percentages or the number of pounds in 100 pounds of either substance:

	Total dry Matter.	Protein.	Fat.	Nitrogen Free Ext.	Crude Fibre.	Ash.
Corn	88.9	16.5	4.8	70.2	1.9	1.5
Acorns with shell (fresh)	49.3	2.2	2.0	34.7	9.4	1.0
Acorns (Dry) without shell	85.6	5.6	4.1	69.2	5.1	1.6
DIFFERENCE IN FAVOR OF CORN.						
First	39.6	8.3	2.8	35.5		
Second	3.3	4.9	7	1.0		

It takes but a glance at the table to show how favorably acorns compare with corn as a fattening food; yet millions of bushels of this choice food product goes to waste in the United States every year.

The school possesses about 400 head of hogs, which are being grown, fattened and prepared for the table, solely on acorns and kitchen slops.

We purchased about 1,000 bushels this fall, and would have doubled the amount had we not feared their keeping qualities when put in bulk, but experience proves that it is only necessary to keep them cool and dry.

Another interesting experiment is the feeding of acorns to milch cows as a grain ration; in other words, instead of giving them corn, and such large quantities of cotton seed and cotton seed meal, acorns are given.

It is impossible at this writing, to give an accurate, well balanced ration, as the exact feeding value of acorns has, as yet, not been determined. We are feeding the following amount with good results: Two quarts at a feed twice per day, making five pounds.

The results have been very gratifying. While the milk did not materially increase as to quantity, it greatly improved in the amount of butter fat, and as far as I am able to determine at this writing, no bad effects on the butter is produced.

This cow has been fed as above described for one month and sixteen days; the amount of fat gained is also quite noticeable, which leads me to believe that beef may be very profitably produced on this natural food product.

Prof. Radusch, late of Harvard University, has quite an extensive poultry ranch on Greycloud Island, Minnesota, and in an interview with him he makes, substantially, the following statement:

> I have tried nearly or quite all of the various poultry foods advertised, and find that acorns, ground up, produce more eggs than anything I have tried; besides keeping the flock healthy.

It is safe to conclude that if fed in too great quantities to laying hens, too much fat would be produced.

The above statement is not at all surprising, if we call certain facts to mind. Many who will read this Bulletin, remember the days of the wild pigeon, when they flew in such great droves as to darken the sun, and remind one of a great moving cloud, indefinite in length, depth and breadth; how they would settle so thickly in their favorite roosting places as to frequently break large limbs several inches in diameter, from the trees upon which they perched.

At this time of the year, which was in the fall, many depended almost wholly on these birds for their supply of meat, and a most excellent dependence it was, as they were easily gotten and the meat of high quality, being so fat that no additional oil was needed in their preparation for the table. Everyone hailed with delight a big crop of acorns, for this meant a great many fat pigeons, and that they would remain a long time.

In feeding acorns there is this precaution necessary; where large quantities are given, plenty of laxative food should be included in the ration, as they are rather binding in their nature and likely to produce harmful results.

Considering the low price of cotton, the high price of corn, the feeding value of acorns and the great quantity produced, the entire South should bend much of their efforts toward saving this valuable natural food. They should be gathered in the fall, put in well ventilated bins, barrels or boxes, and kept as cool as possible to retard the destructive work of the worms and to prevent sprouting. The small varieties are more desirable here than the larger sorts, as they are less liable to sprout in bulk, and are more resistant to worms.

In feeding them to hogs, we find that rather a soft, spongy flesh is produced, with an oily-like lard, that hardens with great difficulty, and frequently not at all. This is readily overcome by feeding corn two or three weeks before butchering, although many hundred pounds of meat go into market, without complaint, that has never been topped off with corn.

A number of persons in this vicinity are feeding them to both cattle and hogs, and in every case report good results.

I know of no wider and more valuable field for experimentation than those of cooking, grinding, steaming, mixing with other foods, digestion experiments, etc; in short, a thorough test of their feeding value, and it is the aim of this Station to perform as many of these experiments as possible, and would like very much to have the co-operation of other Stations.

We hope the days are not far distant when the destruction of our valuable oak forests will cease, and that many of our scrubby pines will give way and be replaced by this useful tree.

The South has much to hope for in the matter of feeding acorns. As the Hon. James Wilson, Secretary of Agriculture, frequently said in addressing our Southern farmers, that the South has the cheapest and best food product in the world, referring to our great output of cotton seed, its by products, with our native and introduced grasses. At present I see no reason why acorns should not be added to the list; in fact, stand at the head in value, alongside of cotton seed meal; the carbohydrates or fat-forming matter coming from the acorns, and the nitrogenous or muscle and milk formers from the cotton seed meal.

The West uses corn largely as the basis of their fat forming foods, as it is produced in such great quantities and so cheaply. The South cannot do this, as the quantity produced is scanty and the cost of production too great; and could it be raised as cheaply as the Western corn, I see no reason why such quantities should be used for fattening purposes, when acorns are so abundant and compare so favorably in fattening value with that of corn, by actual chemical analysis, aside from their dietetic effects, which greatly increases their food value.

The author has spent some time this winter in visiting various acorn districts and observed the following facts: That wherever acorns were plentiful and the cattle and hogs had access to them, they were the fattest and best. I made a number of inquiries, and learned that these cattle and hogs lived solely on acorns and the pipe-stem cane, (Arundinaria Sp.) which remains green all winter, and forms dense patches known as "cane-breaks." Furthermore, I saw the cattle pushing the leaves away with their noses and eating the acorns the same as the hogs.

Horses are very fond of them, and will follow you as quickly for acorns as corn.

In conclusion, we will say that we regret that our knowledge is so meager with reference to experimental data, but as this subject is of such vast importance, not only to Alabama, but to the entire South, and, in a measure, to all acorn-growing districts, that we think it wisdom to present it at this time.

FURTHER READING

Carver, George Washington. Bulletins and Leaflets of the Tuskegee Institute [some may be written by other collaborating authors]:

—. *Feeding Acorns* (1898).

—. *Experiments With Sweet Potatoes* (1898).

—. *Farmer's Almanac* (1899).

—. *Fertilizer Experiments With Cotton* (1901).

—. *The Need of Scientific Agriculture in the South* (1902).

—. *Some Cercosporae of Macon County, Alabama* (1902).

—. *Suggestions for Progressive and Correlative Nature Study* (1902).

—. *Cow Peas* (1903).

—. *How to Build Up Worn Out Soils* (1905).

—. *Cotton Growing on Sandy Upland Soils* (1905).

—. *Successful Yields of Small Grain* (1906).

—. *Saving the Sweet Potato Crop* (1906).

—. *Saving the Wild Plum Crop* (1907).

—. *How to Make Cotton Growing Pay* (1908).

—. *How to Cook Cow Peas* (1908).

—. *Some Ornamental Plants of Macon County, Alabama* (1909).

—. *Nature Study and Gardening for Rural Schools* (1910).

—. *Some Possibilities of the Cow Pea in Macon County, Alabama* (1910).

—. *Possibilities of the Sweet Potato in Macon County, Alabama* (1910).

—. *Cotton Growing for Rural Schools* (1911).

—. *White and Color Washing with Native Clays from Macon County, Alabama* (1911).

—. *The Pickling and Curing of Meat in Hot Weather* (1912).

—. *Dairying in Connection with Farming* (1912).

—. *Poultry Raising in Macon County, Alabama* (1912).

—. *A Study of the Soils of Macon County, Alabama, and Their Adaptibility to Certain Crops* (1913).

—. *A New and Prolific Variety of Cotton* (1915).

—. *When, What and How to Can and Preserve Fruits and Vegetables in the Home* (1915).

—. *Alfalfa: The King of All Fodder Plants, Successfully Grown in Macon County, Ala.* (1915).

—. *How to Raise Pigs With Little Money* (1916).

—. *How to Grow the Peanut and 105 Ways of Preparing It for Human Consumption* (1916).

—. *Three Delicious Meals Every Day for the Farmer* (1916).

—. *How to Dry Fruits and Vegetables* (1917).

—. *43 Ways to Save the Wild Plum Crop* (1917).

—. *How to Grow the Cow Pea and Forty Ways of Preparing It as a Table Delicacy* (1917).

—. *Twelve Ways to Meet the New Economic Conditions in the South* (1917).

—. *How to Grow the Tomato and 115 Ways to Prepare It for the Table* (1918).

—. *How to Make Sweet Potato Flour, Starch, Sugar, Bread and Mock Cocoanut* (1918).

—. *How the Farmer Can Save His Sweet Potatoes and Ways of Preparing Them for the Table* (1922).

—. *How to Make and Save Money on the Farm* (1927).

—. *The Raising of Hogs, One of the Best Ways to Fill the Empty Dinner Pail* (1935).

—. *Can Livestock be Raised Profitably in Alabama?* (1936).

—. *How to Build Up and Maintain the Virgin Fertility of Our Soils* (1936).

—. *Some Choice Wild Vegetables That Make Fine Foods* (1938).

—. *Nature's Garden for Victory and Peace* (1942).

—. *Peanuts to Conserve Meat* (1943).

—. *The Peanut* (1943).

—. "Peanuts to Be Basis for New Foods–Prof. Carver." *Peanut Journal.* April (1921): 11.

Edwards, Ethel. *Carver of Tuskegee.* Cincinnati: Psyche Press, 1971.

Ferleger, Lou , ed. *Agriculture and National Development: Views on the Nineteenth Century.* Ames: Iowa State UP, 1990.

Fletcher, Thomas C. *Scientific Farming Made Easy: or, The Science of Agriculture Reduced to Practice.* London: Routledge, Warne, & Routledge, 1860.

Holt, Rackham. *George Washington Carver: An American Biography.* Garden City: Doubleday and Co., 1943.

Johnston, Sir Harry. *The Negro in the New World.* New York: Macmillan, 1910.

Kremer, Gary R. *George Washington Carver in His Own Words.* Columbia: University of Missouri Press, 1987.

Lloyd, Frederick James. *The Science of Agriculture.* London: Longmans, Green, and Co., 1884.

Macdonald, William. *Makers of Modern Agriculture.* London: Macmillan and Co., Ltd., 1913.

Mackintosh, Barry. "George Washington Carver: The Making of a Myth." *Journal of Southern History* 42 (1976): 525ff.

Marcus, Alan I. *Agricultural Science and the Quest for Legitimacy: Farmers, Agricultural Colleges, and Experiment Stations, 1870-1890.* Ames: Iowa State UP, 1985.

McMurry, Linda O. *George Washington Carver: Scientist and Symbol.* Oxford: Oxford UP, 1981.

Moulton, Forest Ray, ed. *Liebig and After Liebig: A Century of Progress in Agricultural Chemistry.* Washington, DC: American Association for the Advancement of Science, 1942.

Richardson, Clement. "A Man of Many Talents: George W. Carver of Tuskegee." *Southern Workman* 5 (1916): 602ff.

Rodgers, Miles M. *Scientific Agriculture; or, The Elements of Chemistry, Geology, Botany and Meteorology, Applied to Practical Agriculture.* 2nd ed. Rochester: E. Darrow, 1850.

Saloutos, Theodore. *Farmer Movements in the South, 1865–1933.* Berkeley: University of California Press, 1964.

Scott, Roy V. *The Reluctant Farmer: The Rise of Agricultural Extension to 1914.* Urbana: University of Illinois Press, 1970.

Smith, Alvin D. *George Washington Carver: Man of God.* New York, 1954.

Stokes, Anson Phelps. *Tuskegee Institute: The First Fifty Years.* Tuskegee: Tuskegee Institute Press, 1931.

Thrasher, Max Bennett. *Tuskegee: Its Story and Its Work.* Boston: Small, Maynard, 1900.

Washington, Booker T. *My Larger Education: Being Chapters From My Experience.* Garden City: Doubleday, Page, 1911.

36. ALFRED RUSSEL WALLACE

(1823–1913)

Born near Usk in Wales, Wallace's first employment came as a surveyor's apprentice to his brother William in 1837. After an indifferent education, Wallace taught himself botany from popular treatises and field excursions. In 1844 he taught briefly at Collegiate School in Leicester, and began reading works by Darwin, Lyell, and Robert Chambers (*Vestiges of Natural Creation*), who convinced him that species were not static. While at Leiceister, he met the entomologist Henry Walter Bates, who formulated the theory, now widely accepted, of Batesian Mimicry which explains how non-poisonous species imitate the coloration of poisonous or inedible species, thereby protecting themselves from predation.

From 1848–52, Wallace explored the Amazon River in Brazil with Bates; to maximize their efforts, Wallace and Bates took different routes in 1850. Wallace lost most of his specimens, diaries, and sketches, however, when his ship burned and sank on its return voyage across the Atlantic. He was nevertheless able to complete his *A Narrative of Travels on the Amazon and Rio Negro* (1853) after his return to England, establishing himself as a respected naturalist.

Wallace, eager to continue his collecting and to test some of his ideas on animal distribution, sailed to Malaysia and travelled over 14,000 miles throughout the Malay Archipelago from 1854–62. He later wrote the *Malay Archipelago* (1869), frequently reprinted in the nineteenth century, and shipped a huge number of specimens back from the islands, many previously unknown to European naturalists. He began a "Species Notebook" while in Malaysia and also gathered evidence in the late 1850s for "Wallace's Line," a zoogeographic division running through the Malay Archipelago and marking the boundary between the Asian and Australian fauna. Although the exact position of the line is in dispute and was revised by Wallace himself later in life, Wallace's original observation has been generally upheld by recent evidence from plate tectonics and evolutionary theory.

In 1855, Wallace published "On the Law Which Has Regulated the Introduction of New Species," in which he argued that "every species has come into existence coincident both in space and time with a pre-existing closely allied species" (*Contributions* 5). From Malaysia, he later sent Darwin a copy of his second paper on the species question (reprinted below) entitled "On the Tendency of Varieties to Depart Indefinitely from the Original Type" (1858), outlining a theory of species transmutation similar to Darwin's unpublished theory of natural selection, which Darwin had been developing privately since 1839. Darwin's friends Joseph Hooker and Charles Lyell arranged for a reading of Darwin's *Origin* manuscript along with Wallace's essay (both presented in absentia) at the Linnean Society meeting of 1858. Darwin rapidly revised his long essay on species and published it the following year as the *Origin of Species* (1859). Both Wallace and Darwin demonstrated great magnanimity in giving each other credit for the theory of descent with modification.

Like Darwin, Wallace had been influenced by reading Thomas Malthus's "Essay on the

Principle of Population" and concluded that species vary randomly in nature, that useful variations in a species tend to increase, and therefore superior varieties possessed of these variations have a greater chance of surviving and procreating, thus preserving favourable variations. Wallace, however, later split with Darwin over several points of the theory of natural selection including sexual selection by females, the influence of natural selection on human intelligence, and the effect of the use and disuse of parts in speciation. In *Darwinism* (1889), for example, he dismissed Lamarck's and Herbert Spencer's evolutionary hypothesis of the inheritance of acquired characters (chapter 14) through use and disuse, a possible mode of species change endorsed by Darwin in the *Origin*.

From 1870 onwards, Wallace produced a steady stream of books on zoology and evolution, including: 1) *Contributions to the Theory of Natural Selection* (1870)—reprinting his first two papers on natural selection; 2) *The Geographical Distribution of Animals* (1876)—a groundbreaking work of zoogeography in which Wallace studied regions of animal distribution of both extinct and modern genera and families as a key to the mechanisms of distribution; 3) *Tropical Nature and Other Essays* (1878); 4) *Island Life* (1880); and 5) *Darwinism* (1889).

A deep interest in spiritualism in the 1860s led Wallace to the belief that man's brain had evolved to the point where he could transcend the power of natural selection. These views "caused much distress of mind to Darwin" (*My Life* 2.17), but resembled Huxley's final conclusion that ethics could not be explained or regulated through the mechanism of natural selection. Wallace could not convince any of his scientific colleagues, however, to take a serious scientific interest in investigating the supernatural. He wrote two works on spiritualism: *The Scientific Aspect of the Supernatural* (1866), an article which he reprinted and distributed himself, and *Miracles and Modern Spiritualism* (1874).

In *The World of Life* (1910), one of his last books, he surveyed his own thinking on Darwinian evolution and offered the examples of bird feathers and the metamorphosis of insects as "proofs of an organizing and directive life-principle" (309) in the manner of the natural theologians. In 1913, he wrote to James Marchant "the completely materialistic mind of my youth and early manhood has been slowly moulded into the socialistic, spiritualistic, and theistic mind I now exhibit—a mind which is, as my scientific friends think, so weak and credulous in its declining years, as to believe that fruit and flowers, domestic animals, glorious birds and insects, wool, cotton, sugar and rubber, metals and gems, were all foreseen and foreordained for the education and enjoyment of man" (Marchant 413).

In later life, Wallace turned his attention to a variety of social and religious causes including women's liberation, socialism, and the anti-vaccination movement. A selection of his collected articles appeared in two volumes entitled *Studies Scientific and Social* (1900), with an eclectic mix of topics including natural history, land nationalization, public museums, unemployment, and social justice. In his 1903 book *Man's Place in the Universe*, he rejected the possibility of life on other planets, and in *Is Mars Habitable?* (1907), he similarly argued that the chemical makeup of Mars could not support life.

The first selection is Wallace's paper on species which he sent to Darwin from Malaysia. In the second selection from his retrospective book *The Wonderful Century* (1898), Wallace assesses the technological and scientific progress of the nineteenth century, and finds it an extraordinary period which had produced more scientific discoveries than all other previous ages combined. Wallace's lists provide an easily accessible check list of what the nineteenth century saw as its major scientific and technological achievements. Wallace, however, also spoke out bluntly in the same book against European militarism, the neglect of phrenology, social inequality, "the plunder of the earth," and vaccination. Smallpox vaccination, required under British law, became so controversial that in 1896 a special Royal Commission was forced to investigate the practice. Wallace testified before the commission, and later in 1898, using the statistics of the Commission itself (who recommended continuing the use of vaccines) published *Vaccination a Delusion: Its Penal Enforcement a Crime, Proved by the Official Evidence in the Reports of the Royal Commission*, later included in *The Wonderful Century*. His last two publications—*Social Environment and Moral Progress* and *The Revolt of Democracy*—appeared in 1913, the year of his death, and demonstrated his continuing interest in socialism during the final years of his life.

"On the Tendency of Varieties to Depart Indefinitely from the Original Type."

Journal of the Proceedings of the Linnean Society.
Zoology 3 (1859): 53–62.

Instability of Varieties supposed to prove the permanent distinctness of Species.

One of the strongest arguments which have been adduced to prove the original and permanent distinctness of species is, that *varieties* produced in a state of domesticity are more or less unstable, and often have a tendency, if left to themselves, to return to the normal form of the parent species; and this instability is considered to be a distinctive peculiarity of all varieties, even of those occurring among wild animals in a state of nature, and to constitute a provision for preserving unchanged the originally created distinct species.

In the absence or scarcity of facts and observations as to *varieties* occurring among wild animals, this argument has had great weight with naturalists, and has led to a very general and somewhat prejudiced belief in the stability of species. Equally general, however, is the belief in what are called "permanent or true varieties,"—races of animals which continually propagate their like, but which differ so slightly (although constantly) from some other race, that the one is considered to be a *variety* of the other. Which is the *variety* and which the original *species*, there is generally no means of determining, except in those rare cases in which the one race has been known to produce an offspring unlike itself and resembling the other. This, however, would seem quite incompatible with the "permanent invariability of species," but the difficulty is overcome by assuming that such varieties have strict limits, and can never again vary further from the original type, although they may return to it, which, from the analogy of the domesticated animals, is considered to be highly probable, if not certainly proved.

It will be observed that this argument rests entirely on the assumption, that *varieties* occurring in a state of nature are in all respects analogous to or even identical with those of domestic animals, and are governed by the same laws as regards their permanence or further variation. But it is the object of the present paper to show that this assumption is altogether false, that there is a general principle in nature which will cause many *varieties* to survive the parent species, and to give rise to successive variations departing further and further from the original type, and which also produces, in domesticated animals, the tendency of varieties to return to the parent form.

The life of wild animals is a struggle for existence. The full exertion of all their faculties and all their energies is required to preserve their own existence and provide for that of their infant offspring. The possibility of procuring food during the least favourable seasons, and of escaping the attacks of their most dangerous enemies, are the primary conditions which determine the existence both of individuals and of entire species. These conditions will also determine the population of a species; and by a careful consideration of all the circumstances we may be enabled to comprehend, and in some degree to explain, what at first sight appears so inexplicable—the excessive abundance of some species, while others closely allied to them are very rare.

The general proportion that must obtain between certain groups of animals is readily seen. Large animals cannot be so abundant as small ones; the carnivora must be less numerous than the herbivora; eagles and lions can never be so plentiful as pigeons and antelopes; the wild asses of the Tartarian deserts cannot equal in numbers the horses of the more luxuriant prairies and pampas of America. The greater or less fecundity of an animal is often considered to be one of the chief causes of its abundance or scarcity; but a consideration of the facts will show us that it really has little or nothing to do with the matter. Even the least prolific of animals would increase rapidly if unchecked, whereas it is evident that the animal population of the globe must be stationary, or perhaps, through the influence of man, decreasing. Fluctuations there may be; but permanent increase, except in restricted localities, is almost impossible. For example, our own observation must convince us that birds do not go on increasing every year in a geometrical ratio, as they would do, were there not some powerful check to their natural increase. Very few birds produce less than two young ones each year, while many have six, eight, or ten; four will certainly be below the average; and if we suppose that each pair produce young only four times in their life, that will also be below the average, supposing them not to die either by violence or want of food. Yet at this rate how tremendous would be the increase in a few years from a single pair! A simple calculation will show that in fifteen years each pair of birds would have increased to nearly ten millions![1] whereas we have no reason to believe that the number of the birds of any country increases at all in fifteen or in one hundred and fifty years. With such powers of increase the population must have reached its limits, and have become stationary, in a very few years after the origin of each species. It is evident, therefore, that each year an immense number of birds must perish—as many in fact as are born; and as on the lowest calculation the progeny are each year twice as numerous as their parents, it follows that, whatever be the average number of individuals existing in any given country, *twice that number must perish annually*,—a striking result, but one which seems at least highly probable, and is perhaps under rather than over the truth. It would therefore appear that, as far as the continuance of the species and the keeping up the average number of individuals are concerned, large broods are superfluous. On the average all above *one* become food for hawks and kites, wild cats and weasels, or perish of cold and hunger as winter comes on. This is strikingly proved by the case of particular species; for we find that their abundance in individuals bears no relation whatever to their fertility in producing offspring. Perhaps the most remarkable instance of an immense bird population is that of the passenger pigeon of the United States, which lays only one, or at most two eggs, and is said to rear generally but one young one. Why is this bird so extraordinarily abundant, while others producing two or three times as many young are much less plentiful? The explanation is not difficult. The food most congenial to this species, and on which it thrives best, is abundantly distributed over a very extensive region, offering such differences of soil and climate, that in one part or another of the area the supply never fails. The bird is capable of a very rapid and long-continued flight, so that it can pass without fatigue over the whole of the district it inhabits, and as soon as the supply of food begins to fail in one place is able to discover a fresh feeding-ground. This example strikingly shows us that the procuring a constant supply of wholesome food is almost the sole condition requisite for ensuring the rapid increase of a given species, since neither the lim-

1 This is under estimated. The number would really amount to more than two thousand millions! [Wallace's note added to 1870 reprint in *Contributions*.]

ited fecundity, nor the unrestrained attacks of birds of prey and of man are here sufficient to check it. In no other birds are these peculiar circumstances so strikingly combined. Either their food is more liable to failure, or they have not sufficient power of wing to search for it over an extensive area, or during some season of the year it becomes very scarce, and less wholesome substitutes have to be found; and thus, though more fertile in offspring, they can never increase beyond the supply of food in the least favourable seasons. Many birds can only exist by migrating, when their food becomes scarce, to regions possessing a milder, or at least a different climate, though, as these migrating birds are seldom excessively abundant, it is evident that the countries they visit are still deficient in a constant and abundant supply of wholesome food. Those whose organization does not permit them to migrate when their food becomes periodically scarce, can never attain a large population. This is probably the reason why woodpeckers are scarce with us, while in the tropics they are among the most abundant of solitary birds. Thus the house sparrow is more abundant than the redbreast, because its food is more constant and plentiful,—seeds of grasses being preserved during the winter, and our farm-yards and stubble-fields furnishing an almost inexhaustible supply. Why, as a general rule, are aquatic, and especially sea birds, very numerous in individuals? Not because they are more prolific than others, generally the contrary; but because their food never fails, the seashores and river-banks daily swarming with a fresh supply of small mollusca and crustacea. Exactly the same laws will apply to mammals. Wild cats are prolific and have few enemies; why then are they never as abundant as rabbits? The only intelligible answer is, that their supply of food is more precarious. It appears evident, therefore, that so long as a country remains physically unchanged, the numbers of its animal population cannot materially increase. If one species does so, some others requiring the same kind of food must diminish in proportion. The numbers that die annually must be immense; and as the individual existence of each animal depends upon itself, those that die must be the weakest—the very young, the aged, and the diseased,—while those that prolong their existence can only be the most perfect in health and vigour—those who are best able to obtain food regularly, and avoid their numerous enemies. It is, as we commenced by remarking, "a struggle for existence," in which the weakest and least perfectly organized must always succumb.

Now it is clear that what takes place among the individuals of a species must also occur among the several allied species of a group,—viz. that those which are best adapted to obtain a regular supply of food, and to defend themselves against the attacks of their enemies and the vicissitudes of the seasons, must necessarily obtain and preserve a superiority in population; while those species which from some defect of power or organization are the least capable of counteracting the vicissitudes of food, supply, etc., must diminish in numbers, and, in extreme cases, become altogether extinct. Between these extremes the species will present various degrees of capacity for ensuring the means of preserving life; and it is thus we account for the abundance or rarity of species. Our ignorance will generally prevent us from accurately tracing the effects to their causes; but could we become perfectly acquainted with the organization and habits of the various species of animals, and could we measure the capacity of each for performing the different acts necessary to its safety and existence under all the varying circumstances by which it is surrounded, we might be able even to calculate the proportionate abundance of individuals which is the necessary result.

If now we have succeeded in establishing these two points—1st, *that the animal population of a country is generally stationary, being kept down by a periodical deficiency of food, and other checks;* and, 2nd, *that the comparative abundance or scarcity of the individuals of the several species is entirely due to their organization and resulting habits, which, rendering it more difficult to procure a regular supply of food and to provide for their personal safety in some cases than in others, can only be balanced by a difference in the population which have to exist in a given area*—we shall be in a condition to proceed to the consideration of *varieties,* to which the preceding remarks have a direct and very important application.

Most or perhaps all the variations from the typical form of a species must have some definite effect, however slight, on the habits or capacities of the individuals. Even a change of colour might, by rendering them more or less distinguishable, affect their safety; a greater or less development of hair might modify their habits. More important changes, such as an increase in the power or dimensions of the limbs or any of the external organs, would more or less affect their mode of procuring food or the range of country which they inhabit. It is also evident that most changes would affect, either favourably or adversely, the powers of prolonging existence. An antelope with shorter or weaker legs must necessarily suffer more from the attacks of the feline carnivora; the passenger pigeon with less powerful wings would sooner or later be affected in its powers of procuring a regular supply of food; and in both cases the result must necessarily be a diminution of the population of the modified species. If, on the other hand, any species should produce a variety having slightly increased powers of preserving existence, that variety must inevitably in time acquire a superiority in numbers. These results must follow as surely as old age, intemperance, or scarcity of food produce an increased mortality. In both cases there may be many individual exceptions; but on the average the rule will invariably be found to hold good. All varieties will therefore fall into two classes—those which under the same conditions would never reach the population of the parent species, and those which would in time obtain and keep a numerical superiority. Now, let some alteration of physical conditions occur in the district—a long period of drought, a destruction of vegetation by locusts, the irruption of some new carnivorous animal seeking "pastures new"—any change in fact tending to render existence more difficult to the species in question, and tasking its utmost powers to avoid complete extermination; it is evident that, of all the individuals composing the species, those forming the least numerous and most feebly organized variety would suffer first, and, were the pressure severe, must soon become extinct. The same causes continuing in action, the parent species would next suffer, would gradually diminish in numbers, and with a recurrence of similar unfavourable conditions might also become extinct. The superior variety would then alone remain, and on a return to favourable circumstances would rapidly increase in numbers and occupy the place of the extinct species and variety.

The *variety* would now have replaced the *species*, of which it would be a more perfectly developed and more highly organized form. It would be in all respects better adapted to secure its safety, and to prolong its individual existence and that of the race. Such a variety *could not* return to the original form; for that form is an inferior one, and could never compete with it for existence. Granted, therefore, a "tendency" to reproduce the original type of the species, still the variety must ever remain preponderant in numbers, and under adverse physical conditions *again alone survive*. But this new, improved, and populous race might itself, in course of time, give rise to new varieties, exhibiting several diverging modifications of form, any of which, tending to increase the facilities for preserving existence, must, by the same general law, in their turn become predominant. Here, then, we have *progression and continued divergence* deduced from the general laws which regulate the existence of animals in a state of nature, and from the undisputed fact that varieties do frequently occur. It is not, however, contended that this result would be invariable; a change of physical conditions in the district might at times materially modify it, rendering the race which had been the most capable of supporting existence under the former conditions now the least so, and even causing the extinction of the newer and, for a time, superior race, while the old or parent species and its first inferior varieties continued to flourish. Variations in unimportant parts might also occur, having no perceptible effect on the life-preserving powers; and the varieties so furnished might run a course parallel with the parent species, either giving rise to further variations or returning to the former type. All we argue for is, that certain varieties have a tendency to maintain their existence longer than the original species, and this tendency must make itself felt; for though the doctrine of chances or averages can never be trusted to on a limited scale, yet, if applied to high numbers, the results come nearer to what theory demands, and, as we approach to an infinity of examples, become strictly accurate. Now the scale on which nature works is so vast—the numbers of individuals and periods of time with which she deals approach so near to infinity, that any cause, however slight, and however liable to be veiled and counteracted by accidental circumstances, must in the end produce its full legitimate results.

Let us now turn to domesticated animals, and inquire how varieties produced among them are affected by the principles here enunciated. The essential difference in the condition of wild and domestic animals is this,—that among the former, their well-being and very existence depend upon the full exercise and healthy condition of all their senses and physical powers, whereas, among the latter, these are only partially exercised, and in some cases are absolutely unused. A wild animal has to search, and often to labour, for every mouthful of food—to exercise sight, hearing, and smell in seeking it, and in avoiding dangers, in procuring shelter from the inclemency of the seasons, and in providing for the subsistence and safety of its offspring. There is no muscle of its body that is not called into daily and hourly activity; there is no sense or faculty that is not strengthened by continual exercise. The domestic animal, on the other hand, has food provided for it, is sheltered, and often confined, to guard it against the vicissitudes of the seasons, is carefully secured from the attacks of its natural enemies, and seldom even rears its young without human assistance. Half of its senses and faculties are quite useless; and the other half are but occasionally called into feeble exercise, while even its muscular system is only irregularly called into action.

Now when a variety of such an animal occurs, having increased power or capacity in any organ or sense, such increase is totally useless, is never called into action, and may even exist without the animal ever becoming aware of it. In the wild animal, on the contrary, all its faculties and powers being brought into full action for the necessities of existence, any increase becomes immediately available, is strengthened by exercise, and must even slightly modify the food, the habits, and the whole economy of the race. It creates as it were a new animal, one of superior powers, and which will necessarily increase in numbers and outlive those inferior to it.

Again, in the domesticated animal all variations have an equal chance of continuance; and those which would decidedly render a wild animal unable to compete with its fellows and continue its existence are no disadvantage whatever in a state of domesticity. Our quickly fattening pigs, short-legged sheep, pouter pigeons, and poodle dogs could never have come into existence in a state of nature, because the very first step towards such inferior forms would have led to the rapid extinction of the race; still less could they now exist in competition with their wild allies. The great speed but slight endurance of the race horse, the unwieldy strength of the ploughman's team, would both be useless in a state of nature. If turned wild on the pampas, such animals would probably soon become extinct, or under favourable circumstances might each lose those extreme qualities which would never be called into action, and in a few generations would revert to a common type, which must be that in which the various powers and faculties are so proportioned to each other as to be best adapted to procure food and secure safety,—that in which by the full exercise of every part of his organization the animal can alone continue to live. Domestic varieties, when turned wild, *must* return to something near the type of the original wild stock, *or become altogether extinct*.[2]

We see, then, that no inferences as to varieties in a state of nature can be deduced from the observation of those occurring among domestic animals. The two are so much opposed to each other in every circumstance of their existence, that what applies to the one is almost sure not to apply to the other. Domestic animals are abnormal, irregular, artificial; they are subject to vari-

2 That is, they will vary, and the variations which tend to adapt them to the wild state, and therefore approximate them to wild animals, will be preserved. Those individuals which do not vary sufficiently will perish. [Wallace's note added to 1870 reprint in *Contributions*.]

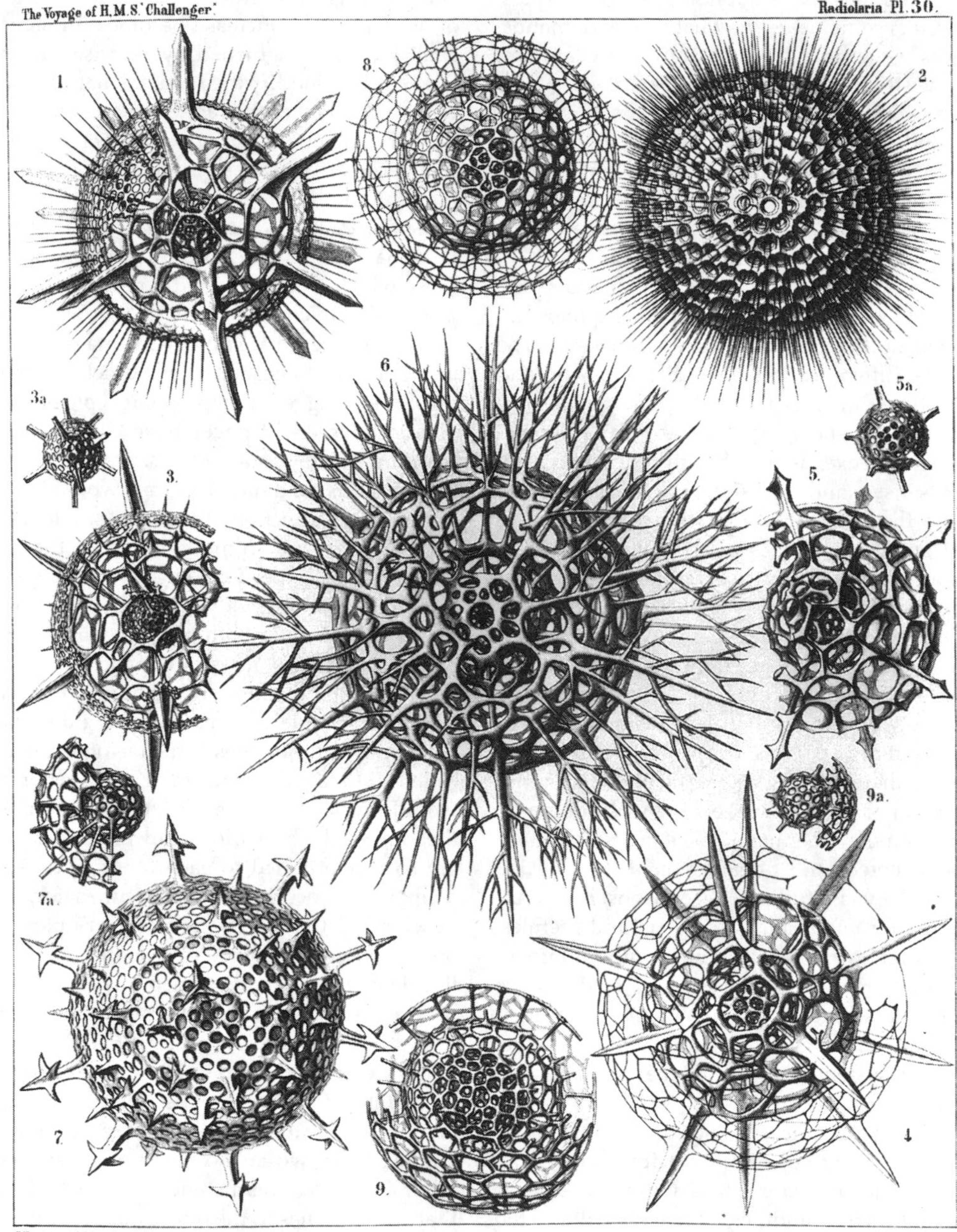

A selection of Radiolarians dredged by the British ship *H.M.S. Challenger* on its scientific voyage through the Atlantic, Pacific, and Indian Oceans. From 1872–76, the ship mapped the world's ocean bottoms and collected over 4,000 previously unknown species of marine life. The plates were drawn by Ernst Haeckel, a German naturalist and friend of Darwin who vigorously promoted Darwin's theories of evolution in Germany.

eties which never occur and never can occur in a state of nature: their very existence depends altogether on human care; so far are many of them removed from that just proportion of faculties, that true balance of organization, by means of which alone an animal left to its own resources can preserve its existence and continue its race.

The hypothesis of Lamarck—that progressive changes in species have been produced by the attempts of animals to increase the development of their own organs, and thus modify their structure and habits—has been repeatedly and easily refuted by all writers on the subject of varieties and species, and it seems to have been considered that when this was done the whole question has been finally settled; but the view here developed renders such an hypothesis quite unnecessary, by showing that similar results must be produced by the action of principles constantly at work in nature. The powerful retractile talons of the falcon- and the cat-tribes have not been produced or increased by the volition of those animals; but among the different varieties which occurred in the earlier and less highly organized forms of these groups, *those always survived longest which had the greatest facilities for seizing their prey*. Neither did the giraffe acquire its long neck by desiring to reach the foliage of the more lofty shrubs, and constantly stretching its neck for the purpose, but because any varieties which occurred among its antitypes with a longer neck than usual *at once secured a fresh range of pasture over the same ground as their shorter-necked companions, and on the first scarcity of food were thereby enabled to outlive them*. Even the peculiar colours of many animals, especially insects, so closely resembling the soil or the leaves or the trunks on which they habitually reside, are explained on the same principle; for though in the course of ages varieties of many tints may have occurred, *yet those races having colours best adapted to concealment from their enemies would inevitably survive the longest*. We have also here an acting cause to account for that balance so often observed in nature,—a deficiency in one set of organs always being compensated by an increased development of some others—powerful wings accompanying weak feet, or great velocity making up for the absence of defensive weapons; for it has been shown that all varieties in which an unbalanced deficiency occurred could not long continue their existence. The action of this principle is exactly like that of the centrifugal governor of the steam engine, which checks and corrects any irregularities almost before they become evident; and in like manner no unbalanced deficiency in the animal kingdom can ever reach any conspicuous magnitude, because it would make itself felt at the very first step, by rendering existence difficult and extinction almost sure soon to follow. An origin such as is here advocated will also agree with the peculiar character of the modifications of form and structure which obtain in organized beings—the many lines of divergence from a central type, the increasing efficiency and power of a particular organ through a succession of allied species, and the remarkable persistence of unimportant parts, such as colour, texture of plumage and hair, form of horns or crests, through a series of species differing considerably in more essential characters. It also furnishes us with a reason for that "more specialized structure" which Professor Owen states to be a characteristic of recent compared with extinct forms, and which would evidently be the result of the progressive modification of any organ applied to a special purpose in the animal economy.

We believe we have now shown that there is a tendency in nature to the continued progression of certain classes of *varieties* further and further from the original type—a progression to which there appears no reason to assign any definite limits—and that the same principle which produces this result in a state of nature will also explain why domestic varieties have a tendency to revert to the original type. This progression, by minute steps, in various directions, but always checked and balanced by the necessary conditions, subject to which alone existence can be preserved, may, it is believed, be followed out so as to agree with all the phenomena presented by organized beings, their extinction and succession in past ages, and all the extraordinary modifications of form, instinct, and habits which they exhibit.

Ternate, February, 1858.

The Wonderful Century: Its Successes and Its Failures.
London: Swan Sonnenschein and Co., Limd., 1898, pp. 150–157.

Estimate of Achievements:—The Nineteenth as Compared with Earlier Centuries

The long crude efforts of society
  In feeble light by feeble reason lead,—
But gleaning, gathering still, effect of cause,
  Cause of effect, in ceaseless sequence fed;
Till, slow developing the eons through,
  The gibbering savage to a Darwin grew—
This hath Time witnessed! Shall his records now,
  The goal attain'd—the end achieved, avow?

—*J.H. Dell.*

Having now completed our sketch of those practical discoveries and striking generalizations of science, which have in so many respects changed the outward forms of our civilization, and will ever render memorable the century now so near its close, we are in a position to sum up its achievements, and compare them with what has gone before.

Taking first those inventions and practical applications of science which are perfectly new departures, and which have also so rapidly developed as to have profoundly affected many of our habits, and even our thoughts and our language, we find them to be thirteen in number.

1. Railways, which have revolutionized land-travel and the distribution of commodities.
2. Steam-navigation, which has done the same thing for ocean travel, and has besides led to the entire reconstruction of the navies of the world.
3. Electric Telegraphs, which have produced an even greater revolution in the communication of thought.
4. The Telephone, which transmits, or rather reproduces, the voice of the speaker at a distance.
5. Friction Matches, which have revolutionized the modes of obtaining fire.
6. Gas-lighting, which enormously improved outdoor and other illumination.
7. Electric-lighting, another advance, now threatening to supersede gas.
8. Photography, an art which is to the external forms of nature what printing is to thought.
9. The Phonograph, which preserves and reproduces sounds as photography preserves and reproduces forms.
10. The Röntgen Rays, which render many opaque objects transparent, and open up a new world to photography.
11. Spectrum-analysis, which so greatly extends our knowledge of the universe, that by its assistance we are able to ascertain the relative heat and chemical constitution of the stars, and ascertain the existence, and measure the rate of motion, of stellar bodies which are entirely invisible.
12. The use of Anaesthetics, rendering the most severe surgical operations painless.
13. The use of Antiseptics in surgical operations, which has still further extended the means of saving life.

Now, if we ask what inventions comparable with these were made during the previous (eighteenth) century, it seems at first doubtful whether there were any. But we may perhaps admit the development of the steam-engine from the rude but still useful machine of Newcomen, to the powerful and economical engines of Boulton and Watt. The principle, however, was known long before, and had been practically applied in the previous century by the Marquis of Worcester and by Savery; and the improvements made by Watt, though very important, had a very limited result. The engines made were almost wholly used in pumping the water out of deep mines, and the bulk of the population knew no more of them, nor derived any more direct benefit from them, than if they had not existed.

In the seventeenth century, the one great and far-reaching invention was that of the Telescope, which, in its immediate results of extending our knowledge of the universe and giving possibilities of future knowledge not yet exhausted, may rank with spectrum-analysis in our own era. The

Barometer and Thermometer are minor discoveries.

In the sixteenth century we have no invention of the first rank, but in the fifteenth we have printing.

The Mariner's Compass was invented early in the fourteenth century, and was of great importance in rendering ocean navigation possible and thus facilitating the discovery of America.

Then, backward to the dawn of history, or rather to prehistoric times, we have the two great engines of knowledge and discovery—the Indian or Arabic numerals, leading to arithmetic and algebra, and, more remote still, the invention of alphabetical writing.

Summing these up, we find only five inventions of the first rank in all preceding time—the telescope, the printing-press, the mariner's compass, Arabic numerals, and alphabetical writing, to which we may add the steam-engine and the barometer, making seven in all, as against thirteen in our single century.

Coming now to the theoretical discoveries of our time, which have extended our knowledge or widened our conceptions of the universe, we find them to be about equal in number, as follows:—

1. The determination of the mechanical equivalent of heat, leading to the great principle of the Conservation of Energy.

2. The Molecular theory of gases.

3. The mode of direct measurement of the Velocity of Light, and the experimental proof of the Earth's Rotation. These are put together, because hardly sufficient alone.

4. The discovery of the function of Dust in nature.

5. The theory of definite and multiple proportions in Chemistry.

6. The nature of Meteors and Comets, leading to the Meteoritic theory of the Universe.

7. The proof of the Glacial Epoch, its vast extent, and its effects upon the earth's surface.

8. The proof of the great Antiquity of Man.

9. The establishment of the theory of Organic Evolution.

10. The Cell theory and the Recapitulation theory in Embryology.

11. The Germ theory of the Zymotic diseases.

12. The discovery of the nature and function of the White Blood-corpuscles.

Turning to the past, in the eighteenth century we may perhaps claim two groups of discoveries:—

1. The foundation of modern Chemistry by Black, Cavendish, Priestley, and Lavoisier; and

2. The foundation of Electrical science by Franklin, Galvani, and Volta.

The seventeenth century is richer in epoch-making discoveries, since we have:—

3. The theory of Gravitation established.

4. The discovery of Kepler's Laws.

5. The invention of Fluxions and the Differential Calculus.

6. Harvey's proof of the circulation of the Blood.

7. Roemer's proof of finite velocity of Light by Jupiter's satellites.

Then, going backward, we can find nothing of the first rank except Euclid's wonderful system of Geometry, derived from earlier Greek and Egyptian sources, and perhaps the most remarkable mental product of the earliest civilizations; to which we may add the introduction of Arabic numerals, and the use of the Alphabet. Thus in all past history we find only eight theories or principles antecedent to the nineteenth century as compared with twelve during that century. It will be well now to give comparative lists of the great inventions and discoveries of the two eras, adding a few others to those above enumerated.

Of course these numbers are not absolute. Either series may be increased or diminished by taking account of other discoveries as of equal importance, or by striking out some which may be considered as below the grade of an important or epoch-making step in science or civilization. But the difference between the two lists is so large, that probably no competent judge would bring them to an equality. Again, it is noteworthy that nothing like a regular gradation is perceptible during the last three or four centuries. The eighteenth century, instead of showing some approximation to the wealth of discovery in our own age, is less remarkable than the seventeenth, having only about half the number of really great advances.

It appears then, that the statement in my first

Of the Nineteenth Century.

1. Railways.
2. Steam-ships.
3. Electric Telegraphs.
4. The Telephone.
5. Lucifer Matches.
6. Gas illumination.
7. Electric lighting.
8. Photography.
9. The Phonograph.
10. Röntgen Rays.
11. Spectrum-analysis.
12. Anaesthetics.
13. Antiseptic Surgery.
14. Conservation of Energy.
15. Molecular theory of Gases.
16. Velocity of Light directly measured, and Earth's Rotation experimentally shown.
17. The uses of Dust.
18. Chemistry, definite proportions.
19. Meteors and the Meteoritic Theory.
20. The Glacial Epoch.
21. The Antiquity of Man.
22. Organic Evolution established.
23. Cell theory and Embryology.
24. Germ theory of disease, and the function of the Leucocytes.

Of All Preceding Ages.

1. The Mariner's Compass.
2. The Steam Engine.
3. The Telescope.
4. The Barometer and Thermometer.
5. Printing.
6. Arabic numerals.
7. Alphabetical writing.
8. Modern Chemistry founded.
9. Electric science founded.
10. Gravitation established.
11. Kepler's Laws.
12. The Differential Calculus.
13. The circulation of the blood.
14. Light proved to have finite velocity.
15. The development of Geometry.

chapter, that to get any adequate comparison with the Nineteenth Century we must take, not any preceding century or group of centuries, but rather the whole preceding epoch of human history, is justified, and more than justified, by the comparative lists now given. And if we take into consideration the change effected in science, in the arts, in all the possibilities of human intercourse, and in the extension of our knowledge, both of our earth and of the whole visible universe, the difference shown by the mere numbers of these advances will have to be considerably increased on account of the marvellous character and vast possibilities of further development of many of our recent discoveries. Both as regards the number and the quality of its onward advances, the age in which we live fully merits the title I have ventured to give it of—THE WONDERFUL CENTURY.

PROGRESS!

Not empanoplied as Pallas, with her spear and Gorgon shield,
But with fair Athene's olive, peaceful Progress takes the field;
Yet that shield is ever ready, and that spear is hers at need,
To protect the field she cultures, and defend the garnered seed;
And the meanest in her legions, marching with a level breast
In unbroken line of duty with her bravest and her best,
Answering only to her watchwords, walking only by her light,
Mustering to her only banner—to the gonfalon of Right;
For that flag's unstained honour—in that flag's unswerving cause—
Knows no other teacher's credo—owns no other leader's laws;
Treads that only Temple's pavement by the feet of Reason trod,
That hath Truth alone for Priestess—Equity alone for God.

—*J. H. Dell.*

FURTHER READING

Beddall, Barbara G. *Wallace and Bates in the Tropics: An Introduction to the Theory of Natural Selection.* London: Macmillan Co., 1969.

Brackman, Arnold C. *A Delicate Arrangement: The Strange Case of Charles Darwin and Alfred Russel Wallace.* New York: Times Books, 1980.

Brooks, J.L. *Just before the Origin: Alfred Russel Wallace's Theory of Evolution.* New York: Columbia UP, 1984.

Clements, Harry. *Alfred Russel Wallace: Biologist and Social Reformer.* London: Hutchinson and Co., 1983.

England, Richard. "Natural Selection Before the Origin: Public Reactions of Some Naturalists to the Darwin-Wallace Papers (Thomas Boyd, Arthur Hussey, and Henry Baker Tristram)." *Journal of the History of Biology* 30 (1997): 267–90.

George, Wilma. *Biologist Philosopher: A Study of the Life and Writings of Alfred Russel Wallace.* London: Abelard–Schuman, 1964.

Loewenberg, Bert James. *Darwin, Wallace and the Theory of Natural Selection.* Cambridge: Arlington Books, 1959.

Marchant, James, ed. *Alfred Russel Wallace: Letters and Reminiscences.* New York: Harper and Brothers Publishers, 1916.

McKinney, H. Lewis. *Wallace and Natural Selection.* New Haven, CT.: Yale UP, 1972.

Severin, Tim. *The Spice Islands Voyage: In Search of Wallace.* Boston: Little, Brown and Co., 1997.

Wallace, A.R. *Alfred Russel Wallace: An Anthology of His Shorter Writings.* Edited by Charles H. Smith. Oxford: Oxford UP, 1992.

—. *Contributions to the Theory of Natural Selection.* London: Macmillan and Co., 1870.

—. *My Life: A Record of Events and Opinions.* 2 vols. London: Chapman and Hall, Ltd., 1905.

—. *Natural Selection and Tropical Nature: Essays on Descriptive and Theoretical Biology.* London: Macmillan and Co., 1891.

Whitmore, T.C., ed. *Wallace's Line and Plate Tectonics.* Oxford: Clarendon Press, 1981.

Index of Names

Index of Topics